Laser Interaction
and Related Plasma Phenomena
Volume 5

Laser Interaction
and Related Plasma Phenomena

Volume 5

Edited by

Helmut J. Schwarz
Professor of Physics
Rensselaer Polytechnic Institute
Troy, New York

Heinrich Hora
Professor of Theoretical Physics
The University of New South Wales
Kensington-Sydney, Australia

M. J. Lubin
Laboratory of Laser Energetics
University of Rochester
Rochester, New York

Presently at
The Standard Oil Co. (OHIO) Research Center
Cleveland, Ohio

and

B. Yaakobi
Laboratory of Laser Energetics
University of Rochester
Rochester, New York

PLENUM PRESS • NEW YORK AND LONDON

Library of Congress Cataloging in Publication Data

Main entry under title:

Laser interaction and related plasma phenomena.

Proceedings of the 1st-   workshops, held at Rensselaer Polytechnic Institute, 1969-

Includes bibliographies.

1. High temperature plasmas—Congresses. 2. Controlled fusion—Congresses. 3. Lasers—Congresses. I. Schwarz, Helmut J., 1915-   ed. II. Hora, Heinrich, ed. III. Rensselaer Polytechnic Institute, Troy, N.Y.

QC718.5.H5L37 1971   530.44   79-135851

ISBN 0-306-40545-8 (v. 5)

Proceedings of the Fifth Workshop on Laser Interactions and Related Plasma Phenomena, held November 5—9, 1979, at the University of Rochester, Rochester, New York

© 1981 Plenum Press, New York
A Division of Plenum Publishing Corporation
233 Spring Street, New York, N.Y. 10013

Printed in the United States of America

PREFACE

The Fifth Workshop on Laser Interaction and Related Plasma
Phenomena was held at the Laboratory for Laser Energetics of
the University of Rochester, November 5-9, 1979. It was a sequel
to the four preceeding workshops held at the Rensselaer Polytechnic Institute, Troy, New York. The Fifth Workshop heard
thirty-seven (37) leading scientists describe the current state-
of-the-art in laser fusion and related laser-matter interaction
phenomena. These proceedings are not a verbatim repetition of
their presentations, but rather an expanded and more detailed
format of the material they covered.

Comprehensive review papers are included, on the following
subjects: new laser drivers, free-electron lasers, light ion
beam drivers, advanced fusion fuels, high compression diagnostics, target fabrication, soliton theory, Rayleigh-Taylor
instability, stimulated scattering, and Stark broadening of
high-Z x-ray lines. Detailed accounts of recent experiments
and current understanding of the field is given by members of
twelve leading laboratories in eight countries active in the
area of laser interaction with matter.

Additional papers were presented on the following subjects:
the first demonstration of nuclear excitation (in ^{235}U) by
resonant electronic transitions in a laser-produced plasma
(essentially the inverse of internal conversion); this work by
a group from Osaka University generated considerable excitement
among participants of the Workshop who speculated on its
potential to isotope separation and x-ray lasing; a detailed
conceptual design of a CO_2 laser fusion reactor (of power
1200 MW) employing magnetically guided liquid lithium blanket,
considerations for the design of laser-induced high pressure
shock waves for the study of equations-of-state, and experiments
leading to x-ray lasing.

The proceedings are grouped into five topical sections.
The first section involves mostly potential drivers for fusion.
Pleasance (Lawrence Livermore Laboratory) describes the current

status and future development of lasers of the type: Group VI, rare gas excimers, metal vapor excimers, rare gas halides and new solid state lasers. Basov et al. (Lebedev Laboratory) describe the DELFIN 216-beam laser system for fusion experiments. Schwarz (Rensselaer Polytechnic Institute) reviews the theory of free-electron lasers and describes a new concept based on two interfering electron beams, whereas Brau (Los-Alamos Laboratory) reviews the status and possible applications of free-electron lasers. Cooperstein (Naval Research Laboratory) describes recent progress in generation, transport and focusing of light-ion beams from pinch-reflex diodes. Deuterium beams of power 0.5 TW were transported over one meter through a discharge channel and focused to a current density of 300 kA/cm^2. A total current of 0.1 MA was inferred from a yield of 10^{12} neutrons emitted when a deuterium beam impinged on a CD_2 target.

The second section, on interaction experiments, contains a wealth of observations on such phenomena as profile steepening, preheat by suprathermal electrons, magnetic field generation, fast ions, x-ray generation, heat transport and Brillouin back-scattering.

In the third section (general considerations on fusion), Kidder (Lawrence Livermore Laboratory) analyzes the conflicting fusion requirements of high corona temperature to counteract heat inhibition, versus low corona temperature to increase the implosion efficiency. He concludes that shorter-wavelength laser drivers would greatly relax these constraints. Miley (University of Illinois) shows a design for a deuterium target with a DT core which breeds its own tritium and would obviate the need for a Lithium blanket.

The fourth section deals with the results and analysis of fusion-related experiments in seven of the major fusion laboratories. Attwood (Lawrence Livermore Laboratory) describes in detail a large variety of diagnostic techniques for present and future high density target implosions. Soures (University of Rochester) shows that symmetric illumination leads to either a higher neutron yield (10^9) or a higher compressed core density ($7 \ gr/cm^3$) for given laser parameters. Kilkenny et al. (Rutherford Laboratory) display a wide variety of transport and compression experiments (involving mostly x-ray spectroscopy and radiography). Afanasiev et al. (Lebedev Laboratory) work in the low irradiance, long laser pulse, ablative regime and attain high densities ($6-8 \ g/cm^3$). Azechi et al. (Osaka, Japan) as well as Bayer et al. (Limeil, France) describe the powerful method of radiography (x-ray back-lighting) and other measurements of target implosions. Stamper et al. (Naval Research Laboratory) describe the ablative, low laser irradiance, acceleration of foils to velocities of 10^7 cm/sec. Finally, Deckman and Halpern

(Exxon Corp.) give a very detailed description of microshell fabrication, characterization, coating, drilling and filling. Development of hemispheres for double-shell targets is also described.

The fifth and last section contains a variety of basic papers on the theory of soliton formation, Rayleigh-Taylor instability, fast electron transport, stimulated Brillouin and Raman scattering and Bohm diffusion in a laser-produced magnetic field. Finally, Griem (University of Maryland) outlines the theory of Stark broadening of x-ray lines of H-like ions in a high density plasma; Hooper (University of Florida) applies the Stark theory to He-like ions.

The fifth workshop was organized at the National Laser Users Facility of the University of Rochester and sponsored in part by the U.S. Department of Energy. The extensive effort of the speakers and their colleagues which went into the preparation of these proceedings is gratefully acknowledged.

We are grateful for the counsel of the Workshop advisors: N.G. Basov (P.N. Lebedev Phsyical Institute of the Academy of Sciences, Moscow, USSR), J.L. Bobin (Commissariat a l'Energie Atomique, France), K. Boyer (Los Alamos Scientific Laboratories, Los Alamos, New Mexico), A.J. DeMaria (United Technologies Research Center, East Hartford, Connecticut), J. Emmett (Lawrence Livermore Laboratory, University of California), A.H. Guenther (Kirtland Air Force Base, New Mexico), P. Harteck (Rensselaer Polytechnic Institute, Troy, New York), R. Hofstadter (Stanford University, California), R.E. Kidder (Lawrence Livermore Laboratory, University of California), D. Pfirsch (Max-Planck-Institute for Plasmaphysics, Garching, Germany), A.M. Prokhorov (P.N. Lebedev Physical Institute of the Academy of Sciences, Moscow, USSR), and Ch. Yamanaka (Osaka University, Japan).

May 1980

Helmut Schwarz
Henrich Hora
Moshe Lubin
Barukh Yaakobi

CONTENTS

I. LASERS AND OTHER DRIVERS

II.   INTERACTION EXPERIMENTS

V. INTERACTION THEORY

ALTERNATE LASER FUSION DRIVERS*

Lyn D. Pleasance

Livermore Laboratory
University of California
Livermore, CA 94550

INTRODUCTION

One objective of research on inertial confinement fusion is the development of a power generating system based on this concept. Realization of this goal will depend on the availability of a suitable laser or other system to drive the power plant. The primary laser systems presently used for laser fusion research, Nd:Glass and CO_2, have characteristics which may preclude their use for this application. Glass lasers are presently perceived to be incapable of operation at sufficiently high average power and the CO_2 laser may be limited by issues associated with target coupling. These general perceptions have encouraged a search for alternatives to the present systems. The search for new lasers has been directed generally towards shorter wavelengths; most of the new lasers discovered in the past few years have been in the visible and ultraviolet region of the spectrum. Virtually all of them have been advocated as the most promising candidate for a fusion driver at one time or another.

Table 1 gives a list of advanced laser candidates which have shown demonstrated performance or have promising characteristics. The wavelength of these lasers range from the mid-infrared to the deep ultraviolet. The degree of development varies widely. The typical energy shown in the second column of Table 1 is an indication of the state of development of each system. The hydrogen fluoride laser has demonstrated large energy output in oscillator configurations. Steady progress is being made on the development of systems at lower energies. The iodine laser is at a stage where it could be configured into a

driver for fusion research experiments. The group VI lasers and
several of the solid state lasers have been demonstrated only at
relatively low energies . The metal vapors have not yet lased.
The rare gas halide lasers are in the first stages of system
development. The rare gas excimer lasers have produced
appreciable energy output but are now receiving less attention
because of intrinsic difficulties with the laser media.

The various candidates can be characterized as either
storage or non-storage lasers. In a conventional energy storage
laser such as Nd:glass or CO_2, electrical input is coupled
directly or indirectly to an excited state population which
stores energy and thereby integrates the power input over a
period long compared to the extraction pulse. For a non-storage
laser, the lifetime of the excited state population is shorter
than or similar to the duration of the extraction pulse. Only
the energy delivered to the laser during the extraction pulse is
available as output. In this case, the laser operates as a
power amplifier rather than an energy integrator. The storage
lifetime of the various fusion candidates are shown in the third
column of Table 1. Several of the lasers can be classified as
storage lasers. However, none of these storage lasers, with the
possible exception of the metal vapors, is excited by direct
electrical input. All others require a separate optical pump
source such as a flashlamp or laser for their excitation. This
has serious implications on the overall efficiency of such
systems. The hydrogen fluoride laser releases chemical energy
for excitation but is initiated by electrical excitation. The
rare gas halide and rare gas excimer lasers are directly excited
by electrical discharges. These are, however, non-storage
lasers. In the past, such non-storage lasers have not normally

Table 1. Lasers Currently under Consideration as
Laser Fusion Drivers

Laser system	Wavelength (μm)	Typical energy (J)	Storage lifetime	Projected laser efficiency
HF	2.6	4000	10^{-8}	5%
Iodine	1.3	700	10^{-6}	< 1%
Group VI	0.5, 0.7	< 0.1	10^{-6}	1%
Solid state	0.3-1.1	< 0.1	10^{-6}-10^{-3}	10%
Metal vapors	0.3-0.7	—	10^{-6}	10%
Rare gas halides	0.2-0.3	300	10^{-8}	5%
Rare gas excimers	0.15-0.18	10	10^{-8}	< 1%

been considered as possible fusion drivers. Efficient excitation of these lasers at the pulse lengths required for fusion applications is incompatible with presently available power conditioning technology. Recently, however, several techniques have been under development which would allow the efficient extraction of energy from non-storage lasers operated at longer pulse lengths. This could have a significant impact on the future of laser drivers for fusion.

The list of candidates is still rather long. Recently, there has been an increasing effort to establish self-consistent criteria for evaluation of laser fusion candidates. The utility of a particular laser system as a fusion driver will ultimately be determined by its ability to produce net power output in a reactor configuration and by its cost. Estimates of the driver requirements for a reactor have varied somewhat over the years. A set of current specifications for a "reactor class" driver are given in Table 2. These specifications apply generally to both laser and heavy ion drivers. The energy and pulse length are determined by the desired pellet gain. Higher input energies produce higher gain. An energy greater than a megajoule will be required to obtain a gain of 100.[1] Shorter wavelength lasers are expected to be more effective at producing acceptable gains. The reactor must be capable of generating sufficient power to excite the laser and to produce net power output. The lower the pellet gain, the higher the required driver efficiency. For a gain of 100, a laser efficiency of 10% or greater is desirable to maintain the recirculating power in the power plant at acceptable levels. Lasers with lower efficiencies could be used with higher pellet gains or with a higher recirculating power. The pulse repetition rate is determined primarily by the desired average power of the reactor and by the response time of the reactor containment vessel and pellet factory. The cost of the laser sub-system must be less than some fraction of the total power plant costs. The current cost of power plants is several billion dollars. The replacement lifetime of major laser subsystem components must be some appreciable fraction of the power plant lifetime for economic operation. A lifetime of 10^8 pulses represents three years of operation at 1 Hz.

The efficiency has been the most restrictive of these specifications for evaluating new laser systems. Lasers with system efficiencies less than 1% can almost be eliminated from consideration. Laser efficiencies greater than 10% are desirable. The projected efficiency of current candidates are shown in the right hand column of Table 1. The iodine laser, the group VI laser, and the rare gas excimer lasers are too inefficient for serious consideration as a fusion driver. This evaluation is based on reasonably hard experimental data. The

Table 2.   Specifications for a Reactor Class
Laser Fusion Driver

Energy	>1 MJ
Pulse length	10 nsec
Efficiency	$>10\%$
Pulse rate	$>1\ H_z$
Cost	<200 M\$
Lifetime	$>10^8$ pulses

systems with highest projected efficiencies have been
demonstrated to lase only at low energy output or have not lased
at all. Two of the more developed systems, the hydrogen
fluoride laser and the rare gas halide lasers, have projected
efficiencies in the intermediate range; less than might be
desired but in the range of interest if high gain targets can be
achieved at moderate energies.

The rare gas halide lasers are of particular importance.
The KrF laser has been demonstrated to have an excited state
production efficiency of greater than 20% and a conversion
efficiency of from energy deposition in the gas to laser output
of greater than 10%. The laser wavelength is 248 nm. This is
probably as short a wavelength as is feasible to use without
damage of conventional optics. Systems studies indicate that
reactor class beam line based on the KrF laser could be
constructed under favorable conditions with a system efficiency
approaching 10%.

One or more of the present advanced laser candidates could
probably be developed into a reactor driver. However, none of
these systems presently satisfy all of the current requirements
for a reactor driver in a compelling fashion. A system which is
more efficient, less complex, and less costly would still be
desirable. Alternately, advances in pellet or reactor designs
which modify some of the requirements would have a significant
impact on the choice of laser subsystem. The discovery and
evaluation of new lasers is continuing. Systems such as the
$O_2(^1\Delta)$:I laser[2] and the free electron laser[3] are
presently under active consideration in the community, for
example. An aggressive search for more efficient lasers is
still needed. However. there is significant lack of compelling
ideas at this time to guide the directions of future research.
The ideas developed in past years have now been investigated and
our understanding of the physics of processes which affect

electronic transition lasers is more extensive. The following sections of this paper describes the status of several of these fusion laser candidates. The emphasis is primarily on the short wavelength lasers and on the issues affecting their utility as a fusion driver.

GROUP VI LASERS

The low atomic number members of the group VIA in the periodic table, oxygen, sulphur, and selenium, have several features which make them suitable as candidates for a storage laser for fusion applications.[4] The energy levels of the selenium atom, shown in Fig. 1, are typical of this group. Sulphur and oxygen have similar energy level structure. The ground state is a ^3P configuration while the first two excited states are singlets. There is a significant energy gap between the ^1S state and the next higher lying levels. Transitions can occur from the ^1S to the ^1D state and to the $^3P_1$ state. These are referred to as the as the auroral and transauroral lines respectively. The atomic oxygen transition at 558 nm is responsible for the familiar green emission from the northern lights. The radiative lifetime of these transitions is long. The stimulated emission cross section for at least one transition in each atom is 5 x 10^{-19} cm^2. This is sufficient to allow energy storage at acceptably low gain. The saturation fluence is well within the damage limits for available optical materials. Deactivation of the ^1S states by most collision partners is relatively

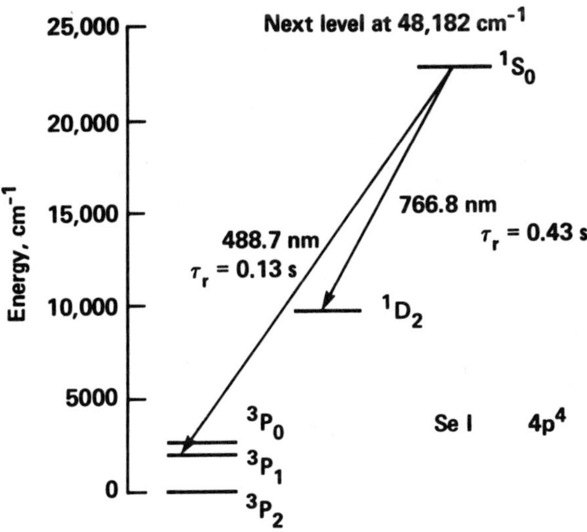

Fig. 1. The energy levels of the selenium atom.

slow. Self-quenching reactions are energetically unfavourable
and the reaction rates are small.

The stimulated emission cross section for transitions from
the ^{1}S state can be increased significantly by collisions with
some gases such as xenon. The collision produces a transient,
slightly bound diatomic molecule, XeO, which radiates on a band
centered to the blue of the atomic emission line.

Group VI atoms do not exist in appreciable quantities as
gases at normal temperatures. However, they can be produced by
electron dissociation or photodissociation of group VI bearing
compounds. In 1974, operation of a XeO laser was demonstrated
using an electron beam to excite a mixture of Xe and O_2.[5]
These experiments confirmed expectations as to the general
characteristics of the group VI lasers. However, two additional
new issues were discovered. Electrons were found to rapidly
deactivate the ^{1}S state, and no source of oxygen atoms was
found which was excited efficiently into the ^{1}S state by
electrons. Production of the corresponding sulphur and selenium
states by electrons has been less effective.

In 1975, it was demonstrated that the ^{1}S state in sulphur
and selenium could be produced with yields close to unity by
photodissociation in the ultraviolet of compounds such as OCS
and OCSe.[6] An external ultraviolet source could be used to
produce the ^{1}S states without the necessity of an electrical
discharge. This approach in principle eliminates the electron
deactivation of the ^{1}S states. Several sources in the
ultraviolet were available for small scale laser experiments.
The rare gas excimer lasers were particularly well suited for
this application. Fig. 2 shows the fluorescence spectrum of a
Xe_2 laser, the absorption spectrum of OCSe,[7] and the
photodissociation yield for OCSe.[6] Similar overlap exists for
OCS and the Kr_2 laser at 142 nm. Recently, an F_2 laser has
been discovered which also matches the dissociation band of
OCS.[8]

Recently, laser action on the auroral and transauroral
lines of Se and the auroural transition in S has been
demonstrated following photodissociation of OCSe and OCS.[9]
The corresponding XeSe laser has also been observed. The output
of an electron beam excited Kr_2 or Xe_2 laser with a energy
of a few Joules was used as a pump source. A population of ^{1}S
excited states of approximately $3 \times 10^{16}/cm^3$ has been
obtained with a fluorescence lifetime of several microseconds.
The conversion efficiency from pump energy absorbed to laser
output was measured to be greater than 30%. The extracted energy
density was greater than 4 J/l.

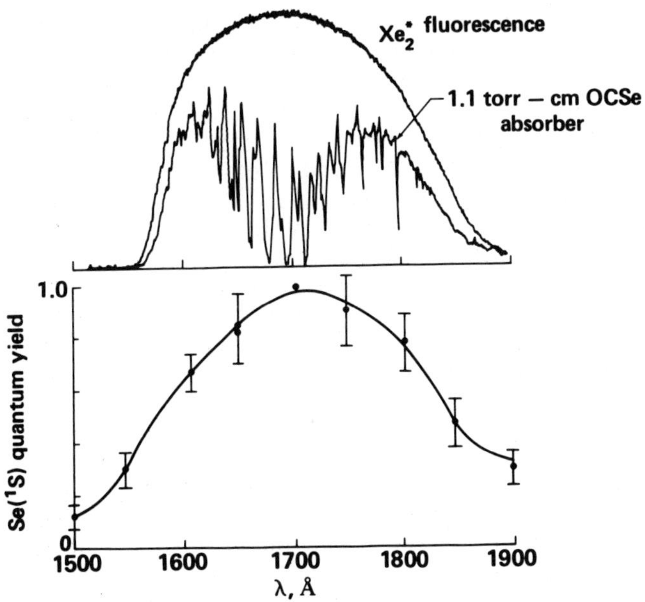

Fig. 2.   The production yield of Se(^{1}S) produced by photodissociation of OCSe, the absorption spectrum for OCSe, and the fluorescence emission spectrum of Xe$_2^*$.

In the laser experiments, the decay rate of the population in the ^{1}S state was observed to increase with increasing pump flux.[10] This effect is caused by electron deactivation of the ^{1}S state. The electrons are present in the laser medium despite the use of an external pump source. The existance of the electrons has been recently confirmed more directly by measurements of microwave absorption.[11] The initial electrons are probably generated by photoionization of the ^{1}S state by the ultraviolet pump flux. These initial electrons are then heated by superelastic collisions with the atomic ^{1}S state and can reach energies sufficient to produce ionization by direct electron impact. This results in the growth of the electron density and the rapid deactivation of the ^{1}S population. Some degree of control of this ionization instability can be achieved by cooling the electrons with collisions in gases with large inelastic collision cross sections such as CO and N_2 and by attachment to SF_6.[10] However, photoionization remains a serious issue for group VI lasers particularly with respect to extension to the excitation of larger volumes and energies.

The development of the group VI laser into a system for fusion applications requires both an efficient storage medium

and an efficient pump source. The efficiency of the rare gas
excimer lasers used as sources for small scale demonstration of
the group VI lasers, generally less than 1%. A group VI laser
is pumped by a narrow band flourescence source. Fluorescence
emission at 175 nm from Xe_2^* has been generated with
efficiencies greater than 40% using electron beam
excitation.[12] However, the overall efficiency in a system
driven by fluorescence is reduced by losses in the geometrical
coupling between source and laser medium. The overall system
efficiency of a Se laser pumped by an Xe_2^* fluorescence
source is approximately 1%.[13] Efficient excitation of the
group VI lasers or any other externally pumped storage medium
will require the development of an efficient laser source.

RARE GAS EXCIMER LASERS

     The excimers have been the most successful class of
electronic transition lasers discovered thus far. These lasers
involve transitions between a bound upper state of a molecule to
an unbound lower state. The production of the excited states is
generally efficient. The absence of population in the lower
state eliminates bottlenecking. In general, the laser
transitions are dipole allowed. The emission bandwidth is broad
giving a relatively low stimulated emission cross section
despite a short radiative lifetime. A large number of excimer
lasers spanning the visible and ultraviolet region of the
spectrum have been discovered. Generally, the excimers are
non-storage lasers and cannnot be used directly as for fusion
applications. Several techniques have been suggested to allow
the use of excimer lasers operated at longer pulse lengths to be
used as sources for a short pulse fusion driver. These include
pulse stacking, Raman pulse compression and pump sources for
other storage lasers. In all of these applications, efficiency
is the parameter of major importance.

     The Xe_2 excimer was the first excimer laser to be
extensively studied. It has been demonstrated that the
conversion efficiency from energy deposited by an electron beam
in xenon gas to photons emitted in fluorescence in the excimer
band at 175 nm can be greater than 40%.[12] Unfortunately,
laser operation has never been obtained with efficiencies
greater than a few tenths of one per cent. The reasons for this
discrepancy have become clearer over the past few years and have
some serious implications in the development of efficient
excimer lasers and visible lasers in general.[14] The kinetic
processes in Xe_2 are characteristic of excimer lasers. An
energy level diagram for the xenon atomic and diatomic species
is shown in Fig. 3. The lowest lying energy levels of the xenon
atom are metastable. Energy entering these levels through
direct excitation or cascade from higher levels is effectively

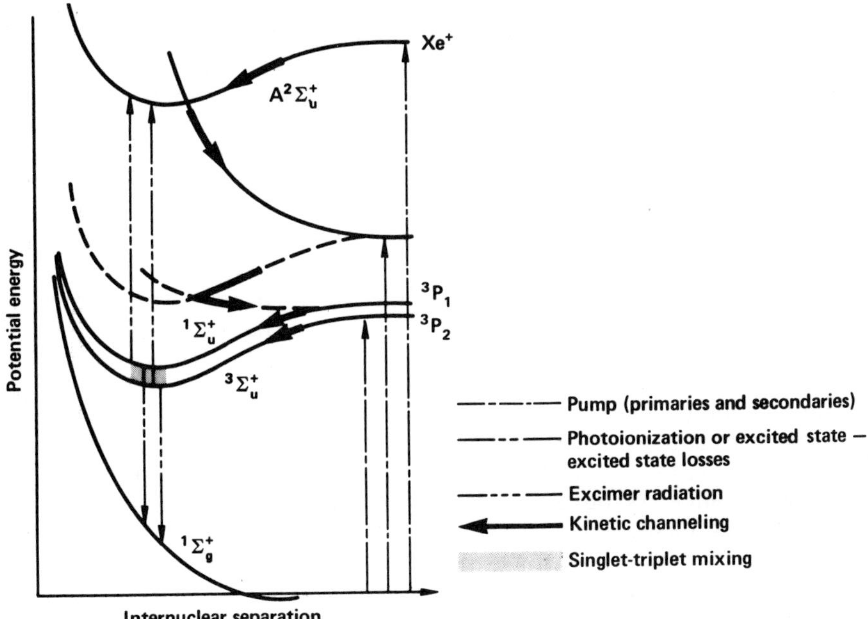

Fig. 3. The energy levels of the Xe and Xe$_2$ showing the
principle processes associated with the electrical
excitation of the Xe$_2^*$ laser.

trapped. This trapped excited state population is efficiently
converted to a bound diatomic excited state Xe$_2^*$ by
collisions. The ground state of the Xe$_2$ is only weakly
bound. The diatomic excited states radiate to the dissociative
ground state. The excited state is inherently inverted and
efficient laser operation should be expected. The gain is
relatively low because the emission bandwith is broad. The
energy channeling from higher lying excited states including the
ion manifold is extremely efficient. However, the Xe$_2$ energy
level structure has two characteristics which account for most
of its inefficiency as a laser. The lowest states in the
Xe$_2^*$ manifold, the $^3\Sigma$, are metastable. The $^1\Sigma$ state,
which radiates, is approximately 0.1 eV higher in energy. The
metastable levels are sufficiently coupled to the $^1\Sigma$ level
that the population in the manifold will ultimately appear as
radiation in fluorescence. However, for laser operation, only
the fraction of the population in the $^1\Sigma$ state is directly
accessible. In addition, laser photons can excite the manifold
of Xe$_2^*$ excited states to higher lying levels. This excited
state absorption has serious implications for all broadband, low
gain lasers. If the absorption cross section is larger than the

stimulated emission cross section, no net gain exists despite
the apparent inversion. High fluorescence efficiency is still
observed if the source is optically thin, i.e., dimensions
smaller than an absorption length. In Xe_2, the absorption
cross section is about 40% of the stimulated emission cross
section;[15] sufficient to allow lasing but insufficient to
allow efficient extraction. In addition, most of the population
resides in the lowest lying metastable level resulting in a low
overall efficiency as a laser.

Several techniques have been suggested to improve the laser
efficiency by increasing the coupling among the excited states.
Electron collisions are one example. In general, this approach
has not been successful. At high densities and temperatures the
electrons tend to photoionize the excimer excited states and to
lower the excitation efficiency by redistributing the population
in the upper states of the Xe and Xe_2. The most efficient
laser performance is obtained by direct pumping with an electron
beam. In this case, energy input is primarily through
ionization and recombination. The use of electron beam
sustainer techniques which heat the electrons and couples
directly to the atomic states has not proved to be effective for
efficient excimer excitation.[16]

The Kr_2^* excimer laser at 142 nm and the Ar_2^* laser
at 128 nm exhibit similar characteristics. The fluorescence
efficiency is generally lower for these lasers than for Xe_2^*.

METAL VAPOR EXCIMERS

The mercury dimer is a second example of an excimer system
which has demonstrated problems associated with excited state
absorption. These excimers are of particular interest because
the lifetime of the emission bands is relatively long and the
mercury atomic metastable levels can be efficiently pumped by
discharges. This raises the possibility of the development of a
discharge pumped excimer storage laser.

The excimer emission bands of mercury were initially
observed in mercury discharges.[17] Two emission bands were
observed, one at 335 nm and one at 480 nm. Until recently, the
origin of these bands was somewhat obscure. There is now
substantial evidence that the 480 nm band originates from a
trimer, Hg_3, formed by collisions with the dimer excited
states. Attempts to obtain laser emission on these bands have
been unsuccessful over the years. The reasons are now fairly
clear. Fig. 4 shows a simplified energy level diagram for
mercury. The level structure in many respects is similar to
that of xenon. The ground state of the dimer is weakly bound.
The excited states of the dimer are bound. As with xenon, the

lowest of the bound dimer states, the 0_g states are
metastable. The 335 nm emission probably originates from the
higher lying 1_u state.

In the mercury system, optical losses at the laser
wavelength are significant. Recently it has been established
that the mercury excimer emission bands are underlain by a
broadband absorption of sufficient magnitude that no net gain is
observed.[18] The specific states responsible for this
absorption have not been completely identified. However, as
with the xenon dimer, absorption from the lowest dimer excited
states is probably responsible. In this case, however, the
absorption cross section is greater than the stimulated emission
cross section and the system does not lase. Further research
may define a regime where losses are sufficiently low to allow
laser operation. The red wing of the 335 nm band where the
losses are lowest has been suggested as one possible
region.[18] However, if the losses remain large, such a system
would not be sufficiently efficient for fusion applications.

Other metal excimer systems have been suggested and
experimentally investigated. The most recent has been
HgCd.[19] An excimer emission band has been observed at
approximately 480 nm. This band consists of two components with

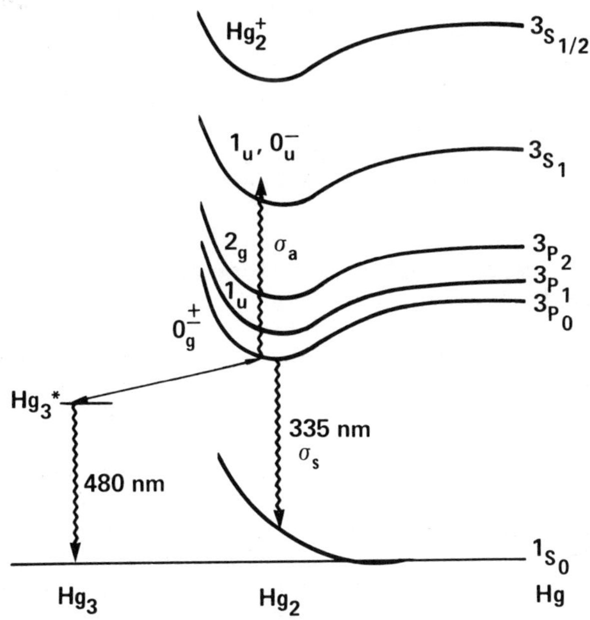

Fig. 4. A simplified energy level diagram for the mercury
         excimer excimers.

lifetimes of a few microseconds. As with Hg_2, however,
measurements show that net absorption exists over the entire
emission band.[19] Other mixed metal systems, Cd_2 and HgZn,
for example, are also possible but the prospects for a metal
vapor excimer storage laser are not encouraging.

RARE GAS HALIDE LASERS

The rare gas halide lasers are the most efficient short
wavelength lasers discovered to date. The KrF laser has been
demonstrated with an efficiency greater than 10% for the
conversion of energy deposited in the gas to laser photons at
248 nm.[20] The XeCl laser at 308 nm, the XeF laser at 353 nm
and the ArF laser at 190 nm have efficiencies which are only
slightly less. The diatomic excited states are produced
efficiently by collisionally stabilized ion-ion recombination in
electron beam generated plasmas. The manifold of excited states
appear to be strongly coupled. No significant optical
absorption from the excited states has been observed. Electron
deactivation is important only in the longer lived xenon halides
and at high pumping rates. The ground states of ArF and KrF are
unbound while the xenon halides have a weakly bound ground state.

The rare gas halide lasers do exhibit some optical loss.
Components of the gas mixture and several of the species
produced in the electrical discharge have broadband absorption
features at the wavelengths of the rare gas halide lasers. Some
of the absorbing species, such as F_- and F_2, are precursors
to the formation of the laser excited states and cannot be
eliminated. The magnitude of the absorption is not
debilitating. However, it does reduce the extraction efficiency
and limit the output intensity which can be obtained from the
laser.

Intrinsic medium absorption has significant implications in
the design of practical laser amplifiers. The basic issues are
illustrated in Fig. 5. The gain in the excimer band is
proportional to the excited state population. The population
and gain decrease as the flux on the laser transition
increases. In the absence of losses, the gain can be reduced to
an arbitrary value by increasing the flux appropriately and the
extraction efficiency approaches unity. In the presence of an
optical loss which does not decrease with increasing flux, the
net gain and extraction efficiency reaches zero at a finite
flux. Application of larger fluxes will drive the medium into
net absorption. The parameter which determines this maximum
flux is g_0/γ , the ratio of small signal gain to loss. At the
limiting flux, the extraction efficiency is zero. In practical
terms, an amplifier would be designed with an output flux less
than this maximum to obtain the optimal extraction efficiency

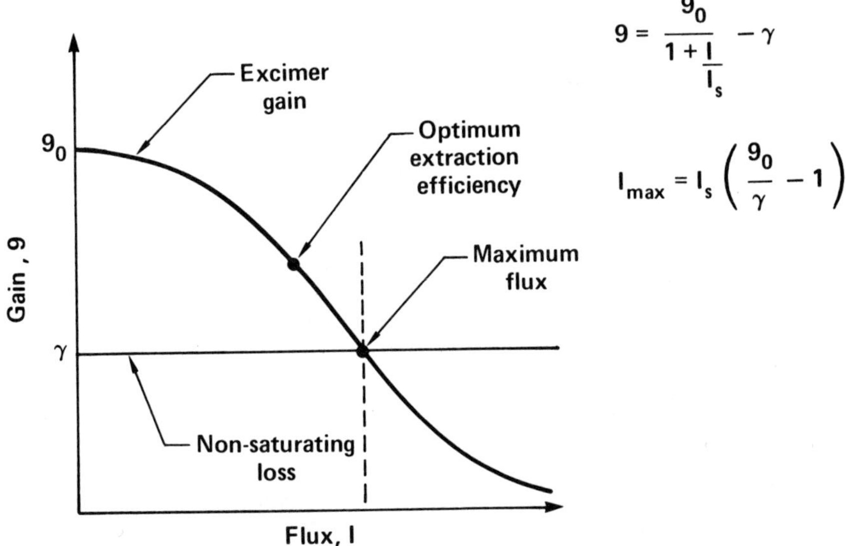

$$9 = \frac{9_0}{1 + \frac{I}{I_s}} - \gamma$$

$$I_{max} = I_s \left(\frac{9_0}{\gamma} - 1 \right)$$

Fig. 5. Non-saturable optical losses in the laser medium limit the maximun intensity which can be obtained from a laser.

which is less than unity. For the KrF laser, a qain to loss ratio of approximately 15 can be obtained under typical conditions. The output flux under these conditions is approximately 5 MW/cm^2, the extraction efficiency is 50% and the intrinsic medium efficiency is greater than 10%.

The radiative lifetime of the rare gas halide lasers is a few nanoseconds. These lasers do not store significant amounts of energy. The efficient use of the laser as a direct power amplifier to provide the required 10-20 ns optical pulse is beyond the capability of current electrical power conditioning technology. In addition, the low output flux density from the laser requires that an unacceptably large optical aperture face a target. Thus, some additional technique for power amplification and time compression is required to allow the laser to be excited over a longer period of time. One approach is to use the rare gas halide lasers to optical pump a storage medium. This approach requires that there be a match between the laser wavelength and the absorption bands of the storage medium and that the extraction efficiency from the storage medium be high. Relatively few systems with this combination have been found. A solid state laser using this approach is described in the next section. The second approach is to develop techniques which are non-resonant and use the lasers

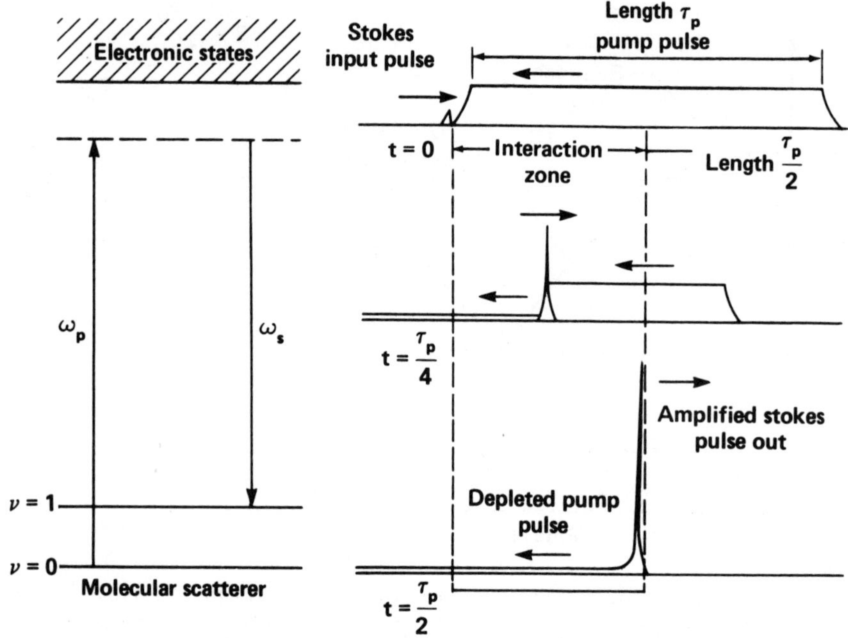

Fig. 6. Pulse compression and amplification by backward wave
         Raman scattering.

more directly. Two techniques of this type are under evaluation
at the present time.

The first is backward wave Raman scattering. The basic
idea is illustrated in Fig. 6.[21] In the presence of an
applied driving field at ω_p, a Raman active medium exhibits
gain at the Stokes frequency, ω_s. Under certain conditions,
this gain can be nominally isotropic. The simultaneous
application of a pump source and an input at the Stokes
frequency will result in the amplification of the Stokes pulse
at the expense of the pump beam. If the Stokes and pump pulse
propagate in the same direction, the Stokes beam will interact
only locally with the pump beam and the energy transfer will be
small. However, if the Stokes and pump pulse counter propagate,
a Stokes pulse of short duration will extract energy from the
entire pump pulse as it propagates through the Raman medium.
The net effect is to compress a long pulse at the pump
wavelength into a short pulse at the Stokes wavelength.

Raman pulse compression has several other advantages. The
Raman process does not require phase coherence in the pump
beam. In principle, beams from multiple sources of relatively
poor spatial quality can be combined in a single Raman cell to

pump a laser with substantially higher beam quality. In
addition, the the optical distortions introduced by thermal
loading of the Raman medium is minimal since the conversion
efficiency is high and the unused pump beam leaves the Raman
medium to be dissipated elsewhere.

Raman pulse compression has some limitations. To maximize
the gain in the backward direction, the linewidth of the pump
source must be less than the homogeneous linewidth of the Raman
scatterer. For methane, the linewidth required is approximately
0.3 cm^{-1} . Linewidths of this magnitude can be generated
efficiently in the rare gas halide lasers. However, this narrow
linewidth may lead to reduced absorption in the fusion target as
a result of Brillouin scattering. In addition, the extraction
efficiency is reduced as a result of the growth of second Stokes
emission. As the intensity of the Stokes pulse increases, it
can itself act as a pump source for amplification at a
wavelength reduced by an additional Stokes shift. Since there
is normally no input at this frequency, the second Stokes
intensity must grow from thermal noise. However, ultimately it
can reach a point where it begins to deplete the original Stokes
pulse. The severity of the second Stokes conversion increases
with the magnitude of the desired pulse compression. Large
pulse compressions can be obtained only at low power gains and
inefficient utilization of the pump beam. For a KrF laser
pumping methane, a power gain of 5 and a pulse compression of 10
should be possible; this is a 50% conversion efficiency. This
is sufficient to allow the use of lasers with 100-200 ns pulse
duration as a pump source for fusion applications. Significant
improvements could be obtained if an absorber for the second
Stokes wavelength were available.

Of the Raman active media, methane appears to have the best
overall performance characteristics. The vapors of metals such
as barium, calcium and lead could also be used as Raman
scatterers. To obtain a large scattering cross section, the
pump wavelength must be in proximity of a specific transition.
This requirement for near resonance reduces the choice of laser
and scatterer. Second Stokes generation is reduced
significantly. However, the Stokes shift in the metal vapors is
large, lowering the efficiency, and parasitic coupling to other
energy levels is more important.

The basic physics of the backward wave Raman process has
been confirmed using pump energies of approximately 1 J.[22]
The gain, the extraction efficiency as a function of pulse
compression, and the onset of second Stokes depletion, are all
predictable. Initial demonstration of the scalability of the
Raman approach requires pump energies of a few tens of Joules
and Raman beams of several square centimeters.[23]

A second approach to pulse compression is illustrated in
Fig. 7. A long pulse can be synthesized from a series of short
pulses. The Fresnel number of a laser amplifier of the size
required for laser fusion is sufficiently large that several
beams can be applied at different angles without serious losses
from diffraction. After amplification, each short pulse
propagates until it is separated from its neighbors. Each beam
is then delayed by propagating through an appropriate distance
and is then directed at a point where all the short pulses
overlap at a common time. For ten nanosecond pulses, the
separation between delay mirrors is approximately 3 meters.
However, for a series of pulses extending to several hundred
nanoseconds, the total separation between the first and last
mirror may reach several hundred meters. The angle of each beam
as it enters and leaves the amplifier must be small to insure
that the volume of the laser amplifier is effectively utilized.
Substantial propagation distances are required for separation of
the individual beams. The generation of the initial beam array
requires similar propagation distances.

An example of an optical stacker design is shown in
Fig. 8.[24] An input array of 16 beams propagating
co-temporally is each delayed by an appropriate amount and
directed towards a large spherical focusing mirror. The sixteen
beams pass sequentially through a two pass amplifier and onto a
second large mirror. The beams then pass through an appropriate
delay generator to regenerate the original array of 16 beams.

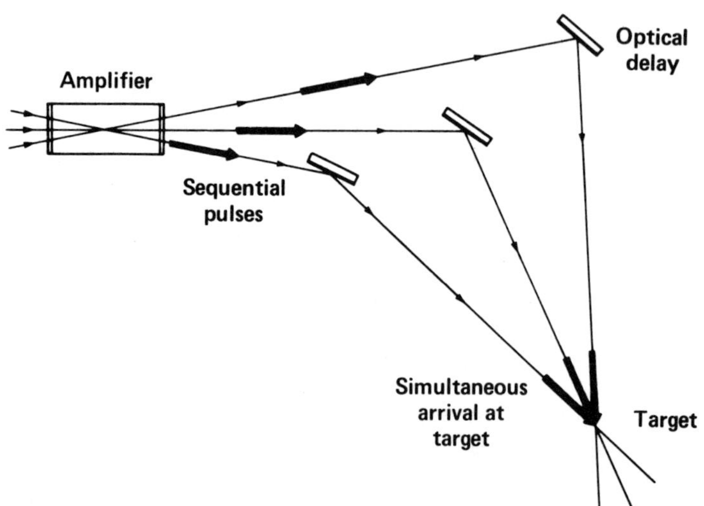

Fig. 7.   Power amplication of short pulses in an amplifier of
          longer pulse duration by sequential pulse stacking.

This array could be used to excite an additional stage of
amplification or be directed towards a target.

     An optical stacker occupies a large amount of space and
uses a large number of optical elements. This approach has the
advantage of allowing the use of long pulse laser sources. In
addition, it gives the system designer some freedom in component
tradeoff against system costs. The stacker shown in Fig. 8 was
designed, for example, using as only flat and spherical optics
to minimize the cost of beam alignment systems and to load the
various optical surfaces maximally.

     Several issues associated with the use of optical stackers
for fusion applications remain to be resolved. Cross talk
between adjacent channels due to scattering from the mirrors and
amplified spontaneous emission from from the amplifier could
produce prepulses which might destroy the target. The severity
of this effect can be minimized by extending the mirrors surfaces
in space at additional cost. The quality of the input beam and
mirror surfaces required for a practical pulse stacker is a significant issue. Beams propagating through different paths will
be subject to aberrations. Sequential propagation of beams
through a saturated gain medium has not yet been demonstrated
satisfactorily. Despite these reservations, it is generally
believed that stackers using between 25 and 50 beams might be

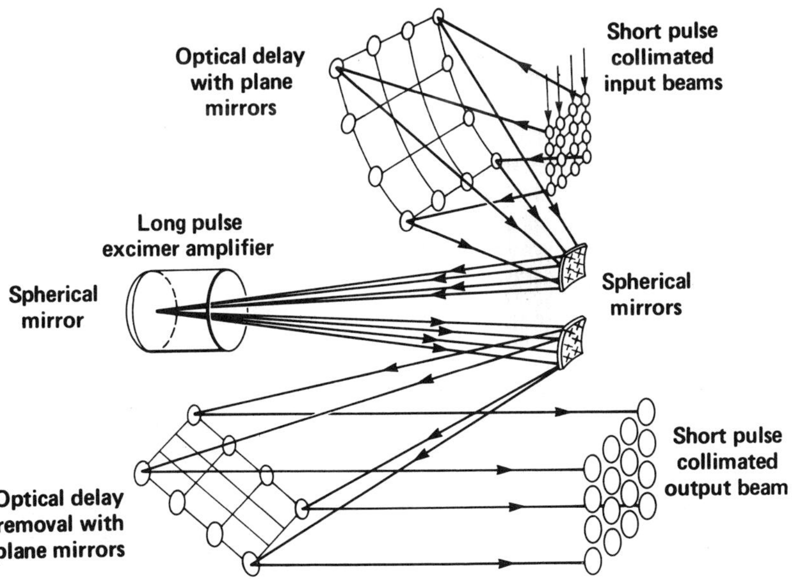

Fig. 8. A 16 beam pulse stacker using a double pass amplifier.

economically and technologically feasible. This would allow the
use of laser sources with a duration of up to 500 ns as fusion
drivers.

Hybrid schemes which use a combination of pulse compression
and Raman compression are also possible. The laser source can be
operated as an optical stacker with the output driving a Raman
compressor of relatively low compression ratio. The beams
through the Raman cell can themselves be stacked for further
time compression. The short pulse output of a Raman compressor
could be used as a pump source for further Raman compression at
a longer wavelength. The tradeoff between these various systems
will be on the basis of technical and economic factors.

Analysis of reactor class systems based on KrF Raman
compression and optical pulse stacking techniques suggest that a
system could ultimately be constructed which could approach the
reactor system requirements given in Table 2.[23] Under optimal
conditions the overall system efficiency could be greater than
5% and could approach 10% if methods could be found to increase
the KrF laser efficiency slightly. Construction of such a
system could not be acheived without substantive improvement in
electrical power conditioning technology and in optical systems
components. To date, the KrF compressor system represents the
most viable short wavelength laser concept for fusion
applications.

NEW SOLID STATE LASERS

The Nd:glass laser system has many excellent attributes as
a fusion laser. The storage lifetime is relatively long.
Flashlamp pumping uses simple power conditioning. The stored
energy in the laser medium density is relatively high. The
technology of laser construction is well developed. These
advantages have encouraged a continuing effort at developing a
solid state laser with higher efficiency, better thermal
properties and shorter wavelength than Nd:glass. These efforts
have taken two general approaches: laser pumping of shorter
lifetime storage media and flashlamp pumping of various ions in
hosts with improved thermal properties.

Laser pumping has several potential advantages over
flashlamp pumping. The narrow laser linewidth allows the
consideration of pumping energy levels which cannot be pumped
efficiently with broadband flashlamps. The higher pumping
intensity allows the consideration of levels with shorter
lifetimes as possible storage media. In addition, much of the
thermal loss in the system occurs in the pump laser medium
rather than in the solid state storage medium. The overall
efficiency cannot be greater than that of the pump laser. This

restricts the choices available for pump lasers to the most
efficient. The pump laser must also match the absorption band
of the solid state laser. Despite these restrictions, several
solid state lasers have been suggested as potential laser pumped
systems. One of these, the Tm^{3+} ion pumped by XeF, has
features which make it of particular interest.

The energy levels of the Tm^{3+} ion are shown in Fig. 9.
Transitions within the 4f manifold would be forbidden except for
the transition moment induced by the fields of the host lattice.
The upper $^1D_2$ level is in coincidence with the 353 nm
emission band of the XeF laser. The $^3H_4$ lower laser level
is 6000 cm^{-1} above the ground state. The relative lifetimes
of the various states are indicated in the table on the right in
Fig. 9. The radiative lifetimes of the upper states are a few
tens of microseconds. This is compatible with efficient
excitation of the XeF laser. The lifetime of the lower level is
substantively longer and upon extraction the laser will
ultimately bottleneck. However, 40% extraction efficiency
should be obtained even with this limitation. The Tm^{3+} laser
has an important additional feature. The corresponding
lifetimes for phonon deactivation are also shown on the right of
Fig. 9. The lifetimes for phonon deactivation for the laser
levels are substantively longer than the radiative lifetime.

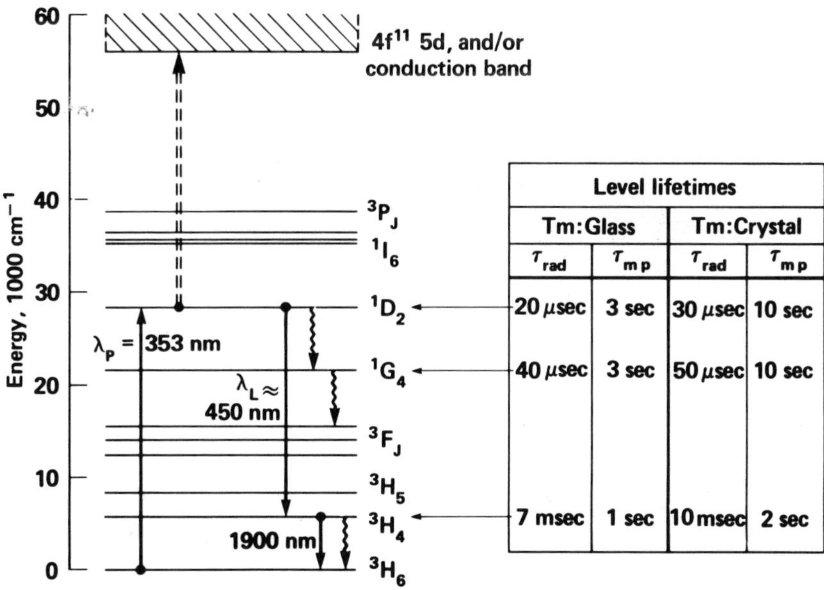

Level lifetimes			
Tm:Glass		Tm:Crystal	
τ_{rad}	τ_{mp}	τ_{rad}	τ_{mp}
20 μsec	3 sec	30 μsec	10 sec
40 μsec	3 sec	50 μsec	10 sec
7 msec	1 sec	10 msec	2 sec

Fig. 9. The energy levels and lifetimes of the Tm^{3+} ion.

Thus, most of the energy deposited in the lower laser level will
be radiated and the heat load on the lattice will be minimal.
This should allow operation of the laser at higher repetition
rates as a result of the lower thermal load on the material.

Recently, the Tm^{3+} laser has been demonstrated with YLF
(Yttrium Lithium Fluoride) as the host.[25] The initial
experiments used small samples and pump energies of a few tens of
millijoules, but larger crystaline samples can be grown. This
development looks promising and will undoubtedly be pursued. The
efficiency of the overall system, however, will ultimately be
limited by the efficiency of the XeF pump laser to a value under
5%.[13]

An alternate approach to solid state lasers is to use ionic
levels which couple efficiently to flashlamp pump in hosts with
good thermal characteristics. No rare earth laser has been identified which is better than Nd:glass. Recently, however, increased attention has been drawn to lasers using the transition
metal ions. Lasing has recently been demonstrated on transitions in the Ni^{2+} ion in MgF_2.[25] Another member of this
group, V^{2+} in MgF_2, may be useful as a fusion laser.

The energy level structure and the corresponding absorption
and emission bands of $V:MgF_2$ are shown in Fig. 10. The medium
is characterized by a compact set of absorption bands which

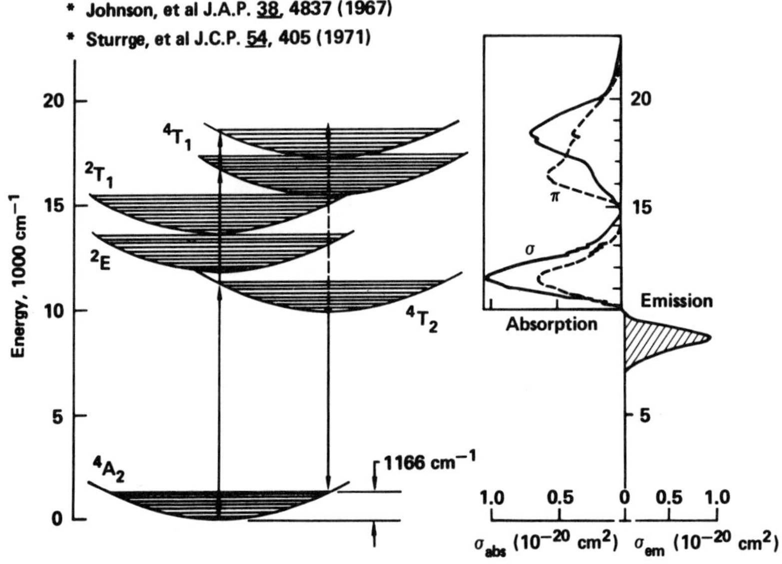

Fig. 10. Energy levels and absorption cross sections for
          vanadium doped magnesium fluoride

rapidly relax to the $^4T_2$ level. The laser transition is to
high lying levels of the ground state manifold. These levels
are rapidly quenched. Operation at cryogenic temperatures would
ensure against bottlenecking. The laser transition is at 1.1 μm.
The lifetime of the $^4T_2$ level is 2.3 msec. The absorption
bands match the emission characteristics of xenon flashlamps.
The MgF_2 host has excellent thermal characteristics and a low
nonlinear index. The MgF_2 is birefringent but it can be grown
in the appropriate orientation. The stimulated emission cross
section is approximately 10^{-20} cm^2 and saturation flux is
approximately $20J/cm^2$. A flashlamp pumped system capable of
repetition rate of several Hertz might be constructed using this
material with an overall efficiency of 10%.[27] Lasing has been
observed previously in $V:MgF_2$ in weakly doped material.[28]
Highly doped material is difficult to grow. Recently, $V:MgF_2$
in laser sized quantities has been grown[29] and are presently
being tested. If the $V:MgF_2$ laser demonstrates its potential,
it would have a significant impact on the development of large
scale solid state lasers capable of operation at high average
powers.

CONCLUSIONS

     Over the past few years, several laser systems have been
considered as possible laser fusion drivers. Recently, there
has been an increasing effort to evaluate these systems in terms
of a reactor driver application. The specifications for such a
system have become firmer and generally more restrictive.
Several of the promising candidates such as the group VI laser,
the metal vapor excimers and some solid state lasers can be
eliminated on the basis of inefficiency. New solid state
systems may impact the long range development of a fusion
driver. Of the short wavelength gas lasers, the KrF laser used
in conjunction with Raman compression and pulse stacking
techniques is the most promising approach. Efficiencies
approaching 10% may be possible with this system. While
technically feasible, these approaches are complex and costly
and are unsatisfying in an aethetic sense. A search for new
lasers with more compelling features is still needed. Based on
present understanding of the processes limiting the performance
of electronic transition lasers, this search may be
unproductive. Ultimately it may be necessary to develop
approaches to laser fusion which allow the use of presently
available laser systems.

ACKNOWLEDGMENT

While the views expressed in this article are those of the author, the material used is from a wide range of sources. The field of advanced lasers is very active and much of the most recent work is as yet published. An emphasis has been placed on data generated by members of the Advanced Laser Group at the Lawrence Livermore Laboratory. However, several other groups have been equally as active. Conversations with colleagues in the field and access to unpublished data from these sources is gratefully acknowledged.

REFERENCES

1.   Target Design, Section 3, Laser Fusion Annual Report, Lawrence Livermore Laboratory, University of California, UCRL-50021-78 (1979).

2.   D. J. Bernard, W. E. McDermott, N. R. Pchelkin and R. R. Bousek, Efficient operation of a 100 W transverse-flow oxygen iodine chemical laser, Appl. Phys. Lett. 34:40 (1979).

3.   John M. J. Madey, Stimulated emission of bremsstrahlung in a periodic magnetic field, J. Appl.Phys. 42:1906 (1971); for proceedings of a recent workshop on free electron lasers see "Novel Sources of Coherent Radiation," Physics of Quantum Electronics, Vol. 6, Stephen F. Jacobs, Murray Sargent III and Marlan O. Scully, ed., Addison Wesley, Reading, MA (1980).

4.   J. R. Murray and C. K. Rhodes, The possibility of high-energy storage lasers using the auroral and transauroral transitions of the column-VI elements, J. Appl. Phys. 47:5041 (1976).

5.   H. T. Powell, J. R. Murray and C. K. Rhodes, Laser oscillation on the green bands of XeO and KrO, Appl. Phys. Lett. 25;730 (1974).

6.   G. R. Black, R. L. Sharpless, T. G. Slanger and D. C. Lorents, Quantum yields for the production of $S(^1S)$ from OCS (1100-1700 A), J. Chem. Phys. 62:4274 (1975); G. R. Black, R. L. Sharpless and T. G. Slanger, Quantum yields for the production of $Se(^1S)$ from OCSe (1100-2000 A), J. Chem. Phys. 64:3785 (1976)

7.   H. T. Powell, private communication.

8.   James K. Rice, A. Kay Hays and Joseph R. Woodworth, VUV emissions from mixtures of F_2 and the noble gases - A molecular F_2 laser at 1575 A, Appl. Phys. Lett. 31:31 (1979).

9.   H. T. Powell and J. J. Ewing, Forbidden transition selenium
        atom photodissociation lasers, Appl. Phys. Lett.
        33:165 (1978); H. T. Powell, D. Proznitz and
        B. R. Schleicher, Sulphur $^1S_0$ - $^1D_2$ laser by OCS
        photodissociation, Appl. Phys. Lett. 34:571 (1979).

10. H. T. Powell and A. U. Hazi, Quenching of high
        concentrations of Se(^1S) produced by photoylsis,
        Chem. Phys. Lett. 59:71 (1978).

11. James K. Rice and Joseph R. Woodworth, High-intensity laser
        photolysis of OCS at 157 nm: S(^1S) production,
        photoionization, and loss, J. Appl. Phys. 50:4415
        (1979).

12. Advanced Quantum Electronics, Section 9, Laser Program
        Annual Report, Lawrence Livermore Labratory,
        University of California, UCRL-50021-75 (1975);
        C. E. Turner, Jr., Near atmospheric pressure xenon
        excimer laser, Appl. Phys. Lett. 31:659 (1977).

13. Advanced Quantum Electronics, Section 8, Laser Fusion
        Annual Report,Lawrence Livermore Laboratory,
        University of California, UCRL-50021-78 (1979).

14. C. W. Werner, E. V. George, P. W. Hoff and C. K. Rhodes,
        Radiative and Kinetic mechanisms in bound-free
        excimer lasers, IEEE J. Quantum Electron. QE13:769
        (1977).

15. Advanced Quantum Electronics, Section 7, Laser Fusion
        Annual Report, Lawrence Livermore Laboratory,
        University of California, UCRL-50021-77 (1977).

16. Ernest E. Huber, Jr., Lawrence R. Jones, Edward V. George
        and Robert M. Lerner, Sustainer enhancement of the
        vuv fluorescence in high-pressure xenon, IEEE J.
        Quantum Electron. QE12:353 (1976).

17. A. O. McCoubrey, The band fluorescence of mercury vapor,
        Phys. Rev. 93:1249 (1954).

18. S. E. Moody and R. E. Center, Optical transmission and
        dissociative recombination in high pressure mercury
        discharges in the Hg_2 excimer system, Technical
        Digest, Topical Meeting on Excimer Lasers,
        Charleston, SC, paper ThB8, Optical Society of
        America, New York (1979).

19. M. W. McGeoch, Cadmium-mercury excimer kinetics and
        transmission measurements, Technical Digest, Topical
        Meeting on Excimer Lasers, Charleston, SC, paper
        ThB7, Optical Society of America, New York (1979).

20. M. Rokni, J. A. Mangano, J. H. Jacob and J. C. Hsia, Rare
        gas fluoride lasers, IEEE J. Quantum Electron.
        QE14:464 (1978).

21. J. R. Murray, Julius Goldhar, David Eimerl and
        Abraham Szoke, Raman pulse compression of excimer
        lasers for application to laser fusion, IEEE J.
        Quantum Electron. QE15:342 (1979).

22. R. R. Jacobs, J. Goldhar, J. R. Murray, D. Eimerl and
     S. C. Brown, Scaled pulse compression experiments at
     248 nm, Technical Digest, Topical Meeting on Excimer
     Lasers, Charleston, SC, paper ThA5, Optical Society
     of America, New York (1979).
23. J. J. Ewing, R. A. Haas, James C. Swingle,
     Edward V. George and W. F. Krupke, Optical pulse
     compressor systems for laser fusion, IEEE J. Quantum
     Electron. QE15:368 (1979).
24. Lynn G. Seppala and Roger A. Haas, Passive optical pulse
     compression systems for rare gas halide fusion
     lasers, (to be published).
25. J. W. Baer, M. G. McKnights and E. P. Chicklis, XeF
     pumped Tm:YLF: An excimer excited storage laser,
     Technical Digest, Topical Meeting on Excimer Lasers,
     Charleston, SC, paper ThA1, Optical Society of
     America, New York (1979).
26. P. F. Moulton, A. Mooradian and T. B. Reed, Efficient cw
     optically pumped $Ni:MgF_2$ laser, Optics Lett. 3:164
     (1978).
27. W. F. Krupke, private communication.
28. L. F. Johnson and H. J. Guggenheim, Phonon-terminated
     coherent emission from V^{2+} ions in MgF_2, J. Appl.
     Phys. 38:4837 (1967).
29. A boule of vanadium doped magnesium fluoride has been
     grown recently by the same group and by the same
     techniques as in reference 26.

* Work performed under the auspices of the U.S. Department of
Energy by the Lawrence Livermore Laboratory under Contract
No. W-7405-Eng-48.

PLASMA HEATING BY RADIATION OF AMPLIFYING MODULE

OF A LASER SYSTEM "DELFIN"

N. G. Basov, A. A. Galichy, A. E. Danilov,
M. P. Kalashnikov, Yu. A. Mikhailov, A. V. Rode,
Yu. V. Senatsky, G. V. Sklizkov, S. I. Fedotov

P.N. Lebedev Physics Institute
Academy of Sciences of the U.S.S.R.
Moscow, U.S.S.R.

Recent progress in laser fusion research has stimulated the development of powerful multi-beam laser installations for heating of microshells. At Lebedev Physical Institute, U.S.S.R., to such laser installations belong "Delfin" (1, 2), "Kalmar" (3, 4), and "UMI-35" (5). In the U.S.A. the installations "Shiva" of Livermore Laboratory (6, 7), 2-beam and 8-beam CO_2 lasers of Los Alamos Laboratory (8), "Zeta" and "Omega" of Laser Energetics Laboratory, Rochester University (9) are in progress and put into operation. In Japan under construction are such Nd-glass lasers as "Gekko-2", "Gekko-4" and "Gekko-12", as well as "Lekko-2" and "Lekko-10" CO_2 lasers (10). Among European installations one should mention a 6-beam Nd-laser developed at Rutherford Laboratory (11), and an 8-beam Nd-glass laser "Octal" designed at Limeil (12, 13). A laser fusion installation includes, as a rule, a laser system itself with means for measuring laser beam parameters, a system for transportation and focusing of beams onto a target, a vacuum target chamber, diagnostic equipment for measuring plasma parameters and for studying the processes of laser interaction with matter, and a system for automation of laser fusion experiments. These installations are notable not only for the type of an active medium, but also for the configuration of active elements and optical schemes of amplifying stages, where the laser radiation gains its energy, and a necessary number of light beams is formed. As for the geometry of amplifying active elements, most widely spread are disc lasers ("Shiva" (6) and "Gekko-4" (10)), lasers on rod active elements ("Delfin" (1), "Kalmar" (3), "Zeta" (9), "Octal" (12)), and plate-type active elements ("UMI-35" (5)). Optical schemes of powerful amplifiers suggest three main variants for the disposition of active elements:

successive, parallel and successive-parallel. In the first case,
a signal is amplified in a chain of successive active elements of
a growing cross-section. An output laser beam is split into beams
of less aperture for forming the focusing scheme. For powerful
amplifying stages this scheme is seldom used because the structure
is complex. It is used, as a rule, in preamplifiers for laser beams
60-80 mm in diameter. The parallel disposition is used when after
preamplification, an initial laser beam is split into a necessary
number of beams, and then each of them is amplified separately.
The "Shiva" installation has 20 parallel amplification stages. The
successive-parallel amplifier scheme used in the "Delfin" setup
provides the following operation.

After preamplification an initial laser beam is split into such
a number of beams that each of them, after a subsequent amplification, becomes equivalent of the initial one. Such splittings and
amplifications are repeated until a necessary number of beams is
obtained to reach preset energy. Such a system has the following
advantages: one can use similar amplifiers in all parts of a laser
installation, which simplifies the operation of a system; besides,
a greater number of output laser beams can be obtained that may be
directed to the target. This would help to improve the symmetry
of target heating.

The laser fusion installation, "Delfin", destined for spherical
heating of targets (2) consists of four independent powerful amplifier modules. Each of them provides a formation of 54 laser beams,
45 mm in diameter, with a possibility of their grouping. The present
paper is studying the operation of an amplifying module, e.g. in
experiments on plasma heating, where the plasma is produced by laser
irradiation of thin foil and shell targets. The automation of physical processes is considered. Note that the amplifying module may
be a basic structural unit for larger laser installations.

The radiation goes to the output after passing through linear
preliminary stages of amplification (LPSA). Figure 1 shows a
scheme of LPSA with a master oscillator and a system of formation
of nanosecond pulses. A Q-switched Nd oscillator (14) is used in
"Delfin" for experiments with subnanosecond pulses. This oscillator, if compared with a conventional mode-locked laser of sub-
nanosecond pulses, possesses a number of advantages. And, first of
all, it has a high synchronization accuracy ($\tau_s \sim 10^{-9}$ sec) of laser
pulses with external recording devices and with laser installation
control apparatus. A nanosecond pulse is generated by a single-
mode master oscillator, whose streak photograph is shown in Fig. 2.
The radiation divergence is diffractional for a ϕ 2 mm light beam.
The required nanosecond duration of pulses with a high contrast
ratio is achieved by using a system of electro-optical Kerr shutters
controlled by a laser-triggered spark gap (15). The streak photograph of a radiation pulse at the LPSA input is given in Fig. 3.

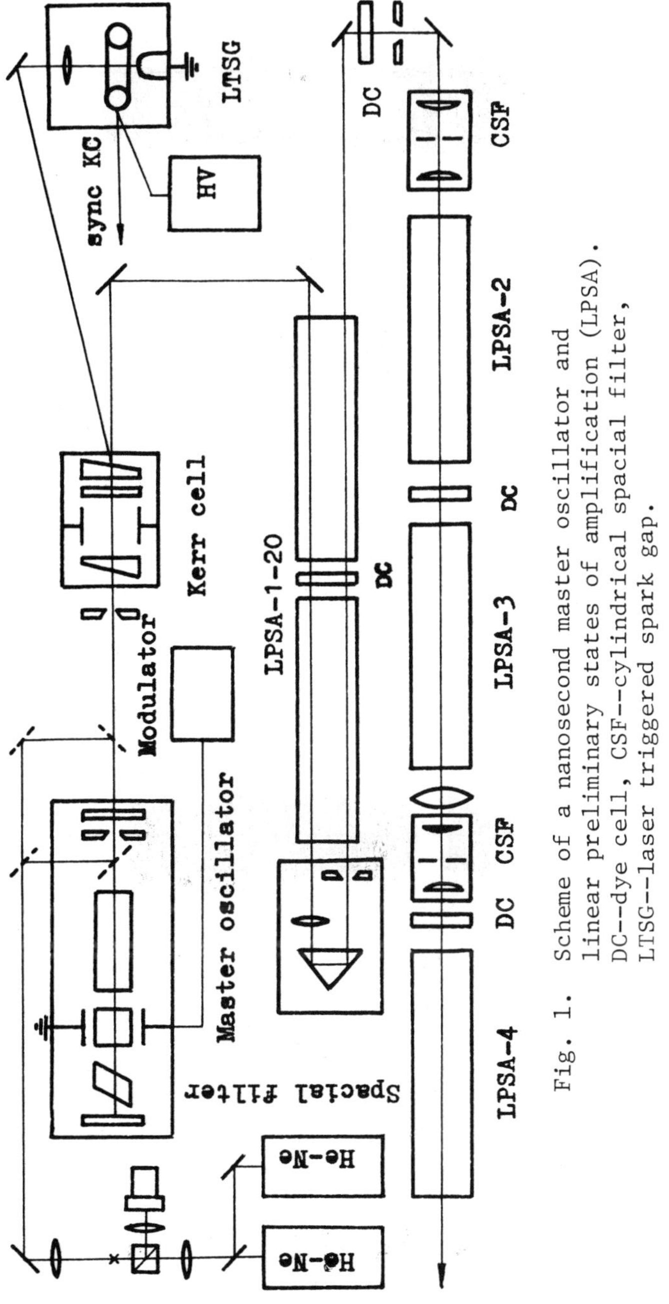

Fig. 1. Scheme of a nanosecond master oscillator and linear preliminary states of amplification (LPSA). DC--dye cell, CSF--cylindrical spacial filter, LTSG--laser triggered spark gap.

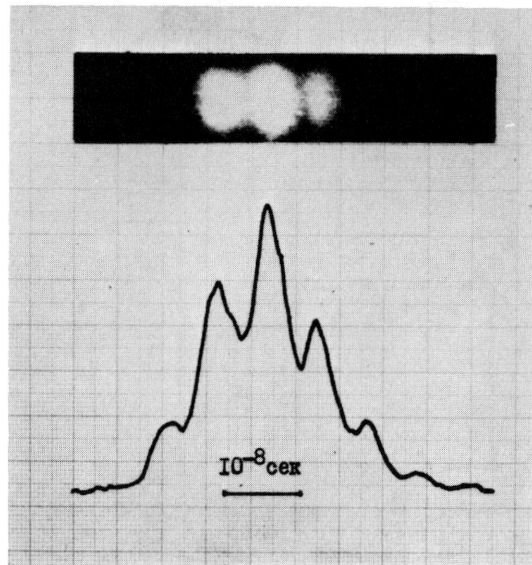

Fig. 2. Master oscillator's pulse.

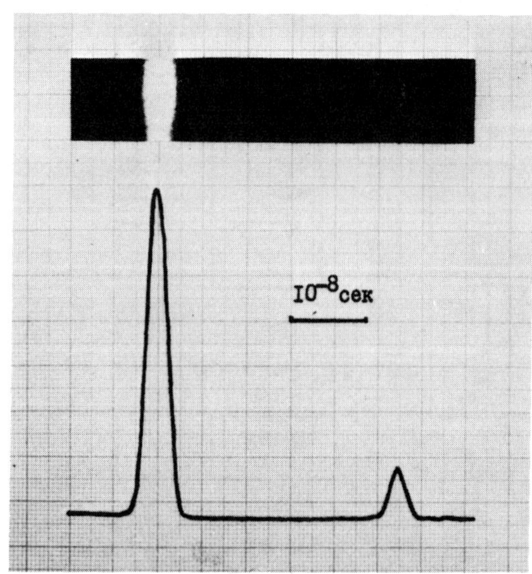

Fig. 3. Shaped pulse.

The pulse duration is $\tau_p \simeq 2.4\times10^{-9}$ sec, the energy, $E \simeq 10^{-2}$ J, the contrast ratio, $K_E \simeq 2.10^5$. A further amplification is accomplished in a five-stage amplifier, whose first two stages operate in a two-passage amplification regime. LPSA amplification stages were isolated by dye filters and spatial filters with slit and circular diaphragms located in cylindric and spherical collimators, respectively (2). At the LPSA output, a laser pulse is formed with the following parameters: the radiation energy, $E_1 \simeq 30-40$ J; pulse duration, $\tau_p \simeq 2.4$ ns; divergence at the level of 80% energy, $\alpha = 1.6\times10^{-4}$ rad; the energy contrast, $K_E \simeq 3.10^6$.

After passing the LPSA output, the laser beam is split into four parts and amplified up to the level of initial energy, and then goes to input of the amplifying module, which is shown schematically in Fig. 4. After a preliminary stage of amplification (PSA), the beam is transformed, by means of a "1 x 3" cylindric collimator, into three beams. Simultaneously, a spatial filtration of a horizontal slit is accomplished. Each of the three transformed beams, after going through dye cells and passing the powerful output stages of amplification (PFSA I) is split into two beams in "3 x 6" cylindric collimators. Then six laser beams are amplified in PFSA II, transmitted through dye cells, and transformed into 18 beams in "6 x 18" collimators. There the beams are spatially filtered in vertical slits which provides, in addition to filtration in "1 x 3"

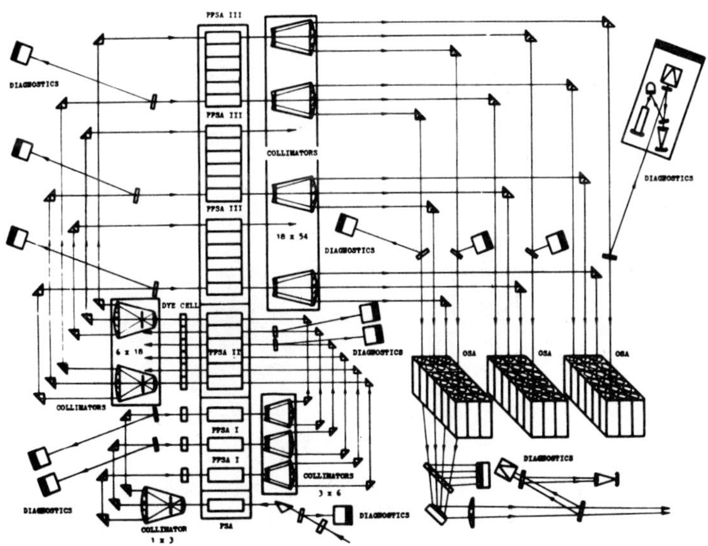

Fig. 4. Optical scheme of preliminary and output powerful stages of amplification.

collimators, a two-dimensional splitting of wavefront components corresponding to the given divergence, or higher than that. Final splitting of laser beams amplified in 18 active elements of PFSA III is accomplished in "18 x 54" collimators. By passing through a prism system of the optical path length compensation, all 54 laser beams go to three output amplification stages, where three composite beams (CB) are formed, each having 19 basic beams.

General view of an amplifying module is shown in Fig. 5. Nominal values of the laser radiation parameters at different points of the amplifying module are listed in Table I.

Table I.

Amplification stage	Number of beams	Beam cross-section cm^2	Gain coefficient	Energy, J	Flux density GW/cm^2	Brightness, $W/cm^2/$ster
PSA	1	15	2.6	45	3.0	
PFSA I	3	45	3.3	120	2.7	
PFSA II	6	90	3.4	320	3.5	2.8×10^{16}
PFSA III	18	270	3.4	880	3.3	
OSA	3x18=54	810	3.6	2500	3.1	4.10^{15}

Fig. 5. General view of the powerful module.

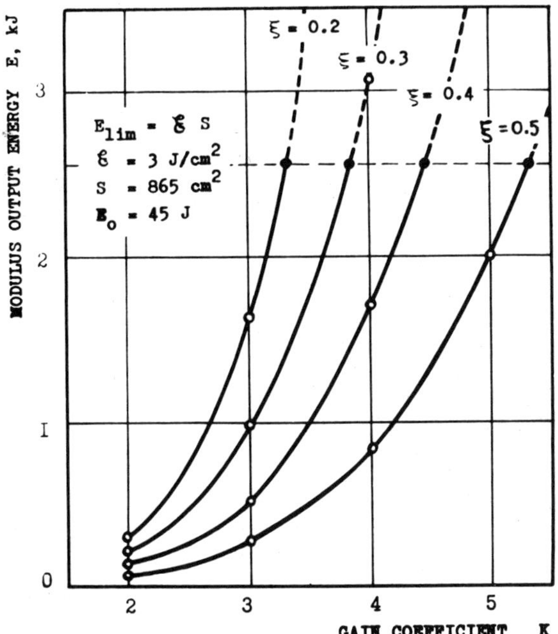

Fig. 6. Dependence of the output energy of the powerful module
upon the gain coefficient of active element.

Figure 6 shows the energy obtained at the output of amplification module versus the gain coefficient K of an active element for
different values of losses in a beam splitting system. The dotted
line (E $\simeq$ 2.7 kJ) corresponds to the ultimate-admissible radiation
energy determined by optical damage threshold of 3 J/cm^2 in a nanosecond interval (16). The output energy is E_O = 45 J. The ultimate
value of output energy at $\sim$ 20% losses is reached when the gain
coefficient is K $\simeq$ 3.3. One should note a sharp dependence of the
achieved energy at K $\geq$ 3 on the value of losses in the splitting
system.

An optical isolation of amplifying stages, and an increase in
the energy contrast of radiation, as well as in LPSA, are accomplished by means of dye cells and slit spatial filters. As dye
cells, we used two types of solutions, 1 and 2.   The curves of dye
transmittance T depending on the flux density q are plotted in Fig.
7. For the dye solution 1, the curves a and b correspond to different values of initial transmittance (2% and 20%, respectively).

The spatial filtration of radiation in amplification module is
illustrated in Figs. 8 and 9. Figure 8 shows the directivity pattern
for the superluminescent radiation of the working active element,

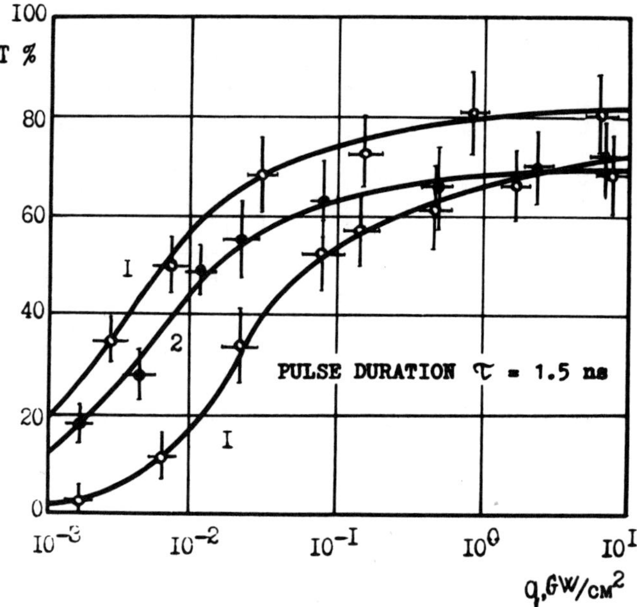

Fig. 7. Transparency of saturable filters vs. incident flux density.

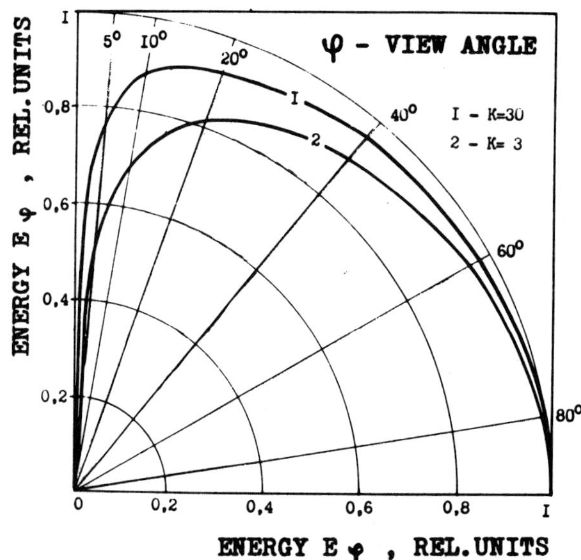

Fig. 8. Angular dependence of superluminosity for different values of an active element gain coefficient K.

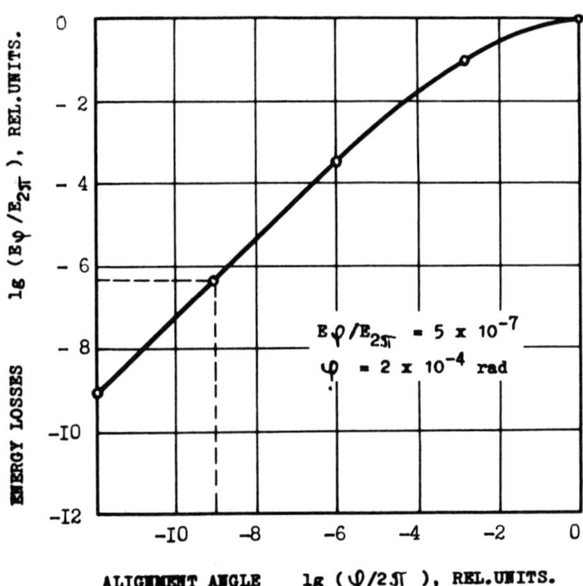

Fig. 9. Energy losses (isolation coefficient) in cylindrical
collimators vs. angle of alignment.

ϕ 45 mm $\times$ 560 mm, for homogeneous pumping. Curve 1 corresponds to
amplification of a weak signal (the gain coefficient, $K \simeq 30$);
curve 2 depicts amplification of the strong signal ($K \simeq 3$). With
growth of K the directivity pattern gets narrower. For the amplification with K = 30 about 80% of the background energy is concen-
trated in the angle $\alpha \simeq 5^{\circ}$, and for K = 3, in $\alpha \simeq 15^{\circ}$. From the
given directivity pattern it follows that the superluminescence
power concentrated in the angle $\Omega = 10^{-2}$ sterad can reach
$p \simeq 10^4 - 10^5$ W, and in the solid angle $\Omega = 10^{-4}$ sterad, p' = 1-10 W.
Thus, the cylindric uncoupling devices can diminish the superlumines-
cense background, which reduces the inverse population in active
elements, and the self-excitation probability of the amplifying system. Figure 9 shows the dependence of relative losses, i.e. the
isolation coefficient, on the angle of tuning. The angle of tuning
is determined by width of a slit placed in focus of the first collimator component.

Laser radiation parameters were measured at different points of
the amplifying module by the following diagnostic means:

(1) Calorimetric measuring of the radiation energy;

(2) Photographic devices for the registration of laser radiation field in the near zone;

(3)   Instruments for measuring the radiation divergence (registrators of radiation field in the far zone);

(4)   Coaxial photoelements for the laser pulse shape recording;

(5)   Semiconductor detectors for measuring the radiation energy contrast.

The laser beam energy was measured both in the amplifying module and at the CB output by means of a 48-channel calorimetric arrangement (17) with glass detectors of the radiation. To record the beam cross-section intensity distribution in the near zone, the beam cross-section was transmitted to the film plane by an optical system. Isodensitograms of the laser beam cross-section in CB, and the intensity distribution by the cross-section of one of the directions are shown in Fig. 10. Figure 11 illustrates the cross-section intensity distribution obtained by a photoprint method in different beams of amplifying stages. The circles denote working zones of beams. Photographs of the near zone of beams not only characterize the adjustment of the system, but make it possible to analyze the growth of perturbations with beam propagation in the amplifying system.

To measure the wavefront structure in the far zone we used special devices of the PIR-1 type (18). They helped us to examine laser radiation in caustics of a long-focus lens. As a result, in the course of the same experiment one can measure angular divergence of radiation, the wavefront curvature radius, and angular dimension of the effective source in a wide energy range of the studied beam. Figure 12 shows the intensity distribution matrix obtained in a far zone of the laser beam with the help of PIR-1. By the horizontal line are the images along the lens caustics with F = 600 mm and 3 mm step; by the vertical line the images with intensity difference of about 4. Besides, Fig. 12 shows the results of processing of the frame corresponding to the caustics minimum. The measured radiation divergence is $\alpha = (1.5 + 1.6) \times 10^{-4}$ rad (one of PSA beams).

The change in the structure of a laser beam when it transmits the amplifying stages is illustrated in Fig. 13. Due to collimator aberrations in the beam splitting system and errors in the preparation of optical surfaces (2), the divergence of radiation increases from 1.6×10^{-4} rad to 3.7×10^{-4} rad. However, the ultimate concentration of laser radiation in the target, in contrast to the possible changes in the illumination intensity distribution of the target surface is more likely to be determined by CB radiation parameters, but not by the divergence of individual beams. The distribution of CB radiation intensity in the far zone is shown in Fig. 14. At the level of 80% energy the divergence reaches α (80) $\simeq 3.9 \times 10^{-4}$ rad. One should note an asymmetry of the focal spot arising from conditions of adjustment and deviations from the symmetry for the individual beams.

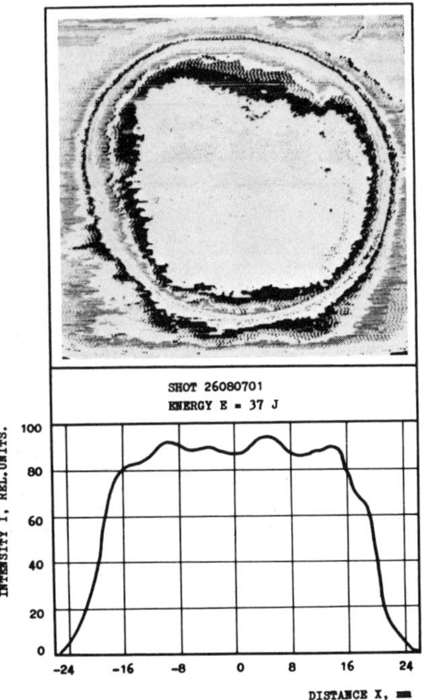

Fig. 10. Near field isodensitogram of a laser beam.

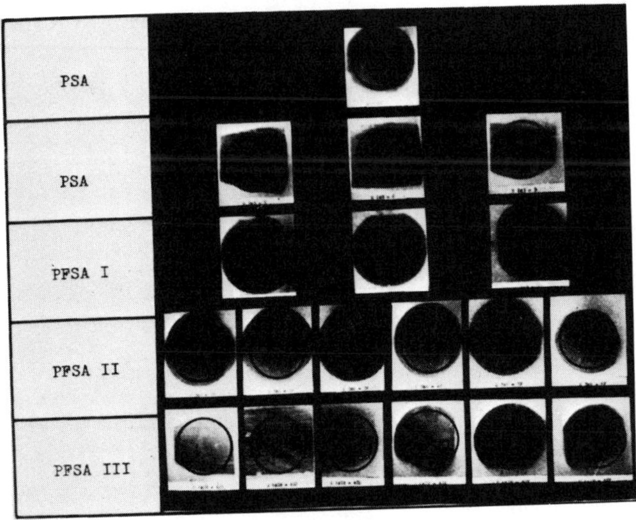

Fig. 11. Near field photographs of laser beams in different
          stages of amplification.

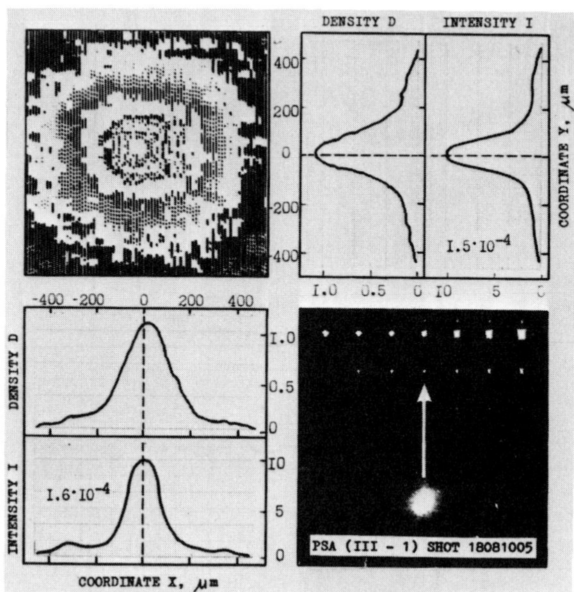

Fig. 12. Matrix of the far field laser beam intensity distribution.

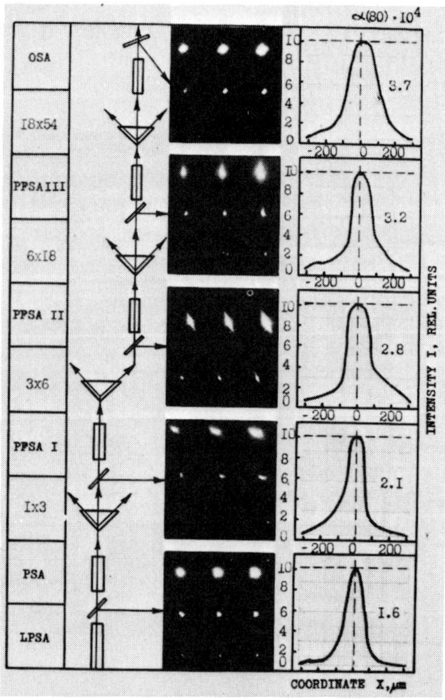

Fig. 13. Far field intensity distribution in different stages of
          amplification.

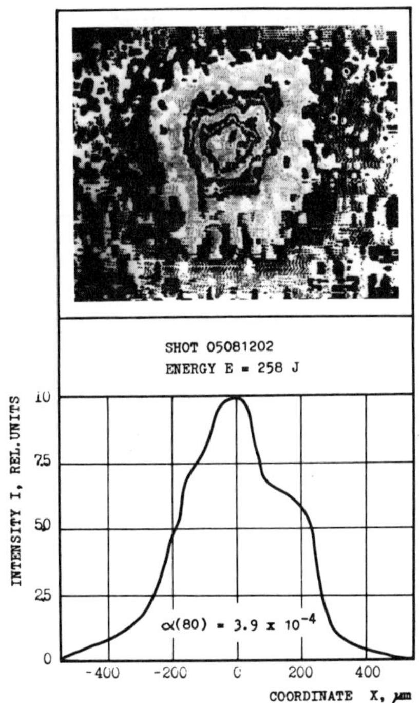

Fig. 14.   Far field intensity distribution of composed beam.

The divergence of the radiation depending on the laser operation regime is illustrated in Figs. 15 and 16. From Fig. 15 one can see a sufficient heightening of the radiation divergence when the energy output from the active amplifying element increases from 20 J to 80 J. Figure 16 illustrates the radiation divergence depending on the interval between laser outbursts.

The interval of $\Delta t \sim 20$ min corresponds to a stationary minimal value of divergence $\alpha = 1.5 \times 10^{-4}$ rad for the given operation regime (pumping energy of the amplifier is 25 kJ, the energy output 45 J). Thus, the time interval of ~ 20 min determines the ultimate operation velocity of the laser system restricted by heat exchange.

In the experiments the laser pulse shape was recorded with the help of coaxial photoelements, as shown in Fig. 3. To measure the contrast ratio of laser radiation in CB we used the procedure described in ref. 19. By applying this method, one can register, with the help of a silicon photodiode, the laser radiation intensity in different time intervals, from hundred microseconds (background of amplifying stages) to tens nanoseconds (Q-switched background), and to nanosecond units. Typical oscillograms of light pulses are given

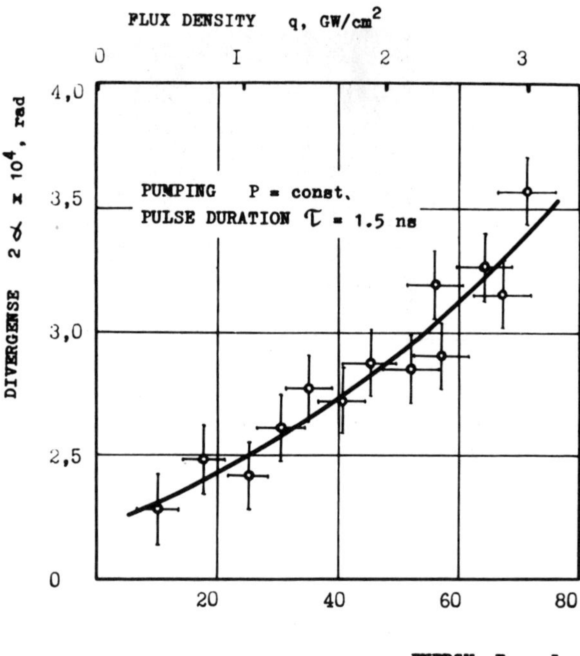

Fig. 15. Divergence of laser radiation vs. output energy of active element.

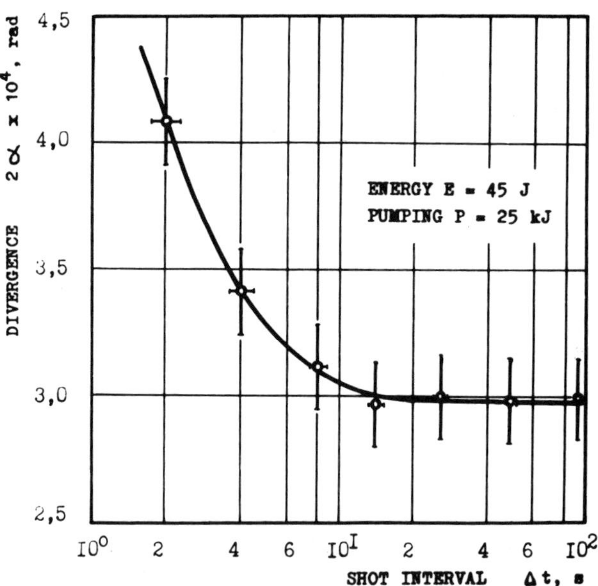

Fig. 16. Dependence of laser radiation divergence (for a single beam) on the time interval between laser shots.

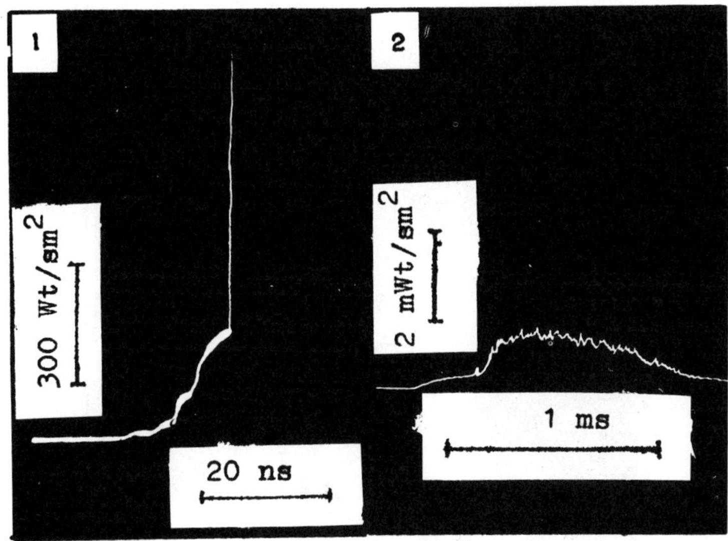

Fig. 17.   Oscillograms of laser prepulses: 1--real shot (master oscillator's prepulse), 2--superluminosity prepulse.

in Fig. 17.   The measured contrast of radiation varied from 3.10^5 to 2.10^6. The main radiation parameters of the amplifying module at the pump energy of 25 kJ in each amplifier are listed in Table II (active element is the rod 45 mm in dia.; the pumping length is 560 mm).

Table II.

Number of beams	Pulse durations sec	Radiation divergence rad	Radiation energy J	Flux density GW/cm^2	Brightness GW/cm^2/strad
18×3=54	2.4×10^{-9}	3.9×10^{-4}	1000	0.6	10^6

The laser beams 45 mm in dia. from each OSA formed CB with parallel composite beams with the help of many-prism mirrors. Then each of CB was introduced into the vacuum target chamber and focused onto the target. General view of the vacuum target chamber with a system of beam convergence and focusing, and the diagnostic devices, is given in Fig. 18.

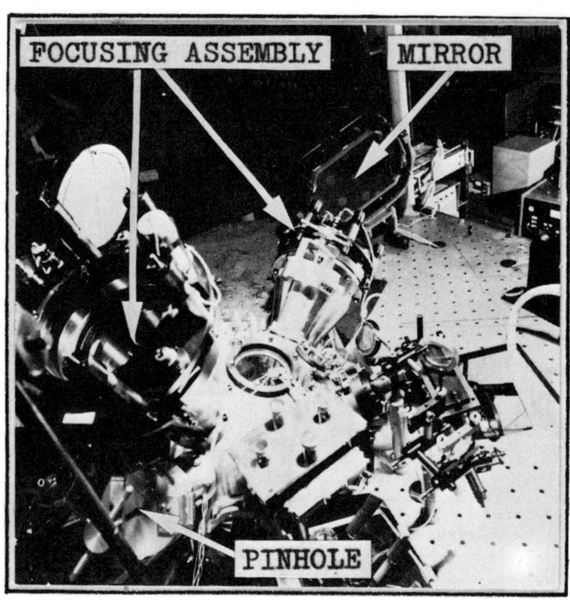

Fig. 18.   General view of the target chamber.

The focusing system of the "Delfin" installation provides the
concentration of laser radiation in the spherical target with initial
inhomogeneity of the surface illumination making 3-10% of the mean
level for the target of 300-1000 μm diameter (20).

In preliminary experiments we used a three-channel variant of
a focusing system. Each of the channels consists of a many-prism
mirror, first component of 300 mm aperture, corrector plate, turning
mirror, second aspherical component of the objective with 180 mm
aperture, and an automatic system of alignment and focusing (fig. 19).

The focusing system of "Delfin" can correct the target surface
illumination that may be performed by each of 216 laser beams. The
automatic guidance focusing system allows one to control the position
of a laser beam, or an entire channel, with the accuracy of ± 5 μm
in the plane perpendicular to the channel axis, and with ± 15 μm
accuracy along the axis (21). The automatic search and guidance
are accomplished in the range of ± 300 μm from the target center.

The focusing system is adjusted by a YAG laser. The densito-
gram of illumination distribution in one of the channels in the
target zone is shown in Fig. 20. The measurements showed that half
of the target-incident energy is contained in a 100 μm spot, and
90%, in the spot of 220 μm diameter. Unlike the measurements taken
by PIR at the laser output, the reported results allow for additional

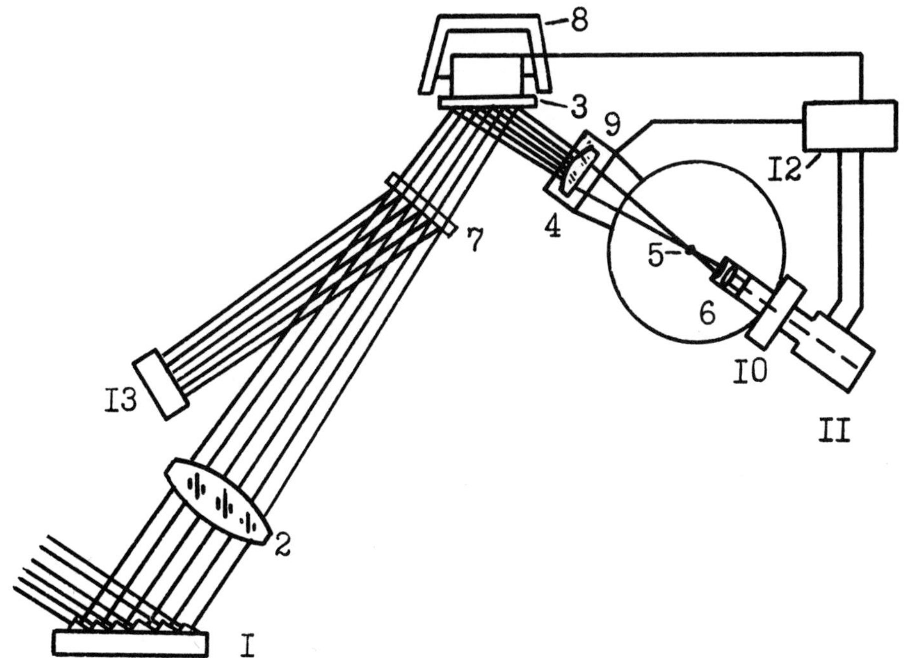

Fig. 19.   Optical scheme of a focusing channel.
1--mirror, 2--1st lens, 3--turning mirror, 4--aspherical
lens, 5--target, 6--object glass, 7--corrector, 8--mirror
setting, 9--lens setting, 10--optical block, 11--receiver,
12--electronics.

filtration of the angular and space laser intensity distribution,
when the laser radiation is transported to the vacuum chamber.

By investigating optical damage threshold of mirror coatings
of 360 x 210 mm aperture we found that the coatings sustain some
tens of shots from the laser system with the flux densities
~ 2 GW/cm^2 at their surface.

The position of the target and the test-target in the center of
a vacuum chamber was visually controlled by TV. Simultaneously,
with the help of adjusting blocks of the focusing system it provided
an accurate position of the spherical target center, of ± 5 μm. The
surface plane of thin foils was set by a control system with accuracy
of ± 10 μm.

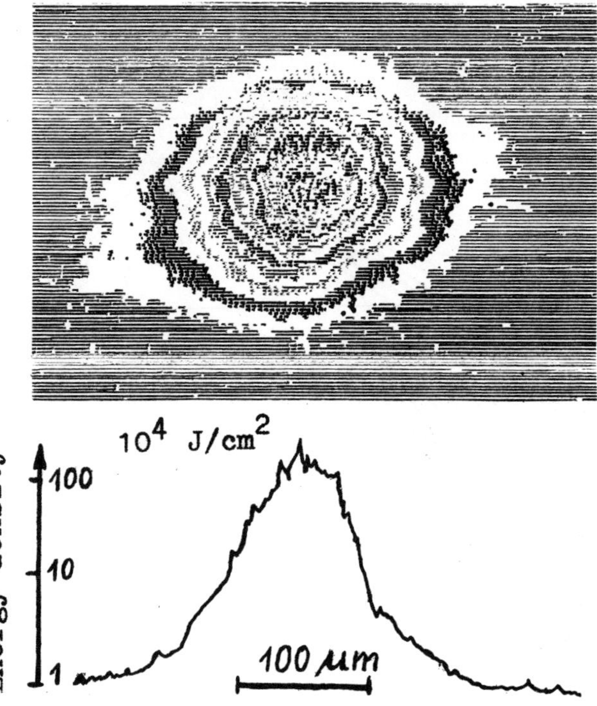

Fig. 20.   Laser beam characteristics in the target plane.

        The plasma heating experiments, we performed, aimed at study-
ing the dependence of plasma target on CB radiation parameters, and
at analyzing the efficiency of thin-wall target heating. As targets,
we used Al foil, 0.5–10 µm thick, and shell spherical targets prepared of (100–300) µm dia. SiO_2 glass with an aspect ratio equal to
$(1-8)\times10^{-2}$. Laser radiation flux density at the target surface
varied as q = $(10^{13}-10^{14})$W/cm^2. The targets were irradiated by one
CB or three CB (for the case of spherical targets only). They did
not provide spherical symmetry of the heating, but allowed one to
model some important conditions of the spherical geometry. The
diagnostics included measurements of plasma-scattered laser radiation at the basic frequency, and the measurements of plasma parameters by X-ray spectroscopic methods and by radiation of multi-
charged [H] – and [He] – like ions. The balance plasma measurements
of radiation scattering were taken by means of coaxial photoelements
calibrated by calorimetric channels. The experiments with Al foils
showed that the laser radiation is efficiently absorbed by 3 µm
foils. The reflected radiation carries about (3–5)% of the incident
pulse energy to the aperture of the focusing objective. And the
plasma-transmitted radiation reaches 10% for the thickness of 2 µm.
When the foil thickness decreases a part of transmitted radiation

grows up to (35–40)% for the thickness of 0.5 μm. The rise of losses
is determined by a correlation between laser pulse duration of 2.4 ns
and the time of plasma expansion below critical value. The amplifying module made it possible to maintain the energy contrast of
radiation at the initial level in the plasma target experiments. To
determine the heating efficiency we measured the temperature and
density of plasma from the spectra of bremsstrahlung and recombination emission (22). A continuous X-ray spectrum was registered by
multi-channel photoemulsion detectors (23) with Be filters of 100 to
2500 μm thickness, and Al filters, 50 μm to 1 mm thick. A typical
photograph of a stepwise attenuator and the results of its procession
are shown in Fig. 21. The measured "integral" plasma electron temperature was $T_e \simeq 300$ eV. When the plasma is heated by CB radiation it
is important to measure temperature distribution in the heated region,
and to determine a correlation of plasma heating with the target
surface irradiation intensity distribution. For that purpose, we
photographed plasma in the intrinsic X-ray emission by 12-channel
pinhole cameras in the spectral intervals with a long-wave cut-off
from $\lambda_1 = 5.5$Å to $\lambda_2 = 2.5$Å at space resolution $\Delta x_o \simeq 20$ μm. Characteristic pinhole photographs are given in Figs. 22 and 23. Figure
22 depicts the convergence of composite beams at one point, and in
Fig. 23 one of the beams is lead out at 1 = 300 μm from the total CB
focal spot. The procession of pinhole photographs shown in Fig. 22
gives a drop of plasma temperature from 800 eV, for the center, to
150 eV for the distance of $\sim$ 350 μm, at the average density of the
most heated part of the plasma, 8.10^{20} cm^{-3}.

Plasma heating was diagnosed from relative intensities of multi-
charged ion components (24, 25) obtained with the help of $10\bar{1}0$ plane
quartz spectrograph (interplane distance, 2d = 8.489 Å) with space
resolution $\Delta x_s \simeq 35$ μm. The quartz crystal we used was previously
calibrated with respect to reflectivity properties, which allowed us
to measure absolute intensities of plasma X-ray emission, as proposed
in ref. 26. The equidensitograms of emission lines corresponding to
most intensive and informative transitions for [H] – and [He] – like
ions of Al in the case of CB energy $E_{CB} = 350$ J and the flux density
at the plane target, $q \simeq 8.10^{13}$ W/cm^2, are shown in Fig. 24. The
same figure shows a two-dimensional plasma image in the resonance,
intercombination, and satellite lines in three cross-sections along
the dispersion direction at different distances from the target surface ($r_1 = 50$ μm, $r_2 = 250$ μm, $r_3 = 450$ μm). The images were recorded
with $\Gamma(\perp) = 2$ magnification perpendicular to the dispersion, and
$\Gamma(||) = 1$ along the dispersion (27). The spectral resolution made
$\Delta\lambda/\lambda \simeq 3.8\times10^{-4}$. The procession results are presented in Fig. 25,
as temperature and density profiles. The absolute intensities of
X-ray plasma emission measured in different experiments showed that
the resonance line of Al XII $1s2p^1P_1 - 1s^2{}^1S_o$ contains from 10^{14} to
5.10^{15} quanta, which makes 20% of energy in the continuous spectrum.
Maximal values of electron temperature in the laser absorption band
make 0.4 keV at the density of 2.10^{20} cm^{-3}. The values of

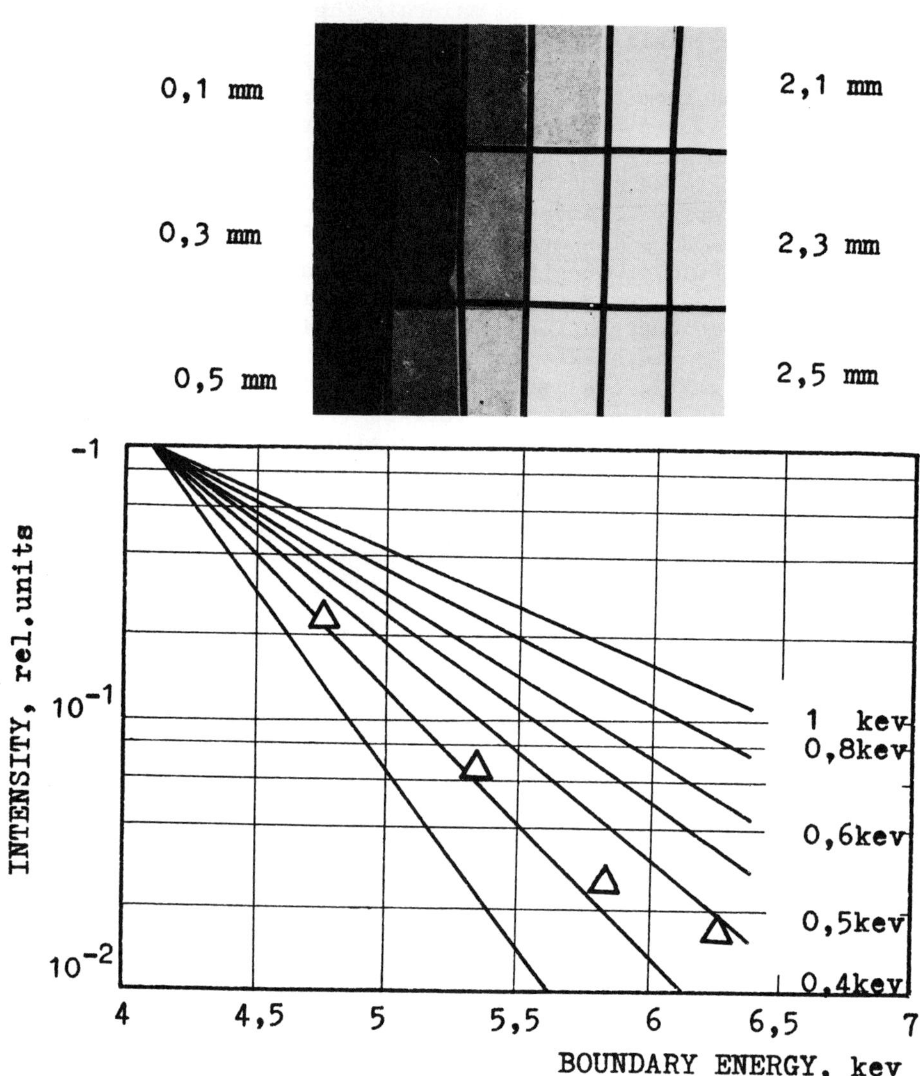

Fig. 21.   Typical photograph of film exposed by X-ray radiation through Be-filters of various thicknesses. Intensity of radiation passed through Be-filters related to intensity through 700 μm thickness filter. Experimental points correspond to electron temperature, $T_e = 0.3$ keV.

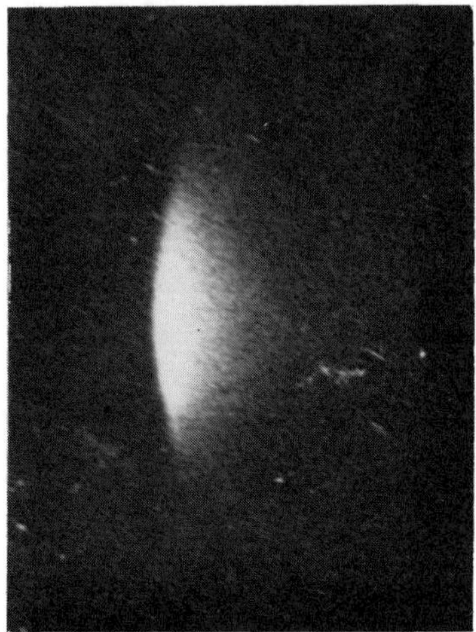

Fig. 22. Typical pinhole photograph of laser plasma (case of
a composed beam).

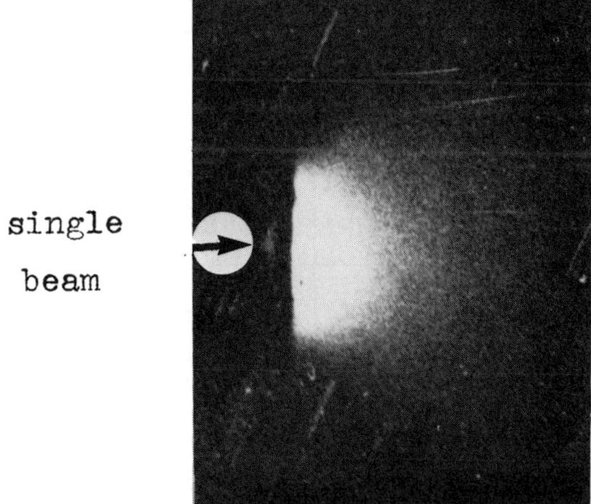

Fig. 23. Pinhole photograph of laser plasma (one beam is removed
from the cluster).

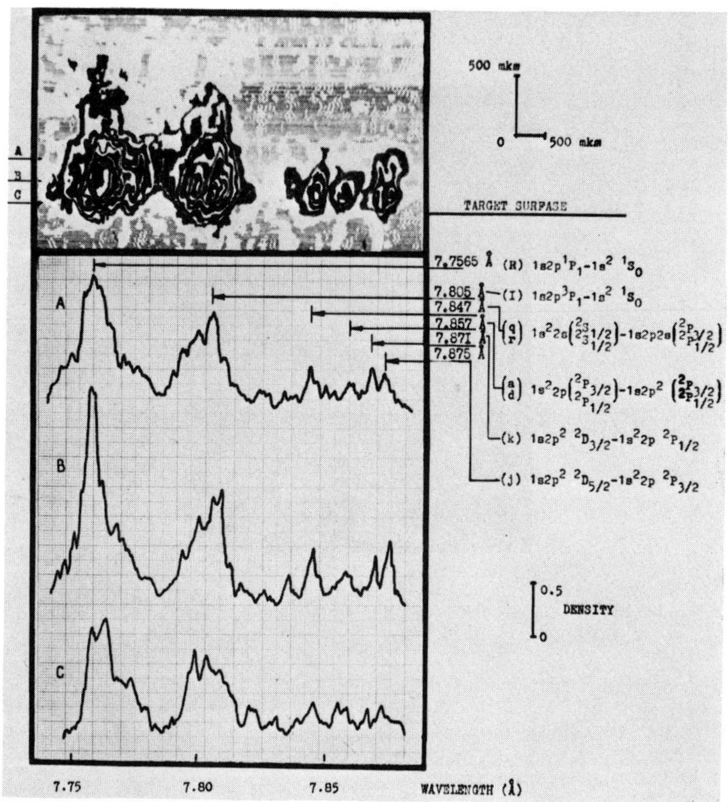

Fig. 24. X-ray radiation of the laser plasma with space resolution
(equidensitogram). Below are shown cross-sections at
the distances from the target surface:
A–450 µm, B–250 µm, C–50 µm.

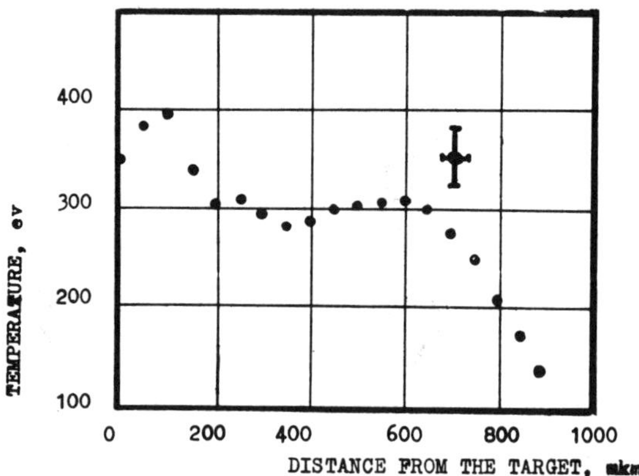

Fig. 25. Plasma electron temperature profile obtained from
x-ray spectra in Fig. 22.

temperature for the reported above heating conditions together with
balance energy measurements both for plane and spherical targets
indicate quite an effective absorption of laser radiation by CB, and
the conversion into the plasma energy.

Requirements put on a system of automation of high power
laser installations in laser fusion experiments indicate a necessity
of constructing such systems on the basis of a minicomputer network
by applying separate processors for implementation of subtasks (subsystems of automation) (28-30). The following functional group complexes of the "Delfin" installation are selected as subsystems of
automation: "feeding," power conditioning of a laser; "alignment,"
alignment of the laser installation proper, pointing and focusing of
laser radiation onto the target, target injection to the focal volume;
"measurement of laser parameters," calorimetry of laser beam, measurement of laser intensity distribution in the far and near fields, etc.;
"plasma parameters diagnostics"--a complex of automated apparatus
for laser plasma diagnostics in optical and X-ray regions of radiation; particle diagnostics and, besides, a series of auxiliary
subsystems, such as "installation readiness," "display representation
of information," etc. The project of a many-computer system of
automation in "Delfin" (29) implies the controlling and computing
complex with central computer PDP 11/70 and a minicomputer network of
PDP 11/04. Controlling and measuring CAMAC apparatus in the form
of many-crate system divided into external independent stations must
be installed in the experimental hall (28, 29).

At present the following apparatus are employed in the installation (31). In the central control room, the operating computer
PDP 11/04 with external terminals and memories is assembled (Fig. 26).
In the experimental hall, peripheral stations are assembled for collecting and preliminary processing of information and controlling
installation units. Now the work is being carried out with subsystems "feeding," "alignment" and "measurement of laser parameters."
A general idea about a peripheral station arrangement is given in
Fig. 27 showing the "feeding" subsystem with crates, interfaces,
transducers (high voltage dividers), and driving devices (commutators
of power circuits) in a power installation of one amplifying module
(54 channels). The connection of peripheral stations with the controlling computer is performed by means of a serial highway connected
with a serial crate's dataway by driver N 3992. Software for func-
tioning of terminals, serial highways and peripheral stations are
developed under the control of the operating system RSX-IIM.

One of the most complex and, at the same time, primary tasks
in automation of the laser thermonuclear experiment is the automation
of powerful laser alignment, a simultaneous pointing and focusing of
a large number of laser beams onto the target, and the automation of
a system of target injection into the focal volume (32). The solution
of these problems gives an opportunity to perform experiments with

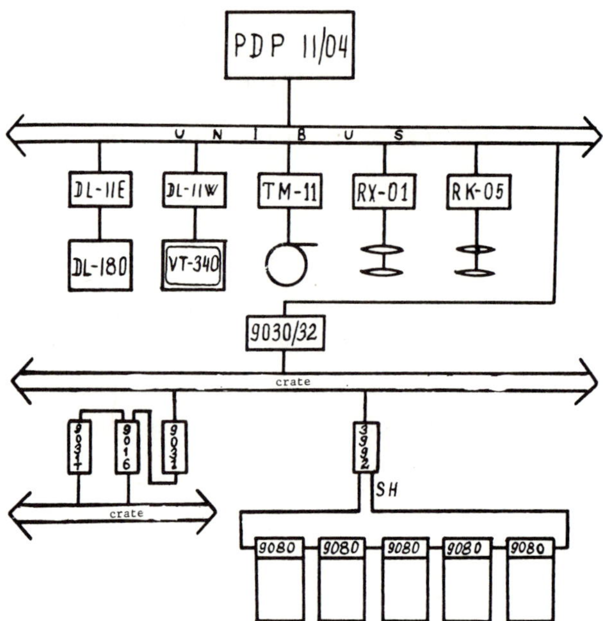

Fig. 26. Block-diagram of master hardware, serial highway and
          peripheral stations: unibus-internal highway of
          computer; DL-11E--asynchronous serial interface;
          DL-11W--asynchronous serial interface and real-time
          clock; TM-11--magnetic tape recorder; RK-05--disc drive
          and controller; RX-11--floppy disc and controller;
          9030/32--master crate controller; 3992--serial highway
          driver; 9031--branch interface; 9016--crate controller;
          9031-T--branch terminator; 9089--serial crate con-
          troller; SH--serial highway.

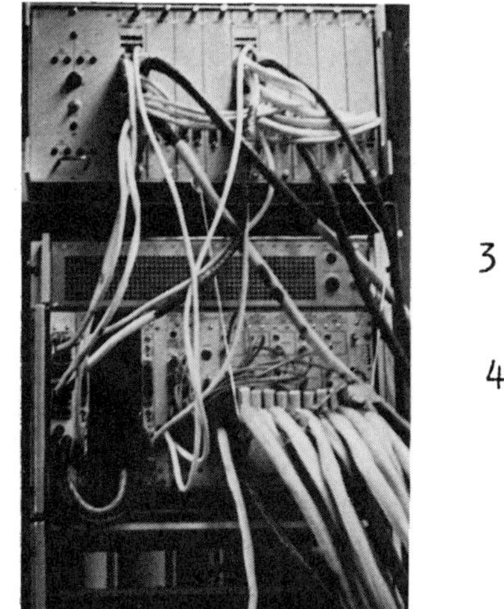

Fig. 27. The peripheral CAMAC station (subsystem "power").
1--rack; 2--block of interfaces; 3--ventilation panel;
4--crate with CAMAC modules; 5--local panel for
controlling power installation.

target heating. To realize automated alignment one should develop a
large complex of driving facilities, gears for different optical
elements, and create a complex of optical-electronical devices coupled
with a computer in order to control the alignment results (32, 33)
(Figs. 19, 28, 29).

A general view of an automatically controlled unit for target
injection into the focal volume is shown in Fig. 28. The moving head
of the mechanism where the working, or the test target, is placed
is oriented towards the center of the chamber. The target moves
along two coordinates with the help of a lever, which has a spherical support. Around this support the level may swing in two planes.
Motion along the third coordinate results from longitudinal displacement of the lever. Stepping motors are used in gears for all three
coordinates. An electronic block of control system ensures not only

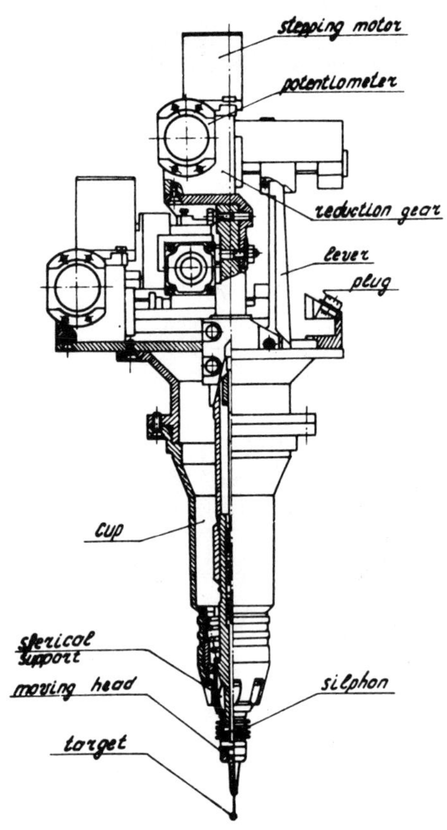

Fig. 28. Total view of the automatically controlled unit of
target injection into the focal volume.

Fig. 29. Laser mirror gimbals and alignment table of parallel
          displacement with stepping motors gears.

the target motion by photodetector signals, but performing of some
auxiliary operations, e.g. distance control. The working- and
test-target position controlling system accomplished visually by
means of a TV system and an automatical system ensures the accuracy
of positioning of spherical targets to ± 5 μm. Figure 29 presents
photographs of laser mirror gimbals with stepping motors gears in
2 angle coordinates, and alignment table with parallel displacement.
The accuracy of motion of these gimbals in the feedback loop system
with a coordinate photodetector reaches 1", and for the tables - 1 μ.

     In the nearest perspective of development of the "Delfin"
automation system, an introduction of additional processors is
planned in the low level of the system (microprocessors "Electronika-
60" for the control of separate tasks in subsystems), and in the
middle level (microcomputers PDP 11/04 and SM-4 for controlling of
the subsystem, and accomplishing intercomputer communications).
For increasing the stability of the automation system to electromagnetic noise the apparatus of the fiber optics communication lines
are under construction. These apparatus will replace the existing
cable links between the central controlling room and the peripheral
stations.

REFERENCES

1. N. G. Basov, O. N. Krokhin, Yu. A. Mikhailov, G.V. Sklizkov
    and S. I. Fedotov, "Laser Interaction and Related Plasma
    Phenomena," H. Schwarz and H. Hora, eds., Plenum Publ.
    Corp., New York (1977), Vol. 4A, 15.
2. N. G. Basov, N. E. Bykovsky, A. E. Danilov, M. P. Kalashnikov,
    O. N. Krokhin, B. V. Kruglov, Yu. A. Mikhailov, V. P.
    Osetrov, N. V. Pletnev, A. V. Rode, Yu. V. Senatsky, G. V.
    Sklizkov, S. I. Fedotov, A. N. Fedorov, Trudy FIAN, 103:3
    (1978)(in Russian).
3. N. G. Basov, O. N. Krokhin, G. V. Sklizkov, S. I. Fedotov,
    A. S. Shikanov, ZhETP, 62:203 (1972)(in Russian).
4. N. G. Basov, A. A. Kologrivov, O. N. Krokhin, A. A. Rupasov,
    G. V. Sklizkov, A. S. Shikanov, Yu. A. Zakharenkov, N. N.
    Zorev, "Laser Interaction and Related Plasma Phenomena,"
    H. Schwarz and H. Hora, eds., Plenum Publ. Corp., New York
    (1977), Vol. 4A, 479.
5. A. M. Prokhorov, S. I. Anisimov, P. P. Pashinin, UFN, 119, 3,
    401 (1976); M. E. Brodov, N. I. Gavrilov, P. I.Ivashkin,
    V. V. Korobkin, V. G. Nikolaevsky, R. V. Serov, Kvantovaya
    Elektronika, 5, 5, 1072 (1978).
6. C. F. Bender, ed., "Laser Program Annual Report-1977," Vol. 1,
    UCRL-50021-77, LLL, Univ. of Calif. (1978).
7. J. A. Glaze, "Recent Results from the Shiva Target Irradiation
    Facility." Report presented at the XII European Conf. on
    Laser Interaction with Matter (Moscow 1978). Abstract
    UCRL 81903.
8. F. Skoberne et al., "Laser Fusion Program at LASL, July 1-
    Dec. 31, 1977," LA-7328-PR Progress Report (1978).
9. Annual Report 1977 of Lab. for Laser Energetics, Vol. 1,
    Univ. of Rochester (1978).
10. Annual Progress Report on Laser Fusion Program, Sept. 1977-
    Aug. 1978, ILE-APR-78, Inst. of Laser Eng., Osaka Univ.
    (1978).
11. Annual Report to the Laser Facility Committee 1979, RL-79-036,
    Science Res. Council, Central Laser Facility, Rutherford
    Lab (1979).
12. M. Andre, J. C. Courteille, J. Y-LeGall, P. Rovati, Octal, a
    new neodymium glass laser built in the Centre d'Etudes de
    Limeil for Laser Fusion Experiments, Report presented at
    the XII European Conf. on Laser Interaction with Matter,
    Moscow (1978).
13. D. Billon, J. C. Couturand, M. Decroisette, P. A. Holstein,
    J. Launspach, M. Louis-Jacquet, C. Paton, J. L. Rocchiocioli,
    A. Saleres, D. Schirmann, Eight beams implosion experiments
    at Limeil, Report presented at XII European Conf. on Laser
    Interaction with Matter, Moscow (1978).
14. N. E. Bykovsky, N. V. Pletnev, Yu. V. Senatsky, Kvantovaya
    Elektronika, 4, 6, 1301 (1977).

15. M. P. Kalashnikov, Yu. A. Mikhailov, G. V. Sklizkov, S. I. Fedotov, A. N. Fedorov, PTE, 6, 124 (1978)(in Russian).

16. O. N. Krokhin, Yu. A. Mikhailov, G. V. Sklizkov, S.I. Fedótov, Kvantovaya Elektronika, 3, 3, 636 (1976).

17. V. N. Bochkarev, B. L. Vasin, G. V. Sklizkov, S. I. Fedotov, L. I. Shishkina, Kvantovaya Elektronika, 5, 3, 710 (1978).

18. A. D. Valuev, B. L. Vasin, B. Yu. Ivanov, N. N. Il'ichev, G. V. Sklizkov, S. I. Fedotov, Preprint FIAN SSSR N239 (1978).

19. A. E. Danilov, Yu. A. Mikhailov, G. V. Sklizkov, S. I. Fedotov, Preprint FIAN SSSR N 73 (1978).

20. A. E. Danilov, N. N. Demchenko, V. B. Rozanov, G. V. Sklizkov, S. I. Fedotov, Preprint FIAN SSSR N 97 (1976).

21. G. Dünnebier, K. Junge, S. Kusch, W. Reinecke, J. Schwieder, R. Ricker, H. Schönnagel, P. Warm, N. G. Basov, A. E. Danilov, G. V. Sklizkov, S. I. Fedotov, Automatic Focusing System of Powerful Laser Installation "Delfin," Report at XII Eur. Conf. on Laser Interaction with Matter, Moscow (1978).

22. Yu. A. Zakharenkov, N. N. Zorev, O. N. Krokhin, Yu. A. Mikhailov, A. A. Rupasov, G. V. Sklizkov, A. S. Shikanov, ZhETP, 70, 2, 547 (1976).
    O. N. Krokhin, Yu. A. Mikhailov, A. A. Rupasov, G. V. Sklizkov, A. S. Shikanov, Yu. A. Zakharenkov, N. N. Zorev, "Proc. of XII Inter. Conf. on Phenomena in Ionized Gases," Eindhaven, Netherlands, 349 (1975).

23. A. A. Kologrivov, Yu. A. Mikhailov, G. V. Sklizkov, S. I. Fedotov, A. S. Shikanov, M. P. Shpolsky, Kvantovaya Elektronika, 2, 10, 2223 (1975).

24. C. P. Bhalla, A. H. Gabriel, L. P. Presnyakov, Monthly Notices of the Royal Astronomical Society, 172, 2359 (1975).

25. A. V. Vinogradov, I. Yu. Skobelev, E. A. Yukov, Preprint FIAN SSSR N 121 (1977).

26. K. Goetz, M. P. Kalashnikov, Yu. A. Mikhailov, A. V. Rode, G. V. Sklizkov, S. I. Fedotov, E. Förster, P. Zaumseil, Preprint FIAN SSSR N 127 (1979).

27. K. Goetz, Yu. A. Mikhailov, S. A. Pikuz, G. V. Sklizkov, A. Ya. Faenov, S. I. Fedotov, E. Förster, P. Zaumseil, PTE, 3, 201 (1978)(in Russian).

28. N. G. Basov, G. V. Sklizkov, Yu. V. Senatsky, Yu. N. Ol'shevsky, S. I. Fedotov, Report N 7-17 at the World Electrotechnical Congress (Moscow, 1977)(in Russian).

29. N. G. Basov, O. N. Krokhin, A. V. Kutsenko, G. V. Sklizkov, L. K. Subbotin, S. I. Fedotov, Trudy FIAN, 112, 13 (1979).

30. P. E. Coyle (ed.), "Laser Program Annual Report-1976," UCRL-50021-76, LLL, Univ. of Calif. (1977).

31. A. P. Allin, Yu. V. Senatsky, G. V. Sklizkov, L. K. Subbotin, A. K. Yakushev, Preprint FIAN SSSR N 156 (1979).

32. N. G. Basov, V. R. Belyan, G. P. Zhilkin, V. S. Zvezdin, V. S.
     Krasovsky, O. N. Krokhin, B. V. Kruglov, A. V. Kutsenko,
     Yu. N. Ol'shevsky, N. V. Pletnev, Yu. V. Senatsky, G. V.
     Sklizkov, S. I. Fedotov, N. N. Sheremetievsky, Trudy FIAN,
     103, 52 (1978).
33. T. V. Koreshkova, N. Pletnev, Yu. Senatsky, G. Sklizkov, L.
     Subbotin, B. Shpilevoi, A. Yuzhakov, A. Yakushev, Preprint
     FIAN SSSR N 64 (1979).

CLASSICAL AND QUANTUM-MECHANICAL VIEWS OF FREE ELECTRON LASERS[*]

Helmut Schwarz[**]

Department of Physics
Rensselaer Polytechnic Institute
Troy, New York 12181

ABSTRACT

About thirty five years ago it had been demonstrated experimentally that a relativistic electron beam properly accelerated and decelerated periodically in a spatially periodic magnetic field could produce radiation in the optical region. But that it is also possible to feed this radiation back into the electron beam had been shown only about ten years ago. It took approximately another five years to combine the two effects and build a free electron laser.

A survey of several concepts of electron interactions with "structures", consisting of spatially periodic magnetic and electric fields, of nonlinear force fields, even of geometrical periodic structures (diffraction gratings) or crystalline ionic structure (channeling radiation) or even of electron wave interference structures will be given.

The light generation based on most of these principles are mainly due to classical electromagnetic theory. However, the

[*]Partially supported by NASA Grant No. NSG 1628.

[**]Also with Departamento de Física, Universidade de Brasília, 70.910 Brasília, DF, Brazil.

production of light from the beating of two electron beams creating
an electron wave interference structure can only be explained quantum-
mechanically.

The two coherent electron waves should have an energy difference
ΔE of a few electron volts to bring the beating frequency $\omega = \Delta E/\hbar$
into the optical region. The electron beams do not have to be relativistic, thus avoiding large linear accelerators. The beating phenomenon that has been demonstrated experimentally with electron
beams converts the very short wave lengths (fractions of Ångstrom)
of the electron beams into optical wave lengths. Electron scattering at the electron wave interference structure and feedback of the
resulting radiation will turn this device in a wave-mechanical free
electron laser.

INTRODUCTION

A free electron laser (FEL) is an electron oscillator and a
feedback amplifier working at optical frequencies. Conventional
lasers (solid state or gas laser) operate, in principle, in the same
way, except that the electrons here are not free but bound to atoms
and molecules or in a crystalline ionic structure; they can have
only specific wave lengths whereas in a FEL the output wave length
could be tuned continuously, in principle, from the infrared up to
the x-ray region. Another advantage of the FEL over conventional
lasers is the fact that light is generated in free space due to
electron oscillations; there is no material, other than electrons,
that could limit the maximum intensity because of material breakdown due to absorption by impurities or self focusing leading to
high electric fields.

Calculations have shown that the overal efficiency of FELs
would surpass that of CO_2 lasers.

FELs can be based on different schemes of electron interactions
with "structures".

The structure could be realized by a periodic electric (G.
Bekefi) or periodic magnetic (H. Motz) field "wiggler" or both.

Even a high intensity laser beam itself could constitute such a
structure for an interacting electron beam, if the energy of the
injected electron beam corresponds to the oscillation energy originating from the nonlinear force of the laser field (H. Hora); under
such condition light amplification will occur provided that the
interacting electron and laser pulses have the same duration.

Also a metallic diffraction grating could serve as an interaction structure whereby an electron beam traverses at grazing

incidence over the grating normal to the grooves (Smith-Purcell-Salisbury effect).

Recently the periodic crystal structure has been used to produce coherent x-rays by channeling electrons and positrons through a well aligned and perfect crystal (M. A. Kumakov).

Also two coherent electron beams (waves) of slightly different energy interfering with each other, would produce such a periodic structure due to quantum "beats" (H. Schwarz).

HISTORICAL REMARKS

Astounding is the fact that up till now approximately 200 papers have been written on FEL, of which there are only a couple dealing with experimental results. E. T. Jaynes[1] has given an extensive survey on the early history of light originating from high energy accelerators. He reports that already in the early 1940's J. Schwinger had made many calculations of radiation from betatrons and synchrotrons, which were only published in part[2] later. The results were somewhat surprising in as much as the frequency turned out to be rather high, reaching a bluish-white light at 80 MeV as also confirmed in 1948 experimentally by Elder, Langmuir, and Pollock[3] with a synchrotron. Such radiation is emitted in pulses of 2×10^{-16}s duration[1], three orders of magnitude shorter than recent mode-locked lasers can produce.

The real birthday of the production of light due to free electrons originating from the modulation of the electron orbits while interacting with an electromagnetic wave of phase velocity equal to or exceeding that of light appears to be indicated in a 1947 patent by Gorn[4], whereas that of light generation due to free electrons not originating from high energy electron accelerators seems to be October 26, 1949 when W. W. Salisbury[5] applied for a patent entitled "Super High-Frequency Electromagnetic Wave Generator". Salisbury proposed the production of "super high" frequency electromagnetic waves (light) of the order of a millimeter or less in wavelength and, in particular, monochromatic light by having an electron beam interact with an optical diffraction grating using a "wave reflector" placed at a small angle above the grating; however, this wave reflector was eliminated by Salisbury in his later experiments.[6] In 1953 independently S. J. Smith and E. M. Purcell[7] reported experiments using a grating of approximately 8000 grooves per cm and an electron beam of 320 kV resulting in light of 5000 Å. Soon after the construction of the first ruby laser, proposals came up to use the Smith-Purcell-Salisbury effect (SPS-effect) for the generation and amplification of coherent radiation, i.e., a free electron laser. For example, H. H. Klinger[8] suggested to consider the SPS-effect as an enery transfer between a backward traveling wave and electrons moving across the grating regarding the device as a slow wave transmission line of periodic structure in the optical region. Since the wave and the group velocities of a backward traveling wave are opposite, an internal feedback occurs and leads to a self-excitation of light waves. Klinger points out that for amplification the light need not be coherent. Such a light wave backward oscillator and traveling wave amplifier had already been described by Klinger[9] in 1954. Also Fox and Smith proposed in 1963 the generation of laser beat frequency radiation in the far infrared by using the SPS effect. G. W. Stroke[11] in his 1967 survey has reported the application of the SPS effect to build coherent radiation sources from microwaves to optical waves and to add mirrors to introduce feedback into the system. The free electron laser had thus been conceived long before the recent work on FEL.

In the 1940's the radiation from accelerators was considered a nuisance and was fortunately not coherent, i.e., the coherence length Δs_c of the electron beam was smaller than the wave length λ of the radiation ($\Delta s_c < \lambda$). Δs_c for the non-relativistic case, $[(eU/m_0c^2) \ll 1]$ is given by

$$\Delta s_c = 2U\lambda_e/\Delta U = 2.453 \times 10^{-9} \sqrt{U}/\Delta U \tag{1}$$

U – average electron energy [in V]; ΔU – energy spread [in V]; λ_e – electron wave length as given by

$$\lambda_e = h/(2em_0U)^{\frac{1}{2}} = 1.226 \times 10^{-9} U^{-\frac{1}{2}} \tag{2}$$

h – Planck's constant; m_0 – electron rest mass; e – its charge; c – speed of light in vacuum[12].

For the relativistic case, Eqs. (1) and (2) read

$$\Delta s_c = \frac{ch}{e\Delta U} (1-\frac{1}{\gamma^2})^{\frac{1}{2}} = 1.240 \times 10^{-6} (1-\frac{1}{\gamma^2})^{\frac{1}{2}}/\Delta U \tag{1'}$$

$$\lambda_e = \frac{h}{m_0c} (\gamma^2-1)^{-\frac{1}{2}} = 2.426 \times 10^{-12}(\gamma^2-1)^{-\frac{1}{2}} \tag{2'}$$

with
$$\gamma = 1 + (eU/m_0c^2) = 1 + 1.957 \times 10^{-6}U$$

For $\Delta s_c \ll \lambda$, the rate of energy loss of N electrons due to radiation increases linearly with N; whereas for $\Delta s_c \gg \lambda$, it increases with N^2.

The latter case is undesirable for a synchotron, but of great value for a free electron laser. Jaynes' prediction of light generation from a Linac lead[1] H. Motz[13] in 1950 to surround the electron beam with a spatially periodic magnetic field, a "wiggler."

Motz' idea had been inspired[14] by R. Kompfner's[15] traveling wave tube: due to the helix, the wave is slowed down to near synchronism with the electron velocity. Based on this model H. Motz doubts[16] whether the FEL should really be called a laser, since in a FEL there is not a population inversion by pumping. A FEL generates synchronously emitted synchroton radiation by tuning the light frequency, thus obtaining synchronism with electrons of an "inverted" part of the electron energy distribution.[16] This theory has recently been expanded by N. M. Kroll.[17]

Two years after his theoretical paper, in 1952, H. Motz et al.[18] could report experimental results detecting light around 4000 Å. Thus, it was shown that the first part of a FEL, viz., the generation of coherent light from free electrons is possible; but the second part, i.e., the feedback of this radiation for amplification had to await its realization, although its possibility had been predicted theoretically for example by J. Schneider[19] in 1959, H. H. Klinger[8,9], A. J. Fox et al.[20], R. H. Pantell[21], and others. However, the first experimental evidence that such a feedback of the generated electromagnetic radiation of optical frequency could occur, was given in 1969 by H. Schwarz and H. Hora[22,23]. The Schwarz-Hora effect (SH effect) as Pantell said[24] "generated interest in the possibility of extending coherent interactions between electro-magnetic waves and electron beams into the optical portion of the spectrum" (verbatim quotation from Ref. 24). The extensive theore-tical work[25] has certainly stimulated the efforts of generating and amplifying coherent light from free electrons, even though the SH effect has been termed "controversial" because of the great experi-mental difficulties which those encountered who tried to reproduce the effect. Most theoreticians are convinced that the SH effect exists[1], as quite a number of theoreticians "were able, without the slightest difficulty, to give quantum-mechanical calculations which predicted virtually everything that Schwarz and Hora reported seeing. It does not speak well for quantum theory if it can so glibly account for nonexistent effects. Anyone who believes that this has happened can hardly avoid asking how many other nonexistent effects it has been predicting all these years; and how much we have been misled in our picture of Nature's workings, as a result." (Verbatim quota-tion from Ref. 1). E. T. Jaynes closes his paper[1] by saying: "It does seem to me that definite proof of its nonexistence would be a considerable embarrassment to quantum theory (or at least to the quantum theorists who found it so easy to account for)."

Although the original Schwarz-Hora experiment has not been re-produced because the few attempts[26-28] to reproduce the experiment lacked the necessary conditions of an extremely good monochromatic-ity, i.e., coherence of the electron beam, the first realization of amplification of radiation from a CO_2 laser[29] was accomplished in 1976 and the first free electron laser at a wave length of 3.4 μm was operated by D.A.G. Deacon et al.[30] in 1977 based on theoretical considerations published by one of the authors[31] of Ref. 30. Since 1977 up to the beginning of 1980 approximately 200 papers have been published on FELs, almost all theoretical; since then no experiments have been reported on free electron lasers at wave lengths shorter than that of the first FEL.[30]

FEL FROM RELATIVISTIC ELECTRON BEAM

C. A. Brau[32] describes in these proceedings almost exclusively the FEL based on relativistic electrons moving in a spatially periodic magnetic "wiggler" field (see Fig. 1 p.   ) which is basically the setup that had been used by H. Motz[18] in 1952 and converted into a FEL by D.A.G. Deacon et al.[30] in 1977 by adding mirrors for feedback and amplification. At low electron densities this feedback of the generated light is based on stimulated Compton scattering whereas at higher electron densities it occurs due to stimulated Raman scattering. The system will only oscillate and amplify these oscillations if the feedback occurs at the right time and the right phase. This means that there are strict coherence conditions to be obeyed.

Stimulated Compton Scattering Region

The wavelength λ of the stimulated radiation is mainly determined by the spatial period λ_q of the magnetic field and the energy E of the electrons accelerated by a voltage U:

$$E = \gamma_o m_o c^2 = eU + m_o c^2 \tag{3}$$

and expressed in terms of

$$\gamma_o = (1 - \frac{v_o^2}{c^2})^{-\frac{1}{2}} = 1 + \frac{eU}{m_o c^2} \tag{4}$$

v_o being the speed of the electrons.

The "wiggler" field as drawn schematically in Fig. 1 of Ref. 32 is circularly polarized and can be represented by

$$\vec{B}_s = B_o [\vec{i} \cos(k_q z) + \vec{j} \sin(k_q z)] \tag{5}$$

whereby z lies in the direction of the axis, i.e., of the electron beam and $k_q = 2\pi/\lambda_q$ is the propagation number of the spatially periodic magnetic field. In all theories it is assumed that B_o is a constant which, as A. Bambini et al.[33] pointed out, can only be valid over a limited region because the field must satisfy the relations

$$\nabla \cdot \vec{B} = 0 \quad \text{and} \quad \nabla \times \vec{B} = 0 \tag{6}$$

However, it can be shown[34] that this assumption of constant B_o is only valid if the electron trajectories remain in a narrow region around the wiggler axis with dimensions much smaller than λ_q. The wavelength of the radiation stimulated by electron scattering is then given[31] by

$$\lambda = \frac{\lambda_q}{2\gamma_o^2} \left(1 + \frac{\epsilon_o \lambda_q^2 r_o B_o^2}{\pi m_o}\right) \tag{7}$$

$$= \frac{\lambda_q}{2\gamma_o^2} (1 + 8.72 \times 10^3 \lambda_q^2 B_o^2) \tag{7a}$$

whereby ε_o = 8.854 x 10^{-12} F/m - permittivity in free space and r_o = = $e^2/(4\pi\varepsilon_o m_o c^2)$ = 2.818 x 10^{-15} m - the classical electron radius.

The gain G in radiation at wavelength λ can be derived, for example, from the solution of the classical Lorentz equation of motion for the relativistic case as this had been done by H. Schwarz and R. Tabensky[35] by adding the spatially periodic magnetic field $\vec{B}_s$ to the magnetic field $\vec{B}$ of the resulting electromagnetic radiation. One arrives then[36-38] at a gain G(t) as a function of interaction duration t defined by the fractional change of radiation energy

$$G(t) = \frac{e^4 B_o^2 n_e \lambda_q}{\pi\varepsilon_o c (\gamma_s m_o \Delta\omega)^3} [1 - \cos(t\Delta\omega) - \tfrac{1}{2}t\Delta\omega\sin(t\Delta\omega)] \tag{8}$$

whereby

$$\Delta\omega = 2\pi \left(\frac{v_o}{\lambda_q} - \frac{c-v_o}{\lambda}\right) \tag{9}$$

is the resonance condition, being the difference between the frequency of acceleration of the electrons due to the spatially periodic magnetic field $\vec{B}_s$ and the frequency at which the simultaneously emitted radiation travels over the electron beam, n_e is the electron density within the electron beam and

$$\gamma_s^2 = \gamma_o^2 \left(1 + \frac{\varepsilon_o \lambda_q^2 r_o B_o^2}{\pi m_o}\right) \tag{10}$$

$$= 2\gamma_o^4 \lambda/\lambda_q \tag{10a}$$

It is interesting to note from Eqs. (9) and (8) that at resonance, $\Delta\omega$ = 0, there will be no gain. In a magnetic field of interaction length L (length of wiggler) the final gain during one passage, which will last t = L/v_o, results from Eq. (8) by introducing a dimensionless resonance parameter $\Omega = \Delta\omega L/v_o$

$$G(\Omega) = \frac{e^4 B_o^2 n_e \lambda_q L^3}{\pi\varepsilon_o c (v_o \Omega \gamma_s m_o)^3} (1 - \cos\Omega - \tfrac{1}{2}\Omega\sin\Omega) \tag{11}$$

The gain function $G(\Omega)$ as represented in Eq. (11) is plotted in Fig.1. One can see that the gain has a maximum for Ω = 2.6062 as indicated also by W. B. Colson[37]. Maximum gain is therefore to be expected, if λ_q and U (the acceleration voltage for the electrons) are chosen in such a way that they fulfill the following two conditions

$$\left(1 + \frac{eU}{m_o c^2}\right)^2 \left\{1 - \left[\lambda\left(\frac{1}{\lambda_q} - \frac{2.6062}{2\pi L}\right) + 1\right]^{-2}\right\} = 1 \tag{12}$$

or

$$(1 + 1.957 \times 10^{-6}U)^2 \{1-[\lambda(\frac{1}{\lambda_q} - \frac{0.4148}{L}) + 1]^{-2}\} = 1 \qquad (12a)$$

and from Eq. (7)

$$2(1 + \frac{eU}{m_o c^2})^2 \lambda = \lambda_q (1 + \frac{\varepsilon_o \lambda_q^2 r_o B_o^2}{\pi m_o}) \qquad (13)$$

or $2(1 + 1.957 \times 10^{-6}U)^2 = \lambda_q (1 + 8.719 \times 10^3 \lambda_q^2 B_o^2) \qquad (13a)$

Figure 1 will give the actual G as a function of the resonance parameter

$$\Omega = 2\pi L(\frac{1}{\lambda_q} + \frac{1-c/v_o}{\lambda}) \qquad (14)$$

if one multiplies the ordinate $f(\Omega)$ with the dimensionless normaliza-
tion factor

$$W = \frac{e^4 B_o^2 n_e \lambda_q L^3}{\pi \varepsilon_o c^4 m_o^3} \{[(1 + \frac{eU}{m_o c^2})^2 - 1][1 + \frac{\varepsilon_o \lambda_q^2 r_o B_o^2}{\pi m_o}]\}^{-3/2} \qquad (15)$$

or

$$W = 3.879 \times 10^{-9} n_e \lambda_q B_o^2 L^3 \{[(1 + 1.957 \times 10^{-6}U)^2 - 1] \times$$

$$\times [1 + 8.719 \times 10^3 \lambda_q^2 B_o^2]\}^{-3/2} \qquad (15a)$$

An interesting classical method to calculate the electrodynamics
of the FEL in the low density stimulated Compton scattering region
has been developed by A. Bambini and A. Renieri[39] by treating the
relativistic movement of the electrons nonrelativistically. This is
accomplished by formulating the equation of movement in a frame that
moves with the fast electrons. Hereby the spatially periodic magnetic field acts on the relativistic electron beam as a radiation
field with wavelength $2\lambda_q$, a "pseudo-radiation" field, according to
W. Heitler[40]. In such a frame[41] the laser radiation field and the
wiggler pseudo-radiation field have the same frequency ω and opposite
wave vectors of equal magnitude k. A more extensive description of
this model is given by A. Bambini, A. Renieri, and S. Stenholm[42].
Hereby singularities in the momentum distribution and their evolution
with intensity and saturation enhancement of the gain were found which
are new features not found in other theoretical studies. Finally A.
Bambini et al.[43] have successfully tried to unify the FEL theories
showing that the quantum mechanical picture[41] as well as the classical picture can well describe the mechanism of a FEL. A. Bambini et
al.[44] have also derived, in the moving frame, equations for pulse
propagation in a FEL which are better suited for the generalization
of single mode theories to multimode operation in order to include
all oscillating modes rather than a Fourier analysis as had been done
by G. Dattoli et al.[45]. A multimode theory for the relativistic FEL
is necessary, since the FEL from an electron beam interacting with a

long spatially periodic magnetic field produces a broad active band
and due to the long laser cavity (12.7 m long)[30] the number of running
modes was large, 10^4 to 10^5.

There have been several proposals to increase the gain of the
relativistic FEL. One is to use a nonuniform, or "tapered" wiggler
as described by C. A. Brau[32,46].

N. M. Kroll and M. N. Rosenbluth have suggested[47] that the gain
could be enhanced by continually decreasing the amplitude B_o of the
spatially periodic magnetic field while keeping its period constant.
Whereas P. Sprangle et al.[47,48] claim that they could reach a single-
pass efficiency[48] greater than 20% by doing the opposite, viz. decreasing the period and increasing the amplitude B_o of the spatially periodic magnetic field. An extensive study of the effect of varying
these parameters B_o and λ_q on the efficiency has been carried out by
P. Sprangle et al.[48], A. T. Lin and J. M. Dawson[49] as well as T. Kwan
and J. M. Dawson[50] suggest to step up the wiggler field intensity
just before the space charge wave saturates; here the conversion
efficiency could be doubled, up to 25%. Along with the efficiency
improvement, the bandwidth could become narrower and the device
itself shorter.

Another proposal published by W. J. Cocke[51] introduces an index
of refraction into the single-mode analysis of the relativistic FEL
with a spatially periodic magnetic field. As one can see from the
gain equation, Eq. 8, decreasing γ_o will increase the gain. If this
decrease is accomplished without changing the other characteristics
of the FEL, a gain increase of one to two orders of magnitude could
be achieved assuming the insertion of a material, for example argon
with a refractive index of only n ≈ 1.0006

$$\delta = 1 - n^{-1} = 6 \times 10^{-4}.$$

which results from the Weizsäcker-Williams approximation

$$\gamma_{o_n}^{-2} = [(\lambda_q/\lambda)^2(\delta-\tfrac{1}{2}\delta^2) + 2(\lambda_q/\lambda)(1-\delta)]/[(\lambda_q/\lambda)+1]^2 \tag{16}$$

which for $\lambda_q/\lambda \gg 1$ and $\delta \ll 1$ simplifies to

$$\gamma_{o_n}^{-2} \simeq \delta + 2\lambda/\lambda_q \tag{16a}$$

One should arrive at more precise results applying the procedure for
the solutions of the laser accelerated charged particles in a medium
of refractive index n as given by H. Schwarz and R. Tabensky[35] after
having added the spatially periodic magnetic $\vec{B}_s$ to the radiation
field. However, the introduction of a gas would broaden the electron
distribution function. This would introduce stricter requirements

on the energy spread of the applied electron beam. For longer wave
lengths FEL (λ > 100 μm) a dielectric wave guide[51] could be used to
increase the gain.

FEL in a uniform magnetic field. W. B. Colson and S. K.
Ride[52,53] have pointed out that laser action can even be expected
from an electron beam passing through a uniform magnetic field, but
only if γ_o >> 1 (i.e., the electrons have to move at relativistic
velocities in relation to the magnetic field) and if the electrons
are injected into the magnetic field $\vec{B}$ slightly off-axis; they must
also oscillate slightly off-resonance.

The energy per unit time which an electron of velocity $\vec{v}$ gains
or loses at a given moment from an electromagnetic field of field
strength $\vec{E}$ can be calculated from the energy equation

$$m_o c^2 \frac{d\gamma}{dt} = e\vec{v}\cdot\vec{E} \tag{17}$$

W. B. Colson and S. K. Ride[52,53] have calculated the gain equation
for the case of electrons moving in a uniform magnetic field $\vec{B}_m =$
$\vec{k}B_o$ ($\vec{k}$ - z-direction of the axis of a solenoid producing the field
B_m). If the beam enters the magnetic field under a small angle it
will spiral at a cyclotron frequency

$$\omega_c = \frac{|e|B_o}{\gamma_o m_o} \tag{18}$$

and it will also give rise to radiation of frequency ω_r leading to
circularly polarized radiation field given by

$$\vec{E}_r = \vec{i}E_o \cos[\omega_r(\frac{z}{c} - t)+\phi] - \vec{j}E_o \sin[\omega_r(\frac{z}{c} - t)+\phi] \tag{19}$$

$$\vec{B}_r = \vec{i}\,\frac{E_o}{c}\sin[\omega_r(\frac{z}{c}-t)+\phi] + \vec{j}\,\frac{E_o}{c}\cos[\omega_r(\frac{z}{c}-t)+\phi] \tag{20}$$

The rotation is on a right-circular helice being in the same sense
as the velocity vector of the electrom beam. Resonance between
the electron beam and the radiation will occur in case the cyclotron frequency ω_c is equal to the frequency at which the radiation
passes over the electron beam, namely

$$\omega_{rz} = (1 - \frac{\vec{v}\cdot\vec{k}}{c})\omega_r \tag{21}$$

which means that

$$\Delta\omega = \omega_c - \omega_{rz} \tag{22}$$

has to be zero for resonance. The relativistic equation of electron
motion can be treated in the same way as before[35] by adding to the
radiation magnetic field $\vec{B}_r$ the uniform axial magnetic field $\vec{B}_m$
and the velocity components v_x, v_y, and v_z can be expressed exactly

in terms of the fields. Use is also made again of the constant[35]

$$\gamma_c = \gamma(1 - \frac{v_z}{c}) \tag{23}$$

Combining Eqs. 17 and 19 leads to

$$\frac{d\gamma}{dt} = \frac{eE_o}{m_o c^2} \{v_x \cos[\omega_r(\frac{z}{c} -t)+\phi]-v_y \sin[\omega_r(\frac{z}{c} -t)+\phi] \tag{24}$$

Continuing the solution process, order by order, as this has been done by W. B. Colson and S. K. Ride[53] one can obtain an expression for the relative change of the electron energy $\Delta\gamma/\gamma$ as a function of time. Taking the average value of this energy change of one electron over the initial phases, i.e.,

$$\Delta E = <\Delta\gamma>m_o c^2 \tag{25}$$

one can then arrive at a gain function G(t) by multiplying Eq. 25 with the number of electrons contained in the interacting volume V and dividing it by the radiation energy

$$E_i = \tfrac{1}{2}\varepsilon_o E_o^2 V \tag{26}$$

initially present in the interaction volume. The result is

$$G(t) = \frac{n_e e^2}{2\varepsilon_o \gamma m_o \Delta\omega^2} \{[\beta_{xo}^2 +2(1-\beta_{zo}^2)][\cos(t\Delta\omega)-1]$$

$$+ \beta_{xo}^2 t\Delta\omega\sin(t\Delta\omega)\} \tag{27}$$

where the usual abbreviations $\beta_{xo} = v_{xo}/c$; $\beta_{zo} = v_{zo}/c$ were introduced.

For the non-relativistic case where the initial velocity components v_{xo} and v_{zo} are much smaller than c and $\gamma \simeq 1$, Eq. 27 reduces to

$$G(t)_{nr} = \frac{n_e e^2}{\varepsilon_o m_o \Delta\omega^2} [\cos(t\Delta\omega)-1] \tag{28}$$

This means that for a non-relativistic electron beam there will never be a positive value for G(t), that is stimulated emission can never occur; a free-electron laser would be impossible. This result is in agreement with Heitler's[40] theory that a system of harmonic oscillators in random phase can only absorb energy from an electromagnetic wave. Only anharmonic oscillators[54] can stimulate emission of radiation from randomly phased harmonic oscillators.

The situation is different for the relativistic case where the last term in Eq. 27 cannot be neglected. Of interest is then the final gain that is the value for Eq. 27 after the electron beam has just left the interaction region of length L; this time is approximately $t = L/v_{zo}$. The final gain can again be expressed in terms of a dimensionless resonance parameter Ω; as this had been done in Eq. 11 for the spatially periodic magnetic field

$$\Omega = \Delta\omega L/v_{zo} = \left[\frac{|e|B_o}{\gamma_o m_o} - (1 - \beta_{zo})\omega_r\right]L/v_{zo} \qquad (29)$$

so that Eq. 27 can be written

$$G_{final}(\Omega) = W_o\{[\tan^2\alpha + \beta_{oz}^{-2}(1-\beta_{oz})](\cos\Omega - 1) + \Omega\tan^2\alpha\sin\Omega\}/\Omega^2$$

$$= 10^{-6}W_o f(\Omega) \qquad (30)$$

where α is the angle between the electron beam's initial direction and the uniform magnetic field $\vec{B}_m = \vec{k}B_o$

$$\tan\alpha = v_{xo}/v_{zo} \qquad (31)$$

and

$$W_o = n_e e^2 L^2/(2\varepsilon_o\gamma m_o v_{zo}^2) \qquad (32)$$

There will be no positive gain at resonance, i.e., for $\Omega = 0$. Also the gain will be negative for the case that the electron beam is injected in the direction of the magnetic field, i.e., $\alpha = 0$, as this follows from Eq. 30 since the term left over has the negative factor $(\cos\Omega - 1)$.

For maximum final positive gain one has to choose such a value for Ω that it fulfills the following two conditions

$$(\Omega^2\tan^2\alpha - 2A)\cos\Omega - \Omega(\tan^2\alpha + A)\sin\Omega + 2A = 0 \qquad (33)$$

and

$$\Omega\cos\Omega(\tan^2\alpha - A) - \tan^2\alpha\sin\Omega(\Omega^2 + 1) < 0 \qquad (34)$$

whereby

$$A = \tan^2\alpha + 2\beta_{zo}^{-2}(1 - \beta_{zo}) \qquad (35)$$

The radial frequency of stimulated radiation that such a free-electron laser will emit is then given by

$$\omega_r = \frac{|e|B_o/(\gamma_o m_o) - v_{zo}\Omega/L}{1 - \beta_{zo}} \qquad (36)$$

As a numerical example, we have chosen an electron beam of energy U = 25 MeV, corresponding to $\gamma = 50$, and an angle $\alpha = 0.027$ rad $\simeq 1.55^o$. This will determine at what value for Ω the gain has a

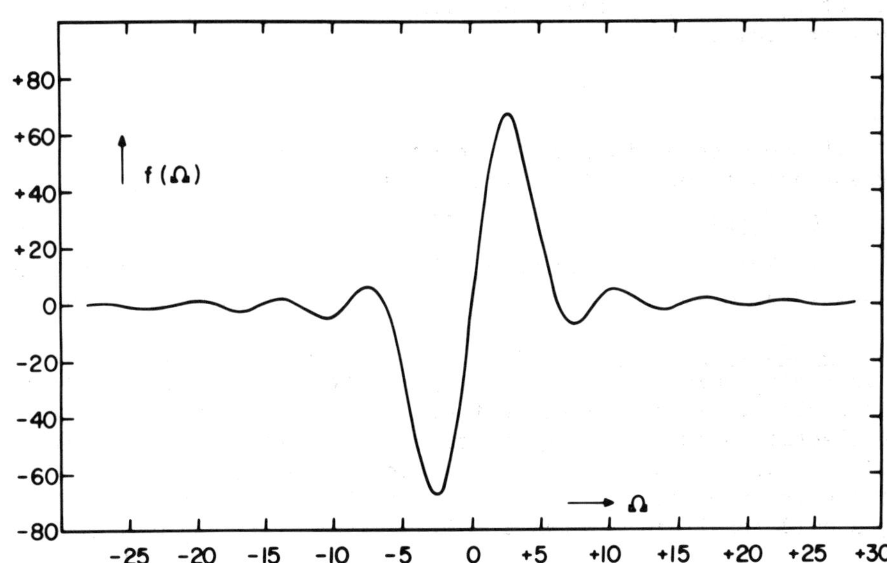

Fig. 1. Normalized gain as a function of resonance parameter Ω for a FEL with static transverse spatially periodic magnetic field ($\Omega = L\Delta\omega/v_o$ from Eq. 14).

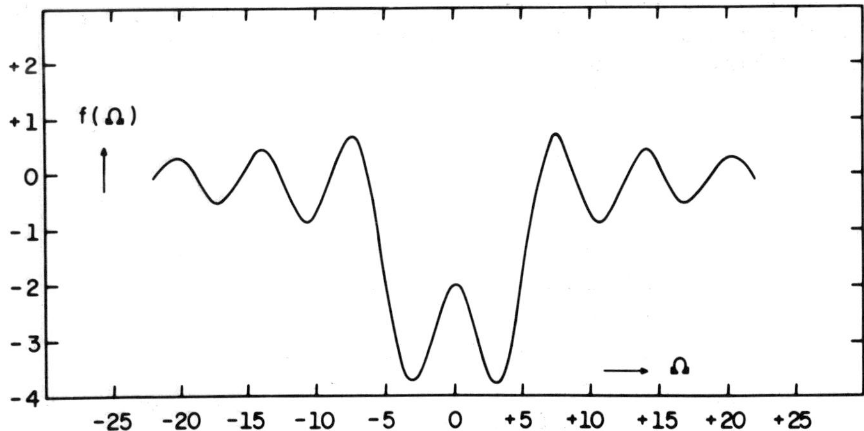

Fig. 2. Normalized gain as a function of resonance parameter Ω for a laser from a relativistic electron beam within a uniform magnetic field ($\alpha = 0.027$ rad; $\gamma_o = 50$) (S.K. Ride and W.B. Colson[52,53]).

maximum. The normalized function $f(\Omega)$ is plotted on Fig. 2 and
shows that with Eqs. 33-35 the maximum for $\gamma = 50$ and $\alpha = 0.027$ rad
lies at $\Omega = \pm 7.5$, so that the final gain for one passage of the
electron beam is given by

$$G_{final} \simeq 7 \times 10^{-5} \, W_o \qquad\qquad (37)$$

If one chooses further an interaction length $L = 5.2$ m, a beam
current density of 5 mA/cm^2 (with $n_e = 10^{12} m^{-3}$), one obtains with
$W_o = 0.01$ a gain of $G_{final} = 7 \times 10^{-7}$. A magnetic field of $B_o =$
200 kG (= 20T) would yield under the above conditions a stimulated
radiation at a wavelength of $\lambda \simeq 15$ μm.

Spatially periodic electric field. The spatially periodic magnetic field could be replaced by spatially periodic transverse electric field[55]. If the electric field is directed at right angle to
the electron beam, the same effect of light generation and amplification should be observable; the electrons cannot distinguish
between the two fields. An electric field with an amplitude $E_o =$
cB_o should result in a similar FEL as a magnetic field of amplitude
B_o.

In the first FEL[30] a magnetic flux density of $B_o = 2.4$ kG =
0.24 T was applied, which means that in order to achieve the same
laser action in a spatially periodic electric field, the peak field
should be $E_o = 720$ kV/cm. G. Bekefi and R. E. Shefer[55] have proposed such a FEL in the Raman range, i.e., for high density relativistic electron beams. The beam is directed through a rippled drift
tube (see Fig. 3). The electric field has a spatial periodicity ℓ

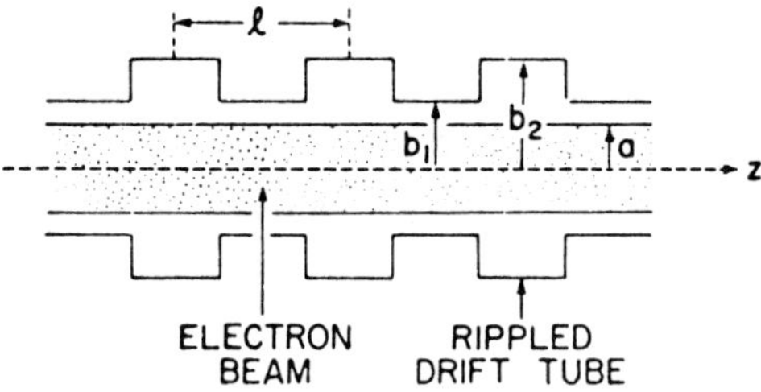

Fig. 3.   Schematics of a Raman FEL applying a rippled electric field
          (G. Bekefi and R. E. Shefer[55]).
          z = electron beam axis; a = e-beam radius
          $b_2 - b_1$ = ripple depth of circular grooves
          ℓ = spatial periodicity

which should be much larger than the ripple depth of the circular
grooves. The wave length of the back scattered wave is[55]

$$\lambda = \ell(1-\beta)/\beta \qquad (38)$$

where $\beta = (1 - \gamma_o^{-2})^{\frac{1}{2}}$ and γ_o as defined by Eq. 4.

The spatial growth rate per unit length for the radiation output
is calculated to be

$$\Gamma = (\frac{\gamma_o^{\frac{1}{2}}\omega_p\ell}{8\pi\beta c})^{\frac{1}{2}} \frac{e}{m_o} \frac{E_o}{\gamma_o c^2} \qquad (39)$$

where

$$\omega_p = [n_e e^2/(m_o\varepsilon_o)]^{\frac{1}{2}} \qquad (40)$$

is the plasma frequency. In the given example[55] the proposed free-
electron Raman laser working at a peak-to-peak electric field of
$2E_o$ = 336 kV/cm with a 1.5 MeV, 20 kA electron beam and tube dimensions of a = b_1 = 0.5 cm and b_2 = 0.75 cm (see Fig. 3) (ripple ℓ=
1.3 cm) would generate a radiation power of 1.2 GW at saturation
and an efficiency of 3.9%. The radiation would have, according to
Eq. 38, a wave length of approximately λ = 500 μm and, according to
Eq. 39, a spatial growth rate of 5.7 [m^{-1}] per length in meter.

G. Bekefi[56] has given some basic thoughts and calculation to
electrically pumped relativistic free electron wave generators including also longitudinal electric fields for high density electron
beams (Raman region).

Stimulated Raman Scattering Region

Whereas in the stimulated Compton scattering region the electrons acquire a velocity modulation transverse to the beam due to
excitation of a single-particle state, in the stimulated Raman
scattering region this is accomplished by the excitation of plas-
mons, i.e., by collective plasma oscillations of the electrons.
Stimulated Compton scattering occurs when the Debye length D_ℓ is in
the same order of magnitude as the wavelength $D_\ell \simeq \lambda$

$$D = (\varepsilon_o kT_p n_e^{-1}e^{-2})^{\frac{1}{2}} \qquad (41)$$

$$\simeq 69\sqrt{T_p/n_e} \qquad (41a)$$

T_p being the temperature of the plasma [°K] and n_e the plasma
(electron) density in [m^{-3}].

In the Raman case the electron density is large so that the
corresponding Debye length is shorter than the radiation wavelength

$(D_\ell < \lambda)$. In this criterion D and λ have to be given in the same
reference frame. Free electron lasers based on stimulated Compton
scattering will have a smaller efficiency.

One can determine a critical radiation wave length λ_{cr} below
which scattering occurs in the stimulated Compton region and above
which it occurs in the stimulated Raman region. This can easily be
shown. Expressing the Debye length as

$$D = 2\pi v_T/\omega_p \tag{42}$$

with v_T being the thermal velocity of the electron plasma related
to the energy spread of the electron beam and ω_p the plasma frequency (Eq. 40). v_T can be determined from the energy spread of the
relativistic electron beam (see for example Ref. 57); its value in
the beam frame is given by $v_T = c\Delta\gamma/\gamma$, which leads to a value for
the Debye length in the beam frame

$$D_b = \frac{2\pi c}{\omega_p} \frac{\Delta\gamma}{\gamma} \tag{43}$$

The critical wavelength measured in the laboratory frame should
be equal to the Debye length, certainly also converted into the
laboratory frame: $D_\ell = D_b/\gamma$, i.e.,

$$\lambda_{cr} = \frac{2\pi c}{\omega_p} \left(\frac{\Delta\gamma}{\gamma}\right)/\gamma \tag{44}$$

or expressing ω_p from Eq. 40 by the current density $J_o = n_e e v_o$
($\simeq n_e ec$ for a relativistic beam) one arrives at

$$\lambda_{cr} = 2\pi \left(\frac{\varepsilon_o m_o c^3}{e}\right)^{\frac{1}{2}} \left(\frac{\Delta\gamma}{\gamma}\right)/ (\gamma J_o^{\frac{1}{2}}) \tag{45}$$

or

$$\lambda_{cr[m]} = 2.31 \times 10^2 \left(\frac{\Delta\gamma}{\gamma}\right)\gamma^{-1} J_o^{-\frac{1}{2}} \tag{45a}$$

Using as a numerical example the data of the first observation of a
FEL[30]: $\Delta\gamma/\gamma = 5 \times 10^{-4}$; $\gamma = 85$; $J_o \simeq 10^5$ am^{-2}, leads to a critical
wavelength of $\lambda_{cr} = 4$ μm. A wavelength of $\lambda = 3.4$ μm had been
measured which means this FEL worked in the stimulated Compton
region.

As pointed out, there is an enhancement of the efficiency, if
higher electron densities are applied; collective interactions take
place whereby the plasma can be excited by parametric instabilities.
However, the frequency of the high electron density FEL is lowered
by the influence of the plasma frequency. T. Kwan and Dawson[50]
have performed computer simulation to investigate a FEL with a
spatially periodic magnetic field. They found the efficiency of

such a laser of frequency ω to be[59]

$$\eta = \frac{[\gamma_o(\gamma_o{}^2-1)]^{\frac{1}{2}}\omega_p}{(\gamma_o-1)(\omega+\omega_q)} \tag{46}$$

where $\omega_q = 2\pi c/\lambda_q$ (λ_q - spatial period of wiggler). Maximum growth rate was found at $\omega_q = 0.105\,\omega_p$ and under these conditions 28% of the beam energy[59] was converted into electromagnetic energy.

The overall efficiency of a FEL can be increased by recycling the electron beam[60,61] through the spatially periodic magnetic field, as proposed already by G.A.G. Deacon et al.[30] The recycling could be done in a storage ring and it should be possible to reach efficiencies higher than 20%. The description of such a storage ring under construction is given by R. Barbini et al.[62] in detail. It is meant for a FEL tunable in the wavelength range between 0.3 and 20 μm with an output average power of the order of 1 kW and an efficiency of 2% light from the electron beam. The storage ring will have an electron beam of 750 MeV energy, a current per bunch in the order of 1 a, and a synchrotron radiation per turn will be 15.5 keV. The overall dimensions will be approximately 44 m for the major axis of the ring and 27 m for the minor axis.

A modified scheme of a FEL using a storage ring has been proposed by L. R. Elias[63] whereby relatively low-energy electron beams are used (E < 5 MeV). Hereby a continuous beam of electrons with a low energy spread interacts with a spatially periodic magnetic structure and generates radiation at a wavelength $\lambda_1 \simeq \lambda_q/2\gamma^2$ (see Eq. 7). The same rippled electron beam interacts with the pump wave and produces laser oscillations at a shorter wavelength $\lambda \simeq \lambda_1/4\gamma^2$. In order to increase efficiency and power, the output electron beam is decelerated and collected at a high negative potential. The maximum small-signal gain for such a laser can be determined by[31,64,65]

$$G_{max} = An_e B_o{}^2\lambda^{3/2}\lambda_1{}^{5/2}N^3 \tag{47}$$

where A is a constant and N the number of pump-wave periods in the interaction region.

NONLINEAR FORCE (FEL) LIGHT AMPLIFIER

B. W. Boreham et al.[66,67] have shown that focusing a laser beam of 4 GW to an intensity of $I = 10^{15}$ W cm^{-2} inside a helium atmosphere of about 10^{-2} -- 10^{-3} Pa generated electrons of 90-100 eV. This energy, as H. Hora[68,69] points out, must originate from the nonlinear force within the strong electromagnetic field of the laser focus setting the electrons in oscillatory motion. H. Hora[70,71] has shown that half of the oscillation energy ε^{osc} is converted in translational kinetic energy ε^{trans} due to nonlinear ponderomotive forces into a direction normal to the laser beam, i.e., along the radial intensity gradient.

$$\varepsilon^{tr} = \varepsilon^{osc}/2 \tag{48}$$

ε^{osc} is given by

$$\varepsilon^{osc} = m_o c^2 (\gamma - 1) \tag{49}$$

which had been evaluated by H. Hora[72] to be approximately:

$$\varepsilon^{osc} = \varepsilon_o E^2/(2n_{ec}n) \tag{49a}$$

where E_o – the peak electric field in the laser, n_{ec} – the critical density, i.e., density which makes the plasma frequency (Eq. 40) equal to the laser frequency, n – the refractive index of the medium. In Eq. 49a it had been assumed that the field strength parameter

$$\alpha = eE_o/(m_o c\omega) \tag{50}$$

smaller than unity which is the case in the above example, viz. $\alpha \simeq 0.02$. The exact value for ε^{osc} is given by introducing into Eq. 49 a more precise expression for γ as derived by H.Schwarz and R. Tabensky[35]. For the case where the refractive index is close to unity they arrive at

$$\gamma = \sqrt{1 + a^2\alpha^2} \tag{51}$$

where the parameter a is unity in case the laser field is circularly polarized; however, for linearly polarized light, which obviously applies in the above experiments[66,67], and $n \simeq 1$, a is given[35] by

$$a^2 = (1 + \frac{9}{16}\alpha^2)/(1 + \frac{1}{2}\alpha^2) \tag{52}$$

a becomes $3\sqrt{2}/4 \simeq 1.06$ for very large α. For $\alpha^2 \simeq 4 \times 10^{-4}$, even with linearly polarized light a^2 is practically unity and γ (Eq.51) can be written approximately as $\gamma \simeq 1 + \alpha^2/2$, so that the oscillation energy , Eq. 49, simplifies, with Eq. 50, and

$$I = n\varepsilon_o c E_o^2/2 \tag{53}$$

to

$$\varepsilon^{osc} = \frac{e^2 I}{n\varepsilon_o c m_o \omega^2} = I/(cnn_{ec}) \tag{54}$$

Considering the experiment[66,67] with an Nd-doped glass laser of I = 10^{19}W/m^2 and $\omega = 1.78 \times 10^{15}$rad/s and for a low density plasma (n=1), one obtains for ε^{tr} a value of 105eV, which is close to what B.W. Boreham et al[66] measured.

Having established experimentally[66,67] and theoretically[70,71]

that half of the oscillation energy converts into translational
kinetic energy, H. Hora[68,69] proposed that the inverse would also be
possible, namely that a beam of electrons having an energy equal to
$\varepsilon^{osc}/2$ and crossing a laser focus of intensity I yielding, according to
Eq. 49, the oscillation energy ε^{osc} would be converted into radiation.
Besides the condition of Eq. 48, a second condition for this energy
transfer to radiation has to be imposed on the duration of the laser
a nd electron beam pulses. Assuming a laser pulse of duration τ_ℓ and
Gaussian intensity distribution, an electron pulse of duration τ_e is
fired radially across this laser focus. τ_ℓ has to be twice as long
as τ_e, so that the electrons are decelerated to zero velocity at the
center of the laser pulse when its intensity is at its maximum. Hora
states[68] that the desired laser amplification is only possible, if
one includes the switching on-and-off of the laser beam, as for example developed by R. Klima et al[73]. If an electron is initially at
rest in the laboratory frame and the laser beam is switched on, the
electron receives its oscillation energy from the optical field and
returns it unchanged after switching off. The laterally injected
electron did not receive its oscillation energy from the optical
field, but from the lateral motion. Switching off leads then to the
desired transfer of the electron energy to the radiation field as an
additional part, resulting thus in an amplification, after which the
electron comes to rest.

The focal diameter is determined by the condition that the
fastest and slowest electron within the electron beam of energy
spread $\Delta\varepsilon^{tr}$ should remain within the laser beam during the interaction time τ

$$r = \tau(2e\Delta U^{tr}/m_o)^{\frac{1}{2}} \tag{55}$$

The electron current must have a density within the laser beam
of

$$j = J/(2r\tau c) \tag{56}$$

since during the interaction duration τ the electron beam will have
traversed a lateral section of the laser beam of width 2r and a
length τc. τ and r are fixed by Eqs. 48-53 and can be expressed in
terms of laser and electron beam energies, E_ℓ and U^{tr} resp. as well
as energy spread ΔU^{tr}:

$$\tau = (4\pi\varepsilon_o c)^{-1/3}\left(\frac{E_\ell}{n\omega^2 U^{tr}\Delta U^{tr}\gamma^{tr}}\right)^{1/3} \tag{57}$$

where $\gamma^{tr} = 1 + eU^{tr}/(m_o c^2) = 1 + 1.957\times10^{-6}U^{tr}$

or

$$\tau = 3.11\left(\frac{E_\ell}{n\omega^2 U^{tr}\Delta U^{tr}\gamma^{tr}}\right)^{1/3} \tag{57a}$$

and r follows with Eq. 57 from Eq. 55. Note that Eq. 57 is an exact
expression using any unrestricted α, contrary to Eqs. 49a and 54.

The amplification A is determined by the ratio of electron beam energy, $E_e = JU^{tr}\tau$, and $E_\ell = I\tau\pi r^2$, which with Eqs. 55-57 is:

$$A = \frac{1}{\pi\varepsilon_o} \left(\frac{e}{2m_o}\right)^{1/2} \frac{j}{n\omega^2(\Delta U^{tr})^{1/2}\gamma^{tr}} \tag{58}$$

or

$$A = 1.07 \times 10^{16} \frac{j}{n\omega^2(\Delta U^{tr})^{1/2}\gamma^{tr}} \tag{58a}$$

An electron source of high current density j_o yielding a large total current J is needed to render optimum amplification. The current density j is either the value calculated from Eqs. 55-57 or the maximally possible current density according to the Langmuir condition:

$$j = j_o\left(1 + \frac{eU^{tr}}{kT}\right) \sin^2\theta \tag{59}$$

whichever is smaller. T is the temperature of the source of current density j_o, θ the half angle subtended by the cone of the electrons at the interaction cross section:

$$A_1 = 2cr\tau \tag{60}$$

The upper limit of θ is half the ratio of the maximum perpendicular electron velocity and parallel velocity – disregarding aberration effects:

$$\theta < \frac{1}{2}(\Delta U^{tr}/U^{tr})^{1/2} \tag{61}$$

so that the current through the interaction region, with $\Delta U^{tr} \ll U^{tr}$, $kT \ll eU^{tr}$, has to obey:

$$J < cr\tau j_o e\Delta U^{tr}/(2kT) \tag{62}$$

As an example H. Hora[68] proposed a CO_2 laser pulse of $E_\ell = 7.85 \times 10^4$J to interact with an electron beam of energy $U^{tr}=0.9$MV, energy spread $\Delta U^{tr}=71$V and current $J=10^5$a (impossible due to Eq. 62). Using Eq. 56 for j, he arrived[68] at $A=2.6 \times 10^{-3}$(which , by the way, should have been 30 times smaller, if Eq. 56 were applicable). However, Eq. 56 cannot be used, as this violates the Langmuir condition, Eq. 59. Equations 59 and 61 have to be applied; they lead, with extreme values of T = 3000K and $j_o=10^5$a/m^2, to an upper limit of $j<7\times10^6$a/m^2 and an amplification $A<10^{-7}$. n has been chosen unity. This value of A is by more than four orders of magnitude lower than claimed for the same numerical example by H. Hora[68].

The proposed scheme is, however, not a FEL, but rather an amplifier which has to start off with a laser of already high intensity to provide for the necessary nonlinear force.

FEL BASED ON SMITH-PURCELL-SALISBURY EFFECT (SPS-EFFECT)

Independently W.W. Salisbury[5,6] and S.J. Smith jointly with E.M. Purcell[7] established that light could be generated from free

electrons which were accelerated to a velocity v_e and passed at
grazing incidence over a metallic diffraction grating (see Fig. 4).
The wavelength λ of the light observed at an angle θ was found to be

$$\lambda = d\left(\frac{c}{v_e} - \cos\theta\right) \tag{63}$$

whereby d is the grating spacing. Smith and Purcell confirmed this
formula experimentally with an electron beam of 320 kV (c/v_e = 1.27),
d = 1/20,000 inches = 1.27 µm when they observed at an angle of θ=
20° light of approximately λ = 4000 Å. Salisbury[6] noticed that in
order to see this light part of the electron current had to strike
the grating and he found[6] that the light intensity was proportional
to this current as well as to the current passing the grating.
Salisbury stated that the light intensity is proportional to the
product of the current reflected from the grating and the undeflected
current passing the grating. This intensity dependence on the current
had been observed for a range of fractions of microampere (light just
visible) up to about 10 ma (light visible in a well-lit room).

The effect has been explained with Maxwell's electrostatic image
theory. As the electrons move over the grating grooves and conducting wires forming peaks, the varying image charge forms a series of
moving dipoles. They, in turn, generate electromagnetic radiation
whose wavelength and intensity can be determined with elementary
calculations[11]. The average radiated power for a dc electron current
J over a total length l_g of the grating pattern will be

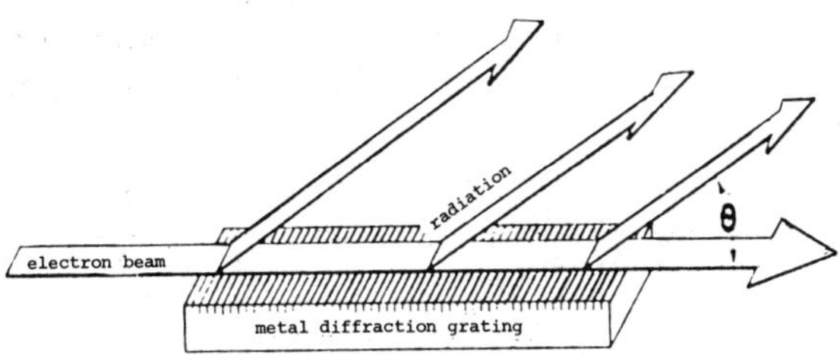

Fig. 4. Schematics of the Smith-Purcell-Salisbury effect[11]

$$\lambda = d\left(\frac{c}{v_e} - \cos\theta\right)$$

d – grating spacing constant; v_e – speed of electrons.

$$<P> = \frac{e\pi^3}{3\varepsilon_o} A^2 L_g \frac{c}{v_e} \frac{J}{\lambda^4} \tag{64}$$

or
$$= 1.87 \times 10^{-7} A^2 L_g \frac{c}{v_c} \frac{J}{\lambda^4} \tag{64a}$$

where A is determined by the dimensions of the dipole which should roughly have the order of magnitude as the grating constant d.

However, in case the electron beam could be "bunched" as, for example, by "beating" two electron beams of slightly different energy as proposed by H. Schwarz[74] (see last paragraph of this paper: "Low Voltage Quantum Mechanical Laser"), the average radiated power would go with the square of the electron beam current J_o, namely

$$<P_b> = \frac{\pi^3}{3\varepsilon_o c} (AL_g)^2 (\frac{c}{v_e})^2 \frac{J_o^2}{\lambda^4} \tag{65}$$

$$= 3.89 \times 10^3 (ALg)^2 (\frac{c}{v_e})^2 \frac{J_o^2}{\lambda^4} \tag{65a}$$

Taking Salisbury's example: $L_g = 2.5 \times 10^{-4}$ m; $c/v = 1.27$; $A \simeq 1.0 \times 10^{-6}$ m; $J = 100$ μa; and $\lambda = 4.5 \times 10^{-7}$ m we obtain from Eq. 64 a value for the average radiation power $<P> = 0.1$ watt, which would go up to 100 watt with the same arrangement if the electron current had been "bunched" (Eq. 65).

A more detailed theory of the SPS effect has been given soon after the experimental results by C. W. Barnes et al.[75] discussing the case of a periodically and an aperiodically modulated electron beam.

G. Toraldo di Francia[76] treated the SPS effect as a special case of Čerenkov radiation and arrived at an expression for the brightness B, i.e., the power per unit solid angle and per apparent emission surface

$$B = \frac{e\delta_m^2}{2\varepsilon_o} \frac{Jm^2}{L_g d^2} \frac{\beta^2 \sin\theta}{(1-\beta\cos\theta)^3} \exp(-4\pi m \frac{a}{d} \frac{\sqrt{1-\beta^2}}{1-\beta\cos\theta}) \tag{66}$$

where L_g, θ, and d have the same significance as in Eqs. 64 and 65; $\beta = v_e/c$; a is the distance at which the electron beam travels over the grating surface. δ_m is a number which depends, as A, on the emitted wavelength; here, it is in the optimum case unity. m gives the order of harmonics.

As mentioned before G. W. Stroke[11] saw already in 1967 the possibility to utilize the SPS effect for a FEL laser. However, only recently has this idea been pursued seriously.

K. Mizumo et al.[77] suggested the realization of the inverse SPS effect. If a light beam of wavelength λ is incident on a metallic

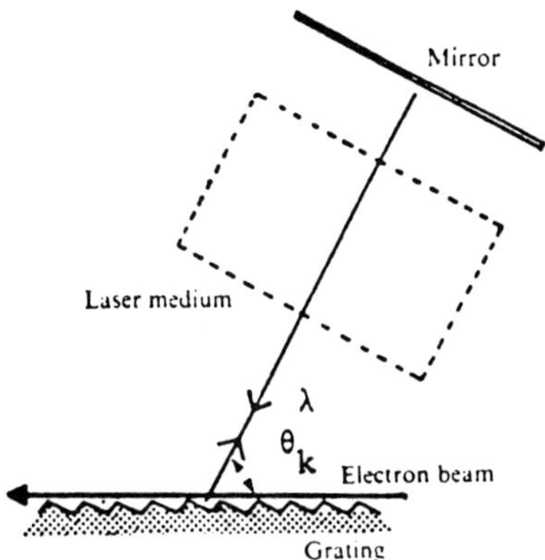

Fig. 5.   Schematics of FEL based on the inverse SPS-effect
according to Mizuno, Ono and Shimoe[77] with Fabry-
Perot resonator.

diffraction grating at an angle θ, so that λ and θ fulfill the SPS
condition, Eq. 63, an electron beam of velocity v_e grazing over the
grating should interact synchronously with the electromagnetic wave,
leading to electron acceleration or deceleration depending on the
electron particle velocity and the phase velocity of the electromagnetic wave. If the electrons are faster than the wave velocity,
they will be decelerated and their lost kinetic energy converts into
light amplification. Mizuno-Ono-Shimoe's scheme of such an SPS
laser is represented in Fig. 5. If the light hits the grating under
an angle θ_k which follows the condition

$$\cos\theta_k = k\lambda/(2d) \tag{67}$$

with k = 0,1,2,...the diffracted beam direction is the same as the
incident beam. K. Mizuno et al.[77] have observed in such a device
with $\theta_k = \pi/2$, millimeter and even submillimeter wave oscillations.

J. M. Wachtel[78] has given detailed design parameters for a FEL
using the SPS effect. He has come to the conclusion that infrared
SPS free electron lasers are feasible, even at beam currents smaller
than 1a in single-mode operation. A power of 100 W may be achieved.

Also A. Gover and A. Yariv[79,80] have discussed the possibility
of constructing a FEL applying the SPS-effect. Their scheme is

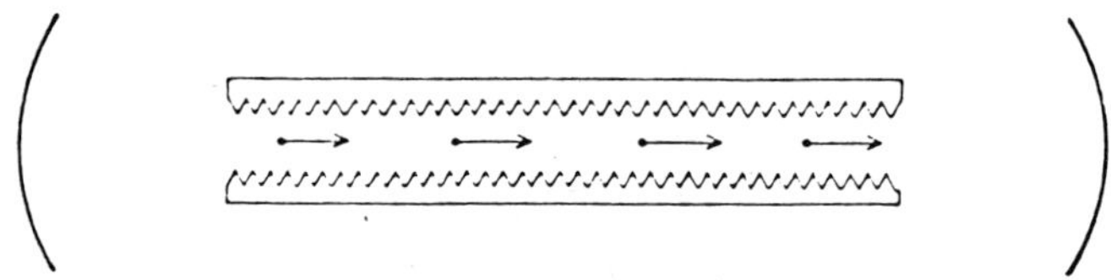

Fig. 6.   Schematics[80] of a FEL based on the Smith-Purcell-
          Salisbury effect.

depicted in Fig. 6.   There is a difference between conventional
lasers where, due to the reflections at both mirrors, gain is
obtained in both directions; but not in the SPS-free electron laser
oscillator, in this case there will be gain only in one direction
of light propagation as indicated on Fig. 6.   The threshold for
lasing is given by an equation similar to the one known for a conventional laser

$$R_1 R_2 e^{(g-\alpha)L} g = 1 \tag{68}$$

where R_1 and R_2 are the reflectivities of the two mirrors; g –
the gain for the SPS – laser; and α – the absorption coefficient
for the electromagnetic energy within the wave guide formed by
the grating surfaces.

CHANNEL RADIATION

     If electrons or positrons or any other sufficiently small
particle beams shine on a crystal they normally do not penetrate
deep into the crystal; their penetration depth follows the usual
exponentially decreasing function.   However, with proper orientation of the particle beam, channels in the crystal will open up,
i.e., between the planes of the ionic structure that make up the
crystal to be a crystal.   In such a case the crystal becomes
practically permeable for the particle beam and one talks about

the channeling effect. M. A. Kumakhov[81] predicted in 1976 that the
channeling of electrons and positrons should generate electromagnetic radiation at energies up to the γ-ray region, whereby the
positrons in their way through the channels will encounter - with a
spatial periodicity of atomic dimensions - positive ions of the
crystal lattice. The mutual repulsion between the positrons and
ions, as the positrons move through the channel, will cause zigzag
motion which can be analyzed with classical electromagnetic theory,
leading to a "classical harmonic oscillator potential". For a given
crystal channel and a given positron energy, a monochromatic highly
directional coherent radiation in the x-ray region should be detectable. If, however, electrons are used a family of lines appears
in the electromagnetic spectrum as this system is more complex, a
quantum mechanical model has to be used to analyze the transitions
between different bound states as they occur in electron channeling.

The first observation of such channeling radiation due to
positrons of 56 MeV traveling along the (110), (111), and (100)
phases and along the <110> axis in silicon had been reported by M.
J. Alguard et al.[82]. X-ray beams of energies with peaks between
10 and 600 keV and beam divergence of 3 mrad x 9 mrad were
observed.[82]

Soon channeling radiation from relativistic electrons could be
measured by R. L. Swent et al.[83,84]. Electrons of 56 MeV and 28 MeV

TABLE I. Observed spectral peaks for planar-channeled electrons.[83]

Beam energy (MeV)	Crystal plane	Photon energy ± 2 keV (keV)	Linewidth ± 2 keV FWHM (keV)	Effective coherence length (µm)
28	(001)	31	4	0.6
	(110)	40	6.5	0.4
		25	5.5	0.5
56	(001)	99	16	0.6
		64	11	0.9
		39	8	1.2
	(110)	128	15	0.7
		94	13	0.8
		68	8	1.2
		52	5	1.9
	(111)	99	20	0.5
		37	10	1.0

energy and currents of 2×10^{-11} a were channeled through silicon
crystals along the same planes and axis. The observed bremsstrahlung x-rays as measured by R. L. Swent et al.[83] are given in Table I.

A quantum mechanical theory of spontaneous and induced radiation of channeled electrons and positrons has been given by V. V.
Beloshitskii and M. A. Kumakov[85] and pertinent basic theories of
transition radiation and transition scattering can be found in a
90-page long report by V. L. Ginzburg and V. N. Tsytovich[86].

It is obvious that, similar to a FEL laser based on the SPS
effect, a cavity could be built for the channeling radiation with
proper feedback systems (see for example: G. C. Baldwin[87]), so that
a laser in the high energy region of the electromagnetic spectrum
(x-rays to γ-rays) might be possible.

LOW VOLTAGE QUANTUM MECHANICAL LASER[74]

If two coherent electron beams originating from the same source
interfere with each other after one of the beams has been accelerated
or decelerated slightly, let us say $\Delta E = 0.2$ --- 100 eV, a beat
phenomenon should occur, i.e., the electron probability density in
the interference region should vary at frequencies $\omega = \Delta E/\hbar = 3 \times 10^{14}$ --- 1.5×10^{17} rad/s.

This would bring the time variation of electron probability
density into the optical region seen from a laboratory frame. Its
wavelength would lie in the region between $\lambda = 10$ nm and 6 μm.
Interaction with an "object", e.g., a "structure" like in the SPS
effect, - which could also be the interference pattern itself -,
would lead to electromagnetic radiation of a frequency according
to the continuously adjustable energy difference ΔE.

This "beat" phenomenon could be achieved by using, for example,
the electron interferometer as developed by Möllenstedt and Duker[88],
see Fig. 7. Hereby, an electron beam is split up into two branches
by a thin wire electrode I at negative potential and led together
by a positively charged wire electrode II. A slightly negatively
charged electrode III decreases the angle between the two branches
before they join and interfere. If one branch has an energy E and
the other a slightly lower energy $E - \Delta E$ the beating frequency
created at the interference region will be $\Delta E/\hbar$. The difference
in energy and momentum of the two branches could be created for
example by a constant time change of the magnetic flux ϕ produced
in a small coil placed inside the electron prism $(\Delta U = d\phi/dt)$.

Let one branch have the energy E and momentum p and the other
the energy $E - \Delta E$ and momentum $p - \Delta p$ then their respective wave
functions will be

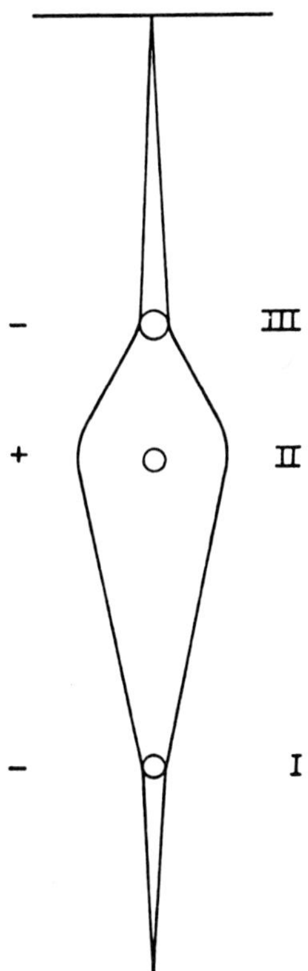

Fig. 7.   Schematics of electron interferometer according to
          G. Möllenstedt[88].

$$\psi_1 = \exp[(\vec{p} \cdot \vec{r} - Et)i/\hbar] \qquad (69)$$

and

$$\psi_2 = \exp\{[(\vec{p} - \Delta\vec{p}) \cdot \vec{r} - (E - \Delta E)t]i/\hbar\} \qquad (70)$$

r – being the length of the electron beam. The wave function of
the electron beam at the interference region is then

$$\psi = a_1\psi_1 + a_2\psi_2 \qquad (71)$$

Its current density in the interference zone is given by

$$\vec{j} = i\hbar e(\psi\nabla\psi^* - \psi^*\nabla\psi)/(2m_o) \qquad (72)$$

Using Eqs. 69-71 for the evaluation of Eq. 72 leads to

$$\vec{j} = [(a_1{}^2+a_2{}^2)\vec{p} - a_2{}^2\Delta\vec{p} + a_1a_2(2\vec{p} - \Delta\vec{p})\cos[(\vec{r} \cdot \Delta\vec{p}-t\Delta E)/\hbar]e/m_o \quad (73)$$

Maximum amplitude of the beat phenomenon will be obtained for $a_1 = a_2 = a$, i.e., if both electron beam branches carry equal current, so that Eq. 73 will become

$$\vec{j} = (2\vec{p} - \Delta\vec{p})\{1 + \cos[(\vec{r} \cdot \Delta\vec{p} - t\Delta E)/\hbar]\}a^2e/m_o \quad (74)$$

Considering only the x-component of the beam along its main axis and the fact that $\Delta E \ll E$, Eq. 74 can be simplified to

$$j = eA\{1 + \cos[\omega(\alpha_o x/v - t)]\} \quad (75)$$

$2\alpha_o$ is a correction factor of almost unity value

$$\alpha_o = (1 + \Delta U/4U)/2 \quad (76)$$

wherein

$$eA = j_o[1 - \alpha_o(\Delta U/U)^{1/2}/2] \quad (77)$$

with

$$j_o = 2a^2e(2eU/m_o)^{1/2} \quad (78)$$

being the current density of the beam before entering the electron prisma, U the acceleration voltage of the undivided beam and $\Delta U \ll U$ the difference between the acceleration voltages of the two branches.

$v_e = (2eU/m_o)^{1/2}$ is the velocity of the electrons before entering the electron interferomenter.

There have been objections that such a system by itself could not radiate[89]; but within a laboratory frame there should be an ac current[90] according to Eq. 75 which with the continuity relationship

$$\nabla\cdot\vec{j} + \frac{\partial\rho_t}{\partial t} = 0 \quad (79)$$

should lead to a time varying charge density

$$\rho_t = \frac{eA\alpha_o}{v_e} \cos[\omega(\alpha_o \frac{x}{v_e} - t)+\phi_o] \quad (80)$$

In order to obtain an estimate of the expected radiation energy, classical calculations were performed considering the problem as the radiation from an ac current (Eq. 75) in the laboratory frame. However, as M. Peshkin[91] stated correctly a transformation into a moving frame would make the time varying term disappear and in such a system the current and charge densities (Eqs. 75 and 80) would be only a function of position in coordinates of the moving frame.

But such a frame would have to move so fast that it would escape
the laboratory space during the experiment[90]. Following a classical treatment[92] of the radiation of an ac current along a path
length L, being the length of the interaction region, at a distance
r >> L and r >> λ, we arrive at an electric field given in spherical coordinates

$$E_r = \frac{eAq\cos\theta}{2\pi\omega\varepsilon_o} L[\frac{k}{r^2} \sin(\alpha-kr) - \frac{1}{r^3} \cos(\alpha-kr)] \tag{81a}$$

$$E_\theta = \frac{eAq\sin\theta}{4\pi\varepsilon_o} L[\frac{k^2}{r} \cos(\alpha-kr) + \frac{k}{r^2} \sin(\alpha-kr) - \frac{1}{r^3} \cos(\alpha-kr) \tag{81b}$$

where k = ω/c the propagation number

$$\alpha = \omega t + \pi/2 - \Delta pL/\hbar \tag{82}$$

There is no azimuthal component of the electric field. However, the
magnetic field turns out to have only an azimuthal component

$$B_\phi = \frac{eAq}{4\pi\varepsilon_o c^2} L \sin\theta[\frac{k}{r} \cos(\alpha-kr) + \frac{1}{r^2} \sin(\alpha-kr)] \tag{83}$$

The time average of radiation into a direction θ (angle between
direction of observation and direction of the interference current)
is given by

$$|(\vec{E}\times\vec{B})\varepsilon_o c^2| \simeq \frac{J_o^2\sin^2\theta}{32\pi^2\varepsilon_o c^3} (\frac{L}{r})^2 \omega^2 \tag{84}$$

assuming
$$\frac{k^3}{r^2} >> \frac{k^2}{r^3} , \frac{k^3}{r^2} >> \frac{k}{r^4} >> \frac{1}{r^5} .$$

Maximum radiation is to be expected at $\theta = \pi/2$, i.e., in a
direction normal to the interference current. Total radiation P
in all directions turns out to be

$$P = \frac{J_o^2}{12\pi\varepsilon_o c^3} L^2\omega^2 \tag{85}$$

or

$$= 1.11 \times 10^{-16} J_o^2\omega^2L^2 \tag{85a}$$

Assuming an initial current J_o = 25 μa, an interaction length
of L = 10^{-7} m and ΔU = 3V, Eq. 85 yields a total radiation power of
10^{-8}W at a radiation frequency of ω = 4.56 $\times$ 10^{15} rad s^{-1}, which is
blue light at a wave length λ = 413 nm. The acceleration voltage,
i.e., the electron velocity v_e has to be large enough, so that the
energy spread remains small, as the coherence length Δs of the
electron beam has to be longer than the interaction length L of the
interference region and also longer than the pulse length v_e/ω of
the expected electromagnetic radiation. These conditions lead to
the inequality

$$\Delta s \simeq \lambda_e^2/\Delta\lambda_e = h v_e/(e\Delta U_{th}) > L > v_e/\omega \tag{86}$$

see Eq. 1; ΔU_{th} being the energy spread of the electron beam related to electron source temperature. If we apply a tungsten filament as an electron source ($\Delta U_{th} \simeq 0.7$ V) and if we accelerate these electrons to $U = 1000$ V ($v_e = 1.88 \times 10^7 ms^{-1}$), we arrive at a coherence length of $\Delta s = 1.1 \times 10^{-7}$m and a light pulse length $v_e/\omega = 4.1 \times 10^{-9}$m, which remains within the conditions required by inequality[86].

The gain of the proposed quantum mechanical light generator during one passage is

$$G = (12\pi\epsilon_o c^3)^{-1} J_o L^2 \omega^2/U \tag{87}$$

or
$$= 1.11 \times 10^{-16} J_o L^2 \omega^2/U \tag{87a}$$

For the given example the gain would be only 6×10^{-7}. But it is interesting to note that it increases with the square of the frequency of radiation and of the interaction length, which most other free electron light generators do not do.

The theoretical treatment of this light generator should be done completely quantum-mechanically as the original idea has been based on quantum mechanical thoughts. Two material waves of slightly different wave lengths interfere and result in a material wave whose beat frequency is given by the difference of the two beating material waves bringing thus the material wavelength into the optical region. In the interference zone there will be then a wave function of two superimposed waves with two different energy levels of electrons which, due to scattering at the interference pattern, jump from one level to the other radiating thus energy of a frequency corresponding to the energy difference of the two states. Energy and momentum will be conserved by scattering at the interference pattern or at a harmonic oscillator potential created for example by a metallic diffraction grating as has been used in the Smith-Purcell-Salisbury effect. This latter idea of combining the SPS effect with the beating phenomenon had been suggested by K. Mizumo, S. Ono, and O. Shimoe[93].

The beating phenomenon has been established experimentally by G. Möllenstedt et al.[94] when they measured an ac current due to the interference of two electron waves that had been generated by an interferometer similar to the one depicted in Fig. 7. One branch had been accelerated "mechanically" by reflecting it from a piezo-electric crystal being subjected to continuously increasing voltage; thus moving its surface at a small rate causing a minute velocity difference ($\Delta v_e \simeq 2$ cm/s) of the electrons of this branch in relation to the other. This led to a beat frequency of 1 Hz which was exactly the frequency of the measured ac current.

Recently O. Schwerzer[95] has worked out a theory of the wave mechanical beating and the Doppler effect of two electron waves. The importance of the coherence length of the electron wave is stressed.

The success of building FELs depends very much on this aspect of electron-optical technology. Unfortunately increasing the electron current decreases the coherence length. The electron source should be as cold as possible; therefore, field emission cathodes should be used.

Also, in the case of the wave mechanical light generator in this chapter, recycling of the electron beam should increase the gain substantially. Certainly a cavity with reflecting mirrors would turn this wave mechanical light generator into an oscillator and amplifier. The gain would be comparable with relativistic high power FEL.

Its advantage over such FEL would be the increasing gain with the square of the frequency and its low voltage electron generator avoiding the relativistic linear accelerators.

REFERENCES

1. E. T. Jaynes, "Ancient History of Free-Electron Devices" in "Novel Sources of Coherent Radiation:, Edited by S. F. Jacobs, M. Sargent III, and M. O. Scully (Addison-Wesley Publishing Co., Reading, Mass. 1978) pp. 1-39.
2. J. Schwinger, Phys. Rev. 70, 798 (1946); 75, 1912 (1949).
3. F. R. Elder, R. V. Langmuir, and H. C. Pollock, Phys. Rev. 74, 52 (1948).
4. E. J. Gorn, U.S. Patent No. 2,591,350 applied April 26, 1947, granted April 1, 1952.
5. W. W. Salisbury, U.S. patent No. 2634372, applied October 26, 1949, granted April 7, 1953.
6. W. W. Salisbury, Science 154, 386 (1966).
7. S. J. Smith and E. M. Purcell, Phys. Rev. 92, 1069 (1953).
8. H. H. Klinger, IEEE Proc. 51, 1367 (1963).
9. H. H. Klinger, "Funk und Ton" (in German) 8, 509 (1954).
10. A. J. Fox and N.W.W. Smith, IEEE Proc. 52, 429 (1964).
11. G. W. Stroke, in "Handbuch der Physik", Edited by S. Flügge (Springer Verlag 1967) Vol. XXIX, Article: "Diffraction Gratings" pp. 440-443 (in English).
12. All formulae in this paper are written in such a way that the SI system of units (Système Internationale) can be used which is identical with the rationalized MKSA system.
13. H. Motz, J. Appl. Phys. 22, 527 (1951).
14. H. Motz, Physics Today 32, 15 and 78 (Dec. 1979).

15. R. Kompfner, IRE Proc. $\underline{35}$, 124 (1947).

16. H. Motz, Phys. Lett. $\underline{71A}$, 41 (1979).

17. N. M. Kroll, "The Free Electron Laser as a Traveling Wave Amplifier" In "Novel Sources of Coherent Radiation" Edited by S. F. Jacobs, M. Sargent III, and M. O. Scully. (Addison-Wesley Publishing Co., Reading, Mass. 1978), pp. 115-156.

18. H. Motz, W. Thon, and R. N. Whitehurst, J. Appl. Phys. $\underline{24}$, 826 (1953).

19. J. Schneider, Phys. Rev. Lett. $\underline{2}$, 504 (1959).

20. A. J. Fox and N. W. W. Smith, IEEE Proc. $\underline{52}$, 429 (1964).

21. R. H. Pantell, G. Soncini, and H. E. Puthoff, IEEE J. Quantum Electron $\underline{4}$, 905 (1968).

22. H. Schwarz and H. Hora, Appl. Phys. Lett. $\underline{15}$. 349 (1969).

23. H. Schwarz, Bull. Am. Phys. Soc. $\underline{13}$, 897 (1968); Trans. N.Y. Acad. Sci., $\underline{33}$, 150 (1971); Proceedings 2nd International Conf. on Light Scattering in Solids, Paris 1971; edited by M. Balkanski (Flammarion-Sciences-Press) pp. 123-127; Appl. Phys. Lett. $\underline{19}$, 148 (1971); ibid. $\underline{20}$, 148 (1972); Proceedings of 1st Conference on Interaction of Electrons with Strong Electromagnetic Field, Balatonfüred, Hungary 1972, (edited by J. Bakos), pp. 39-57; and Proceedings of 2nd Workshop on Laser Interaction and Related Plasma Phenomena (edited by H. Schwarz and H. Hora)(Plenum Press, N. Y. 1972) pp. 209-225; Bull. Am. Phys. Soc. $\underline{19}$, 642 (1974); Proceedings 2nd Conference on Interaction of Electrons with Strong Electromagnetic Field, Budapest, 6-10 Oct. 1975, pp. 292-341 (Central Research Institute, Budapest, Sept. 1977).

24. C. K. Chen, J. C. Sheppard, M. A. Piestrup, and R. H. Pantell, J. Appl. Phys. $\underline{49}$, 41 (1978).

25. A. Salat, J. Phys. C $\underline{3}$, 2509 (1970); R. L. Harris and R. F. Smith, Nature $\underline{225}$, 502 (1970); A. R. Hutson, Appl. Phys. Letters $\underline{17}$, 343 (1970); L. L. van Zandt and J. W. Meyer, J. Appl. Phys. $\underline{41}$, 4470 (1970); E. E. Bergman, Nuovo Cimento $\underline{143}$, 243 (1973); L. A. Bolshov, A. M. Dykhne, and V. A. Ryakov, Phys. Lett. 42A, 259 (1973); C. S. Chang and P. Stehle, Phys. Rev. A $\underline{5}$, 1928 (1972); M. I. Dyakonov and D. A. Varshalovich, Phys. Lett. $\underline{35A}$, 277 (1971); J. A. Elliot, Journ. Phys. C. $\underline{5}$, 1976 (1972); L. D. Favro, D. M. Fradkin, and P. K. Kuo, Lett. Nuovo Cimento, $\underline{4}$, 1147 (1970); Phys. Rev. D $\underline{3}$, 2934 (1971); L. D. Favro, D. M. Fradkin, P. K. Kuo, and W. B. Rollnik, Appl. Phys. Lett., $\underline{19}$, 378 (1971); Bull. Am. Phys. Soc. $\underline{16}$, 25 (1971); J. D. Gibbon and R. K. Bullough, Journ. Phys. C, $\underline{5}$, L80 (1972); G. R. Hadley, J. J. Stanek, and R. H. Good, Jr., Journ. Appl. Phys. $\underline{43}$, 144 (1972); V. M. Horoutmian and A. K. Avetissian, Phys. Lett. $\underline{44A}$ 281 (1973); A. G. M. Janner and P. L. LaFleur, Phys.

Lett. 36A, 109 (1971); Yu. S. Korobockko, B. D. Grachev,
and W. I. Muney, Zurn. Tekh. Fiz. 42, 2422 (1972)1 H. J.
Lipkin, Journ. Appl. Phys. 43, 3011 (1972)1 G. B. Lubkin,
Physics Today (June 1971), p. 17, I. M. Makhrila, Sov.
Phys. Nucl. Phys. 15, 305 (1972); H. Masakazu, German
Patent 214 86 471 (1972); B. M. Oliver and L. S. Cutler,
Phys. Rev. Lett. 25, 273 (1970); L. A. Rivlin, JETP
Lett. 13, 257 (1971); K. C. Rogers, Science Year, The
World Book of Science (Chicago, Ill. 1972), p. 355; P.
L. Rubin, JETP Lett. 11, 239 (1970); R. W. Schmieder,
Bull. Amer. Phys. Soc. 16, 833 (1971); Lawrence Berkeley
Labs. Rpt. LBL-252 (September 1971), Appl. Phys. Lett.
20, 516 (1972); K. P. Sinha, Current Science 41, 124
(1972); L. L. vanZandt, Appl. Phys. Lett. 17, 345 (1970);
R. M. Bevensee, Lawrence Radiation Laboratory Report
94550, TID-4500, UC-34 Physics UCRL 51050, May 1971;
G. D. Ward, Diss. Univ. Maryland (1971); D. A.
Varshalovich and M. I. Dyakonov, JETP Lett. 11, 411
(1970); Zurn, Eksp. Teor. Fiz. 60, 90 (1971); Review/
Ideen der exakten Wissenschaften, Stuttgart (May 1972)
(No. 5), p. 299; P. S. Farago and R. M. Sillitto, Proc.
Roy. Soc. Edinburgh A 71, 305 (1974); H. Hora, Phys.
stat. sol. (b) 80, 143 (1977). C. Becchi and G.
Morpurgo, Phys. Rev. D 4, 288 (1971); Appl. Phys. Lett.
21, 123 (1972); J. Kondo, J. Appl. Phys. 42, 4458 (1971).
D. Marcuse, J. Appl. Phys. 42, 2255 and 2259 (1971);
P. L. LaFleur, Lett. Nuovo Cimento 8, 520 (1973), and
unpublished; P. W. Hawkes, "Coherence in Electron Optics"
Section 2. "The Schwarz-Hora effect" in "Advances in
Optical and Electron Microscopy" Edited by V. E. Cosslett
and R. Barer (Academic Press 1978) pp. 162-184.
26. R. Hadley, D. W. Lynch, E. Stanek, and E. A. Roasauer,
    Appl. Phys. Lett. 19, 145 (1971); H. Schwarz, ibid. 19,
    148 (1971).
27. L. Pfeiffer, D. L. Rousseau, and A. R. Hutson, Appl. Phys.
    Lett. 20, 147 (1972); H. Schwarz, ibid. 20, 148 (1972).
28. G. Gould, D. B. Scarl, and L. Silverstein, Annual Summary of
    Research in Electronics at Polytechnic Institute of
    Brooklyn, Microwave Research Institute, New York, 1972,
    p. 141
29. L. R. Elias, W. M. Fairbank, J. M. J. Madey, H. A. Schwettman,
    and T. I. Smith, Phys. Rev. Lett. 36, 717 (1976).
30. D. A. G. Deacon, L. R. Elias, J. M. J. Madey, G. J. Ramian,
    H. A. Schwettman, and T. I. Smith, Phys. Rev. Lett. 38,
    892 (1977).
31. J. M. J. Madey, J. Appl. Phys. 42, 1906 (1971).
32. C. A. Brau, These Proceedings, pp.
33. A. Bambini, A. Renieri, and S. Stenholm, Phys. Rev. A19,
    2013 (1979).

34. L. R. Elias et al., Stanford University Synchrotron Radiation
    Project Report No. 77/05, 1977.
35. H. Schwarz and R. Tabensky, in Laser Interaction and Related
    Plasma Phenomena, Editors: H. Schwarz and H. Hora (Plenum
    Press, New York 1977) Vol. 4B, pp. 961.
36. W. B. Colson, Phys. Lett. 59A, 187 (1976).
37. W. B. Colson, Phys. Lett. 64A, 190 (1977).
38. W. B. Colson, Ph.D. Thesis submitted to Stanford University,
    Physics Department, Sept. 1977.
39. A. Bambini and A. Renieri, Lett. Al Nuovo Cimento 21, 399 (1978).
40. W. Heitler, The Quantum Theory of Radiation (Oxford 1960).
41. A. Bambini and S. Stenholm, Preprint Series in Theoret. Physics,
    Research Institute for Theoret. Phys. University of
    Helsinki, Finland, No. HU-TFT-79-20. 9 pages.
42. A. Bambini, A. Renieri, and S. Stenholm, Phys. Rev. A 19,
    2013 (1979).
43. A. Bambini and S. Stenholm, Preprint Series in Theoretical
    Physics, Research Institute for Theoretical Physics,
    University of Helsinki, Finland, No. HU-TFT-79-24. 31 pages.
44. A. Bambini, R. Bonifacio, and S. Stenholm, Prepring Series in
    Theoretical Physics, Research Institute for Theoretical
    Physics, University of Helsinski, Finland, No. HU-TFT-79-33.
    8 pages.
45. G. Dattoli and A. Renieri, Lett. Nuovo Cimento 24, 121 (1979).
46. C. A. Brau, IEEE J. Quantum Electron., QE - 16, 335 (1980).
47. P. Sprangle, Cha-Mei Tang, and W. M. Manheimer, Phys. Rev.
    Lett. 43, 1932 (1979).
48. P. Sprangle, Cha-Mei Tang, and W. M. Manheimer, Phys. Rev.
    A 21, 302 (1980).
49. A. T. Lin and J. M. Dawson, Phys. Rev. Lett. 42, 1670 (1979).
50. T. Kwan and J. M. Dawson, Phys. Fluids 22, 1089 (1979).
51. W. J. Cocke, Optics Commun. 28, 123 (1979).
52. W. B. Colson and S. K. Ride, Preprint HEPL 815 Physics Depart-
    ment, Stanford University, March 1978.
53. S. K. Ride and W. B. Colson, Appl. Phys. (Germany) 20, 41 (1979).
54. M. Borenstein and W. E. Lamb, Jr., Phys. Rev. A 5, 1298 (1972).
55. G. Bekefi and R. E. Shefer, J. Appl. Phys. 50, 5158 (1979).
56. G. Bekefi, Preprint PFc/Ja-79-13 Plasma Research Report MIT,
    Plasma Fusion Center, Cambridge, MA 02139 (PRR 79/21,
    October 1979).
57. A. Hasegawa, Bell Syst. Tech. J. 57, 3069 (1978).
58. T. Kwan, J. M. Dawson, and A. T. Lin, Phys. Fluids 20, 581
    (1977).
59. T. Kwan and B. B. Godfrey, IEEE Trans. Nucl. Sci. NS-26, 3833
    (1979).
60. L. R. Elias et al., Stanford High Energy Physics Laboratory
    Report No. HEPL-812, 1976; Appl. Phys. (Germany) 19, 97
    (1979); Phys. Rev. Lett. 38, 892 (1977); Appl. Phys. (Germany)
    19, 295 (1979); Errata: 20, 196 (1979).
61. R. H. Helm, IEEE Trans. Nucl. Sci. NS-26, 3824 (1979).

62. R. Barbini, G. Dattoli, T. Letardi, A. Marino, A. Renieri, and
     G. Vignola, IEEE Trans. Nucl. Sci. NS-26, 3836 (1979).

63. L. R. Elias, Phys. Rev. Lett. 42, 977 (1979).

64. F. A. Hopf, P. Meystre, M. O. Scully, and N. H. Louisell, Phys.
     Rev. Lett. 37, 1215 (1976).

65. N. M. Kroll and W. A. McMullin, Phys. Rev. A 17, 300 (1978).

66. B. W. Boreham and H. Hora, Phys. Rev. Lett. 42, 776 (1979).

67. B. W. Boreham and B. Luther-Davies, J. Appl. Phys. 50, 2533
     (1979).

68. H. Hora, Proceedings of 2nd Internat. Conf. Energy Storage,
     Compression and Switching, Venice, December 5-8, 1978.
     (Ed.: H. Sahling, Plenum Press, New York 1979).

69. H. Hora, B. W. Boreham, and J. L. Hughes, Sov. J. Quantum
     Electron. 9, 464 (1979).

70. H. Hora, Laser Plasmas and Nuclear Energy, (Plenum Press,
     New York, 1975).

71. H. Hora, E. L. Kane, and J. L. Hughes, J. Appl. Phys. 49,
     923 (1978).

72. H. Hora, Z. Phys. 226, 156 (1969).

73. R. Klima and V. A. Petrzilka, Czech J. Phys. B 22, 896 (1972).

74. H. Schwarz, Phys. Rev. Lett. 42, 1141 (1979); Errata ibid.
     43, 238 (1979).

75. C. W. Barnes and K. G. Dedrick, J. Appl. Phys. 37, 411 (1966).

76. G. Toraldo di Francia, Nuovo Cimento 16, 61 (1960).

77. K. Mizuno, S. Ono, and O. Shimoe, Nature 253, 184 (1975);
     O. S. Heavens, Nature 253, 157 (1975).

78. J. M. Wachtel, J. Appl. Phys. 50, 49 (1979).

79. A. Gover and A. Yariv, Appl. Phys. (Germany) 16, 121 (1978).

80. A. Gover and A. Yariv, "Collective and Single Electron Inter-
     actions and Cerenkov-Smith-Purcell Free Electron Lasers"
     in "Novel Sources of Coherent Radiation" Edited by S. F.
     Jacobs, M. Sargent III, and M. O. Scully (Addison-Wesley
     Publishing Co., Reading, Mass. 1978) pp. 197-240.

81. M. A. Kumakhov, Phys. Lett. A 57, 17 (1976); Phys. stat. sol.
     (b) 84, 41 (1977); R. L. Walker, B. L. Berman, and S. D.
     Bloom, Phys. Rev. A 11, 736 (1975). R. W. Terhune and
     R. H. Pantell, Appl. Phys. Lett. 30, 265 (1977); R. H.
     Pantell and M. J. Alguard, J. Appl. Phys. 50, 798 (1979).

82. M. J. Alguard, R. L. Swent, R. H. Pantell, B. L. Berman, S.
     D. Bloom, and S. Datz, IEEE Trans. Nucl. Sci. NS-26, 3865
     (1979).

83. R. L. Swent, R. H. Pantell, M. J. Alguard, B. L. Berman, S.
     D. Bloom, and S. Datz, Phys. Rev. Lett. 43, 1723 (1979).

84. Editor: Science News 116, 389 (1979).

85. V. V. Beloshitskii and M. A. Kumakhov, Sov. Phys. JETP 47,
     652 (1978).

86. V. L. Ginzburg and V. N. Tsytovich, Physics Reports 49,
     No. 1, 1-89 (1979).

87. G. C. Baldwin, These Proceedings, Vol. 3B, pp. 875.

88. G. Möllenstedt and H. Düker, Z. Physik 145, 377 (1956); W.Bayh,
       ibid., 169, 492 (1962).
89. A. Peres, Phys. Rev. A 20, 2627 (1979).
90. H. Schwarz, Phys. Rev. A 20, 2628 (1979).
91. M. Peshkin, Phys. Rev. A 20, 2629 (1979).
92. C. A. Holt, Introduction to Electromagnetic Fields and Waves
       (John Wiley & Sons, New York 1963) pp. 276.
93. K. Mizuno, S. Ono, and O. Shimoe, Private Communication,
       August 1979.
94. G. Möllenstedt and H. Lichte, Proceedings of the Ninth Inter-
       national Congress on Electron Microscopy, Toronto, 1978
       (Electron Microscopy Society of America, New Castle County,
       Delaware, 1979), Vol. I, pp. 178-179.
95. O. Scherzer, Optik 54, 315 (1979).

COMMENTS BY H. HORA TO THE REMARKS FOLLOWING EQ. (62)

H. Schwarz reviews my preceeding publications on the free electron
laser amplifier based on the nonlinear force. I based my calculations on electron beams of MeV energy and current densities less
than the 10^{11} Amp/m^2 realised by Yonas et al.[1]. If there are discrepancies to the Langmuir condition, this should not surprise.
There was the controversy between Ready et al.[2] and Honig[3] whether
electron currents from a laser irradiated target should be some
milliampère according to the Langmuir condition or some 100 ampère
as observed. Later measured kiloampère[4] were definitely non-classical. The oberserved higher currents are just explaining the difference in the orders of magnitudes. My calculations were based on
observations. Why classical theories are nonvalid or non-applica-
ble is not necessary to be answered in such a case.

1. G. Yonas, et al., Plasma Physics and Controlled Nuclear
       Fusion, (IAEA, Vienna, 1979), Vol. III, p. 129.
2. D. Lichtman and J.F. Ready, Phys. Rev. Lett. 10, 342 (1963)
3. R.E. Honig, Appl. Phys. Lett. 3, 8 (1963).
4. G. Siller, et al., These Proceedings Series, Vol. 2, p.254
       (1972).

RESPONSE OF H. SCHWARZ TO THE ABOVE COMMENTS

In his comments H. Hora quotes correctly measurements of laser-
induced electron emissions. However, in all those measurements
and his remarks there is nothing stated about the energy spread,
whereas this is definitely an important factor in making his
interesting scheme work. The energy spread determines the divergence angle θ (see Eq. 61), which enters in Eq. (59) for the
determination of the effective current density. Hereby the
current density measured near the cathode is reduced by a factor
smaller than $\sin^2\theta$ (= 2×10^{-5}, using H. Hora's example).

FREE ELECTRON LASERS

Charles A. Brau

Los Alamos Scientific Laboratory

Los Alamos, NM 87545

Free electron lasers represent a new and exciting class of
potentially high power lasers. Although they are too new for their
problems and limitations to have been adequately explored, they have
the potential to become tunable, efficient, powerful sources of
coherent radiation. The purpose of the present talk will be to
describe, in a simple way, how free electron lasers operate and to
review the experiments which have been performed. In addition,
several of the proposed technological approaches to high power and
efficiency will be examined, and their possible applications briefly
discussed.

Fundamentals

Although the operation of a free electron laser was first
predicted by means of quantum mechanics,[1] there is nothing inherently quantum mechanical about a free electron laser. In fact, $\hbar$
appears nowhere in the final formulas, at least for photon energies
small compared with the electron energy (many MeV). Accordingly, it
is more convenient and powerful to adopt a classical point of
view,[2,3] which we shall do here.

Basically, a free electron laser consists of a relativistic
electron beam, a spatially periodic, magneto-static "wiggler" field,
and an optical radiation field propagating in the direction of the
electron beam, as shown in Fig. 1. The wiggler field is aligned
normal to the electron beam and alternates its sign periodically
along the direction of propagation of the electron beam. As the

                                                                C. A. BRAU

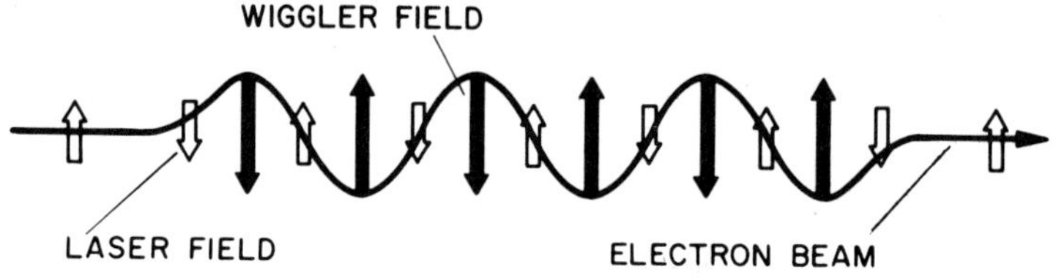

Fig. 1.   Schematic diagram illustrating the basic principle of a free electron laser. As the electrons pass through the free electron laser, the periodically varying force of the wiggler magnet (solid arrows) acts on the electrons, forcing them to oscillate transversely. The oscillating field of the laser (open arrows) does work on the oscillating electrons to retard (as shown) or accelerate them, depending on the relative phase of the electron and laser oscillations.

electrons pass through the wiggler field they undergo forced transverse oscillations, as shown. As they oscillate the electrons spontaneously emit "magnetic bremsstrahlung" peaked in the forward direction. To obtain stimulated emission, a laser field is superimposed on the oscillating electrons, as shown. Since the electric field of the laser radiation is transverse, it is capable of doing work on the transversely oscillating electrons. However, inasmuch as the electric field is rather weak, many transverse oscillations are needed to accelerate or decelerate the electrons very much. Nevertheless, when the laser frequency is nearly resonant with the electron oscillations, significant amounts of energy can be exchanged between the electrons and the laser field. The condition for this resonance is given by the expression (in MKS units)

$$\gamma_r^2 = \frac{\lambda_W}{2\lambda_L} \left(1 + \frac{e^2 B^2 \lambda_W^2}{4\pi^2 m^2 c^2} \right) , \tag{1}$$

where γ_r is the energy of a resonant electron divided by its rest energy, λ_W is the wiggler period, λ_L the laser wavelength, e the electron charge, B the rms magnetic induction of the wiggler, m the electron rest mass and c the velocity of light. The term in parentheses represents a correction (generally of the order of unity) for the change in path length caused by the electron wiggles in large magnetic fields. The allowable detuning of the laser from the resonance condition, the "gain bandwidth" of the free electron laser amplifier, is given by the approximate expression

$$\frac{\Delta\lambda}{\lambda} \sim \frac{1}{2N} \qquad (2)$$

where N is the number of magnet periods along the length of the wiggler.

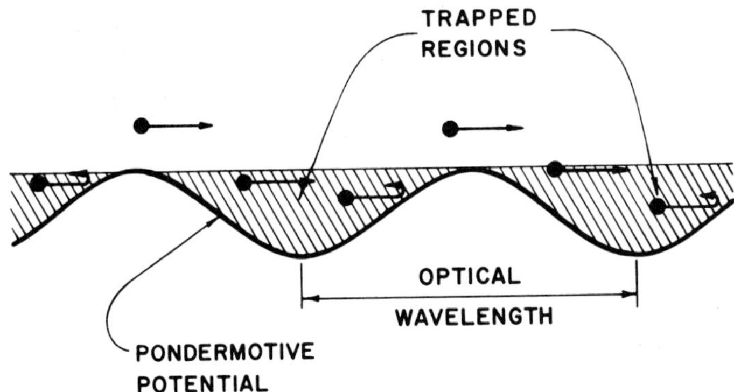

Fig. 2.　Schematic diagram of the electrons moving in the pondero-motive potential. Depending on the phase of the electrons with respect to the optical field, the electrons are accelerated or decelerated. The electrons in the trapped regions oscillate longitudinally at the synchrotron frequency and, after many periods, lose all memory of their initial velocity.

Whether an electron is accelerated or decelerated by the laser radiation depends on the phase of the optical field with respect to the electron's oscillation. Some electrons are accelerated, while those half an optical wavelength ahead or behind are decelerated. To first order, as many electrons are accelerated as are decelerated, and there is no net energy exchange between the electron beam and the laser beam. The electrons cannot, of course, exchange energy with the static magnetic wiggler field. However, as the accelerated electrons catch up to those which have been decelerated, they form into bunches at the optical wavelength. Once bunched, they can give up (or absorb) a net amount of radiation. To see how this happens, it is useful to think of the electrons as moving in a periodic potential (called the "ponderomotive" potential) which moves at a velocity corresponding to the resonant electron energy γ_r, given by Eq. (1). The periodic acceleration and deceleration of the electrons in this potential is analogous to that of marbles rolling on a corrugated roof, as shown in Fig. 2. The period of the corrugations corresponds to the laser wavelength, while the depth of the corrugations depends on the strength of the laser and wiggler

fields. Electrons entering the wiggler with the resonant velocity
correspond to marbles situated on the roof with no initial velocity.
Clearly, they never get anywhere, on the average, which is why there
is no net energy exchange (and therefore no gain or loss) for resonant electrons. Electrons entering the wiggler with a velocity
greater than the resonant value correspond to particles situated on
the roof with an initial velocity to the right. These nonresonant
electrons experience some average deceleration, which amounts to net
gain in the laser. This is clearly true for electrons close enough
to resonance to be trapped in the ponderomotive potential; after
many so-called "synchrotron" oscillations in the "buckets" of the
ponderomotive potential their average velocity decreases from the
initial velocity to zero. There is also a small initial deceleration
for electrons too far from resonance to be trapped. However, this
averages out after several periods of the pondermotive potential.
We see from this analogy that the limit for extracting energy from
the electrons in such a laser is set by the difference between the
initial velocity and the resonant velocity. Since the maximum non-
resonance for which gain is observed is given by Eq. (2), the
saturated energy extraction efficiency is given by the expression

$$\eta_X \; \lesssim \; \frac{1}{2N} \; . \tag{3}$$

In general, the width of the final electron energy distribution
function, corresponding to the rattling around of the electrons in
the buckets, will be comparable to the depth of the ponderomotive
potential. For a strongly saturated laser, in which most of the
electrons are trapped, this spread can be much larger than the
energy extraction. We will see the importance of this later.

    Fortunately, it may be possible to extract more energy than
this by using techniques related to those used in rf accelerators.[4,5]
Once the electrons are trapped in buckets moving at the velocity
corresponding to γ_r, the buckets - with the electrons in them - can
be slowed down. This is accomplished by using a nonuniform, or
"tapered" wiggler in which the bucket velocity (γ_r) becomes progressively smaller as the wiggler period is reduced (see Eq. (1)).
The principle difficulty with this approach is that to trap and
decelerate a significant fraction of the electrons requires very

---

*The synchrotron frequency corresponding to longitudinal oscilla-
tions in the ponderomotive "buckets" should not be confused with the
frequency of the transverse oscillations which occur at the optical
frequency. This is clear from the fact that the longitudinal accel-
erations represent the cumulative work of the transverse electric
field after many oscillations. The synchrotron frequency, corresponding to the depth of the ponderomotive buckets, depends on the
strength of the optical and wiggler fields. A typical laser having
hundreds of transverse oscillations (wiggler periods) along its
length may have a few synchrotron oscillations.

large (GW) laser intensities. The faster the buckets are decelerated, the more the electrons tend to slosh forward in the buckets.
and spill out, becoming detrapped. For initially resonant electrons,
the optimum balance between deceleration and trapping corresponds to
about 40% capture.[6]   A further difficulty of tapered wigglers is
their low small-signal gain.[7]

   Up to this point the discussion has addressed the behavior of
individual electrons in the laser and wiggler fields. When this is
a satisfactory approximation, the free electron laser is said to be
in the "Compton" regime. At sufficiently high electron-beam current
densities, electron-electron interactions must also be taken into
account.[8,9] These take the form of plasma oscillations which may
become excited under conditions when the plasma frequency is comparable to the transit time of the electrons through the wiggler, or
larger. For a cold (monoenergetic) electron beam this condition may
be expressed

$$\frac{\omega_p \tau}{\gamma^{3/2}} \gtrsim 1 \; , \tag{4}$$

where $\omega_p/\gamma^{3/2}$ is the (Lorents transformed) plasma frequency and τ is
the time for the beam to traverse the length of the wiggler. Under
conditions when Eq. (4) is satisfied, negative energy plasma waves
can become strongly excited in the beam by a parametric instability.
The resonant laser frequency for this interaction is shifted downward from the value given by Eq. (1) by the plasma frequency. Both
the small signal gain and the energy extraction efficiency are
greatly enhanced by collective interactions, even with uniform
period wigglers. The saturated extraction efficiency in the presence
of collective interactions is given by the formula

$$\eta_X \sim \frac{1}{2N} \frac{\omega_p \tau}{\pi \gamma^{3/2}} \; . \tag{5}$$

Comparing this with Eq. (3), we see that the saturated extraction
efficiency is enhanced by the factor $\omega_p \tau / \pi \gamma^{3/2}$, which may be quite
large. Extraction efficiencies in excess of 20% are possible when
collective effects are used to enhance the performance of tapered
wigglers. Because of the factor $\gamma^{3/2}$ in the denominator of Eq. (4),
collective effects are most important at relatively low energies,
typically below 10 MeV, and correspondingly long wavelengths, generally in the far infrared region and beyond.

Experiments

   At the present time, the experimental data on free electron
lasers are rather meager. The theory has progressed rather rapidly

since the electrons and fields in vacuum which make up a free
electron laser provide a tractable system. On the other hand, the
experiments require powerful electron accelerators with unique
capabilities. Nevertheless, data have been obtained in both the
Compton (single-particle) and Raman (collective interaction) regimes.
In general, the agreement with theoretical predictions has been
good, which is most encouraging for the future of free electron
lasers.

Up to the present time, all the experiments in the single
particle regime have been carried out by Madey and his coworkers
using the Stanford superconducting linac.[10,11] This machine has
unusually good energy resolution (a few parts in 10^4) and collima-
tion, which are achieved at the expense of relatively low peak
current and short "microbunch" length. As in all rf linacs, the
electrons are accelerated only during a brief interval in the rf
cycle during which the accelerating fields reach their maximum
value. The electron microbunches which emerge at the peak of every
rf cycle (or subharmonic thereof) have a very short duration, about
4 ps (1.3 mm) in the Stanford experiments. In the first series of
experiments, the linac was adjusted to produce a 24 MeV electron
beam with a peak current of 70 mA. The wiggler had a length of
5.2 m, a period of 3.2 cm, and a magnetic field of 0.24 T. The gain
at 10.6 μm was measured with a 100 kW CO_2 laser. The observed gain
was 7% per pass, in good agreement with theory. This corresponds to
0.25% energy extraction from the electrons, which is close to the
theoretical saturated value of 0.31% given by Eq. (3). In a second
series of experiments, the linac was adjusted to produce 43 MeV
electrons with a peak current of 2.6 A. An optical cavity was used
to provide feedback and oscillation was observed at 3.4 μm. The
peak power was 40 kW, corresponding to 0.14% efficiency. Again,
this is close to the theoretical saturated efficiency given by Eq.
(3). Because of the microbunch structure of the electron beam, the
laser output emerged in correspondingly short pulses, modelocked by
the gain medium. Therefore, it was necessary to tune the length of
the cavity so that the returning optical pulse entered the wiggler
coincident with an electron microbunch. It is interesting to note
that the laser linewidth in these experiments (7 cm^{-1}) was apparently
just the transform width of the 4 ps microbunches.

Further experiments on free electron lasers in the Compton
(single particle) regime are planned at several laboratories in the
US and in Europe. These experiments will make use of both rf electron linacs and storage rings. The principle thrust of most of these
experiments will be to see if the efficiency can be improved by
means of tapered wigglers.

Another set of experiments has been undertaken by groups at
TRW,[12] Columbia University,[13] and the Naval Research Laboratory[14] to
explore free electron lasers operating in the Raman (collective

interaction) regime. In a joint experiment[15] using a 1.2 MeV, 25 kA electron beam, the Columbia and NRL groups produced a peak output of about 1 MW at a wavelength of 400 μm. The laser bandwidth, about 2%, was much narrower than the spontaneous emission bandwidth. The observed efficiency was about 0.02%, which is much less than the theoretical saturated efficiency of a few percent predicted by Eq. (5). The discrepancy is probably due to the inadequate beam uniformity and the fact that the laser did not reach saturation during the 30 ns electron beam pulse. Improved experiments are now being undertaken which should reach saturation and produce much improved efficiency. In addition, tapered wigglers will be tested to determine if even higher efficiency can be obtained.

Technology
<u>Technology</u>

     Basically, four configurations have been proposed for high power free electron lasers. These are illustrated schematically in Fig. 3. In the single pass approach, an accelerator is used to form the electron beam, which is passed once through the free electron laser and then discarded. The overall efficiency of such a device will be the product of the accelerator efficiency and the laser extraction efficiency, so that both must be high. In order to achieve high laser extraction efficiency, it is necessary to trap and strongly decelerate the electrons in the ponderomotive field. As discussed above, this requires enormous laser intensities. This approach has been proposed for producing short, powerful laser pulses for inertial confinement fusion, since the necessary high intensity is inherent in the application.[16] Alternatively, it should be possible to achieve extraction efficiencies in excess of 20% from relatively modest optical fields by using collective effects, as discussed above.[8] This should be a fruitful approach for infrared and longer wavelengths. To achieve shorter wavelengths at the low electron beam energies where collective interactions are strongest, an rf field propagating in opposition to the electron beam could be substituted for the magneto-static field to act as a short wavelength wiggler.[14,17,18] This requires rather high rf power, of the order of 1 Gigawatt or more, to provide the wiggler.

     Other approaches to improved efficiency make use of the energy remaining in the electron beam when it emerges from the wiggler. In the storage ring approach, shown in Fig. 3, the electrons lose a small fraction of their energy as they pass through the free electron laser. This energy is replaced by the rf accelerator structure in the ring as the electrons circulate around to re-enter the free electron laser. Very high efficiency is possible, at least in principle, by using this approach. The difficulty is that the energy spread induced by the laser action may exceed the energy extracted, as discussed earlier. This spread must be removed by synchrotron radiation damping as the electrons circulate around the ring before further laser energy may be extracted from the electron

beam. The laser power in such a system is limited by the rate at
which the synchrotron radiation can damp out the momentum spread,
and by the momentum spread acceptable to the free electron laser and
to the storage ring. Additionally, the efficiency is reduced by the
synchrotron radiation required to damp the momentum spread. For a
uniform wiggler, the necessary synchrotron losses reduce the efficiency to that of a single pass laser.[19] The use of a "gain-expanded"
wiggler has been proposed to increase the momentum spread acceptable
to the wiggler.[20] Alternatively, it may be possible to decrease the
energy spread induced in the electron beam by the free electron
laser by using a so-called "phase displacement" wiggler.[4] This idea
is related to techniques used in rf accelerators, and requires a
very slowly tapered wiggler. Despite the difficulties facing storage
rings, they are attractive for very short wavelengths - possibly as
short as the vacuum ultraviolet - because they operate at very high
electron energies with extremely well collimated (low emittance)
electron beams.

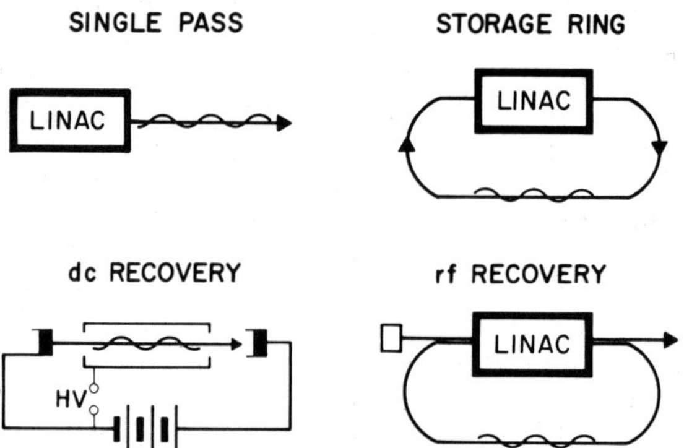

Fig. 3.   Schematic diagram of four configurations proposed for high
           power free electron lasers. The single-pass approach is
           simplest, but depends on high free electron laser extrac-
           tion efficiency to achieve high overall efficiency. The
           storage ring re-accelerates the electrons from the free
           electron laser to their original energy and re-uses them.
           The dc recovery system collects the spent electrons at a
           high potential to recover their energy. The rf recovery
           system decelerates the electrons in the accelerator to
           recover their energy before they are discarded.

The problems introduced by the momentum spread of the emerging
beam can be largely circumvented by recovering the energy of the
electrons without re-using the electrons themselves. The simplest
approach, at least conceptually, makes use of an electrostatic
accelerator, such as a Van de Graff accelerator, to to accelerate
the electrons for insertion into the free electron laser.[17] The
emerging electrons are then decelerated electrostatically and collected at the highest potential they can reach with their remaining
energy. They are then returned to the original potential by means
of a (relatively) small power supply, as indicated in Fig. 3. Very
high overall efficiencies should be possible with this approach,
even though the extraction efficiency of the free electron laser may
be much smaller. Due to the difficulty of building electrostatic
accelerators much beyond 25 MeV, this approach should find its
principal application at mid- to far-infrared wavelengths.

For shorter wavelengths, in and near the visible portion of the
spectrum, higher energy electrons (50-100 MeV) are required. These
are most conveniently generated with rf electron linacs. To recover
the energy of the electrons emerging from the free electron laser,
they are brought around and re-inserted into the linac as shown in
Figs. 3 and 4. It must be remembered that the electrons accelerated
by the linac emerge in short (< 100 ps) "microbunches" synchronized
to the phase of the rf fields in the accelerator cavity structure at
which maximum acceleration occurs. After being brought around, the
electron microbunches are inserted into the linac 180° out of phase
from the electrons being accelerated, so that the rf fields have
reversed. Thus, instead of being further accelerated, as would be
the case in a microtron, the electrons are decelerated and their
energy returned to the rf fields, as in a klystron. This energy is
then available for accelerating fresh electrons. Studies carried
out at Los Alamos indicate that it should be possible in this way to
build free electron lasers with as much as 25% overall efficiency,
operating at more than 100-kW average power. The short-wavelength
limit depends on the emittance of the electron beam, that is, the
degree to which it can be collimated and focussed.[6] Visible wavelengths should be possible with present technology, and improved
injector designs should extend this into the near ultraviolet.

Applications

Free electron lasers offer several potential advantages over
conventional lasers. These include continuous tunability, efficiency, power, beam quality and cost. Wavelength tuning is achieved
by varying the electron beam energy and/or the wiggler period and
magnetic field, as indicated by Eq. (1). Wavelengths from the
millimeter region to the ultraviolet appear possible. High overall
efficiency, as much as 40% or more, may be achieved by using tapered
wigglers, collective effects (in some regimes), and by recovering

the energy of the electrons emerging from the free electron laser,
as outlined above. High power is possible because high power, high
reliability electron accelerators already exist. SLAC, for example,
has an average beam power of 200 kW. By avoiding a gaseous medium
with its associated inhomogenieties, good optical beam quality can
be achieved. Also, the flow system and pump power are avoided.
Finally, low capital and operating costs should be possible. Since
the capital cost of large lasers is driven by the power supply and
cooling costs, which vary in proportion to the input power, high
efficiency should make possible low capital cost (per output Watt).
Studies by Los Alamos indicate that costs of the order of $50/Watt
(output) are possible for very large systems. Low operating cost
depends on efficiency and reliability. When the projected costs of
amortization and operation are added together, they amount to a few
cents per mole of photons at visible wavelengths.

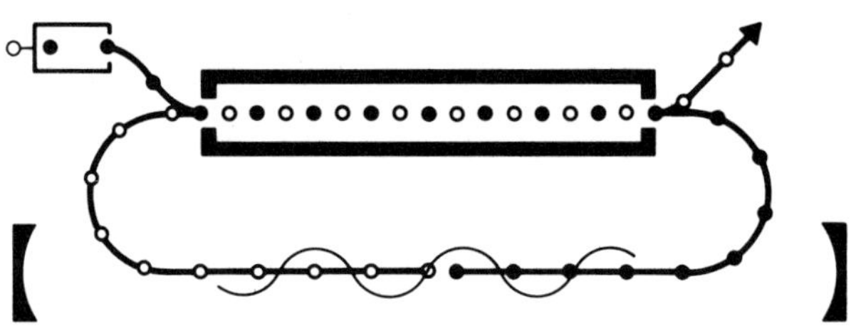

ENERGY ~100 MeV	**EFFICIENCY:**
POWER ~100 kW	LASER/e-BEAM ~10%
WAVELENGTH ~1 μm	LASER/rf ~40%
	LASER/dc ~25%

Fig. 4.   Schematic diagram of the rf racetrack energy recovery
            system. To decelerate the electrons emerging from the
            free electron laser they are re-inserted into the accel-
            erator 180° out of phase from the accelerating electrons
            so that the rf fields have reversed. High power and high
            overall efficiency can be achieved this way.

     A number of applications have been suggested for high power
lasers. Besides military applications (communications, radar,
weapons), industrial processing, such as welding, metalworking, and
especially chemical processing, looks very promising, at least

superficially. For chemical processing, tunability, power and, above all, cost are the important factors. In Fig. 5 are displayed the unit cost and annual production of a variety of chemicals. Also displayed are the unit costs of rare gas halide laser photons and free electron laser photons. The cost per pound for photons is projected on the basis of one near ultraviolet photon for each product molecule, presumed to have a molecular weight of 100. Evidently, all but the cheapest (and most widespread) chemicals would be accesible to photochemical processing on this basis. However, the most important applications of laser photochemistry will probably involve considerable leverage, requiring only one photon for many molecules of product. Laser purification of feed-stocks - removing a few impurities from many product molecules - and laser cross-linking of polymers are examples of highly leveraged processes. Such leveraging would bring the laser costs well below those indicated in Fig. 5, and make it worthwhile to think about large scale photochemistry.

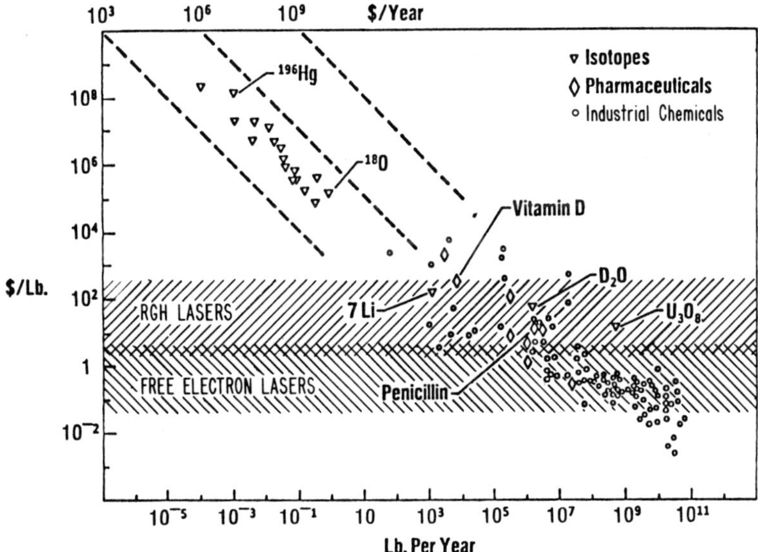

Fig. 5.   Chart showing cost vs. consumption of chemicals together with projected cost of photons from rare gas halide (RGH) and free electron lasers. One photon is presumed to be required for each product molecule of molecular weight 100.

Laser fusion, power transmission to satellites, laser propulsion for maneuvering in space and even transmission of power from a solar space station back to earth are other "far-out", high power applications which have been speculated about. Surely, it will be a long time before free electron lasers make a useful contribution to these areas. But even in the near term, free electron lasers will be a useful source of coherent radiation for spectroscopy in wavelength regions not otherwise accessible, such as the far infrared.[21] Free electron lasers have a long way to go yet, having barely gotten started. But if progress is made as rapidly as some hope, some of the speculation may become reasonable.

References

1. J. M. J. Madey, J. Appl. Phys. $\underline{42}$, 1906 (1971).
2. F. A. Hopf, P. Meystre, M. O. Scully and W. H. Louisell, Opt. Commun. $\underline{18}$, 4 (1976).
3. W. B. Colson, Phys. Lett. $\underline{59A}$, 187 (1976).
4. N. M. Kroll, P. L. Morton and M. N. Rosenbluth, "Variable Parameter Free Electron Laser," Workshop on Free Electron Generators of Coherent Radiation, Telluride, Colorado, August 13-17, 1979.
5. N. M. Kroll, P. L. Morton and M. N. Rosenbluth, "Enhanced Energy Extraction in Free Electron Lasers by Means of Adiabatic Decrease of Resonant Energy", Workshop on Free Electron Generators of Coherent Radiation, Telluride, Colorado, August 13-17, 1979.
6. C. A. Brau and R. K. Cooper, "Variable Wiggler Optimization," Workshop on Free Electron Generators of Coherent Radiation, Telluride, Colorado, August 13-17, 1979.
7. C. A. Brau, IEEE J. Quant. Electron. (in press).
8. T. Kwan, J. M. Dawson and A. T. Lin, Phys. Fluids $\underline{20}$, 581 (1977).
9. N. M. Kroll and W. A. McMullin, Phys. Rev. $\underline{A17}$, 300 (1978).
10. L. R. Elias, W. M. Fairbank, J. M. J. Madey, H. A. Schwettman and T. I. Smith, Phys. Rev. Lett. $\underline{36}$, 717 (1976).
11. D. A. G. Deacon, L. R. Elias, J. M. J. Madey, G. J. Ramian, H. A. Swettman and T. I. Smith, Phys. Rev. Lett. $\underline{38}$, 892 (1977).
12. M. Zales Caponi, J. Munch and H. Boehmer, "Optimized Operation of a Free Electron Laser, Spanning the Single Particle and Collective Regimes", Workshop on Free electron Generators of Coherent Radiation, Telluride, Colorado, August 13-17, 1979.
13. T. C. Marshall, S. Talmage and P. Efthimion, Appl. Phys. Lett. $\underline{31}$, 320 (1977).
14. V. L. Granatstein, S. P. Schlesinger, M. Herndon, R. K. Parker and J. A. Pasour, Appl. Phys. Lett. $\underline{30}$, 384 (1977).
15. D. B. McDermott, T. C. Marshall, S. P. Schlesinger, R. K. Parker and V. L. Granatstein, Phys. Rev. Lett. $\underline{41}$, 1368 (1978).

16. A. Szoke, V. K. Neil and D. Prosnitz, "A Tutorial Summary of Variable Wiggler Free Electron Lasers as Well as a Summary of Some Proposed Experiments", Workshop on Free Electron Generators of Coherent Radiation, Telluride, Colorado, August 13-17, 1979.

17. L. R. Elias, Phys. Rev. Lett. $\underline{42}$, 977 (1979).

18. D. A. Reilly, M. S. Tekula, R. M. Patrick, "Beam Heating Constraints on a Low Voltage Free-Electron Laser with Visible Output", Workshop on Free Electron Generators of Coherent Radiation, Telluride, Colorado, August 13-17, 1979.

19. C. Pellegrini, "Synchrotron Radiation Problems in Storage Ring Version of FEL", Workshop on Free-Electron Generators of Coherent Radiation, Telluride, Colorado, August 13-17, 1979.

20. T. I. Smith, J. M. J. Madey, L. R. Elias and D. A. G. Deacon, J. Appl. Phys. $\underline{50}$, 4580 (1979).

21. E. D. Shaw and C. K. N. Patel, "Theoretical Considerations for FEL's in the Far Infrared", Workshop on Free Electron Generators of Coherent Radiation, Telluride, Colorado, August 13-17, 1979.

NRL LIGHT ION BEAM RESEARCH FOR INERTIAL CONFINEMENT FUSION[†]

G. Cooperstein, Shyke A. Goldstein[*], D. Mosher
R. J. Barker[*], J. R. Boller, D. G. Colombant,
A. Drobot[**], R. A. Meger[*], W. F. Oliphant,
P. F. Ottinger[*], F. L. Sandel[*], S. J. Stephanakis,
and F. C. Young

Naval Research Laboratory
Washington, DC 20375

ABSTRACT

There is presently great interest in using light ion beams to drive thermonuclear pellets. Terrawatt-level ion beams have been efficiently produced using conventional pulsed power generators at Sandia Laboratory with magnetically-insulated ion diodes and at the Naval Research Laboratory with pinch-reflex ion diodes. Both laboratories have recently focused ion beams to pellet dimensions. This paper reviews recent advances made at NRL in the area of ion production with pinch-reflex diodes, and in the areas of beam focusing and transport. In addition, modular generator and beam requirements for pellet ignition systems are reviewed and compared with the latest experimental results. These results include the following: (1) production of $\geq$ 100 kJ proton and deuteron beams with peak ion powers approaching 2 TW on the PITHON generator in collaboration with Physics International Co., (2) focusing of 0.5 TW deuteron beams produced on the NRL Gamble II generator to current densities of about 300 kA/cm^2, and (3) efficient transport of 100 kA level ion beams over 1 meter distances using Z-discharge plasma channels.

[†] Work supported by Defense Nuclear Agency, Washington, DC 20305 and Department of Energy, Washington, DC 20545
[*] JAYCOR, 205 S. Whiting Street, Suite 500, Alexandria, VA 22304
[**] Science Applications, Inc., 8400 Westpark Dr., McLean, VA 22101

1. INTRODUCTION

The use of light ion beams (protons, deuterons, etc.) for inertial confinement fusion has been seriously pursued following the theoretical prediction[1-3] and the experimental documentation[4], that small-area diodes ($\sim$ 100 cm^2) can be used to generate ion beams efficiently at the megavolt-megampere level. Soon after this, results of ballistic beam focusing experiments were presented[5] and a technique for transporting ions in a plasma Z-discharge was introduced[6]. Recently, terrawatt-level ion beams with focused ion current densities in excess of 100 kA/cm^2 have been produced using water-dielectric transmission-line generators at Sandia Laboratories (SANDIA) with magnetically-insulated ion diodes[7] and at the Naval Research Laboratory (NRL) with pinch-reflex ion diodes[8].

Target designs for light ion beams[9] call for delivery of about 2 MJ to an $\sim$ 1 cm diameter pellet in an $\sim$ 10 ns time scale in order to achieve high-gain thermonuclear ignition. Present pulsed power technology provides up to 10 TW single generator modules from which up to 300 kJ of ions can be extracted in 50-100 ns. Thus, a large number of modules and the means to transport energy from these modules onto the pellet are needed. Additionally, compression of the pulse to the pellet-implosion time-scale is required.

Two different approaches addressing these problem areas have emerged. An approach researched by SANDIA[10] involves pulsed-power techniques for shortening the accelerating-voltage pulse, and self-magnetically insulated flow of electromagnetic energy in vacuum transmission lines which terminate in small ion diodes close to the pellet. Packing of transmission lines near the target, coupling of diodes to the lines, and ion-beam focusability will be investigated with the 36 short pulse, 1 TW PBFA I modules now in construction at SANDIA. For pellet ignition, this approach will require on the order of 100 modules.

A second approach, researched by NRL, involves the extraction and focusing of beams from self-insulated pinch-reflex ion diodes, coupled with transport of these focused beams in $\sim$ 1 cm diameter Z-discharge transport channels. Pulse compression during transport to the pellet-implosion time-scale is achieved by increasing the accelerating voltage with time. These techniques are appropriate for modules operating up to the 10 TW level with a 50-100 ns pulse duration. The number of modules required for this approach is on the order of 10.

In the present report, major results of a combined experimental and theoretical study of the second approach are reviewed with concentration on recent advances made at NRL in the area of ion production using pinch-reflex diodes, and in the areas of beam focusing and transport.

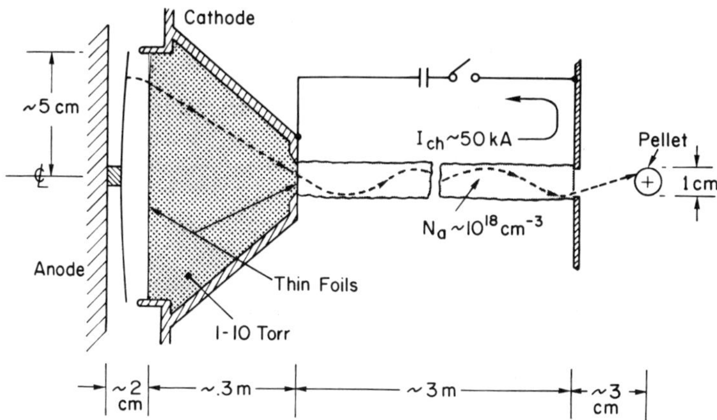

Fig. 1.   Conceptual schematic of NRL Light-Ion Inertial Confinement Fusion Approach.

A conceptual schematic of the NRL approach is shown in Fig. 1. The diode and transport system together with the pulsed power generator (not shown) is one module of a multi-module pellet-ignition system.   A pinch-reflex ion diode ($\sim$ 10 cm in diameter) produces the ions and properly aims them towards the transport system entrance aperture.   The ion beam then free streams in a gas-filled chamber towards an $\sim$ 1 cm diameter focus.   The gas in the drift region allows the ion beam to be highly charge- and current-neutralized.   Inside the channel, the ions are confined radially by the azimuthal magnetic field produced by the discharge of an external capacitor bank.   The $\sim$ 50 kA current in the channel is sufficient to provide radial confinement for ions entering the channel with transverse velocities below about 15% of their axial velocity.   The plasma density in the channel must be sufficiently high to provide inertial resistance to channel expansion forces during beam transit and sufficiently low to prevent excessive energy loss of the beam during transport.   Beam power multiplication of about a factor-of-five is achieved during transport to the pellet by ramping the accelerator voltage in time. Beams emerging from the channel propagate the last few centimeters to the pellet with low divergence because of the small transverse velocity.   Several ion beams can then be overlapped onto the pellet.

The second section of this paper discusses ion production in pinch-reflex ion diodes including recent experiments[11] in collaboration with Physics International Co. on the PITHON generator.   In these experiments, greater than 100 kJ of protons and deuterons were produced with peak ion powers approaching 2 TW.   In Sec. 3, the self-focusing of ion beams with planar diode geometries is reviewed

and ion focusing experiments with curved geometries are reviewed in
Sec. 4.   In these experiments, 0.5 TW deuteron beams produced on the
NRL Gamble II generator were focused to current densities of about
300 kA/cm^2.   In Sec. 5, the numerical simulation of pinch-reflex
ion diodes is discussed.   Theoretical work on ion orbits in the
transport channel is reviewed in Sec. 6, and in Sec. 7 results of
recent transport experiments are discussed.   In these experiments,
high current ion beams have been efficiently transported 1 meter
distances using Z-discharge plasma channels.   The MHD response of
the channel induced by beam passage is then considered in Sec. 8.
In Sec. 9, bunching of ion beams in transport channels is briefly
discussed.   Finally, the results of this research are used to determine a range of system parameters which are appropriate for driving
high-gain pellets with proton or deuteron beams.

2.   ION PRODUCTION

Self-pinched electron flow and laminar ion flow in a large-
aspect-ratio electron-beam diode are conceptually illustrated in
Fig. 2.   After an initial phase leading to self-pinched flow[12], the
electrons primarily originate from the edge of the cathode and flow
under the dominant influence of the self-magnetic fields towards
the diode axis ending up in a tight pinch at the center of the anode
plane.   The ions, which are primarily protons originating from the
desorbed gases making up the anode plasma, are only slightly bent

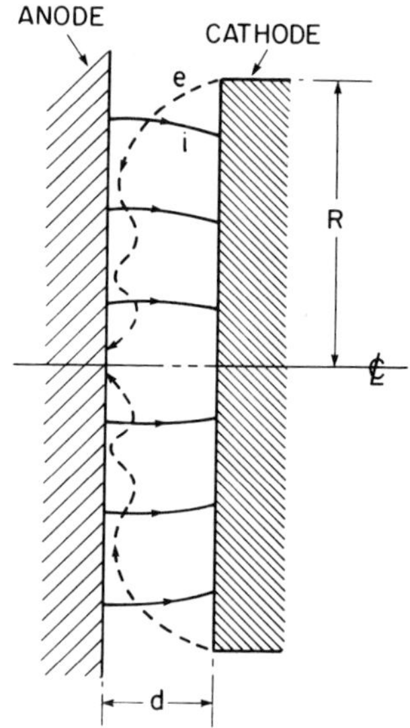

Fig. 2.   Conceptual schematic of electron and ion flow in large-
         aspect-ratio (R >> d) vacuum diode.

by the diode's self-magnetic fields due to their heavier mass. Thus
ions flow in almost straight lines toward the cathode as they accelerate up to the diode voltage. The enhanced ion currents come about
because of the additional electron space charge in the diode caused
by the large electron path length (R) relative to the ion path length
(d). For the simple diode configuration shown in Fig. 2 analytical
theory predicts[1] and numerical simulation confirms[13] an ion to electron current ratio of

$$\frac{I_i}{I_e} > \frac{R}{d}\left(\frac{eV}{2m_i c^2}\right)^{\frac{1}{2}}$$

where R is the diode radius, d is the anode-cathode gap spacing,
V is the diode voltage and m_i is the ion mass.

Early experimental results[14] with R $\simeq$ 6 cm and d $\simeq$ 0.4 cm
giving $(I_i/I_e) \sim 1/2$ were in excellent agreement with this theory.
These demonstrated 0.8 MeV proton currents of up to 200 kA. Using
spherical section electrode structures, focused deuteron current
densities of up to 70 kA/cm^2 were obtained[15]. These early results
were obtained at 0.5 TW with the Gamble II generator operated for
the first time in positive polarity. In positive-polarity operation,
the cathode is mounted on the door of the generator so that ions
accelerated through the diode potential can be injected into a drift
tube region through a thin cathode transmission foil.

In the most recent positive polarity experiments[8] on upgraded[16]
Gamble II at 1.5 TW, $\sim$ 60% conversion efficiency from generator
power to ion power was achieved using the refined diode design
illustrated in Fig. 3. A thin CH_2 foil stretched across a dielectric ring is electrically connected to the positive electrode by a
1 cm-diameter rod on the diode axis. Plasma formation due to surface
flashover causes the plastic foil to become a conductor early in the
pulse and provides a source of ions. This geometry enhances ion
emission by increasing the electron path length (and therefore lifetime) relative to that of ions by forcing the electrons to reflex
through the thin foil as they pinch in radially. This reflexing is
not due to spatial charge building up in the vacuum gap behind the
foil because this is quickly neutralized by ions from the anode
plasma. Rather, it is due to the azimuthal self-magnetic field
caused by the return current flow through the center conductor.
Thus, electron reflexing is magnetically induced behind the foil.
In front of the foil, the electrons reflex in the self-consistent
diode fields. The ions that are produced on the back of the anode
foil do not carry diode current because the back plate is at anode
potential. Results of the numerical simulation of this diode
configuration will be presented in Sec. 5.

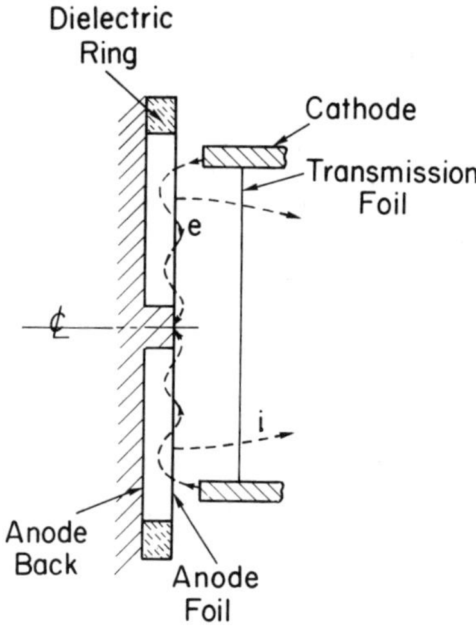

Fig. 3. Pinch-reflex diode schematic.

Typical diode electrical characteristics are shown in Fig. 4a
for a 6 cm cathode radius, a 0.4 cm anode-cathode gap spacing, and
a 0.5 cm vacuum gap behind the 0.01 cm CH_2 anode foil. Ion currents
of 500 kA have been measured with 1 MA total diode currents. This
corresponds to an average source ion current density of 5 kA/cm^2.
After passing through a phase associated with establishing the
charged-particle flow illustrated in Fig. 3, the diode impedance
remains well matched to the 1.5 ohm output impedance of the Gamble
II generator. Closure of the anode-cathode gap due to electrode
plasma motion usually occurs near the end of the electrical power
pulse. The diode impedance can be adjusted by varying the cathode
emitting area, the anode-cathode gap spacing, or the applied voltage.
The natural voltage ramp which occurs demonstrates that accelerating
voltages appropriate for ion bunching during transport to fusion
pellets can be achieved with these generators. The net current
signal (the current entering the cathode foil) is interpreted as the
total ion current flowing in the anode cathode gap. The location
of the Rogowski coil used to make this measurement will be illustrated later in Fig. 6. On a few selected shots, the measured net
current exceeded 700 kA at 1.3 MV out of 1.1 MA of total diode
current. Carbon activation[17] by proton beams indicates that greater
than 50% of the net-current signal is due to protons. The interpretation of these results is limited by an unknown correction for
deuteron activation. When the anode is coated with CD_2, neutron
time-of-flight measurements show that greater than 50% of the net
current takes the form of deuterons. Up to 10^{12} neutrons have
been measured on Gamble II. This data is consistent with ion

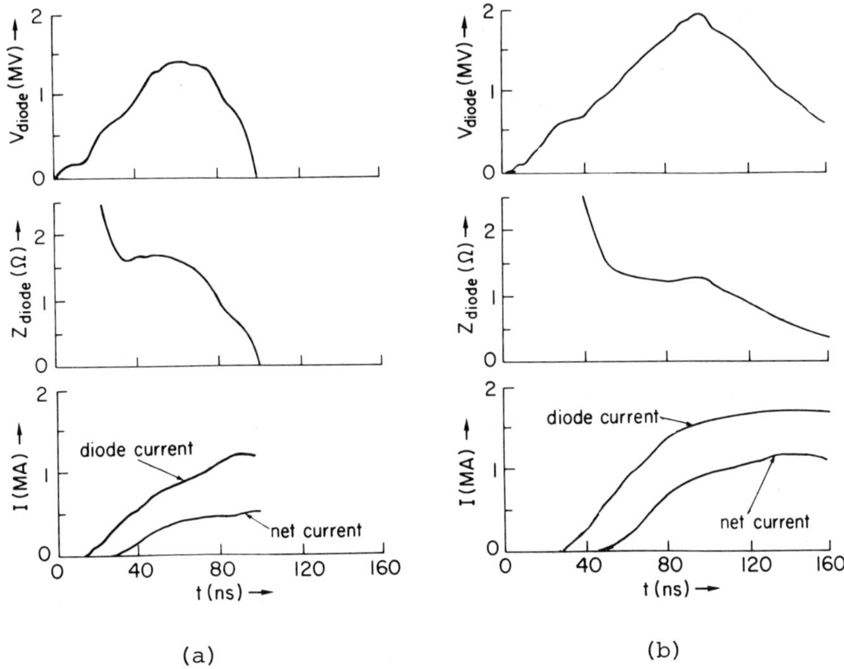

Fig. 4.   Diode electrical characteristics for (a) Gamble II and
          (b) PITHON.

energies of at most 300 kV lower than the diode voltage of 1.3 MV.
Several mechanisms are being investigated as a source of this loss.

In the most recent ion production experiments[11], a pinch-reflex
ion diode, similar to that used on Gamble II (Fig. 3), was successfully
operated at 3 TW on the Physics International PITHON generator. The
main difference between the geometry used on Gamble II and that used
on PITHON was that PITHON presently only operates in the normal
negative polarity mode. Because the cathode was now at a 2 MV potential, this required special techniques to be able to both read out
the ion current monitors and to fill the drift region behind the
cathode with gas. Deuteron current measurements were made in the
conventional manner by coating the anode foil with CD_2 and using
both nuclear activation and time-of-flight neutron detectors to
detect the neutrons produced from a (greater than one range thick)
CD_2 target placed behind the cathode.

Typical diode electrical characteristics are shown in Fig. 4b
for a 6 cm cathode radius, a 0.3 cm anode-cathode gap spacing, and
a 0.5 cm vacuum gap behind the 0.01 cm CH_2 anode foil. Peak diode
voltages of 2 MV at 1.5 MA diode current were obtained with flat
impedance behavior even though the pulse length was almost twice as
long as that on Gamble II. Ion currents of up to 1 MA were obtained

corresponding to peak ion powers approaching 2 TW. Over 100 kJ of
ions were produced out of just over 200 kJ of electrical energy
delivered to the diode. Impedance control and electrical reproducibility were excellent. When deuterons were produced, the neutron
yield approached 10^{13} from the D-D and D-carbon reactions occurring
in the CD_2 target.

As will be discussed in the last section, the voltage and ion
currents produced on PITHON are in the range for a single module of
a multi-module pellet ignition system. The long impedance lifetimes
and high reproducibility associated with the $\sim$ 1 cm gaps between the
anode and cathode transmission foil indicates that programmed-voltage
waveforms for beam bunching during transport can be employed. Focus
perturbing effects associated with time varying B_θ fields in the
diode region can be controlled for these $\sim$ 1 cm diode gap by electrode shaping and naturally occurring gap closure as discussed in
the next section.

3.   ION FOCUSING IN PLANAR GEOMETRY

The processes believed to be important in ion focusing are schematically illustrated in Fig. 5. Ions traversing the vacuum gap
are radially accelerated inward by the azimuthal magnetic field of
the diode current flow. Particle-in-cell codes predict that the
predominant electron flow in the diode is close to the anode surface
where most of the potential drop across the diode takes place. Thus,
the electron current contributes to the angular deflection of ions
only for a small fraction of the vacuum space. The ions are radially

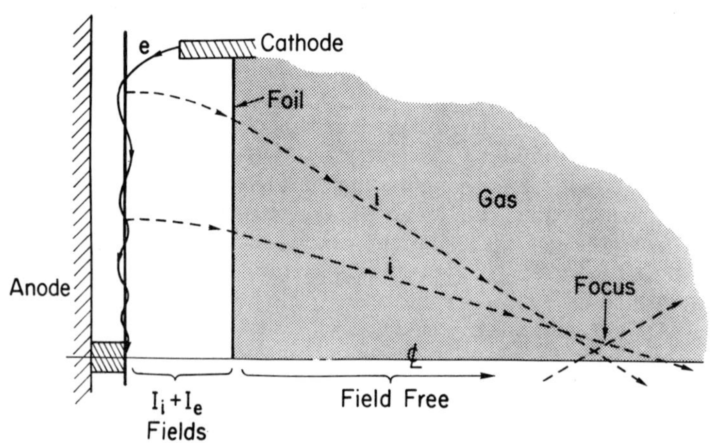

Fig. 5. Light ion focusing schematic.

accelerated primarily by their own current. If the ion current density were approximately uniform over the anode surface, then the resulting linear $B_\theta(r)$ would cause the ions to be pointed towards an approximately-common focus as they enter the cathode transmission foil. Scattering during passage through the 1.8 μm polycarbonate (KIMFOL) foil is about 20 mrad for 1 MeV protons. Beam induced breakdown of the low pressure gas filling the drift region to the right of the cathode foil allows beam current neutralization. The magnetic field acting on the ions is reduced by a factor of 50 by this neutralization. The ions have therefore nearly straight-line trajectories in this region and free stream towards the focus.

One major limitation to tight focusing is asymmetry in power and charged particle flow in the diode. This can be overcome by careful experimental design which results in well-centered electron pinches and centered symmetric ion focal spots. Another major limitation is the actual radial distribution of ion current which may direct ions emitted at different radii to different foci. In principle, for a given diode voltage and current, the shapes of the anode and cathode surfaces can be adjusted to correct for focus-perturbing effects of non-uniform ion current distributions. For instance, the launch angle can be changed by curving the anode sur-face. The radial deflection due to diode magnetic fields can also be changed by varying the axial position of the cathode foil as a function of radius. The anode-foil thickness can also be varied as a function of radius to vary the degree of electron reflexing and thus control the emitted ion current profile. A final major limita-tion to tight focusing is the time variation in the focusing magnetic fields which cause a change in focus location with time. This effect can be partially offset by the natural reduction of the anode-cathode (AK) gap due to electrode plasma motion. Thus, as the diode current increases, the increasing radial acceleration of the ions can be offset by the reduction of AK gap with time. In principle, both the initial AK gap spacing and the cathode foil material could be chosen to minimize this time variation effect.

Long focal lengths (i.e. large f-numbers) are achieved by mag-netic self-focusing acting alone in a flat-anode geometry. This configuration is employed to inject ions into transport channels since larger-angle injection requires excessive transport-channel currents. Spherically-curved anodes in conjunction with self-magnetic forces are used to obtain shorter focal-lengths appropriate for target studies with the highest-focused current densities.

Three sets of focusing experiments in planar geometry were performed on Gamble II with electrical characteristics similar to those of Fig. 4a.

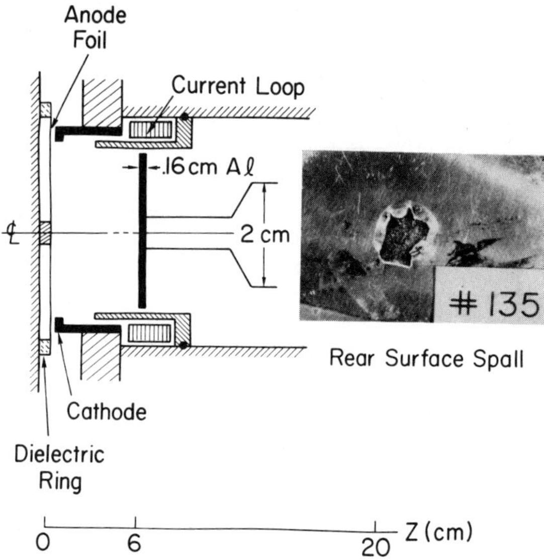

Fig. 6. Self-focusing in planar geometry with no cathode foil.

The first self-focusing experiments were performed with the
planar diode geometry illustrated in Fig. 6 where a 0.01 cm thick
CH_2 anode foil stretched across a dielectric ring was used with a
12 cm diameter hollow cathode without a cathode transmission foil.
The ion beam was observed to self-focus 6 cm from the anode as
indicated by the 1 cm^2 rear surface spall produced on a 0.16 cm
thick aluminum witness plate. No spall was observed when the witness
plate was placed a few centimeters to either side of this position.
The focus location was in agreement with self-consistent analytical
and numerical calculations for the self-focusing of a 500 kA, 1.3 MV
proton beam. These calculations consist of solving ion orbit equations in the diode electromagnetic fields (determined from current
and voltage measurements) and in the focusing/drift region magnetic
fields (determined from net current measurements).

In the second set of self-focusing experiments, the 1.8 μm
thick cathode foil was placed at 4 mm from the cathode face as shown
in Fig. 7, and the drift tube was evacuated. A Rogowski coil in the
evacuated drift tube measured a net current of about ½ the total ion
current. The ion self-focus, as determined from the rear surface
spall of the witness plate, moved from 6 cm to about 12 cm from the
anode plane, again consistent with the predicted location for this
reduced net current in the drift section. The origin and nature of
this net current has been discussed[18].

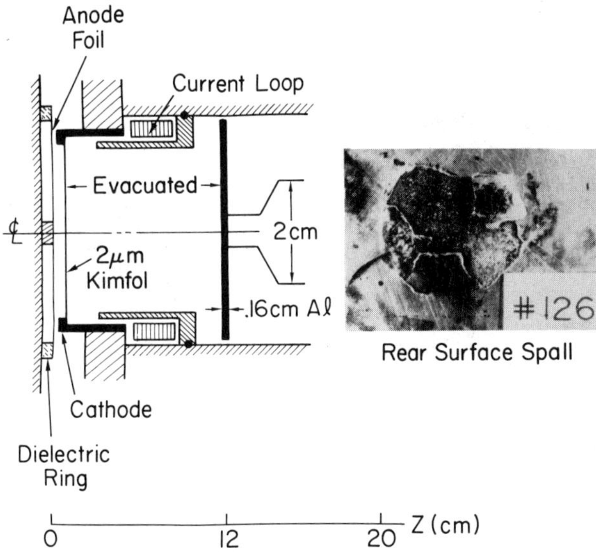

Fig. 7. Self-focusing in planar geometry with cathode foil and
evacuated drift tube.

The final self-focusing experiments were performed with the
cathode foil 2.8 cm from the anode and with the drift tube region
behind the foil filled with 2 Torr of air. The large separation
between the anode and the cathode foil insured that negligible electron current flowed from the cathode foil to the anode. This experiment is illustrated in Fig. 8. The cathode foil was bowed by the
air pressure in the drift tube. The Rogowski coil surrounding the
cathode foil support structure monitored the total current flowing
into the cathode foil. The second Rogowski coil in the drift tube
read less than 2% of the current read by the first for drift
tube pressures between 0.2 to 3 Torr. With gas in the drift
section, an aluminum witness plate placed at the 20 cm predicted
focus location showed a 3 cm diameter rear surface spall. Computer
analysis predicting ion orbits indicates that the 2-3 cm diameter
focus spot size is consistent with axial motion of the best-focus
location due to the time variation of diode current and voltage.
That is, although the current and voltage values may be appropriate
for tight focusing at peak power, the beam is under-focused early
and over-focused late in time. No spall occurred a few centimeters
to either side of this position and moving the cathode foil moved
the focus location as predicted. It is important to note that
damage patterns associated with focusing in the neutralizing gas
backgrounds of this last set of experiments were typically well
centered and symmetrical, whereas, focal spots obtained in vacuum
display azimuthal irregulatiries and beam filamentation. Aluminum

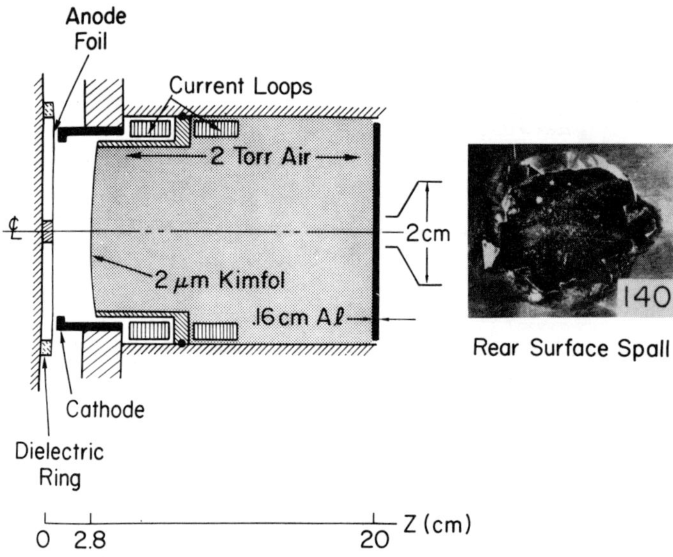

Fig. 8.   Self-focusing in planar geometry with cathode foil and
          gas in drift tube.

K_α x-ray pinhole photographs[7] showed a peak current density of about
50 kA/cm^2.

4.   ION FOCUSING IN CURVED GEOMETRY

     In order to achieve high ion current densities for beam target
experiments, a 0.025 cm thick spherical-section plastic anode foil
with a 12.7 cm radius of curvature was used. As shown in Fig. 9,
the same cathode geometry previously shown was employed except for
the closer proximity of the thin cathode transmission foil to the
cathode surface. In this short-focal-length geometry, the ion
launching angle dominates over the magnetic deflection in the diode.
Focal spot broadening due to radial and temporal field effects is
thus minimized. Aluminum witness plates placed at the best focus
location exhibit smaller area spall patterns (1-1.5 cm diameter)
than in flat geometry with gas. When the anode was coated with
CD_2, the observed focus was close to the geometric focus because of
the small magnetic deflection of deuterons. By using small CD_2
targets and measuring neutrons from D-D and D-carbon reactions with
activation counters and time-of-flight detectors, peak deuteron
current densities of about 300 kA/cm^2 over 0.5-1 cm^2 were inferred.
Typical neutron yields were a fraction of 10^{12} neutrons.

     Important new experimental results for focusing are provided
by a pinhole shadowbox diagnostic shown in Fig. 10. Small ion

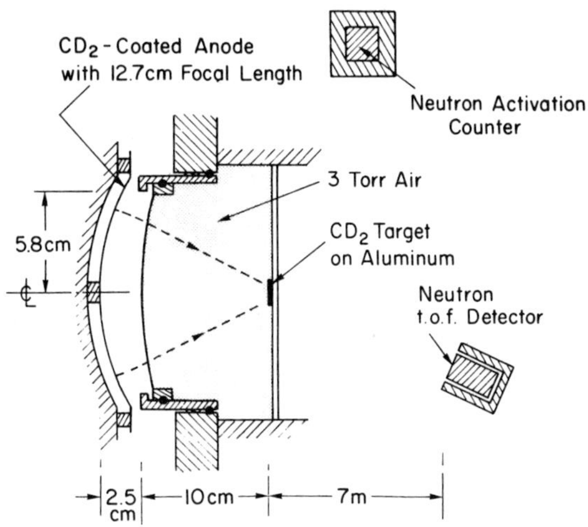

Fig. 9.   Geometric focus experiments.

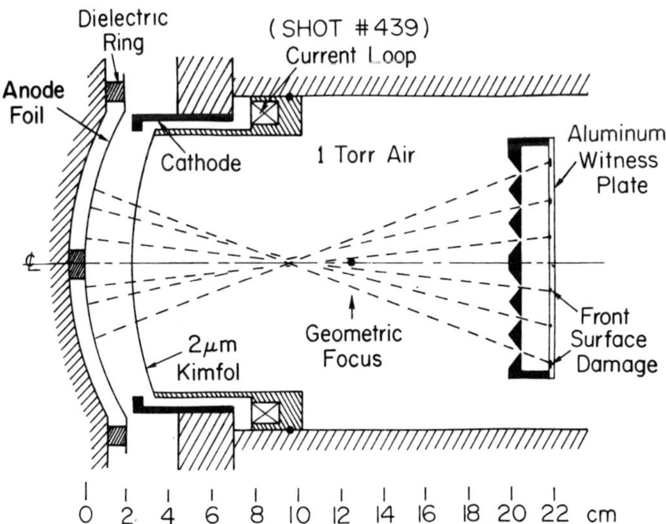

Fig. 10.   Ion pinhole shadowbox.

beamlets are produced by masking off all of the beam drifting
in the neutralizing background except for that portion emerging
through 1 mm diameter holes placed along two diameters at right
angles. Damage patterns produced on a aluminum witness plate 1 cm
behind the pinhole array (shown as solid spots in the figure) allows
one to determine ballistic orbits for ions entering pinholes at
various radii. When the shadowbox is placed further from the diode
than the center of the ion focus, these orbits can be projected
from the damage spots back through the pinhole to the focus. These
are shown as dashed lines in the figure. In this manner, the best
focus location and the radial dependence of ion deflection angle can
be determined. Additionally, the radial extent (typically a few
millimeters) of each damage pattern reflects the time-variation of
deflection angle due to time-varying diode fields. The preliminary
Gamble II results with protons demonstrate that radial deflections
of damage patterns (shown in Figure 10) are consistent with the
measured focus location. Diode deflection angles were found to be
roughly proportional to radius. Also, the significantly smaller
observed azimuthal extent of damage compared with the radial extent
of damage suggests that with proper aspheric anode and cathode
shapes, focusing of ion beams to areas very much less than the
present 1-3 cm^2 should be possible.

5.   NUMERICAL SIMULATION OF PINCH-REFLEX ION DIODES

Numerical simulations of electron and ion flow in pinch-reflex
diodes were conducted for the configuration shown in Fig. 11 using
the NRL DIODE2D simulation code.[19] The initial equipotential lines
are also shown in the figure. The diode dimensions modeled in the
simulation had an R/d=20 aspect-ratio where r=6.8 cm was the outer
radius of the hollow cathode and d=0.32 cm was the separation of the
cathode face from the anode foil. The thickness of the cathode
shank was 0.8 cm and the scattering foil was assumed to be 125 μm
thick polyethylene. The distance between the reflexing foil and
the solid anode was 0.29 cm. This space was assumed to be filled
with ions which charge neutralize the reflexing electrons, making
it unnecessary to solve for the electrostatic fields in that region.
The distance from the reflexing foil back to the ion transmission
foil inside the hollow cathode was 0.7 cm. These new simulations
were carried out on a finer 64×40 mesh rather than the 32×40 mesh
used in previously reported results.[20] The simulations also
assumed azimuthal symmetry and were done in (R,Z) geometry with
a time step of 0.25 psec.

The time evolution of the electron and ion flow in the diode
is shown in Fig. 12 for a simulation done at 1.5 MV. The positions
of the electrons and ions are shown in (R,Z) configuration space at
four different times. Early in time (step 500), the electrons
(one in 10 is plotted on the figure) had transited the diode and

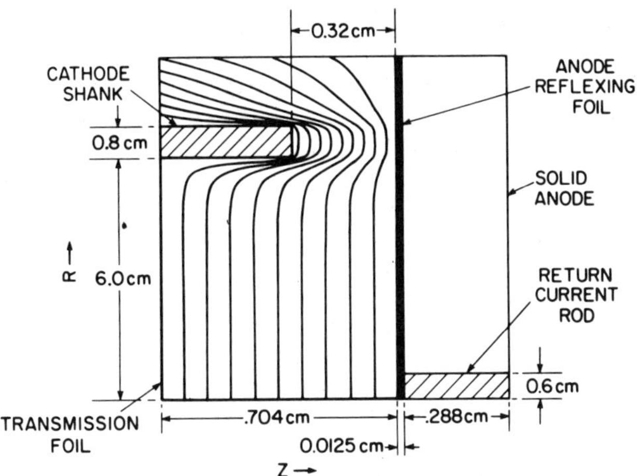

Fig. 11. Configuration used for pinch-reflex diode simulation showing equipotential lines due to applied voltage.

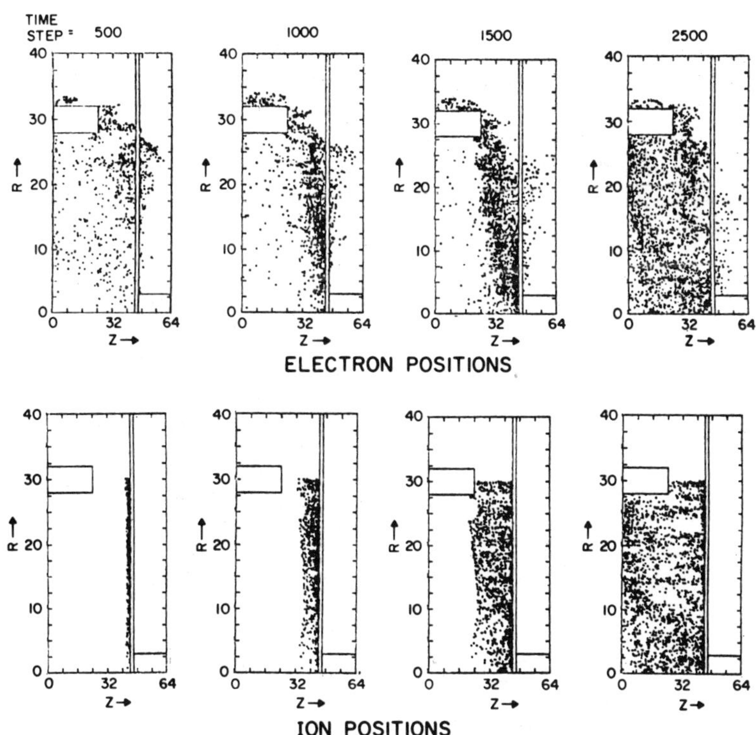

Fig. 12. Time history of the electron and ion positions in the diode.

reflexed several times through the foil. The ions had not yet
crossed the diode gap and were clumped along the anode foil. By
time step 1000, the electrons had established the pinched flow
along the diode side of the reflexing foil. At step 1500, the
pinched electron flow was well established and the electrons filled
the diode gap. However, electrons did not flow along the foil except
at low radii because of the tendency for electrons to follow ions
at large radii to provide space charge neutralization. At step 2500,
all the ions had now crossed the gap and a quasi-steady state was
established. The flow pattern shown at step 2500 persisted for the
remainder of the run which was taken to step 4000.

In the quasi-steady state the equipotential lines are piled
up near the anode foil so that most of the potential drop occurs
over a distance much less than the 0.32 cm gap between the cathode
face and the anode foil. This justifies the assumption of Sec. 3
that the ion bending between the anode and cathode foil is primarily
due to the self-magnetic field of the ions. The current density
profile of the ions arriving at the cathode transmission foil is
shown in Fig. 13. This current profile, which is the result of
averaging the arriving ion current for 500 time steps, shows a
1/R dependence as predicted by analytic theory[2] for planar diodes.
The simulation for this case assumed that ion emission occurred
along the reflexing foil up to a radius of 6.4 cm. In this case,
the net ion current was 590 kA and the electron current 330 kA.
In order to determine if the average ion current density could be
increased by suppressing ion emission at large radii, a case where
emission was allowed only up to a radius of 3.0 cm was also

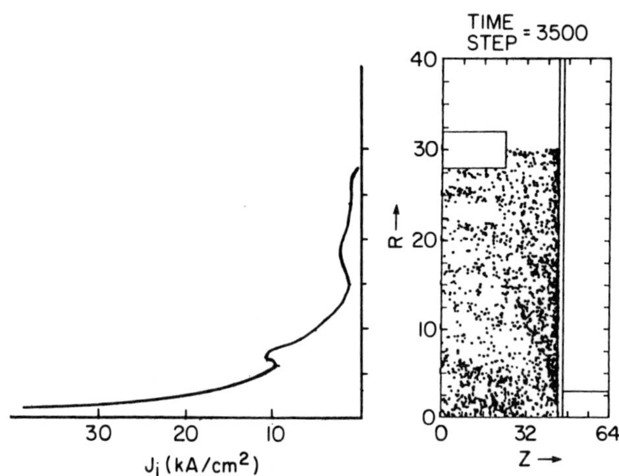

Fig. 13.   Ion current density profile at the cathode and ion
           positions in the quasi-steady state.

simulated. In this case, the total diode current remained close
to 900 kA and the total ion current decreased by about a factor
of 2.

The simulation results are in reasonable agreement with the
NRL experimental data with respect to total ion and electron
currents. The geometry was different enough so that it is difficult
to make a more quantitative statement. The transmission foil is
further removed from the cathode face in the present Gamble II
experiments than in the simulation and this may affect the diode
behavior. Further simulations that are closer to the actual experimental configurations are being conducted so that a more direct
comparison with experiments can be made.

6.   ION ORBITS IN Z-DISCHARGE CHANNELS

As shown in Fig. 14, ions enter the discharge channel with a
range of injection angles up to θ_M determined by the anode radius
and distance to the focus. The externally-applied current flowing
through the channel must be sufficient to confine ions with maximum
transverse kinetic energy, that is, ions which enter the channel
with the maximum injection angle at the maximum injection radius,
r_s. The required discharge current can be determined from conservation of ion energy and P_z conical momentum

$$V_r^2 + V_z^2 = V_o^2 \;\; ; \qquad V_z = V_o \cos\theta + \frac{q}{m} \int_{r_o}^{r} B_\theta(r')dr' \quad ,$$

where $\frac{1}{2}mV_o^2$ is the energy of an ion with charge q, θ is the injection
angle and r_o is the injection radius. In order for the magnetic

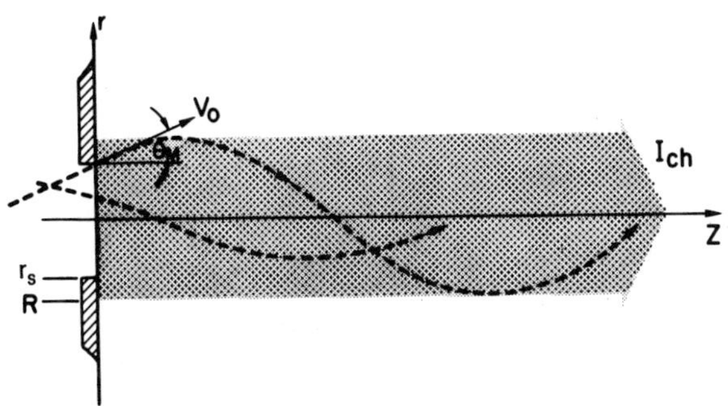

Fig. 14. Ion orbits in a Z-discharge channel.

field to be just sufficient to confine a maximum-transvere-energy
ion to the channel radius (R), V_r must vanish when $r = R$ for initial
conditions $\theta = \theta_M$ and $r_o = r_s$. Substituting into the previous
equation, results in the expression

$$q \int_{r_s}^{R} B_\theta dr = mV_o (1-\cos\theta_M) . \tag{1}$$

Defining an ideal channel as one carrying a uniform discharge current
so that $B_\theta \sim r$ when $r \leq R$, Eq. (1) takes the form

$$I_{ch}(A) = \frac{10^{-3}V_o(1-\cos\theta_M)}{1 - (r_s/R)^2} \tag{2}$$

for protons where I_{ch} is in amperes when V_o is in cm/s. This result
indicates that megampere discharge currents are required to confine
ions injected with large angles. Protons of 5 MeV with $\theta_M = 0.2$
radians require about 120 kA for confinement within the channel
when $r_s/R = 2/3$.

Analytical and computational techniques have been used to
determine the effects of various forms of nonideal channel behavior
on ion confinement and beam bunching.[21] Axial electric fields
associated with plasma currents and expansion disrupt confinement
and reduce beam power multiplication due to bunching only if large
enough to strongly slow the beam during transport. Radially non-
uniform net currents develop as the beam passes through the background because of beam-induced MHD expansion of the channel. The
effects of the resulting B_θ fields on beam confinement are discussed
in Sec. 8. Channels which taper to smaller radius as they approach
the pellet were investigated as a means of increasing the beam
current density. A WKB-like analysis for the ion motion shows that
only a small improvement in current density due to radial beam
compression can be achieved before disruption of axial compression
by bunching. Both analytic and numerical calculations were performed
to determine ion confinement in a channel with a large amplitude
m = 0 sausage instability. Fig. 15 shows one result of this work.
The channel B_θ was modeled as $\sim r$ inside the channel radius R(z),
and $\sim 1/r$ outside. With every change in axial location $\Delta z = \lambda$,
the value of R changed abruptly and randomly to simulate the nonlinear development of the instability. A large number of ion orbits
were numerically determined for a range of injection angles up to
θ_M and injection radii up to r_s. The figure shows a cross-section
of beam ion density at various axial locations along the channel.
The case shown corresponds to $r_s = 0.4$ cm, $R = 0.6$ cm, $E_i = 5$ MeV,
$\theta_M = 0.2$ radians, $I_{ch} = 120$ kA, $\lambda = 3.8$ cm, $\Delta R_{rms}/\bar{R} = 1$. This high

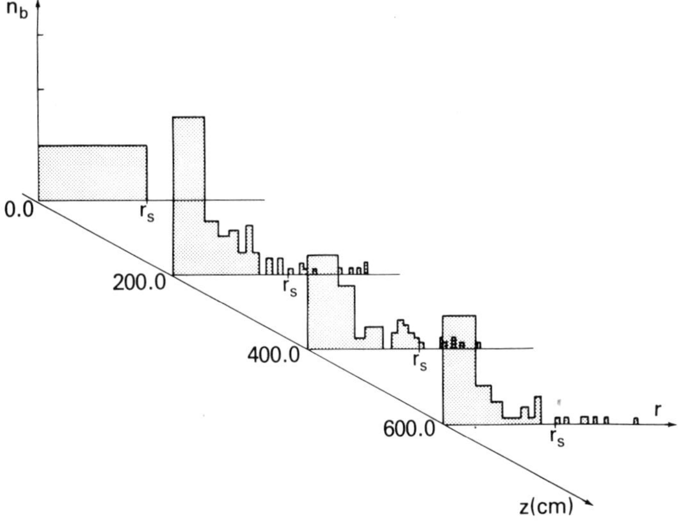

Fig. 15.   Beam density profile at various axial locations along
           a sausage unstable channel.

degree of fluctuation is required for radial loss of half the beam
ions after transporting 6 m. When ΔR_{rms} is decreased by about a
factor of 2, only about 9% of the beam is lost at 6 m. Good radial
confinement is predicted for all cases in which the axial wave
length of the instability is much smaller than the betatron wavelength of the ion in the channel. Electrostatic and electromagnetic
microinstabilities driven by the relative streaming between the
beam ions and the channel plasma have been discussed previously.[22]

7.   ION TRANSPORT EXPERIMENTS

     Figure 16 illustrates the experimental configuration for
transport experiments on Gamble II. The planar, pinch-reflex diode
configuration shown previously in Fig. 8 was used to bring a 0.5 MA,
1.4 MV proton beam to a narrow-angle focus ($\sim$ 50 kA/cm^2) 20-30 cm
downstream from the diode. The transport section consisted of a
copper pipe containing an insulating ceramic liner filled to the
0.2-2 Torr air background pressure of the focusing-drift section.
The inside diameter of the ceramic liner defines the diameter of
the wall-stabilized discharge. A discharge channel diameter of
4.5 cm was employed. The discharge current was provided by a
20 kV, 10 kJ capacitor bank. The current rose to about 50 kA in
15 μs after which time the ion beam was injected into the channel.
This current satisfies Eq. (2) for the injection angles employed
in the experiment. Measurements of transport efficiency were
provided by diagnosis of 6 MeV gamma rays produced by the interaction of beam protons with fluorine[23]. Collimated scintillator-

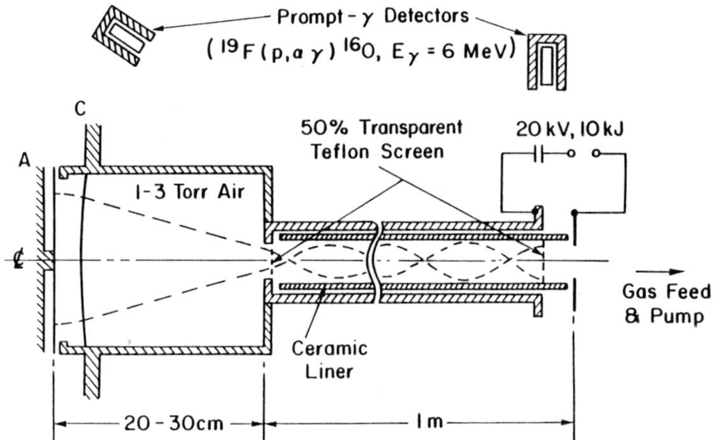

Fig. 16.   Ion transport experiment schematic.

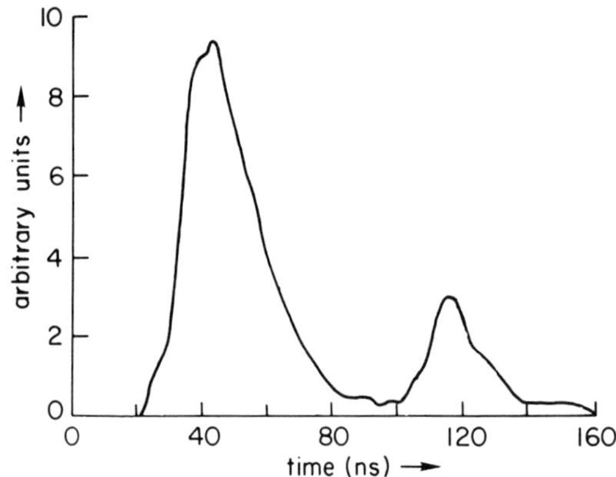

Fig. 17.   Prompt-γ signal from two 50% transparent teflon targets
placed on meter apart in the channel.

photomultiplier detectors were used to detect the gamma rays produced in 50% transparent teflon screens placed at the entrance and exit apertures of the transport system.

The signal shown in Fig. 17 was obtained with a single detector placed about 4 m from both screens and behind about 0.5 m of concrete. Signals from the two screens are separated by the ion time-of-flight through the 1 m discharge channel. Because the teflon screens are 50% transparent, the second signal must be multiplied by a factor of 2 in order for comparison with the signal from the entrance aperture. Since the gamma-ray production cross-section is a strong function of the proton energy, the transport efficiency could not be immediately inferred from the ratio of signals associated with the two screens. Assuming no ion energy loss during transport, experimental results indicated a lower limit of 50% transport efficiency. If an energy loss of 200 kV is assumed, which is consistent with the relative timing between the prompt-γ signals shown in Fig. 17, then the particle transport efficiency in the channel is about 80%. A more concrete example of the ability to transport ion beams is provided by witness plates at the exit aperture. Aluminum plates 0.16 cm thick exhibited rear surface spalls with widths less than the channel diameter. Null experiments, measurements without channel current, did not show evidence of ion transport.

In more recent transport experiments a smaller 1.6 cm diameter channel was used with a much faster risetime (0.5 μs) capacitor bank than previously employed. Preliminary results indicate that proton beams with current densities of several tens of kA/cm^2 were efficiently transported over 1 m distances. Unlike the large diameter channel transport experiments, shots were taken with the ion beam injected into the channel while the channel plasma was still radially imploding. On these shots, the shapes and relative timing of the input and output prompt-γ signals suggest that the protons may have gained rather than lost energy in the channel. This observation would be consistent with recent theoretical predictions[24] that the transported ion beam could gain energy from the fields associated with a radially imploding Z-pinch plasma. This phenomenum is presently under investigation.

8.   RESPONSE OF THE CHANNEL TO BEAM TRANSIT

As the focused ion beam transits the transport channel, it heats the background plasma by collisional deposition and induces plasma-electron return currents which resistively heat the plasma. Since the electromagnetic fields induced by the beam can vary substantially from those assumed to confine the beam ions, it is important to determine how the self-consistent fields alter beam confinement and energy loss during transport. To answer these questions, a 1-D radial, time-dependent MHD code has been developed[25]. The code solves the usual conservation equations for particle number, momentum and energy. The $\vec{\nabla} \times \vec{B}$ and $\vec{\nabla} \times \vec{E}$ Maxwell's relations, and electron transport equations (using classical coefficients for electrical and thermal conduction) close the set of equations. The beam is specified by its current density in space and time and the particle energy as a function of time. Collisional beam deposition and plasma-electron return-current heating are included as is bremsstrahlung radiation cooling. Development of radial profiles in time for background density, electron and ion temperature, azimuthal magnetic field and radial velocity are shown in Fig. 18 for a 50 kA discharge in deuterium carrying 600 kA of 5 MeV protons in a 50 ns pulse. Since the plasma is initially at rest, no radial velocity profile is shown in Fig. 18a. The initial temperature distribution is thus determined by an equilibrium pressure balance condition. Once the beam enters the channel, the plasma pressure and $\vec{j} \times \vec{B}$ forces acting on the plasma push material out of the channel interior and pile it up against the steep density gradient at the channel edge. The initial magnetic field profile corresponds to uniform current density arising from the long times and the low conductivities associated with channel creation. Once the beam heats the channel (Fig. 18b), the conductivity rises quickly and magnetic field lines become frozen in the plasma. The magnetic field is then convected out of the center of the channel along with the plasma flow.

A primary criterion for maintenance of beam confinement is suggested by Eq. (1) and is given by the integral extended from r_0 to R. If the value of the integral is reduced, ions injected with large r_0 or θ_0 can reach the high density edge of the channel and be lost. For low beam current cases, the plasma channel response to the beam is sufficiently gentle not to disrupt confinement. However, in the high-current cases, ions with larger transverse energy are lost.

The evolution of the confinement integral with time does not depend sensitively on the ion energy. This behavior can be understood qualitatively by examining the plasma equation of motion.

$$\rho \, \frac{dV_r}{dt} = - \frac{\partial p}{\partial r} - j_z B_\theta \approx j_b B_\theta \qquad (3)$$

The final expression on the right hand side of Eq. (3) results from the dominance of the $\vec{j} \times \vec{B}$ term over the pressure term and good beam current neutralization. For a maximum allowed acceleration given by $2R/\tau^2$, where τ is the beam duration, Eq. (3) yields

$$I_b I_{ch} < 300 \, \rho \, R^4/\tau^2 \qquad (4)$$

for maintenance of beam confinement. In Eq. (4), the currents are

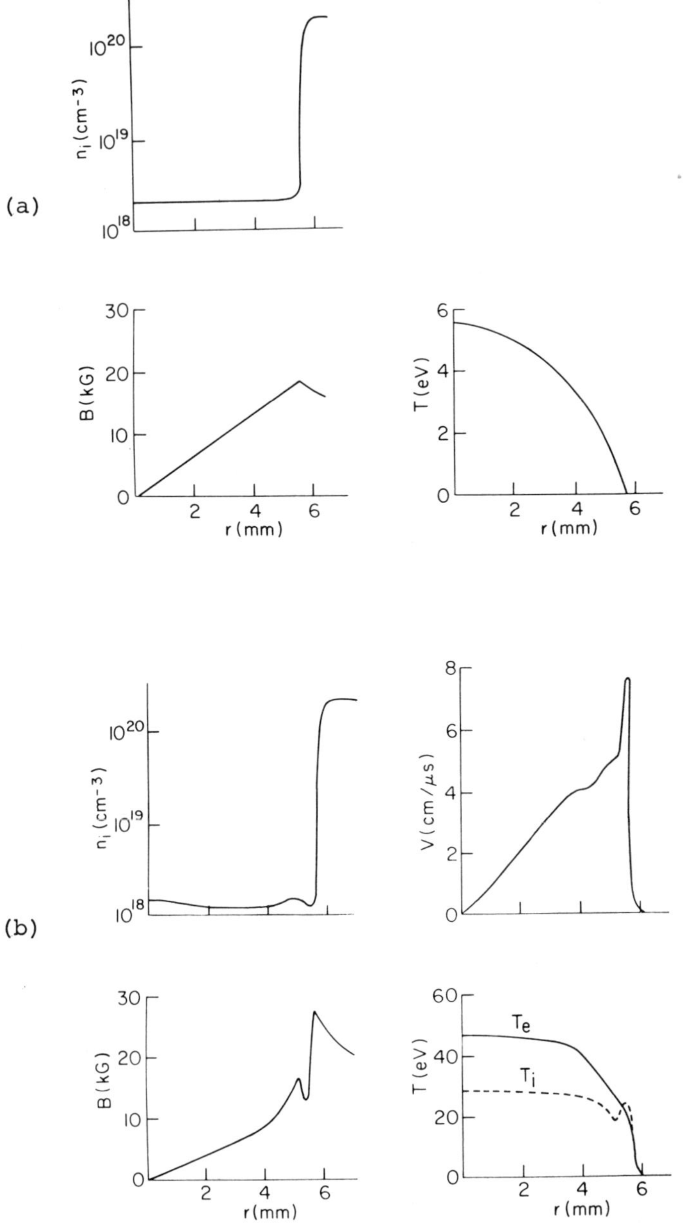

Fig. 18. Initial (a) and final (b) radial profiles in time for background plasma density, azimuthal magnetic field, electron and ion temperature ($T_i = T_e$ initially), and radial plasma velocity ($V=0$ initially).

in amperes and the terms of the right-hand-side are in cgs units.
This condition is in qualitative agreement with the results of the
MHD code. Because of the greater primary beam deposition, lower-
voltage ion beams require lower background density. Thus, the
maximum allowed value of $I_b I_{ch}$ is reduced with lower ion voltage.
This indicates that less power can be transported in channels
carrying lower-voltage ions unless the channel current can be
significantly reduced. However, the channel current is determined
primarily by the spread of injection angles and θ_M is unlikely to
be reduced for less-stiff lower-energy ions. Additionally, Eq. (4)
is easier to satisfy with the beam in the bunched state since $I_b \tau$
remains constant as τ is reduced. If a final focusing stage after
transport can be employed[26], then the beam can be transported with
a radius larger than the pellet size. In this case the constraints
of Eq. (4) can be considerably relaxed because of the strong R^4
dependence.

A second important consideration for beam transport channels
is the total energy loss suffered by the beam during transport. The
energy loss consists of both that due to primary deposition and that
due to slowing in the electric field produced by return currents and
plasma expansion. For the example considered in Fig. 18, the 5 MV
protons lose about 200 kV/m. This energy loss could be reduced
or possibly turned into an energy gain[24] if the channel plasma during
beam injection is imploding with a velocity of $\sim 10^7$ cm/sec using
a 100 kA channel current.

9.   BEAM BUNCHING DURING TRANSPORT

Since the time of flight of the ions through the channel to the
pellet usually exceeds the beam pulse duration, applying an accelerating voltage ramp at the diode can result in power multiplication
(PM) by beam bunching[27]. Since the pulsed-power technology employed
results in ion emission times of 50-100 ns and the desired
irradiation time on the pellet is about 20 ns, as much as a factor-
of-5 bunching is desirable. For purely axial motion, ions accelerated by the ideal voltage waveform

$$\Phi(t) = \frac{\Phi_o}{(1-t/t_a)^2} \; ; \; 0 \leq t \leq \tau < t_a \tag{5}$$

all arrive at the target at the same time t_a. In reality, errors
in voltage shape, $\delta\Phi$, and the spread in ion injection angles contribute to a spread in arrival times, δt_a, thus limiting the degree
of power multiplication to a factor $\tau/\delta t_a$. For a desired PM factor
of 5, analysis indicates that $\delta\Phi/\Phi \leq 0.15$ and $\theta_M \leq 0.3$ radians are
sufficient for parameters of interest.

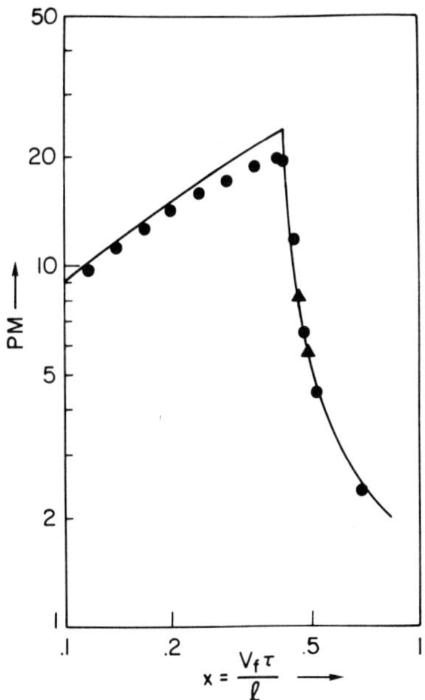

Fig. 19.   Theoretically achievable power multiplication predicted
by analytic modeling (solid curve) and macroparticle code.

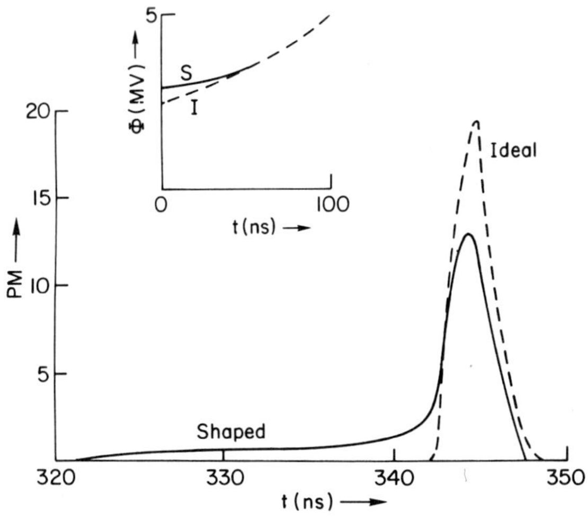

Fig. 20.   Bunched pulse shaping.

Figure 19 compares analytic modeling of achievable power multiplication with a macroparticle code, where ℓ is the distance to the target and V_f is the ion speed at $t = \tau$. Below $x = 0.414$, PM is limited by the spread in injection angles $\theta_M = 0.2$ radians. Above this value, PM is determined by a maximum voltage ramp $\Phi_f/\Phi_o = 2$ in order to control the energy spectrum of ions depositing in the pellet.

Figure 20 compares the time variation of power on target using two voltage waveforms and the parameter set $r_s = 0.4$, $R = 0.6$ cm, $\ell = 7.5$ m, $\theta_M = 0.2$ radians. Peak power multiplication is achieved by employing the ideal voltage. In that case, the power multiplication is limited by the dispersion in injection angles. When the accelerating voltage is higher than the ideal early in time, some ions arrive early on target. This results in a shaped foot which may be desirable for some inertial-confinement-fusion target configurations.

10.   SYSTEM REQUIREMENTS FOR INERTIAL CONFINEMENT FUSION

Pulsed power, ion focusing, transport, bunching and beam-target interaction considerations can be combined in order to determine an acceptable set of system parameters for the ignition of a high-gain fusion pellet. High voltage beams are easily focused because of their relative insensitivity to time-varying magnetic fields in the diode. However, their long range in materials leads to inefficient energy coupling to fusion targets. Low voltage beams suffer from excessive energy loss in the transport channel and difficulty in focusing. For any voltage, channel hydrodynamic response calculations indicate difficulty in transporting ion beams much in excess of 1 MA without employing a final focusing stage. Thus, for a fixed power and energy requirement delivered to the target, MHD considerations establish a lower limit on the number of channels which can be used to transport ion beams to the target. An upper limit on the number of channels which can be used is set by considerations that include channel creation energy and beam overlap problems. These points are considered in Fig. 21 which defines an acceptable parameter range for a 2 MJ ion beam incident on a high-gain pellet. The SANDIA PBFA concept has design objectives in the lower right-hand corner of the window. Research at NRL concentrates primarily on parameters of the upper left. Table 1 summarizes an acceptable range of system parameters based on the considerations discussed above. Parameters based on two pellet requirements and two ion pulse durations are shown. The stored energy is an estimate of the total electrical energy stored in the Marx generators required to provide the total ion beam energy. The shorter-duration beams require higher powers in the diode but also require less bunching in order to achieve power on target. The focused power represents the total focused ion beam power for all the modules into the entrance of all the channels.

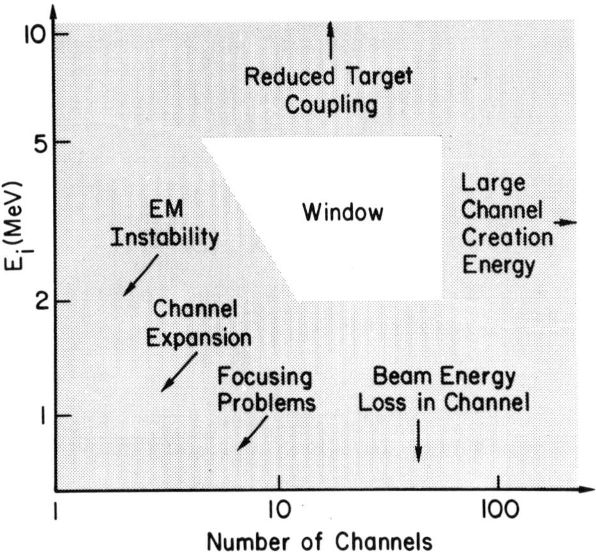

Fig. 21. Channel number considerations for protons in hydrogen.

Table 1. Ignition System Parameter

ENERGY ON PELLET (MJ)	3		2	
POWER ON PELLET (TW)	200		100	
STORED ENERGY (MJ)	24		16	
PULSE DURATION (ns)	100	50	100	50
FOCUSED POWER (TW)	35	70	23	45
POWER MULTIPLICATION	6	3	4.5	2
FOCUSED CURRENT (MA)				
5 MeV PEAK	7	14	4.5	9
2 MeV PEAK	18	35	12	23
MIN. BUNCHING LENGTH (m)				
5 MeV PROTONS	6	2.5	6	2
2 MeV DEUTERONS	3	1	2.5	0.8

Focused current represents the total current transported by all
channels. Using 1 MA as the current which can be transported in any
given channel, the focused current numbers represent an estimate
of the minimum number of channels required to transport the total
beam. The minimum channel length (i.e. stand-off distance) is
arrived at by determining the smallest transport length which can
achieve the desired degree of power multiplication (PM) using an
accelerating voltage which varies by no more than a factor of 2.
This bunching length may be estimated from[21]

$$\frac{V_f T}{\ell} = \frac{0.414}{1-1/PM} \quad .$$

The scaling of ℓ with PM results from the fact that smaller lengths
than necessary for optimal bunching are sufficient for moderate
power multiplication factors.

Extension of these results to ion species other than $Z = 1$
requires further investigation of transport properties. Higher
voltage beams can transport at lower beam current to achieve beam
powers comparable to $Z = 1$. Channel currents are uncertain. From
Eq. (1), I_{ch} is proportional to the larger values of m/q. However,
it tends to be reduced by the smaller values of θ_M permitted by
better focusing control.

In the near future scaling of present results will be attempted
in the following areas. Diode behavior at the 3-5 TW level on
PITHON will be studied in order to check ion beam efficiency and
focusability. The physics of higher voltage (up to 10 MV) diodes
operating at impedances up to 20 ohms will be studied on the
Harry Diamond Laboratory Aurora generator.[28] Transport of ion
beams will be performed in longer channels where beam bunching may
be investigated. Larger diameter channels with a final focusing
stage will be studied in order to achieve transport of higher
current, longer duration ion beams. The effect of imploding plasma
discharges on the energy of the transported ion beam will be
investigated. These experiments will provide the MA/cm^2 level
fluxes required for relevant ICF beam target studies.

ACKNOWLEDGMENT

The authors are indebted to R. Genuario and J. Maenchen
for their essential contribution to the ion beam experiments on
PITHON.

REFERENCES

1. Shyke A. Goldstein and R. Lee, Phys. Rev. Lett. 35, 1079 (1975).
2. J. W. Poukey, J. R. Freeman, M. J. Clauser, and G. Yonas,
   Phys. Rev. Lett. 35, 1806 (1975).
3. J. W. Poukey, J. Vac. Sci. Technol. 12, 1214 (1975).
4. A. E. Blaugrund, G. Cooperstein, J. R. Boller and Shyke A.
   Goldstein, Bull. Am. Phys. Soc. 20, 1252 (1975).
5. P. A. Miller, C. W. Mendel, D. W. Swain, and S. A. Goldstein,
   in Proceedings of the International Topical Conference on
   Electron Beam Research and Technology, Albuquerque, New Mexico
   (1975), p. 619; G. Cooperstein, S. J. Stephanakis, J. R. Boller
   R. Lee, and Shyke A. Goldstein, in Proceedings of the 1976
   IEEE International Conference on Plasma Science, Austin, Texas
   (1976), (IEEE Cat. 76CH1083-5-NPS), p. 126.
6. Shyke A. Goldstein, D. P. Bacon, D. Mosher and G. Cooperstein,
   Proceedings of the Second International Topical Conference on
   High Power Electron and Ion Beam Research and Technology,
   Ithaca, New York (1977) p. 71.
7. D. J. Johnson, G. W. Kuswa, A. V. Farnsworth, Jr., J. P. Quintenz,
   R. J. Leeper, E. J. T. Burns, and S. Humphries, Jr., Phys.
   Rev. Lett. 42, 610 (1979); D. J. Johnson, Bull. Am. Phys. Soc.
   24, 925 (1979).
8. G. Cooperstein, Shyke A. Goldstein, D. Mosher, F. W. Oliphant,
   F. L. Sandel, S. J. Stephanakis and F. C. Young in Proceedings
   3rd International Topical Conference on High Power Electron
   and Ion Beam Research and Technology, Novosibirsk, USSR (1979).
9. R. O. Bangerter and D. J. Meeker, in Proceedings of the Second
   International Topical Conference on High Power Electron and
   Ion Beam Research and Technology, Ithaca, New York (1977), p.183;
   J. H. Nuckolls, in Proceedings of the Topical Meeting on
   Inertial Confinement Fusion, San Diego, CA (1978), (Optical
   Society of America 78CH1310-2QEA).
10. G. Yonas, Sci. Am. 239, No. 5, 50 (1978); P. A. Miller,
    D. J. Johnson, T. P. Wright and G. W. Kuswa, Comments on
    Plasma Physics and Controlled Fusion, 5, 95 (1979).
11. S. J. Stephanakis, J. R. Boller, G. Cooperstein, Shyke A.
    Goldstein, D. D. Hinshelwood, D. Mosher, W. F. Oliphant,
    F. C. Young, R. D. Genuario and J. E. Maenchen, Bull. Am.
    Phys. Soc. 24, 1031 (1979).
12. A. E. Blaugrund, G. Cooperstein, and Shyke A. Goldstein,
    Phys. Fluids 20, 1185 (1977).
13. J. W. Poukey, in Proceedings of the International Topical
    Conference on Electron Beam Research and Technology,
    Albuquerque, New Mexico (1975), p. 247.

14. S. J. Stephanakis, D. Mosher, G. Cooperstein, J. R. Boller,
     J. Golden, and Shyke A. Goldstein, Phys. Rev. Lett. $\underline{37}$,
     1543 (1976).
15. Shyke A. Goldstein, G. Cooperstein, Roswell Lee, D. Mosher and
     S. J. Stephanakis, Phys. Rev. Lett. 40, 1504 (1978).
16. J. R. Boller, J. K. Burton, and J. D. Shipman, Jr., in Pro-
     ceedings of the IEEE Second International Pulsed Power Confer-
     ence, Lubbock, Texas (1979).
17. F. C. Young, J. Golden, and C. A. Kapetanakos, Rev. Sci. Instrum.
     $\underline{48}$, 432 (1977).
18. Shyke A. Goldstein, A. T. Drobot, in Proceedings of 1979 IEEE
     International Conference on Plasma Science, Quebec (1979), (IEEE
     Cat. 79CH1410-ONPS), p. 161.
19. R. Lee and Shyke A. Goldstein, NRL Memorandum Report 3702 (1978);
     D. Mosher, G. Cooperstein and Shyke A. Goldstein, NRL Memorandum
     Report 4130 (1979).
20. A. T. Drobot, R. J. Barker, R. Lee, A. Sternlieb, D. Mosher and
     Shyke A. Goldstein, in Proceedings of the 3rd International
     Topical Conference on High Power Electron and Ion Beam Research and
     Technology, Novosibirsk, USSR (1979).
21. P. F. Ottinger, D. Mosher, Shyke A. Goldstein, in Proceedings of
     the 1979 IEEE International Conference on Plasma Science, Quebec
     (1979), (IEEE Cat. 79CH1410-ONPS), p. 105, and to be published
     in Phys. Fluids (1980).
22. P. F. Ottinger, D. Mosher and Shyke A. Goldstein, Phys. Fluids
     $\underline{22}$, 332 (1979); P. F. Ottinger, D. Mosher and Shyke A.
     Goldstein, NRL Memorandum Report 4088 (1979), and submitted to
     Phys. Fluids.
23. J. Golden, R. A. Mahaffey, J. A. Pasour, F. C. Young, C. A.
     Kapetanakos, Rev. Sci. Instrum. $\underline{49}$, 1384 (1978).
24. Shyke A. Goldstein and D. A. Tidman, private communication.
25. D. G. Colombant, Shyke A. Goldstein, D. Mosher, in Proceedings
     of the 1979 IEEE International Conference on Plasma Science,
     Quebec (1979), (IEEE Cat. 79CH1410-ONPS), p. 105.
26. P. F. Ottinger, Shyke A. Goldstein and D. Mosher, private
     communication.
27. D. Mosher, Shyke A. Goldstein, Bull. Am. Phys. Soc. $\underline{23}$, 800
     (1978).
28. B. Bernstein, I. Smith, IEEE Trans. Nucl. Sci. $\underline{20}$, 294 (1973).

SHORT WAVELENGTH POPULATION INVERSIONS ASSOCIATED WITH

CHARGE TRANSFER IN LASER-PRODUCED PLASMAS*†

R. C. Elton, T. N. Lee, R. H. Dixon,
J. D. Hedden, and J. F. Seely

Naval Research Laboratory
Washington, D. C. 20375

INTRODUCTION

An important current requirement for achieving lasing at
extremely short wavelengths in highly-stripped atoms is the obtain-
ment of a significant degree of population density inversion between
the lasing states[1]. We have previously reported the measurement of
such inversions between the upper $n=4,5,6$ and the lower $n=3$ excited
levels in C^{4+} and C^{5+} ions for potential quasi-cw lasing in the
$\lambda=35-76$ nm wavelength range[2,3]. These results were obtained from
soft x-ray spectroscopic measurements of intensity distributions
within both the first and second series spectra, using a laser-
produced carbon plasma expanding from a graphite target into a 1-2
torr gaseous background. Most spectra were spatially-resolved
with intensity inversions occurring at a distance of 10-20 mm from
the target surface, where the line emission is viewed through an
optically thin layer. Free electrons from the gas at a measured[2]
density of $5 \times 10^{16} cm^{-3}$ assured collisional equilibrium among
sublevels[3]. The excitation mechanism for the observed anomalies
was associated mainly with electron capture into the $n=4$ level
through a resonant charge transfer process, involving neutral
carbon atoms as electron donors. Overpopulation of nearby levels
was accounted for by free-electron capture and by electron-
collisional mixing. A numerical rate equation analysis resulted in

---

†Supported in part by the U. S. Department of Energy.

a self consistent pumping rate for the resonant charge transfer
process of approximately 5×10^5 sec^{-1}.

This charge transfer pumping process requires a sufficient
density of neutral atoms in the path of the carbon ions emerging
from the heated target. Since the results were independent of
the background gas used, the charge transfer was associated with
neutral carbon atoms, the presence of which was detected prior to
the arrival of the ion-dominated blast wave. These earlier neutrals
have recently been shown[4] to originate at the target, and are
thought to be formed rapidly by charge transfer interactions between
fast target ions and neutral gas **atoms**. These interactions may occur
within a distance of $\lesssim 6$ mm from the target surface, ahead of the
detonation wave which is driven by target energy.

An estimate of the density of early neutral carbon atoms has
been obtained[5] from a time-resolved measurement of the emission at
274.9 nm. This measurement was performed with a monochromator
calibrated using a deuterium standard lamp. A density magnitude
of $N_A \sim 3 \times 10^{13}$ cm^{-3} was deduced for the carbon atoms, with some
uncertainty due to the sensitivity of the excited state population
to the estimated temperature of $T_e \approx 1.4$ eV. With this density and
a measured[2] ion velocity of 5×10^6 cm/sec, a cross section of
$\sigma \sim 3 \times 10^{-15}$ cm^2 was estimated, using the relation $N_A \langle \sigma v \rangle =$
5×10^5 sec^{-1}. While theoretical calculations are not available
for the (C, C^{n+}) reaction, accurate calculations[6-8] for the
(H, C^{n+}) reaction favoring the n=4 excited state are in general
agreement with this estimate from experimental data. Thus, while
the neutral atom density is relative low, the charge tranfer cross
section is very large, and the measurements completed so far support
the feasibility of resonance charge transfer as the dominant pumping
process for the observed population density anomalies and inversions.

CURRENT RESULTS

In order to more fully understand and optimize the population
inversion at higher densities for gain experiments, it is important
to extend the measurements to similar ionic species of other
elements. The highest species obtainable with the 5 Joule, 15
nanosecond neodymium laser is N^{5+}. However, similar one- and two-
electron species in lithium and boron are available (because of
toxicity beryllium would require very special handling). The
results of initial tests with targets of boron nitride and lithium
are reported for the first time at this workshop. The boron and
nitrogen spectroscopic results support the previous conclusions for
carbon, and the lithium results are even more definitive in
supporting the resonant charge transfer pumping process. It should
be emphasized that these data are preliminary and obtained under
less-than completely consistent conditions. Varying ion conditions
and instrumentation will be noted where appropriate.

Carbon

The spectra of C^{4+} and C^{5+} ions are traced in Fig. 1, both at a 19 mm distance from the target where the inversion occurs and also for a more normal spectral series distribution near the target surface. Again, intensity inversion implies population density inversions of magnitude 3-4X. Also sketched in Fig. 1 are the relevant ion and neutral energy levels[9], showing the typically 10-15 eV energy defect expected for the exothermic charge transfer process.

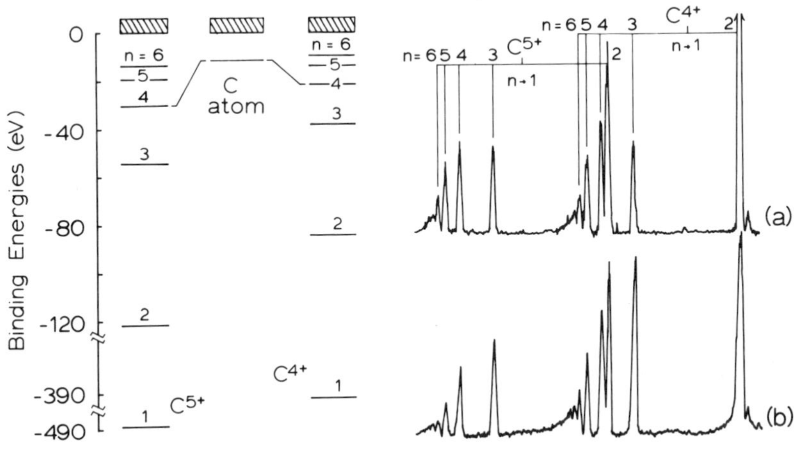

Fig. 1 Binding energies for C,C^{4+},C^{5+} and retraced spectra (λ=2.6-4.0 nm)[9] at: (a) 19mm above and (b) at the graphite target region. Obtained with a 2 Torr hydrogen atmosphere.

Boron-Nitrogen

The energy levels drawn in Fig. 2 for the boron ion created from a boron nitride (BN) target again indicate a preferential population of the n=4 level over n=3, particularly for the B^{4+} hydrogenic ion (4-3, $\lambda \approx$ 75 nm), and a possible additional population enhancement of the n=3 level of the B^{3+} helium-like ion (3-2, $\lambda \approx$ 40 nm). This is again supported by the spectral traces included in Fig. 2 at 15 mm and at the target. For nitrogen from this target, only the spectrum from N^{5+} was observed at 15 mm and again showed enhanced n=4, 5 population (4,5-3, $\lambda \approx$ 35, 52 nm) compared to the spectrum at the target, as expected from the level diagram included in Fig. 2.

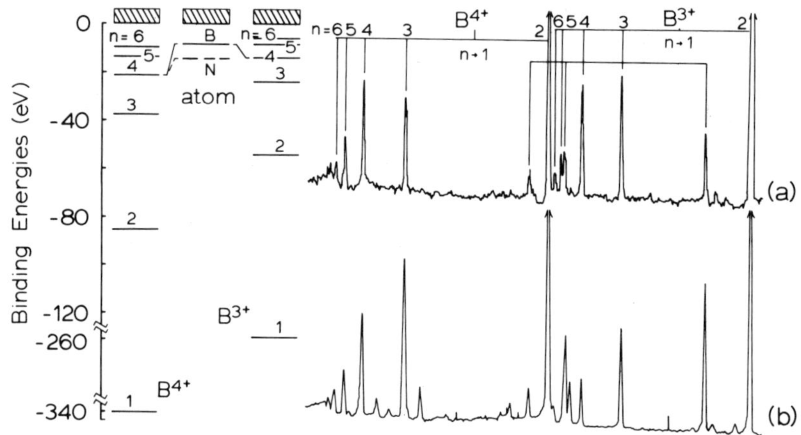

Fig. 2.   Binding energies for B^{3+}, B^{4+} ions and B,N atoms, and
retraced spectra (λ=3.7–6.0 nm)[9] at : (a) 15 mm above and
(b) at the boron–nitride target region. Obtained with a
1 Torr helium atmosphere.

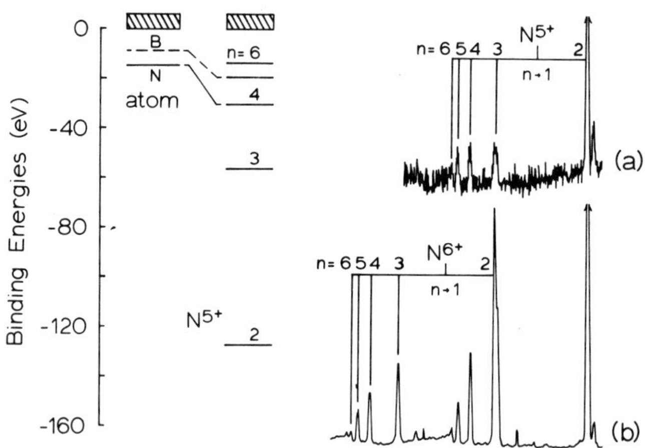

Fig. 3.   Binding energies for the N^{5+} ion and B,N atoms, and
retraced spectra (λ=1.9–2.9 nm)[9] at: (a) 15 mm above and
(b) at the boron–nitride target region. Obtained with
a 1 Torr helium atmosphere. The N^{6+} spectrum was only
observed near the target.

<u>Lithium</u>

From the energy levels shown in Fig. 4, the lithium ion-
atom combination appears to offer a particularly definitive test
of the resonance charge transfer pumping process in this experiment. As indicated, the n=3 level is preferred in hydrogenic
Li^{2+} (3-2, λ=73 nm), while n=2 population is preferred instead of
n=3 in helium-like Li^+. Indeed, the spectral tracing included
in Fig. 4 supports both of these expectations when the data
obtained at 6 mm are compared to those at the target. These data
for lithium offer the most definitive evidence obtained so far in
support of the resonance charge transfer pumping hypothesis,
orginally proposed[2] as the dominant pumping mechanism for these
ionic species. Since these particular inversions were obtained in
vacuum at a distant of 6 mm from the target, a cavity experiment
at a wavelength of 73 nm appears to now be a realistic possibility.

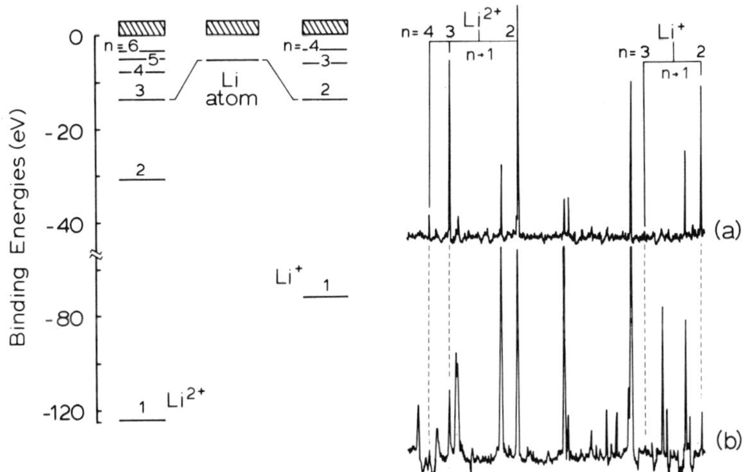

Fig. 4   Binding energies for the Li^+, Li^{2+} ions and Li atoms, and
          retraced spectra (λ=10 - 20 nm)[9] at: (a) 6 mm above and
          (b) at the lithium target region. Obtained in vacuum. The
          unidentified lines are from oxygen ions.

OTHER EXPERIMENTS

There are also data obtained at the University of Rochester
on similar 1- and 2-electron species of aluminum (and magnesium)
which show similar intensity analomies for the n=4 level in x-ray
resonant spectra[10-12]. The measurements are performed in vacuum
with higher power lasers and again inversions were only observed

when the expansion region was perturbed, in this case with a thin
(100 μm) magnesium foil barrier placed 100 μm from the target at
the edge of the laser beam. Another similarity to the NRL data
on lighter elements is the maintenance of a "normal" spectrum
near the target surface. In aluminum 1- and 2-electron ions, the
level binding energies are larger so that resonant charge transfer
pumping of the n=4 level would only occur in ion-ion collisions
such as indicated in the diagram shown in Fig. 5. While the ion-
ion resonance charge transfer process in aluminum, with potential-
curve crossings and with six 2p-donor electrons available, is
expected to have a lower cross section than that for ion-atom
reactions by 1-2 orders of magnitude[13], compensation may be present
in the aluminum plasma due to the higher density and temperature
expected to be present near the target. Although little is known
concerning the required energy defects for ion-ion collisions of
this type, the analogies of the aluminum data with those of the
lighter ion suggest that the neon-like species Al^{3+} or Mg^{2+} might
be involved. A reaction involving these closed-shell species is
particularly interesting since they are expected[14] to "linger" in
a transient plasma, even in the presence of higher species. Indeed,
it is vital to this proposed process that they dominate over other
higher-ionized species which could similarly fill lower levels, thus
depleting the net inversion achieved*.

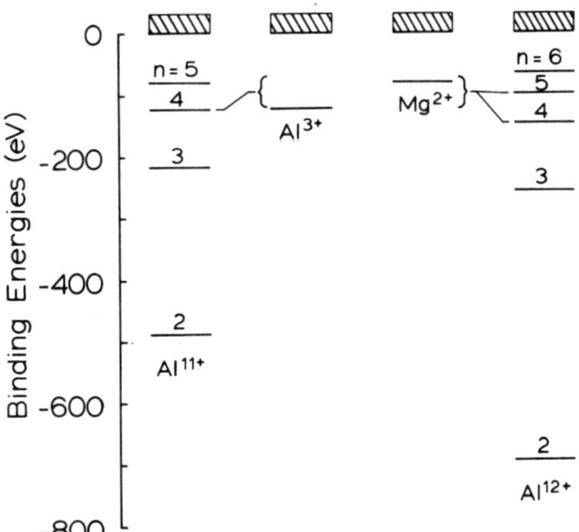

Fig. 5.   Binding energies for Al^{11+}, Al^{12+} product ions and
          Ml^{3+}, Mg^{2+} neon-like donor ions for a suggested ion-ion
          resonance charge transfer pumping mechanism.

---

*From useful discussions with Y. Contourie and J. Forsyth, Univ.
of Rochester, during this workshop.

Extending this suggested correspondence of the light ion results
to the data on aluminum and magnesium, we plot in Fig. 6, for
various atomic numbers Z, the ground state binding energies of the
neon-like donor species. Also plotted are the n=3,4,5 levels of the
1- and 2-electron product ions for the same element (identical
elements are not a necessary requirement). Regions of possibly
favorable energy defect for an exothermal reaction are shaded in
Fig. 6. This graph suggests that for a lighter element such as
sodium the n=5 level may be preferentially populated rather than
n=4; and also that the n=4 observed anamolous populations may
disappear for Z $\gtrsim$14 until n=3 enhancement becomes likely as Z
approaches 20. If, observed, this would be analogous to the
definitive test described above with lithium for the atom-ion
reaction at low Z. It is known that the necessary stripping for
elements as heavy as titanium at Z=22 can be achieved with present
technology. For example, n=3 enhancement for Z=20 (calcium) is suggested (Fig. 6), which if sufficient could conceiveably lead to 3-2
population inversion on a 1.6 nm transition in the x-ray region.

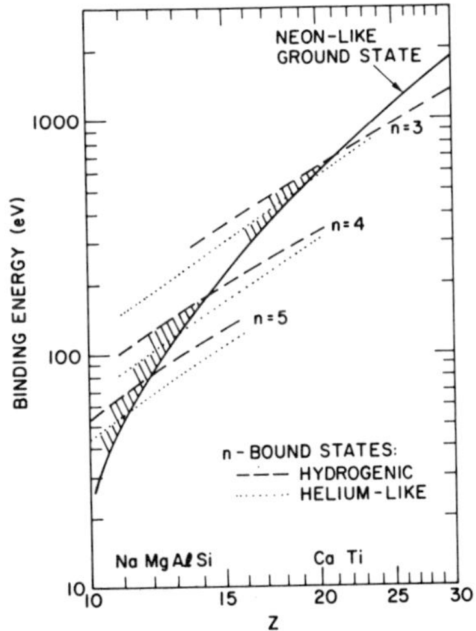

Fig. 6. Binding energies for ground state of neon-like ions and
n=3,4,5 of 1- and 2-electron species for nuclear charge Z.
Shaded areas indicate possibly favorable ion-ion resonance
charge transfer combinations.

SUMMARY

Earlier reported evidence of population inversions in highly-
stripped ions associated with resonance charge transfer pumping is
now supported both by consistent measurements of sufficient neutral

atom densities and most recently by similar but more definitive
measurements in other light ions, particularly those of lithium.
The wavelengths for potential gain experiments with such light
ions span the 35–73 nm range, in which cavity experiments on low-
gain systems are rapidly becoming a reality. Although still
preliminary in nature it is projected that the present light-ion
results may be relevant to some heavier ion data on aluminum which
would require an ion-ion charge transfer process. The promise of
a unified pumping process of high probability, such as offered by
resonance charge transfer, for elements extending from lithium to
calcium could be the production of amplified spontaneous emission to
wavelengths as short as 1.6 nm, and perhaps coherent cavity lasing in
the longer-wavelength vacuum-ultraviolet region in the not too
distance future.

REFERENCES

1. R. W. Waynant and R. C. Elton, Review of short wavelength laser
       research, Proc. IEEE 64:1059 (1976).
2. R. H. Dixon and R. C. Elton, Resonance Charge transfer and
       population inversion following C^{5+} and C^{6+} interactions with
       carbon atoms in a laser-generated plasma, Phys. Rev. Letters
       38:1072 (1977).
3. R. H. Dixon, J. F. Seely, and R. C. Elton, Intensity inversion in
       the Balmer spectrum of C^{5+}, Phys. Rev. Letters 40:122 (1978).
4. R. H. Dixon, J. F. Seely, and R. C. Elton, Observation of a
       fast neutral component from a laser-produced carbon plasma
       Bull. Am. Phys. Soc. 24:765 (1979).
5. R. H. Dixon, J. F. Seely, and R. C. Elton, Measured charge
       transfer cross-section for carbon ions in a plasma, Bull.
       Am. Phys. Soc. 24:765 (1979).

6. A. Salop and R. E. Olson, Charge exchange between H(1s) and
       fully stripped heavy ions at low-keV impact energies,
       Phys. Rev. A 13:1312 (1976).
7. R. E. Olson and A. Salop, Charge transfer and impact-
       ionization cross sections for fully and partially stripped
       positive ions colliding with atomic hydrogen, Phys. Rev. A
       16:531 (1977).
8. A. Salop, The distribution of excitation resulting from
       electron capture in stripped-ion-hydrogen-atom collisions,
       J. Phys. B; Atomic Molecular Physics 12:919 (1979).
9. R. L. Kelly and L. J. Palumbo, "Atomic and Ionic Emission Lines
       Below 2000 Ångstroms, Naval Research Laboratory Report 7599",
       U. S. Gov. Printing Office, Washington, DC (1973).
10. V. A. Bhagavatula and B. Yaakobi, Direct observation of
       population inversion between Al^{+11} levels in a laser-
       produced plasma, Optics Comm. 24:331 (1978).

11. V. A. Bhagavatula, Experimental evidence for soft x-ray population
     inversion by resonant photoexcitation in multicomponent laser
     plasmas, Appl. Phys. Letters 33:726 (1978).
12. Y. Contourie and J. M. Forsyth, Euv and x-ray spectroscopy
     of an inverted laser-produced aluminum plasma (abstract),
     J. Opt. Soc. Am. 69:1410 (1979).
13. V. P. Zhdanov, Resonance charge exchange in ion-ion collisions,
     Sov. Phys. Tech. Phys. 21:117 (1976).
14. C. R. Stumpfel, J. L. Robitaille, and H. Z. Kunze, Initial
     appearance of highly charged ions from magnesium surfaces
     irradiated by Q-switches ruby pulses, J. Appl. Phys.
     43:962 (1972).

THE INTERACTION OF INTENSE NANOSECOND CO_2 LASER RADIATION WITH

LIMITED MASS TARGETS

M.C. Richardson, M.D.J. Burgess, N.H. Burnett,
N.A. Ebrahim, G.D. Enright, R. Fedosejevs, P.A.
Jaanimagi, C. Joshi, R.S. Marjoribanks, and
D.M. Villeneuve

National Research Council of Canada
Division of Physics
Ottawa, Canada

ABSTRACT

A review is presented of a study of the interaction of
intense nanosecond CO_2 laser radiation with spherical and microdisc targets. These investigations on limited mass targets have
concentrated on, elucidating the principal characteristics of
physical phenomena existing in the critical density region during
the interaction, and, identifying the gross features of superthermal particle behaviour. The spatial distribution of the
electron density in the region of n_c and up to 40 n_c has been
characterised. The formation of a steepened profile, and its
subsequent development into an overdense bump at intensities too
low for significant radiation pressure effects are discussed in
terms of a model having a sharp temperature transition within the
interaction region. Measurements of the fractional absorption on
spherical targets are described and some deductions made on the
partition of energy into fast ions, based on analysis of the
density profiles and quantitative fast ion spectrometry. The
lateral transport of energy by fast electrons has been investigated
with a number of diagnostic approaches and has led to a clearer
understanding of the role of fast electrons in establishing a
plasma sheath about the target and in providing an avenue for
effective transport of energy to remote regions of microdisc
targets.

INTRODUCTION

In this paper we review our recent studies of the interaction
of intense ($\sim 10^{14}$ W cm^{-2}), nanosecond CO_2 laser radiation with
spherical and microdisc targets, an area of primary importance
to current laser fusion investigations. In particular emphasis
has been placed on defining the conditions existing in the interaction region and investigating the characteristics and behaviour
of the superthermal ion and electron emission from such targets.
From these investigations has come a clearer understanding of the
interaction process, as well as the identification of a number of
features of this interaction hitherto not specifically studied.

Experimental Conditions

These studies were performed with the use of the COCO-II
laser facility, detailed characteristics of which were described
fully at the previous workshop[1]. The present single beam
experiments were made at energy levels up to 40 J focussed onto
the target by f/2.5, 20 cm focal length, off-axis parabolic optics
producing focal spot intensities of up to 5×10^{14} W cm^{-2}. The
intensity distribution at and about focus was limited by surface imperfections in the focussing mirror, Fig. 1(a), and had a
minimum half intensity diameter of 75 µm, and half-energy diameter
of ~ 110 µm with prepulse radiation levels <10^8 W cm^{-2}. Under
these conditions, picosecond visible-light interferometry established that no detectable plasma ($n_e < 10^{18}$ cm^{-3}) existed on the
surface of the target prior to the arrival of the pulse. Diagnosis
of the output pulse shape with the aid of picosecond upconversion
techniques[2], and with the use of a fast pyroelectric detector
– 1 GHz oscilloscope combination[3] showed that the pulse shape
can be approximately characterised by a rising linear ramp of
~ 300 ps risetime and an exponential fall time of between 800 ps
and 1 ns (Figs. 1(b) and 1(c)). The targets, in general simple
selected unfilled glass microballoons or planar microdiscs, were
mounted on thin (<10 µm diameter) glass or metal stalks, and
aligned in the focal plane to an accuracy of ~ 30 µm.

Primary Features of Nanosecond 10 µm Interaction

Previous studies of short pulse 10 µm interaction primarily
with massive planar targets, at NRC and elsewhere, have helped
to clarify the principal physical processes. Early interferometric
studies showed strong profile steepening in the critical density
region with the clear formation of upper and lower density pla-
teaux[4], the scale length L at n_c having a value of $\sim \lambda_o$. The
existence of such density profiles, together with the measured
values of absorption[5] and electron temperature[6] T_e imply that

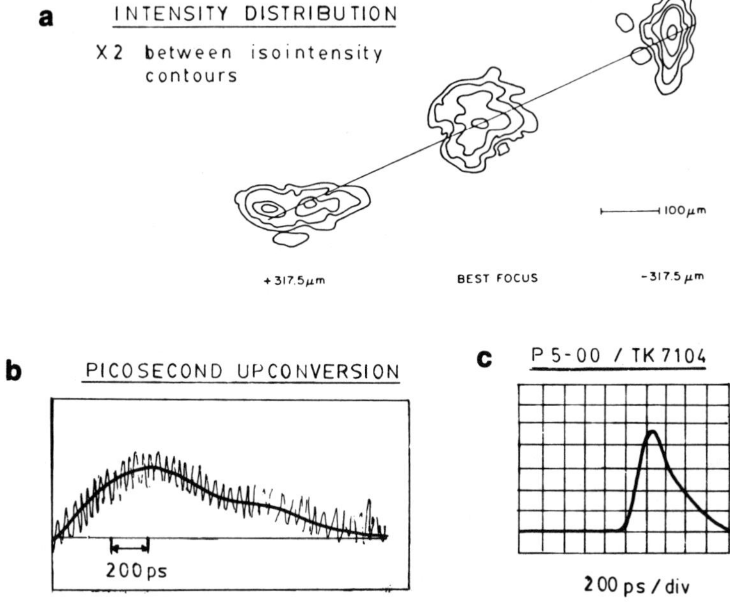

Fig. 1   (a) Iso-intensity contours of the beam distribution in the focal region; (b) and (c) Typical laser pulse shape as determined by picosecond upconversion techniques and a fast pyroelectric detector, respectively. The fast modulation on the upconverted signal in (b) results from mode-beating in the Nd:YAG laser used as a pump source, and conveniently provides temporal calibration.

classical inverse Bremsstrahlung absorption is not effective. Moreover, experiments on massive planar targets have established a distinct dependence of the fractional absorption on laser beam polarisation and angle of incidence[5], whereas little dependence on intensity or target Z was observed. The absorption for p-polarised radiation was significantly greater than for s-polarised light, and decreased markedly with increasing incidence angle, having a maximum value $\sim$50% at $\lesssim$20°. Similar observations have been found in short pulse (typically $\sim$100 ps) 1.0 μm experiments[7-9] where a distinct peak in the angular dependence for p-polarised radiation has been found. These results together with observations of the fast ion emission[10] are widely interpreted as establishing the importance of resonance absorption[11] as the main interaction mechanism.

The existence of strong resonant electrostatic fields in the critical region can also be inferred from the observation on massive planar targets, in both 10 μm[12] and 1 μm experiments[13], of a complete series of integral harmonics in the backscattered and sidescattered emission from the plasma. In the former case, the identification of the harmonics of up to 11 ω_o have been reported, and more recently[14] the 12 ω_o harmonic has been identified. These harmonics all possess similar spectral characteristics and have been shown to display a dependence on laser intensity of $I_{n\omega_o} \sim I^n$ up to a value of $I \sim 10^{14}$ W cm^{-2} where a saturation effect becomes evident[15]. Ultrafast temporal analysis of these harmonics has indicated there is considerable picosecond substructure in their temporal development[16] at high intensities ($>10^{14}$ W cm^{-2}) which may well be related to preliminary evidence for the existence of similar structures in the hard x-ray emission from such plasmas, observed with the aid of a picosecond x-ray streak camera[17].

Further evidence for the dominance of resonance absorption in the interaction can be found from the characterisation of the superthermal electron distribution produced by this process. Determination of the characteristic temperature T_H of this distribution from the x-ray continuum emission from the plasma has found it to follow a dependence on $(I\lambda_o^2)$ of the form $T_H \sim (I\lambda_o^2)^\sigma$ where σ has a value of 0.25 - 0.40[18], in qualitative agreement with numerical simulations of resonance absorption on a steepened density profile[19-22].

Although these basic features of the interaction are largely accepted, the partitioning of energy into the target, and into the fast ion component, and the lateral flow of energy are not well understood. The following sections outline investigations of these features made on microballoon and microdisc targets at NRC during the past year.

Picosecond Interferometry

Interferometric measurements were made, primarily on glass microballoon targets, of the development of the electron density profile during and following irradiation using a folded-wavefront interferometer[23]. The 0.53 μm, 50-150 ps probe pulse used to illuminate the latter was derived from the output of an actively mode-locked Nd:phosphate glass laser[24] which was synchronised to the CO_2 laser pulse to an accuracy of $\sim$200 ps. Electron densities were obtained by numerical Abel inversion[25] of the fringe shift data. Effective spatial resolution in the axial direction was dependent upon fringe separation, varying from 8 μm in the underdense region to 2 μm at the highest densities, $\sim$40 n_c with a spatial resolution in the transverse direction of 20 μm.

Profile Steepening

A representative sample of results taken on 140 μm diameter microballoons, at irradiation intensities of $\sim$0.5 × 10^{13} W cm^{-2} can be seen in Fig. 2 which illustrates the development of the profile from an initial sharp step and short plateau through the formation of an overdense bump to its subsequent decay into an extended plateau. In many interferograms the plasma expanded preferentially towards the laser beam while only occasionally at higher intensities and at later times did cratering of the profile occur as previously observed on planar targets[26,27]. The scale length decreased with increasing intensity up to 10^{13} W cm^{-2} where it reached a minimum of 3.4 μm. At these lower intensities the value of scale length and its dependence on incident intensity was in fair agreement with the L $\sim$ I$^{-0.48}$ prediction of Estabrook and Kruer[19] if an effective electron temperature at n_c of T_{eff} $\sim$ 1.3 keV was assumed[28]. However this temperature is considerably higher than the time and space averaged cold electron temperature T_e $\sim$ 250 eV deduced from x-ray line emission and continuum measurements[6,29,30] for nanosecond CO$_2$ laser pulses at 10^{13} W cm^{-2}. At higher incident energies, and in general later probe times, the scale length increased with intensity, rising to $\sim$5.7 μm at I $\sim$ 10^{14} W cm^{-2}.

Density Step Height

Some of the axial density profiles show clear evidence for the existence of an overdense bump having characteristics similar to the compressional shock transition from supersonic to subsonic flow predicted by Max and McKee[31] and Virmont et al.[32]. A typical example is shown in Fig. 2(f). The plateau height N_p, defined as the peak value of the density bump when observed or the point of minimum gradient for an inclined shelf when it was not, is plotted as a function of I in Fig. 3. A least square fit to the data gives

$$N_p = 13.6 \ (10^{-13} \ I)^{0.39}$$

where I is in W cm^{-2}. These values are in disagreement with those predicted by the isothermal model of Lee et al.[33] for T_e $\sim$ 250 eV. Such a large discrepancy has not yet been found in results obtained with 1.06 μm radiation[26,34] and indicates that for 10.6 μm the effect of the superthermal plasma component on the plasma profile must be large. Steepening of the profile with significant step heights has been predicted even at intensities where the radiation pressure is less than the thermal plasma pressure[35,36]. The effects of the superthermal component may be seen by employing a model which includes a sharp temperature step at n_c, as well as a density step[28]. This results in an upper to lower density ratio

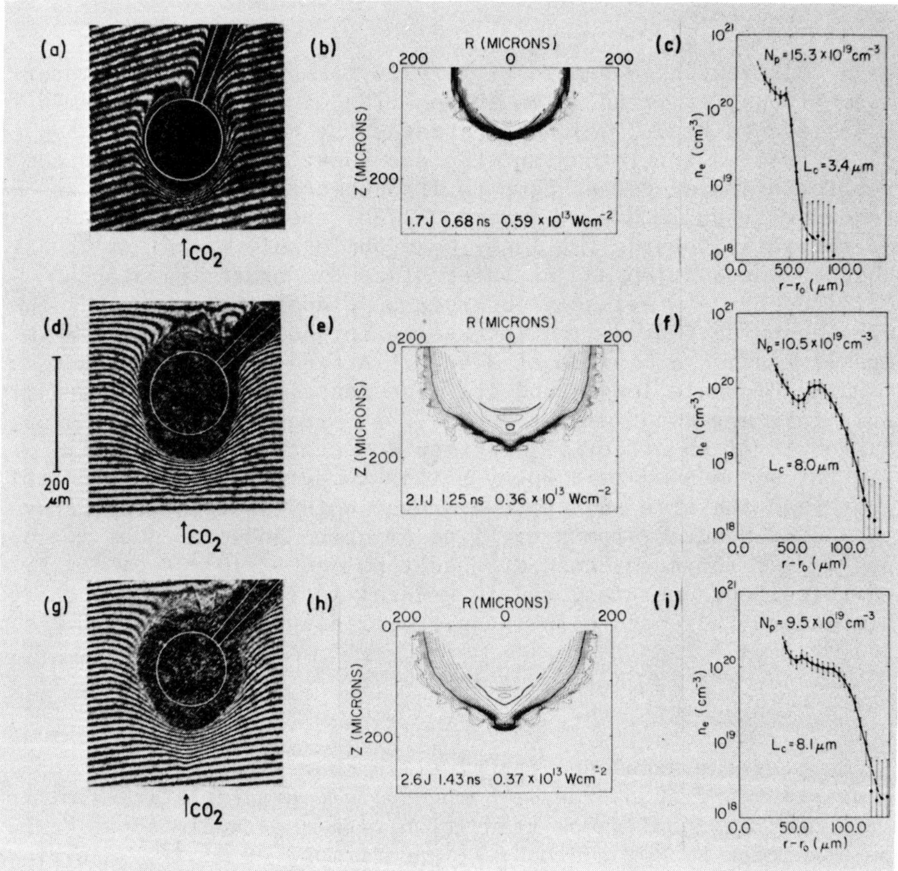

Fig. 2   Interferograms, (a), (d) and (g), electron density contour
plots, (b), (e) and (h) corresponding axial electron density profiles, and (c), (f) and (i), of the plasma produced
on 140 μm diameter microballoon targets at similar energies
for times of t = 0.68 ns, 1.25 ns and 1.43 ns. The CO_2
laser beam is incident from the direction of the arrow.
The energy, probe time, incident intensity, upper plateau
height and scale length at critical density are given for
each shot on the axial density profile. Logarithmic
contour intervals, ten per decade, are displayed with dark
contours at 10^{19} and 10^{20} cm^{-3} and dotted contours at
multiples of 2.5 and 5. (a) and (g) were taken probing
the plasma perpendicular and parallel to the polarisation
of the CO_2 laser respectively while (d) was taken with an
elliptically polarised CO_2 laser pulse. Time t = 0 ns
corresponds to 20% of the leading edge of the CO_2 laser
pulse.

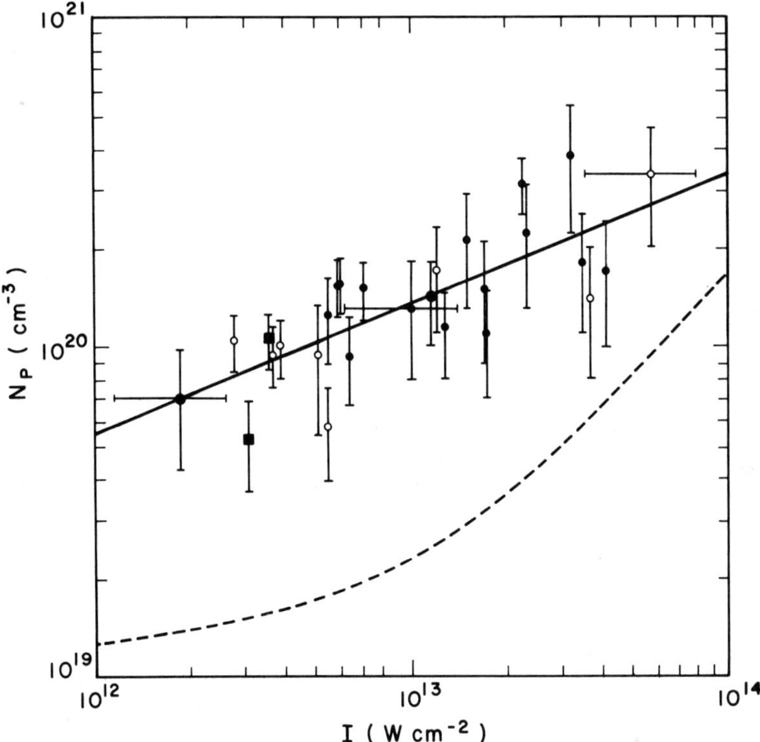

Fig. 3   Upper density plateau height versus incident intensity.
Different symbols represent different polarisations of
the CO$_2$ laser beam relative to the probe direction:
perpendicular (closed circles), parallel (open circles)
and elliptically polarised (closed squares). The dotted
line indicates the value of N_p predicted by the isothermal
model for a $T_e \sim 250$ eV.

$$\frac{n_u}{n_1} = \frac{P_\ell + (P_\ell^2 - 4v_\ell^2 c_u^2)^{\frac{1}{2}}}{2c_u^2}$$

where P_ℓ is related to the radiation pressure P_R as

$$P_\ell = v_\ell^2 + c_\ell^2 + \frac{2P_R}{Mn_\ell}$$

and where v_ℓ is the lower shelf velocity and c_u, c_ℓ are the ion
acoustic velocities of the upper and lower shelves respectively.
Assuming the temperatures of the upper and lower shelves to be
the cold electron temperature, $T_e = 250$ eV, and the hot electron

temperature[18,29], $T_H \sim 5$ keV, respectively, and taking values of $v_\ell \sim 10^8$ cm s^{-1} and $n_\ell \sim 10^{18}$ cm^{-3}, then an upper shelf density of ~ 12 n_c is predicted for an intensity of 10^{13} W cm^{-2}. This interpretation implies the existence of a transition layer in the vicinity of n_c, separating a cold dense plateau from the hot tenuous blow-off. This view is supported by measurements of T_e and T_H derived from x-ray and fast ion measurements made on 10.6 μm irradiated microballoons[29]. These indicate that a two-temperature electron distribution, with values for T_H/T_e of 20-50 existed for values of $I\lambda^2$ of 10^{15}-10^{16} W·μm^2·cm^{-2}. For values of $T_H/T_e > 5 + \sqrt{24}$ a rarefaction shock has been predicted to occur in an expanding two temperature plasma[35,37], giving rise to a density discontinuity at $n_e \sim n_H$. If, for the present experimental conditions, the region below n_c is predominantly hot electrons, then this density discontinuity would be located at the steepened surface and would then contribute to the total step height through n_c. The fast rarefaction would then extend from the lower shelf density to where quasi-neutrality breaks down as the hot electron debye length exceeds the density scale length.

Magnetic Field Generation

Strong evidence now exists for the generation of megagauss fields in plasmas produced off planar targets by intense short duration 1.06 μm pulses[38,39]. As yet no such direct evidence exists in plasmas produced by 10.6 μm pulses at similar values of $I\lambda^2$. Because Faraday rotation detection of magnetic fields with optical probes is dependent on density, their measurement in the region of n_c for 10 μm is limited in practice to fields $\gtrsim 1$ MG.

Nonetheless, several features of plasmas created from solid targets by intense ($>10^{13}$ W cm^{-2}) nanosecond CO_2 laser pulses are consistent with the presence of large magnetic fields. Several years ago, magnetic fields of plasmas created off solid targets in a vacuum were detected with fast magnetic probes[40]. More recently spatial characterisation of the plasma with the aid of XUV spectroheliography and interferometry add greater credibility to their existence[41]. The spectroheliograph[42], incorporating a toroidal grating, produces two-dimensional images of the plasma with a spatial resolution of ~ 60 μm, in discrete XUV lines in the 200-500 Å region. A typical spectrohelogram of the plasma produced off an Al microdisc target is shown in Fig. 4. The individual images of the plasma are in emission lines from transitions of ions of AlV through AlX. As can be seen the images show the creation of similar plasmas on the front and rear sides of the target, a phenomenon which will be discussed in detail later. However, an additional feature of the images is their highly elongated form, which may be several millimeters in length

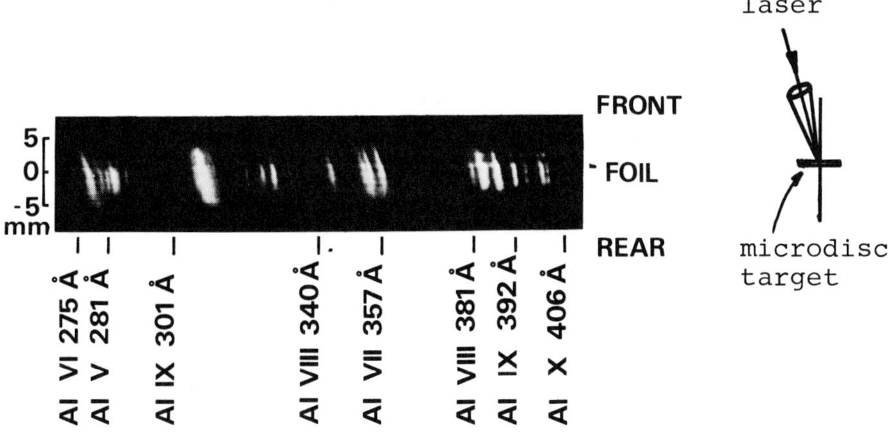

Fig. 4   Spectroheliogram of plasma produced off 12.5 μm thick,
         500 μm diameter disc target.

while $\sim$500 μm in diameter at the target surface. It has been
suggested[41] that this effect, and the plasma-constriction observed
interferometrically could be the result of large scale azimuthal
magnetic fields existing in the underdense region. It is clear
that MHD effects will directly influence the plasma dynamics if
the ratio of kinetic to magnetic pressure in the plasma $\beta = 2\ nT_e/$
$B^2 < 1$. For a 300 eV plasma at $n = 10^{19}$ cm^{-3}, $\beta = 1$ for a value
of $B = 600$ kG. After the laser pulse, as the plasma cools and
expands, smaller fields would be sufficient to confine the underdense region of the plasma.

The axial constriction of the expanding plasma produced on
empty glass microballoons irradiated by 10.6 μm pulses also
suggests the presence of azimuthal fields. An illustration of
this phenomenon is shown in Fig. 5. As can be seen, the radial
density profiles up to and above 10 n_c are axially constricted,
perhaps implying fields of $\sim$1 MG for $T_e \sim 250$ eV.

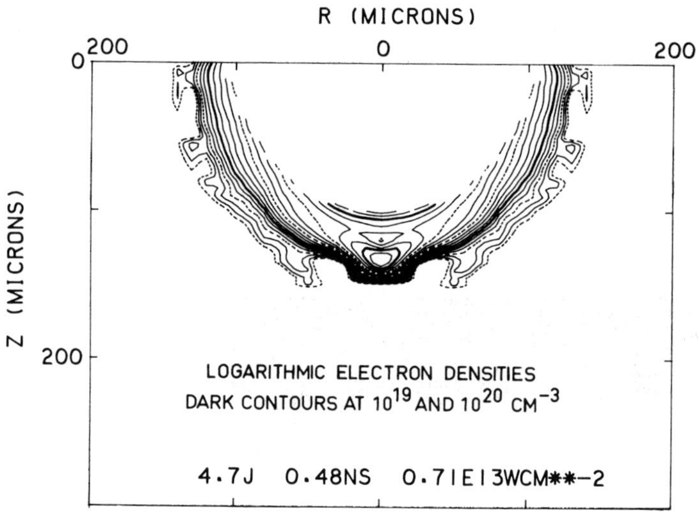

Fig. 5   Two-dimensional electron density profile of plasma created
off 140 μm diameter microballoon, obtained 480 ps into
the laser pulse when the intensity was 7 10^{12} W cm^{-2},
showing radial constriction of the plasma expansion.

<u>Energy Partition</u>

X-ray photography, together with interferometry provides
a means of roughly estimating the local electron temperature.
A typical pinhole photograph of a single side irradiated empty
glass microballoon supported on fine glass stalk is shown in Fig.
6. The lack of exposure from regions of the microballoon outside
the focal zone where electron densities >10^{21} cm^{-3} are known to
exist during the laser pulse, implies a T_e < 100 eV in these
regions. A better estimate of T_e can be deduced from analysis
of the development of the electron density profile in these regions.
Assuming that these regions may be characterised by an isothermal
blow-off, then the variation of the scale length L with time is
used to determine the isothermal sound speed. Reduction of interferograms similar to those shown in Fig. 2 then imply a temperature
of ∿30 eV for empty microballoons irradiated with single 30 J pulses.

Taking into account the kinetic energy of ion expansion
acquired during the laser pulse by a plasma of this temperature,
and assuming that the whole shell is uniformly heated, requires
an energy of ∿1.5 J. However detailed measurements of the plasma

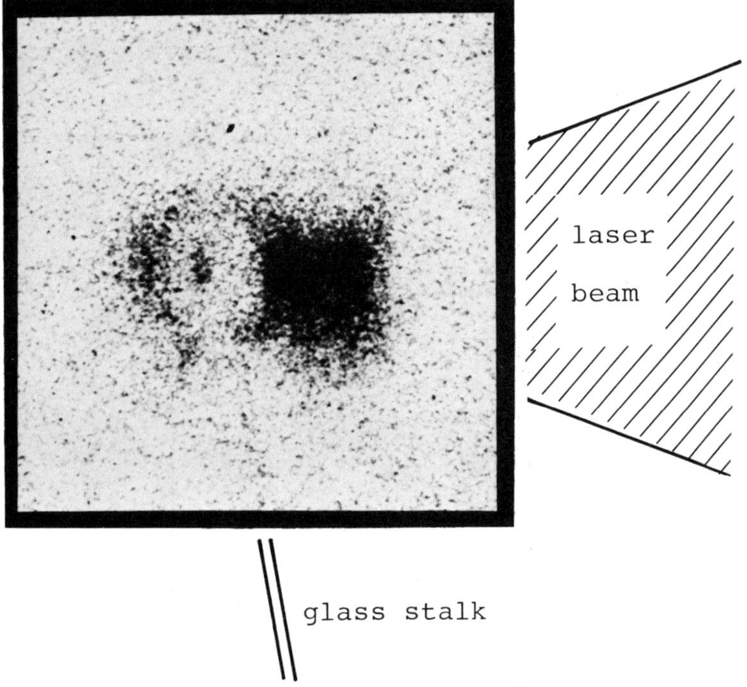

Fig. 6   X-ray pinhole photograph of 140 μm diameter empty glass
         microballoon irradiated with single, nanosecond, 25 J
         CO$_2$ laser pulse.

blow-off distribution about the target with a 4π array of 20
calorimeters typically produces energy distributions of the form
shown in Fig. 7, where for 30 J incident laser energy, the total
fractional absorption, was ∿20%. Thus, from the above, ∿4.5 J,
or ∿75% of the absorbed energy remains unaccounted for in the
plasma. It would thus appear from these arguments that as much
as 75% of the laser light absorbed in the plasma created off
these thin shell targets, is coupled into the fast ion component.

Fast Ions Off Microballoon Targets

     Quantitative measurements of the fast ion flux from micro-
balloon targets have been made at specific angles from the
irradiation axis[29]. Although no attempt has yet been made to
directly estimate the fractional partition of the absorbed energy
into the fast ion component, it was found that the species of
fast ions detected, and their energy distribution were dependent
on the angle of observation. Moreover, a significant flux of
fast ions was observed at an angle of 110° from the irradiation
axis[29]. A number of mechanisms could give rise to this broad
angular distribution of fast ions. These include, (i) ion acceleration by fast electrons isotropised by electron-ion scattering

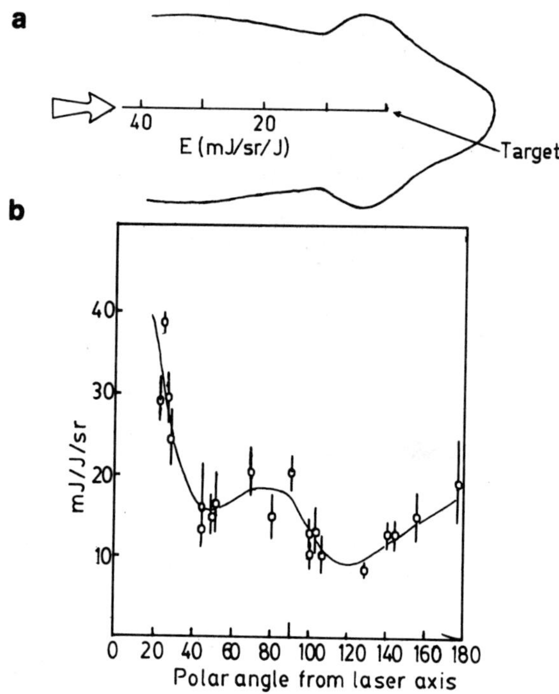

Fig. 7   Energy deposition in the plasma blow-off from a single-
         sided irradiated glass microballoon, as determined from a
         4π array of 20 particle calorimeters, plotted as (a) a
         polar distribution about the axis of irradiation and
         (b) the fractional energy per steradian as a function of
         polar angle. All the data points have been normalised
         to the incident laser energy.

in the shell, (ii) large angle deflection of fast ions from the
interaction region by strong magnetic fields, (iii) acceleration
of a thin surface layer of ions by the electrostatic field surrounding the target and (iv) lateral spreading of a given neutral
plasma sheath comprising the fast rarefaction. The latter process
will be discussed in more detail in subsequent sections. If this
ion flux were emitted uniformly from all unirradiated regions of
the microballoon, then it could contain as much, if not more,
energy than is observed in the fast ion plume directed toward
the laser beam.

Lateral Energy Transport Through Fast Electrons

It is now firmly established that at these irradiances, energy coupling to the plasma occurs through collisionless processes which generate high energy electrons. These electrons directly influence many of the features already discussed. However, at the present time the transport of energy into the target by these electrons, a subject of considerable importance to laser fusion, is not well understood[43]. Much of the experimental data can be interpreted by assuming that classical heat conduction is inhibited and multigroup flux-limited diffusion treatments of electron transport have been included in numerical simulations of the interaction. Possible physical mechanisms leading to such inhibited energy flow have received much attention in the past few years, and in particular self-generated magnetic fields and ion acoustic turbulence have been modeled in computer simulations. Alternatively, if classical thermal conduction is assumed, without the inclusion of other inhibiting mechanisms, agreement between one dimensional simulations and experimental data can apparently only be effected by a large reduction in the fractional absorbed energy[44]. It would therefore appear that with this assumption, physical mechanisms must be involved to account for the effective loss of absorbed energy. However, in all these studies the behaviour of the electrons has been treated one-dimensionally and it has only been very recently that two-dimensional aspects of superthermal electron transport have been considered[45,46]. Consideration of the lateral motion of the superthermal electrons would then lead to a possible energy loss mechanism accounting for the reduced thermal transport. To investigate these effects in more detail, we have recently performed a series of experiments on microdisc targets.

One-sided Irradiation of Microdisc Targets

In experiments in which microdisc targets of diameter 0.5 – 10 mm and thicknesses in the range 12.5 μm to 150 μm were irradiated with nanosecond 10.6 μm pulses at intensities of $\sim 10^{14}$ W cm^{-2}, we have found that a significant fraction of the energy is coupled to the rear of the target, resulting in the formation of a hot plasma[47]. Figure 8 shows an x-ray photograph of such an irradiated target, the pinhole camera recording x-ray emission of energy in excess of 0.6 keV. Whereas an intense emitting region of ~ 300 μm diameter exists on the front surface of the 500 μm diameter target, a thin intense x-ray region also extends over a large part of the rear surface of the target. The existence of comparable temperatures on the front and rear surfaces of the target can also be inferred from charge collector data[47], and from the spectra recorded by the spectroheliograph. As can be seen from Fig. 4, the spectroheliogram of a 500 μm diameter,

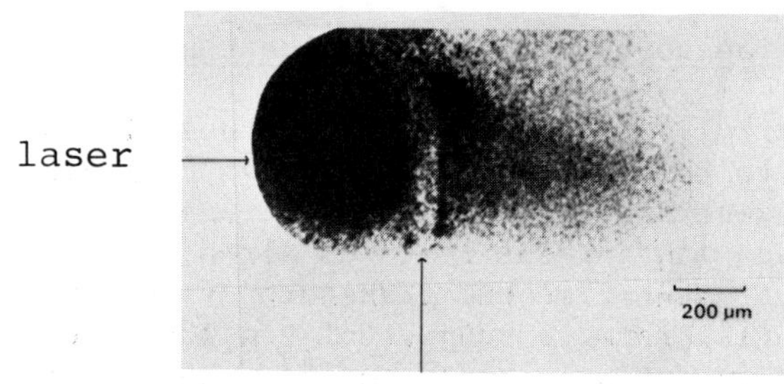

Fig. 8   X-ray photograph of 150 µm thick, 500 µm diameter Al
         disc target irradiated from one side by a single
         nanosecond 25 J CO_2 laser pulse.

12.5 µm thick target, identical highly charged ions exist in the
blow-off plasmas from both sides of the target. Other diagnostics
indicate that this high temperature plasma is created on the rear
side of the target during irradiation. Visible picosecond streak
photography of the expanding plasmas on both sides of the target
show not only that the plasma on the rear side of the target is
created within 1 ns of that on the front side, but also that the
front and rear plasmas have comparable expansion velocities,
(Fig. 9(a)), implying similar thermal temperatures[50]. Moreover,
the fast ion component observed from the rear surface contains
fully stripped C ions, which must have been created at early times
since they are only accelerated during the laser pulse.

Fractional Partition of Energy to the Rear of Microdisc Targets

     Measurements of the fractional absorption, made with an array
of 20 plasma calorimeters[48] deployed about the target have been
determined from the plasma blow-off distribution on both sides of
the target. Symmetrised polar distributions of the plasma blow-off
for 25 µm thick Au targets are shown in Fig. 10. Whereas 22% of
the total absorbed energy ($E_{abs} \sim 0.3 \, E_{laser}$) is emitted towards
the rear for a 500 µm diameter target, this value is reduced to
9% for a target of 5 mm diameter. Moreover in the latter case
the front emission is several times more intense along the target
normal, suggesting a predominantly one-dimensional expansion from
the target surface. The dependence of the fractional absorption,
and the fractional energy partition between front and rear plasmas
as a function of target thickness and target diameter are shown
in Fig. 11.

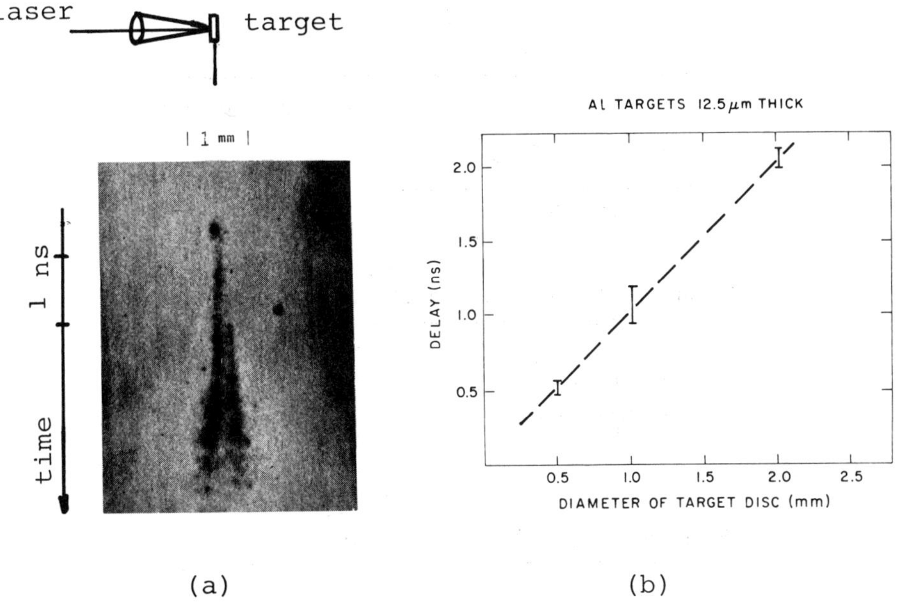

Fig. 9   Time resolved infrared photography of plasma expansion
         from microdisc targets. (a) streak photograph of front
         and rear plasma expansion from 150 μm thick, 500 μm dia-
         meter Al disc target; (b) variation of delay in onset
         of rear plasma expansion as a function of target diameter.

Whereas the total integrated absorption is found to be
independent of target conditions, a large variation of the partition of energy to the rear hemisphere plasma blow-off with target
diameter is observed. The lack of any dependence of this partition on target thickness eliminates any mechanisms dependent on
bulk absorption or the transport of energy through the target by
fast electrons.

Variation of Hard X-ray Continuum with Target Diameter

Measurements of the x-ray continuum observed in a direction
45° to the front target normal also show a dependence on target
diameter. Figure 12(a) shows the dependence of x-ray intensity
of the superthermal continuum emission in the 5-30 keV range as
a function of target diameter. This would suggest that a fraction
of this emission may be emanating from the periphery of the front
surface of the target, and is consequentially reduced as the target diameter is reduced. This view is perhaps supported by the
observation that the x-ray continuum spectrum in this energy range
is also dependent on target diameter. Figure 12(b) shows the
spectrum obtained for two targets of diameter 500 μm and 7 mm. A
significant reduction in the superthermal component is observed in

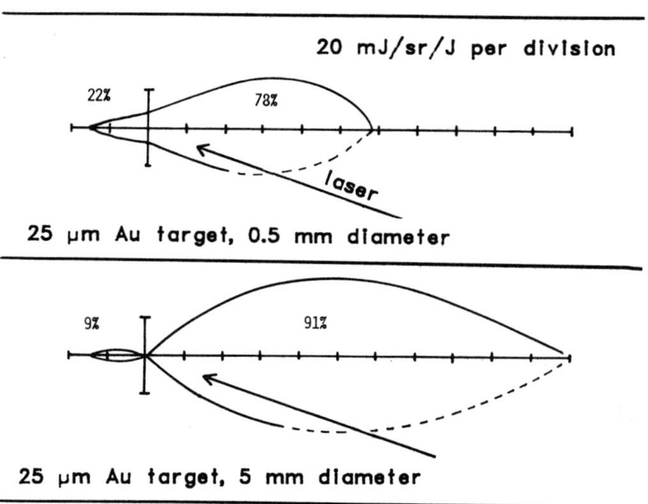

Fig. 10   Plasma blow-off distributions from front and rear plasmas
produced off single-sided irradiated Au discs. The energy
flux has been normalised to the incident laser energy.

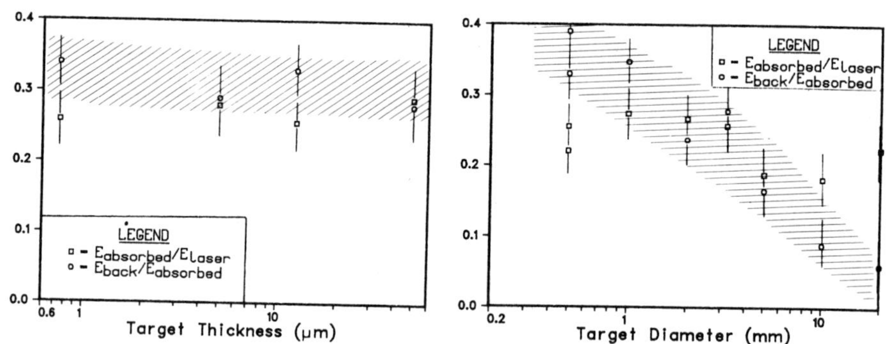

Fig. 11   Variation of the total fractional absorption, and the
fractional energy deposition on the rear of microdisc
targets as a function of (a) target thickness for a 500
μm diameter disc and (b) target diameter for a 12 μm
thick disc.

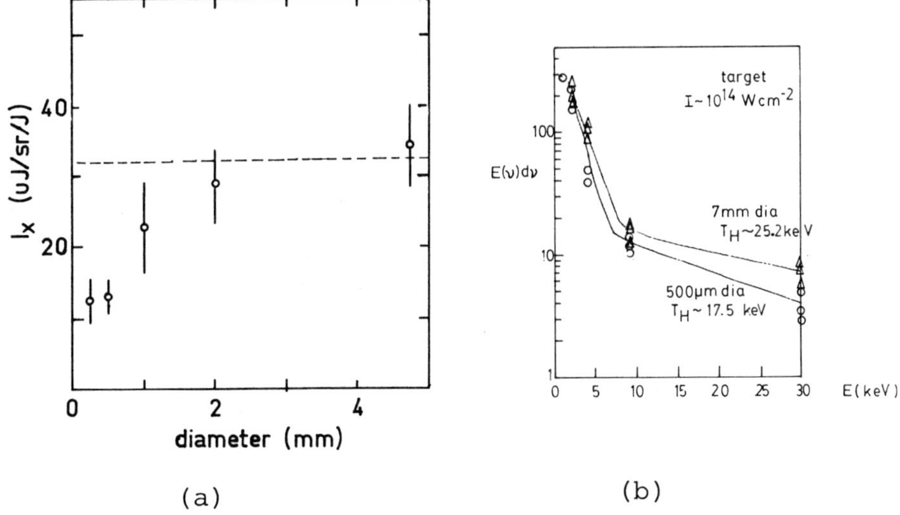

(a)   (b)

Fig. 12   Dependence of hard x-ray emission on target characteris-
          tics. (a) shows variation of integrated x-ray emission
          above 5 keV as a function of target diameter. (b) shows
          the dependence of the emitted x-ray spectrum on target
          diameter. The target thickness in all cases was 12 μm.

the former case, suggesting that caution should be exercised in
determining absolute values of T$_H$ in the interaction region from
such measurements.

Evidence for Fast Electron Bombardment of Rear Surface of Target

It has been suggested that the plasma formed on the rear
surface of the target results primarily from the transport of
superthermal electrons around the target[47]. The effects of such
electrons on the rear surface of the target has been observed in
both the back streaming fast ion distribution, and in the observation of K$_\alpha$ emission from the rear side of the target. The fast
ion emission on both sides of the target has been diagnosed with
the aid of Thomson parabola spectrometers, utilising cellulose
nitrate recording film, deployed normal to the target surfaces.
A summary of the ion species observed on both sides of the 500 μm
diameter, 150 μm thick Al target is shown in Table I. On the
irradiated side of the target fast ions of charge species Al^{3+}
through Al^{+11} are detected as well as protons and carbon ions
probably originating from a thin contamination layer on the surface
of the target. The maximum cut-off energy for Al ions was 2.5 MeV,
suggesting the emission of electrons of energy ∿200 keV. Electrons
in the range 50-600 keV have been detected in the forward direction
with the aid of small 180° electron spectrometers[49]. The fast ion
emission observed from the rear side of the plasma only consisted

TABLE I

	Charge Species	E_{max} (keV)	T_H (keV)
Front	$Al^{+11} \rightarrow Al^{+3}$	2500	25
	$C^{+6} \rightarrow C^{+1}, H^+$	1400	
Rear	$C^{+5} \rightarrow C^{+1}, H^+$	600	

of ions originating from the hydrocarbon contamination layer; the
C ions having maximum energies of 0.6 MeV, suggesting that only a
thin layer, formed the fast rarefaction. Since the hot electron
temperature, as determined both from the slope of fast ion energy
spectrum in the isothermal limit and from the x-ray continuum
spectrum, is $\sim$20 keV, it is unlikely that these ions are accelerated by electrons passing through the target (the penetration
depth of 20 keV electron in Al is $\sim$3 μm).

Moreover direct evidence of heating of the rear surface of
the target exists in the observation of K_α emission on the rear
side of the target. These measurements were made with 500 μm
diameter, 150 μm thick Al disc targets. With the aid of an R.A.P.
crystal spectograph, the relative K_α flux from the front surface
of the target compared to that from the rear surface was found to
be $\sim$1.5. Since, under these conditions, the effect of radiation
pumping of K_α on the rear surface of the target can be discounted,
this observation provides a measure of the level of electron bombardment on the rear surface of the target.

<u>Formation of Plasma Sheath</u>

These observations lead to the view that considerable lateral
transport of energy by fast electrons occurs in nanosecond CO_2
laser interaction with limited mass targets. This most probably
occurs through the early establishment of a space charge sheath
about the target resulting from electrons making large amplitude
orbits from the target but confined by its potential[51], and from
electrons escaping completely from the target field. In this
pure electron gas region ahead of the quasi-neutral sheath, the
electron trajectories are determined by space charge repulsion
and the self-consistent magnetic field. These electrons could
follow trajectories of length comparable to the target and thus
may well be able to induce significant heating of the rear surface.
Once a plasma sheath is formed around the target, lower energy
hot electrons with large angular momentum may transport energy

to the rear surface by reflection within this region. Additional
evidence to support this interpretation can be drawn from time-
resolved photography of twin disc targets, separated by $\sim$300 μm
and, supported on fine, 2 mm long, glass stalks, of the type shown
in Fig. 13(a)[50]. As can be seen from the streak photograph, Fig.
13(b), of the visible light from the two targets, plasma forms on
both sides of the surrogate target, situated 300 μm behind the irradiated target, within 1 ns. Moreover, the velocity of expansion
of the plasma from the second target is comparable to that of the
plasma from the irradiated target, suggesting the creation of
similar temperatures. The x-ray photograph of a similar target,
Fig. 13(c), clearly indicates that the plasma formed on the front
surface of the surrogate could have a temperature comparable to
that of the plasma irradiated by the laser. Since the total stalk
length between targets is >5 mm, it is clear that surface return
current effects[51] cannot be responsible for the formation of the
plasma on the surrogate target at about the same time as the
plasma on the rear side of the irradiated target is formed. Moreover, the time delay of the onset of visible emission from the
surrogate target is comparable to the time delay of the onset of
emission from the rear side of the front target. Studies of this
latter delay on single disc targets suggested a propagation velocity of $\sim$10^8 cm s^{-1}, Fig. 9(b). These measurements confirm
earlier estimates of this propagation velocity[52]. These were made
from interferograms of irradiated microballoons, where an ionisation front was observed to propagate across the microballoon and
along its thin glass support with a velocity of $\sim$10^8 cm s^{-1}. Thus
the formation of plasma on both sides of the surrogate target is
most probably due to the progressive expansion of the superthermal
plasma, with a velocity $\sim$10^{8}cm s^{-1}, enveloping the second target
and permitting the effective transfer of energy between the two
targets by fast electrons.

Conclusions

These studies of microtargets irradiated by nanosecond 10 μm
radiation in the intensity range 10^{13}–10^{14} W cm^{-2} have isolated a
number of effects not clearly observed in previous studies. A
common denominator of all these effects is the behaviour and
characteristics of the hot electrons generated in the critical
region. Although a clearer view of the critical density region
and of the lateral transport of energy is evident from these
studies, they pose as many questions as they answer. In particular the form and development of the potential sheath about the
target and its concurrent effects on the fast ion distribution
require much further study. The partition of energy into the
fast ions, an estimate of which is made in this study, is still
not firmly established. Although the form of the density profile
at and above n_c has been clearly determined, a totally consistent
interpretation of its two-dimensional character is still lacking.

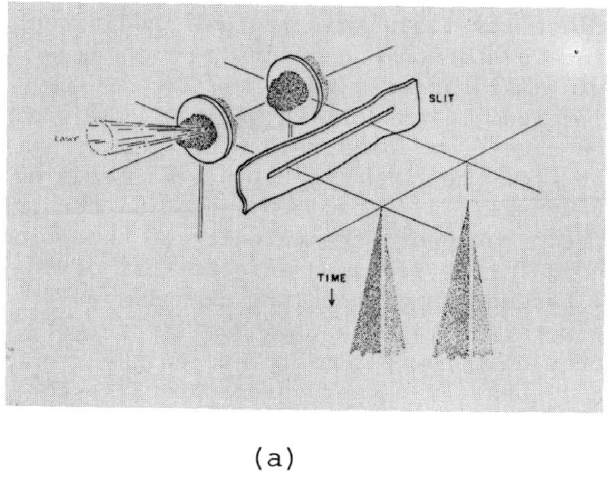

(a)

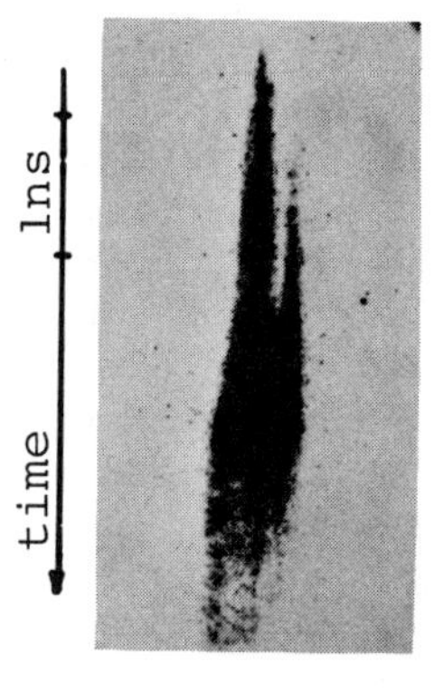

(b)

(c)

Fig. 13 Irradiation of twin disc targets (a) schematic arrangement
for streak photography of double disc target, each 500 μm
diameter, 150 μm thick, supported on 2 mm long glass
stalks, separated by ∿300 μm. (b) typical infrared streak
photograph of plasma expansion from front and rear surfaces of both disc targets. (c) x-ray pinhole photograph
of similar target showing plasma produced off surrogate
target.

The role of mechanisms potentially giving rise to thermal flux inhibition is not clearly understood, and the existence of magnetic fields in the interaction region remains to be investigated. We hope to be able to address some of these problems in the near future.

Acknowledgements

The authors wish to acknowledge with gratitude the help and advice of many colleagues particularly U. Feldman and N.R. Isenor, and the continuing technical support of P. Burtyn, G. Berry, Y. Lupien, K. McKee, W.J. Orr, and R. Sancton throughout these investigations.

References

1. M.C. Richardson, H.H. Burnett, H.A. Baldis, G.D. Enright, R. Fedosejevs, N.R. Isenor, and I.V. Tomov, Laser Interaction and Related Plasma Phenomena, Vol. 4A, p. 161, Plenum Press (1977).

2. P.A. Jaanimagi, M.C. Richardson, and N.R. Isenor, Opt. Lett. 4, 45 (1979).

3. Molectron, P-500 pyroelectric detector in combination with Tektronix 7104 Oscilloscope.

4. R. Fedosejevs, I.V. Tomov, N.H. Burnett, G.D. Enright, and M.C. Richardson, Phys. Rev. Lett. 39, 932 (1977).

5. D.M. Villeneuve, G.D. Enright, M.C. Richardson, and N.R. Isenor, J. Appl. Phys. 50, 3921 (1979).

6. G.D. Enright, N.H. Burnett, and M.C. Richardson, Appl. Phys. Lett. 31, 494 (1977).

7. K.R. Manes, V.C. Rupert, J.M. Auerbach, P. Lee, and J.E. Swain, Phys. Rev. Lett. 39, 281 (1977).

8. R.P. Godwin, C.G.M. van Kessel, J.N. Olsen, P. Sachsenaier, R. Sigel, and K. Eidmann, Z. Naturforsch. A32, 1100 (1977).

9. A.S. Maaswinkel, K. Eidmann, and R. Sigel, Phys. Rev. Lett. 42, 1625 (1979).

10. C. Joshi, G.D. Enright, and M.C. Richardson, Appl. Phys. Lett. 34, 625 (1979).

11. V.L. Ginzburg, "The Propagation of Electromagnetic Waves in Plasmas", (Pergamon, New York, 1964) Chap. 4.

12. N.H. Burnett, H.A. Baldis, G.D. Enright, and M.C. Richardson, Appl. Phys. Lett. 31, 173 (1977).

13. E.A. McClean, J.A. Stamper, B.H. Ripin, H.R. Greim, J.M. McMahon, and S.E. Bodner, Appl. Phys. Lett. 31, 825 (1977).

14. R.M. Marjoribanks, P.A. Jaanimagi, G.D. Enright, and M.C. Richardson (unpublished).

15. H.A. Baldis, N.H. Burnett, G.D. Enright, and M.C. Richardson, Appl. Phys. Lett. 34 (1979).

16. P.A. Jaanimagi, G.D. Enright, and M.C. Richardson, IEEE Trans. Plas. Sci. PS-7, 166 (1979).

17. P.A. Jaanimagi, R.S. Marjoribanks, R.W. Sancton, G.D. Enright and M.C. Richardson Proc. 13th Int. Cong. on High Speed Photography, Tokyo. S.P.I.E. (1978) p. 546.
18. G.D. Enright, M.C. Richardson, and N.H. Burnett, J. Appl. Phys. 50, 3909 (1979).
19. K. Estabrook and W.L. Kruer, Phys. Rev. Lett. 40, 42 (1978).
20. D.W. Forslund, J.M. Kindel, K. Lee, E.L. Lindman, and R.L. Morse, Phys. Rev. A11, 679 (1975).
21. K. Estabrook, E.J. Valeo, and W.L. Kruer, Phys. Fluids 18, 1151 (1975).
22. D.W. Forslund, J.M. Kindel, and K. Lee, Phys. Rev. Lett. 39, 284 (1977).
23. R. Fedosejevs, I.V. Tomov, N.H. Burnett, and M.C. Richardson, Proc. 12th Int. Cong. High Speed Photography, 1976, p. 401, pub. S.P.I.E. (1977).
24. R. Fedosejevs and M.C. Richardson (to be published).
25. K. Bockasten, J. Opt. Soc. Am. 51, 9, 943 (1961).
26. D.T. Attwood, D.W. Sweeney, J.M. Auerbach and P.H.Y. Lee, Phys. Rev. Lett. 40, 184 (1978).
27. R. Fedosejevs, M.D.J. Burgess, G.D. Enright, and M.C. Richardson, Bull. Am. Phys. Soc. 23, 7, 768 (1978).
28. R. Fedosejevs, M.D.J. Burgess, G.D. Enright, and M.C. Richardson, Phys. Rev. Lett. 43, 1664, (1979).
29. C. Joshi, N.A. Ebrahim, and M.C. Richardson (to be published).
30. H. Pepin, B. Grek, F. Rheault, and D.J. Nagel, J. Appl. Phys. 48, 312 (1977).
31. C.E. Max and C.F. McKee, Phys. Rev. Lett. 39, 1336 (1977).
32. J. Virmont, R. Pellat, and A. Mora, Phys. Fluids 21, 567 (1978).
33. K. Lee, D.W. Forslund, J.M. Kindel, and E.L. Lindman, Phys. Fluids 20, 51 (1977).
34. A. Raven and O. Willi, Phys. Rev. Lett. 43, 278 (1979).
35. B. Bezzerides, D.W. Forslund, and E.L. Lindman, Phys. Fluids 21, 2179 (1978).
36. F.J. Mayer, R.L. Berger, and C.E. Max, KMSF Internal Report, KMSF-U878 (1979).
37. L.M. Wickens, J.E. Allen, and P.T. Rumsby, Phys. Rev. Lett. 41, 554 (1978), L.M. Wickens and J.E. Allen, J. Plas. Phys. 27, 162 (1979).
38. J.A. Stamper, E.A. McClean, and B.H. Ripin, Phys. Rev. Lett. 40, 1177 (1978).
39. A. Raven, O. Willi, and P.T. Rumsby, Phys. Rev. Lett. 41, 554 (1978).
40. R. Serov and M.C. Richardson, Appl. Phys. Lett. 28, 115 (1975).
41. N.A. Ebrahim, M.C. Richardson, R. Fedosejevs, and U. Feldman, Appl. Phys. Lett. 35, 106 (1979).
42. U. Feldman, G.A. Doschek, D.K. Prinz, and D.J. Nagel, J. Appl. Phys. 47, 1341 (1976). N.A. Ebrahim, M.C. Richardson, U. Feldman, and G.A. Doscheck, J. Appl. Phys. (to be published).

43. W.L. Kruer, UCRL Report 82585 Lawrence Livermore Laboratory (1979).
44. S.J. Gitomer and D.B. Henderson, Phys. Fluids 22, 365 (1979).
45. J.R. Albritton, I.B. Bernstein, E.J. Valeo and E.A. Williams, Phys. Rev. Lett. 39, 1536 (1977).
46. K.A. Brueckner, Nucl. Fusion 17, 1257 (1977) and K.A. Brueckner and Y.T. Lee, Nucl. Fusion (to be published).
47. N.A. Ebrahim, C. Joshi, D.M. Villeneuve, N.H. Burnett, and M.C. Richardson, Phys. Rev. Lett. (to be published).
48. D.M. Villeneuve and M.C. Richardson, Rev. Sci. Instr. (to be published).
49. C. Joshi and N.A. Ebrahim, Bull. A.P.S. 24 990 (1979).
50. R.S. Marjoribanks, G.D. Enright, and M.C. Richardson (unpublished).
51. R.F. Benjamin, G.H. McCall, and A.W. Ehler, Phys. Rev. Lett. 42, 890 (1979).
52. R. Fedosejevs, M.D.J. Burgess, G.D. Enright, and M.C. Richardson (unpublished).

POLARIZATION AND ANGLE-OF-INCIDENCE DEPENDANCE OF SCATTERING AND

ABSORPTION IN SHORT LASER PULSE INTERACTION EXPERIMENTS

C. Gouedard, A. Saleres, M. Decroisette

C.E.A. - Centre d'Etudes de Limeil

B.P. n° . 27, 94190 Villeneuve-Saint-Georges, FRANCE

I. INTRODUCTION

Photon absorption in short pulse laser-matter experiments
mainly results from collisionnal and resonant absorption. P polarization increases absorption, enhancing the cold electron temperature through thermalization of an important fraction of resonantly accelerated electrons; meanwhile, a few ions are accelerated
by non-thermal electrons seeing a nearly constant intense localized
electrostatic field driven by the incident electromagnetic field
near the critical density n_c. Resonant absorption was experimentally studied through X-ray emission[1], ion distribution function[2]
or more directly by photon calorimetry[3]. Other processes such as
parametric instabilities[4], scattering processes, steepening and rippling of the critical surface[5] appeared to be invoked to account for
experimental absorption measurements.

Extensive theoretical and experimental investigations have
dealt with scattering processes, which appear to be strongly dependant on target (atomic number, surface smoothness, geometry)
and irradiation (intensity, pulse duration, contrast ratio). In
ultra short low Z experiments, important diffuse scattering was
observed[3] ; backscattering into the incident light cone, attributed
to stimulated Brillouin scattering, has been studied versus intensity and pulse duration[6,7,8,9].

In this paper, we describe experimental results obtained with
a multichannel calorimeter allowing to perform both precise energy
balance measurements, and angular distribution studies of the

scattered light in and out the incident light cone. Scattering
anisotropy and resonant absorption has been measured with planar
D_2 target irradiated with 80 ps 1,06 µm pulses in S and P polarizations, versus the angle of incidence α. Angular separation of
specular reflection and backscattering, which occurs for $\alpha > 15°$
allowed spectral observations on these two components.

II. SCATTERING AND ABSORPTION MEASUREMENTS

The photon energy balance in laser produced plasma experiment
can be performed by two means :

First, the most precise one, is to measure the whole amount of
non-absorbed energy by means of an integral calorimeter. This technic
provides a high accuracy by using a single calorimeter, spatial
integration of lobes being readily obtained.

However it doesn't give any information on mechanisms which can
only be inferred from a close examination of spatial distribution
of non-absorbed light. In fact, it appears that scattering centered
on laser axis contains a large amount of scattered light (up to 60 %)
and seems to be a different mechanism from pure Brillouin backscattering. This has been observed in réf. 3 even with 30 ps pulses and
0.53 µm irradiation.

This scattering is greatly reduced when rotating the target,
and it must be separated out to study specific effects of resonant
absorption.

A second fruitful way of investigation is therefore to study the
angular distribution of scattered light. This technic requires many
calorimeters distributed in the half space in front of the plane
target surface, as in the experiment we present.

This method has been used by many authors with a limited number
of detectors placed in most of cases in the plane perpendicular to
the incident polarization vector. Doing that keeps from recording
information on azimuthal anisotropy.

As an example, experiments on microdots with large aperture
lens showed a maximum of scattering in the plane perpendicular to
the polarization. This result has been interpreted as an effect
of resonant absorption but also of Brillouin side scattering[10,11]

The experimental arrangement consists in the focalization of
a Nd-glass laser beam,80 ps duration, up to 10 Joule, by an f/9
lens onto the surface of a planar solid D_2 target.

Use of solid D_2 was justified by simultaneous study of plasma expansion with a well defined Z number.

Many indications developed in the following allow us to estimate that a good planarity of the target is obtained. A polarizor was placed in front of the chamber in order to achieve a well defined polarization on the target. A quartz plate allowed a $90°$ rotation of this polarization. All reported shots were chosen to lay in the energy range 3 - 5 J on target, with a mean focal spot size of about 80 µm measured by $2\omega_0$ imaging of the plasma. This focal spot size was not very sensitive to focusing conditions due to the small aperture f/9 of the focusing lens.

Thus, flux on the target was in all shots of about 10^{15} W cm^2. The power contrast ratio was measured in a range between 10^4-10^6. Noise consisted in prepulses located 500 ps to 2 ns before the main pulse. It was found that prepulses, at levels lower than 10^{-4}, did not play an important role on the overall energy balance.

The scattered light detection device in interaction chamber consisted in an array of twenty volume absorbing glass calorimeters; these calorimeters were calibrated relatively to that devoted to incident energy measurement ; hence, a relative precision on sensivity of 2 % was achieved.

The integrated energy was obtained by analytical fitting of experimental data and an accuracy of 15 % was estimated on lobes integration. The calorimeters were situated on an half sphere in the interaction chamber, each of them viewing the plasma under a solid angle of 6 10^{-3} rd.

Their position in the space in front of the target was defined by θ and φ as shown in Fig. 1.

Typical   results of angular scattering with $22°$ rotated target are shown on Fig. 2 in the horizontal plane.

This angular scattering exhibits several typical features :

It appears that scattering is contained mainly in two lobes. First of all, a specular lobe is observed in the direction 2α from the laser axis where α is the inclinaison of the target. This feature is an "a posteriori" proof of the planarity of the plasma. This lobe has a mean half width at half maximum of $15°$. This is rather large in comparison with the incident beam, due to diffuse reflection and/or refraction on the plasma.

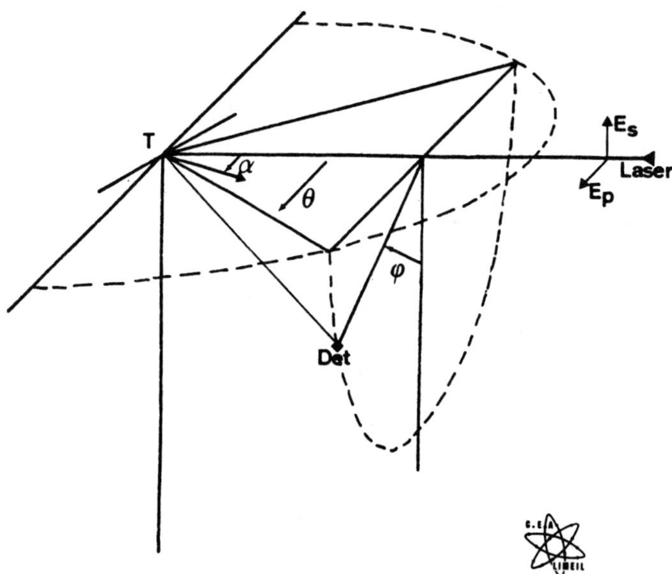

Fig. 1 Orientation of detectors in absolute coordinates. θ is the angle between the line-of-sight of the detector and the laser axis. φ is the azimuthal angle in the vertical plane containing it. α is the angle of incidence on the plane target.

However, this specular lobe is much less broad than the other, scattering lobe, centered on the laser axis.

This scattering lobe exhibits a mean half width at half maximum of 30°, much more than for specular lobe.

Another feature concerning the scattering lobe is its azimuthal anisotropy in respect to φ defined by fig. 1.

Fig. 3 shows this anisotropy for S and P polarization. In both cases, the diffusion is maximum in the plane perpendicular to the polarization vector.

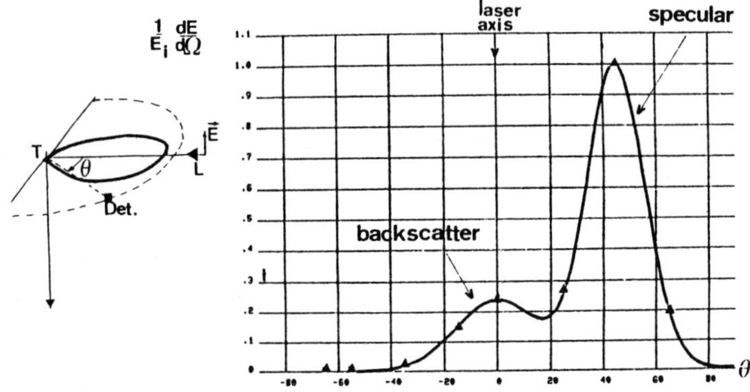

Fig. 2 : Example of lobes measured in the horizontal plane. They are well discriminated here for $\alpha = 22°.5$

The asymmetry is increasing with increasing θ and is well reproduced by an anisotropy factor $f_s = 1 - \sin^2\theta\cos^2\varphi$

and $f_p = 1 + tg^2\theta\cos^2\varphi$ for both polarization.

Analytical fitting of experimental data is shown in Fig. 4 in P polarization for nearly normal incidence.

Brillouin side scattering mechanism can explain such an anisotropy due to an overall polarization factor $|\vec{e}_o \cdot \vec{e}_d|^2$ present in the expression of growing rates of instable waves, where $\vec{e}_o$ ($\vec{e}_d$) are polarizations of incident (scattered) waves.[12,13]. This factor, averaged on scattered wave polarization reproduces exactly the anisotropy factor defined above.

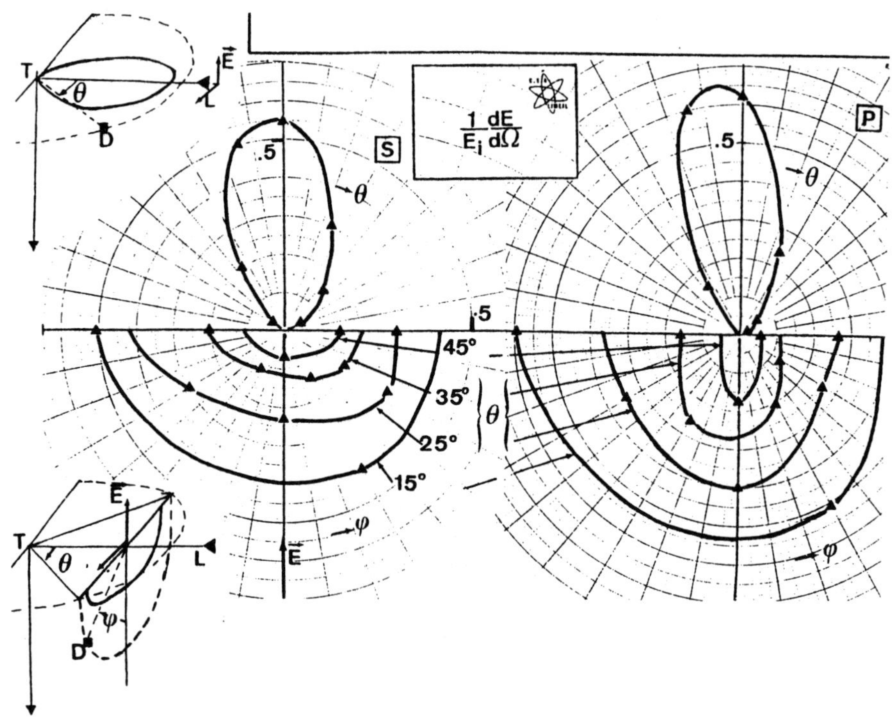

Fig. 3 : Experimental polar data for scattering in both S (left) and P (right) polarization. The scattering is maximum in the plane perpendicular to the polarization vector. This asymmetry increases for increasing θ.

As a remark, it has to be noted that such an anisotropy is characteristic of all scattering processes, stimulated or not [11].

On an another hand, small aperture of the focused beam prevented from a possible anisotropy due to resonantly absorbed rays (incidence is always defined at $\pm$ 3 %).

Not taking in account this asymetry in scattering lobes leads to a mean error (depending on angular spreading) of about 10 % on the integrated energy.

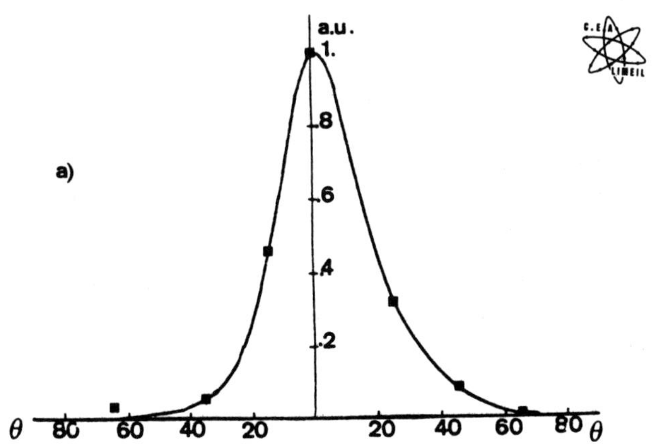

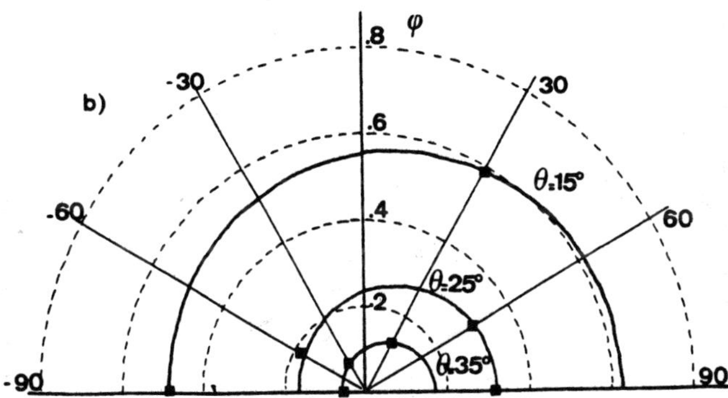

Fig. 4 : Analytical fit of experimental data (squares) in horizontal plane (a) and out of it (b). Both lobes are fitted by a function $e^{-k\theta^2}$ and the backscattering anisotropy is fitted by a factor $1 + tg^2\theta \cos^2\varphi$ for "P" polarization.

Another important cause of error is the broad angular spreading of diffusion lobe. In fact, the backscattering spreads far outside the solid angle of the focusing lens. The fraction of scattered energy collected in this solid angle is about 30 % of the whole lobe, for $f/2$ aperture lens, coherent with Ripin's result[6]. Accordingly it

appears that pure Brillouin backscattering is smeared out in a
great amount of scattered light. So, prepulse effect on scattered
light proportion is very poor in our experiment provided that this
prepulse didn't exceed 10^{-4}.

As a consequence such a prepulse effect on absorption is very
weak, at least for planar target experiments.

In Fig. 5 is plotted the fraction of incident energy backscattered
towards the laser (Not only limited at the focusing lens solid angle)
versus angle of incidence α.

It can be seen that scattered light represents up to 60 % of
incident energy for normal incidence, and is the decisive parameter
for non-absorption.

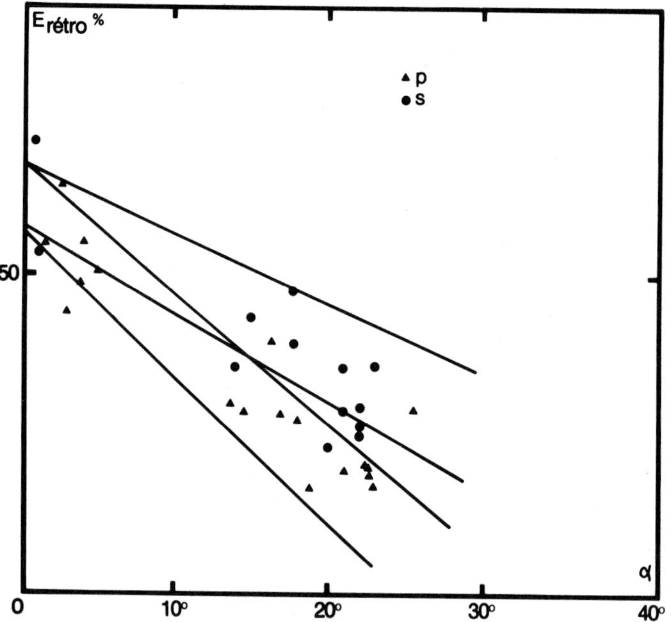

Fig. 5 : Plot of backscattered energy versus angle of incidence α
for S or P polarization. Dramatic effect at normal incidence (60 %)
leads to very poor absorption(see Fig. 7).

Such a dramatic scattering may be partly due to lightness of
D_2 ions, the gradient scale length being proportional to $(m_i)^{-1/2}$.

It can be seen also that scattering is slightly decreased for P polarization at angles where resonant absorption takes place.

As a proof of resonant absorption mechanism, Fig. 6 represents
the fraction of energy specularly reflected from the target.
Reflection is strongly reduced in P polarization at angle of about
15° although it is monotonically increasing with α in S polarization.

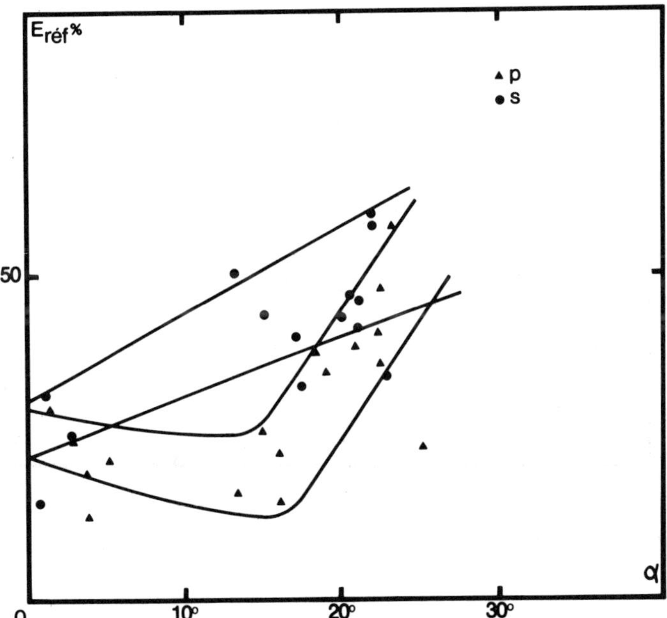

Fig. 6 : Specularly reflected light versus angle of incidence.
Sensitive decreasing for P polarization at an angle $\alpha \simeq 15°$ gives
evidence of resonant absorption, leading to a gradient scale of
about 3 μm.

Lastly in Fig. 7 is plotted the absorption in the case of S or P polarizations.

Effect of polarization is evident, due to both effects of resonant absorption and slight decrease of scattering in P polarization.

It has to be noted the rather great dispersion of data, due to strong interaction induced rippling of the plasma surface leading to a mixing of polarization effects.

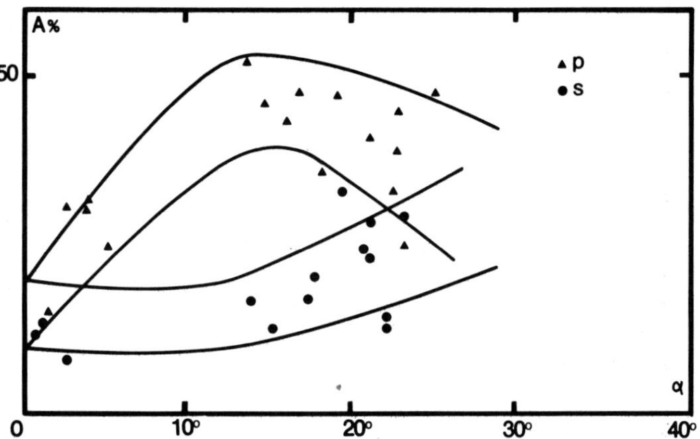

Fig . 7 : Absorption versus angle of incidence.

A 10 - 20 % absorption at normal incidence is rather small, corresponding to the very large amount of scattering in these conditions.

III. SPECTRAL OBSERVATIONS

Results described above show that specular reflected light could be separated from scattered one, provided the angle of incidence on target was greater than 15°. Thus, it was possible to perform spectral analysis of both components. In normal incidence, scattered light was discriminated from reflected one owing their different angular speading. Here, we shall see that distinction appears also on spectral energy distribution.

Spectral recordings were performed with two high resolution grating spectrographs : the first was devoted to backwards observations, with a dispersion of 2 Å/ mm with a practical resolution of 0.3 Å ; the second one presented a same order dispersion, and a resolution of 1 Å, and was used for specular analysis.

Spectral observations were performed by means of two Reticon photodiode multichannel systems allowing direct scope recordings. Elementary diode size and spacing ($\simeq$ 20 μm) preserved the initial resolution.

Experiments were performed always with the same power contrast ratio (> 10^{+5}) in S and P polarization for three angles of incidence : 0°, 15°, 25°.
Typical spectra recordings are presented on Fig. 8. First, no significative differences appeared between S and P cases, except slightly more perturbed spectral profiles in P polarization.

At normal incidence (a) two groups of lines are observed, respectively at 0 Å and $\left[\,-10 \text{ to } - 20\,\right]$Å, the minus sign standing for a blue shift. For increasing angles on incidence α, the blue shifted lines disappear in backwards observations (b,c) ; meanwhile, in specular observation (c) only a blue shifted component is present. Thus in normal incidence the blue shifted lines can be coherently attributed to the scattered light component.

The following qualitative interpretation for normal incidence can be presented : the blue shift of the specular light is representative of the turning point region velocity (Vc) and wave number (k_c) according to $\dfrac{\Delta\lambda}{\lambda_0} = - 2 \dfrac{Vc}{G} \dfrac{k_c}{k_0}$. Concerning the scattering component, the spectral shift results from competition between Doppler effect and acoustic shifts , according to $\dfrac{\Delta\lambda}{\lambda_0} = - 2 \dfrac{k_e}{k_0} \dfrac{Ve - Cs}{C}$ where

k_e and V_e are wave number and velocity of the scattering zone.

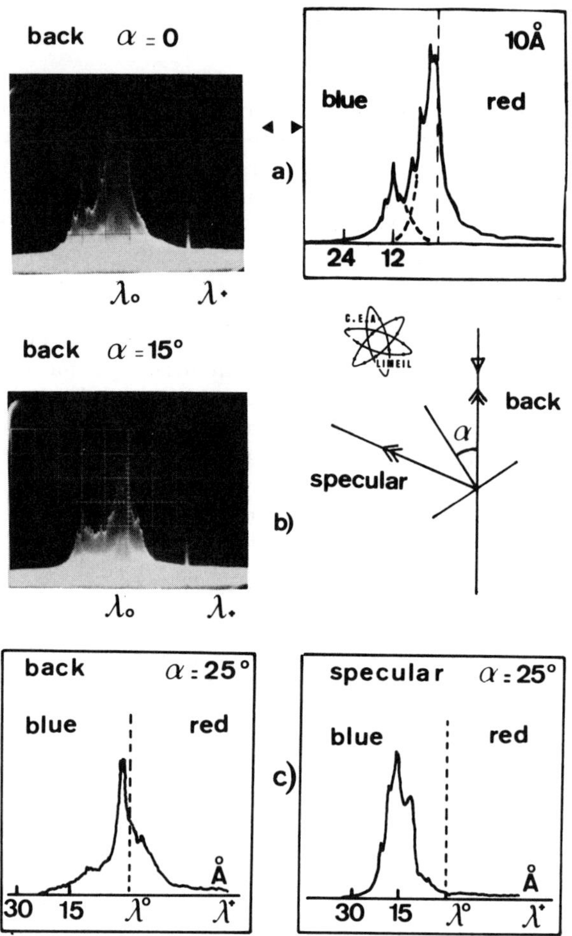

Fig. 8 : Spectral separation of the two components of reflected light
at λ_0.
At normal incidence - a) backscattered and specular reflected light
are both present in backwards observation ; b) is an intermediate
regime for $\alpha = 15°$
At 25° incidence -c) blue shifted specular component is absent on
backwards observation but appears alone in specular observation ;
contrast ratio is always less than 10^{-5}.

In the particular cases we have studied $\Delta\lambda = 0$ means the scattering zone is slightly supersonic.

As a remark, it has to be noted that respective spectral component intensities of two groups of lines are not directly correlated to calorimetric measurements, as aperture of backreflection calorimeter and spectrograph are different (respectively $f/3$ and $f/12$; in the last case indeed, pure Brillouin scattering may be more important.

IV- CONCLUSION

Interaction experiments have been performed with short 1,06 µm laser pulses and planar D_2 target, in P or S polarization.

Both calorimetric and spectroscopic measurements have allowed us to discriminate stimulated backscattering and specular reflection. It appears that limitation in absorption results mainly from an important stimulated scattering, characterized by a very large angular distribution maximum in a plane perpendicular to the incident electric field vector; in normal incidence its lobes contain up to 60% of the incoming energy.

Moreover, resonant absorption was observed in P polarization, at an optimum angle of incidence of 15°, corresponding to a mean gradient scale length of 3 µm.

At last the backscattered light presents a nearly unshifted spectrum, signifying the scattering zone is slightly supersonic.

Authors wish to thank the laser team for P 102 management, and B. Aveneau, G. Faucheux, J.L. Larcade, and D. Schiess for technical assistance.

REFERENCES

1. J.E.Balmer, J.P. Donaldson, Phys. Rev. Lett. 39 1084 (1977).
2. J.S. Pearlmann and M.K. Matzen, Phys. Rev. Lett., 39 140 (1977).
3. A.G.M. Maaswinkel, K. Eidmann, R. Sigel, Phys. Rev. Lett., 42, 1625 (1979).
4. K.R. Manes, V.C. Rupert, J.M. Auerback,P. Lee, J.E. Swain, Phys. Rev. Lett., 39 281 (1977).
5. D.W. Phillion, R.A. Lerche, V.C. Rupert, R.A. Haas, M.S. Boyle, Phys. Fluids, 20 1892 (1977).
6. B.H. Ripin, F.C. Young, J.A. Stamper, C.M. Armstrong, M. Decoste E.A.Mc Lean, S.E. Bodner, Phys. Rev. Lett., 39 611 (1977).
7. B.H. Ripin, A. Phys. Rev. Lett., 30 134 (1977).
8. D.W. Phillion, W.L. Kruer, V.C. Rupert, Phys. Rev. Lett., 39 1529 (1977).

9.   A.A. Offenberger, M.R. Cervenan, A.M.Yam, A.W. Pasternak,
       Appl. Phys. $\underline{47}$ 1451 (1976).
10. J.J. Thomson, C.E.Max, J. Erkkila, J.E. Tull, Phys. Rev. Lett.,
       $\underline{37}$ 1052 (1976).
11. R.A. Haas, W.C. Mead, W.L. Kruer, D.W. Phillion, H.N. Kornblum,
       J.D.Lindl, D. Mc Quigg, V.C. Rupert, K.G. Tirsell,
       Phys. Fluids, $\underline{20}$ 322 (1977).
12. C.S. Liu, M.N. Rosenbluth, R.B. White, Phys. Fluids, $\underline{17}$ 1211
       (1974).
13. I.M. Gorbunov, A.N. Polyanichev, Sov. Phys., JETP, $\underline{47}$ 290
       (1978).

STIMULATED BRILLOUIN SCATTERING EXPERIMENTS[*]

D.C. Slater, R.L. Berger, Geo. Busch, C.M. Kinzer,
F.J. Mayer, L.V. Powers and D.J. Tanner

KMS Fusion, Inc.
Ann Arbor, Michigan 48106

INTRODUCTION

High intensity laser light incident on a plasma can excite
collective modes which may enhance either absorption or reflection
of the laser energy. The stimulated Brillouin scattering process
(SBS), in which the incident light wave decays into a backward-
traveling wave and an ion wave, has long been a subject of concern
for laser fusion because the stimulated reflection can prevent
light from reaching the critical density surface where most absorption processes are strongest. In recent years, experiments have
confirmed that SBS is present for plasma and laser parameters of
interest to laser fusion[1,2].

This report describes two experiments in which SBS would be
expected to play an important role. In the first experiment, we
find a clear signature of the Brillouin backscatter of a short
(100 psec) pulse from a long ($\sim$50 μm) gradient length gas target
plasma. The second experiment used much longer ($\sim$1 nsec) pulses
on spherical glass shell targets. These experiments were done with
both narrow (<5Å) and broad (>30Å) bandwidth laser light. Using
one-dimensional, spherically symmetric fluid simulations, we have
attempted to model many of the laser-plasma interaction processes
which combine to determine the amount of absorbed energy in the
long-pulse experiments. These simulations indicate that modest
laser bandwidths are successful in reducing the level of SBS at
the irradiances (<10^{15} W/cm^2) used in these experiments.

*This research is supported by the United States Department of
Energy under Contract number DE-AC08-780P40030.

GAS TARGET EXPERIMENTS

Laser plasma interaction experiments using CO_2 lasers and gas targets[3,4] have produced clear evidence of Brillouin backscatter since a direct measure of the ion acoustic frequency shift was obtained. We were motivated by these experiments to examine the backscatter from a gas target using the Nd:glass laser which, operating with a short pulse, could perhaps minimize plasma motion effects e.g., Doppler shift in short pulse solid target experiments, and ionization wave effects in long pulse CO_2 laser gas target experiments. We were particularly interested in 1) obtaining a clear signature of Brillouin backscatter in spectral measurements, 2) observing saturation of the backscattered energy with laser power, and 3) observing any systematic spectral changes with increasing laser power.

Our experimental set-up, shown in Figure 1, is quite simple. We have made use of a puffed nitrogen gas jet as the laser target. High pressure nitrogen (about 10 atmospheres) is allowed to leak through a 100 μm diameter pinhole producing supersonic nozzle flow into a vacuum chamber. The gas density falls off approximately as $\rho \sim \rho_0 (r_a/r)^2$ in this free expansion[6]. Here r_a is the pinhole radius; ρ_0 was high enough to produce a critical density surface if the

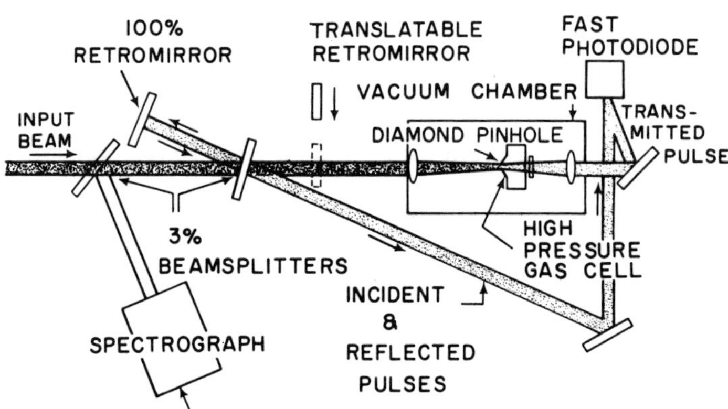

Fig. 1. Experimental configuration for the gas target reflection and spectral measurements.

nitrogen was partially ionized. The laser beam was directed on axis, up the density gradient, and the focal position could be shifted axially. The electron density at the lens focus was not measured, and is only crudely estimated at about $10^{21} cm^{-3}$. The experiments were performed at a focal position which maximized backscatter.

The Nd:glass laser had an approximately square pulse shape of 100 picosecond duration, maximum energy of 700 millijoules and a beam diameter of 40 millimeters. The laser energy was focussed onto the gas jet with an f/3.5 lens. We made measurements of pre-pulses (less than 10^{-6} of the main pulse) and measured both the incident and reflected beam energy with the same fast calibrated photodiode. A retro mirror provided reflection calibration of the photodiode. Spectral measurements of the backscattered light were made with a Czerny-Turner grating spectrograph having 1.2 Å resolution. Far-field focal spot measurements showed that 90% of the energy was contained within a 20 μm diameter which gave a peak focused intensity of 2×10^{15} W/cm^2.

Figures 2a-2d show the backscattered energy spectrum at increasing incident powers (I) from $I = 2 \times 10^{14}$ W/cm^2 to $I = 2 \times 10^{15}$ W/cm^2. The backscattered energy in Figure 2a is seen to be red shifted by about 4.5Å with no energy at the incident frequency ω_0. This shift is typical of all the backscattered spectra even at higher incident powers. The red shift clearly indicates the Brillouin backscatter mechanism, and is in qualitative agreement with the measurements of Offenberger[3,4].

The measured linewidths and frequency shifts are plotted in Figure 3 as a function of laser power. At the lower laser powers, the ratio of width-to-shift is approximately one-third, but increases to unity at higher power. The spectra in Figure 2 are typical of the data included in Figure 3. Some energy appears at ω_0, presumably reflected from a nearly stationary critical density surface. As the laser power increases, the linewidth is increased but the shift remains constant. The backscattered energy fraction increases as the incident power increases up to 2×10^{15} W/cm^2, where it saturates at ~40%.

It is tempting to relate the linewidth to $\overline{Z}T_e/T_i$ by attributing the width to ion Landau damping. (Here $\overline{Z}$ is the mean ionization state; T_e and T_i are the electron and ion temperatures.) Using an analysis similar to that of Offenberger, the width-to-shift ration of one-third implies $\overline{Z}T_e/T_i = 1.3$. For this very small ratio, stimulated scattering from ions, rather than ion-acoustic

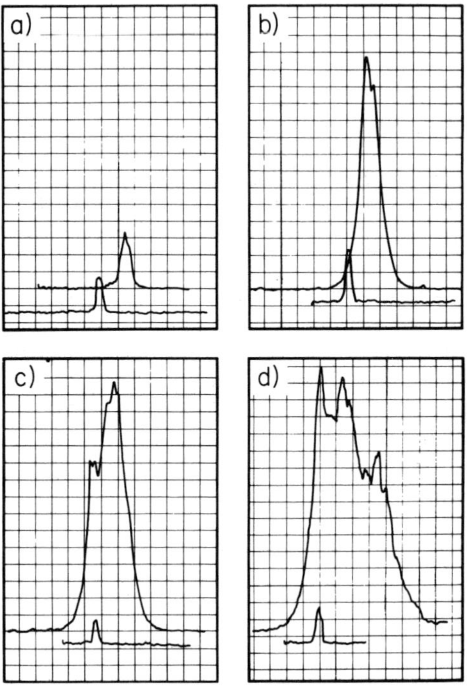

Fig. 2. Microdensitometer traces of backreflected spectra. Wavelength increases to the right. Lower traces indicate the incident 1.06 μm laser energy. The incident power increases from 2×10^{14} W/cm^2 in a) to 2×10^{15} W/cm^2 in d).

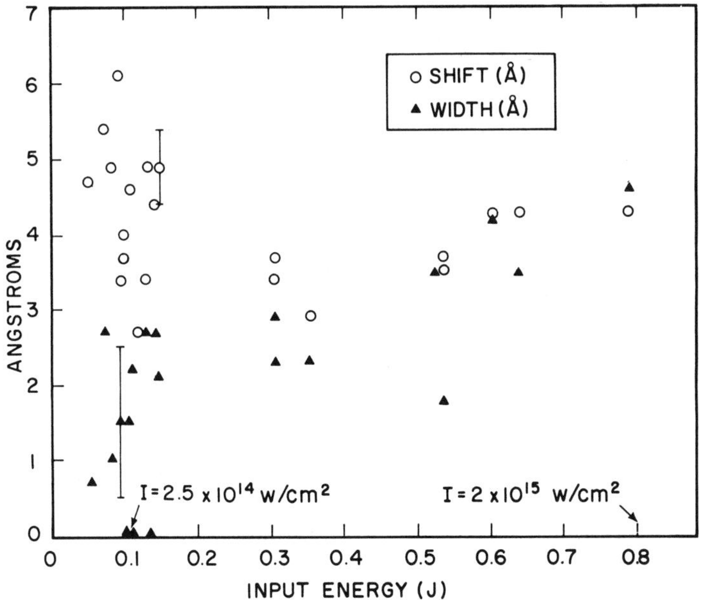

Fig. 3.   Measured Brillouin redshift and line width (FWHM) as a
function of incident power.

waves, dominates. In this circumstance the frequency shift is

$$\frac{\Delta\omega}{\omega_o} = 2\left[(1-n/n_c)T_i/m_i c^2\right]^{1/2} \tag{1}$$

where m_i is the ion mass and c is the velocity of light. Using
Eq. (1) and the measured shift, we obtain T_i = 800 eV for n/n_c =
0.2. This temperature is implausibly large. A more reasonable
explanation is that $\overline{Z}T_e/T_i \gg 1$. Then Brillouin scattering off
ion acoustic modes is expected and Eq. (1) is replaced by

$$\frac{\omega_s}{\omega_o} = \frac{2k_o C_s}{\omega_o} = 2\left[(1-n/n_c)(\overline{Z}T_e+3T_i)/m_i c^2\right]^{1/2} \tag{2}$$

The measured shift then determines $\overline{Z}T_e+3T_i \approx 800$ eV so that $T_e \approx$
100-200 eV. The linewidth in this case is not due to ion Landau

damping alone, but could result from a combination of ion damping, time dependence of the electron temperature and ionization state, and plasma inhomogeneity. The power dependence of all these quantities would lead one to expect linewidth changes as the power is increased.

It is quite interesting that energy is reflected from the critical surface, because inverse bremsstrahlung, in our low temperature plasma, ought to be nearly 100% efficient in absorbing the energy which is not Brillouin backscattered. We believe this reflected energy at ω_0 is not stray light because in some shots at similar laser power and much reduced gas pressure (where there was insufficient gas density to produce a critical surface) we found no energy at ω_0. Also, experiments with no puffed gas showed no reflection. The appearance at the higher powers of the reflected light at ω_0 may be due to high-field inverse bremsstrahlung[7] reducing the absorption at ω_0. From our measured backscatter shifts, we estimate $(V_{os}/V_{th})^2 \sim 2$ at the higher powers which is sufficient to produce reduced inverse bremsstrahlung absorption. Further experiments will be required to quantify this result.

In some shots, such as the one shown in Figure 2d, we observed a nearly periodic modulation structure on the long-wavelength side of the spectrum. A qualitatively similar result has been obtained in particle simulation studies of SBS[8], but has not been previously observed experimentally. In these simulations the cascade of lines on the red side of ω_0 and separated by the ion acoustic frequency is caused by multiple Brillouin reflections in the underdense plasma[9]. The significance of multiple scattering is: (1) for each scattering the light has another chance to be absorbed by inverse bremsstrahlung, (2) each scattering off the critical surface gives the light another chance to be absorbed there, and (3) each scattering heats the ions.

We believe these long gradient length, short pulse experiments are effective in isolating the physics of the stimulated Brillouin scattering from the hydrodynamic flow questions associated with long pulse experiments on solid targets. Similar gas target experiments might also be suitable for Raman scattering and filamentation studies.

SPHERICAL TARGET EXPERIMENTS

Spherical glass shell targets ranging in diameter from 109 to 290 μm and in wall thickness from 0.8 to 2.5 μm were irradiated with 1.06 μm laser light. The nominal laser irradiance (laser power divided by target surface area) was varied from 2×10^{13} to 5×10^{14} W/cm^2. The laser pulse was formed by stacking 22 individual

40 psec FWHM equal intensity pulses with 40 psec peak-to-peak spacing. The resulting pulse shape was essentially square in time, with intensity variations of about 2:1 during the 850 psec pulselength. A streak camera record of the temporal pulse shape was taken for almost every shot. The target illumination system consisting of two lens-ellipsoidal mirror pairs provides almost uniform illumination at near-normal incidence over the entire target surface.

The laser was operated in two different modes for these experiments. In our standard configuration, a plasma spatial filter[10] is included in the amplifier chain. This device acts to smooth the time-integrated laser intensity distribution both in the near-field and at the focal position of the target surface. It also serves to broaden the time-integrated laser bandwidth from a few angstroms to a few tens of angstroms. Measured bandwidths and spectral shapes varied somewhat for these shots, with 50Å FWHM typical, and 30Å FWHM as the smallest observed. Several shots were taken with the plasma spatial filter removed. In this case the laser bandwidths were measured to be approximately 3Å. The intensity modulation at a focal plane equivalent to the target surface was recorded on several shots. With the plasma filter in place, intensity variations of 2:1 peak-to-valley ratio were typical. No focal plane data is available without the plasma filter for these long pulses, but shots at ~300 psec showed variations of 8:1.

The energy absorbed by each target was measured with two thermo-pile differential calorimeters which sample the plasma blowoff energy. Two uncovered TLD chips at a third location provided an additional measurement which in a few cases indicated an asymmetric energy expansion pattern. The measured energy absorption fraction as a function of irradiance is plotted in Figure 4. Each point represents one target shot. The error bars show the standard deviation of a weighted average of the three measurements. The solid-circle data were taken with the plasma filter included in the laser system. The open-circle data, taken with the plasma filter removed, show lower absorption at all irradiances. We believe the improved absorption with the plasma filter is due to the increase in laser bandwidth, which reduces the amount of SBS. This hypothesis is discussed below. The alternative explanation is that the smoother intensity distribution when the PSF is used leads to better absorption. We discount this explanation on the basis of shorter (~200 psec) pulse experiments done at comparable irradiance. In those experiments, absorption averaged 21±4% with the plasma filter and 20±3% without it[11]. The differences in absorption with and without the PSF are much greater in the long-pulse data of Figure 4.

Wavelength spectra of the backreflected light were obtained for several of the long-pulse shots with narrowband incident light. The spectra show a two-component structure. One peak is slightly

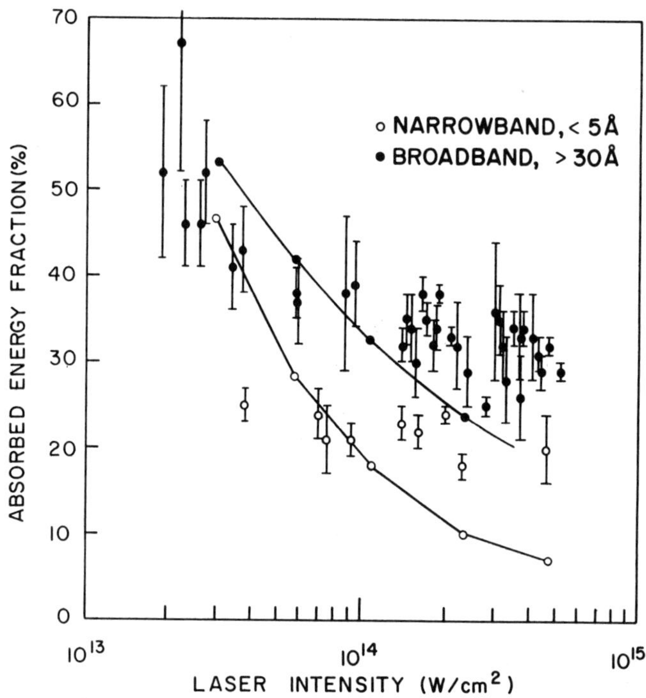

Fig. 4.   Absorbed energy fraction versus on-target irradiance for
          narrowband (PSF out) and broadband (PSF in) light. Solid
          lines are calculated values.

blue-shifted from the incident laser wavelength and is typical of
spectra taken on short pulse shots. The second component is red
shifted, and is not observed in short pulse experiments. Typical
spectra at two irradiances are shown in Figure 5. At the higher
irradiance, the peaks broaden and merge but the two-component
structure is still evident.

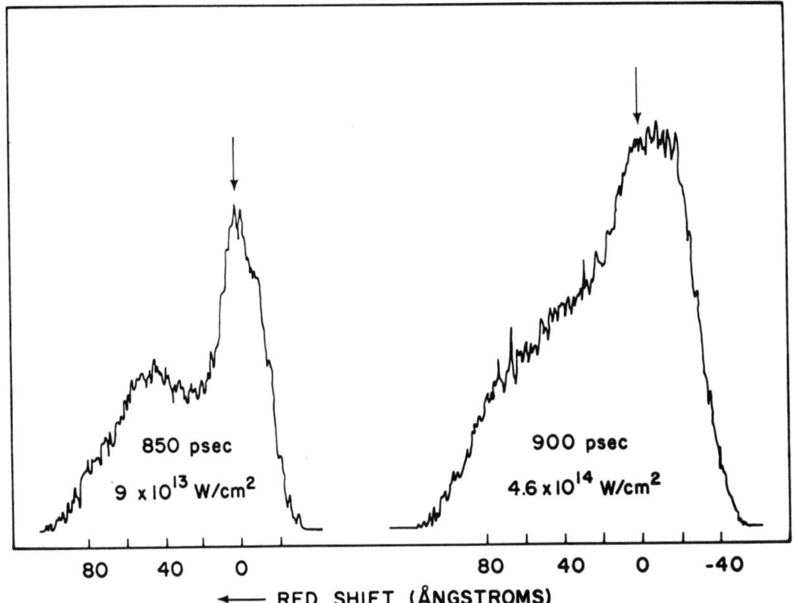

Fig. 5.   Backreflected light spectra. The arrows indicate the
position of the incident light.

SBS MODELING AND SIMULATION STUDIES

The absorption of high intensity laser light by a plasma is
determined by several processes, including but not limited to SBS,
inverse bremsstrahlung absorption and resonance absorption. The
detailed mix of mechanisms depends on plasma variables such as the
density profile and the electron and ion temperatures and also
depends on laser parameters such as wavelength, intensity, bandwidth,
and focusing conditions. Simulations of laser-target interaction
experiments with fluid hydrodynamics codes generally require considerable freedom to adjust the absorbed energy fraction to match
an experimentally determined value. In order to understand the
absorption data of Figure 4, we have attempted to model the relevant
physics with a minimum of adjustable parameters.

We begin with a one-dimensional, spherically symmetric fluid
hydrodynamics simulation code, with inverse bremsstrahlung absorption and flux-limited thermal transport characterized by the flux-
limit parameter f. To this we add a model of stimulated Brillouin

scattering which explicitly includes the laser bandwidth. A number of simplifying assumptions built into the model are discussed here. Since the growth time for the instability is short compared to hydrodynamic time scales, a steady-state model of the interaction is used. The model includes plasma inhomogeneity, laser bandwidth, pump depletion, and ion heating. The damping time of ion acoustic waves is assumed short compared to the acoustic wave transit time across the plasma scale length and to the growth time of the instability. The ion Landau damping is computed for a single-temperature, single-species Maxwellian distribution. The percentage of laser light reflected by SBS depends on the noise level from which the reflected wave grows and on the integral

$$z = \int_{r}^{r_o} dr' \frac{\gamma_0(r')}{v_g(r')}$$

where $\gamma_0(r')$ and $v_g(r')$ are the growth rate of the instability and the group velocity of the light waves respectively.[2] The percentage reflection as a function of the factor z for various noise levels is shown in Figure 6.

Both ion heating and laser bandwidth can reduce the amount of SBS. Due to our assumptions about ion damping rates, the ion wave energy is lost to the plasma immediately. Because the ion thermal conductivity is small, the ions can be heated to a level comparable to the electron temperature, which leads to a reduction in the maximum growth rate. When the laser bandwidth ν exceeds the ion-wave damping rate, the Brillouin growth rate γ_0 is reduced to the value. γ_0^2/ν, thus reducing the amount of scattered light[12]. Recent microwave experiments have measured a reduction in backscatter with increased bandwidth[13]. The SBS model in our simulation code includes the effects of ion heating, plasma inhomogeneity and laser bandwidth in the calculation of z.

The absorbed energies calculated with this model are shown as solid curves in Figure 4. Two features are evident in these results: 1) the differences between calculated broad- and narrowband absorption are similar to the experimental differences, and 2) the intensity dependence of the calculated absorption does not match the data.

Two improvements to the model described above are being investigated in some detail. At the lowest intensities in Figure 4, a difference exists in the absorption data of narrowband and broadband illumination, even though very little Brillouin scattering and little profile steepening is predicted. A possible explanation is offered in some recent work by Randall et. al.[14] where it is shown that doppler-shifted light reflected from the moving critical surface can act as the "noise" source for backscattering the incident light. For small values of $z(z<1)$ and weak absorption, the light transmitted to the critical surface is reduced by a factor $(z+1)^{-1}$. In one

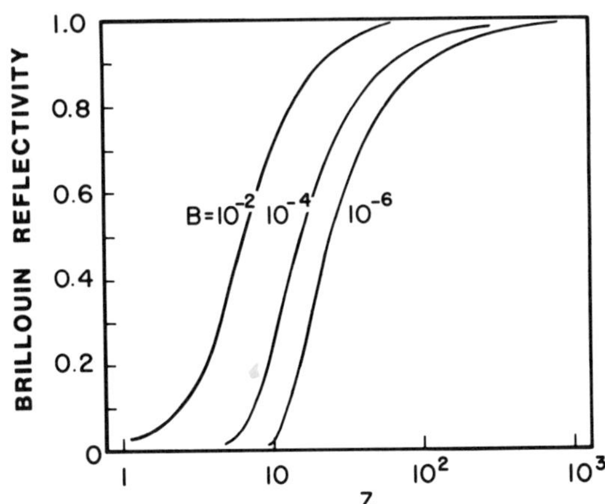

Fig. 6.   Calculated SBS reflection fraction as a function of z for
several background noise levels.

simulation, at 3×10^{13} W/cm^2 the Brillouin-scattered light fraction
increased to 78% in the narrowband case using the Randall model,
while the inverse bremsstrahlung fraction fell from 45% without this
effect to 14% with it. It is therefore possible that substantial
amounts of SBS can be present even at relatively low incident intensity.

The second improvement is the ability to resolve small scale
density features. It is well known that the density profile calculated from hydrodynamics and heat flow alone is modified around
the critical density surface by the laser-plasma interaction. This
modification - generally a steepening - affects both inverse bremsstrahlung and SBS. In most cases however, it is not possible to
resolve these changes in Lagrangian fluid codes because of zone
size limitations. To demonstrate the effect of profile steepening
we ran a test case simulation of a 169 joule, 0.9 nsec narrowband
laser pulse incident on a 114 μm diameter target (4.6×10^{14} W/cm^2).

When profile steepening was neglected, 90% of the energy was back-scattered and 8% absorbed by inverse bremsstrahlung. The same problem was run in a special code which automatically adds fine-scale zones around the critical density. In this case the Brillouin-scattered fraction fell to ~15% while the absorption grew to 20%, much closer to the experimental value.

The model calculates the growth of SBS light from noise at only one frequency - that with the highest growth rate - and therefore would not necessarily reproduce the spectrum of backscattered light. The fastest growth usually arises in a region moving at the ion sound speed. Since the doppler shift from this motion effectively cancels the SBS red shift, our model predicts only a slight blue shift in the time-integrated spectrum. This appears in contradiction to the experimentally observed red-shifts of Figure 5.

We can summarize the simulations in the following ways: 1) a standard, flux-limited hydro code simulation with the standard model of SBS and inverse bremsstrahlung absorption predicts too much absorption at low intensity, too little absorption at high intensity, but a laser bandwidth dependence similar to the data; 2) adding a mechanism in which light reflected from critical enhances SBS will reduce the absorption at low intensity; and 3) allowing for profile modification can substantially reduce SBS at moderate intensities.

The strong coupling between SBS, inverse bremsstrahlung and profile steepening emphasizes the subtleties involved in trying to account for overall absorption fractions.

REFERENCES

1. B.H. Ripin, F.C. Young, J.A. Stamper, C.M. Armstrong, R. Decoste, E.A. McLean, S.E. Bodner, Phys. Rev. Lett. $\underline{39}$, 611 (1977).
2. D.W. Phillion, W.L. Kruer, V.C. Rupert, Phys. Rev. Lett. $\underline{39}$, 1529 (1977).
3. A.A. Offenberger, et al., J. Appl. Phys. $\underline{47}$, 1451 (1976).
4. A. Ng, et al., Phys. Fluids $\underline{42}$. 307 (1979); N.H. Burnett, Anomalous Absorption Conf., Tucson, Ariz. 1978.
5. F.J. Mayer, G.E. Busch, C.M. Kinzer, K.G. Estabrook, KMSF Report U-904 (1979).
6. F.P. Boynton, AIAA Journal $\underline{5}$, 1703 (1967).
7. V.P. Silin, Zh. Eksp. Teor. Fiz. $\underline{47}$, 2254 (1964); [Sov. Phys. JETP $\underline{20}$, 1510 (1965)]; P.J. Catto and T. Speziale, Phys. Fluids, $\underline{20}$, 167 (1977).
8. K. Estabrook, Bull. Am. Phys. Soc. $\underline{21}$, 1067 (1976).
9. T. Speziale, J.F. McGrath, R.L. Berger, KMSF Report U-947 (1979).
10. N.K. Moncur, Appl. Opt. $\underline{16}$, 1449 (1977).

11. KMSF 1978 Annual Report on Laser Fusion Research, page 2-51.

12. J.J. Thomson, Nucl. Fusion $\underline{15}$, 237 (1975).

13. A. Mase, H. Huey, M. Rhodes, N.C. Luhmann, Bull. Am. Phys. Soc. $\underline{24}$, 980 (1979).

14. C.J. Randall, J.J. Thomson, K.G. Estabrook, Phys. Rev. Lett. $\underline{43}$, 924 (1978).

PULSE-LENGTH, POLARIZATION, AND Z DEPENDENT PROPERTIES OF LASER

PRODUCED PLASMAS AT HIGH IRRADIANCES

S. Jackel, B. Arad, S. Eliezer, Y. Gazit, A.D. Krumbein,
H.M. Loebenstein, Y. Paiss, I. Pelah, D. Salzmann,
N. Spector, A. Zigler, H. Zmora and S. Zweigenbaum

Soreq Nuclear Research Centre, Yavne, Israel

A series of investigations was conducted at high irradiances
with 1.06μm laser light pulses produced by either a mode-locked or
a Q-switched Nd:YAG oscillator driving the Soreq High Irradiance
Laser System. Both short (60 psec) and long (2.5 nsec) pulses were
used and the range of irradiances was from 10^{14} to 10^{16} W/cm^2. Low
and high atomic number targets were irradiated at normal and oblique
incidence using, in some cases π and σ polarized light.

Target momentum measurements were made with a torsion pendulum
for aluminum slab targets using both π and σ polarizations. Clear
evidence of resonant absorption was obtained for 60 psec pulses
while no light polarization or angle of incidence dependent resonant absorption effects were detected for long pulse irradiations.

Calorimetry and spectroscopy of ω and 2ω light reflected back
into the focusing lens was performed on a number of target materials
of varying atomic number with 2.5 nsec duration pulses. Results
show a dramatic reduction in the specularly reflected light from
high atomic number targets. Brillouin backscatter was observed
throughout the range $10^{14} - 10^{16}$ W/cm^2. Spectra of the 2ω light
indicated the presence of an electron-ion decay instability with a
threshold at 2×10^{15} W/cm^2.

X-ray spectra in the range of 4.8 - 8 $\overset{\circ}{A}$ emitted from laser
produced plasmas of Hf, Ta, W, Re, and Pt were obtained. Lines
belonging to the Ni-like isoelectronic sequence were identified and
the interpretation of some previously measured lines of this
sequence were revised. These results are of interest for plasma
diagnostics such as impurity studies in various plasma machines,
or in research into the dynamics of laser produced plasmas.

197

1.   THE LASER SYSTEM

The laser at Soreq has been designed for single beam, high
irradiance , 1.06µm wavelength interaction studies (see Fig. 1).
In order to perform experiments in the 1 - 10 nsec pulse duration
regime, of interest in current laser-fusion research, a new
Q-switched Nd:YAG oscillator has been developed. Laser pulse
durations were previously limited to 30-500 psec pulses obtained
with an etalon tuned, active-passive mode-locked, Nd:YAG oscillator.

A schematic of the passively Q-switched oscillator is shown in
Fig. 2. For stable nanosecond operation it was found that a short
Q-switched oscillator cavity (25 cm) and a relatively short flashlamp pulse (150µsec) were important. The pulse duration was
controlled by varying the concentration of the Q-switching Kodak
Nickel Dye. Figure 3 shows the dependence of the pulse length
on the unbleached dye transmission.

Stable operation resulted when the flashlamp voltage was set
at 5 V above that required for the lasing threshold. Statistics
taken over 100 shots showed no misfires and a pulse height variation of less than 2%. It should be noted that longer flashlamp
pulses, longer cavity lengths, and flashlamp voltages far above
threshold only resulted in longer, structured pulses and in
multiple pulses. (Additional details on the Q-switched oscillator
may be obtained from reference 1.)

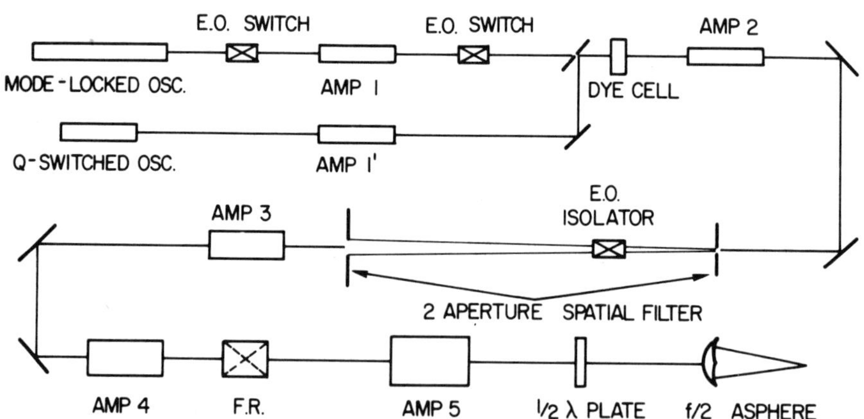

Fig. 1.   Schematic of the Soreq High Power Laser System (E.O.
          switch, E.O. isolator = Pockels cell + two polarizers,
          F.R. isolator = Faraday rotator + two polarizers).
          Output beam is focused in an f/5 cone.

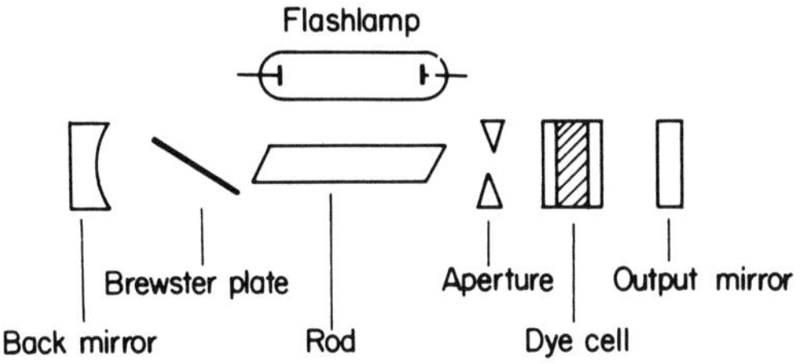

Fig. 2. Q-switched oscillator schematic

The amplifier chain is divided into two sections: a preamplifier section composed of two amplifiers and an amplifier section composed of three amplifiers, the largest of which is 50mm in diameter. The two sections are separated by a two aperture spatial filter. The first aperture, 2mm in diameter, selects out the central portion of the preamplifier section output. The second

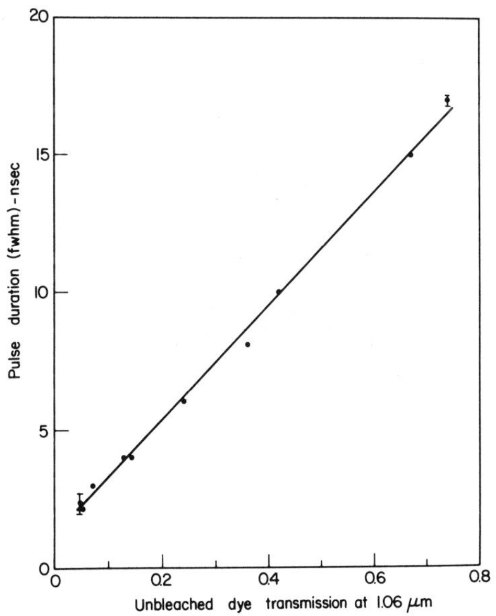

Fig. 3. Dependence of the laser pulse duration on the
unbleached dye transmission at 1.06μm

aperture is located in the Frauenhofer zone of the first aperture (15m distance) and has a diameter such that the aperture edge falls on the first zero of the Frauenhofer diffraction pattern. The resultant profile has a smooth bell shape and is propagated through the amplifier section without any beam breakup.

Pre-pulse suppression, target isolation, and backscatter protection are obtained with dye cell, Pockels cell and Faraday rotation units. The dye cell gives a contrast ratio of 10, while the Pockels cell and Faraday rotation units each have contrast ratios of about 300. An additional two Pockels cell units are used with the mode-locked oscillator for pulse selection.

Typical output power is 10 G watt in a diffraction limited beam. Full power far-field measurements indicate a 5µm fwhm focal spot (80% of the energy within 10µm)^2, so that interaction experiments can be performed at the 10^{16} Watt/cm^2 level (Fig. 4). The stability of the laser system output has been held to the oscillator output stability of 2%. The system repetition rate is nine minutes.

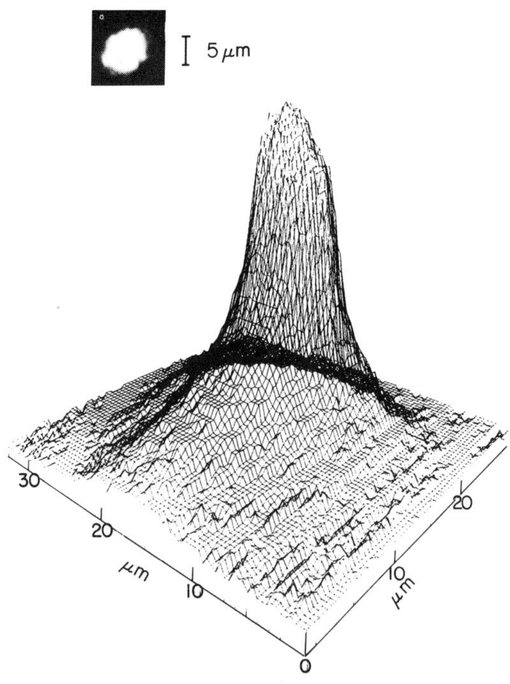

Fig. 4. Laser beam focal spot as determined from a full power
far-field measurement (1µrad = .15µm)

To summarize, we can obtain from our system a maximum energy
of 20J in a 2.5 nsec pulse, a maximum irradiance on target of
2×10^{16} W/cm^2 , laser pulse durations which can be varied from
50 psec to 15 nsec and a laser focal spot diameter as small as 5μm.

2. RESONANT ABSORPTION FOR SHORT AND LONG LASER PULSES

Necessary for successful laser fusion is the efficient
absorption of laser light in the coronal plasma surrounding the
target. Equally important, however, is the coupling of this
absorbed energy to the dense target material so that inward directed momentum will be imparted to compress and heat the fusion
fuel.

Efficient laser light absorption is believed to occur through
the process of resonant absorption[3]. In experiments conducted with
short duration laser pulses ($\tau \lesssim 100$ psec), increased laser light
absorption[4] and momentum transfer[5] have been observed when conditions were optimized for the resonant absorption process. Current
and planned laser fusion experiments, however, require much longer
laser pulses ($\tau = 1 - 10$ nsec) and it is important to determine in
this regime whether or not resonant absorption exists and if it
will improve the efficiency of momentum transfer to the target.

We have determined experimentally the resonant absorption
contribution to the momentum imparted to targets irradiated with
short (60 psec) or long (2.5 nsec) duration laser pulses. The
targets, polished aluminum slabs, were irradiated at oblique
incidence with high focused intensity ($10^{14} - 10^{16}$ W/cm^2), π and σ
polarized laser light. (The target was always placed within the
5μm fwhm diameter diffraction limited focal spot of the f/5
convergent cone of light[2].) The primary diagnostic was a torsion
pendulum which measured the momentum imparted to the target[6].
Supporting data was provided by charge collectors (Faraday cups).

Principal conclusions are that for 60 psec pulses, π polarized
light resulted in 35% more momentum transfer to the target, in
agreement with resonant absorption experiments performed elsewhere[4].
For 2.5 nsec pulses, however, no difference was observed in the
target momentum between irradiations with π and σ polarized light.
Furthermore, for short pulse irradiations the momentum coupling
efficiencies for π and σ polarized light, which are characterized
by momentum/energy on target, were both independent of intensity
over the range $10^{14} - 10^{16}$ W/cm^2. For long pulse irradiations the
momentum coupling efficiency decreased by more than a factor of two
over the same intensity range.

Figure 5 shows the experimental results. It can be seen that
for the 60 psec duration pulses, the momentum transfer for π

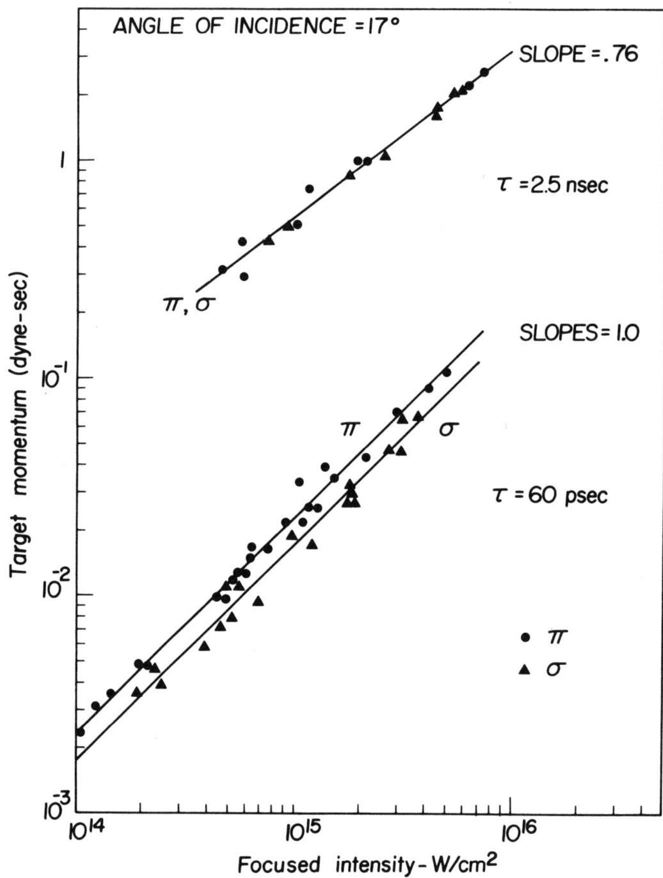

Fig. 5.   Plot of target momentum vs. focused laser intensity using
π and σ polarized laser light for laser pulse durations of
60 psec and 2.5 nsec

polarization was 35% greater than for σ polarization. For the
2.5 nsec duration pulses, however, no difference was seen between
the π and σ polarizations. For both the short and long duration
pulses, the momentum increased with focused intensity. The scaling
was, however, stronger for the 60 psec pulses ($P{\sim}I^{1.0}$) than for the
2.5 nsec pulses ($P{\sim}I^{.76}$). It may be noted that the long pulse
scaling was similar to the scaling previously observed for 0.5 nsec
pulses over the range $10^{13} - 10^{14}$ W/cm² [6].

    Plotting the ratio of momentum to incident energy (P/E) allows
the momentum coupling efficiencies for short and long duration
pulses to be compared. Figure 6 shows plots of the specific momentum curves obtained from the least squares curve fits of Fig. 5.
For 60 psec pulses, the momentum coupling efficiency was 35%
greater for π than for σ polarization. The coupling efficiencies
were independent of focused intensity. For the 2.5 nsec pulses
there was no difference between π and σ polarization and the momentum coupling efficiency decreased with increasing intensity. For
focused intensities greater than 3×10^{14} W/cm^2 the coupling efficiency for long pulses was less than for short pulses.

    Optimum resonant absorption occurs when $(\ell/\lambda_0)^{2/3}\sin^2\theta=0.15$
where λ_0 is the laser light wavelength, ℓ is the density scale-
length, and θ is the angle of incidence[3]. If ℓ were larger for
long pulses than for short pulses then the resonant absorption
peak would occur at a smaller angle of incidence. Spatially
resolved measurements made with an X-ray pinhole camera, an X-ray
spectrometer, and a 2ω telemicroscope, all indicated a plasma size
of 30µm. This gives an upper limit for the density scalelength,
i.e. $\ell \lesssim$ 30µm, and thus a lower limit to the angle for optimum
resonant absorption $\theta \gtrsim 7^\circ$. A series of shots (60) were made at
constant energy with the angle of incidence varying from 0° to 17°.
The result is shown in Fig. 7. At no angle was there any difference

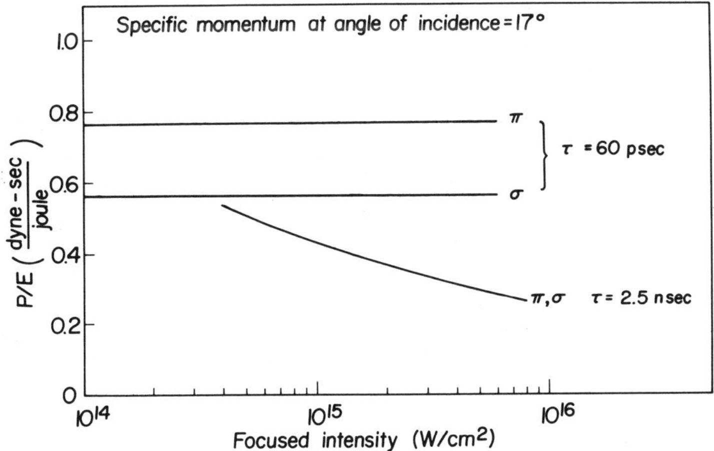

Fig. 6. Ratio of target momentum to the incident energy, P/E
        (specific momentum) plotted as a function of focused
        intensity for π and σ polarized laser light incident
        at an angle of 17° and pulse durations of 60 psec and
        2.5 nsec.

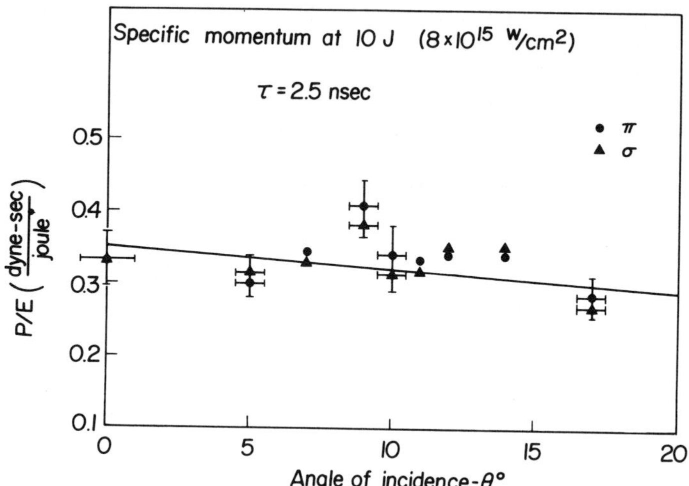

Fig. 7.   Plot of specific momentum (P/E) as a function of the
angle of incidence for both π and σ polarized light at
a focused intensity of 8×10^{15} Watt/cm^2 and for a pulse
duration of 2.5 nsec

between π and σ polarizations. The momentum coupling efficiency
decreased slowly with increasing angle of incidence. Some fine
structure may exist.

Since the target momentum equals the blow-off momentum, additional information can be obtained from charge collector data. A
typical charge collector trace is shown in Fig. 8. The observed
ion spectrum is divided into separate "fast" and "thermal" ion
groups. The fast ions, although they may contain 50% of the
observed ion energy, account for less than 10% of the observed ion
mass. The momentum of a group of ions P is related to the mass of
the group $\tilde{m}$ and the energy E_i by $P = \sqrt{2\tilde{m}E_i}$. Thus, the greater part
($\gtrsim 80\%$) of the momentum should be contained in the thermal group.

The momentum, P, of the thermal group can be related to a
characteristic ion current, $\tilde{i}$ in the charge collector by the
relation: $P \simeq [m_i r/(e\, Z)]\tilde{i}$ where r is the target-charge collector
distance, m_i is the ion mass, Z is the ionization, and e is the
electron charge. The momentum of the blow-off is, thus, proportional to the characteristic current in the ion collector, as
observed experimentally (see Fig. 8).

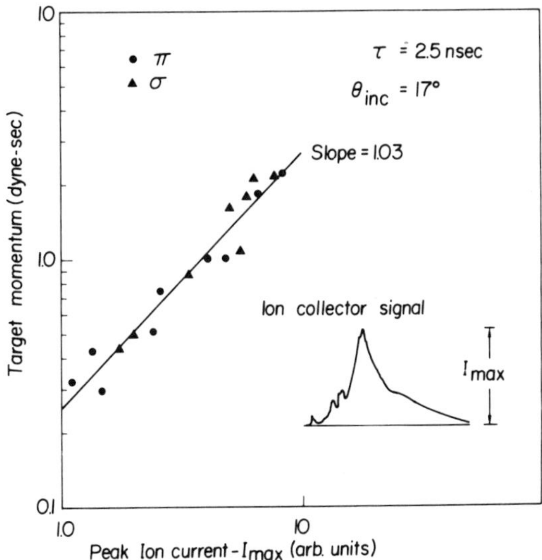

Fig. 8.   Plot of the target momentum as a function of the peak
ion current (I_{max}) measured in the ion charge collectors
for both π and σ polarizations. The pulse duration is
2.5 nsec and the angle of incidence is 17°. In the insert
is a sketch of a typical ion collector signal. The charge
collector was located in the plane defined by the optical
axis and the target normal, at an angle of 40° to the
optical axis.

At this point it should be noted that since the targets were
slabs, it is possible that vaporized material generated with heat
conducted away from the laser-plasma interaction region, may have
contributed to the observed target momentum ("late effect").
Experiments were conducted with thin foils mounted on a "double
pendulum"[6]. The thin foils had less material surrounding the
interaction region and should have had less of a "late effect".
The results indicated that for the aluminum targets used in the
experiment, the "late effect" contributed a maximum of 20% to the
total momentum.

A characteristic feature of resonant absorption is a larger
velocity for the fast ion group[4]. Such a behavior was observed by
us in the short pulse experiments. The ratio of the velocities of
the fast to thermal ion groups was found to be about 3 for π
polarized light and 2 for σ polarization with a thermal ion velocity
of about 4×10^7 cm/sec at a laser intensity of 10^{15} Watt/cm^2. In
the long pulse regime, however, the fast to thermal ion velocity
ratio equals to 2 for both π and σ polarizations with the same
thermal ion velocity and at the same light intensity. Thus the

fast ion data for long laser pulses also shows no evidence of
polarization dependent resonant absorption.

The disappearance of polarization dependence during long pulse
irradiations might possibly result from: (a) spherical expansion of
the plasma, (b) long scale lengths, (c) ineffective coupling of
suprathermal electrons to the dense target material and (d) small
scale critical surface rippling or plasma turbulence.
(a) Reflectivity measurements made during the experiments[7] indicated a clear laser light back reflection dependence on the angle
of incidence. This implies a planar geometry.
(b) 2ω light is emitted from points on the critical surface
reached by laser light. 2ω was observed during the experiments
and it may be concluded that laser light did reach the critical
surface[7]. Furthermore, plasma dimensions were measured from
spatially resolved 2ω light and X-ray emissions, and an upper limit
for the density scalelength of only $30\mu m$ was obtained, which is not
large enough to explain the data.
(c) The resonantly absorbed energy deposited in suprathermal
electrons, might not couple to the dense target material but might
produce fast ions in the corona. Charge collector data does not,
however, show significant fast ion dependence on polarization. It,
therefore, does not appear that resonant absorption without corona-
to-target coupling can explain the results.
(d) Small scale rippling or turbulence at the critical surface
would tend to smear out any angle of incidence or polarization
dependences. It would, therefore, be impossible to distinguish
between resonant absorption and other absorption mechanisms. It
would also mean that there may always be some resonant absorption.
Small scale rippling has been observed for short pulses[8] but might
have a growth-rate slow enough to affect only the absorption of
long pulses. Small scale rippling or turbulence is not inconsistent
with the experimental data.

The decrease of the momentum coupling efficiency with intensity
for long laser pulses (see Fig. 6) may be attributed to two effects,
an increase in the plasma temperature and a decrease in the fractional absorption. The plasma momentum, P, can be expressed by
$P = 2\eta E/\tilde{v}$, where η is the fraction of energy absorbed and $\tilde{v}$ is a
characteristic ion blow-off velocity. For short pulses, our experi-
ments indicate that both P/E and $\tilde{v}$ are constant as a function of
laser intensity, thus implying η independent of laser intensity.
For fixed focusing geometry and light polarization, η has indeed
been observed to be independent of intensity for short laser
pulses[2]. For long pulses, however, our results show that for an
angle of incidence of 17°, P/E varies as $E^{-0.24}$, while $\tilde{v}$ (as
measured by the charge collectors) varies as $E^{0.14}$, thus implying
that the fractional absorption decreases as $E^{-0.10}$. For normal
incidence $P/E{\sim}E^{-0.14}$ and $\tilde{v}{\sim}E^{0.17}$ so that $\eta{\sim}E^{0.03}$. It would,

therefore, appear that the decrease of P/E with laser intensity for long pulses is due mainly to an increase in the plasma temperature (at the expense of increased mass ablated) rather than to a pronounced decrease in the fractional absorption.

Up to this point, only energy deposition effects were treated in computing the momentum imparted to the target. As pointed out[9] by Hora and Lindl and Kaw, ponderomotive forces may also impart a net momentum to the target. Their results indicate that the net momentum to the target is given by

$$P_\pi \simeq \frac{3E}{2\omega\ell} \frac{\omega^2}{\nu^2} e^{-2\frac{\omega\ell}{c}\frac{\nu}{\omega}} \quad \text{and} \quad P_\sigma \simeq \frac{E}{2c}\left[\left(\frac{\omega}{\nu}\right)^{\frac{1}{2}} + \left(\frac{\nu}{\omega}\right)^{\frac{1}{2}}\right]$$

where ω is the laser frequency, ℓ is the plasma density scalelength, and ν is the effective collision frequency. Note that in these expressions the momentum has an explicit linear dependence on the laser energy, but that the momentum also depends on ν and ℓ. If ν and ℓ are independent of E then the momentum transfer due to ponderomotive forces will conform with the experimentally observed scaling for 60 psec pulses. Note that for $\nu/\omega \lesssim 1$, as ν/ω increases, P_π and P_σ decrease.

A lower limit to the value of ν/ω can be obtained from the short pulse experimental data by assuming that all of the additional momentum observed with π polarization was due to ponderomotive forces. ν/ω is then calculated using the above equation for P_π. (If only a fraction of the additional momentum was due to ponderomotive forces then ν/ω would be larger.) From the short pulse results we have that for E = 100 mJ, the addition in momentum transfer to the target was 1.8×10^{-2} dyne-sec. Using the expression for P_π and taking[8] $\ell = 1\mu m$ we obtain $\nu/\omega \simeq 0.02$. For this value of ν/ω, $P_\sigma \simeq 10^{-4}$ dyne-sec and so was negligible. Since the momentum contributed by ponderomotive forces was overestimated, $\nu/\omega \gtrsim 0.02$.

To summarize this section, the contribution of resonant absorption to the momentum imparted to targets irradiated with high intensity laser light was measured for laser pulses of 60 psec and 2.5 nsec duration. Resonant absorption contributions were clearly observed for the short pulses. No polarization dependent effects were observed for the long pulses. The disappearances of $\pi-\sigma$ and angle of incidence dependences for long pulses may be the result of small scale rippling or turbulence at the critical surface. The momentum coupling efficiency was found to be independent of intensity for short pulses and was found to decrease for long pulses. This behavior appears to be a consequence of corona temperature variations with incident light

intensity. A lower limit for the effective collision frequency
was obtained ($\nu/\omega_{\zeta}.02$).

3. Z DEPENDENCE OF BACKSCATTERING IN LASER PRODUCED PLASMAS

In this section we present the results of back reflection
experiments performed with 2.5 nsec duration pulses at focused
intensities ranging from 4×10^{14} to 2×10^{16} W/cm^2. The focused light
was contained within a diffraction limited f/5 cone. (The focusing
lens was an f/2 asphere.) The targets were polished slabs of both
low (Al,Mg) and high (Pt,Au) atomic number elements. The diagnostics consisted of incident and backscatter calorimetry and backscatter spectroscopy at the fundamental and second harmonic frequencies. Spatially resolved 2ω photography[10]provided critical
surface excursions. Spatially resolved X-ray line spectra
revealed plasma dimensions[11] and ionization states while spectral
line measurements and ion collector data were used to obtain plasma
temperatures.

The targets, situated in the laser beam focal spot, were
irradiated at both normal and oblique incidences (17º). At normal
incidence both specularly reflected and stimulated backscatter
light was collected through the focusing lens. At oblique incidence, however, only the stimulated backscatter light was collected.
It was thus possible to distinguish between the two components of
the reflected light.

Figure 9 shows the results of the calorimetric measurements.
It can be seen that at normal incidence, the reflectivity from the
low atomic number targets remained constant at about 20% except
at the highest intensities. This represents the sum of the
specular reflection and the stimulated backscatter. At oblique
incidence, however, the reflectivity which consisted only of
stimulated backscatter was significantly lower but increased with
intensity. The difference between these two curves is approximately
equal to the specularly reflected light. This difference indicates
that the specular reflectivity decreased with intensity or in other
words that the absorption efficiency of that component of the light
that reached the critical surface increased with intensity.

Consider next the results from the high atomic number targets.
The oblique incidence results were approximately the same as for
low Z elements. Thus, the stimulated backscatter reflectivity was
not significantly effected by the target material. The normal
incidence data from high Z elements, was, however, significantly
different. The reflectance dropped down to approximately the level
of the oblique incidence stimulated backscatter component. This is
indicative of an almost complete absence of specular reflection
and of improved absorption.

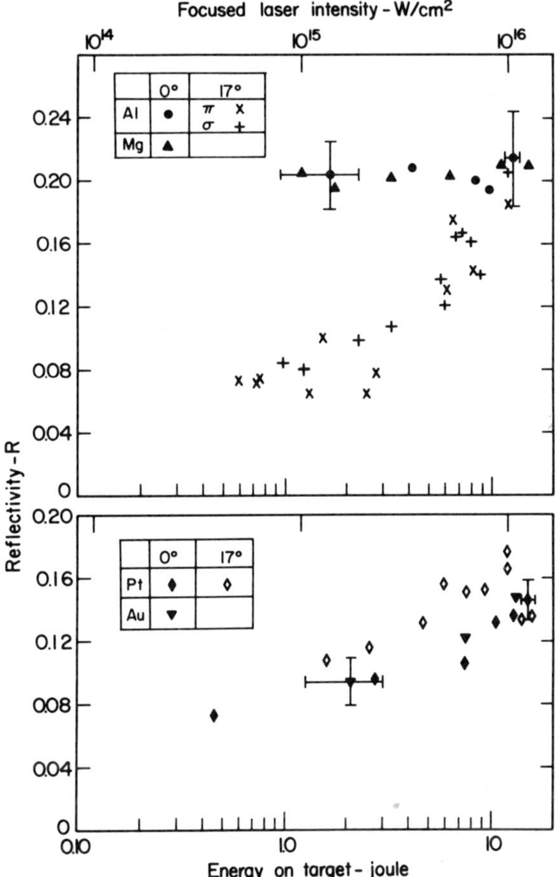

Fig. 9.   Reflectivity of low and high Z targets. For 0° Al, Pt
and Au sufficient data was obtained to allow averaging
within discrete energy bins. Error bars give standard
deviations. Note that for 17° Al, data was collected
for π and σ polarizations. 17° Pt and Au data was
with 45° polarization.

Summarizing the results presented in Fig. 9 we note the
following features: (a) No difference in stimulated backscatter
when the incident light is either π or σ polarized. (b) From
the difference in reflectivity when the angle of incidence is 0°
or 17°, one can deduce the specular reflection. (c) Back reflection increases with irradiance while specular reflection appears
to decrease. (d) Stimulated backscatter is the same for low and
high Z targets. (e) For high Z targets no specular reflection
was observed.

Note that the normal incidence measurement of stimulated plus specular backreflection was, for low Z targets, independent of intensity. This is consistent with the long pulse momentum measurements on Al from which it was deduced that the fractional absorption did not vary with intensity. This constant reflectivity is the result of two opposing processes: increased stimulated backscatter in the underdense plasma, and increased absorption of the light that reaches the critical density regime.

Inverse bremsstrahlung absorption alone cannot explain all of the specular reflection results. Although it does predict lower specular reflectivity for high Z plasmas, it also predicts diminished absorption and increased specular reflection for increased laser intensity and corona temperature[12]. The experimental results indicate an opposite intensity dependence. Improved absorption can be explained by an additional absorption mechanism such as an electron-ion decay instability, i.e. decay of a photon into plasma waves[13], resonant absorption in a turbulent plasma, i.e. coupling of longitudinal oscillations in the resonant region to plasma waves via ion-acoustic turbulence[14], or turbulent heating i.e., enhanced scattering of electrons oscillating in the laser light's electric field by turbulent structures[15].

Evidence of parametric instabilities or resonant absorption in a turbulent plasma is seen in the 2ω spectra (Fig. 10). There are two components to these spectra. At low intensities a single peaked distribution within several angstroms of $\lambda_0/2$ is observed. At higher intensities a red shifted peak is observed with a threshold at approximately 2×10^{15} W/cm^2. This red shifted component quickly increases in amplitude and obscures the original unshifted spectra. (Both low and high Z materials behave similarly.)

There are three possible generation mechanisms for the 2ω light (p_2). The first involves the second order current generated by the electric field associated with resonant absorption[16]. The second involves the reverse parametric coalescence of Langmuir waves (1) and photons (p) at frequency ω i.e., $1+p \rightarrow p_2$ or $1+1 \rightarrow p_2$ with the Langmuir wave produced by a parametric instability[17]. The third mechanism also involves the reverse parametric coalescence of Langmuir waves and photons but in this case the Langmuir waves are produced by the interaction of plasma oscillations, produced by resonant absorption, with ion-acoustic turbulence[14].

The red-shifted 2ω component has characteristics of both a parametric instability and a source of resonant field-turbulence. Experiments on low (Mg, Al) and high (Pt, Au) atomic number targets

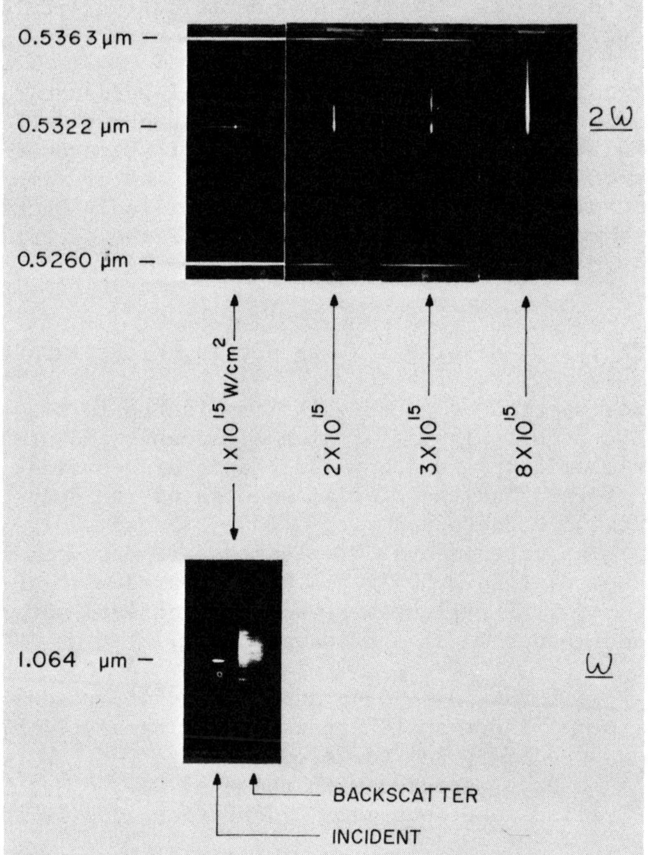

Fig. 10. 2ω and ω spectroscopy results for low Z targets.
Narrow lines superimposed on 2ω spectra are calibration
lines from a Rb vapor lamp. ω spectra consist of
incident and reflected light taken simultaneously. The
bandwidth of the incident laser light is less than the
.5Å instrumental resolution

show a Z/A dependence [Z/A = 1/2 → Δλ = 11Å $\pm$ 3Å ,   [14,18]
Z/A = 1/4 → Δλ = 5Å $\pm$ 2Å]. This is what the theories predict
The magnitudes of the red-shift also agree with the theoretical
prediction[7,18]. The sharp threshold behavior of the red-shifted
component agrees with the threshold behavior characteristic of
parametric instabilities[17]. It is not clear how resonant fields
(generated via a "linear process") coupled to turbulence (whose
strength increases gradually with laser intensity) can explain

the sharp 2ω threshold. Thus, a parametric instability source
mechanism best explains the data.

In conclusion, we have seen experimental evidence for strongly
Z-dependent absorption. Inverse bremsstrahlung may be dominant
at low intensities but the results indicate that parametric instabilities (aperiodic or parametric decay) exist and are important
coupling mechanisms at high intensities. Brillouin backscatter
was evident throughout the intensity range of the experiment rising
from 7% at 4×10^{14} W/cm^2 to about 20% at 2×10^{16} W/cm^2.

4. X-RAY SPECTRA OF Ni-LIKE IONS OF Hf, Ta, W, Re AND Pt

The X-ray spectra of the Ni-like isoelectronic sequence are
characteristic of hot plasmas of high Z elements and are of
interest for atomic spectroscopy and plasma diagnostics. We have
recorded and identified the Ni-like spectra of the heavy elements
Hf, Ta, W, Re and Pt,obtained with a 15J, 2.5 nsec Nd:glass laser
irradiation. The experimental wavelengths are measured with an
accuracy of better than 0.005Å, and are compared with ab-initio
calculations[20,21]. Excellent agreement is achieved between the
calculated and measured value of wavelengths.

Together with the $3d^{10}-3d^94p$ and $3d^{10}-3d^94f$ "resonance"
transitions, some "inner-shell" transitions, namely, $3p^63d^{10}-$
$3p^53d^{10}4s$ and $3p^63d^{10}-3p^53d^{10}4d$ were also observed. Moreover by
comparing the observed spectra with those of Sm, Gd and Dy[22] we
were able to revise some doubtful   identifications in these
elements.

The X-ray spectra in the 5-8Å range were analyzed with two
flat crystal spectrometers and recorded on KODAK SC7 films, behind
a 15μm thick Be filter. KAP (2d=26.6Å) and Gypsum (2d=15.15Å)
crystals were employed. The spectral resolution of the spectrometers was better than 0.005Å. Each spectrum of the high Z
elements was obtained with two laser shots. For wavelength calibration, we used Al, Si and P H- and He-like spectra, as well as,
the K_α lines of these elements (superimposed on the unknown
spectrum). These spectra have a wide enough wavelength range so
as to span the unknown X-ray lines in the observed spectra of the
high Z elements. The wavelengths of the calibration lines were
taken from Refs. 23 and 24.

The continuum portion of the emitted X-ray spectrum was
recorded with an array of PIN detectors, with different filters in
front of them. The electron temperature deduced from these
measurements was in the range of 1-1.5 keV.

The identification of the lines was performed by means of comparison with ab-initio computations[20],[21]. The parametric potential method in its relativistic version[25] was used for the calculation of energy levels and transition probabilities. The angular coefficients in j-j coupling were obtained using the program of I.P. Grant[26]. Configuration interactions within the complexes were taken into account, but their effect on the energy values was found to be negligible. In general, the coupling was nearly pure j-j (97%) except for two levels of the $3d^9 4f$ configuration, which were found to be only 75% pure. Furthermore, the calculation of the energy of the $3d^9(3/2)4f_{5/2}$ (J=1) levels was slightly sensitive to mixing with the $3d^9 5f$ configuration.

The parametric potentials were optimized separately for the ground state $3d^{10}$ on the one hand, and for all the excited states on the other hand. This "adiabatic" procedure[22] was found to yield excellent results for calculations of X-ray spectra[22].

A typical spectrum of Hf recorded with a microphotometer is shown in Fig. 11, together with the wavelengths and classifications of those lines whose identifications are established. In Table 1, we give the observed wavelengths and intensities of the classified lines of Hf XLV. The experimental intensities given in the table are corrected for the Be filter absorption, the photographic film[25] and the microphotometer response. As can be seen, the agreement between the calculated and measured wavelengths is excellent. The r.m.s difference is about 0.004Å, which equals the experimental error. A more complete listing of the observed lines in the five elements studied can be found in refs. 28 and 29.

All transitions in Table 1 are from levels of J=1 to the J=0 ground states. Besides the "resonance" $3d^{10}$-$3d^9 4f$ and $3d^{10}$-$3d^9 4p$ transitions, quite strong $3p^6 3d^{10}$-$3p^5 3d^{10} 4s$ "inner shell" lines were also observed in the spectra. In the case of Hf XLV, $3p^6 3d^{10}$-$3p^5 3d^{10} 4d$ transitions were also observed. For the other elements the corresponding wavelengths were outside the range of the crystal spectrometer. As can be seen from Table 1, the intensities of these lines are comparable to those of the "resonance" lines, in particular, if one compares them with the $3d^{10}$-$3d^9(5/2)$ $4f_{5/2}$ line, which moreover is blended with a broad "band" (composed probably of many lower ionization satellites. This observation led us to reinvestigate some of the corresponding weak lines in Sm, Gd and Dy which were tentatively identified as belonging to the $3d^{10}$-$3d^9(5/2)4f_{5/2}$ transition by Burkhalter et al.[20] We propose to classify these measured transitions as belonging to the array $3p^6 3d^{10}$-$3p^5 3d^{10} 4s$. Figure 12 shows a plot of the reduced transition energy E/Z as a function of Z along the isoelectronic sequence. It is clear that with this revised identification, all the experimental points fall on straight lines. Moreover, we have

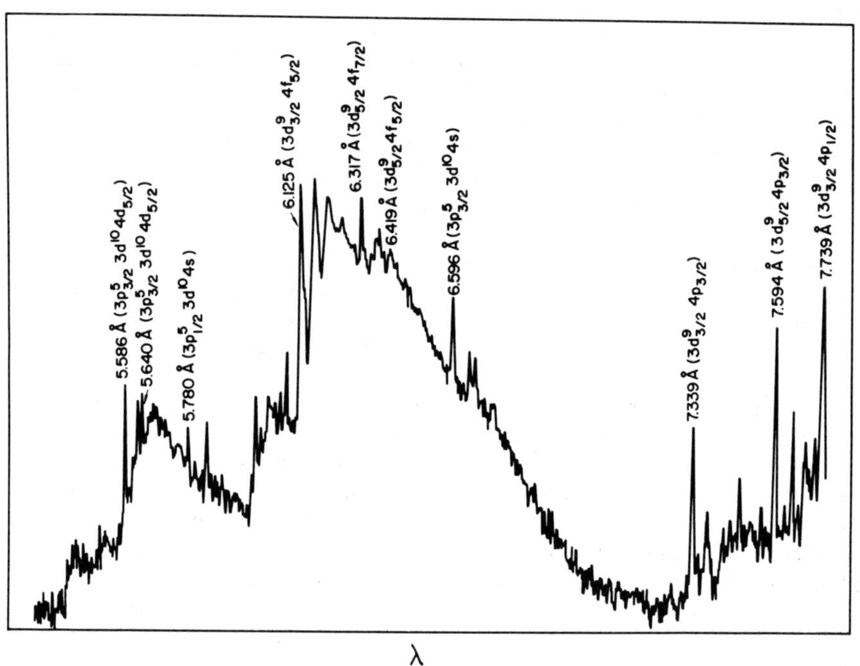

Fig. 11.   X-ray spectrum of Hf.   All indicated transitions are to the $3p^6 3d^{10}$ ground state of Hf XLV

computed the expected wavelengths of these transitions and obtained a good agreement with the measured wavelengths as given in Ref.22. On the other hand, the calculated values for the transitions $3d^{10}-3d^9(5/2)4f_{5/2}$ fall on a straight line with the experimental points of the heavy elements measured in the present work.

In conclusion, strong lines of Hf, Ta, W, and Pt belonging to the Ni-like sequence of these elements were observed.   The general features of the spectra obtained from all the elements studied are preserved, with the exception of the disappearance of the "inner shell" lines ($3p^6 3d^{10}-3p^5 3d^{10}4s$ and $3p^6 3d^{10}-3p^6 3d^{10}4d$ transitions) in the Pt spectrum.   The spectra are composed of a band type continuum emission together with the line spectra.   It

Table 1. Experimental and Calculated Results for Ni-Like Hf

Transition	Experiment		Calculated	
	$\lambda(\text{Å})$ [a]	Intensity	$\lambda(\text{A})$	gf
$3p^6 3d^{10} - 3p^6 3d^9 (3/2) 4p_{1/2}$	7.739	80	7.744	0.136
$-3p^6 3d^9 (5/2) 4p_{3/2}$	7.594	85	7.599	0.255
$-3p^6 3d^9 (3/2) 4p_{3/2}$	7.339	40	7.346	0.027
$-3p^5 (3/2) 3d^{10} 4s$	6.596	80	6.596	0.347
$-3p^6 3d^9 (5/2) 4f_{5/2}$	6.419	15 [b]	6.411	0.007
$-3p^6 3d^9 (5/2) 4f_{7/2}$	6.317	30	6.322	1.67
$-3p^6 3d^9 (3/2) 4f_{5/2}$	6.125	100	6.122	6.14
$-3p^5 (1/2) 3d^{10} 4s$	5.780	30	5.776	0.052
$-3p^5 (3/2) 3d^{10} 4d_{3/2}$	5.640	30	5.632	0.052
$-3p^5 (3/2) 3d^{10} 4d_{5/2}$	5.586	85	5.579	1.20
$3p^5 (1/2) 3d^{10} 4d_{3/2}$			5.014	0.436

a) ± 0.005Å   b) blended line

is worth noting that upon going from Hf^{+44} to Pt^{+50}, the relative
intensity of the line spectrum decreases relatively to the continuum. In the gold spectrum, the line spectrum was already too
weak to be analysed quantitatively.

These kinds of transitions in the Ni-like isoelectronic
sequence are generally quite strong in hot plasmas of medium and
high Z elements. Indeed, these ions have a closed 3d shell in
their ground state and their spectra is characterized by a small
number of emission lines that are prominent in a rather wide
temperature range. Therefore, the results of this work is of
interest for plasma diagnostics, such as impurity studies in
various plasma machines, or research into the dynamics of laser
produced plasma[30].

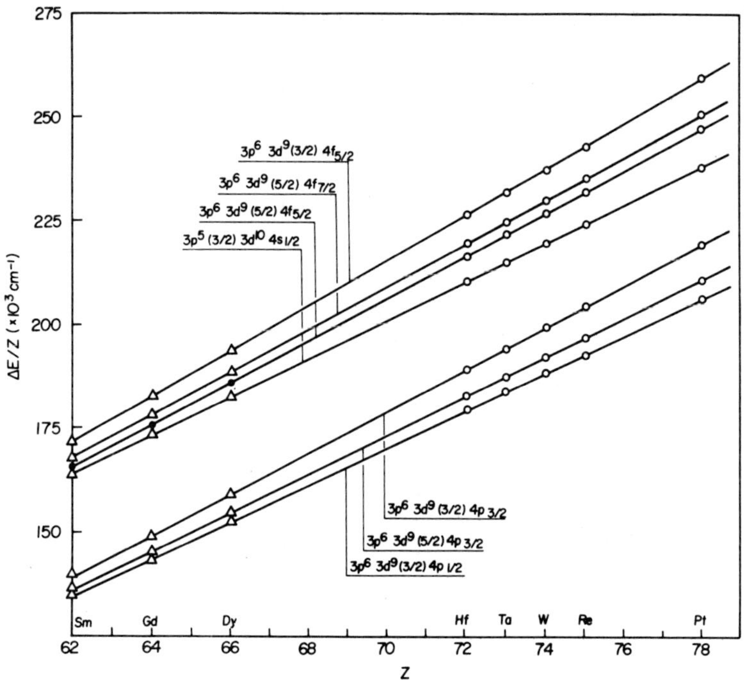

Fig. 12.   Plot of $\Delta E/Z$ versus Z for the observed and classified
transitions in the Ni-like isoelectronic sequence for
Z=62 through Z=75.   All transitions are to the $3p^6 3d^{10}$
ground state.
0 - present work, experimental; ● - present work,
theoretical; Δ - Ref. 22 experimental

REFERENCES

1. S. Jackel, H.M. Loebenstein, A. Zigler, H. Zmora and
    S. Zweigenbaum, "Soreq Plasma Physics Dept. Report: A Stable
    Time-Tunable Q-Switched Oscillator for Operation in the Nano-
    second Regime" (1979).
2. H. Szichman and S. Zweigenbaum, J. Phys. E 12, 87 (1979).
3. D. Forsland, J. Kindl, K. Lee, E. Lindman and R. Morse, Phys.
    Rev. Lett. 11, 679 (1975).
4. J. Pearlman, J. Thomson and C. Max, Phys. Rev. Lett. 38.
    1397 (1977).
    K. Manes, V. Rupert, J. Auerbach, P. Lee and J. Swain, Phys.
    Rev. Lett. 39, 281 (1977).

R. Godwin, R. Sachsenmaier and R. Sigel, Phys. Rev. Lett. 39, 198 (1977).

5. B. Arad, S. Eliezer, S. Jackel, H.M. Loebenstein, I. Pelah, A. Zigler, H. Zmora and S. Zweigenbaum, Conf. Record - Abstracts of 1979 IEEE International Conference on Plasma Sciences, IEEE Catalog # 79 CH 1410-0 NPS.

6. S. Zweigenbaum, Y. Gazit and Y. Komet, Plasma Phys. 19, 1035 (1975).
    S. Zweigenbaum, Y. Gazit and Y. Paiss, J. Phys. E: Sci. Instrum. 11, 831 (1978).

7. S. Jackel, H.M. Loebenstein, A. Zigler, H. Zmora and S. Zweigenbaum, Appl. Phys. Lett. (in press).

8. D. Atwood, D. Sweeney, J. Auerbach and P. Lee, Phys. Rev. Lett. 40, 184 (1978).

9. H. Hora, Phys. Fluids 12, 182 (1969).
    J. Lindl and P. Kaw, Phys. Fluids 14, 371 (1971).

10. S. Jackel, J. Albritton and E. Goldman, Phys. Rev. Lett. 35, 514 (1975).

11. A. Zigler, H. Zmora and Y. Komet, Phys. Lett. 60A, 319 (1977).

12. P. Catto and T. Speziale, Phys. Fluids 20, 167 (1977).

13. Basov, Kroknin, Pustovalov, Rupasov. Silin, Sklizkov, Tikhonchuk, Shikanov, Sov. Phys. JETP 40, 61 (1975).

14. R. Cairns, private communication.

15. R. Fraehl and W. Kruer, Phys. Fluids 20, 55 (1977).
     W. Manheimer, D. Colombant and B. Ripin, Phys. Rev. Lett. 38, 1135 (1977).

16. N. Erokhin, V. Zakharov and S. Moiseev, Sov. Physics JETP 29, 101 (1969).

17. A. Vinogradov and V. Pustovalov, Sov. Phys. JETP 36, 492 (1973).

18. O. Krokhin, V. Pustovalov, A. Rupasov, V. Silin, G. Sklizkov, A. Starodub, V. Tikhonchuk and A. Shikanov, JETP 22, 21 (1975).

19. M. Born and E. Wolf, "Principles of Optics", Fourth Ed. (Pergamon Press 1970) p.441.

20. M. Klapisch, A. Bar-Shalom, J.L. Schwob, B.S. Fraenkel, C. Breton, C. De Michelis, M. Finkenthal and M. Mattioli, Phys. Lett. 69A, 34 (1978).

21. M. Klapisch, J.L. Schwob, M. Finkenthal, B.S. Fraenkel, S. Egert, A. Bar-Shalom, C. Breton, C. De Michelis and M. Mattioli, Phys. Rev. Lett. 41, 403 (1978).

22. P.G. Burkhalter, D.J. Nagel and R.R. Whitlock, Phys. Rev. A9, 2331 (1974).

23. F.V. Aglitzki, V.A. Boiko, A.V. Vinogradov, Kvant Elektron 4, 908 (1974).

24. U. Feldman, G.A. Doschek, D.J. Nagel, R.D. Cowan and R.R. Whitlock, Astrophys. J. 192, 213 (1974).

25. M. Klapisch, J.L. Schwob, B.S. Fraenkel and J. Oreg, J. Opt. Soc. Am. 67, 148 (1977).

26. I.P. Grant, Comput. Phys. Comm. 11, 397 (1976).

27. K. Eidman, M.H. Key and R. Sigel, J. Appl. Phys. 47, 2402 (1976).

28. A. Zigler, H. Zmora, N. Spector, M. Klapisch, J.L. Schwob and
     A. Bar-Shalom, J. Opt. Soc. Am. (in press).
29. A. Zigler, H. Zmora, N. Spector, M. Klapisch, J.L. Schwob and
     A. Bar-Shalom, "Nickel-like Spectrum of Platinum Emitted from
     Laser Produced Plasma", Soreq Internal Report (1979).
30. A. Zigler, H. Zmora, H.M. Loebenstein and J.L. Schwob, J. Appl.
     Phys. $\underline{50}$, 165 (1979).

EXPERIMENTAL OBSERVATIONS OF NON-LINEAR

FORCE EFFECTS IN LASER-PLASMA INTERACTIONS

B. Luther-Davies, V. Del Pizzo, B.W. Boreham

Laser Physics Laboratory
Department of Engineering Physics
Research School of Physical Sciences
The Australian National University
Canberra, ACT, 2600, Australia

ABSTRACT

Non-linear forces associated with strong electric field
gradients in laser-produced plasmas are expected to have a strong
influence on the laser-plasma interaction especially at high laser
intensities. Two components these field-gradient forces must be
considered. Firstly, a component transverse to the propagation
direction of the beam can arise due to the non-uniformity in the
laser beam irradiating the plasma. The second component exists
because of the dielectric swelling of the incident field as the
plasma density approaches the cut-off value. These components can
lead respectively to (1) rippling and curvature of the critical
density surface, beam filamentation and self-focussing, and (2)
plasma acceleration and profile modification.

We report results of experimental studies of some of these
phenomena. Firstly, the process of whole beam self-focussing is
considered and experiments described in which plasma and laser
conditions were established in which our computations predicted it
should occur and strong evidence of self-focussing obtained.
Secondly, results from an experimental study of electron energies
ejected from gases ionized by intense laser radiation are discussed
which confirm the existence and correct magnitude of the radial
non-linear force component. Thirdly, indirect measurements of
plasma density scale lengths inferred from measurement of the
optimum angle for resonance absorption in a laser-produced plasma
are discussed.

INTRODUCTION

A substantial body of evidence exists to suggest that the non-linear force associated with strong gradients of the electric field of an intense laser beam plays an important role in determining the nature of the laser-plasma interaction. In simplified form the non-linear force, f_{NL}, acting as the electrons in the plasma may be written [1]

$$f_{NL} = \frac{-e^2}{4m_e \omega^2} \nabla \langle E^2 \rangle \tag{1}$$

where $\langle \rangle$ denotes averaging over one optical cycle (NB: the force on the ions is a factor of m_e/M_i smaller than that on the electrons and is usually neglected). The non-linear force is non-zero when there exists gradients of $\langle E^2 \rangle$ in the plasma and in the general situation of a non-uniform laser beam propagating up the density gradient of a non-uniform plasma there are two components of that force which must be considered. The first arises due to the spatial non-uniformity of the laser beam transverse to its propagation direction and causes ejection of electrons (and ions which are electrostatically coupled to the electrons when the Debye length, $\lambda_D \ll D$, the beam diameter), from the highest intensity regions of the beam. This can result in rippling and curvature of the critical density surface [23]. Such density modifications effect the local value of the plasma refractive index which can result in filamentation or self-focussing [4-8] of the incident laser beam. The second component of the force arises due to the dielectric swelling of the incident laser field in the region near the critical density surface where the value of the plasma dielectric constant becomes small. In this case the non-linear force acts to expel the underdense plasma down the plasma density gradient from regions adjacent to the critical surface and, as has been demonstrated in computer simulations [9-13] this can lead to modification of the plasma density and velocity profiles so that density jumps from sub- to super-critical values occur in only a few tens of Debye lengths. A plasma with its density profile modified in this manner would be expected to show lower levels of absorption by collisional processes and an enhancement of resonance absorption in comparison with a plasma whose profile is determined by hydrodynamic expansion alone. A growing body of experimental evidence exists which demonstrates effects attributable to the non-linear forces in laser-plasma interactions particularly when short duration high intensity laser pulses are used.

Experimental research within the Laser Physics Laboratory at The Australian National Univeristy has, for some time, concentrated on studies of some effects associated with non-linear forces. The three areas which will be discussed in this paper are:

(a) self-focussing of laser beams in laser-produced plasmas;

(b) acceleration of electrons produced by ionization of gases by intense laser pulses;

(c) profile modification.

(a) SELF-FOCUSSING

   1. Computational Results on Self-focussing

There have been extensive theoretical studies of self-focussing and filamentation of a laser beam propagating in a high density plasma [4-8] and some experimental evidence exists to support the theory [14-18]. In our research we have aimed to make a parallel theoretical and experimental study of the process with the specific aim of determining the conditions in which whole beam self-focussing could be expected to occur with the resultant concentration of a large fraction of the laser power into a single narrow filament. The in-plasma enhancement of the laser flux density by such a means could lead to a false interpretation of data from laser-plasma interaction experiments were it to go undetected.

Our theoretical treatment of self-focussing is based upon the Transient Self-focussing Numerical Code (TRSF) developed by Siegrist [7,19] which studies propagation of a laser beam which is Gaussian in time and space in a multiply-ionized, absorbing plasma with arbitrary values of plasma density, temperature and thickness, and laser pulse duration, beam diameter and intensity. Two self-focussing mechanisms are applicable for laser pulse durations in the short time regime ($<$ 10 nsec) relevant to most fusion-orientated experiments, those being ponderomotive force [4] and relativistic self-focussing [6,20] mechanisms respectively. TRSF has been used to predict the thresholds for efficient self-focussing to occur (ESF) for a wide range of plasma and laser conditions. The definition of efficient self-focussing is an arbitrary one chosen so that thresholds are calculated so that the in-filament intensity exceeds the peak input intensity in the laser beam by one order of magnitude - this criterion being chosen with the aim of producing an observable effect (eg. X-ray hot spots) in a given experimental situation.

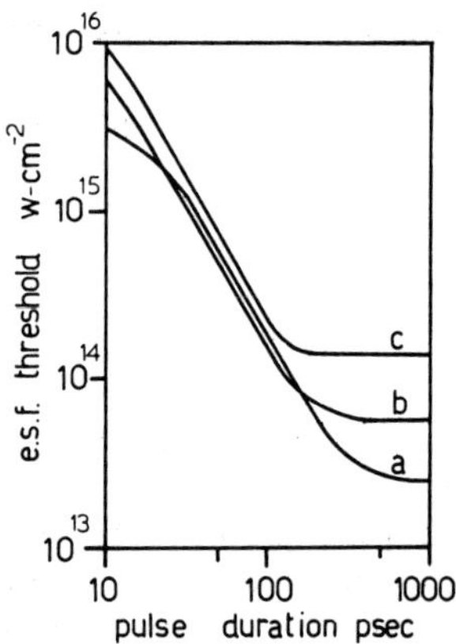

Fig.1. ESF threshold as a function of pulse duration for different
       degrees of ionization for an aluminium plasma, (a: $q = 1$;
       b: $q = 7$; c: $q = 13$), $N = 8 \times 10^{20}$ cm^{-3}, $T_e = 1$ keV, plasma
       thickness $= 180$ µm, beam diameter $= 36$ µm.

       Some of the results of the TRSF computer calculations are
summarised below [8].

(i) Transient Behaviour

       The minimum threshold for relativistic self-focussing were
found to be typically two to three orders of magnitude higher than
the minimum values for ponderomotive force self-focussing.
However, since it takes a finite time for the plasma to respond to
the ponderomotive force, the mechanism is fully effective only for
relatively long pulses (> 200 psec) [8]. In comparison the
relativistic mechanism acts essentially instantaneously and hence
for short duration pulses self-focussing is dominated by that
mechanism. The result is that the ESF threshold is dependent on
laser pulse duration as shown in Figure 1.   Typically the
steady-state conditions where the ponderomotive force mechanism
dominates are obtained for pulses longer than about 200 psec. In
the steady-state the plasma response time is short compared with
the laser pulse duration and the plasma density becomes reduced in
the vicinity of the laser beam according to the simple relationship:

$$N = N_o \exp \left\{ - \frac{q}{q+1} \frac{e^2}{4m\omega^2 kT_e} \langle E^2 \rangle \right\}$$

(2)

$$= N_o \exp\{- \beta \langle E^2 \rangle\}$$

where N_o is the initial plasma density. The refractive index profile in the plasma can then be calculated using the expression

$$\eta = \sqrt{1 - N/N_{cr}}$$

(3)

which taken in conjunction with the wave equation describing propagation of the E–M wave in the plasma can be used to calculate the ESF threshold. For the intermediate duration pulses the threshold rises with decreasing pulse duration (t_p) until for $t_p < 10$ psec a second pulse-duration-independent region is just apparent corresponding to dominance of relativistic self-focussing.

(ii) Absorption

In extended plasmas where the self-focussing lengths are large, absorption by collisional processes cannot be neglected and has a strong effect on the ESF thresholds. For a wide range of conditions the ESF threshold decreases with increasing plasma temperature (T_e) because of the reduced absorption levels. In contrast, for ponderomotive force self-focussing it is normally expected that as T_e increases the ESF threshold rises due to increased plasma pressure. Using TRSF it has been found that plasma pressure effects only dominate for rather high temperatures as shown in Figure 2.

Inverse Bremsstrahlung absorption also increases rapidly with increasing plasma density and hence the highest density plasmas do not necessarily have the lowest self-focussing thresholds as would be expected from refractive index considerations alone. In typical conditions the lowest thresholds occur for $N \simeq 0.3$-0.6 N_{cr} with only small changes of threshold occurring for densities between 0.05 and 0.85 N_{cr}.

(iii) Stability of the Beam to Filamentation

As would be expected from studies of self-focussing in solids and liquids the conditions in which whole beam self-focussing would

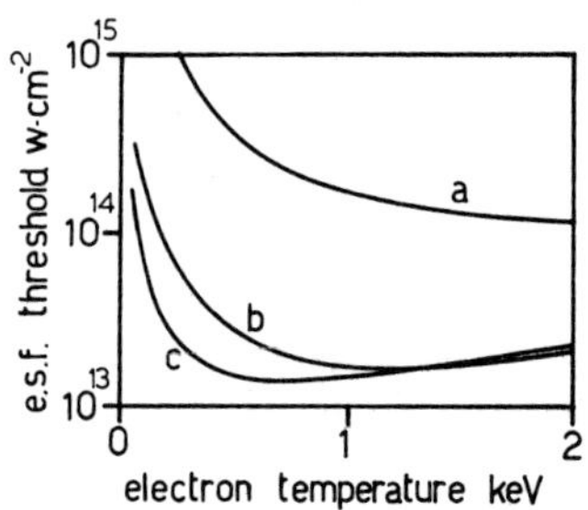

Fig.2. ESF threshold versus T_e and pulse duration (t_p) for a Deuterium plasma, (a: t_p = 20 psec; b: t_p = 100 psec; c: t_p = 5 nsec), N = 9 x 10^{20} cm^{-3}, plasma thickness = 180 µm, beam diameter = 36 µm.

be observed are rather restricted with the more likely occurrence being break-up of the beam into multiple filaments. Using the familiar results from instability theory of Bespalov and Talanov [21] a laser beam with average intensity I propagating in a non-linear medium with an intensity-dependent refractive index (η) given by the expression

$$\eta = \eta_o + \gamma I \qquad (4)$$

is unstable to break-up into multiple filaments due to the presence of small modulations of intensity on the beam. The growth rate for a ripple with a spatial frequency K is given by:

$$g = K \left\{ \frac{\gamma I}{\eta_o} - \frac{K^2 \lambda^2}{16 \pi^2} \right\}^{\frac{1}{2}} \qquad (5)$$

which has a maximum value at a spatial frequency

$$K_{max} = \frac{2\pi}{\lambda} \left\{ \frac{2\gamma I}{\eta_o} \right\}^{\frac{1}{2}} \qquad (6)$$

Filamentation would be expected for $K_{max} \gg 2\pi/D$, where D is the initial beam diameter. The existence of the optimum ripple diameter is simply a result of competition between self-focussing and diffraction.

For a laser beam undergoing self-focussing due to the ponderomotive force mechanism in steady-state conditions we may write the refractive index approximately as

$$\eta = \eta_o + \frac{N\beta\langle E^2\rangle}{2N_{cr}} = \eta_o + \gamma I \tag{7}$$

Considering the specific case of self-focussing in a Deuterium plasma with a temperature of 250 eV, density $0.3\ N_{cr}$ and thickness 100 μm and initial ripple amplitude $\Delta I/I = 0.01$ the variation of filamentation threshold with ripple size can be calculated. This is shown in Figure 3 which indicates the optimum ripple diameter is about 8-9 μm and this value decreases with increasing laser intensity in accordance with equation 6. Inserting the value for

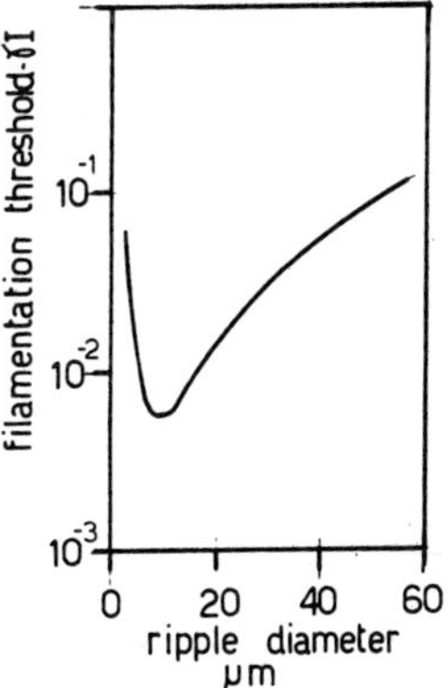

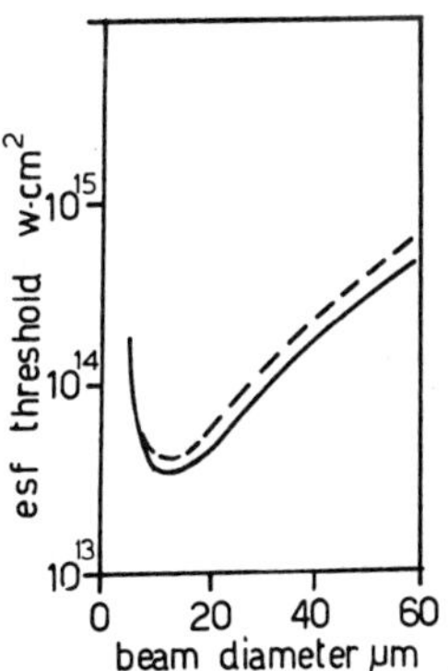

Fig.3. Steady state filamentation threshold versus ripple size for a Deuterium plasma. $N = 3 \times 10^{20}$ cm^{-3}, $T_e = 250$ eV, $\Delta I/I - 0.05$.

Fig.4 Steady state ESF threshold versus beam diameter for a Gaussian beam. Plasma parameters are as in Figure 3.

γ from equation 7 indicates that at intensities of 10^{15}–10^{16} W/cm^6 the optimum ripple diameter is only a few microns at hence above about 10^{15} W/cm^2 in plasmas with moderate temperatures we can expect a laser beam with a diameter of greater than about 20 microns to be unstable to filamentary break-up and whole beam self-focussing is unlikely to be observed.

As a further check on the beam size dependence of the self-focussing threshold TRSF has been used to calculate self-focussing thresholds for a Gaussian beam in a plasma with the same conditions as for Figure 3. The results are shown in Figure 4 which shows only minor differences from the form obtained from the simple instability theory. Figure 5 shows extensive calculations of ESF thresholds for steady-state ponderomotive force self-focussing. The region accessible for an experimental study corresponds to beam radii above 10 microns. From Figure 5 we can deduce that to avoid filamentation the propagation distance in the plasma must be large (at least 100 μm), the beam diameter as small as possible and the laser intensity very close to the threshold value since even small increases in laser intensity cause a rapid decrease in self-focussing length (plasma thickness) for the smaller sized beams. For each laser intensity there exists a particular optimum Gaussian beam radius (R_{opt}) and corresponding self-focussing length (Z_f) for which the ESF threshold is a minimum. We find empirical relationships for Z_f and R_{opt} as below.

$$Z_f = 10^{17} I^{-1.1}$$
$$R_{opt} = 1.7 \times 10^6 I^{-0.4} \tag{8}$$
$$Z_f, R_{opt} \ (\mu m) \ , \ I(W/cm^2)$$

[cf. Figure 1 in reference 22 where $Z_f \alpha I$ and $R_{opt} \alpha I^{-0.5}$].

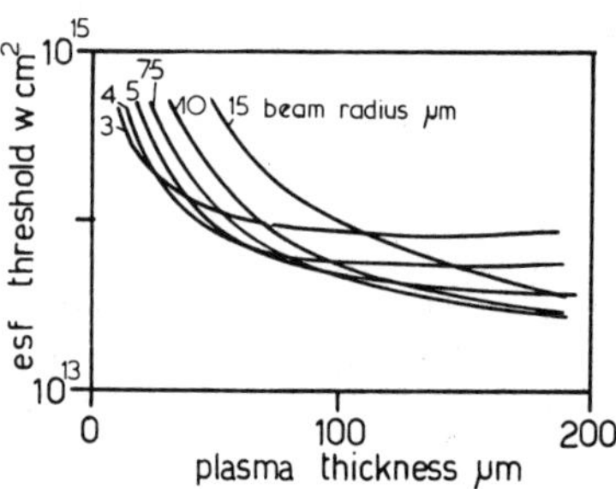

Fig.5. Steady-state ESF thresholds for Deuterium plasma as a function of plasma thickness and beam radius. T_e = 250 eV, N = 3 x 10^{20} cm^{-3}.

From the broad comments above a number of deductions can be made to determine the optimum conditions for whole beam self-focussing to be observed. The propagation distance in the plasma must be large (100–200 μm), the beam diameter small (10–20 μm) and the laser intensity rather close to the threshold value to avoid filamentation. The plasma temperature must be relatively high and the average plasma density (and atomic number) low to avoid threshold increases and reduction of effective plasma thickness due to absorption. The laser pulse duration should be greater than about 200 psec to avoid an increase in threshold due to transient effects and probably greater than several nsec to allow time for the plasma to expand into vacuum to provide the required plasma propagation length. Specific calculations for three different low–Z target materials are shown in Figure 6. Plasma densities were 0.6 N_{cr}, plasma thickness 180 μm and average ionization state q of the plasmas calculated from the empirical relationship given by Shearer and Barnes [42].

$$q = 26 \left\{ \frac{T_e}{1 + (26/Z)^2 T_e} \right\}^{\frac{1}{2}} \tag{9}$$

The electron temperature values required to obtain thresholds within a factor of two of the minimum values range from 100 to 750 eV for D_2 up to 300 to 5 keV for aluminium with predicted thresholds in the range $10^{13} - 2 \times 10^{14}$ W/cm^2.

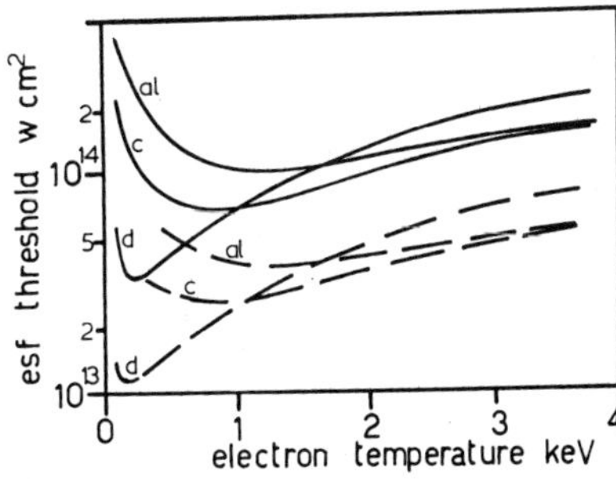

Fig.6. Steady-state ESF thresholds versus target material, plasma temperature and beam diameter, D (D = 36 μm solid line, D = 18 μm hatched line), N = 6 x 10^{20}, plasma thickness = 180 μm.

2. Experimental Studies of Self-focussing

An experiment was established in order to try and achieve the conditions required for self-focussing as predicted in Figure 6. Laser pulses were supplied by the ANU SPL system using a Q-switched Nd:YAG oscillator producing about 3 joules in a 5 nsec duration pulse (FWHM). The pulses were focussed into an evacuated target chamber using an F = 1.5 aspheric lens and the intensity distribution within the focal region measured by ablating a thin metal film off a glass substrate with the attenuated beam at various different power levels. In the focus the beam was approximately Gaussian with a diameter of 13±3 μm, leading to peak on-axis beam intensities of about 2×10^{14} W/cm^2.

The basic apparatus is shown in Figure 7 and the following diagnostics were used:

(1) A six channel X-ray spectrometer using P-1-N diode detectors for determination of T_e.

(2) A low resolution (20 μm) twin pin-hole X-ray camera for determination of the time average axial variation of plasma density and temperature using the two foil technique.

(3) A high resolution (3 μm or 10 μm) X-ray pin-hole camera.

(4) Various diagnostics for studying the light transmitted through the targets.

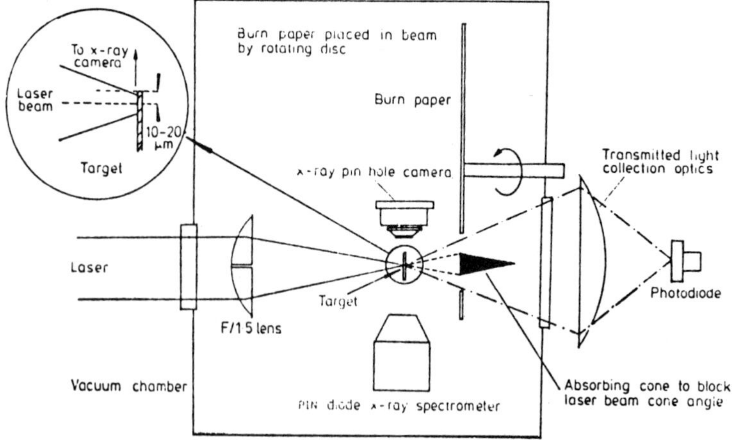

Fig. 7. Experimental apparatus used to study self-focussing.

It was hoped to obtain indications of self-focussing using two diagnostics. Firstly, by monitoring the laser light transmitted through the plasma, self-focussing would be expected to be detectable by an increase in the divergence of the transmitted light in a similar way to self-focussing experiments in gases [17]. Secondly, high resolution X-ray images of the plasmas could be expected to show hot spot of filamentary structures if self-focussing occurred. In order to use the transmitted light diagnostic thin foil targets were used (5μm aluminium) which became underdense and showed large transmission (60-70%) for the latter half of the pulse at intensities above about 2-3 x 10^{13} W/cm^2.

Measurements of the spatial and laser intensity dependence of the plasma density and temperature showed that the plasmas produced had approximately the right paramters for self-focussing. The plasma temperatures were in the range of 300 eV at intensities above 3 x 10^{13} W/cm^2 and the plasma density dropped with a scale length of 50-60 μm (to half density) in front and behind the target, with the plasma expansion being approximately isothermal. The large scale length provided useful propagation distance of 100-200 μm as required.

3.   Evidence of Self-focussing

At low intensities 7 x 10^{13} W/cm^2 the high resolution X-ray images of the plasmas showed little evidence of substructure, the emission being relatively uniform over the emitting region in front and behind the target. At intensities above 10^{14} W/cm^2, however, filamentary structures were observed in about 45% of laser shots with the remaining 55% showing confused 'hot spot' structures but no distinct filaments. Two examples of the phenomenon are shown in Figure 8. The laser beam was incident from the right and the target position is evident by the dark 'waist' region. The filaments were evident emerging from the over-exposed region in front of the target, visible in the region partially obscured by the target edge and passed through the plasma at the rear. The filaments were about the same width (10 μm and 6 μm for Figure 8) for a distance of greater than 60 μm, the width being close to the camera resolution in both cases.

Microdensitometer traces show the presence of the filament which is clearly evident in the Abel inversions in Figure 9. At highest resolution the filament appeared 'hollow' as shown in Figures 9.2a and 9.2c. We suggest that these structures are caused by self-focussing of part of the incoming laser beam, the filament first forming in the high density region 20-40 μm in front of the target and propagating for a distance of about 60-100 μm before breaking up in the region of decreasing plasma density behind the target. Depletion of the plasma density on the filament axis due to the strong radial ponderomotive forces combined with local

(a)

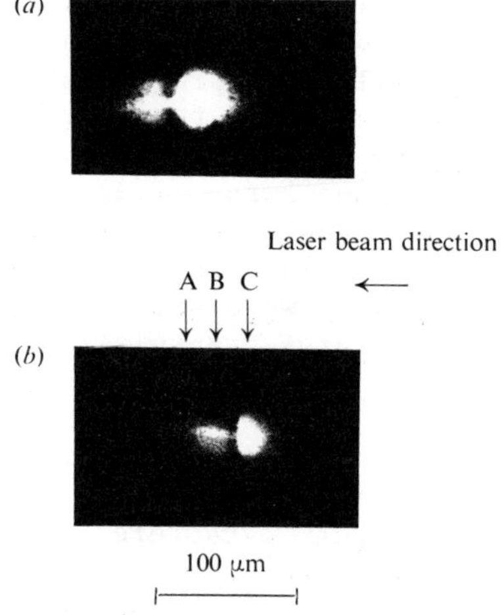

Fig. 8. X-ray photographs obtained through 1.5 μm thick aluminium foils showing evidence of self-focussing. Spatial resolution was 10 μm in (a) and 4 μm in (b). The laser beam was incident from the right.

heating due to concentration of laser energy in the filament could result in X-ray emission as shown in Figures 8 and 9 [23,24]. In accordance with this model tests also showed that the filament was hotter than the surrounding plasma.

The transmitted light diagnostics also provided indications of self-focussing. At intensities above about 5×10^{13} W/cm^2 the transmitted light showed increased divergence, and a smoother 'Gaussian' profile in comparison with the light distribution at lower intensities and with no target. Measurements with 'burn paper' were consistent with an abrupt increase in transmitted beam divergence over the value with no target present at an intensity of 5×10^{13}W/cm^2. Since at these higher intensities the transmitted light was close to Gaussian the waist diameter in the plasma could be estimated from the beam divergence measurements – a value of 3–4 μm was obtained – a factor 3–5 smaller than the measured laser beam diameter.

In summary having established laser and plasma parameters close to the values predicted for self-focussing using TRSF we have observed behaviour in our experiments which are consistent with self-focussing having occurred. The observation of threshold behaviour, hot X-ray filaments and increased transmitted beam divergence are perhaps not entirely unambiguous but suggest that self-focussing may occur when long laser pulses propagate in extended plasmas.

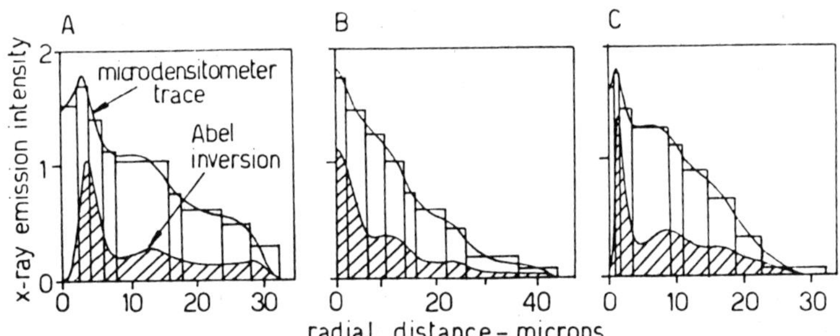

Fig.9. Microdensitometer traces and corresponding Abel inversions
of radial emission intensity distributions from the scans
of Figure 8b in positions A, B, and C.

(b)   ACCELERATION OF ELECTRONS PRODUCED BY IONIZATION OF GASES

The radial component of the non-linear force which is
responsible for ponderomotive force self-focussing has been studied
in an experimental situation which is much simpler than that
encountered in experiments in high density plasmas where many other
competing effects are operative.

This situation is one where a low pressure gas is ionized by
intense laser radiation to form a limited volume containing charged
particles which are then accelerated by the non-linear force out of
the laser beam. At low gas pressures collective effects can be
neglected and hence the energy gained by the electrons and ions is
essentially determined by a single particle/field interaction. By
careful measurement of the emitted energy spectrum of the particles
and the various laser parameters the validity of equation 1 can be
verified.

Typical values of electron energies obtained by non-linear
force acceleration can be calculated by integration of equation 1
over the time and space history of the local field intensity
experienced by the charged particle. Two limiting cases can be
treated [25]. Firstly, the so-called high intensity limit where
the particle moves out of the beam in a time short compared with
the laser pulse duration, here the energy is simply equal to the
work done in removing it from the beam, ie.

$$\varepsilon = \int_{R}^{\infty} f_{NL}\,dr \tag{10}$$

where the particle is initially of radial position R. For
neodymium laser radiation the electron energy is

$$\varepsilon_e = 10^{-13}\,I(R) \tag{11}$$

In the second case, the low intensity limit, the particle receives
an impulse but does not move significantly out of the beam during
the pulse and the electron energy is given by

$$\varepsilon_e = \int_{-\infty}^{\infty} f_{NL} \, dt \tag{12}$$

The transition between the two regimes occurs when $r_o/v_e \simeq t_p$ with
r_o the beam radius, v_e the electron velocity and t_p the laser pulse
duration. For an electron energy of 1 eV, and $t_p = 25$ psec, $r_o \simeq$
15 μm the transition from the low to high intensity limit occurs
for $I \simeq 10^{13}$ W/cm^2. Large electron energies of 100–1000 eV can be
expected for peak laser intensities in the region of 10^{15}–10^{16}
W/cm^2 with the form of the spectrum related to the intensity
distribution of focussed laser beam by equation 11.

A test of the validity of equation 1 in predicting the
magnitude of the non-linear force was therefore obtained by
measurement of the electron energy spectrum from laser induced
ionization of gases at high intensities ($> 10^{14}$ W/cm^2) where
acceleration by the non-linear force would be the only mechanism
capable of producing high electron energies.

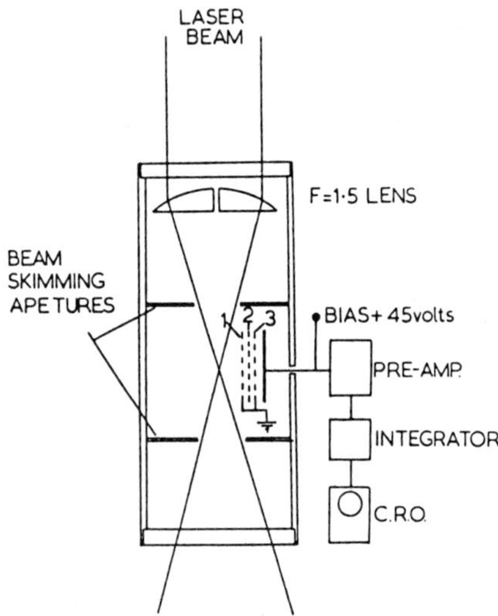

Fig. 10. Experimental apparatus for measurement of electron energies
emitted in ionization of gases. Grid 2 of the retarding
grid analyser was biassed negatively to prevent electrons
with energies less than the retarding voltage from reaching
the collector.

The apparatus is shown in Figure 10. Single pulses from the ANU SPL laser using a passively mode-locked Nd:YAG oscillator were focussed into low pressure helium gas ($\sim 10^{-4}$ torr, a value chosen to avoid collective effects) using an aspheric F = 1.5 lens. Peak powers of up to 70 GW were available focussed to a spot diameter of 13 ± 3 μm. The experiment described here was performed at lower power with a maximum focussed beam intensity of 3 ± 1 x 10^{15} W/cm^2 in a 25 psec duration pulse.

Electron energies were measured using the retarding field electron energy analyser. Helium gas was chosen because of its low atomic number and high ionization potential hence only first ionization state electrons would be expected to be generated at this intensity. Using the Keldysh [26] formula for tunnelling the calculated ionization threshold was 3 x 10^{14} W/cm^2. The intensity distribution within the focal volume was carefully mapped to allow I(R) to be determined accurately and hence the expected spectrum of energies calculated. Experimental values of electron energy and the calculated integrated energy spectrum using the I(R) data and equation 11 are shown in Figure 11. Clearly a very close correlation was obtained confirming the correct magnitude and existence of the radial non-linear force component [27].

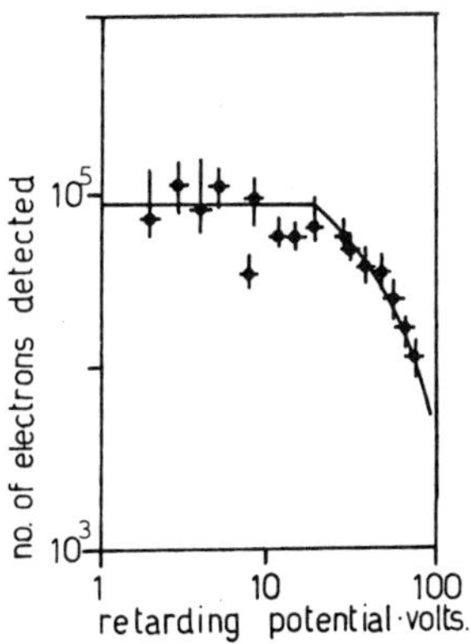

Fig. 11. The variation of the number of electrons detected as a function of the retarding potential for Helium at 10^{-4} torr. The solid curve represents the calculated 'spectrum' for electron acceleration by the non-linear force.

(c) PROFILE MODIFICATION

Recently there have been several measurements of density distributions within laser-produced plasmas using interferometry which have shown that the density scale length in the vicinity of the critical density surface is much smaller than expected for hydrodynamic expansion of the plasma [28-31]. The explanation of these observations involves modification of the density profiles by non-linear forces associated with the strong electric field gradients in that region. Such profile modification has been demonstrated in computer simulations where density jumps from sub- to super-critical values in only a few tens of Debye lengths are predicted. The best spatial resolution obtained using interferometry is about 1 micron and this value is insufficient to determine whether the sub-micron scale lengths predicted in the computer simulations do occur in an experimental situation.

An important prediction of the model of profile modification by the non-linear force is that the plasma density scale length should decrease with increasing laser intensity. This behaviour affects the temperature of resonantly-heated suprathermal electrons within the plasma since according to Friedberg et al [32] the average energy of these electrons $\langle \varepsilon \rangle$ is given by:

$$\langle \varepsilon \rangle = (m_e \omega_o \phi_{abs} L/N_{cr})^{\frac{1}{2}} \tag{13}$$

and is sensitive, therefore, to laser intensity both through ϕ_{abs}, the absorbed laser flux density, and L, the density scale length. The intensity dependence of $\langle \varepsilon \rangle$ has been deduced from high energy X-ray Bremsstrahlung emission from laser-produced plasmas [11,33] and is in good agreement with the functional dependence obtained in simulations of resonance absorption in a profile steepened plasma reported by several workers but departs significantly from the $\sqrt{I}$ variation expected from equation 13.

We have established that resonance absorption can be detected in plasmas generated by focussing 25 psec duration pulses from a Neodymium laser onto planar targets by measuring the angle and polarisation dependence of the high energy X-ray Bremsstrahlung emission [34]. Similar resonant behaviour has also been reported in measurements of the electron temperature [35,36], the plasma reflectivity [37] and plasma absorption [38]. According to Ginzberg [39] and Friedberg et al. [32] this resonance angle can be related to the plasma density scale length L by the expression:

$$Sin\Theta_m \simeq 0.8(k_o L)^{-\frac{1}{3}} \tag{14}$$

for $k_o L > 1$.

In principle, therefore, measurement of Θ_m gives a measure of the scale length L. Some variation in Θ_m can be found from values reported in the literature with an indication that Θ_m increases with increasing laser intensity, corresponding to a reduction in L. However, since the experiments were performed in widely varying conditions it is difficult to make a meaningful comparison of the data.

In our experiments we have concentrated on the determination of Θ_m from resonance in high energy X-ray Bremsstrahlung emission from polished, planar targets. The 25 psec duration laser pulses were focussed onto either planar 50μm thick copper targets or 30μm aluminium using F = 2 aspheric optics. Target surfaces were hand polished but generally contained some micron-sized surface scratches within the beam area. The X-ray emission was monitored using a seven-channel silicon p-i-n diode K-edge spectrometer with channels spanning the spectral range from 1.4 to 56 keV.   The individual channel signals were recorded and the data processed by computer by modelling the emission by a two temperature Maxwellian distribution. Estimates of the plasma temperature and suprathermal electron temperature were obtained from the slopes of the spectrum in high and low photon energy regions. The detector array viewed the target at 90° to the incoming beam direction along its electric field vector.   Measurements were made to check that the X-ray emission was isotopic particularly in the high energy channels.

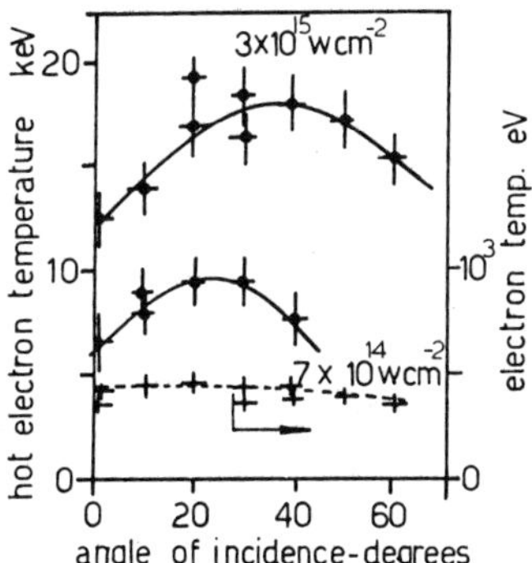

Fig. 12. Two examples of resonance behaviour in T_{hot} versus angle of incidence for p-polarised light incident upon 30 μm thick aluminium targets. Also shown are values for T_e.

Laser energy and pulse duration were monitored for each laser shot and the laser spot diameter was 35±5 μm in the focal plane and, therefore, intensities between 3×10^{14} W/cm^2 and 10^{16} W/cm^2 were available. A search was made for resonance behaviour in the suprathermal electron temperature (T_{hot}) and electron temperature (T_e) to determine Θ_m for several different laser intensities. At large angles of incidence slight increases were made in the laser power to hold the on-target intensities constant.

Two examples of resonance behaviour in T_{hot} are shown in Figure 12 for intensities of 7×10^{14} W/cm^2 and 3×10^{15} W/cm^2 respectively. A clear difference in Θ_m is apparent with Θ_m increasing from 25° to 35° with the increase in laser intensity. This implies from equation 14 a reduction in L from $\sim$ 1 μm down to $\sim$ 0.4 μm. The maximum values of the hot electron temperature were 10 keV and 18 keV in the two cases whilst the electron temperature remained constant at 400 ±50 eV. No resonance behaviour was detected in T_e in contrast to results published by other workers. We note that in comparison with the data from Balmer and Donaldson our data would give an apparent resonant rise would electron temperature if a two foil technique was used to measure T_e since the suprathermal tail affected the spectra down to photon energies in the 5-10 keV range. Electron temperature values were found to be essentially independent of laser intensity throughout the whole range of measurements above $\sim$ 2×10^{14} W/cm^2.

The plasma scale lengths implied from equation 14 are plotted as a function of laser intensity in Figure 13 together with values deduced from data from other workers. The curve shows that the

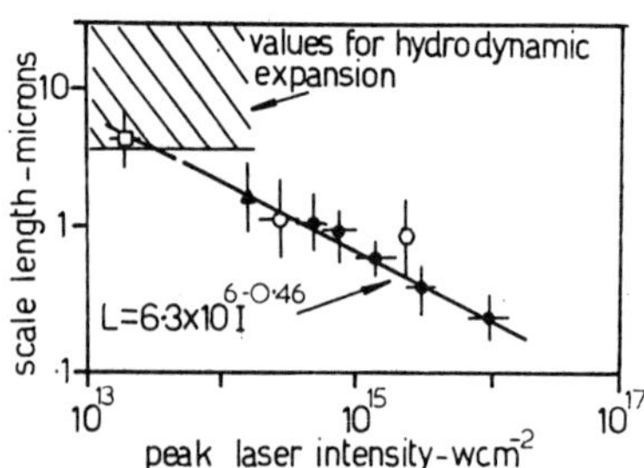

Fig. 13. Inferred values of the plasma density scale length calculated from the resonance angle data. Values from other workers are also shown as follows: □ ref 36; Δ ref 35, ○ ref 37; ⬡ ref 38.

scale length decreases monotonically with increasing laser intensity and is best represented by the relationship:

$$L(\mu m) = 6.3 \times 10^6 \ I(W/cm^2)^{-.45 \pm .05} \tag{15}$$

Using the measured value of T_e this result can be compared with the results of computer simulations by Estabrook and Kruer [11]. Inserting the value for the Debye length into their expression gives $L = 3 \times 10^6 \ I^{-.48 \pm .07}$ which only differs from the experimental result by a factor of two numerically possibly due to inherent time and space averaging in the experimental situation.

SUMMARY AND CONCLUSIONS

Three different experiments have been described which provide indications of the importance of non-linear forces in laser plasma interactions in widely different laser and plasma conditions. Strong evidence of ponderomotive force induced self-focussing of long laser pulses in extended plasmas has been obtained. Further experiments are planned to search for self-focussing in the transient regime and an investigation of filamentation seem worthwhile as well as a search for correlations between other plasma measurements (eg. T_{hot}) and the observation of self-focussing. Existence of the radial component of the non-linear force has been confirmed from the experiments on gases. This type of experiment has subsequently proved useful in determining the ionization thresholds for multiply-ionized gas species as a means of making tests of the validity of the theoretical models for the ionization process at high laser intensities [40,41].

Finally, resonance behaviour in the suprathermal X-ray emission observed for short duration pulses has also provided good support to the concept of profile modification by the non-linear force. Confirmation of the scale length reduction with increasing intensity awaits studies with very high resolution interferometers.

ACKNOWLEDGEMENTS

The authors would like to thank Professor H Hora, Professor R Mavaddat and Dr J L Hughes for valuable discussions of this work and also all members of our technical staff who contributed to the projects.

REFERENCES

1. See for example, H Hora, Phys Fluids 12, 182 (1969).
2. E J Valeo, K G Estabrook, Phys Rev Letts 34, 1008 (1978).
3. C Randell, J S DeGroot, Phys Rev Letts 42, 179 (1979).
4. H Hora, Z Physik 226, 156 (1969).
5. P Kaw, G Schmidt, T Wilcox, Phys Fluids 16, 1522 (1973).
6. C E Max, J Arons, Phys Rev Letts 33, 209 (1974).
7. M R Siegrist, Optics Comms 16, 402 (1976).
8. V Del Pizzo, B Luther-Davies, M R Siegrist, Appl Phys 14, 381 (1977), Appl Phys 18, 199 (1979).
9. D W Forslund, J M Kindel, K Lee, E L Lindman, R L Morse, Phys Rev A11, 679 (1975).
10. K Estabrook, E J Valeo, W L Kruer, Phys Fluids 18, 1151 (1975).
11. K Estabrook, W L Kruer, Phys Rev Letts 40, 42 (1978).
12. P Dragila, J Krepelka, Jrnl de Physique 39, 617 (1878).
13. K Baumgartel, K Sauer, Phys Letts 70A, 107 (1979).
14. C Yamanaka, T Yamanaka, J Mizui, N Yamaguchi, Phys Rev A11, 2138 (1973).
15. A J Alcock, C DeMichelis, V V Korobkin, M C Richardson, Appl Phys Letts 14, 145 (1969).
16. R A Haas, M J Boyle, K R Manes, J E Swain, J Appl Phys 47, 1318 (1976).
17. A J Alcock, in Laser Interaction and Related Plasma Phenomena, Vol 2, 155, Plenum Press, New York (1972).
18. T P Donaldson, I J Spalding, Phys Rev Letts 36, 467 (1976).
19. M R Siegrist, B Luther-Davies, J L Hughes, Optics Comms 18, 603 (1976).
20. H Hora, J Opt Soc Am 65, 882 (1975).
21. Y I Bespalov, V J Talanov, JETP Letts 3, 307 (1966).
22. J W Shearer, J L Eddleman, Phys Fluids 16, 1753 (1973).
23. V Del Pizzo, B Luther-Davies, J Phys D 12, 1261 (1979).
24. M R Siegrist, J Appl Phys 48, 1378 (1977).
25. M J Hollis, Optics Comms 25, 395 (1978).
26. L V Keldysh, Sov Phys JETP 20, 1307 (1966).
27. B W Boreham, B Luther-Davies, J Appl Phys 50, 2533 (1979).
28. R Fedosejevs, I V Tomov, N H Burnett, G D Enright, M C Richardson, Phys Rev Letts 39, 932 (1977).
29. H Azechi, S Oda, K Tanaka, T Norimatsu, T Sasaki, T Yamanaka, C Yamanaka, Phys Rev Letts 39, 1144 (1977).
30. D T Attwood, D W Sweeney, J M Auerbach, P H Y Lee, Phys Rev Letts 40, 184 (1978).
31. A Raven, D Willi, Phys Rev Letts 43, 278 (1979).
32. J P Friedberg, R N Mitchell, R L Morse, L I Rudsinski, Phys Rev Letts 28, 795 (1972).
33. B Luther-Davies, Optics Comms 23, 98 (1977).
34. B Luther-Davies, Appl Phys Letts 32, 309 (1978).
35. J L Pearlman, M K Matzen, Phys Rev Letts 39, 140 (1978).

35.   J L Pearlman, M K Matzen, Phys Rev Letts 39, 140 (1978).

36.   J E Balmer, T P Donaldson, Phys Rev Letts 39, 1084 (1978).

37.   R P Godwin, R Sachsenmaier, R Sigel, Phys Rev Letts 39, 1198 (1977).

38.   K R Manes, V C Rupert, J M Auerbach, P Lee, J E Swain, Phys Rev Letts 39, 281 (1977).

39.   V L Ginzberg, in The Propagation of Electromagnetic Waves in Plasmas, Pergamon Press, New York (1959).

40.   B W Boreham, Europhysics Conference on Multiphoton Processes, Benodet, France, June 1979.

41.   K G H Baldwin, Laser Induced Ionization of Argon, MSc Thesis, The Australian National University, 1979.

42.   J W Shearer, W S Barnes in "Laser Interaction and Related Plasma Phenomena,' Vol 1, p 314 (Plenum Press, New York 1971).

RECENT STUDIES AT INRS-ENERGIE ON CO_2 LASER

BEAM INTERACTION WITH PLANE TARGETS

T. W. Johnston, H. Pépin, F. Martin, B. Grek, P. Church,
J. Geoffrion, J. C. Kieffer, J. P. Matte, G.R. Mitchel

INRS-Energie
Varennes, Québec, Canada, JOL 2PO

and

R. Décoste

Institut de Recherche de l'Hydro-Québec
Varennes, Québec, Canada, JOL 2PO

At INRS-Energie we have been studying laser-target interaction
with short pulses of 10 μm radiation incident on various large
targets[1], with a special interest in the transition from modest to
strong profile modification occurring at $I\lambda^2$ values of about
10^{15} watt/cm^2 × (μm)^2.

EXPERIMENTAL

The laser can provide up to 10 J in a pulse of duration 1.2 to
1.5 nsec, focused to produce a time-space average of 3×10^{13} watts/
cm^2 inside the 130 μm diameter half-energy area. The targets are
ribbons of various kinds, polyethylene, several metals (e.g. aluminum, lead) usually 125 μm thick, and thin films of polyethylene,
either on a glass substrate (microscope slide)(with polyethylene
thicknesses down to 0.04 μm) or free-standing (minimum thickness
0.08 μ). The 7.5cm diameter vertical laser beam is focused onto the
horizontal targets with a vertical off-axis f/1.5 parabola and an
effective angle of incidence of 28°, in S polarization.

Low-energy X-rays (1–10 keV) are detected by PIN Si diodes.
The PIN detectors have a 250 μm active thickness. Filters providing
various X-ray energy responses are placed at the end of a shielded
tube 10 cm in front of the active Si layer. The diameter of the

aperture placed in front of the filter and the distance to the plasma
specify the detector solid angle. The whole assembly is covered with
lead. Electron sweeping magnets, consisting of two horseshoe magnets
positioned along the detector axis cm ahead of the aperture, provided
a field of 600 G for a distance of about 38 mm. Along the axis directly normal to the target it was necessary to use three such magnets to
eliminate the electrons.

High-energy X-rays (10–50 keV) are detected by a combination of
a 1 mm thick NaI scintillator and a photomultiplier tube (RCA 8575).
High-Z X-ray absorption filters are located 10 cm in front of the
NaI crystal. A thick lead aperture is used to define the solid
angles. It is very important to shield the aluminum cylinder containing the whole system with a thick layer of lead, particularly in the
area around the scintillator. Electron sweeping magnets are also
used 10 cm in front of the absorption filters.

The various filters used with the corresponding detector are
presented in Table I.

An extended dynamic range for the collection of the X-ray data
is ensured by the use of variable apertures in front of the X-ray
filters and of neutral density filters placed on the front end of the
photomultipliers. The scintillator-photomultiplier combinations
were calibrated with radioactive sources (Cd and Am).

In order to measure the X-ray spectrum from 1 to 50 keV, eight
detectors were used simultaneously: six diodes and three scintilla-
tor-PM assemblies. The six silicon diodes were arranged in two
groups of three. The detectors viewed the plasma at an angle of
45° with respect to the target normal.

Table 1.   Characteristics of the Multichannel Spectrometer

Channel	Detector	Filter	Energy Range (keV)
1	Si diode 250 μm	25 μm Be	1–1.5
2	Si diode 250 μm	100 μm Be	1.4–2.1
3	Si diode 250 μm	25 μm Ti	3–5
4	Si diode 250 μm	25 μm Ni	6–8.3
5	Si diode 250 μm	50 μm Zn	7–9.6
6	NaI 1 mm	100 μm Mo	12–20
7	NaI 1 mm	250 μm Sn	20–30
8	NaI 1 mm	650 μm Ga	35–50

<u>Interferograms</u> were taken with a 20 psec 15 mJ ruby laser pulse that was synchronized to the CO$_2$ laser through the use of active mode lockers, driven from a common source[2]. The jitter between the two lasers is less than we can measure and certainly less than 200 psec. This well-synchronized ruby pulse, that is considerably shorter than the main CO$_2$ pulse, allows us to measure the temporal evolution of the plasma. The interferometer used is a grating interferometer chosen for ease of alignment in nearly white light. (The probe laser correlation time may well be less than 3 psec.)

The method of <u>recording images at 10.6 μm</u>[3] is a two-step process whereby the infrared radiation is used to temporarily sensitize silver halide film to a subsequent pulse of visible light that actually exposes the film. The dynamic range of the technique is at least 40 (50 mJ/cm^2 for nanosecond 10 μm pulses). High-contrast, fine-grain films can be chosen such that the spatial resolution is limited only by diffraction effects at the infrared wavelengths.

RESULTS

The basic overall result is that we see the onset of a change in behaviour of many of the diagnostic signals (or parameters deduced from them) at irradiance somewhat less than 10^{13} watts/ sq. cm, i.e. I$\lambda^2 \sim 5 \times 10^{14}$ watts/cm$^2 \times (\mu$m$)^2$. This is the intensity range where v_{osc}/v_e (the ratio of oscillating electron velocity in the transverse field of the laser to the cold thermal electron velocity) goes from being definitely less than one to definitely more than one, i.e. from moderate profile modification, due only to resonance fields, to strong profile modification, due to both resonance fields and electromagnetic fields.

Let us examine the various pieces of evidence.

The most direct (but the most difficult to use for numerical results) is interferometry. At 4×10^{12} W/cm^2 at 1.3 nsec after the beginning of the laser pulse we have what can be termed a "normal interferogram," Fig. 1. At 3×10^{13} W/cm^2, with the same timing, we have clear evidence in an interferogram (see Fig. 2) indicating profile modification so strong that we cannot Abel-invert the interferogram with confidence, since varying the smoothing procedure used results in large changes in the electron density in the middle. Nonetheless it is clear that the profile modification is very strong in the Abel inversions of Figs. 1 and 2, as presented in Figs. 3 and 4 respectively.

Energetic electrons are produced in significant amounts as laser pulse irradiance is increased above 5×10^{12} W/cm^2. This is inferred from the change in the signal of the lowest energy X-ray channel (25 μm Be) when the permanent magnets which were mentioned above and which protect against energetic electrons, are removed[1].

INTERFEROGRAMS FOR I (CO_2) INTENSITIES

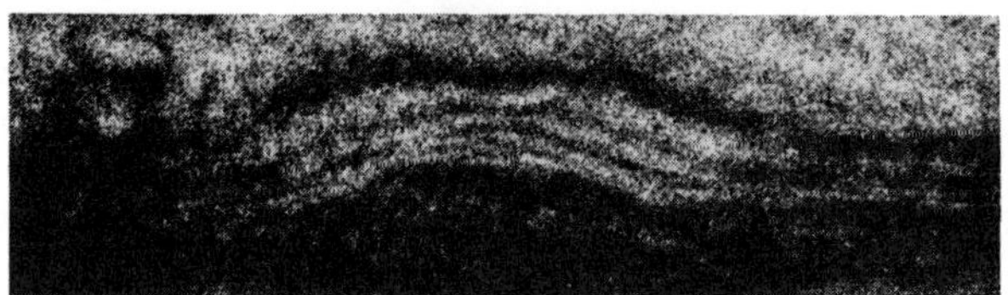

Fig. 3.   Abel inversion of Fig. 1.

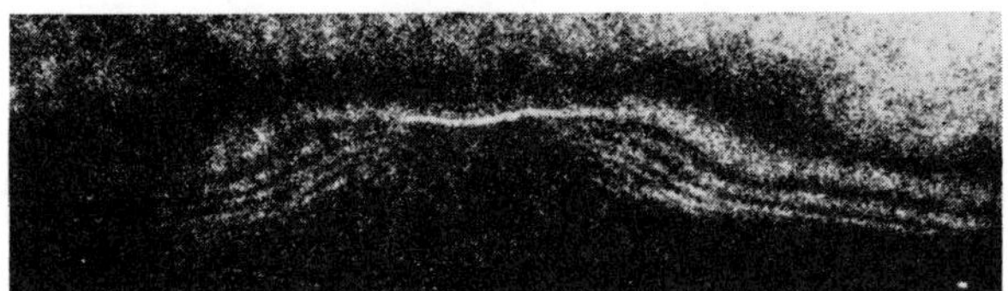

Fig. 2.   (3×10^{13} W/cm^2).

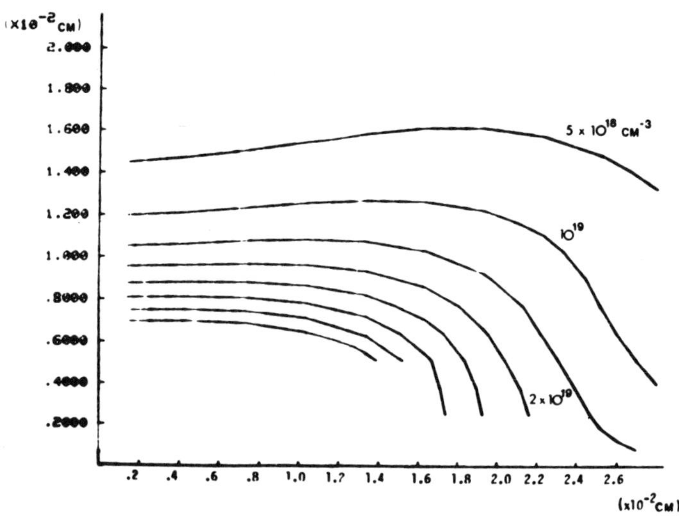

Fig. 1.   (4×10^{12} W/cm^2).

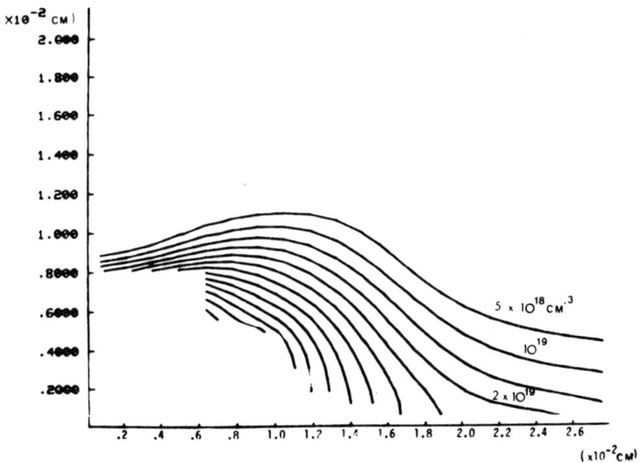

Fig. 4.   Abel inversion of Fig. 2.

At lower laser energies no signal change is seen when the magnets are removed, while at higher energies the ratio between the signals with and without increases rapidly (I ≃ 5 x 10^{12}).

The power law index for soft X-ray intensity, (25 μm Be filter channel) versus laser energy changes suddenly (from 2.25 to 1.2) as the flux passes the critical flux value (Fig. 5).

While low energy X-rays ·(25 μm Be filter channel) with a polar angle with respect to the target normal greater than 40° are hardly affected, at near-normal angles (θ < 40°) a large increase in soft X-ray anisotropy is seen as the critical intensity level is passed. The low energy X-ray emission then becomes strongly peaked in the target normal direction.

The power law index for the hot temperature (inferred for the hot electrons from the higher X-ray channels) decreases suddenly from 0.6 ∿ 0.7 to 0.22 ∿ 0.35 as the laser intensity passes the critical value (Fig. 6), in agreement with results from many other laboratories[4,5].

Times of arrival of ion peaks decrease suddenly and significantly, indicating a sudden increase in ion average drift velocity, again at the critical intensity value.

In general, the 10 μm backscatter image (formed as discussed elsewhere[3]) begins to show an increasing number of spots as the intensity increases. (These spots seem to have phase structure and may be linked to filamentation.) While there is no clear-cut threshold, spots are not seen below 5 x 10^{12} watts/cm^2 (Fig. 7).

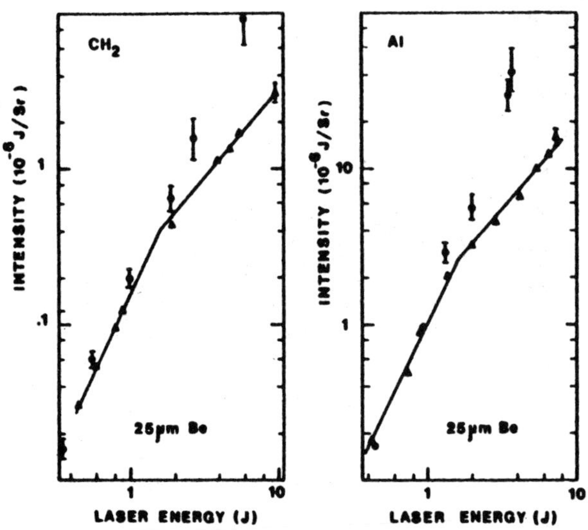

Fig. 5.   X-ray intensity (through 25 µm Be filter) versus laser energy for CH_2 and Al targets.   (▲ magnet in front, • no magnet).

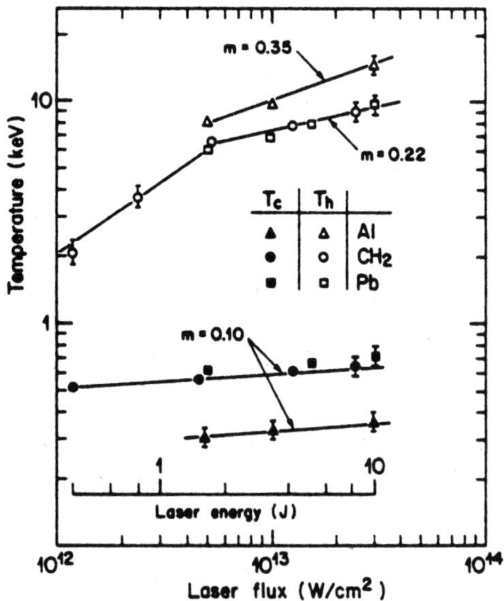

Fig. 6.   Cold and hot temperatures inferred from X-ray spectra versus laser intensity and energy.

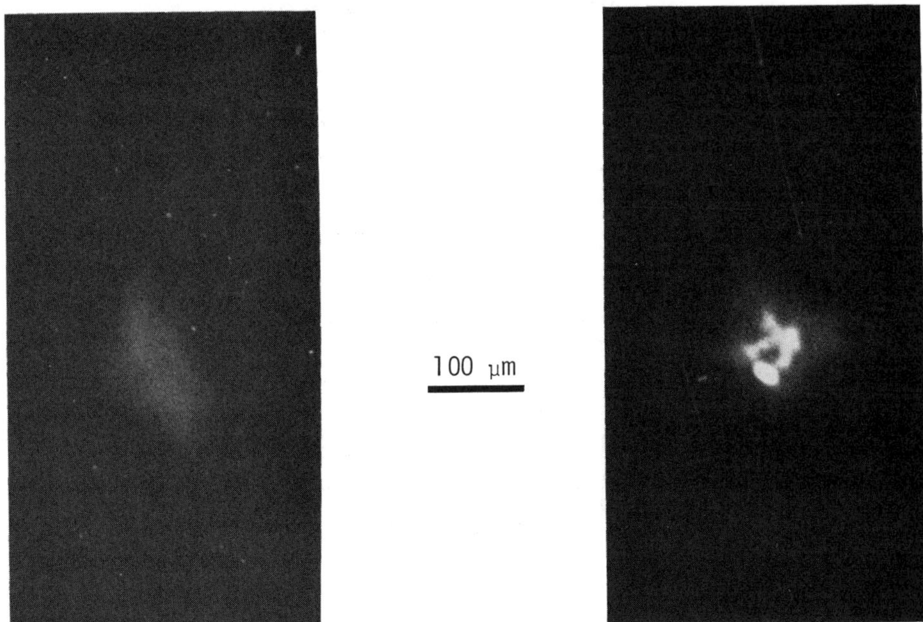

Fig. 7. Backscatter images of 10.6 µm light at mean intensities of (left) 7.5×10^{12} W/cm^2, (right) 1.5×10^{13} W/cm^2.

Finally there is the evidence given by the analysis of the X-ray yield from targets consisting of polyethylene layers on a glass substrate (a microscope slide). The inferred penetration depth of the relatively cold electrons appears to undergo a <u>very significant change</u> near 2J (I $\sim 6 \times 10^{12}$ watts/cm^{-2}), which indicates strongly the presence of inhibition of electron energy flow into the target. Since this data presentation is novel, we will now discuss this result in more detail. The basic concept, following Young et al.[6], is the use of a substrate of significantly higher Z than polyethylene, in this case glass ($\sim$ SiO$_2$) and the study of the <u>ratio</u> of X-ray intensity in a detector system to that of the substrate alone as target, as a function of the thickness of a polyethylene layer on the substrate and as a function of the energy of the laser pulse. Fig. 8 shows typical behaviour of this ratio as the polyethylene layer thickness is increased for a given laser pulse energy. The X-ray intensity ratio decreases with polyethylene thickness until it reaches a value after which any rate of decrease with thickness is much smaller. The curves demonstrate a well-developed knee (resembling two straightline segments more than, say, an exponential transition). The value of polyethylene thickness at which this knee occurs is what we call the <u>penetration depth</u>. For thicknesses greater than this value, little of the X-ray emission

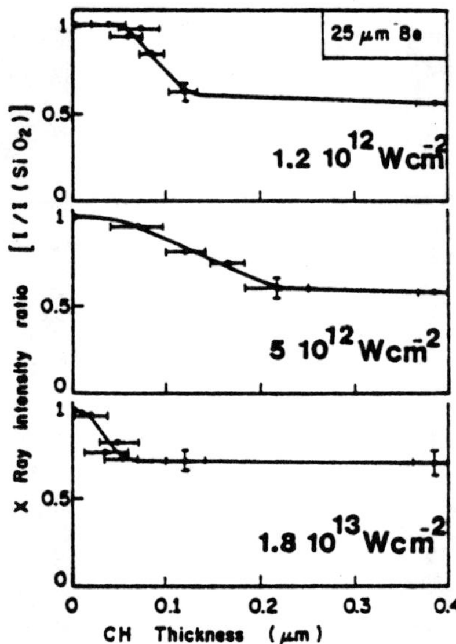

Fig. 8.   X-ray emission ratios (with substrate/without) for the
          1-1.5 kV channel as a function of CH thickness
          (up to 0.6 μm).

in the detector is coming from the substrate. The penetration depth
associated with the lower X-ray energies (and hence related to lower
energy electrons) increases with laser irradiance up to 5×10^{12}
watts/cm^2, at which point it decreases abruptly, from 0.3 μm to
less than 0.1 μm. This behaviour is plotted in Fig. 9a. Given that
the lower electron temperature T_c hardly changes with laser irradiance, the implication of further strong reduction of transport from
the interaction region by factors of 3 or 4 seems inescapable.

There is another way in which this behaviour can be verified.
That is by using no substrate and plotting the ratio with a given
thickness to that of a thick film at the same laser pulse irradiance.
However, the results of foil experiments should be taken with caution,
since plasma expansion before the foil becomes subcritical may
affect the values measured. Still, there is fairly good agreement
between the behaviour of the penetration depth obtained from layered
targets and the burnthrough depth inferred from foils.

Closer study of thick films revealed that the X-ray intensity
ratio (for thicknesses rather greater than the penetration depth
previously) was still increasing but much more slowly, on a scale
of microns rather than tenths of microns (Fig. 9).

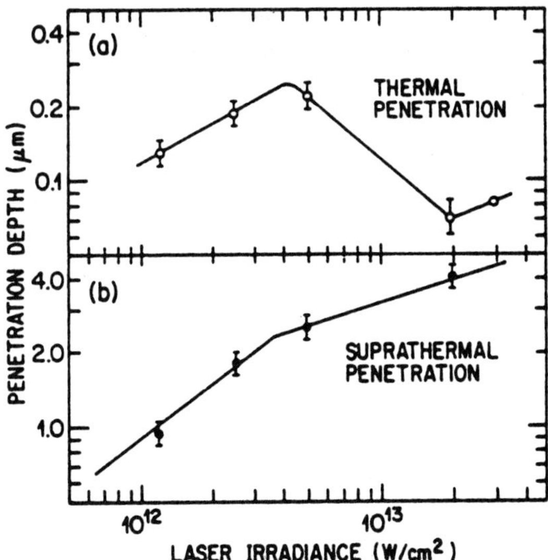

Fig. 9a,b. Inferred penetration depths as a function of intensity: (a) thermal depth, (b) fast electron depth (note different scales).

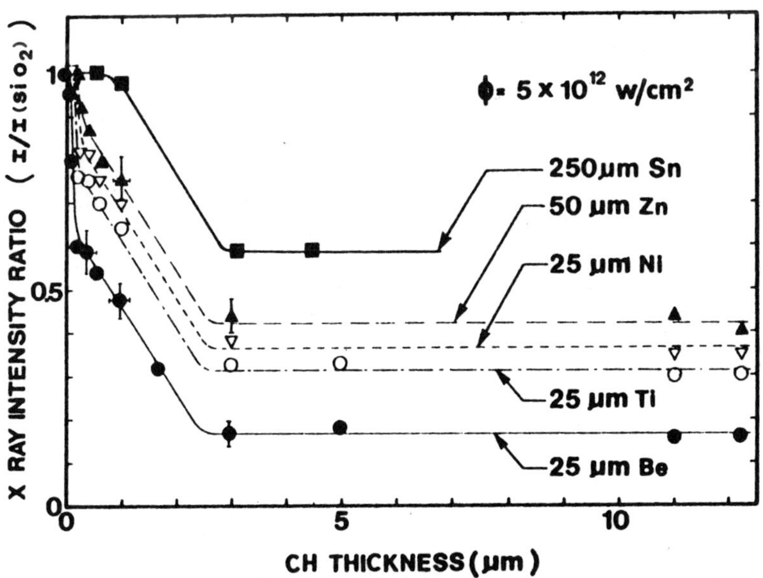

Fig. 10. X-ray emission ratios in several channels for various CH thickness (up to 12 μm).

This defines, in a fashion similar to the above, a deeper penetration depth which we associate with the hot electrons. This penetration depth varies with energy in a manner reminiscent of the hot temperature (Fig. 9b). This faster electron penetration behaviour can be seen with other filters as well (25 μm Ti, 50 μm Zn).

The penetration depth behaviour for lower X-ray energies is not only a clear evidence for the onset of some drastic change in behaviour as $v_{osc}/v_e \sim 1$ is passed, but is also a clear demonstration that an effective reduction of thermal energy flow into the target is taking place, beginning at $I\lambda^2$ values of about 6×10^{14} W/cm$^2 \times (\mu$m$)^2$, since the fractional absorption is independent of the laser irradiance at about 30%.

CONCLUSIONS

The behaviour of T_{HOT}, the appearance of fast electrons in unshielded X-ray detectors, the ion data and the interferometer results are all consistent with the model previously advanced by Forslund, Kindel and Lee[4]. The behaviour of the deeper penetration ($\sim$ 3 μm) depth also fits that concept--at least to the extent that suprathermal electron penetration increases weakly with irradiance. On the other hand the dramatic behaviour of shallower penetration depth ($\sim$ 0.3 μm) does not seem to fit the model in any simple way and seems at first glance in contradiction with the "no flux limit necessary" assertion of Gitomer and Henderson[7]. The results could indicate strong heat transport inhibition above the critical irradiance. Another possible explanation, which might still fit the "no flux limit thesis," is that the dramatic effect might be due to a striking change in plasma potential due to escaping electrons and the effect of that potential on the ablation rate. Again, the ponderomotive steepening of the density gradient coupled with the resonant absorption mechanism could be such that suddenly less absorbed energy is available to thermal electrons and correspondingly more for energetic electron production.

Determining which of these three possibilities is correct will be essential to a clear understanding of the rate of thermal transport at fluxes higher than critical. It is likely, however, that the energy delivery is then actually dominated by the superthermals.

REFERENCES

1. H. Pépin, F. Martin, B. Grek, T. W. Johnston, J. C. Kieffer and
    G. R. Mitchel, Evidence of nonlinear processes from X-ray
    spectra of CO_2 laser-irradiated targets, J. Appl. Phys. 50:
    6784 (1979).
2. F. Martin, B. Grek, H. Pépin, A synchronized actively mode-
    locked ruby laser, Rev. Sci. Inst. (to be published).

3. G. R. Mitchel, B. Grek, T. W. Johnston, F. Martin and H. Pépin, Nanosecond photography at 10.6 μm using silver halide film, Appl. Optics, 18:2422 (1979).

4. D. W. Forslund, J. M. Kindel and K. Lee, Theory of hot-electron spectra at high laser intensity, Phys. Rev. Lett. 39:284 (1977).

5. K. Nishikawa, Summary of inertial confinement fusion, Nucl. Fusion, 19:137 (1979).

6. F. C. Young, R. R. Whitlock, R. Décoste, B. H. Ripin, D. J. Nagel, J. A. Stamper, J. MacMahon and S. E. Bodner, Laser-produced-plasma energy transport through plastic films, Appl. Phys. Lett. 30:45 (1977).

7. S. J. Gitomer and D. B. Henderson, Re-examination of strongly flux-limited conduction in laser-produced plasmas, Phys. Fluids, 22:364 (1979).

LASER-DRIVEN SHOCKWAVE EXPERIMENTS AT EXTREME HIGH PRESSURES*

Richard M. More

University of California Lawrence Livermore Laboratory

Livermore, California 94550

ABSTRACT

Laser-driven shockwave experiments have been proposed for
accurate determination of equation of state data in the multi-
megabar pressure range. This paper gives a quantitative analysis of the prospects for such experiments.

In order to unambiguously interpret shockwave data, one
requires a clean shock--that is, a planar, steady shock wave
entering cold material (without significant preheat perturbation). We examine the problems of attaining sufficiently clean
shocks at high pressure and develop scaling relations which
relate the pressure achieved to laser intensity, pulse energy,
etc. It is shown that significantly higher pressures can be
achieved when structured (layered) targets are used.

1. INTRODUCTION

High power pulsed laser irradiation of small targets appears
capable of generating remarkably high pressures. Experiments
recently conducted at the JANUS laser system in Livermore find
shock velocities compatible with pressures of 20 Mbar (2
TPa).[1] Calculations predict that larger lasers can generate
shockwave pressures up to 100 Mbar (10 TPa)--extending by two
orders of magnitude the range of pressure (and temperature) now
available in the scientific laboratory.[2,3]

---

*Work performed under the auspices of the U.S. Department of
Energy by Lawrence Livermore Laboratory under contract
#W-7405-Eng-48.

How far can the laser technique extend? Can a laser generate steady shock waves or is there a limitation to impulsive
blast waves? How harmful is fast-electron preheat? Can one
usefully exploit the preheat for novel types of hydrodynamic
experiments? Does the generation of steady shocks (without
preheat) require unreasonable laser energies?

This paper reports a parameter study aimed at answering
these questions.[4] The objective is to develop quantitative
information to assist in planning experiments, designing shock-
wave targets and selecting priorities for laser development.

To be successful, a laser-driven shockwave experiment must
achieve (1) a high pressure, (2) a good contrast between the
shock itself and any preheat from fast electrons generated by
the laser, and (3) a steady shock which moves at constant
velocity. These objectives conflict, and therefore it is necessary to examine them closely to identify favorable compromises.
In every case, greater laser energy yields better results
although this may not represent a practical solution in the
immediate future. Additional constraints are imposed on laser-
shock experiments by the need to accurately measure both shock
and particle speeds; however, this paper will concentrate on the
fundamental physical limitations involved in production of clean
shocks by laser irradiation.

There is little doubt that lasers can generate high pressures. Heavily diagnosed laser fusion implosion experiments
performed on the JANUS laser (including space- and time-resolved
x-ray detectors, x-ray spectra, alpha particle and neutron counters and plasma calorimetry) show good agreement with computer
calculations which indicate pressures of $\sim$ 200 Mbar (20 TPa) in
the imploded DT fuel.[5] Larger lasers built in recent years
(ARGUS, SHIVA) can achieve still higher pressures; however, the
spherical implosion geometry does not lend itself to accurate
measurement of material properties.

A survey of available equation-of-state theories shows that
experiments must have accuracies of 15% or better to produce
useful information.[6] With few exceptions[7,8] the existing
experiments are confined to a pressure range below 5 Mbar. Thus
it appears entirely reasonable to compromise on the peak pressure in order to achieve conditions better suited to accurate
measurement.

The most serious problem for laser-generated shock studies
is caused by hot electron preheat. Laser absorption, especially
at high intensities, leads to copious production of superthermal
electrons having energies of 1-100 keV.[9] These electrons
traverse a thin target much faster than a shock-wave and deposit

their energy as hot electron preheat in the material ahead of
the shock. This preheat interferes with the standard Hugoniot
measurement technique which determines the thermodynamic state
of the shocked material by measuring the particle (material) and
shock (wave) speeds.[10] The Hugoniot analysis assumes that
both temperature and density are known for the (cold) material
ahead of the shock. Experiments of this type can be analyzed
with confidence only if the hot electron preheat temperature is
small. Therefore, a large part of our work is aimed at quantifying the preheat and locating conditions in which it will be
negligible or unimportant.

The calculation of laser-matter interaction is so complicated that it contains substantial uncertainties. It is very
important to understand that the uncertainties do not imply
correspondingly large error-bars in an equation-of-state experiment. If a shockwave is traversing cold material (of known
temperature and density), and if the shock speed and particle
speed are accurately measured, then the pressure, density and
energy of the heated material are uniquely determined by conservation laws.[10] The error or uncertainty in the inferred pressure and energy is controlled exclusively by uncertainties in
the measured velocities or in the initial density and temperature. The calculation of laser-matter interaction can be used
to design an experiment but is not used in the data-reduction.
The direct and simple interpretation of shock-wave data is the
best reason for choosing Hugoniot experiments as the first step
of a general attack on high pressure physics with lasers.

2. CALCULATIONAL MODEL

The calculations reported here are performed with several
computer codes (LASNEX) developed by George Zimmerman and his
collaborators.[4] Our calculations use a tabular theoretical
equation of state; at very high pressures this is based mainly
on the Thomas-Fermi-Kirzhnitz model.[11,12]

The computational model contains several empirical parameters which provide flexible description of otherwise difficult
physical processes. We have selected parameter settings typical
of recent laser fusion target calculations while recognizing
that the laser-shock experiments may lead to refinement of these
parameter choices.

The laser absorption process is calculated as follows: an
incident laser ray is traced by geometrical optics up to the
critical density surface, and its intensity is reduced by
inverse bremsstrahlung absorption which heats the thermal electrons.[4] Near the critical surface, a fixed fraction (30% in
our calculations) of the energy is deposited with generation of

- Laser intensity I

- Pulse duration τ

- Depth to observation point d

- Target material, structure

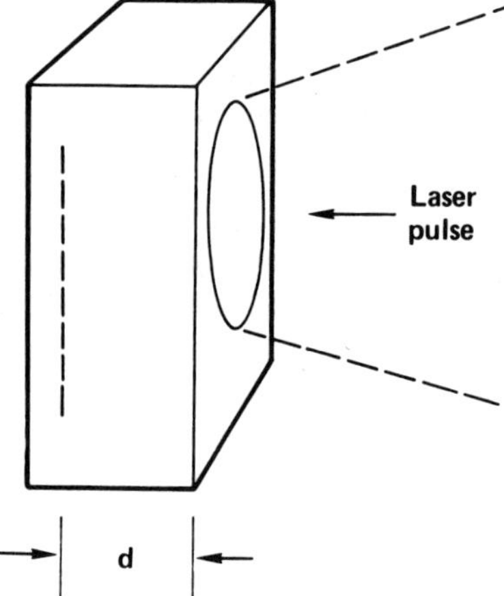

Fig. 1   There are four main design variables in a planar shock-wave
         experiment. Representative values are intensity $I = 2 \times 10^{14}$
         watt/cm, pulse length $\tau = 300$ psec, target thickness
         $d = 25$ μm, target material = aluminum foil.

superthermal electrons. The remainder of the laser light is
reflected and again attenuated by inverse bremsstrahlung. It is
necessary to emphasize the arbitrariness of this prescription,
because it predicts an absorption of $\sim$ 60% at 10^{14} watt/
cm^2. The experimental absorption is more likely to be
40-50%. We have not altered the absorption model because the
discrepancy mainly involves energy absorbed to inverse bremsstrahlung which plays a comparatively weak role in shock generation and preheat formation.

The energy absorbed by superthermal electrons is assumed to
generate a Maxwellian electron spectrum with temperature[13]

$$kT_{HOT} = 2.6 \text{ keV } (I\lambda^2)^{0.425}$$

where I is the laser intensity (units of 10^{14} watt/cm^2) and
λ the laser wavelength in microns (fixed at unity). The formula
actually employed also includes a weak dependence on the thermal
electron temperature.

Thermal electron transport is inhibited by a model which
invokes ion turbulence limiting the electron heat current to
flow with a drift velocity proportional to the ion sound
speed.[14] It is this transport inhibition which reduces the
importance of the inverse bremsstrahlung absorption fraction.

We have performed a series of calculations which alter
zoning, electron and photon group structure and other parameters; typically these changes produce less than a 10% altera-
tion in the computed results. However, the superthermal electron generation and transport constitute the most uncertain
aspect of our calculations, and we look forward to early experimental test of the preheat predictions.

The atomic ionization state is calculated by a method based
on the Saha equation; this predicts excessive recombination at
densities above the solid state.[15] We have adjusted the minimum electron temperature used by the Saha equation to force
reasonable initial ionization of the cold solid target material.

3. AN EXAMPLE AT 10 MBar (1 TPa)

Laser-driven shock experiments may be characterized by four
design variables (Figure 1):

a.   The peak <u>laser intensity</u>·I in the focal spot (a value
      of 2×10^{14} watts/cm^2 will be taken as typical of
      recent experiments on the Livermore JANUS laser).

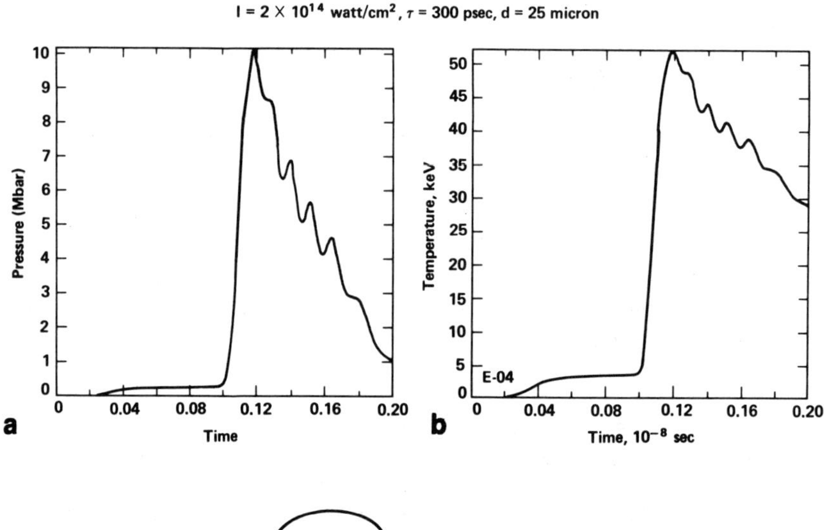

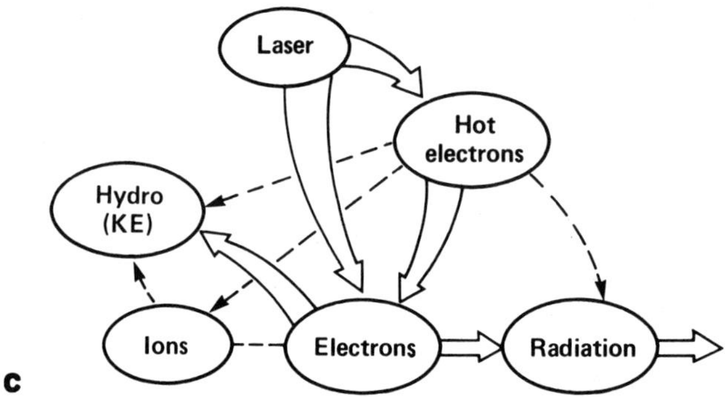

Fig. 2   (a) Calculated pressure versus time at 25 μm depth in aluminum foil irradiated as described in Fig. 1. The time is in units of 10^{-8} sec so that the shock arrives at $\sim$ 1 nsec. (b) Temperature (units are 10^{-4} keV) versus time; the 1/2 eV early signal is preheat which appears during the laser pulse. (c) Energy transfers which occur during the calculation; numerical values are indicated in the text.

b. The _pulse duration_ τ (300 psec for recent experiments at JANUS). For short pulse calculations, the laser intensity is assumed to rise linearly to its peak in 300 psec and then to decline to zero in another 300 psec (this will be referred to as a 300 psec pulse). For longer pulse lengths, the laser pulse is held constant at the peak power from 300 psec to time τ, and drops to zero at τ + 300 psec.

c. The _depth_ d to the observation point on the rear surface of the target (typically, d = 25 microns). To prevent rear-surface perturbation by evaporation caused by preheat, the computational targets are usually more than 25 microns thick (e.g., 35 μ).

d. The _target material_ (typically aluminum foil).

We are free to adjust these variables to achieve favorable shock pressure, preheat contrast and shock steadiness.

A sample calculation is given in Figures 2a, b which show pressure and temperature at a depth d = 25 μ in an aluminum foil target irradiated by a 300 psec pulse at I = 2×10^{14} watt/ cm^2. The shock arrives at $\sim$ 1.05 nsec and reaches a pressure of 10 Mbar and temperature of 5.4 eV. These conditions represent a significant extension of the pressure range available in the laboratory.

Unfortunately, the calculated preheat temperature is about 1/2 eV, corresponding to a 10% perturbation of the shock state, and the pressure relaxes rapidly after the shock has passed, implying that a rarefaction has overtaken the shock.

For these reasons, the conditions reached in the calculation are not quite satisfactory for a scientific measurement at high precision.

We have examined the calculated energy transfers in this example, in order to learn which interactions are most important (Figure 2c). With the absorption model described in section 2, some 35 Joules of energy are absorbed by the target. Of this, 14 Joules are directly absorbed by superthermal electrons, 21 by the thermal electrons. During the calculation, 11 Joules are transferred from superthermals to thermals. Smaller amounts of energy are transferred from superthermals to ion thermal energy (1.5 J), hydrodynamic motion (this is an electrostatic coupling, which transfers 0.8 J) and radiation (0.006 J).

The thermal electrons lose approximately 5.5 Joules to recombination radiation and transfer about 21.2 Joules to the hydrodynamic kinetic energy. The electrons continuously share energy with the ion thermal distribution.

At the end of the calculation, the thermal ion energy is 1/2 Joule, the electron energy is 6.4 Joules, and 22.8 Joules have been converted to hydrodynamic energy. Most of this hydro-dynamic energy resides in a small number of ions expanding toward the laser at high velocities.

The hydrodynamic efficiency of shock generation can be defined as the final kinetic energy of the rear quarter of the target divided by the total laser energy. Defined in this manner, the hydrodynamic efficiency is approximately one percent.

Similar percentage efficiencies were inferred from laser implosion experiments by Pelah, et al.[16] The reason for the low efficiency is largely the unfavorable partition of energy between the small number of high-energy blowoff ions and the remainder of the target. Because of momentum conservation, it would be very useful to reduce the velocities of the ablated ions.[17] This idea is investigated in section 7. below.

4. DEPENDENCE ON LASER PULSE CONDITIONS

This section will describe a series of calculations done at various laser pulse lengths and focal spot intensities. In order to directly compare these different cases, pressure and temperature are examined at a fixed depth of 25 μm below the front (laser) surface. The target is 35 μm thick so that rear-surface release does not affect the conditions reached at the 25 μm depth.

As is evident from figures (3a, b) and (4a, b), the preheat is very severe for high laser intensities. For this reason, the 25 micron position is much too close to the laser absorption region for meaningful Hugoniot experiments at high intensities (i.e., beyond 5×10^{14} watt/cm^2). Even though these condi-tions are not suitable for principal Hugoniot measurements, it is interesting to examine these calculations for the following reasons:

a) It is desirable to perform experiments on thin targets in which the preheat temperature is observed in order to verify or correct the computer calculation of pre-heat. The preheat temperatures calculated here are large enough (at high laser intensity) to produce easily visible signals in an optical streak camera. Experiments of this type, which observe and resolve

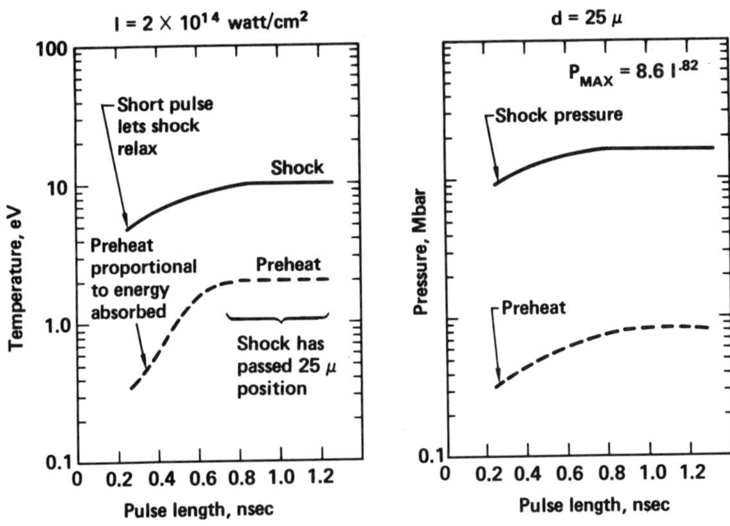

Fig. 3   (a) Shock and preheat temperature as functions of laser pulse length τ for fixed intensity (I = 2x10^{14} watt/cm^2) and fixed depth (d = 25 μm).   (b) Shock and preheat pressures as functions of laser pulse length τ.

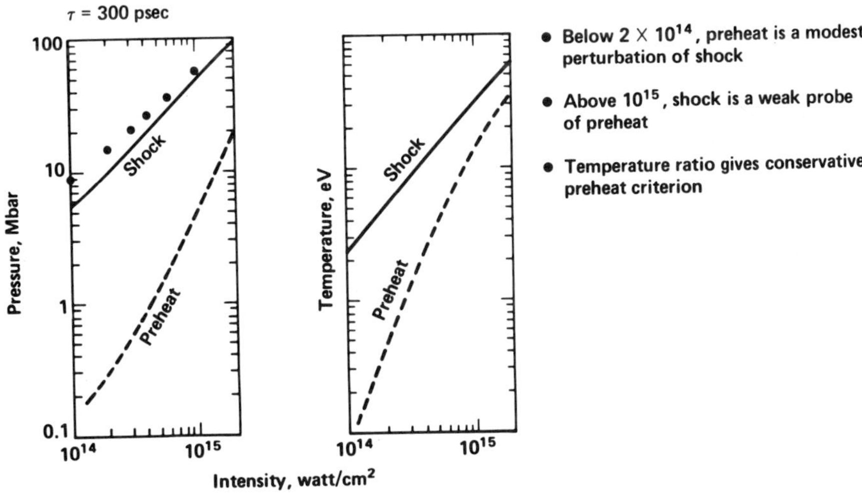

Fig. 4   (a) Shock and preheat pressure as functions of the laser intensity for fixed pulse length (τ = 300 psec) and fixed target thickness (d = 25 m). The dots are the pressures attained in long-pulse irradiation.   (b) Shock and preheat temperatures.

preheat and shock luminosity on the same shot, have
recently been performed at LLL by N. C. Holmes and R.
J. Trainor.[18]

b)   The calculations give guidance for design of preheat
     hydrodynamic experiments in which hot electrons give
     rapid isochoric (constant-volume) heating of a thin
     target. The subsequent expansion can be observed and
     compared with hydrodynamic calculations.

c)   Even if a thick target is selected to avoid hot elec-
     tron preheat of the rear surface, we want to understand
     conditions in the front portion of the target, because
     that is where the shock wave is generated.

Figures 3a, b show the shock temperature (pressure) and preheat temperature (pressure) as a function of laser pulse length
for fixed laser intensity ($I = 2 \times 10^{14}$ watt/cm^2) and fixed
depth (25 μ) in an aluminum target.

For long pulses, the shock pressure achieved at the 25
micron position ($\sim$ 16 Mbars in Figure 3b) is about 60% higher
than the pressure produced in short-pulse calculations. A close
examination of the short-pulse case shows that an equally high
pressure was generated near the front surface, but that the high
pressure is attenuated by $\sim$ 60% during propagation to the 25 μm
position. This attenuation occurs because a release wave
rapidly overtakes the shock in the short-pulse calculations.

After examining many calculations like those plotted in
Figure 3b one can extract the useful approximate scaling law

$$P_{max} = 8.6 \ I^{.82} \tag{1}$$

where P_{max} is the pressure (Mbars) generated in long pulse
irradiation and I is the laser intensity divided by 10^{14}
watts/cm^2. This scaling law gives the pressure at 25 microns
for long pulse cases; the same equation gives the maximum pressure (produced near the front surface) for shorter pulses.
Equation (1) differs only slightly from an analytic result
obtained by Kidder.[19]

The hot electron preheat temperature, shown in Figure 3a,
rises in time proportional to the absorbed laser energy until
one reaches long pulse conditions ($\tau \geq 0.9$ nsec). This preheat
temperature is defined as the temperature existing at the 25 μm
position just before the shock wave arrives. The preheat temperature does not increase for pulses longer than 0.9 nsec
because that is the shock arrival time under long-pulse conditions.

It is clear from Figures 3a, b that the ratio of shock and preheat temperatures is more favorable for short pulses. However, in the short pulse case, a release wave overtakes the shock before it crosses the sample. This means there is a conflict between the goals of a steady shock and a preheat-free target environment.

Figures 4a, b show the dependence of shock conditions on laser <u>intensity</u> (I) for fixed depth (25 microns) for a 300 psec pulse. These curves clearly illustrate the difficulty of doing clean experiments at very high pressures.

Figure 4a shows that good contrast between shock and preheat is obtained for laser intensities below 2×10^{14} W/cm^2, corresponding to pressures less than 15 Mbars. (The dots in Figure 4a indicate the shock pressure achieved in long pulse irradiation.)

Figures 4a, b show that pressures up to 100 Mbar are obtained with laser intensities greater than 10^{15} watts/cm^2, but that these pressures are generated in the presence of severe preheat. For $I \geq 10^{15}$ W/cm^2, the preheat temperature is 50% of the shock temperature. In this case, the shock is a weak probe of the preheated material.

Experiments at these conditions would be interesting integral hydrodynamic experiments but could not be analyzed through the usual Hugoniot technique without a direct, specific measurement of the preheat temperature.

It is useful to develop a quantitative measure of the importance of preheat. Comparison of Figures 4a, b shows that the temperature ratio (shock to preheat) provides a more conservative criterion than the pressure ratio. The next section gives an analytic formula for the perturbation of the Hugoniot adiabat produced by a specified level of preheat.

5. CONSTRAINTS ON LASER SHOCK EXPERIMENT

This section identifies the constraints on target thickness, area and laser pulse length which assure generation of a steady shock entering cold material. The combined effect of these constraints is to limit the range of pressures accessible for a given laser energy.

A first constraint is required to assure that edge rarefactions do not interfere with the high pressure region in the shock front (see Figure 5). If one wishes to observe a uniform shock wave arriving over an area $A_g = \pi R_g^2$ on the rear surface of the target, it is necessary that the laser light

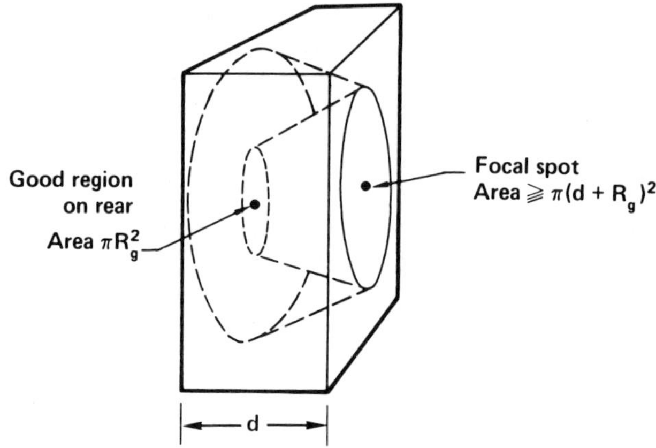

Fig. 5   It is necessary to select a large focal spot area in order
to prevent edge rarefactions from reducing the shock pressure observed on the target rear surface.

- Short pulse allows pressure release
- Possible decoupling of critical surface
- Prepulse can destroy front surface
- Energy loss to sides weakens shock

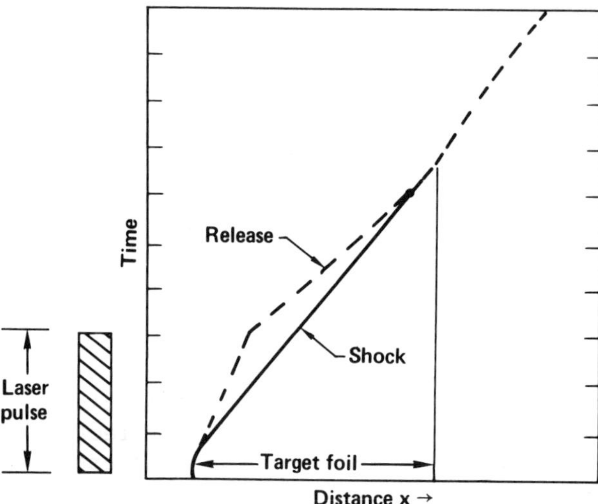

Fig. 6   When the laser pulse ends, a release wave (dotted) is
launched; it rapidly moves to overtake the shock and reduces the pressure. If the critical-density surface becomes decoupled from the ablation region, this problem
might occur during the laser pulse.

cover a larger area $A = \pi (d+R_g)^2$ on the front surface. (This estimate assumes that the edge rarefaction enters at an angle of approximately 45°, but whatever the angle the focal spot area A is proportional to d^2 for thick targets.) If this condition is met, edge rarefactions do not perturb the pressure in the central part of the shock wave.

This condition becomes a difficult constraint for thick targets because it requires coverage of a large focal spot. In addition, it is necessary that the laser beam give spatially uniform irradiation over the focal spot in order to produce a planar shock.[20] Thermal conduction in the laser absorption region is able to reduce modest spatial variations of the laser intensity.

The second constraint is required to obtain a steady shock whose velocity remains constant as it traverses the sample. The space-time graph of Figure 6 shows the trajectory of the release wave (rarefaction) which begins at the front surface when the laser intensity drops. This rarefaction travels at the sound speed in the shocked material; i.e., faster than the shock. Once it overtakes the shock, the peak pressure begins to decline resulting in an unsteady shock. An accurate measurement of the shock speed is only possible if the rarefaction does not overtake the shock before it reaches the rear surface of the sample.

A simple estimate (including the particle velocity of the shocked material) shows that the release wave does not overtake the shock at depth d if the driving pressure persists for a time $\tau > d/2D$. For strong shocks, the shock speed D is approximately proportional to the square root of the pressure P. The inequality $\tau > d/2D$ requires a long laser pulse for a thick target.

There may be problems in achieving a continuous (steady) driving pressure even if the laser is able to produce a long pulse. Potential troubles include fluctuations (spiking) in the laser output, decoupling of the critical surface, and other energy loss mechanisms which may remove energy from the focal spot during the laser pulse.

The third constraint, most difficult of all, is imposed by the need to avoid hot electron preheat and its deleterious effects (Figure 7). The hot electrons are generated in the process of laser absorption with energies between 2 keV and 200 keV. A fraction of these electrons penetrate into the sample at speeds much greater than the shock speed. These electrons raise the sample temperature almost simultaneously with the rise of the laser pulse; at any point in the sample the preheat temperature is proportional to the total laser energy absorbed. The manner in which the preheat temperature decreases with depth

- **Hugoniot displaced by ΔT, $\Delta\rho$**
- **Average shock speed is reduced slightly**
- **Shock traversing preheat is not steady**
- **Rear surface destroyed (thin target)**
- **Preheat launches precurser shock (thick target)**

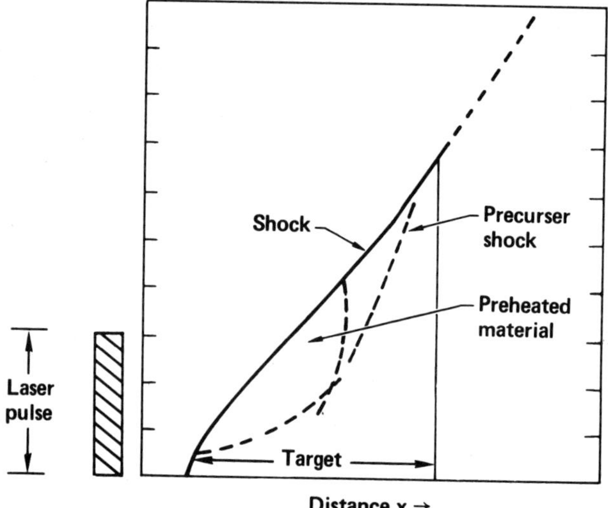

Fig. 7   Space-time plot showing region ahead of shock which is affected by hot-electron preheat. In a very thick target, there is the possibility of a precursor shock launched by preheat.

into the sample is controlled by the range-energy relation for
the superthermal electrons.

It is useful to set a criterion for the severity of preheat. One might be tempted to compare the preheat temperature
with the initial temperature of the sample, but this condition
is much too severe. It is shown below that the correct criterion is obtained from the ratio $T_{preheat}/T_{shock}$ of preheat
temperature to shock temperature.

The problems produced by preheat include the following:

a. The preheat displaces the Hugoniot locus; i.e., the
density and pressure produced differ from those expected from a shock of the same pressure entering cold
material. An approximate formula for this displacement
is given below.

b. The speed of a shock traversing preheated material is
different from that of a shock (at the same pressure)
traversing cold material. However, this velocity
change is small; i.e., second order in the temperature
ratio $T_{preheat}/T_{shock}$.

c. The preheat is formed with a spatial variation arising
from the energy loss of the original hot electrons.
The shock traverses this spatially non-uniform layer
and leaves behind material with corresponding non-
uniform density and temperature. The material behind
the shock oscillates until its inhomogeneities are
smoothed. If these oscillations are severe, they
interfere with generation of a steady shock.

d. For relatively thin targets, the preheat temperature
may exceed the melting temperature of the target material. The rear surface of the target is then subject to
hydrodynamic disassembly. If the rear surface expands,
the observed shock may have properties (arrival time,
temperature, or pressure) significantly different from
those it had in the bulk.

e. For experiments at very high pressures ($\sim$ 100 Mbar),
where thick targets are necessary to avoid major preheat perturbation, the preheated material is raised to
a high temperature ($\sim$ 50-100 eV) and can launch a
precursor shock ahead of the main shock wave.

In a well-designed experiment, conditions are adjusted so
that preheat is a small perturbation on the Hugoniot measurement. The size of the perturbation resulting from a specified

level of preheat can be estimated as follows: assume that preheat raises the temperature of the target without changing its
density, and assume further that the shock pressure is fixed.
It is straightforward to apply the Hugoniot relations to connect
the final (shocked) density or energy to the preheat temperature. The relation which results is (see Figure 8)

$$\frac{\Delta E}{\Delta E_0} = \frac{1+\frac{1}{2}(\rho-\rho_0)\ (\partial P/\partial E_0)/(\rho_0\rho)}{1+\frac{1}{2}(p_s+p_0)\ (\partial P/\partial E)/(\rho^2\ \partial P/\partial\rho)} \tag{2}$$

where ΔE is the change in final state energy (with respect to a
shock of the same pressure entering cold material) and ΔE_0 is
the energy per gram associated with the preheat; p and p_0 are
the shocked and initial pressures. The righthand side of this
equation is 1.5 - 2 for tabular equations of state or for ideal
gases. In effect, 1/2 eV of preheat simply raises the shock
wave temperature by 1/2 eV. In order to perform experiments
with an accuracy of a few percent at pressures above 10 Mbars,
the preheat must be kept below about 1/10 of an eV. If this
condition is satisfied, rear surface disassembly is also reduced
to a minor problem.

The computer code[4] provides numerical calculation of hot
electron generation, transport energy loss and heating of the
target. The code also simulates the hydrodynamic disruption (d.
or e. above) produced by the preheat. It is useful to extract
crude analytic formulas in order to establish the trends which
govern shock-wave target performance. For this purpose, we can
combine the computed scaling of hot electron temperature with
well-known range-energy relations (Fig. 9).

The hot electron temperature T_{hot} (keV) for laser intensity I (in units of 10^{14} watts per cm^2) may be written (the
expression given applies to Nd-glass laser radiation):

$$T_{hot} \stackrel{\sim}{=} (2.6\ keV)\ (I)^{.425} \tag{3}$$

While the hot electron temperature is given by the above
equation, the hot electrons responsible for the preheat in the
rear portion of the target have energies about 5 kT_{hot}. (This
conclusion follows from numerical analysis of the preheat.) The
calculated preheat is approximately described by an equation for
the preheat temperature T_e

$$T_e = \eta\ \frac{I\cdot t}{C_v}\ \phi(x) \tag{4}$$

where η is the hot electron production efficiency, I is the peak
laser intensity, t is the time that the laser has been on, C_v
is a material specific heat (constant density) and $\phi(x)$ is
approximately

Assume: Fixed initial density ρ_0
          Fixed shock pressure p_s
          $p = p(\rho, E)$

$$\frac{\Delta E}{\Delta E_0} = \frac{1 + \frac{1}{2}(\rho - \rho_0)\left(\frac{\partial p}{\partial E_0}\right)\bigg/\rho_0\rho}{1 + \frac{1}{2}(p_s + p_0)\left(\frac{\partial p}{\partial E}\right)\bigg/\left(\rho^2\frac{\partial p}{\partial \rho}\right)}$$

≈ 2 for ideal gas

≈ 1.5 for aluminum tabular EOS

½ eV of preheat raises T_s by ½ eV

Fig. 8   Equation for final specific energy with (ΔE) and without (ΔE_0) preheat.

Range $R(E) \simeq 0.6\,\mu\,(E/10\text{ keV})^2$ for 10-100 keV

Hot electron temp $T_{HOT} \sim 2.6$ keV $(I\lambda^2)^{.425}$

Preheat temp $T_e = \eta\,\dfrac{I \cdot t}{C_v}\,\phi(x)$

η = hot electron production efficiency

$\phi(x)$ is larger than $\exp(-x/R(T_{HOT}))$

Fig. 9   Simple approximate formulas for scaling of hot-electron preheat produced by a laser pulse of constant intensity I which has been applied for a time t.

$$\phi(x) \sim e^{-x/R}$$

The range R extracted from the computer calculations is then approximately the range of electrons having energy $\sim 5\ kT_{hot}$. Close to the front surface, the typical hot electron energy is smaller ($\sim kT_{hot}$), and at greater depths in the sample, very low preheat temperatures are produced by the few electrons having still higher energies.

In the next section this scaling of preheat is combined with other constraints to identify the laser energy requirement for strong shock experiments.

6. SCALING LAWS FOR HOMOGENEOUS TARGETS

We have seen that three interrelated problems must be overcome to generate clean, strong shocks. In each case a solution is possible by raising laser energy.

Hot electron preheat preceding the shock causes the measured Hugoniot to be displaced from the desired principal Hugoniot. The straightforward solution to this problem is to absorb the preheat in a thicker target. The required target thickness d must exceed R(E), the hot electron range, for electron energies $E \sim 5k\ T_{hot}$. The range-energy relation,[21] combined with the dependence of T_{hot} on laser intensity I then results in

$$d \gg R(E) \sim T_{hot}^{2} \sim I^{0.85} \sim p \tag{5}$$

where the last proportionality is deduced from the observed scaling of peak pressure with laser intensity (section 3). In order to avoid preheat, the target thickness must rise proportional to the desired pressure.

The second problem is the release wave which begins at the end of the laser pulse and moves to overtake the shock wave. The straightforward solution to this problem is to require the laser pulse length τ to be long enough so that

$$\tau \overset{>}{=} d/2D \tag{6}$$

where d is the target thickness and D is the shock speed, proportional to the square root of the pressure. For high pressures, d is large and Eq. (6) requires a long pulse and therefore a large laser energy.

Finally, the focal spot area A must increase (proportional to d^2) in order to prevent edge rarefactions from reducing the pressure.

The total energy the laser must supply is determined by the peak intensity I, area A and pulse length τ according to

$$E = I \cdot A \cdot \tau \qquad\qquad (7)$$

Combining the above scalings, one finds (Fig. 10)

$$E \propto p^4 \qquad (8)$$

To produce twice the pressure, one requires 16 times the laser energy. This is not the energy required to produce the high pressure _per se_; the scaling law gives the energy required to produce the pressure in the form of a clean, steady shock wave entering cold material.

Although three successive LLL lasers have delivered energies 10 times larger than their predecessors, the scaling expressed by the relation (8) shows that even the largest laser cannot freely explore the realm of very high pressures when used with homogeneous targets. (See section 7 below.)

Figure 11 shows a calculation for a thick target (50 μm of aluminum) which supports the conclusions drawn above. The quantity plotted is the temperature as a function of time for 9 positions in the target (the positions are initially 5 μm apart). The laser remains at an intensity of 2×10^{14} W/cm^2 for the time interval from 300 psec to 900 psec; the laser intensity has dropped to zero at 120 nsec.

It is easy to distinguish the linear temperature rise caused by preheat in the first few zones (x < 25 μm) from the sharp temperature rise associated with the shock itself. Two features are immediately apparent:

1)   The contrast between shock and preheat is much better at the 45 μm position near the rear surface, and

2)   The shock jump is roughly constant as it rides over the preheat, so that the post-shock temperature is the sum of the preheat temperature and this constant. (This illustrates the rule of thumb given in section 5.)

It is interesting to observe that the temperature continues to rise for the forward zones (x < 25 μm) even after the shock has passed. This temperature rise can be described as "post-heat", i.e., deposition of energy by superthermal electrons generated after the shock passes the zone in question. The rear zones do not experience post-heating because the laser intensity falls after 900 psec.

Energy = $I \cdot A \cdot \tau$

Intensity   Area   Pulse length

$$\begin{cases} I \sim p^{1.22} \\ A \gtrsim d^2 \\ \tau \geqslant d/2D \qquad D \sim \sqrt{p} \\ d \gg R(E) \sim T_{hot}^2 \sim I^{.85} \sim p \end{cases}$$

$$\boxed{\text{Gives} \quad E \sim p^4}$$

**Additional problem at high pressure from precursor shock
launched by preheat**

Fig. 10   The energy required to produce a clean steady shock, without preheat perturbation, rises proportional to the fourth power of the shock pressure.

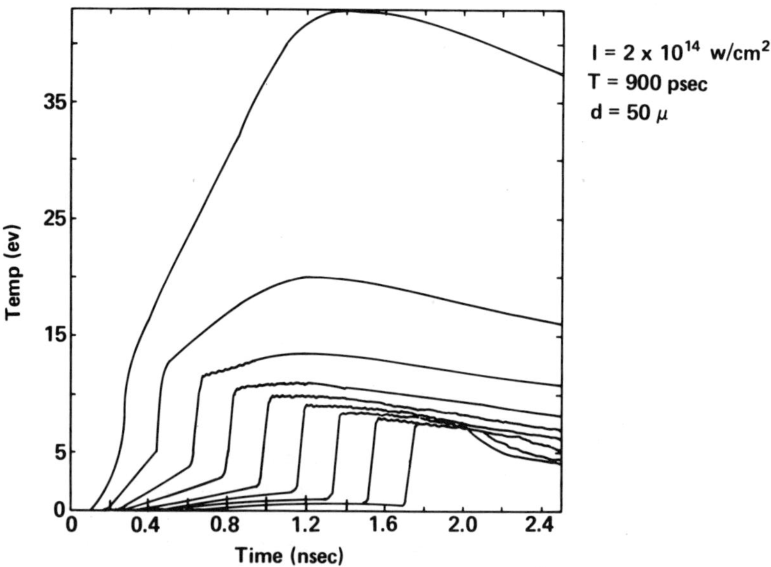

Fig. 11   Time-dependent temperature in zones which are 5, 10, ..., 40 and 45 microns into a 50 m aluminum target. For the rear surface zones, the preheat perturbation is relatively small.

Detailed inspection of this calculation shows that the rear part of the target experiences a 15 Mbar shock with less than 1/2 eV of preheat. These conditions appear to be comparatively favorable.

7. RESULTS WITH STRUCTURED TARGETS

In this section we examine the results obtained with a more elaborate target (Fig. 12). It will be shown that the presence of a tungsten layer leads to a dramatic improvement in target performance. Our initial examination of targets of this type was based on the hope of raising the hydrodynamic efficiency by ablating heavier material; in the final analysis, that is not the most important advantage of layered targets.

The targets considered here consist of thick aluminum foils. The side facing the laser has a thin layer (3 microns) of tungsten, covered with another two microns of aluminum. The outer layer is present in order to have the laser impinge on aluminum, for which the hot-electron generation is comparatively well-understood.

The computational analysis of targets in this category shows:

- The tungsten layer significantly reduces hot-electron preheat penetration.

- The massive tungsten acquires enough momentum to sustain the shock pressure for several hundred picoseconds after the end of the laser pulse. Energy deposition in the tungsten causes it to expand and exert pressure on the aluminum even after the first rarefaction enters the tungsten.

- One expects that material having so many electrons per atom will remain colder than aluminum and thereby cool the critical surface. This effect would (presumably) enhance the inverse bremsstrahlung absorption; however, no temperature difference is actually observed in the calculations.[22]

- It is also a plausible conjecture that the heavier atoms ablate at lower velocities which should raise the hydrodynamic efficiency. This gain could not be detected in the computer output, probably because of the aluminum overlayer.

- The most significant result is that the target with a tungsten pusher achieves good shock-to-preheat contrast

- Impedes preheat penetration
- Acts as pusher to maintain shock
- Could reduce temperature near critical density
- Could increase hydrodynamic efficiency
- Thinner target implies smaller spot size and shorter pulse

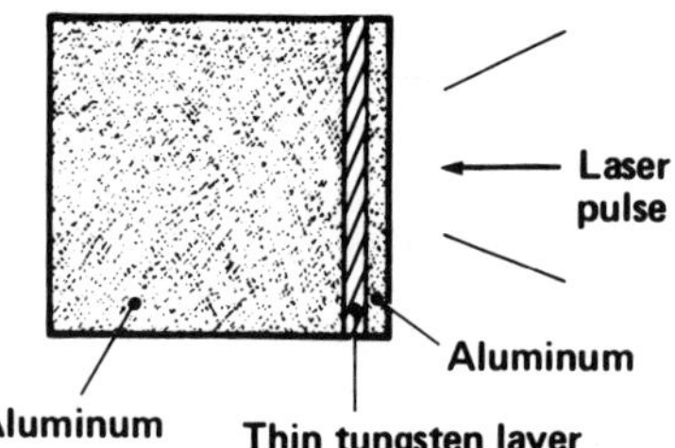

Fig. 12   Conceptual design for a structured laser shock-wave target.

in about <u>half the thickness</u> of the all-aluminum target. The implication of this result, according to Eqs. (5, 8) is a dramatic reduction in the energy required to achieve a given shock pressure.

To expand on this point, we observe that a factor of two reduction in target thickness d significantly reduces both the required pulse length and focal-spot area, and this leads to almost an order-of-magnitude reduction in the required laser energy.

This gain has been obtained mainly because the tungsten layer impedes penetration of hot electrons. However the tungsten also functions as a very effective pulse-stretcher; the computed shock pressures are extremely stable for as much as a nanosecond after the shock enters the aluminum. This is presumably caused by momentum storage in the tungsten as well as by the high pressure generated in the tungsten by electron energy deposition.

It is our belief that future laser-driven shock-wave experiments will find the advantages of layered targets to be irresistable.[23] However, there are significant practical difficulties in bonding flat thin layers of material with the exceptional tolerances required to conduct accurate measurements on the scale of tens of microns and tens of picoseconds. More generally, a high level of technical artistry is obviously necessary to perform accurate scientific studies of tiny specks of matter which have been heated to astrophysical conditions.

ACKNOWLEDGEMENTS

The work of this paper has been significantly assisted by many scientists including Minao Kamegai, who helped perform the numerical calculations; Robert J. Trainor and Neil C. Holmes, who are performing laser shock experiments at LLL; and George B. Zimmerman, David Bailey, David Kershaw, and Judith Harte, who developed computer codes used in the calculations.

REFERENCES

1. R. J. Trainor, J. W. Shaner, J. M. Auerbach and N. C. Holmes, Phys. Rev. Letters <u>42</u>, 1154 (1979); L. Veeser and J. Solem, Phys. Rev. Lett. <u>40</u>, 1391 (1978).
2. Portions of this work have previously appeared in Bull. A.P.S. <u>24</u>, 720 (1979).
3. As a personal note, I recall that informal discussions on the use of large lasers for shockwave experiments occurred here in Rochester as early as 1975 (M. Lubin, E. Goldman, L.

   Goldman, E. Thorsos and R. More). Then as now, the critical
   question was the feasibility of obtaining accurate scientific data from laser experiments.

 4. The calculations reported here were performed with several
    computer codes developed by G. Zimmerman and his collabora-
    tors. Partial description of these codes is given in the
    Laser Program Annual Reports for 1975 through 1978 (Lawrence
    Livermore Laboratory Rept. UCRL-50021-75, UCRL-50021-76,
    etc.) See also G. B. Zimmerman and W. L. Kruer, Comments on
    Plasma Physics $\underline{2}$, 51 (1975).

 5. J. Larsen, unpublished.

 6. R. M. More, Phys. Rev. A$\underline{19}$, 1234 (1979).

 7. C. E. Ragan III, M. G. Silbert, and B. C. Diven, J. Appl.
    Phys. $\underline{48}$, 2860 (1977) and more recent work (unpublished).

 8. L. V. Al'tshuler, B. N. Moiseev, L. V. Popov, G. V. Simakov,
    and R. F. Trunin, Soviet Physics JETP $\underline{27}$, 420 (1968).

 9. H. G. Ahlstrom, Lawrence Livermore Laboratory Rept.
    UCRL-79819 (1977) gives a thorough review of laser absorp-
    tion phenomena.

10. G. E. Duvall, in <u>Physics of High Energy Density</u>, edited by
    P. Caldirola, Academic Press, New York, 1971.

11. S. I. McCarthy, Lawrence Livermore Laboratory Rept.
    UCRL-19364 (unpublished).

12. N. N. Kalitkin, Soviet Physics JETP $\underline{11}$, 1106 (1960).

13. Lawrence Livermore Laboratory Rept. UCRL-50021-77 (page
    6-18)(1978).

14. Lawrence Livermore Laboratory Rept. UCRL-50021-76 (section
    4-7.3)(1977).

15. G. B. Zimmerman and R. M. More, Lawrence Livermore Labora-
    tory Rept. UCRL-81336 (1978).

16. I. Pelah, E. Thorsos and E. Goldman, unpublished.

17. Dr. C. Cranfill drew the author's attention to this possi-
    bility (private communication).

18. N. C. Holmes, R. J. Trainor and R. M. More, to be presented
    at the APS Plasma Physics meeting in November 1979.

19. R. Kidder, in <u>Physics of High Energy Density</u>, edited by P.
    Caldirola, Academic Press, New York, 1971.

20. Failure to achieve sufficient spatial homogeneity can dras-
    tically reduce the shock pressure attained (R. Schultz and
    L. Veeser, personal communication).

21. R. D. Evans, <u>The Atomic Nucleus</u> (Mc-Graw-Hill Book Co., New
    York, 1955), p. 624.

22. This idea was suggested to the author by M. Rosen.

23. Efforts to fabricate layered targets are underway in
    H-Division at the Lawrence Livermore Laboratory.

SPACE-DEPENDENT SHIFT OF SPECTRAL LINES IN LASER-PRODUCED PLASMAS

A. Carillon, P. Jaeglé, G. Jamelot and C. Wehenkel

GRECO du CNRS "I. L. M." Ecole Polytechnique
91120 Palaiseau, France
Laboratoire de Spectroscopie Atomique et Ionique
Université Paris-Sud, Bât. 350, 91405 Orsay, France

ABSTRACT

A space-dependent red-shift of extreme ultraviolet lines of aluminum laser-plasma is observed on a distance covering more than $100\,\mu$ in a single-shot experiment. For radiation coming from highest density plasma, the shift can exceed the broadening of the lines. In the first approximation, this shift does not depend on plasma optical thickness. Various aspects of Stark effect are discussed in order to account for these observations.

INTRODUCTION

On account of the role of ion spectra for studying the properties of dense plasmas produced from laser-irradiated target,
the understanding of the relation between spectral emission and
plasma parameters has to be improved. The fact that density and
temperature exhibit strong variations along the normal to target
surface, and possibly along directions parallel to the surface,
is of great importance for physical processes such as generation
of magnetic fields (1, 2) , emission of fast particles (3) , and
also because it modifies appreciably the observed features of
core emission in the X-ray and extreme ultraviolet (EUV) ranges
(4, 5). Therefore a large spatial resolution is to be desired in
the spectral studies. Here we report the observation of a space-
resolved red-shift of spectral lines, measured for a single laser-
shot on a distance covering more than $100\,\mu$ along the focusing

axis of the laser-beam. Such an observation of large systematic shift of ion lines is a new fact in dense plasma diagnostics which, until now, have been centered rather on line broadening (6, 7, 8) .

I. EXPERIMENTAL SET UP ON TOROIDAL MIRROR AND SPATIAL RESOLUTION.

The experimental arrangement provides a spatial resolution with a luminosity large enough for recording the EUV spectrum from a single laser shot ; toroidal mirror, working in an unusual way, is used to focus the plasma radiation in the plane of the entrance slit of a grazing incidence spectrograph. Let R = OM and r be the torus radii and u the glancing angle of the radiation on the mirror (figure 1). After reflexion on a toroidal mirror, a light beam emitted by a pin-point source leans on two perpendicular focal lines T' and S', the positions of which being given by the well known relations

$$\frac{1}{OT'} = \frac{1}{OS} + \frac{2}{R \sin u} \tag{1}$$

$$\frac{1}{OS'} = \frac{1}{OS} + \frac{2 \sin u}{r} \tag{2}$$

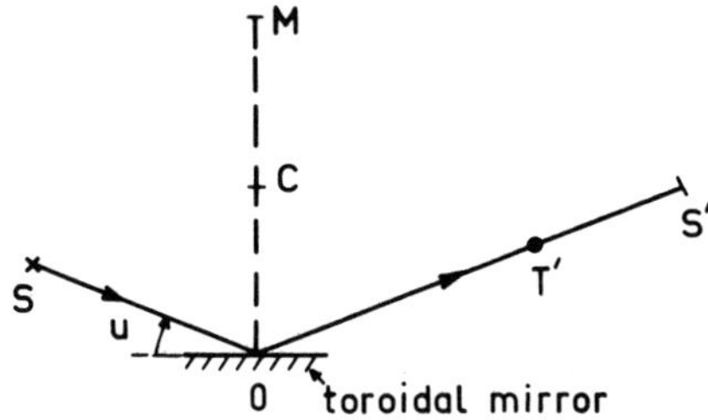

Figure 1 : focal lines given by a toroidal mirror from a pin-point source.

Under the condition OS = - R sin u, the source and the tangential focal line T' are on the focusing circle of the mirror, the radius of which being R/2 . Instead of satisfying the stigmatic condition $r/R = \sin^2 u$, under which the two focal lines are coplanar are

the image of a pin-point source is an approximate point, we have
chosen the radii such that

$$\frac{r}{R} = 2 \sin^2 u \qquad (3)$$

Then the sagittal focal line is transferred to infinity, and all
rays emitted by the pin-point source S become parallel to the
plane of focusing mirror after being reflected. So, if the entrance
slit of the spectrograph is set on the tangential focal line T', the
distribution of radiation along the entrance slit is fairly well
duplicated on the detector.

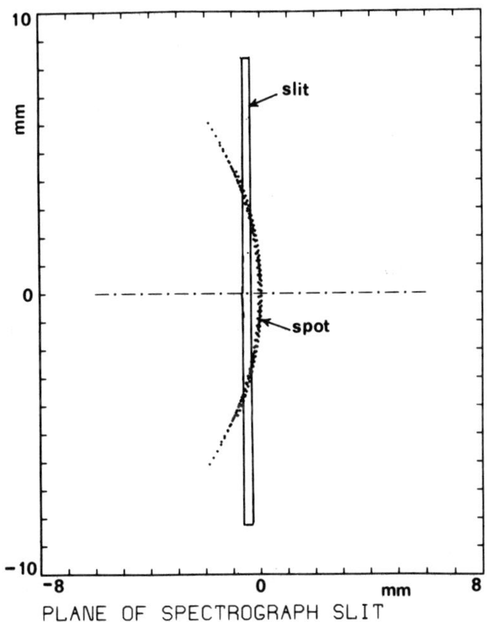

Figure 2 : Calculated illumination at the spectrograph
entrance for a point source in the plasma ; a toroidal
mirror projecting the horizontal focal line to infinity is used.

A computer calculation shows that the illumination of the
entrance slit plane, from a pin-point source near the focusing
circle, consists of a curved spot as plotted on figure 2 for
R = 200 cm, r = 2 cm and u = 4°. Thus a pin-point source yields
two symmetric lighted spots on the slit itself, at the crossing of
the slit with the curve.
When the source point moves along the radius of the focusing

circle from A to B, that is to say along the laser axis in our
experiment, as shown on left side of figure 3, the curved spot
runs horizontally in the slit plane from A' to B' (figure 3, right
side). The two crossing points of a spot with the slit slide
symmetrically along the slit when the spot is shifted in the plane.

As their separation varies as the square of the shifting of the
source, this system amplifies, on the detector, the distance between the emitting points in the plasma.

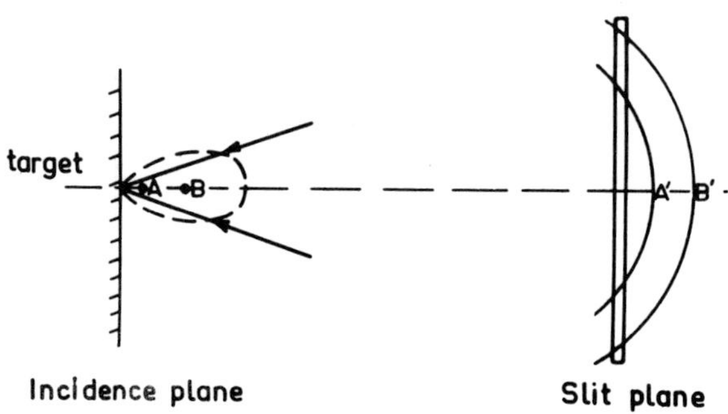

Figure 3 : principle of spatial resolution

If an emitted line is submitted to a space-dependent spectral
displacement, the dispersion by the spectrograph turns the spatial
resolution in to a distortion of the line shape, as this is shown
on the figure 4. The central part of the spectrum corresponds to
the zone of the plasma near the focusing circle the mirror, and
the upper and the lower part to external zones of the plasma.
The shape of the lines depends on the position of the target and
shows a white zone if the target is external to the focusing circle
($\Delta x < o$).

The calculated limit of the spatial resolution is as large as
5μ . A computer simulation , performed for a 5μ wide slit,
leads to the curve on figure 5 which gives the correspondence
between emitting points along the laser axis in the plasma, 0
being the abscisse of the focusing circle and printed positions on
the photographic plate.

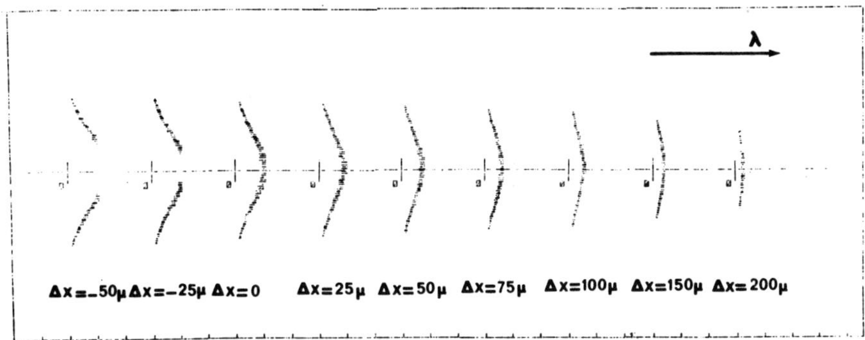

Figure 4 : calculated shapes of a space dependent shifted line for several distances along a radius Δ x between the target and the focusing circle of the mirror. (Δx $>$ o target inside the circle Δx $<$ o target outside).

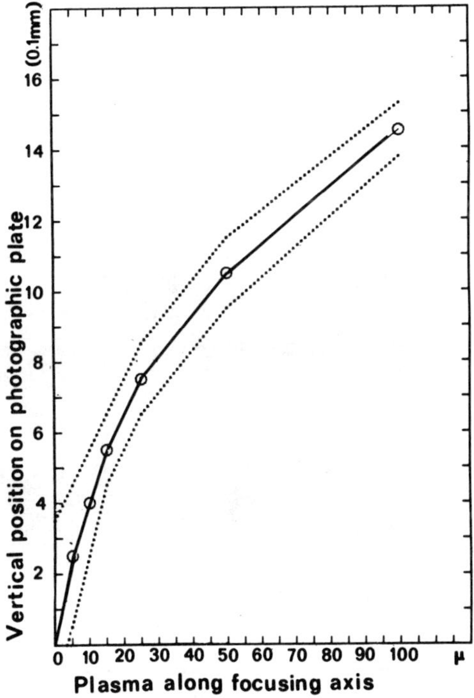

Figure 5 : Location of plasma emitting points versus printed positions on photographic plate.

II. EXPERIMENTAL RESULTS

Using this system, the spectra presented on figure 6 have
been recorded each from a single laser-shot of a Nd-laser focused
on an aluminum target. The pulse width was 2. 5 ns and the
power density 5 x 10^{14} W/cm^2. From the line intensities of highly
charged aluminum ions, a temperature of about 400 eV is assumed
for the plasma. The EUV spectrograph used a 2400 g/mm grating
with a curvature radius of 1 m. Most of lines appearing in the
spectra are due to transitions between levels of quantum number
2 and 3 of ions Al^{10+}, Al^{9+} and Al^{8+}.

The first spectrum (a) exhibits a large central white band, as
it can be seen on the calculated line shapes of figure 4 for Δx $<$o.
This band corresponds to the 25μ long non emissing zone included between the focusing circle and the target. As shown by the
pictures of figure 6, the lines exhibit a band which reveals a
space dependent shift. The upper and lower ends of the lines
correspond to an emitting zone far from the target (more then
200μ) where the plasma is cold and underdense, and so are
unshifted. The central zone of the lines, corresponding to the
hot and dense layers of the plasma near the target, appears
shifted towards the large wavelengths. The densitometer traces
on figure 7 which correspond to the (b) spectrum of figure 6

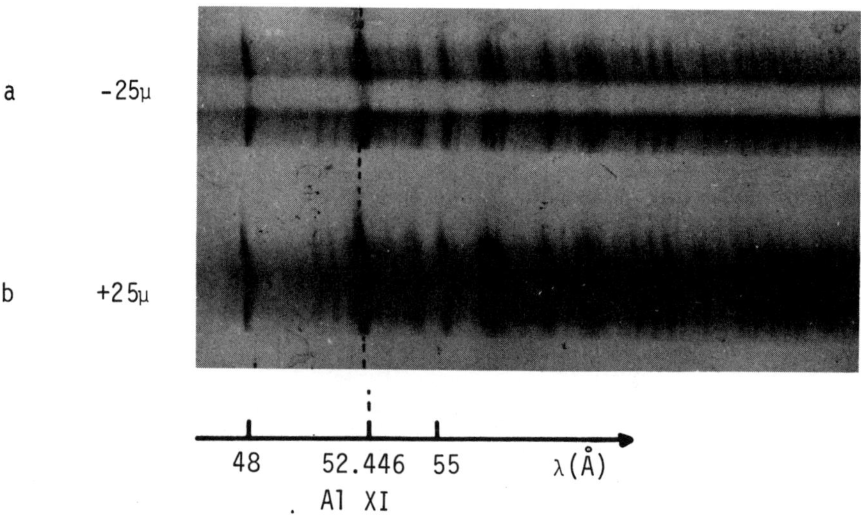

Figure 6

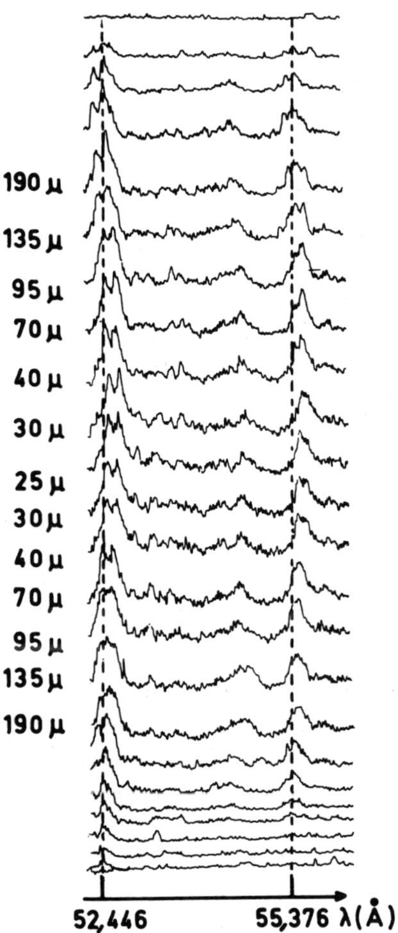

Figure 7 : Densitometer traces showing the space-dependent red shift of the lines. The distance of the plasma emitting points to the target surface are plotted on the left (in microns) for every spectrum.

show the experimental shift according to the distance between
the target surface and the emitting plasma point, for lines of
ions Al^{10+} (52.446 Å and 52.299 Å) and Al^{9+} (55.376 Å and
55.272 Å). The maximum shift is of about 0.75 Å. This value,
which varies little from one line to another in the spectral range
displayed on figure 6, exceeds the width of many lines.

III. DISCUSSION

The reabsorption in an optically thick plasma is known to
yield broadening of spectral lines. This effect has been extensively calculated in our previous works (4, 9), especially in cases
where the Doppler shift due to ion expansion velocity yields line
asymmetry. Asymmetry results from the opacity of external
expanding plasma shell and it appears as a blue-shift of self-
reversed profile which allows the top of the line to be practically
unshifted. Such features depend strongly on line intensity what
is not the case of the shifts observed here. Thus, in the present
work, only the line widths are possibly affected by plasma opacity. As to the line shift, Stark effect must be considered for
explaining it.

Recent works suggest that an electric field other than the
ion microfield can be generated in plasma as a consequence of
suprathermal electron production (10, 11, 12) ; it is a possible
cause of Stark displacement. However, in our time-integrated
experiment, it may contribute only a small part in producing
the line redshift because its peak value does not exceed 10 % of
the ion microfield. Only in a time-resolved experiment, the
measurement of the shift could give valuable information on this
field wich is to be considered in studying transport of energy in
the plasma.

The intense field of the Nd-laser can also be invoked as a
source of Stark effect. If it would be the case in the experiment
reported here, the shift must be sensitive, for instance, to
small variations of the focusing distance used for producing the
plasma. In fact, the observed shifts are practically constant
when this distance varies of more than $\pm$ 100 μ , what must
modify significantly the laser intensity in the considered zone
of the plasma. Furthermore, the laser field effect decreases
rapidly with the main quantum number of the transition upper-
level and thus it is not expected to have a large part in the
perturbation regarding the present lines.

The perturbation of ion levels by the next electrically

charged particles has thus the main part in the observed features.

A red-shift of lines is generally expected from the quadratic quasistatic Stark effect (13) while the impact approximation predicts red or blue shifts (14, 15) . We do not know any attempt for calculating the full profiles of non-hydrogenic lines for plasma density as large as 10^{21} particles per cubic centimeter in folding electron and ion contribution, not only in broadening but also in shifting the lines. However it is possible to ensure that our experimental result is consistent with the experienced values of plasma density in using the following rough approximation. First, on account of the ratio between the electronic temperature and the energy of the transitions under consideration, electron-ion collisions have mostly inelastic effect which will results in broadening rather than in shifting the lines (16) . Neglecting this contribution to the line shift leads to a fairly small overestimation of the ion density. As far as we want merely verify that this density does not exceed the solid target density, this is of no consequence. Second, the order of magnitude of ion density required for producing the observed shift can be deduced from the mean value of the Holtsmark field strength F and from an appropriate expression for the quadratic Stark effect (17) :

$$\delta E = - n^6 (a_0^3 / Z^4) F^2 \qquad (4)$$

leading to :

$$N = 2.54 \times 10^{22} \frac{Z^{3/2}}{n^{9/2}} \delta E^{3/4} \qquad (5)$$

where N is the perturber density in cm^{-3}, δE the shift in eV, Z the perturber charge and n the quantum number of the upper level. For Z = 10 n = 3 and δE = 0.75 eV, the maximum value of shift, the density must be :

$$N = 5 \times 10^{21} \ cm^{-3}$$

for ions, so that the electronic density should be $5 \times 10^{22} \ cm^{-3}$. These values are quite credible at short distance from the target. However, it must be pointed out that, on account of the charge Z and the electronic density, one cannot exclude a contribution to the displacement from the so-called polarization shift (18) .

It is possible to reproduce fairly well the general feature of the lines shape by simulation as shows it the figure 8. The density is assumed to decrease exponentially with the distance to the target, with a scalelength of 200μ . The line width is

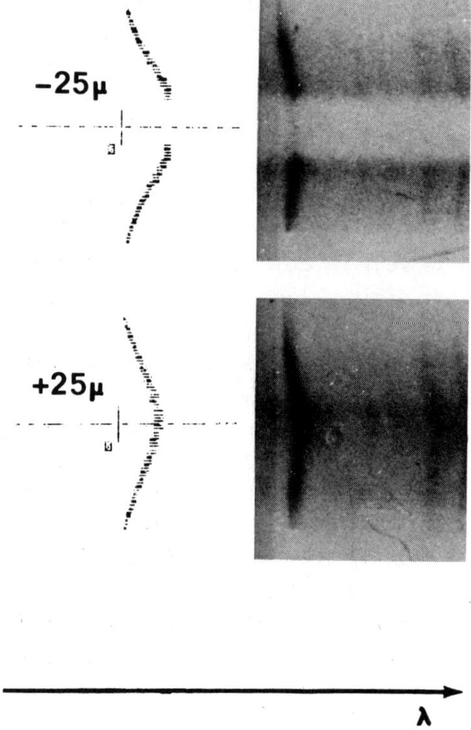

Figure 8 : Simulation of lines

assumed to be proportional to the density and the red shift
proportional to the 4/3 power of the density, according to (5).

In conclusion, since reabsorption does not affect significantly
the observed effect, the measurement of the line shift will likely
be an efficient tool for laser-plasma studies above the critical
density, for which the laser light frequency is equal to the plasma frequency, provided that further improvement of Stark shift
calculation allows an accurate derivation of density from the
experimental results.

REFERENCES

1   J. A. Stamper, "Laser Interaction and Related Plasma
     Phenomena", ed. by H. J. Schwarz and H. Hora
     (Plenum Press, N. Y.), vol. 4B, p. 721 (1977)

2   H. G. Ahlstrom, J. F. Holzricher, K. R. Manes, E. K. Storm,
     M. J. Boyle, K. M. Brooks, R. A. Haas, D. W. Phillion,
     V. C. Rupert, ref. (1), vol. 4A, p. 437

3   E. Yablonovitch, réf. (1), vol. 4A, p. 367

4   P. Jaeglé, G. Jamelot, A. Carillon, A. Sureau, J. Physi-
     que Colloq. 39, C4-75 (1978)

5   J. P. Apruzese, J. Davis, K. G. Witney, J. Quant. Spec-
     trosc. Radiat. Transfer, 17, 557 (1977)

6   B. Yaakobi, D. Steel, E. Thorsos, A. Hauer B. Perry,
     Phys. Rev. Lett. 39, 1526 (1977)

7   R. J. Tighe, F. C. Hooper, Phys. Rev. A 15 1773 (1977)

8   Hoe Nguyen, J. Grumberg, M. Caby, E. Leboucher,
     C. Couland, GRECO ILM   Internal Report n°6 (1978)

9   G. Jamelot, P. Jaeglé, A. Carillon C. Wehenkel, J.
     Physique Colloq. 40, C1-91, (1979)

10   D. Shwarz, J. Virmont, C. Jablon, Bull. Am. Phys. Soc.
     23, 809 (1978)

11   J. S. Pearlman, G. H. Dahlabaka, Appl. Phys. Lett. 31,
     414 (1977)

12   R. F. Benjamin, G. H. McCall, A. W. Ehler, Phys. Rev.
     Lett. 42, 890 (1979)

13   H. A. Bethe, E. E. Salpeter, Quantum Mecanics of one-and
     two-electron Systems (Springer Verlag, Berlin and New
     York), p. 233 (1957)

14   H. R. Griem, Spectral Line Broadening by Plasmas
     (Academic Press, New York and London) (1974)

15   N. Konjevic, W. L. Wiese, J. Phys. Chem. Ref. Data 5
     259 (1976)

16   M. Baranger, Atomic and Molecular Processes, Ed. by
     D. R. Bates (Academic Press, New York and London)
     p. 511 (1962)

17   H. R. Griem, ref. 14 p. 25

18   H. R. Griem, ref. 14 p. 146

NUCLEAR EXCITATION OF ^{235}U BY ELECTRON TRANSITION

IN LASER PRODUCED URANIUM PLASMA

Y. Izawa, H. Otani and C. Yamanaka

Institute of Laser Engineering
Osaka University
Suita, Osaka 565, Japan

ABSTRACT

Production of the 26 min isomer of ^{235}U has been observed in
a laser produced uranium plasma by detecting internal conversion
electrons from ^{235}U^m. It can be attributed to nuclear excitation
by electron transition (NEET). The cross section and the efficiency of NEET are also discussed.

I. INTRODUCTION

In 1973, Morita[1] pointed out that in the deexcitation of the
orbital electron system, a third mechanism named "nuclear excitation by electron transition (NEET)" is theoretically possible besides X-ray and Auger electron emission. He also discussed the
applicability of NEET to isotope separation, especially to that
of ^{235}U. The existence of NEET has been verified experimentally
in ^{189}Os by Otozai et al[2].
Recently Okamoto[3] proposed a new method to improve the efficiency of NEET for uranium isotope separation, in which a scheme
similar to the pellet implosion in inertial confinement fusion is
utilized. A pellet or a hollow shell of natural uranium is bom-
barded by lasers or charged particle beams. The electrons produced by the incident beam collide with the uranium atoms and
ionize them to give rise to NEET. Here we present the experimental
production[4] of a ^{235}U isomer by the NEET process in a laser produced plasma.

II. NEET PROCESS IN ^{235}U

The deexcitation mechanisms of the electron system are shown in Fig. 1. The lines with circles refer to the electronic levels and the open circles indicate the electron holes. In the initial state, there is an electron hole in an inner shell. There are three final states; X-ray emission, Auger electron emission and nuclear excitation. In order to realise the nuclear excitation, two conditions must be satisfied. The first is that the difference between the excitation energy E_N and $(E_2 - E_1)$ should be small compared with the interaction energy of nuclear and elctronic states. The second is that the spin and parity of the resultant system should be conserved.

Nuclear and atomic energy levels in ^{235}U are shown in Fig. 2. The nucleus of ^{235}U has a first excited isomeric level, which decays to the ground state with a half-life of 26 min. This transition is not accompanied by γ-rays but only by very soft internal conversion electrons. The transition energy was determined by Neve de Mevergnies to be 73 ± 5 eV.[5],[6]

If an electron hole is produced at the $5d_3/_2$ level in ^{235}U and the electron transition A occurs from $6p_3/_2$ to $5d_3/_2$, an isomer ^{235}U^m may be produced by the nuclear excitation B. The combination of deexcitation A and excitation B satisfies the required conditions for NEET, because these transitions have the common E3 components and the approximately equal transition energies.

In order to produce the electron hole at the $5d_3/_2$ level of U, we used a laser produced plasma. The reason for this is that by laser irradiation a plasma with electrons of high density $(n_e \gtrsim 10^{19}$ cm$^{-3})$ and moderate temperature (several tens to a few hundreds of eV) is easily obtained.

III. EXPERIMENTAL APPARATUS

Experimental apparatus is shown in Fig. 3. A TEA CO_2 laser beam, whose energy is 1 J in 100 ns at 10.6 μm, was focused by a Ge lens on a metallic natural uranium target in a vacuum chamber. Uranium ions in the laser produced blow-off plasma were deflected by an external electric field and collected on a stainless steel plate. After 100 laser pulses on the target with a repetition rate of 0.5 pps, the collector plate was transferred to the detector which is similar to the one described by Neve de Mevergnies.[5] We used a channel electron multiplier (Murata type EMW6081B), and a control grid (made in stainless steel, 150 mesh, and 70 % open area) located between the collector plate and the multiplier. A variable retarding voltage for the electrons was applied between the plate and the grid. The grid and the multiplier input was kept to be the same potential. The output from the multiplier was counted through an amplifier. Counting was started within 2 minutes after the laser irradiation.

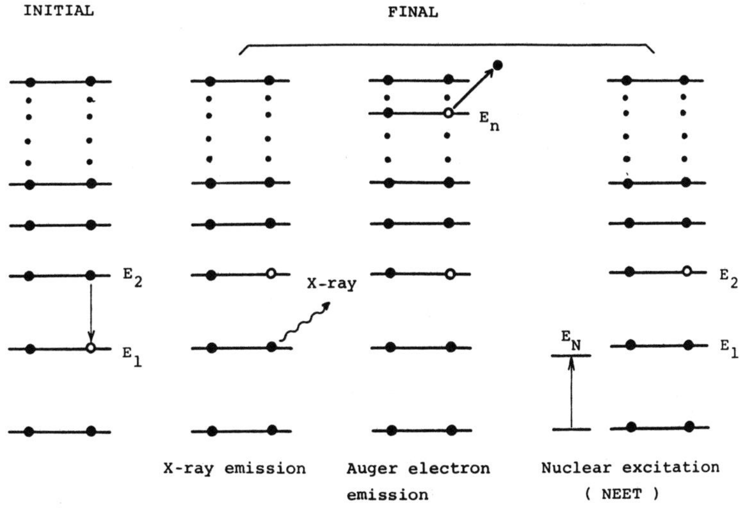

Fig. 1. Deexcitation processes of the electron system.
Initial state (left) and three final states
are shown.

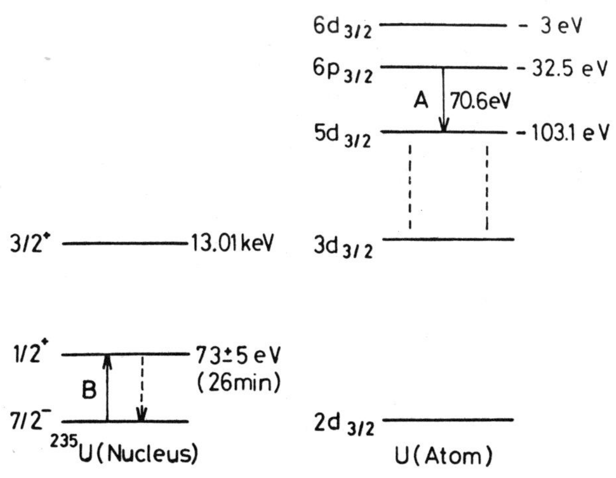

Fig. 2. Nuclear levels in ^{235}U and atomic levels in U.

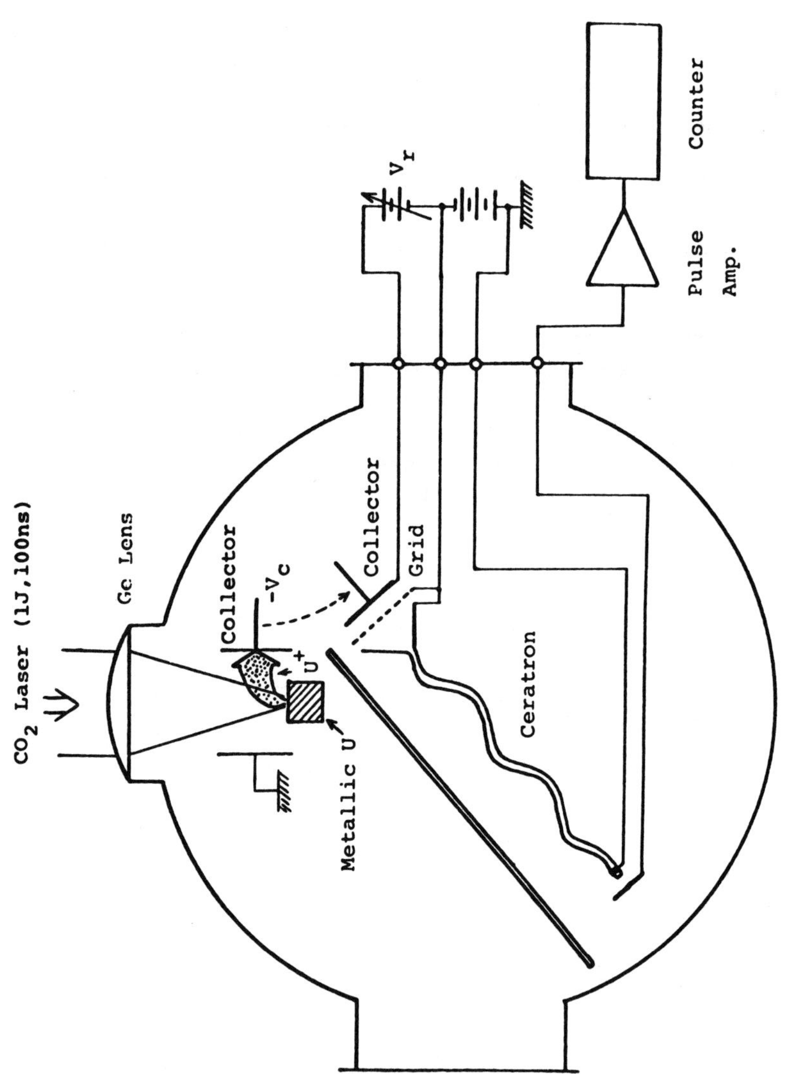

Fig. 3. Experimental set-up of the nuclear excitation.

IV. EXPERIMENTAL RESULTS AND DISCUSSIONS

The decay curve of the counting rate without retarding voltage is shown in Fig. 4 by the open circles. The decay consists of three components, which are a faster decay, a slower decay and a time-independent component. The decaying components are also shown in Fig. 4 by the solid circles, which were obtained by subtracting the time-independent component from the total counting rate. Half-lives of the decays were estimated from least-squre fits to be 1.0 ± 0.1 min for the faster decay and 25.7 ± 0.4 min for the slower decay.

The decay curve with the retarding voltage of 90 V is shown in Fig. 5. In this case, only the faster decay and the time-independent component are observed but the slower decay is not observed. . This shows that the slower decay is caused from the electrons, the energies of which are lower than 90 eV. Therefore it is concluded that the slower decay is attributed to the internal conversion electrons from the isomers ^{235}U^{m}, which were produced by NEET in the laser produced plasma.

The faster decay was observed, whether the collector plate was transferred to the detector or not. But the time-independent component disappeared when the plate was not transferred.

When a high power laser beam is focused on a target, vaporization of the target material occurs at the surface, accompanied by ejection of material. This material consists of neutral atoms, ions and electrons. In our experiments, almost all of the ions and electrons ejected are deflected by the external electric field and collected on the plates. But the neutral atoms expand into the chamber without being deflected. Finally they are pumped out by the vacuum system.

The origin of the faster decay is supposed to be α emission from the neutral ^{234}U and ^{238}U atoms evaporated from the metallic uranium target. The time-independent component seems to be due to α emission from the ^{234}U and ^{238}U ions collected on the plate.

It was tried to roughly estimate the NEET cross section. The number of isomers produced by NEET in our experimental scheme[3] is described by

$$N = n_e \, n \, V < \sigma_N v > \Delta t, \qquad (1)$$

where n_e is the electron density of the laser produced plasma, n is the density of ^{235}U in the plasma, V is the volume of the plasma, σ_N is the NEET cross section, v is the electron velocity and Δt is the laser pulse width. The bracket indicates the average over the Maxwellian distribution of the electron velocities.

From the counting rate for the slower decay and the efficiency

                                                     Y. IZAWA ET AL.

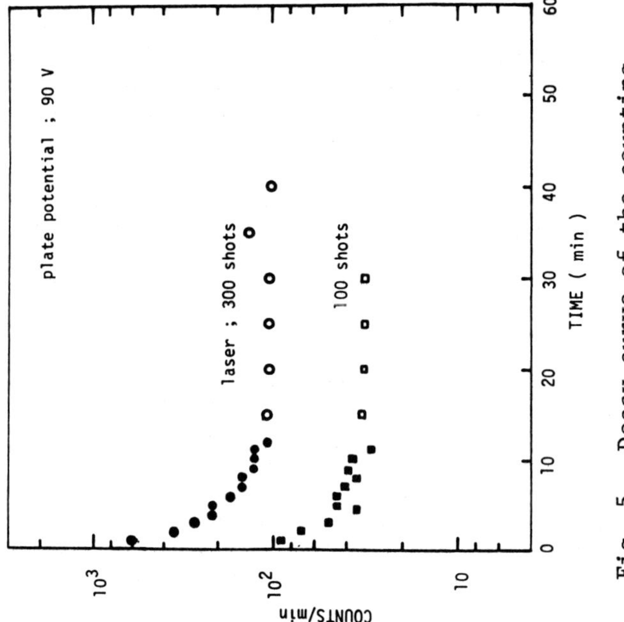

Fig. 5. Decay curve of the counting rate with the retarding voltage of 90 V between the collector plate and the grid.

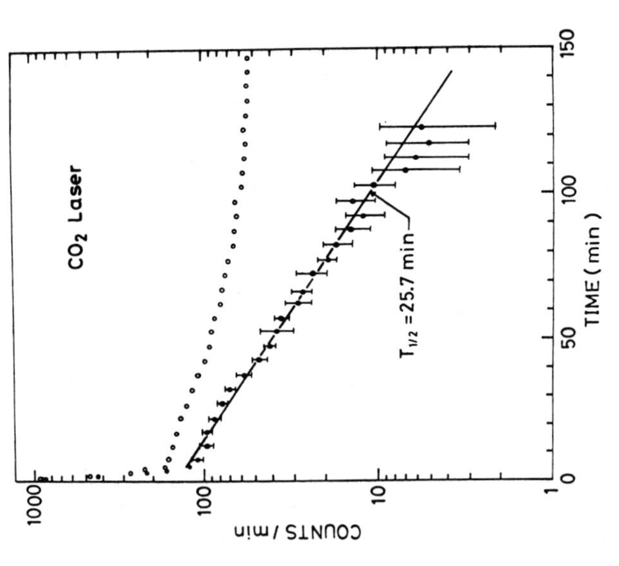

Fig. 4. Decay curve of the counting rate. The open circles are the total counting rate in one minute. The solid circles show the decaying components, which were obtained by subtracting the time-independent component from the total counting rate.

of counting system, $N \sim 10^3$ nuclei/laser pulse was estimated.
In the laser produced plasma, the electron density varies rapidly
with space from the solid density downwards. Here we assume
$n_e = 10^{19}$ cm^{-3}, which is the cut-off density for the CO_2 laser.
This assumption is not so improbable, because at the higher
density region in the plasma, the electron temperature is too low
to produce the electron holes at the $5d_{3/2}$ level and in the underdense region, the electron density is too low to produce enough
NEET.

A value of $<\sigma_N v> = 1.4 \times 10^{-20}$ cm$^3 \cdot$ sec^{-1} was obtained by
using $n = 10^{19}/140$ cm^{-3}, $V = 10^{-6}$ cm^3, and $\Delta t = 100$ ns. This value
is one order of magnitude smaller than that calculated by Okamoto.[3]

Possible competitive processes with NEET are the Coulomb excitation due to inelastic electron scattering by the nucleus and
the photoexcitation of nucleus by X-ray emission from the laser
produced plasma. The cross sections for these processes are far
below comparing with the NEET process.

We define $R = C_{iso}/C_{ion}$, where C_{iso} and C_{ion} are the counting
rate for the faster decay due to the internal conversion electrons
from the isomers at the time $t = 0$ and that for the time-independent
component due to α emission from the uranium ions, respectively.
C_{iso} is proportional to the number of isomers.

$$C_{iso} \propto N \propto n_e \, n \, <\sigma_N v> V \propto n_e^{\,2} <\sigma_N v> V$$

On the other hand, C_{ion} is proportional to the total number of
the uranium ions produced by the laser.

$$C_{ion} \propto n_e V$$

As mentioned above, it is assumed that $n_e = n_c$ (cut-off density
of the laser beam). Therefore R is described by

$$R = C_{iso}/C_{ion} = n_c <\sigma_N v> \qquad (2)$$

and R gives a measure of the NEET efficiency. In order to get the
higher efficiency the laser with shorter wavelength is desirable,
because the cut-off density is inversely proportional to the
square of the laser wavelength. The experimental results for the
CO_2 and Nd : YAG laser irradiations are compared in Table 1.
It is clear that R ($= 2.0 \times 10^2$) at 1.06 μm is two orders of
magnitude greater than R ($= 2.9$) at 10.6 μm.

$<\sigma_N v>$ is a function of the electron temperature of plasma,

Table 1

Measured value of $R = C_{iso}/C_{ion}$ for the CO_2

and Nd:YAG laser irradiations.

	Wavelength (μm)	Cut-off density (cm^{-3})	$R = C_{iso}/C_{ion}$
CO_2 Laser	10.6	10^{19}	2.9 ± 0.2
Nd:YAG Laser	1.06	10^{21}	(2.0 ± 0.3) × 10^2

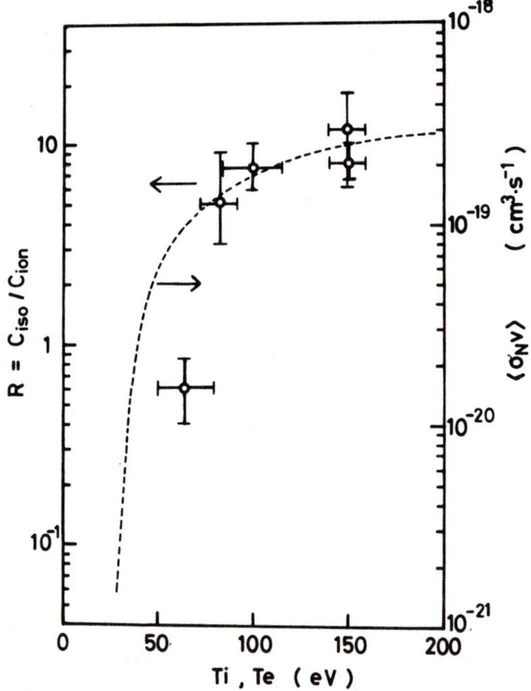

Fig. 6.   Temperature dependence of R for the 1.06 μm irradiations. The dotted line shows the theoretical calcu-
lation of $\langle \sigma_N v \rangle$ as a function of the electron temperature. The vertical scale of the curve was chosen so
as to best fit to the measured values.

which has been calculated by Okamoto[3]. The temperature dependence
of the NEET efficiency was measured by varying the focusing condition of the laser beam to the uranium target. Measured values
of R for 1.06 μm irradiation are shown in Fig. 6 as a function of
the ion energy. In the region where the ion energy is lower than
50 eV, the production of isomer was not observed. In the figure,
the dotted line indicates the theoretical calculation of $< \sigma_N v >$
as a function of the electron temperature. The vertical scale
of the right hand side was chosen so as to best fit to the measured values.

Experimentally obtained dependency of R on the ion energy is
fairly in good agreement with the theoretical calculation except
for the low temperature region.

V. CONCLUSION

The ^{235}U^m isomers were produced by the NEET process in the
laser produced uranium plasma, which was confirmed by detecting
the internal conversion electrons from the isomeric level of ^{235}U.

The temperature dependence of the excitation rate to the
isomeric level in the plasma showed a fairly good agreement with
the theoretical calculation. The efficiency of NEET increased
linearly with the electron density of the plasma, which shows the
availability of the pellet compression to improve the production
rate of isomers.

REFERENCES

1. M. Morita, Prog. Theor. Phys. <u>49</u> (1973) 1574.
2. K. Otozai, R. Arakawa and M. Morita, Prog. Theor. Phys. <u>50</u>
    (1973) 1771.
3. K. Okamoto, J. Nucl. Scie. Tech. <u>14</u> (1977) 762.
4. Y. Izawa and C. Yamanaka, submitted to Phys. Letters B.
5. M. Neve de Mevergnies, Phys. Letters <u>32B</u> (1970) 482.
6. M. R. Schmorak, Nuclear Data Sheets <u>21</u> (1977) 117.

ESTIMATE OF THE EXCITATION RATE OF THE ISOMERIC

LEVEL OF ^{235}U BY LASERS

K. Okamoto

Dept. of Applied Mathematics
University of New South Wales
Kensington, N.S.W., Australia, 2033

ABSTRACT

Izawa and Yamanaka have succeeded in exciting the first isomeric level of ^{235}U by nuclear excitation by electron transitions
(NEET) using lasers. For comparison with this experiment the
estimate of the NEET probability averaging over the Maxwellian distribution of the electron velocity is made. The experimental value
is lower than the theoretical value by about one order of magnitude,
but considering large uncertainties both in theory and experiment
the agreement can be said to be reasonable.

I   INTRODUCTION

When an atom is excited by electrons, photons or ions, and an
atomic electron is ionized, the atomic excitation energy E_A is usually released in the form of X-ray or Auger electrons.   However
Morita[1] has pointed out the third mechanism called nuclear excitation
by electron transitions, abbreviated as NEET, in which E_A is transferred to the nucleus to raise it to its excited state if there
happens to be a level whose energy E_N is close to E_A. The detailed
explanation of NEET is already given in the last workshop.[2]

Morita has developed the theory of NEET, and applied it to the
^{235}U isotope separation, but found that the NEET probability P was
too small for mass production of ^{235}U. However the present author
has shown[3] that the method, which is similar to that of the pellet
bombardment used in inertial confinement nuclear fusion, could

improve the efficiency of NEET by orders of magnitude. The possibility of application of NEET to grasers was discussed in the last workshop.[2] The detailed calculations of P have also been made recently.[4]

In the experimental side Otozai et al.[5] have succeeded in exciting the 6h isomeric level of ^{189}Os by NEET. They have repeated their
experiment and have reported their more detailed results later.[6]

Very recently Izawa and Yamanaka[7] have succeeded in exciting the
26 min. isomeric level of ^{235}U by NEET using CO_2 and YAG lasers, thus
adding the second example of the nucleus excited by this new
mechanism.

Their experiment is particularly important in the following sense.
Since the lasers used produce light of the wave length $1 \sim 10\mu$, their
experiment shows that a nucleus has been excited by infra-red light
for the first time throughout the history of nuclear physics of 70
years. In the history of laser technology, too, this is the first
example of the nuclear excitation by lasers. In this paper the
theoretical estimate of the NEET excitation rate of this ^{235}U isomeric
level averaged over the Maxwellian distribution of the electron
velocity is given and is compared with experiment.

II   EXCITATION RATE

The NEET cross section σ_N is given by

$$\sigma_N = P\sigma_A \tag{1}$$

where P is the NEET probability and σ_A is the cross section to ionize
the electron in question.

The 26 min. isomeric level in question is the first excited
state of ^{235}U, for which E_N is 73 ± 5eV.[8] Thus for atomic transitions to excite this level E_A must also be about 70eV. The relevant
transitions are $P_3O_4(6p_{3/2} - 5d_{3/2}; E_A = 70.6eV)$, $P_3O_5(6p_{3/2} - 5d_{5/2}; E_A = 60.3eV)$, and $O_4O_3(5d_{3/2} - 5p_{3/2}; E_A = 86.7eV)$, of which P_3O_4
is the most important. The values of P for these transitions are
7.22×10^{-11}, 5.05×10^{-13} and 1.44×10^{-11} for the P_3O_4, P_3O_5 and
O_4O_3 respectively, as already shown in Ref. 3. The details of the
calculations are explained in Ref. 4.

However to compare the theoretical results with experiment these
values of P must be converted into σ_N and averaged over the Maxwellian distribution of the electron velocity v.

Such an average, denoted as $< \sigma_N v >$, can be calculated as follows.
Since P is independent of v, $< \sigma_N v > = P < \sigma_A v >$.   Then

$$\langle \sigma_N v \rangle = \frac{2P}{\sqrt{\pi}} \frac{1}{(kT)^{3/2}} \int_1^{\infty} e^{-\frac{E}{kT}} \sqrt{E}\, \sigma_A(E) v(E)\, dE \tag{2}$$

where I is the ionization energy of the electron in question (in the present case $5d_{3/2}$, $5d_{5/2}$, $5p_{3/2}$). After substituting numerical values (2) becomes

$$\langle \sigma_N v \rangle = \frac{6.693 \times 10^7}{(kT)^{3/2}} \cdot I^2 P \int_1^{\infty} \sigma_A(x) e^{-\alpha x}\, x\, dx \tag{3}$$

where $\alpha = I/kT \quad x = E/I$.

To evaluate the integral $\sigma_A(x)$ must be assumed. Unfortunately there is no experimental data on ionization cross sections of outer electrons of uranium, and so the so-called Thomson formula is assumed

$$\sigma_A(x) = 3.519 n \left(\frac{I_H}{I}\right)^2 \frac{1}{x} \left(1 - \frac{1}{x}\right) \times 10^{-16} \; \text{cm}^2 \tag{4}$$

where n is the number of electrons in the shell and I_H is the ionization energy of Hydrogen, 13.6eV. Then $\langle \sigma_N v \rangle$ is given by

$$\langle \sigma_N v \rangle = 2.355 \times 10^{-8} \; n \, P \left(\frac{I_H}{I}\right)^2 (kT)^{\frac{1}{2}} H(\alpha) \tag{5}$$

where

$$H(\alpha) = \alpha [e^{-\alpha} - \alpha \int_{\alpha}^{\infty} \frac{e^{-x}}{x}\, dx] \tag{6}$$

and kT is in eV, $\langle \dot{\sigma}_N v \rangle$ is in cm^3/sec.

Numerical calculations are carried out for the temperature of 100 ~ 200eV, which seems reasonable for the experiment of Ref.7. The resultant values of $\langle \sigma_N v \rangle$ are 20 ~ 30 $\times$ $10^{-20}\text{cm}^3/\text{sec.}$, the contrubution from the P_3O_4 transition being more than 90%, wheras the experimental value is 1.4 $\times$ $10^{-20}\text{cm}^3/\text{sec.}$[7] Hence the experimental value is lower than the theoretical value by about an order of magnitude, as already mentioned in Ref.7. However considering the uncertainties of both theory and experiment, the agreement can be said to be fair.

The author would like to thank Dr. Y Izawa of Osaka University for sending him the experimental data before publication.

REFERENCES

1. M. Morita, Progr. Theor. Phys. 49(1973)1574.
2. K. Okamoto, Laser Interaction and Related Plasma Phenomena
    Vol. 4A, ed. by H.J. Schwarz and H. Hora (Plenum Publishing
    Corporation, 1977) p283.
3. K. Okamoto, J. Nuc. Sci. Tech.14(1977)762.
4. K. Okamoto, submitted to Nuc. Phys.
5. K. Otozai, R. Arakawa and M. Morita, Progr. Theor. Phys. 50(1973)
    1771.
6. K. Otozai, R. Arakawa and T. Saito, Nuc. Phys. A297(1978)97.
7. Y. Izawa and C. Yamanaka, submitted to Phys. Letters and also the
    paper submitted to this workshop.
8. Neve de Mevergnies, Phys. Letters 32B(1970)482.

CONSEQUENCES OF INTENSITY CONSTRAINTS

ON INERTIAL CONFINEMENT FUSION

Ray E. Kidder

Lawrence Livermore Laboratory
University of California
Livermore, California 94550

ABSTRACT

It is shown that the conflicting requirements of high implosion efficiency (*low* corona temperature) and adequate energy
transport (*high* corona temperature) can, together with other
effects, limit useful infrared light intensities to values on the
order of 100 Tw/cm^2. Increased interest in ultraviolet lasers,
for which this intensity constraint is expected to be less severe,
and the entry of charged-particle drivers in the inertial confinement fusion (ICF) competition are consequences of this limitation.

Analytical results based on a simple model are presented
which show how the gain of an ICF target is modified by the existence of an arbitrary intensity constraint.

INTRODUCTION

In 1962 it was estimated that DT ignition by laser-driven
implosion would require a pulse energy of <u>100 kJ</u> (at 0.69 μ) and
an intensity on target of 500 Tw/cm^2 [1]. These requirements,
though far beyond the state of the art at that time, were deemed
feasible, with the result that a research program with this
objective in view was begun at Livermore.

This work was performed under the auspices of the U. S. Department
of Energy by the Lawrence Livermore Laboratory under contract
number W-7405-ENG-48.

Progress in laser development led to the availability in 1967 of focused intensities as high as 10^5 Tw/cm^2 (at 1.06 μ) [2]. Theoretical calculations predicted that intensities in the range 10-10^5 Tw/cm^2 would be capable of exciting a large variety of light-driven plasma instabilities and other nonlinear effects, including the generation of anomalously energetic suprathermal or "hot" electrons [3-7].

In 1971, experiments at Livermore indicated that a hot-electron component with an apparent temperature of 40-50 keV was being generated in a 1-2 keV CD_2 plasma at an intensity of 200 Tw/cm^2 (at 1.06 μ) [8-9]. These experiments also suggested that stimulated Brillouin scattering, for which the threshold intensity was estimated to be approximately 10 Tw/cm^2 [9], was responsible for the observed back-reflection of the incident laser light.

Although the prospects of utilizing such high light intensities did not seem favorable in view of the existing theoretical and experimental picture, it was nonetheless proposed in 1972 that intensities on the order of 10^5 Tw/cm^2 might be used to achieve 10,000-fold compression and central ignition of a bare droplet of liquid DT [10]. The authors of this proposal advised that "the Basov (Soviet) team is likely to reach the so-called break-even point within the next year, slightly ahead of the Americans" [11]. KMS Fusion Inc. advised its shareholders: "Your company predicted to the U.S. Atomic Energy Commission that it would reach break-even in energy before December 31, 1973" [12].

At the same time that these sanguine predictions were being broadcast, it was pointed out that the transport of energy within the laser-heated plasma might be limited to a value nearly two orders of magnitude less than had been previously assumed [13]. This effect, which has been observed experimentally, together with the generation of hot electrons, corona-core decoupling [14], and the excitation of various plasma instabilities, acts to limit the focused intensity that can be effectively used to drive implosions. A result of these limitations is that the "breakeven" predicted with a kilojoule of light in 1973 is now thought to require at least 100 kilojoules, focused to an intensity of a few hundred terawatts/cm^2, values remarkably similar to those initially estimated in 1962.

The focused light intensity that can be effectively used to drive implosions is clearly a matter of primary importance to laser-driven fusion. In this paper we shall consider the factors that limit that intensity, the limits they impose, and the consequences of these limits for inertial confinement fusion generally.

INTENSITY CONSTRAINTS

The least laser pulse energy W_L (joules) required to achieve a "pellet gain" G_p by imploding and centrally igniting a DT target is given by [15]

$$W_L = (G_p/115)^{12/5} \alpha^3 / \varepsilon_{LW}^{17/5} \quad \text{joules,} \tag{1.1}$$

where α is the ratio of the internal energy of the compressed DT to that which would result from isentropic compression at minimum entropy, and ε_{LW} is the ratio of internal energy W_F (of compressed DT) to laser pulse energy. The "pellet gain" is defined as the ratio of the thermonuclear energy yield W_{TN} to the energy of the laser pulse W_L.

The property of this result to which we wish to draw attention is its sensitivity to the efficiency ε_{LW} with which the laser pulse energy is converted to internal energy W_F of the DT fuel. A factor of two reduction in the efficiency ε_{LW} results in a <u>ten-fold</u> increase in the laser pulse energy required to achieve the same gain. It is therefore important that the implosion be efficient.

An estimate of the efficiency of a laser-driven implosion is provided by the efficiency ε of a "rocket" ablatively propelled by a steady, planar isothermal expansion. This efficiency is easily found to be

$$\varepsilon = (\beta^2/2)/(e^\beta - 1), \quad \beta = u/2c_i \quad , \tag{1.2}$$

$$c_i = (2.2 \times 10^7) \sqrt{T_e(\text{keV})} \ \text{cm/sec} \ (Z = 2A \gg 1) \quad , \tag{1.3}$$

where u is the velocity of the "rocket", i.e., imploding shell, and c_i is the exhaust velocity, i.e., isothermal sound speed. The isothermal expansion is actually spherical rather than planar, and for this reason the efficiency will be overestimated by Eq. (1.2). This overestimate can be severe if a substantial part of the absorbed laser light is wasted in maintaining the isothermality of a divergent supersonic flow.

If the efficiency ε as given by Eq. (1.2) is to exceed 10%, for example, then

$$c_i \leq (2.23)u \quad , \quad (\varepsilon \geq 0.1) \quad . \tag{1.4}$$

If we assume a typical implosion velocity u of 3×10^7 cm/sec, it
follows from Eqs. (1.3) and 1.4) that

$$T_e < 9 \text{ keV} \quad , \qquad (\varepsilon \geq 0.1, \quad u = 3 \times 10^7 \text{ cm/sec}) \quad , \qquad (1.5)$$

that is, if we are to achieve adequate propulsive efficiency, the
temperature of the laser-heated corona must be limited.

The maximum power density that can be transported by the
electrons of a plasma is the saturated intensity

$$F_{sat} = n_e [2(kT_e)^3/\pi m]^{\frac{1}{2}} \quad . \tag{1.6}$$

Bickerton [13] first pointed out that this result may overestimate
the maximum achieveable power density by a factor ~ 60 $(= \sqrt{M/Zm})$
if the velocity of thermal energy transport is limited to the ion
acoustic velocity by the unstable excitation of ion sound, a
possibility that arises when $ZT_e \gg T_i$ as is often the case.
Recent experimental results are consistent with transport inhibition factors of this magnitude [16].

If we assume a transport inhibition factor of 60, then the
maximum power density F that can be transported to the ablation
front from the critical surface is

$$F = (1/60)F_{sat}(n_e = n_{ec}) = (3.15)T_e^{3/2}(\text{keV})/\lambda_L^2(\mu) \tag{1.7}$$

$$< 85 \text{ Tw/cm}^2 ; \quad T_e < 9 \text{ keV} \quad , \quad \lambda_L > 1\mu \quad ,$$

where the critical electron density n_{ec} is given by

$$n_{ec} = \pi/r_o \lambda_L^2 \qquad (r_o = e^2/mc^2) \quad , \tag{1.8}$$

for light of wavelength λ_L.

The efficiency ε will therefore be less than 10% if the
absorbed infrared ($> 1\mu$) light intensity is as great as 85 Tw/cm^2,
a limit based on the conflicting requirements of adequate energy
transport and low "rocket" exhaust temperature. The wavelength
dependence exhibited in Eq. (1.7) suggests that the same 10%
efficiency might be achieved at a 10-fold greater intensity if
ultraviolet light (1/3 μ) were used instead of infrared light
(1.06 μ).

This intensity constraint based on the need for efficient
ablative implosion, together with other intensity constraints
imposed by plasma instabilities, hot electron generation, self-
focusing, corona-core decoupling, etc., suggests that light

intensities useful for driving implosions may be limited to values less than ~ 100 Tw/cm^2 (at 1.06 μ or greater) [17]. More generally, we shall assume that the maximum useful light intensity is I_L, and examine the consequences of this assumption.

LEAST MECHANICAL POWER DENSITY NEEDED TO ACCELERATE A THIN
SHELL TO A SPECIFIED VELOCITY

We consider a thin, hollow, spherical shell of solid DT fuel with an initial density ρ_o (=0.2 g/cm^3) and aspect ratio ζ (= $R_o/\Delta R_o$ >> 1). The mechanical power density I_F transmitted into this shell is the product pu of pressure and velocity at the outer surface of the shell. The least peak mechanical power density needed to implode the shell to a desired velocity u is achieved when I_F is constant in time. An elementary calculation shows this least value to be

$$I_F = \rho_o u^3/\zeta \qquad . \qquad (3.1)$$

The efficiency with which the incident light intensity I_L is converted into mechanical intensity I_F supplied to the fuel will be denoted by ε_{LI}. This efficiency will usually be comparable to the efficiency ε_{LW} with which light energy W_L is converted into internal energy of the compressed fuel, i.e.,

$$I_F = \varepsilon_{LI}I_L \quad , \quad W_F = \varepsilon_{LW}W_L \quad , \quad (\varepsilon_{LI} \sim \varepsilon_{LW}). \qquad (3.2)$$

For example, if $\varepsilon_{LW} = 0.05$ (half the incident light is abosrbed and the implosive efficiency is 10%), then a laser intensity I_L of 400 Tw/cm^2 is required to accelerate a solid DT shell, having an aspect ratio of 27, to an implosion velocity of 3×10^7 cm/sec.

LEAST PULSE ENERGY NEEDED TO ACHIEVE IGNITION

The mean internal energy (per gram of fuel) of the imploded fuel at ignition time is the sum of the energy w_S needed to create the central ignition "spark" and the energy w_C required to compress the fuel. This energy is approximately equal to the maximum specific kinetic energy attained by the shell during implosion, i.e.,

$$w = w_S + w_C \simeq u^2/2 \qquad . \qquad (4.1)$$

The energy of compression w_C is, for DT,

$$w_C = \alpha w_o \eta^{2/3} \quad , \qquad (\eta = \rho/\rho_o \gg 1) \quad , \tag{4.2}$$

where the constant w_o equals 0.178 MJ/g for DT, and α is defined as in Eq. (1.1).

The ignition "spark" energy w_S is

$$w_S = f_S e_S \quad , \qquad f_S = M_S/M = (H_S/H)^3 \quad , \tag{4.3}$$

where e_S (= 116 T_S (keV) MJ/g for DT) is the internal energy of the spark, f_S is the fraction of the total fuel mass M encompassed by the spark, and H and H_S denote ρR and ρR_S of the fuel and spark, respectively.

The total internal energy W_F in the compressed fuel can be written as follows in terms of H^3

$$W_F = (4\pi w/3\rho^2)H^3$$

$$\quad = (4\pi w/3\rho_o^2)(\alpha w_o H/w_C)^3 \quad . \tag{4.4}$$

The quantities w and w_C can be eliminated from Eq. (4.4) using the relations

$$w_C = w - (e_S H_S^3)/H^3 \quad , \tag{4.5}$$

$$w = (\zeta I_F/\rho_o)^{2/3}/2 \quad , \tag{4.6}$$

so that W_F is expressed in terms of H, the ρR of the compressed fuel.

The least energy $(W_F)_{MIN}$ needed to achieve fuel-ignition is obtained by minimizing $W_F(H)$, given by Eqs. (4.4–4.6), with respect to H. The result obtained is that

$$(W_F)_{MIN} = 2\pi(8/3)^4(\alpha w_o)^3 e_S H_S^3/(\zeta I_F)^2 \tag{4.7}$$

when H equals H_{MIN}

$$H_{MIN} = 2(e_S H_S^3)^{1/3}(\rho_o/\zeta I_F)^{2/9} \quad . \tag{4.8}$$

$W_F(H)$ also has the property

$$W_F \xrightarrow[H \to H_\infty]{} \infty \quad , \qquad H_\infty = H_{MIN}/2^{2/3} \quad , \qquad (4.9)$$

which implies that the conditions specified cannot be simultaneously satisfied at any finite energy W_F if $H \le H_\infty$.

The properties of the ignition spark are contained in the factor $e_S H_S^3$ appearing in Eqs. (4.7) and (4.8). To ensure ignition, this factor should not be less than $\sim 6 \times 10^{14}$ cgsu, corresponding to a spark temperature T_S of 20 keV and spark H_S of 0.3 g/cm^2 [15].

We note that the least fuel-energy needed to achieve ignition decreases as the square of the mechanical intensity I_F delivered to the fuel, according to Eq. (4.7). A restriction to lower light intensities therefore implies a requirement for strongly increased light-pulse energy to achieve ignition and burn.

We also note that $(W_F)_{MIN}$ decreases with the square of the aspect ratio ζ. The ratio α, however, tends to increase with ζ, the least values achieved in a series of WAZER computer program [18] runs, spanning a large range of aspect ratios, being given by [19]

$$\zeta \simeq 3\alpha^3 \quad .$$

If this effect is accounted for, $(W_F)_{MIN}$ still decreases with increasing ζ, but only as the first power.

INFLUENCE OF INTENSITY CONSTRAINTS ON PELLET GAIN

The "fuel gain" G_F is shown in Fig. 1 as a function of H, the ρR of the compressed fuel. G_F is related to the "pellet gain" G_P according to

$$G_p = \varepsilon_{LW} G_F \quad , \qquad (5.1)$$

the "pellet gain" denoting the ratio of the thermonuclear energy yield to the energy of the incident laser pulse. The solid curves show G_F versus H for a sequence of values of the fuel energy W_F, and are taken directly from Fig. 2 of Ref. [15] where the details of their derivation are described. The curves apply to the case in which the parameter α, defined in Section 2 above, is assumed equal to 2. The solid straight line appearing in Fig. 1 represents the limit of G_F vs H as W_F tends to infinity.

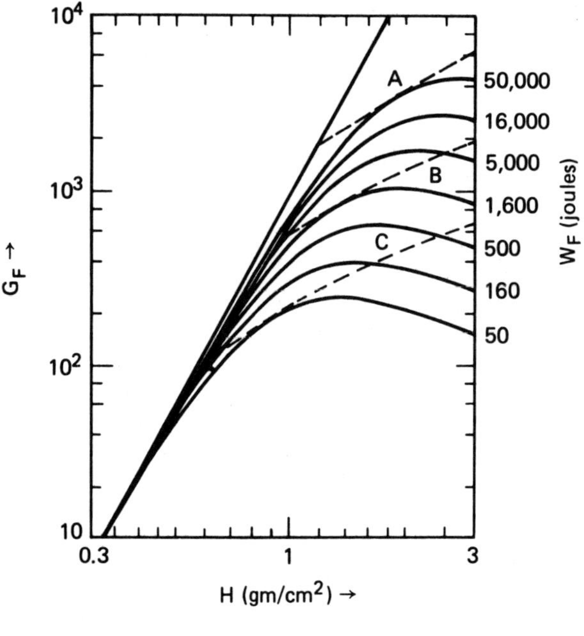

Fig. 1

The dashed curves shown in Fig. 1 show G_F vs H for three values
of the mechanical intensity I_F. They are obtained by substituting
$W_F(I_F)$ given by Eqs. (4.4)-(4.6) into the relations defining
$G_F(H;W_F)$ in Ref. [15], and apply to the case in which the aspect
ratio ζ of the fuel-shell equals 24. The three values of I_F were
chosen so that $(W_F)_{MIN}$ equals 50, 1600, and 50,000 joules, with
the result that the dashed curves are tangent to the solid curves
having these values of W_F.

To translate the results shown in Fig. 1 into laser pulse
energy W_L and intensity I_L, it is necessary to specify the conver-
sion efficiencies ε_{LW} and ε_{LI}. If we assume 50% light absorption
and 10% efficiency in converting absorbed light energy into fuel
internal energy, then ε_{LW} equals 0.05. The energies W_F listed in
Fig. 1 should therefore be multiplied 20-fold to obtain the corresponding laser pulse energies W_L, which then range between one kilo-
joule and one megajoule. If ε_{LI} equals ε_{LW}, then the listed values
of I_F should similarly be multiplied 20-fold to obtain the light
intensities I_L that apply to the dashed curves (A), (B), and (C),
namely

$$(A) \quad 3.5 \times 10^{14}$$

$$(B) \quad 2.0 \times 10^{15} \quad (w/cm^2)$$

$$(C) \quad 1.1 \times 10^{16}$$

We see that if central ignition and burn are to be accomplished with as little as one kilojoule of laser light the light intensity used must exceed 10^{16} w/cm^2 (with $\zeta = 24$), and the maximum pellet gain G_p achievable at this energy is approximately 10. These conditions are reminiscent of those postulated to provide "breakeven at a kilojoule" in the early 70's.

On the other hand, if effects such as those considered in Section 2 should limit the useful light intensity to less than 3×10^{14} w/cm^2, then at least one megajoule of laser light will be needed to achieve ignition and burn under the conditions we have specified. However, the pellet gain that is required for commercial electricity generation via pure-fusion ($G_p > 200$) then becomes potentially possible.

SUMMARY AND CONCLUSIONS

Limitations on useable light intensity (at 1.06 μ or greater wavelength) have resulted in increased light-pulse energy requirements for laser-driven fusion. Breakeven is no longer thought possible at a kilojoule. Indeed, current estimates are that breakeven will require pulse energies in the range 100-400 kilojoules, to be provided by the SHIVA-NOVA facility in the early 1980's. These conditions have led to increased interest in shorter wavelength lasers, such as KrF (0.250 μ) and XeCℓ (0.306 μ) for which focused intensity constraints are expected to be less severe.

The limitation of useable light intensity to a few hundred terawatts/cm^2 has allowed charged-particle drivers, especially ion beams, to enter the ICF competition, since it is projected that such intensities may be achievable with these drivers. These drivers can also more readily provide the multi-megajoule pulses that appear to be needed for commercial ICF, have high energy-conversion efficiency, demonstrated pulse repetition rate capability (in the case of heavy ion drivers), and appear to enjoy markedly superior target-coupling characteristics.

Heavy-ion accelerators are projected to be capable of achieving sufficiently low emittance at high power that their beams, like laser beams, can be focused to high intensity in vacuum at the stand-off distances required for ICF reactor survival. They are late entries in the ICF race, but at the moment seem to show considerable promise for the commercial production of ICF power, and for the breeding of fissile fuels [20].

REFERENCES

[1]   R. E. Kidder, <u>LRL Laser Research Program</u>, COTM-63-7, 1963
      (unpublished).
[2]   M. Cox et al., *Laser Focus* <u>3</u>, 21 (1967).
[3]   D. F. DuBois and M. V. Goldman, *Phys. Rev. Letters* <u>14</u>, 544
      (1965).
[4]   K. Nishikawa, *J. Phys. Soc. Japan* <u>24</u>, 1152 (1968).
[5]   P. K. Kaw and J. M. Dawson, *Phys. Fluids* <u>12</u>, 2586 (1969).
[6]   R. E. Kidder, "Interaction of Intense Photon and Electron
      Beams with Plasmas," in <u>Physics of High Energy Density</u>,
      edited by P. Caldirola and H. Knoepfel (Academic Press,
      New York, 1971).
[7]   A. J. Palmer, *Phys. Fluids* <u>14</u>, 2714 (1971).
[8]   S. W. Mead et al., *App. Opt.* <u>11</u>, 345 (1972).
[9]   J. W. Shearer et al., *Phys. Rev.* <u>6A</u>, 764 (1972).
[10] J. Nuckolls et al., *Nature* <u>239</u>, 139 (1972).
[11] Newsweek, January 31, 1972.
[12] KMS Industries Inc. Shareowners Report, Second Quarter 1973,
      August 10, 1973.
[13] R. J. Bickerton, *Nuclear Fusion* <u>13</u>, 457 (1973).
[14] R. E. Kidder and J. W. Zink, *Nuclear Fusion* <u>12</u>, 325 (1972).
[15] R. E. Kidder, *Nuclear Fusion* <u>16</u>, 405 (1976).
[16] P. H. Lee and M. D. Rosen, *Phys. Rev. Letters* <u>42</u>, 236 (1979).
[17] B. H. Ripin et al., *Phys. Rev. Letters* <u>43</u>, 350 (1979).
[18] R. E. Kidder and W. S. Barnes, <u>WAZER, A One-Dimensional, Two-
      Temperature Hydrodynamic Code</u>, Lawrence Livermore Laboratory
      Report UCRL-50583, January 31, 1969.
[19] R. E. Kidder, *Nuclear Fusion* <u>16</u>, 3 (1976).
[20] H. A. Bethe, *Physics Today* 32, 44 (May 1979).

THE POTENTIAL ROLE OF ADVANCED FUELS IN

INERTIAL CONFINEMENT FUSION

George H. Miley

Fusion Studies Laboratory
Nuclear Engineering Program
University of Illinois
Urbana, IL 61801

ABSTRACT

The potential importance of developing advanced (non D-T)
fuel pellets for inertial confinement is discussed. Reduced radioactivity due to low tritium involvement and less neutron activation, improved blanket flexibility with the removal of tritium
breeding requirements, and improved mating of the output energy
spectrum with non-electrical applications such as synthetic fuel
production could lead to technical advantages and earlier public
acceptance. As a possible first step to advanced-fuel pellets,
the A-FLINT concept of a D-T core ignited, deuterium pellet is
proposed which would offer tritium self-sufficiency. A design is
described that uses 0.1-MJ internal energy in a $\rho R \sim 7$ gm/cm^2
compressed pellet, giving a tritium breeding ratio of ~ 1.0 and an
internal pellet gain of ~ 700. While not optimized, these results
appear encouraging, especially for efficient drivers such as
heavy-ion-beam accelerators.

INTRODUCTION: ADVANCE FUELS - THE MOTIVATION

Advanced fuels (AFs) have been defined as any fusion fuel
other than "conventional" D-T. Indeed, if elements through boron
are considered, several dozen reactions have non-negligible fusion
cross sections below 1 MeV.[1] However only a few have been
identified that combine a reasonable cross section, the possibility of ignition, and natural availability. Thus another way to
view these fuels, indicated in Fig. 1, is to divide them into
"low-temperature" deuterium-based fuels that can ignite below 50

- LOW TEMPERATURE (D-based) ⎫ ignite
    cat.-D, D-^{3}He, D-D   ⎭ < 50 keV
- HIGH TEMPERATURE
    (p-based)
        p-^{6}Li, p-^{11}B   ⎫ minimum
    (D-based)   ⎬ neutrons
        D-^{6}Li, ...   ⎭

Figure 1.   Classification and characteristics of advanced
            fuels.

keV and "high-temperature" fuels where ignition, if possible,
requires temperatures above ~200 keV. Of particular interest are
the proton-based fuels since, without the presence of deuterium
(except for some produced in side reactions), neutron production is
minimized, leading to the popular designation "neutronless" fuels.
Of the low temperature fuels D-^{3}He is the closest competitor in
this respect, but unfortunately the ^{3}He must be bred, e.g. via a
semi-catalyzed-D reactor[2].

There are two key reasons to consider AF fusion in spite of
the increased difficulties associated with the higher temperature
required. First, this offers a way to reduce the tritium inventory and handling, plus eliminating the need for a lithium
breeding zone in the blanket[3]. Second, in principle, the fuel
can be selected to provide an output energy split best suited to
the application.[4] Thus, if "free" neutrons (i.e., not required for tritium breeding) are desired, catalyzed-D is an ideal
fuel.* If photon radiation is to be emphasized, p-^{11}B is a good
choice.

For the above reasons AF fuel potentially offers some important advantages. First, and perhaps paramount, is quicker public
acceptance due to the improved safety, easier maintenance (due to
reduced neutron activation) and higher efficiency (assuming advantage is taken of the increased charged-particle energy-fraction
to employ direct conversion). Second, the potential ability to
better adapt fusion to a variety of uses such as synthetic fuel
production could heighten interest in this energy source. Even

---

*Catalyzed-D (or Cat.-D) refers to a plasma where the products T
and ^{3}He from D-D reactions thermalize in the plasma and react
with deuterium at the same rate they are produced [1].

for electrical production new avenues may be opened. For example,
the development of small, relatively "clean" satellite reactors
operating on D-^{3}He fuel has been suggested.[2] These satellites could be located close to the user while larger semi-
catalyzed-D reactors supplying the ^{3}He would be more remotely
sited. This ability to create a combination of small satellites
simultaneously with central power stations could potentially
provide a necessary and attractive approach to our expanding
energy needs in the twenty-first century.

TIMING AND DEVELOPMENT SCENARIOS

     It may be thought that AF fusion, due to the higher tempera-
ture plasma required, is well behind D-T in time. This view however, may exaggerate the difficulty of heating to higher
temperatures once we learn how to generate fusion grade D-T plasmas. Since both catalyzed-D and D-^{3}He fuels ignite below 50
keV[2], these fuels could provide the stepping stone to
other AFs. Indeed, while D-T seems destined to be the choice for
breakeven and early demonstration experiments, the possibility of
developing catalyzed-D for first generation commercial plants can
not (and should not) be ruled out.

     This line of reasoning raises two questions: What confine-
ment approach is best suited to burning AFs? What developmental
scenario can be envisioned? Let us briefly consider each question
in turn.

     As illustrated in Fig. 2, both inertial and magnetic confine-
ment have advantages and disadvantages relative to advanced fuels.
A key factor in favor of inertial confinement is the elimination
of cyclotron radiation (due to the absence of confining magnetic
fields). Indeed, cyclotron losses become increasingly crucial at
the higher plasma temperatures required by AFs. The need for
pulsed operation, however, introduces problems. With the very
large investment of energy required to achieve a burn with AFs,
the recirculating power fraction increases, placing increased
emphasis on the development of efficient, relatively low-cost,
drivers (i.e., advanced lasers, accelerators, etc.). Both confinement approaches, however, must face the issue of how to
develop adequate heating methods to achieve the higher temperatures (indeed, as discussed later, for inertial confinement, this
appears to mandate the use of a D-T spark ignition process).
Finally, to take advantage of the AF, simplified blanket designs
are desired. Elimination of hazards associated with liquid metals
appears to be a desireable part of this goal.[5] While this
is relative straightforward for magnetic systems (once tritium
breeding is eliminated)[6], inertial confinement presents a
more complex problem. Various schemes for protection of the first

Approach	Component			
	Cyclotron Radiation	Recirculating Power	Heating	Blanket
Inertial	+		?	
Magnetic		+	?	+

Figure 2.   Comparison of confinement approaches for
advanced fuels.   A (+) indicates a positive
advantage.

wall against both radiation and shock damage have involved liquid
metals in the form of "water falls", ablative films, etc.[7]
Alternative approaches such as gas or magnetic protection appear
less well developed but would be most desirable for alternate
fuels.[8]   Thus, preliminary studies where the exhaust
plasma is guided to a MHD conversion channel illustrate the
direction needed for efficient recovery of the fusion energy
contained in this plasma.[9]

If advanced fuels are deemed attractive, serious thought must
be given to ways to introduce them as early as practical in the
development of fusion power.   A rough example of a possible
approach for inertial confinement is shown in Fig. 3.   The
Engineering Test Facility (ETF), presently slated to follow
scientific feasibility by DOE[10], could be designed with
sufficient driver capability to both conduct engineering tests
with D-T pellets and establish scientific feasibility for a
catalyzed-D pellet (termed A-FLINT as described later).   Depending
on the results of these tests, the catalyzed-D approach could then
be thrown into competition with D-T for development as an
electrical generating plant, and (or), as discussed earlier, be
developed for non-electrical applications.

<u>Difficulties and the Challenge</u>

Cross sections for selected fusion reactions are shown in
Fig. 4.   The lower reactivities (i.e., $\langle\sigma v\rangle$ values) for the
advanced fuels, combined with the fact that $\langle\sigma v\rangle$ peaks at higher
temperatures, introduce a basic obstacle to effective use of these
fuels.   This is most easily illustrated for magnetic confinement.
Assuming that these systems are limited by the magnetic field
strength, then it is easily shown using pressure balance that the
plasma power density is proportional to $\langle\sigma v\rangle/T^2$.   Consequently,
as seen from Fig. 5, the maximum power densities for even the

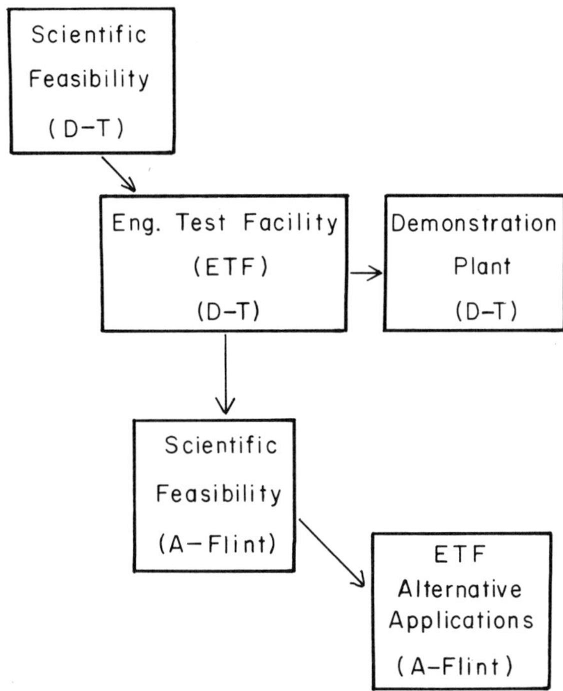

Figure 3. A possible strategy for introducing AFs through the early introduction of a deuterium burning A-FLINT type pellet.

deuterium-based AFs are approximately two orders of magnitude lower than for D-T. Further, both the ignition temperature and the temperature corresponding to the maximum power density are higher than for D-T by a factor of four or more.

Based on these observations, it might be thought that AFs could never be competitive. This overlooks some other factors, however. For example, studies show that as higher β devices are developed (e.g. flux-conserving tokamaks, field-reversed mirrors, etc.) the power densities in D-T devices will be restricted by wall-loading limitations to values below the maximum suggested by Fig. 5. (Neutron damage limits seem most restrictive for D-T while, in contrast, the surface heat load due to radiation limits AF power densities). Consequently, under these conditions, AFs could have power densities comparable to D-T[11-13] and this, combined with their other advantages, would make AFs the preferred approach.

It might be anticipated that analogous considerations apply to AFs for inertial confinement. A very simple model can be used

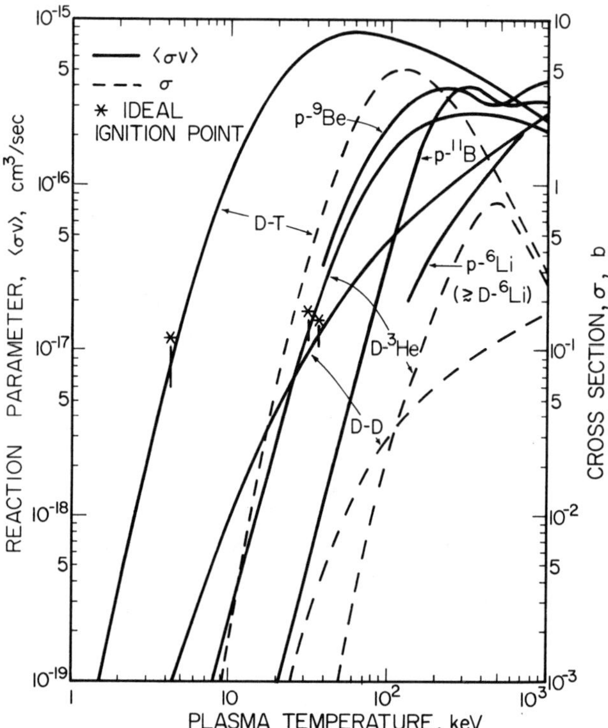

Figure 4.   Cross-section (σ) and reactivity ($\langle\sigma v\rangle$) for key fusion
reactions. Although the peak $\langle\sigma v\rangle$ for reactions like
D-^{3}He and p-^{11}B come within a factor of 2-3 of the peak
D-T reactivity, their peaks occur at much higher temperatures (200-400 keV vs ~ 60 keV for D-T). From Ref. 1.

to illustrate this. We assume a homogeneous pellet uniformly
heated to equal ion (i) and electron (e) temperatures (Θ) during
the burn (B):

$$\Theta_i \sim \Theta_e \sim \Theta_B \quad . \tag{1}$$

While this model neglects important effects such as spark heating,
since we are only interested in relative trends, it is sufficient
for present purposes. Then the pellet energy gain G_c, defined
as the total fusion energy produced divided by the energy invested
in heating the pellet, is simply

$$G_c \sim K_0 \; \frac{\langle\sigma v\rangle \, E_f}{E_c} \, n\tau_B \quad , \tag{2}$$

where E_f is the fusion energy per reaction, n is the compressed
pellet ion density, τ_B is the burn time, E_c is the heating
energy per gram, and K_0 is an appropriate numerical factor.

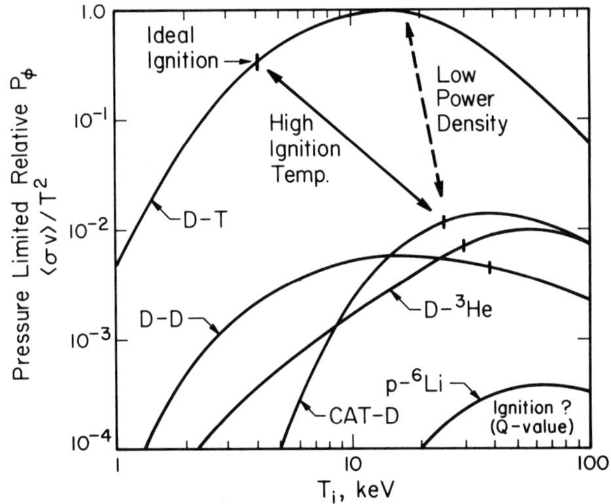

Figure 5.   Relative power densities achieveable with various
            fuels confined in a pressure limited (i.e., magnetic
            field strength limited) system. From Ref. 1.

With the conventional assumptions of a sonic time limit for the
burn, adiabatic compression, and neglecting degeneracy effects, we
can use the following relations

$$E_c \sim K_1\, \Theta_B \tag{3a}$$

$$n\tau_B \sim K_c\, R/\Theta_B^{1/2} \tag{3b}$$

$$\Theta_B \sim K_3 \rho^{2/3} \quad , \tag{3c}$$

where ρ and R are the compressed density and radius respectively,
and the K's are appropriate numerical constants. The total
heating energy required, E_c, is then easily found to be

$$E_c \sim K_4\, R^3 \Theta_B^{5/2} \quad . \tag{3d}$$

If it is assumed that we are[11] driver limited, i.e., E_c
is limited to a maximum value independent of the pellet or fuel,
Eqns. (3a)-(3d) can be substituted into Eqn. (2) to give:

$$G_c \sim K_5\, \langle \sigma v \rangle\, E_f/\Theta_B^{5/6} \quad . \tag{4}$$

In other words, according to this model, the energy gain for AFs
suffers a disadvantage factor proportional to $\langle \sigma v \rangle\, E_f/\Theta_B^{5/6}$

(analogous to the $\langle \sigma v \rangle\, E_f/T^2$ disadvantage for magnetic confinement systems).

A slight variation might be considered whereby the pellet would be heated to Θ_c to achieve ignition while bootstrapping, the burn would occur at a higher temperature Θ_B. This gives

$$G_c \sim K_3\, \langle \sigma v \rangle\, E_f/\Theta_B^{1/2}\, \Theta_c^{1/3} \quad . \tag{5}$$

The precise relation between the temperatures depends on specific conditions. However, for present illustration, we assume that Θ_B/Θ_c varies in proportion to the ratio of $\langle \sigma v \rangle^{1/2}$ for various fuels.

Results are illustrated in Table 1 where the ratios of the gain for D-T to that for various AFs are shown. Here the burn temperature is set to roughly correspond to the peak $\langle \sigma v \rangle$ value for all fuels except D-D and p-^{6}Li where 500 keV is arbitrarily selected as a maximum value.

As seen from this table, the various AFs suffer a disadvantage factor ranging between one and two orders of magnitude compared to D-T. While this seems quite severe, it might be properly questioned whether or not moderating factors may exist as in the case of magnetic confinement described earlier. Indeed, the situation is not too clear. For example, one factor may be that spark ignition will be more difficult to achieve in the smaller D-T pellets. An approach stressed here that can reduce the disadvantage is to use a D-T core that is spark ignited and then burn propagates into a surrounding AF layer. Despite this, however, the AF pellet gain will probably remain well below that for D-T. This places a strong stress on finding an efficient driver so that the larger recirculation fraction involved does not place an undue penalty on

Table 1. Gain Deficit

	Θ_B (keV)	$\langle \sigma v \rangle / \langle \sigma v \rangle_{DT}$	Gain Ratio	
			$\sim\ \langle \sigma v \rangle / \Theta_B^{5/6}$	$\sim \langle \sigma v \rangle / \Theta_B^{1/2}\Theta_c^{1/3}$
D-T	60	1.0	1.0	1.0
D-D	(500)	0.25	15–59	10–42
D-^{3}He	300	0.38	10	7
p-^{6}Li	(500)	0.25	80	60
p-^{11}B	300	0.38	20	15

the AF plant. If this is possible, then the advantages of reduced
tritium and activation would swing the scale in favor of AFs.

An alternative view of these considerations is instructive at
this point. An intermediate form of Eqn. (4) is:

$$G_c \sim K_6 \langle \sigma v \rangle E_f R \qquad\qquad\qquad (6a)$$

$$E_c \sim K_7 R^3 \Theta_B^{5/2} \quad . \qquad\qquad\qquad (6b)$$

In other words, the gain scales directly with the pellet radius R.
However, as the pellet radius increases, so does its volume, hence
mass. This results in a rapidly increasing input energy requirement E_c, so that if a upper limit is set on E_c, the maximum
radius, hence gain is fixed. Again the disadvantage suffered by
AFs due to a lower $\langle \sigma v \rangle$ and higher Θ_B is evident.

These observations can be summed up as follows. Competitive
AF pellets will require a large ρR pellet. However, regardless of
the compression scheme employed, there will be a limit (ρ_{max})
to the density achievable. Then to maximize ρR, a large R, hence
large mass, is desirable. This will be limited, however, by the
input energy E_c available for heating.

The key questions then boil down to: Do the benefits of AF
pellets outweigh the complication of a potentially lower gain? Can
the high compressions and large E_c required be achieved in practice? Can D-T core ignition and burn propagation be used to reduce
the constraints on burning AF pellets? Will the latter processes,
yet to be demonstrated experimentally, eventually work? While the
above questions were posed for AF pellet development, it is important to realize that the basic physics of high compression, core
ignition and burn propagation are also crucial to the attainment of
high gain D-T pellets. Thus success with D-T represents a strong
step towards AFs.

The calculations described in the remainder of this paper
were designed to explore the potential of D-T core ignition and
burn propagation in some detail.

Prior AF Pellet Studies

Earlier AF pellet studies have considered the extreme cases
of a deuterium and a p-^{11}B pellet. We will review the latter
first. While the p-^{11}B approach was first proposed for inertial
confinement by Weaver, et al.[14], they did not present specific
results. However, later calculations by Hora[15] and by
Moses[16] are summarized in Table 2. These calculations

Table 2. p-^{11}B Pellet Studies

	E_o (MJ)	ρR (g/cm^2)	G_c
H. Hora, Ref. 15	0.1	(a)	1
	1.0	(b)	20
G. Moses, Ref. 16	(c)	500	170

(a) Compression ratio, 10^3
(b) Compression ratio, 10^4
(c) Implosion velocity, ~3 x 10^8cm/sec; spark ignition "micro core" of ρR ~ 50g/cm^2.

were performed prior to revisions of the "best estimate" for the p-^{11}B cross section which have made prospects for ignition seem questionable. At any rate, the results of Table 2 show that the operation with such pellets will be very demanding on technology. Consequently, the present studies have concentrated on deuterium (D) pellets since they appear to represent the best prospect for a follow-on to D-T pellet operation.

Prior D-pellet studies are summarized in Table 3. Wood[17] first disclosed the possibility of a catalyzed-D type pellet that would be tritium self-sufficient. While details were not presented, he indicated that ignition would only require about three times the input energy needed for an "equivalent" D-T pellet. Subsequently, Nuckolls, et al.[18] included a similar pellet (called a deuterium burner) as one of four reference designs for heavy-ion beam studies. Again, details were not given, but as indicated in Table 4, the competitive use of such pellets in place of a "standard" D-T design would require an increase in accelerator efficiency from 10 to 50% along with a higher repetition rate (20 vs 2 pulses per sec). While more demanding, such improvements in performance do not appear beyond reach, especially since the input energy requirement remains at 10 MJ as per the Standard D-T pellet.

Other D-pellet studies have been reported by Moses[16] and by Skupsky[19]. While Moses used spark ignition, he assumed pure deuterium. Since ignition requirements for pure deuterium are almost an order of magnitude more severe than for the more energetic cat.-D fuel[20], it is perhaps not surprising that he found such large energy input and ρR values were necessary. Skupsky likewise considered spark ignition, but in his case, a D-T spark core was used along with an outer deuterium layer seeded with up to 10% tritium. Still, his reference design involved a relatively large ρR and, while giving a reasonable gain, it did not achieve tritium self-sufficiency.

Table 3. D-Based Pellet Studies

REFERENCE	TYPE	E (MJ)	ε (J/gm)	ρR (gm-cm^{-2})	TBR	G_c
WOOD, Ref. 17	Cat.D Burner (T; ^{3}He Seed)	3 x DT	?	?	$\gtrsim$ 1.0	1/2 x DT
NUCKOLLS, Ref. 18	Cat.D Burner (T; ^{3}He Seed)	10	3×10^7	?	$\gtrsim$ 1.0	?
MOSES, Ref. 16	Pure D Spark	> 100	1.5×10^9	40-80	–	200-300
SKUPSKY, Ref. 19	50/50 D-T Spark; 90/10 Outside	0.16 abs.	1.6×10^8	25	~ 0.4	580
1978 A-FLINT, Refs. 21-23	50/50 D-T Spark; Pure D Outside	1.8 abs.	9.7×10^7	13	~1.1	1700
Present A-FLINT		0.1 abs.	5.87×10^7	6.8	~1.0	700

The present studies were initiated in 1978[21-23] in an attempt
to achieve a tritium self-sufficient pellet design with minimum input requirements. The approach taken utilized spark ignition with
a D-T core (radius equal or larger than the spark region -- see
Fig. 6) which results in burn propagation into an outside deuterium layer. As indicated in Fig. 7, tritium left unburned after the
implosion-burn would be extracted in the burn-chamber exhaust,
chemically separated from other species, and then returned to the
pellet "factory" for use in manufacture of subsequent D-T micro-
cores. Assuming losses of $\lesssim$ 10%, a *Tritium Breeding Ratio* (*TBR*) of
about 1.1 should provide a closed cycle which avoids the need to
breed tritium in a lithium blanket. (Here we define TBR as the
amount of tritium contained in the chamber exhaust divided by that

Table 4.   Target Requirements*

	E(MJ)	P(TW)	(%)	RR(s^{-1})	GeV	Confid.
Standard	10	600	10	2	100	Hi
Adv.	1	100	10	20	40	Mod.
D	10	600	50	20	100	Mod.
non-cryo.	10	600	50	10	100	Mod.

*From Ref. 18

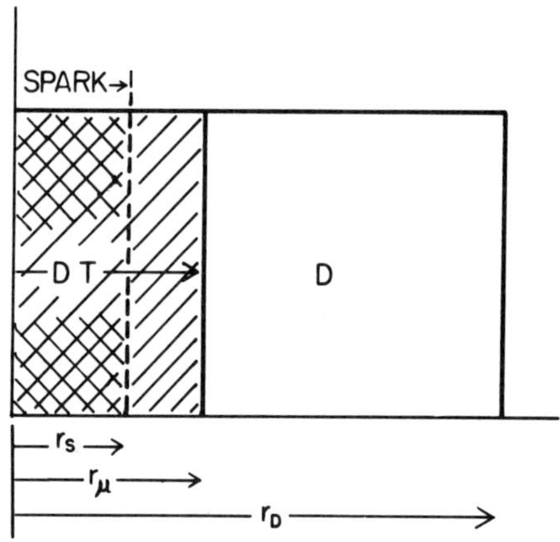

Figure 6.   Conceptual layout of the A-FLINT pellet. Shock heating
             of a spark core occurs in a region of radius r_s, the
             D-T micro-core region extends to radius r_μ (which could
             equal r_s), while an outer region of deuterium is
             possibly seeded with small amounts of tritium. Finally,
             a tamper (not shown) region can surround the deuterium.

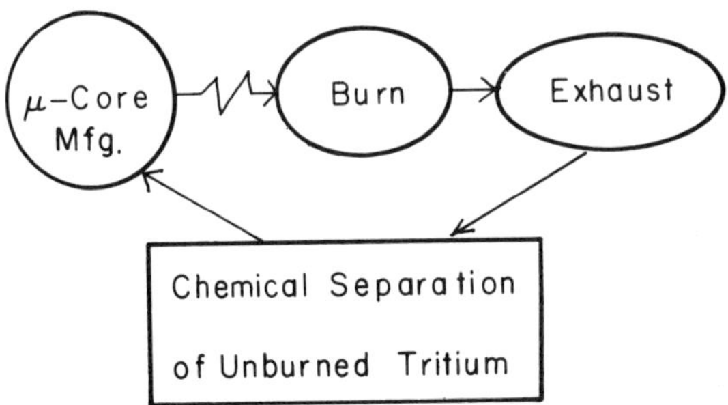

Figure 7.   Flow diagram for tritium whereby tritium produced by
D-D reactions in the burn is used to manufacture
subsequent D-T micro-cores. A pellet tritium breeding
ratio (TBR) slightly greater than 1.0 is required for
self-sufficiency, i.e., to eliminate the need for a
lithium breeding blanket.

required to manufacture the original pellet, including the D-T
core.) Consequently, this pellet concept was named as A-FLINT,
signifying Advanced-Fuel Layered Ignitor/pellet Nurturing Tritium.

As indicated in Table 3, the original A-FLINT calculations[21-23] assumed a heavy-ion driver and rather large energy
inputs ($\sim$ 1.8 MJ absorbed). In contrast, the studies described
here were designed to study A-FLINT with smaller input energies.

The basic restrictions placed on the present A-FLINT studies
are listed in Table 5. If a 10% coupling efficiency can be
achieved, the 100-kJ absorbed energy could be obtained with a 1-MJ
driver, not a unreasonable goal, especially for heavy-ion beam approaches. In order to assure tritium self-sufficiency, stress is
placed on obtaining an acceptable TBR. The pellet gain must then
be optimized within these limitations.

To implement such a strategy, it is necessary to study the
key issues listed in Table 6. Some questions (such as the minimum
practical spark radius, r_s, and maximum compressed density,

Table 5. Strategy

- 100 kJ max. absorbed energy

- min. TBR ~ 1.1

- optimize gain

ρ_{max}) ultimately require experimental verification. In this
sense, the present studies must be viewed as quite preliminary.

In the present studies, a maximum compression ratio of
5×10^3 was selected somewhat arbitrarily. While still high compared to anticipated near-term performance, it is an order of
magnitude lower than used in many early studies such as Skupsky's
calculation.[19] The minimum compression is set both by
the density necessary to maintain burn propagation and by energy
input limitations (as discussed earlier, for constant ρR with decreasing ρ, the pellet mass must increase). The compression of
5×10^3, although not optimum, appears "comfortable" in these
respects for present purposes.

Calculational Model -- Ashes

Calculations employed an improved version of the burn program
Ashes that was briefly described in Ref. 21. Basically, this program starts from initial compressed density-temperature profiles
that are either assumed or taken from an implosion code. In the
present case, an ideal adiabatic compression is assumed along with
uniform shock-heated spark core.

Program Ashes then treats the burn phase using explicit
fusion-product transport superimposed on a one temperature-one
fluid model of the background plasma. The pellet is zoned by
concentric shells, and for tracking fusion products, is broken up
into angular bins (Fig. 8). It has been written in a modular form

Table 6. Key Questions

- min. achievable spark radius

- max. achievable compression

- optimum r_μ/r_D

- degree of tamping possible

ANGULAR BINS

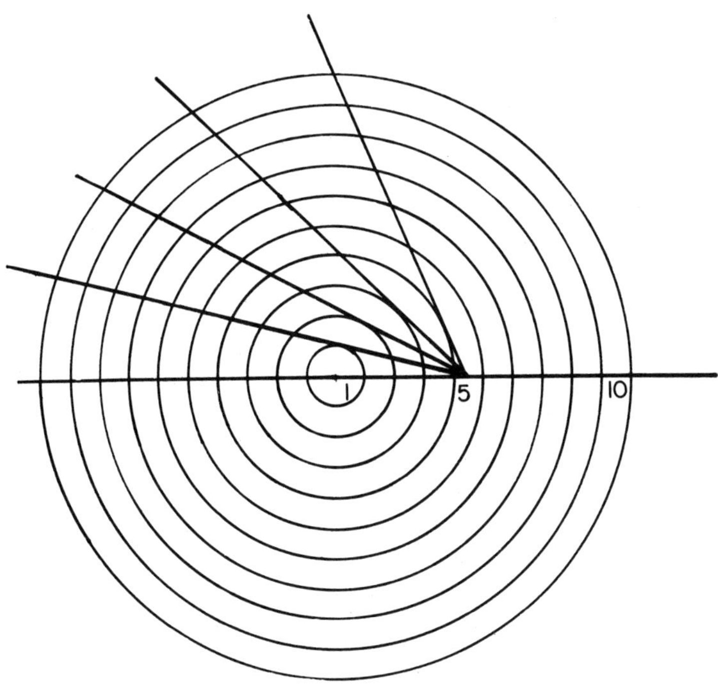

Figure 8.   Program Ashes uses a grid of angular bins superimposed
            on radial zones to track fusion products, Bremsstrah-
            lung radiation, and neutrons. Depending on the birth
            point of the particles, bins can originate from any
            zone.

so that each subroutine serves one main function (Fig. 9). An
effort has been made to name all variables, arrays, and subrou-
tines to describe their prospective functions.

Once data have been read in, and inital conditions set,
subroutine TRACK calculates the length that fusion products will
travel from their zone of birth till when they leave each zone,
for all angular bins. TRACK also computes the fraction of fusion
products that travel in each angular bin, for every zone.

Logical function TIMEUP is used to determine when the burn is
over. The burn time is equal to the radius of the pellet divided
by the sound speed, calculated for each zone, for each time step
Δt, depending on the position of the outward moving sound wave.

PROGRAM ASHES

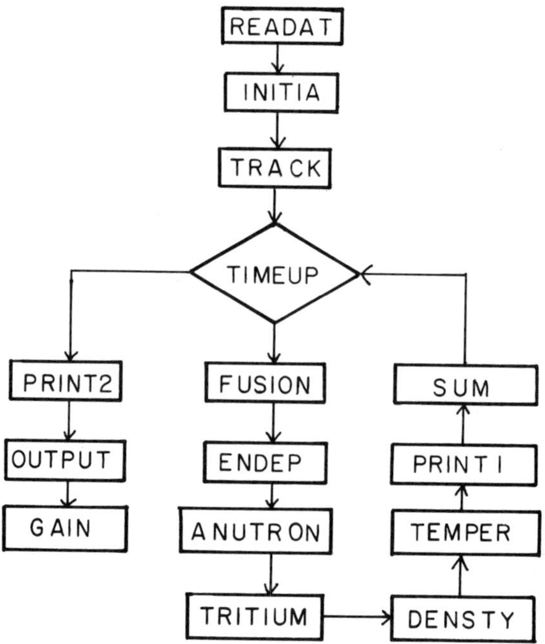

Figure 9.   Flow diagram for the computer code Ashes used for
            pellet burn calculations.

        If the time is not up, the program goes through another loop
(Fig. 9) that calculates the number of fusions (D-T and D-D) in
each zone.

        Next, subroutine ENDEP calculates the energy deposited by
alpha, proton, and ^{3}He fusion products using the lengths
calculated from the ray tracing routine, TRACK. Tritium is
tracked separately because of its large reaction cross section and
our interest in tritium self sufficiency. A "unified theory" is
used to model the slowing down of the fusion products. (24,25)

        Subroutine ANUTRON calculates the heating due to the 14.1-MeV
D-T and the 2.5-MeV D-D neutrons, using the lengths calculated in
TRACK and assuming a single-collision elastic scattering model.
The energy lost is then deposited in the corresponding zone.

Subroutine DENSTY revises the density of deuterium and tritium in each zone, for each time step Gains are due to thermalization of fusion products, while losses are due to burn-up.

Subroutine TEMPER recalculates the temperature of each zone for each time step. Thermal conduction losses from inner zones to adjacent outer zones are computed using an average Spitzer thermal conductivity. This causes a smoothing of the tempreature profile as ignited zones adjust to the same temperature. Bremsstrahlung losses are calculated for the average temperature of the zone over the particular time step. Gains are due to the energy deposited by fusion products and by neutron heating. PRINT1 prints information of interest for each zone, for each time step, while SUM totals quantities such as the number of D-T reactions. The program then returns to TIMEUP, to check if the pellet has disassembled. If not, it goes on to calculate the fusions for the next time step. When the outward-propagating sound wave reaches the outer radius, the program goes to PRINT2 to print both global and zone-wise data.

Comparison Calculations

As a test of program Ashes, it was checked against the D-T pellet calculations by Skupsky[19] noted in Table 7 and Fig. 10. Results are for spark ignited D-T pellets with a compression ratio of 5.6×10^4, spark to outer radius ratio of 1/3, a uniform density profile, and ρR values ranging from 5 to 25 gm/cm^2. In view of the many differences inherent in the calculational techniques, the agreement is considered good.

Table 7. Comparison Calculations

	Skupsky, Ref. 19			ASHES		
$\rho R(gm/cm)$	5	10	25	5.21	10.4	26.2
E (kJ)	1.3	10.0	160	1.4	11.0	175
Yield (kJ)	596	5640	1.1×10^5	589	6040	1.3×10^5
Gain	466	553	699	424	549	720
Deuterium Burnup	0.36	0.45	0.60	0.34	0.44	0.60

Note: Pure D-T Pellet Compression of 5×10^4 (cf. Ref. 19).

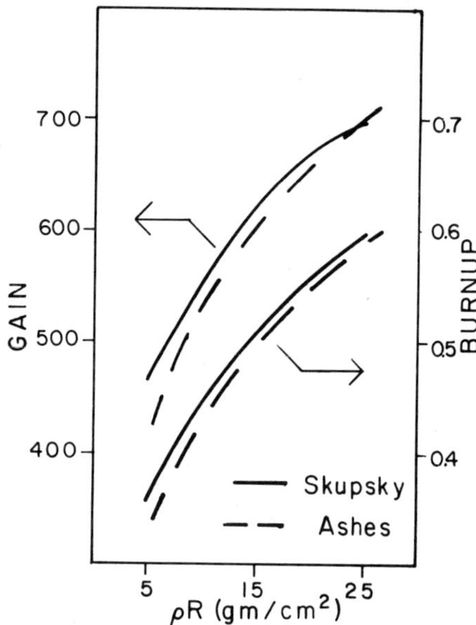

Figure 10.   Comparison of homogeneous D-T pellet calculations
using program Ashes with prior calculations by
Skupsky, Ref. 19.   Results are for large compression
ratios ($\sim 5 \times 10^4$).

Results from a more demanding test, namely, Skupsky's $\rho R = 25$ gm/cm^2 layered pellet with 10% tritium in the outer region, are shown in Table 8.   Again, the agreement is satisfactory, lending further confidence to the accuracy of program Ashes.

<u>0.1 MJ A-FLINT Results</u>

A variety of approaches to D-based pellets with a tritium breeding capability can be envisioned as indicated in Table 9. While some of the multiple zone designs might offer better performance, the basic A-FLINT design (cf. Fig. 6) was selected for the present study largely because it potentially offers a good performance with minimum complications, both relative to calculations and, more importantly, relative to ease of manufacturing.   Further, while some improvement in performance appears possible by shaping the density profile (e.g. use of a 1/r density profile in the outer layer as done in Ref. 22), the benefits from this were not thought to be large enough to warrant the added complexity.   Consequently, all results presented here employ uniform density profiles.

Table 8.   Layered-D Pellet* Comparison

	Skupsky, Ref. 19	ASHES
ρR (gm/cm^2)	25	23.5
E_{in} (kJ)	160	175
Yield (kJ)	9.25×10^4	6.70×10^4
Gain	578	382
Deuterium Burnup	.38	.33
Tritium Breeding Ratio	–	.296

*50/50 D/T in microcore; 90/10 D/T in outer core; compression ratio 5×10^4 (cf. Ref. 19).

Table 9.   Approaches to Tritium Breeding Pellets

- pure D
- D-uniformly seeded with T, T & ^{3}He
- Zoned designs

     DT core‖D

     DT core‖D, T

     DT core‖D, T, ^{3}He

     Multiple zones

- Seeding with other materials

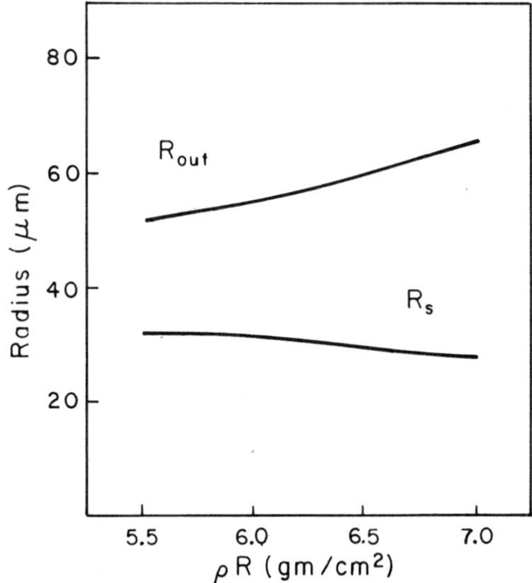

Figure 11.   Spark and outer layer radii for present A-FLINT
pellets.

     With the constraints of Table 5, the two key remaining vari-
ables are ρR and the spark temperature T_s. The latter was
optimized through parametric studies where the spark radius R_s
was varied. Then overall energy balance requirements determine
the outer compressed radius R as a function of ρR corresponding to
Fig. 11 (as shown later, the range $6 \lesssim \rho R$ $(gm/cm^2) \lesssim 7$ turns out
to be of interest here). Thus pellets considered here have a nom-
inal compressed radius of ~ 60 μm.

     The TBR, as might be anticipated, turns out to be a strong
function of the percent tritium in the outer layer. As shown in
Fig. 12, the TBR decreases rapidly with increasing initial
("seed") tritium. In fact, for a TBR ~ 1.0, a pure deuterium
layer is required. The problem is that the volume of the outer
layer is so large that even a few percent seed tritium represents,
in absolute terms, an amount roughly equal to that in the spark
core. Consequently, the deuterium burn fraction must be roughly
doubled to even maintain the same TBR. It might be though that
the increase in gain with added seed tritium would make it attractive to maintain some seed material. However, as seen from Fig.
12, the increase in gain is not so marked as the precipitous drop
in TBR. Consequently, due to the continued stress on TBR, present
studies were restricted to pure deuterium outer layers.

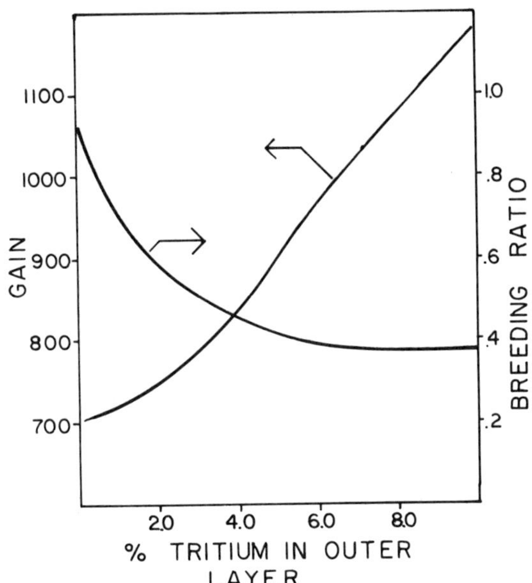

Figure 12.   Variation of gain and TBR with percent tritium seed in outer deuterium layer.

Final results showing the calculated TBR and pellet gain vs ρR in the range of 5.5 to 7 gm/cm^2 are presented in Fig. 13. It is apparent that the "best" pellet would have $\rho R \sim 7$ gm/cm^2, giving a TBR ~ 1.0 with a gain G_c, of ~ 700. Under these conditions, a deuterium burn up fraction of ~30% is obtained. As shown earlier in Table 3, this 0.1-MJ pellet remains fairly competitive with earlier designs requiring much larger energy inputs and, of the designs cited, has the smallest compression ratio (5×10^3) and smallest ρR (~7 gm/cm^2). Indeed, other than the "cat.-D-burner" pellets where little data is available, it is the only design (except for the 1978 A-FLINT pellet) having a satisfactory TBR.

Several characteristics of these pellets are of interest. As shown in Fig. 14, of the tritium produced by D-D reactions, only 30% is recovered at $\rho R \sim 7$ gm/cm^2. This is still adequate to achieve the TBR of ~1.0 cited earlier, and the fact that a reasonable amount burns is important to the maintenance of the burn propagation and gain. In other words, the outer layer acts as a semi-catalyzed-D fuel which, as noted earlier, has a much lower

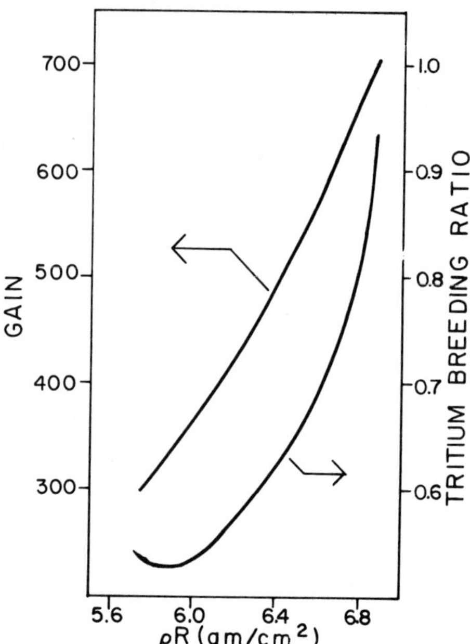

Figure 13.   Gain and TBR calculated for present A-FLINT pellet.
For $\rho R \sim 6.8$ gm/cm^2, a TBR ~ 1.0 is obtained with a
gain of ~ 700.

ignition temperature than pure deuterium. The tritium that burns
is largely composed of thermalized tritium that reacts with deuterium in "Maxwellian" fusion reactions (vs. fusion while slowing
down). Much of that recovered comes from thermalized tritium in
the outermost layers that have not yet heated to very high temperatures prior to disassembly. Some comes from high-energy tritium
that escapes and thermalizes in the outer tamper region.

Burn propagation is obviously crucial to the performance of
these pellets. Time evolution of the temperature profile is shown
in Fig. 15. The profile increases quite rapidly in outer region
so that the heating appears somewhat similar to the uniform ignition process reported in Ref. 15. A distinct burn point is
discernable, however. Thus, as shown in Fig. 16, a propagation
velocity of 2.3 x 10^7 cm/sec can be associated with the

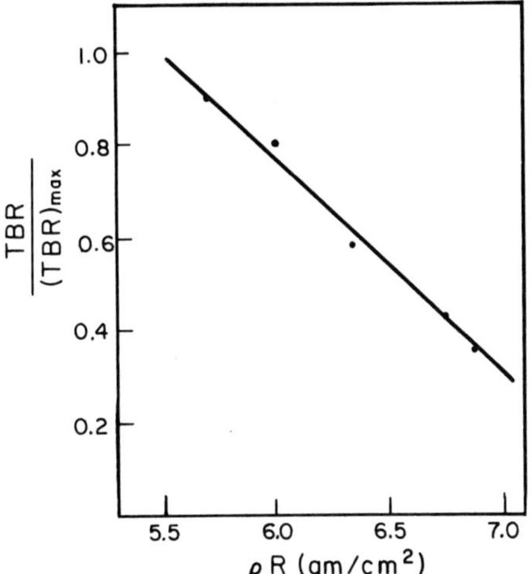

Figure 14.   Variation of the ratio of tritium recovered to that
originally produced [$\sim$ TBR/(TBR)$_{max}$ where (TBR)$_{max}$
implies 100% recovery] for the present A-FLINT pellet.

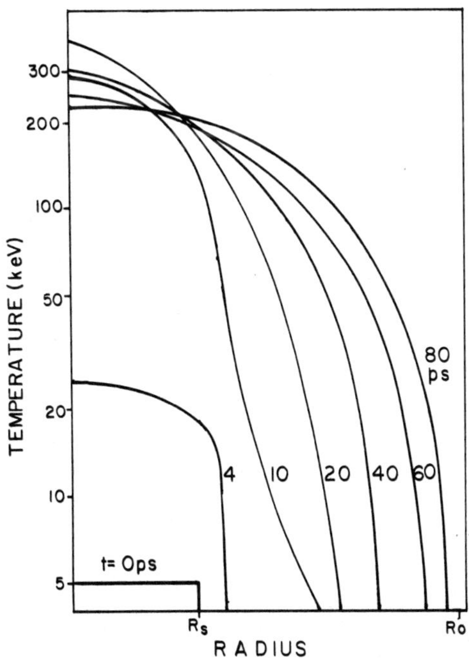

Figure 15.   Temperature-radius contours as a function of time
starting from a 5-keV spark region.

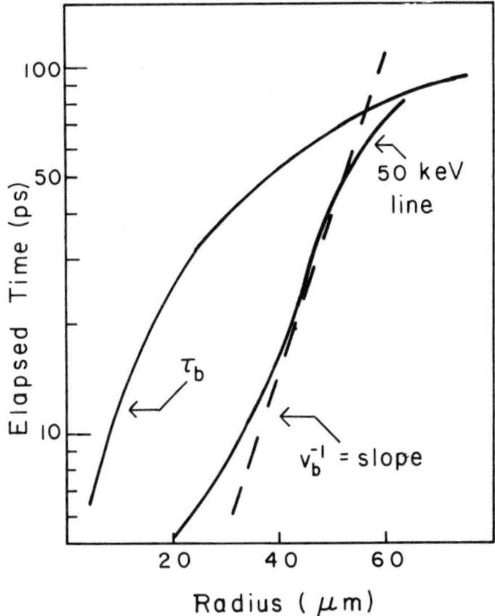

Figure 16.   A plot of the time to react 50-keV vs radius suggests
effective burn front propagation velocity (v_b) of
2×10^7 cm/sec. For comparison, the propagation time
(τ_b) for a sonic wave evaluated at the initial
temperature is also shown.

slope of the line defining the time to reach 50 keV vs radius.
Another way of viewing this is shown in Fig. 17, where the location of the region receiving the maximum energy deposition from
fusion products is plotted as a function of the time step. Again,
a relatively constant velocity of ~ 2×10^7 cm/sec is observed.

SUMMARY

     As outlined in the introduction, there is a strong motivation
to seek ways to develop AF pellets for application in inertial
confinement. The A-FLINT approach appears to represent the eas-
iest route to a relatively near term AF pellet. This pellet would
not offer the flexibility possible with an array of pellets based
on different fuels, but it would have the very significant advantage of eliminating the need for breeding tritium in the blanket.
Indeed the results presented here are quite encouraging. With an
investment of 0.1 MJ of absorbed energy, reasonable A-FLINT opera-

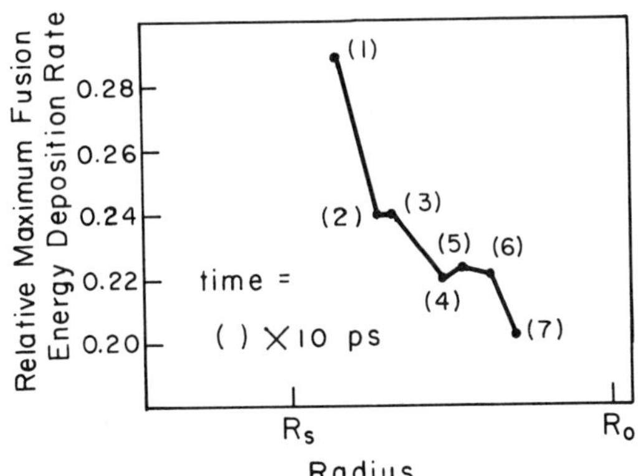

Figure 17.   Propagation of fusion-product energy deposition front.
Location of maximum energy deposition rate is shown as
a function of time. While the absolute magnitude decreases with propagation, a characteristic burn front
is evident.

tion is predicted. While the calculations are simplified and contain optimistic assumptions, the results are sufficiently encouraging to make more detailed studies worthwhile. Ultimately, experimental data are needed, but many of the fundamental questions
such as spark ignition and burn propagation are equally important
to D-T pellet development. Thus this data should become should
become available in the near future. Consequently, a decision
about the amount of effort that should be expended on the A-FLINT
concept should be possible in time to allow such pellets to be a
contender for commercial application.

Acknowledgments

This work represents a joint effort with Dr. Chan Choi, Dr.
Tom Blue, Chris Powell, and David Harris. The calculations presented were largely done by Chris and David. Discussions with
Drs. S. Skupsky, R. Bangerter, and H. Hora plus vital suggestions
by Dr. C. Choi are gratefully acknowledged. This research was
partly supported by the Electric Power Research Institute.

REFERENCES

1. G. H. Miley, Fusion Energy Conversion, Am. Nucl. Soc.,
    La Grange Park, IL, 1976. Chapter 2.
2. G. H. Miley, et al., "Studies of Catalyzed-D and D-^{3}He
    Fusion Reactor Systems," Proc. 2nd Int. ANS Topical Mtg.
    on the Technology of Controlled Nuclear Fusion, Richland,
    WA, (1976) Vol. I, p. 119.
3. G. H. Miley, "Advanced-Fuel Fusion Concepts," Atomkernergie
    (ATKE) Bd. 32 (1978) Lfg. 1, pp. 12-18. also see "Advanced
    Fuel Furor--The Alternative to DT Fusion," 14th Fusion
    Power Coordinating Committee, Oak Ridge Nat. Laboratory,
    Oak Ridge, TN, August 1978.
4. Workshop on Synthetic Fuels from Fusion, EPRI ER 439-SR,
    Electric Power Research Institute, Palo Alto, CA, April
    1977.
5. J. P. Holdren, "Safety and Environmental Aspects of Fusion
    Reactors," UCRL-78759, Lawrence Livermore Laboratory,
    Livermore, CA, Oct. 1976.
6. J. A. Fillo and J. R. Powell, Fusion Blankets for Cat.-D and
    D-^{3}He Reactors, in Proc. Review Mtg. on Adv. Fuel
    Fusion (C. Choi, ed.) EPRI ER-536-SR, Electric Power
    Research Institute, Palo Alto, CA. Sept. 1977, pp. 57-68.
7. M. Monsler, et al., Electric Power From Laser Fusion: The
    HYLIFE Concept, Proc. 13th Intersociety Energy Conversion
    Engineering Conf., San Diego, CA., August 1978. Also see
    UCRL-81259.
8. John H. Nuckolls, "Target and Reactor Design," in "ERDA Summer
    Study of Heavy Ions for Inertial Fusion," R. W.
    Bangerter, W. B. Herrmannsfeldt, D. L. Judd, and L.
    Smith (eds.), LBL-5543, Lawrence Berkeley Lab., Berkeley,
    CA, 1976.
9. Sec. 4-1.4.c, Ref. 1.
10. R. Schreiver, Overview of the U. S. Inertial Confinement
    Program, Minicourse on Inertial Conf. Fusion, IEEE
    International Conference on Plasma Science, Montreal,
    Canada, June 1979.
11. J. A. Fillo, et al., "Exploratory Studies of High-Efficiency
    Advanced-Fuel Fusion Reactors," EPRI ER-581 (1977); and
    EPRI ER-919, Electric Power Research Institute, Palo
    Also, CA, (1978).
12. C. C. Baker, et al., "Fusion Reactor Technology Impact of
    Alternate Fusion Fuels, 8th IEEE Sym. Eng. Problems of
    Fusion Research, San Francisco, CA, November 1979.
13. J. R. Roth and H. C. Roland, "The Effect of Wall Loading
    Limitations and Choice of Fuel Cycle on the Feasibility
    of High Beta Fusion Reactors, 8th IEEE Sym. Eng. Problems
    of Fusion Research, San Francisco, CA, November 1979.

14. T. Weaver, G. Zimmerman, and L. Wood, "Exotic CTR Fuels; Non-Thermal Effects and Laser Fusion Applications," UCRL-74938, Lawrence Livermore Laboratory, Livermore, CA (1973).

15. H. Hora, et al., "Calculations of Inertial Confinement Fusion Gains Using a Collective Model for Reheat, Bremsstrahlung and Fuel Depletion for High-Efficient Electrodynamic Laser Compressions," Plasma Physics and Controlled Nuclear Fusion Research 1978, (IAEA, Vienna), Vol. III, p.237.

16. G. Moses, "Advanced Fuels" in "SOLASE, A Conceptual Laser Fusion Reactor Design," Report UWFDM-220, Univ. of Wisc., Madison, Wisc. Dec. 17. 1977, pp. III-F-1 to F-10.

17. L. Wood, "Advanced Topics in Inertial CTR: A Status Report," Proc. First IEEE Minicourse on Fusion, Chapter 12, Austin, TX, (1976)

18. Ref. 8; also see Jack Hovingh, "Ion-Beam Reactor First-Wall Design," in Ref. 8.

19. S. Skupsky, Nucl. Fusion, 18, 6 (1978).

20. T. Chu and G. H. Miley, "D-D Plasma Fusion Power-Balance Calculations," Technology of Controlled Thermonuclear Fusion Experiments and the Engineering Aspects of Fusion Reactors, AEC Symposium Series 31, TIC, Oak Ridge, TN, p. 322-346 (1972).

21. G. H. Miley, C. K. Choi, and D. Lee, "Ion Beam Implosions Using Layered Pellets of Advanced Fuels," Proc. of 2nd Int. Topical Conf. on High Power Electron and Ion Beam Research and Technology, Oct. 3-5 (1977) Vol. I, p. 243.

22. C. K. Choi, G. H. Miley, D. Lee and C. Powell, "D-T Seeded, Deuterium-Layered, Tritium Self-Sufficient Pellets, Third ANS Topical Meeting on Fusion, Santa Fe, NM, May, 1978, p. 405.

23. C. K. Choi, Thomas E. Blue, and George H. Miley, "Advanced-Fuel Pellet Approaches to Inertial Fusion," Energy, 4, 157 (1979).

24. C. K. Choi, W. R. Sutton, and G. H. Miley, Trans. Am. Nuc. Soc., 33, 34 (1979).

25. C. K. Choi and W. R. Sutton, "Non-Classical Nature of the High-Energy Fusion-Product Slowing in Dense Plasmas," IEEE International Conference on Plasma Science, 1B10, p. 119, Montreal, Canada (1979).

CHARGED PARTICLE ENERGY LOSS RATES

AND RANGES IN PLASMA

Jiri Stepanek

Swiss Federal Institute for Reactor Research

5303 Würenlingen, Switzerland

ABSTRACT

The energy loss and the penetration length of charged particles
in a plasma are calculated using the exact binary collision theory
in which the collective effects are included only through the use of
Debye screening in the Coulomb cross sections. The results are
compared with different analytical approximations of the binary collision theory such as Kammash's, Perkins', Ligou's, Rose's, Ray, Hora's
and to Sigmar-Joyce's more rigorous treatment based on the Balescu-
Lenard kinetic equations (in which the collective effects are included in the normal way through the plasma dielectric coefficient).
It is shown that in mirror-confined plasmas, in Tokamak-like plasmas
and in many other cases, the collective effects are small and therefore the differences found in the recent comparisons between the
binary collision theory and the solutions with more rigorously
included collective effects were often caused by the approximations
made in the analytical solutions of the binary collision theory.
Kammash's approximation gives the best agreement for mirror confined
plasma or for Tokamak-like plasmas. Whereas this approximation
breaks down for highly compressed D-T plasma or for low temperature
D-T plasma, the approximation pointed out by McNally and Sharp and
used by Perkins gives the best average agreement in all considered
cases.

For laser compressed plasma with densities $\geq 10^{22}$ cm^{-3} and
temperatures below 10 keV large discrepancies occur between the
exact and all approximate solutions of the binary collision theory.

On the other hand, the collective effects of plasma begin to
be too strong in this range of temperatures and densities, so that
the binary collision theory is no longer satisfied.

1. INTRODUCTION

The additional heating of the high temperature plasmas for
thermonuclear reactions by charged energetic particles is one of the
important problems of nuclear fusion. The effects of the additional
heating of the hot plasma were studied in recent papers for magnetically confined plasma as well as for laser compressed plasmas [1-13].

One of the most important processes for additional heating is
the Coulomb scattering of the reaction products, such as the 3.5 MeV
alpha particles, for example. During the slowing down of these
particles the lost energy is transferred to the plasma electrons and
ions, which increase their energy to values where fusion cross sections are greater.

The exact calculations of the energy loss rate is therefore
important. The most widely used method is the binary collision
theory which leads to simple equations for the energy loss of the test
particle. Different simple analytical solutions to this equation
were used in recent papers[1, 2, 9-13]. Some of these methods do not
give good results in all ranges of densities and temperatures or
break down with increasing plasma density or with decreasing temperature.

For seven different cases we calculate the energy loss rates,
time dependence of the test particle energy, thermalization time
and range. The results of the recent analytical solutions of the
binary collision theory are compared to its exact solution.

Because the collective plasma effects which are important
especially at high densities are included in the binary collision
theory only through the Debye length, the results of the binary collision theory for some of these cases are also compared to the theory
of Sigmar and Joyce[14], which is based on the solution of the Balescu-
Lenard kinetic equations, where the collective effects are considered
more rigorously through the plasma dielectric coefficient.

In seven cases, the energy loss rates of deuterons in electron-
plasma, energy losses and time dependence of the energy of 3.5 MeV
alpha-particles during their slowing-down in the mirror confined
plasma, the thermalization times of protons in Tokamak-like plasma,
energy loss rates and ranges of 3.5 MeV alpha-particles in solid and
compressed D-T plasma and the ranges of 14.7 MeV protons in solid and
compressed D-T plasma are calculated.

It is shown that for mirror confined plasmas and for Tokamak-
like plasmas, the exact solution of the binary collision theory is
in very good agreement with the Sigmar-Joyce results based on the
solution of the Balescu-Lenard equations with more rigorously
treated collective plasma effects. The differences found in the

recent papers between the approximative binary collision theory and the results of Sigmar and Joyce were often caused by the approximative nature of the analytical solution than by the collective effects of the plasma. From the analytical approximations, the Kammash's approximation is the best for the mirror confined plasma as well as for a Tokamak-like plasma. This approximation breaks down for high density or low temperature of D-T plasma. The approximation used by McNally and Sharp or by Perkins gives the best average agreement in all cases considered in this paper.

Large discrepancies between the exact solution of the binary collision theory and all its approximative analytical solutions can be seen especially by the calculation of the 3.5 MeV alpha particles and 14.7 MeV proton ranges in the temperature range $kT \leq 10$ keV.

Unfortunately no results based on Balescu-Lenard equations were available at the present time but a method based on this equation is in development. The alpha particle energy loss rates and ranges and proton ranges in laser compressed plasma will be presented in the near future.

2. BINARY COLLISION THEORY

The binary collision theory is described in detail by Trubnikov[1], Shkarofsky, et al.[2], Kammash[9], and by others.

One considers as in ref. 9 a test particle of mass m and charge number Z and a field particle of mass M_j and charge number Z_j. Their velocities before and after collision are in the laboratory coordinate system $\bar{v}$, $\bar{v}_j$ and $\bar{v}'$, $\bar{v}_j'$, respectively. One considers also that there are J different particles in the field with Maxwellian distribution of the number density $F(V,\Omega(\psi))$:

$$F(V_j,\Omega(\psi)) = 2\pi \left(\frac{M_j}{2\pi kT_j}\right)^{3/2} V_j^2 e^{-M_j V_j^2/2kT_j}, \tag{1}$$

where V_j is the magnitude of particle velocity with direction $\Omega(\psi)$ and kT_j is the temperature.

The average scattering cross section is a function of the angle between critical and final velocities in the C.M. coordinate system and also the function of the relative speed V_r between both particles. Therefore, the number of collisions per unit time experienced by a test particle with a field particle j is:

$$\bar{\sigma}_s(v) = \frac{2\pi}{v^2} \left(\frac{M_j}{2\pi kT_j}\right)^{3/2} \int_0^\infty V_j e^{-\frac{M_j V_j^2}{2kT_j}} \int_{|v_j-v|}^{V_j+v} V_{rj}^2 \int_0^\pi \int_0^{\pi/2} \sigma_s(V_{rj},\theta)\sin\chi\, d\chi\, dv\, dV_{rj}\, dV_j, \tag{2}$$

where $\overset{*}{v}$ is the magnitude of velocity $\overline{v}$, θ is the angle between initial and final velocities of the test particle in C.M. coordinate system and V_{rj} is the magnitude of the relative velocity $\overline{V}_r'$ between the test and field particle given by

$$V_{rj}^2 = v^2 + V_j^2 - 2\, vV_j \cos\psi \, , \tag{3}$$

where ψ is the angle between velocities $\overline{v}$ and $\overline{V}$ and the angles χ and ν define the solid angle into which the test particle is scattered (in C.M.S.):

$$d\Omega' = \sin\chi\, d\chi\, d\nu \, .$$

The scattering cross section for both unshielded and shielded Coulomb potentials can be expressed[9,15] as:

$$\sigma_s(V_r,\theta)_j = \left(\frac{e^2 ZZ_j}{2\mu_j V_{rj}^2} \right)^2 \frac{1}{(\sin^2(\theta/2) + 1/4\theta_o^2(V_{rj}))^2}, \tag{4}$$

where μ_j is the reduced mass

$$\mu_j = \frac{mM_j}{(m+M_j)} \, . \, ,$$

and the angle θ_o is defined by either the classical or quantum approach as

$$\theta_o(V_{rj}) = \frac{2e^2 ZZ_j}{\mu_j \lambda_D V_{rj}^2} \quad \text{for} \quad 0 \le V_{rj} \le \frac{2e^2 ZZ_j}{\hbar} \, ,$$

$$\tag{5}$$

$$\frac{\hbar}{\mu_j \lambda_D V_{rj}} \quad \text{for} \quad \frac{2e^2 ZZ_j}{\hbar} < V_{rj} < \infty \, ,$$

where both forms are equal for $V_{rj} = 2e^2 ZZ_j/\hbar$, and λ_D is the Debye length given by[2]:

$$\lambda_D^2 = \frac{1}{4\P \sum\limits_{j=1}^{J} \frac{n_j e^2 Z_j^2}{kT_j}} \, , \tag{6}$$

The Debye length takes into account the shielding of the field particle j by other particles in the field. n_j is the number of test particles j per unit volume, e is the electron charge and z and z_j are the charge number of the test and field particle, respectively.

Inserting the energy loss $\frac{m}{2}(v^2 - v'^2)$ of the test particle per collision into Eq. (2) one gets the expression for the average energy loss rate:

$$\frac{\partial E}{\partial t} = - N_j \frac{\pi m}{v} \left(\frac{M_j}{2\pi kT_j}\right)^{3/2}$$

$$\int_0^\infty V_j e^{-\frac{M_j V_j^2}{2kT_j}} \int_{|V_j - v|}^{V_j + v} V_{rj}^2 \int_0^\pi \int_0^{\pi/2} (v^2 - v'^2)\sigma_s(V_{rj},\theta)\sin\chi d\chi d\nu dV_{rj}dV_j, \tag{7}$$

using the formula (5) for $\sigma_s(V_{rj},\theta)$ and the following substitutions

$$x = \frac{M_j}{2kT_j} V_{rj}^2 \tag{8}$$

$$R_j^2 = \frac{M_j}{2kT_j} v^2, \tag{9}$$

$$\alpha_j = \frac{M_j}{2kT_j} \left(\frac{2e^2 ZZ_j}{\hbar}\right)^2, \tag{10}$$

$$\beta_j = \frac{M_j}{2kT_j} \left(\frac{e^2 ZZ_j}{\mu_j \lambda_D}\right), \tag{11}$$

$$\gamma_j = \frac{M_j}{2kT_j} \left(\frac{\hbar}{2\mu_j \lambda_D}\right)^2, \tag{12}$$

one gets after some integrations of Eq. (7):

$$\left(\frac{\partial E}{\partial t}\right)_j = -n_j \frac{\sqrt{\pi}(e^2 Z Z_j)^2}{v\mu_j} \left\{ \int_0^{\alpha_j} \left[\ln\left(\frac{\beta_j^2+x^2}{\beta_j^2}\right) - \frac{x^2}{\beta_j^2+x^2} \right] f(x)dx \right.$$

$$\left. + \int_{\alpha_j}^{\infty} \left[\ln\left(\frac{\gamma_j+x}{\gamma_j}\right) - \frac{x}{\gamma_j+x} \right] f(x)dx \right\} , \tag{13}$$

where

$$\tag{13a}$$
$$f(x) = x^{-3/2} e^{-(R_j^2+x)} \left[2R_j x^{1/2} \cosh(2R_j x^{1/2}) - (1+\frac{2M_j x}{m+M_j}) \sinh(2R_j x^{1/2}) \right].$$

<u>Analytical expressions for $(\partial E/\partial t)$</u>

a.   Kammash[9] makes the assumptions

$$\ln\left(\frac{\beta_j^2+x^2}{\beta_j^2}\right) - \frac{x^2}{\beta_j^2+x^2} \sim \ln(1/\beta_j^2), \text{ for } 0 \leq x \leq \alpha_j,$$

and
$$\tag{14}$$
$$\ln\frac{\gamma_j+x}{\gamma_j} - \frac{x}{\gamma_j+x} \sim \ln(1/\gamma_j), \text{ for } \alpha_j \leq x \leq \infty,$$

and gets after analytical integration of Eq. (13):

$$\left(\frac{\partial E}{\partial t}\right)_j \sim -M_j \frac{\pi(e^2 Z Z_j)^2}{M_j} \sqrt{\frac{2m}{E}} \left[(f_1(R_j) - f_2(R_j,\alpha_J))\ln(1/\beta_j^2) \right.$$

$$\left. + f_2(R_j,\alpha_j) \ln(\gamma_j) \right] . \tag{15}$$

where

$$f_1(R_j) = \mathrm{erf}(R_j) - (1 + \frac{M_j}{m}) R_j \frac{2}{\sqrt{\pi}} e^{-R_j^2} , \tag{16}$$

and

$$f_2(R_j, \alpha_j) = \frac{1}{2} \left[\mathrm{erf}(R_j + \sqrt{\alpha_j}) + \mathrm{erf}(R_j - \sqrt{\alpha_j}) \right.$$

$$\left. (1 + \frac{M_j}{m}) \frac{1}{\sqrt{\pi \alpha_j}} \left(e^{-(R_j + \sqrt{\alpha_j})^2} - e^{-(R_j - \sqrt{\alpha_j})^2} \right) \right] , \tag{17}$$

with

$$\mathrm{erf}(R_j) = \frac{2}{\sqrt{\pi}} \int_0^{R_j} dt\, e^{-t^2} . \tag{18}$$

The "classical" limit of Eq. (15) is

$$\lim_{\alpha \to \infty} \frac{\partial E}{\partial t} = -n_j\, \frac{\pi (e^2 Z Z_j)^2}{M_j} \sqrt{\frac{2m}{E}}\, f_1(R_j)\, \ln(1/\beta_j^2), \tag{19}$$

whereas the "quantum" limit is

$$\lim_{\alpha \to o} \frac{\partial E}{\partial t} = -n_j\, \frac{\pi (e^2 Z Z_j)^2}{M_j} \sqrt{\frac{2m}{E}}\, f_1(R_j)\, \ln(1/\gamma_j). \tag{20}$$

b. Perkins[12] uses for the ions the approximation suggested by McNally and Sharp[11]

$$\frac{\partial E}{\partial t} \approx - n_j\, \frac{\pi (e^2 Z Z_j)^2}{M_j} \sqrt{\frac{2m}{E}}\, f_1(R_j)\, \ln(1 + (1/\beta_j^p)^2), \tag{21}$$

where for electrons

$$\beta_j^p = \frac{(e^2 Z Z_j)}{\mu_j \bar{V}_{r_j}^2 \lambda_D} , \quad \text{for } 0 \leq \bar{V}_{r_j} \leq \frac{2 e^2 Z Z_j}{\hbar} , \tag{22}$$

and

$$\beta_j^p = \frac{h}{2\mu_j \bar{V}_{r_j} \lambda_D} \quad , \quad \text{for} \quad \frac{2e^2 ZZ_j}{\hbar} < \bar{V}_{r_j} < \infty \, , \, j = e, \tag{23}$$

where $\bar{V}_{r_j}$ is an average relative velocity (see below), the Debye length is defined by Eq. (6) and the function f_1 by Eq. (16).

For the ions the classical approximation (22) is used for the whole range of the relative velocity $0 \leq \bar{V}_{r_j} \leq \infty$. This is in agreement with the suggestion made by Fraley[3] and referenced by McNally, and Sharp[11].

Already Fermi and Teller[17] have pointed out that the conventional approximation which includes only the term $1/\beta_j^2$ (see Eq. (15) for example) becomes negative with a density increase or temperature decrease. In contrast to this, Eq. (21) approaches zero asymptotically with high density or low temperature.

The approximation (21) can be obtained using in Eq. (13) the approximation

$$\ln\left[\frac{\beta_j^2 + x^2}{\beta_j^2}\right] - \frac{x^2}{\beta_j^2 + x^2} \simeq \ln\left[1 + \left(\frac{\bar{x}}{\beta_j}\right)^2\right] \, , \tag{24}$$

and

$$\ln\left[\frac{\gamma_j + x}{\gamma_j}\right] - \frac{x}{\gamma_j + x} \simeq \ln\left[1 + \left(\frac{\bar{x}}{\gamma_j}\right)\right] \, , \tag{25}$$

where an average value of $\bar{x}$ or relative velocity (see Eqs. (8) and (27)) is used.

Additionally, Eq. (15) is replaced by its limits $\alpha \to \infty$ for

$$0 \leq \bar{V}_{r_j} \leq \frac{2e^2 ZZ_j}{\hbar} \quad \text{and} \quad \alpha \to 0 \quad \text{for} \quad \frac{2e^2 ZZ_j}{\hbar} < \bar{V}_{r_j} \leq \infty \, .$$

To calculate the average value of the relative velocity $\bar{V}_{r_j}$ in Eq. (21) is a complicated process. Therefore, the simple formula

$$\bar{V}_{r_j}^2 = v^2 + v_j^2 \, , \tag{26}$$

is used. It leads to

$$\bar{v}_{rj}^2 = \frac{1}{mM_j} (2M_j E + 3mkT_j) ,\qquad (27)$$

where the value $3kT_j$ was used for the field particle energy.

c. Ligou[12] is using the same approximation as Perkins[12] Eq. (21) but with the difference that the classical approximation for β_j is used in all ranges of relative velocities for electrons as well as for the ions.

In Ligou's original formula the factor $(1 + \frac{M_j}{m})$ in formula (16) for the function f_1 was omitted. In our calculations it is shown that this leads to poor agreement with the other results. Therefore, we have included this factor and the calculations referenced as Ligou[corr].

d. Ray and Hora[9] have solved the energy loss rate of charged particles on electrons in plasma on the basis of the Fokker-Planck formalism.

Ray and Hora's formula for the energy loss rate on electrons is:

$$(\frac{\partial E}{\partial t})_e = - n_j \frac{\pi (e^2 Z)^2}{M_j} \sqrt{\frac{2m}{E}} \ f_1(R_j) \ \ln(1/\beta_j^{R1})^2),\qquad (28)$$

where

$$\beta_j^{R1} = \frac{1}{\frac{3}{2} kT_j} (\frac{e^2 Z}{\lambda_D}) ,\qquad (29)$$

and the function $f_1(R_j)$ is defined by Eq. (16). Note, that Eq. (28) is the same as the "classical" approximation of binary collision theory (Eq. (19)), except that $2kT_j$ in β_j (Eq. (11)) is replaced by $\frac{3}{2}kT_j$ and only the electrons are considered in the Debye length (Eq. (6)).

Because formula (28) breaks down for small values of $1/\beta_j$, Ray and Hora have developed a "collective" interaction model for the energy loss rate. Considering the collective effects formulated by Gasiorowicz et al.[15], Ray and Hora have derived for the energy loss rate:

$$\frac{\partial E}{\partial t} = - n_j \frac{\pi (e^2 Z)^2}{M_j} \sqrt{\frac{2m}{E}} \ \ln(1/\beta_j^{R2})^2 ,\qquad (30)$$

where

$$\beta_j^{R2} = \frac{e^2 Z}{M_j v^2 \lambda_D}$$

This Eq. can be obtained by approximation $\ln(\beta^2+x^2)/\beta^2-x^2/(\beta^2+x^2) \rightarrow \ln(R^2/\beta)^2$, using the classical limit (19) of the analytical solution of B.C. (Eq. (13)) and replacing the function f_1 by its value for $R \rightarrow \infty$, $f_1 \equiv 1$. This last approximation can be used (see Eq. (16)) for high energy or low temperature.

3. CALCULATIONS

Seven different cases are calculated in order to show the suitability of the analytical approximations of the binary collision theory in comparison to their exact solution and additionally to the results of Sigmar and Joyce[14] based on the solution of the Balescu-Lenard equations in which the plasma collective effects are accounted for more exactly through the plasma dielectric coefficient.

3.1. In the first problem (see Fig. 1) taken from (9), the time dependent energy loss rate of deuterons on electrons with density of 10^4 per cm^3 and temperature of 1.0 eV is calculated using the exact numerical solution of the binary collision theory (Eq. (13)) and their different approximative analytical solutions used by Kammash, Perkins and Ray and Hora.

All the analytical solutions differ considerably from the exact numerical solution for medium energies. Only at high energy do the approximations of Perkins and of Ray and Hora (collective) follow closely the exact solution. The approximation of Ray and Hora breaks down completely for low energies. The boundary between the quantum ($\alpha \rightarrow 0$) and classical ($\alpha \rightarrow \infty$) approximation is at 200 keV.

The maximal energy loss rate is in the range of equal velocities of the test and field particles. For the effective temperature of the field particles 2 kT the corresponding energy is about 8 keV.

3.3. In the second case we calculate the time dependence of the energy losses of 3.5 MeV alpha-particles in mirror confined D-T plasma with electron density of $n_e = 10^{14}$ cm^{-3}, deuteron and triton density of $n_d = n_t = 0.5$ n_e, deuteron and triton temperature of $kT_D = kT_t = 100$ keV and electron temperatures of $kT_e = 20$, 40 and 80 keV (see Fig. 2). The values obtained using the exact binary collision theory are compared to Kammash's and Perkins' analytical approximation, to Rose's[4] high energy approximation of the binary collision theory and to Sigmar-Joyce[13] results obtained solving the Balescu-Lenard equation.

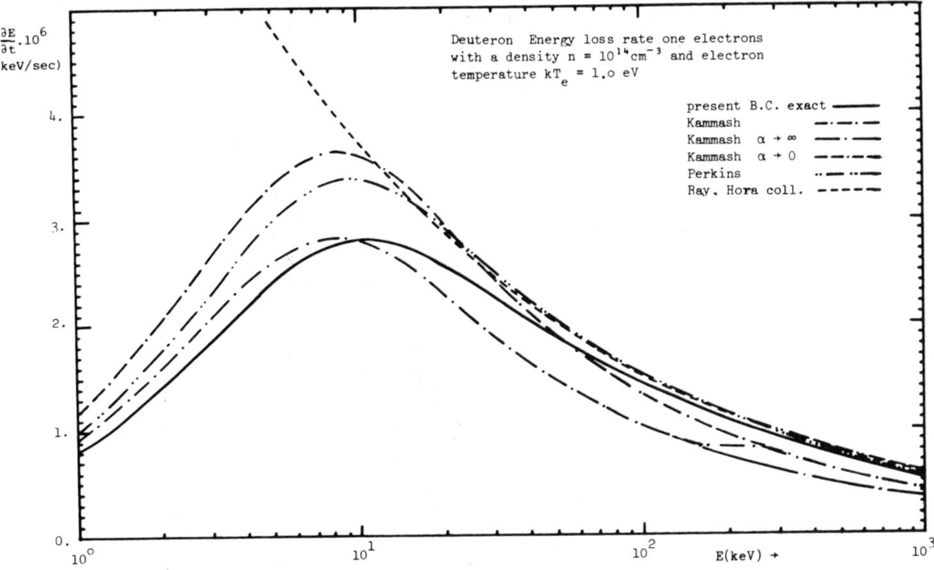

Fig. 1. Deuteron energy loss rate versus energy.

For each temperature the upper curves belong to electrons and ions whereas the lower curves belong to electrons only.

Whereas the Rose approximation breaks down for low energies because the test particle velocity is considered to be much greater than the velocity of the target particle, the other approximations agree well with the Sigmar, Joyce results for low energies.

The best agreement with the Sigmar, Joyce results, for high energies as well, is obtained by the exact numerical solution. The differences between the analytical solutions of the binary collision theory and the Sigmar, Joyce results are therefore not caused by the collective effects as suggested by Kammash[9] but by the approximation itself.

For higher energies the energy loss rate to electrons becomes more and more important.

The maximal energy loss rate to ions is at about 500 keV whereas the maximal energy loss to electrons for the three values of temperature occurs at the energy of $3.2 \, 10^5$, $6.5 \, 10^5$ and $12.8 \, 10^5$ keV, resp. These are the energies at which the curves in Fig. 2 would reach their maximum.

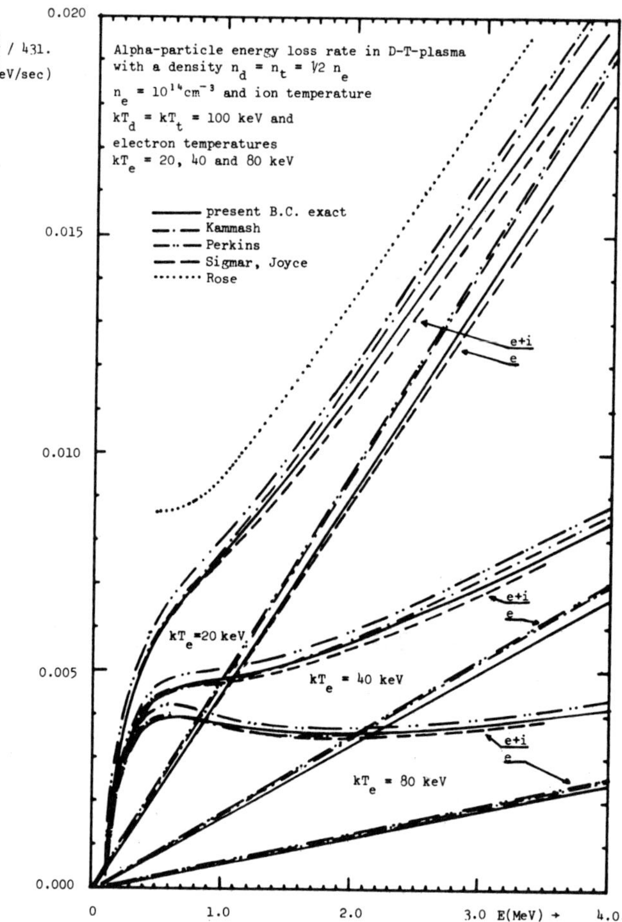

Fig. 2. Alpha-particle energy loss rate versus energy.

In Fig. 3 the alpha particle of energy as a function of time is plotted. The energy E of the particle is defined by

$$t(E) = \int_{E}^{E_o} dE' \ / \ \frac{\partial E'}{\partial t}.$$

Very good agreement is obtained between the exact binary collision theory and the Sigmar, Joyce results. Then the Kammash's, Perkins' and Ligou's corr. approximations follow. The difference between Perkins' and Ligou's results is caused by using the quantum approximation for higher energies and electron by Perkins, whereas Ligou is using the classical approximation.

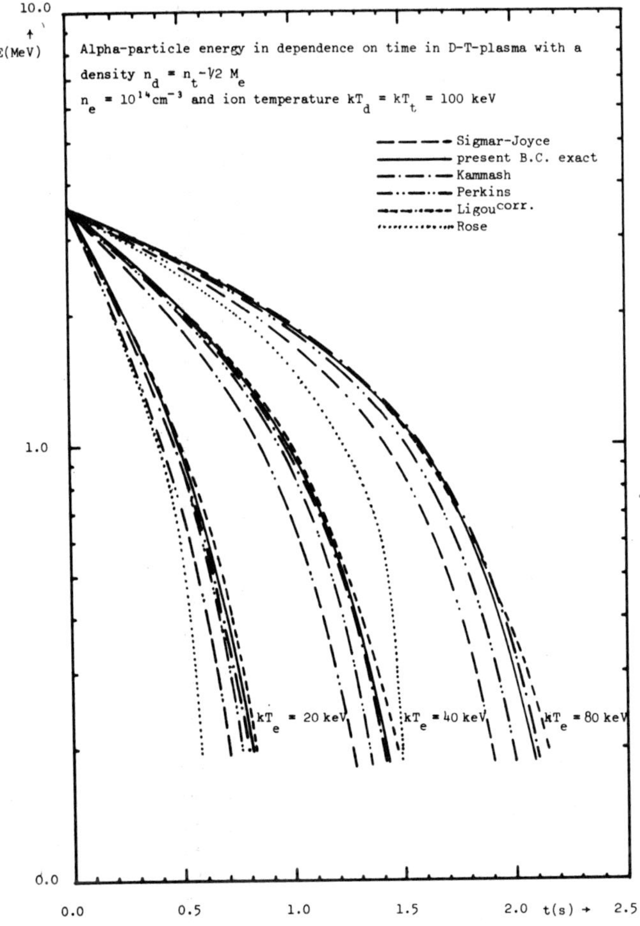

Fig. 3.   Alpha particle energy versus time.

The approximation of Rose leads to large deviations at low energies due to its approximation described in the previous case.

3.3. Fig. 4 shows the thermalization time of protons in the Tokamak-like electron-proton plasma of equal density of electrons and protons $n = 5.10^{13}$ cm^{-3} and temperatures $kT_e = 1$ keV and $kT_p = 0.5$ keV. The thermalization time is defined by

$$T_{th} = \int_{2kT_e}^{E_o} dE' \ / \ \frac{\partial E'}{\partial t}$$

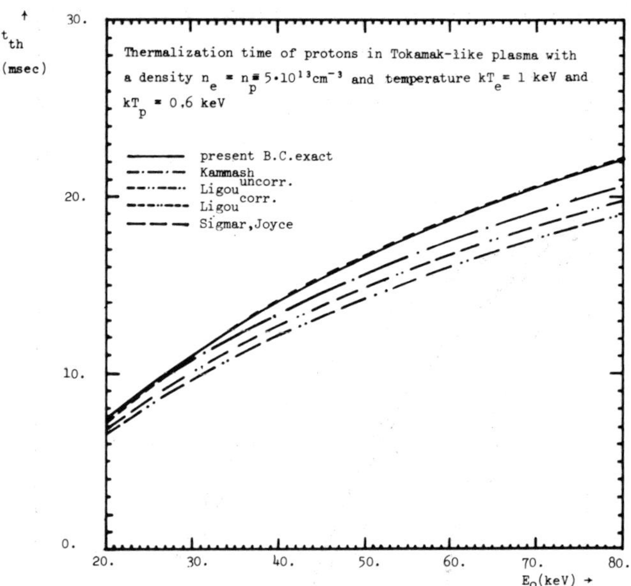

Fig. 4. Thermalization time of ptorons versus injection energy.

Excellent agreement between the exact binary collision theory and the Sigmar-Joyce theory is achieved. Again the Kammash's approximation seems to be the best. The omitted factor $1 + \frac{M_j}{m}$ in Ligou's theory leads to underestimation of the thermalization time.

<u>3.4.</u> In the fifth case, the loss rate of alpha-particles in D-T plasma is calculated. The plasma densities are $n_d = n_t = 1/2k_e$ and $n_e = 4.5\ 10^{22}$ and the plasma temperatures $kT_e = kT_d = kT_t = 1$ eV, 100 eV and 10 keV.

In Fig. 5 the values of $\frac{\partial E}{\partial t}$ computed using the exact solution of the binary collision theory (Eq. (13)) and the values obtained by Perkins are plotted.

Whereas for the temperatures $kT_e = 0.1$ and 10 keV good agreement is achieved, large discrepancies can be seen for very low temperatures (1 eV). The Kammash's approximation, which is not shown in Fig. 5, breaks down completely for kT = 1 keV.

Fig. 6 shows the 3.5 MeV alpha-particle ranges as function of plasma temperature in the D-T plasma described earlier, but for the electron densities $n_e = 4.5\ 10^{22}$, $4.5\ 10^{24}$ and $4.5\ 10^{26}$ cm^{-3}.

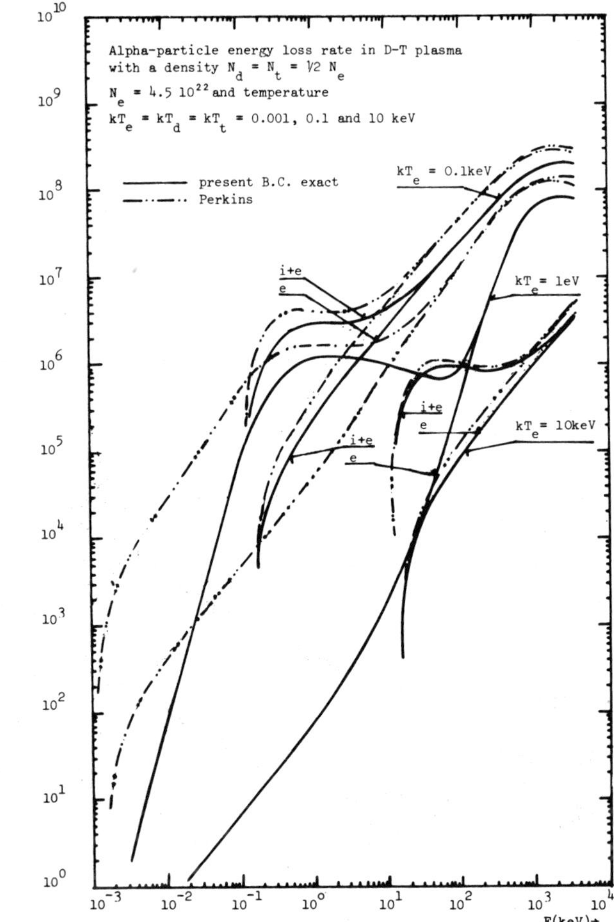

Fig. 5. Alpha-particle energy loss rate versus energy.

The range is defined by

$$R(cm) = \int_{3/2\ kT_e}^{E_o} dE' \Big/ \frac{\partial E'}{\partial x},$$

so that the particle slowing down is followed up to the energy $\frac{3}{2} kT_e$.

The exact solution of the binary collision theory is compared to Kammash's, Perkins', Ligou's and Ray and Hora's results.

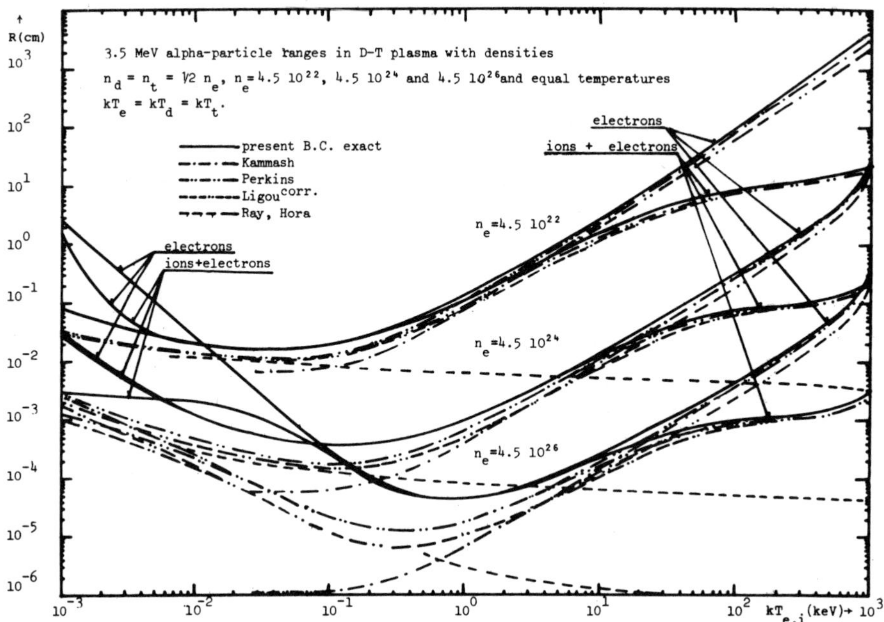

Fig. 6.   3.5 MeV alpha-particle ranges versus D-T plasma
temperature.

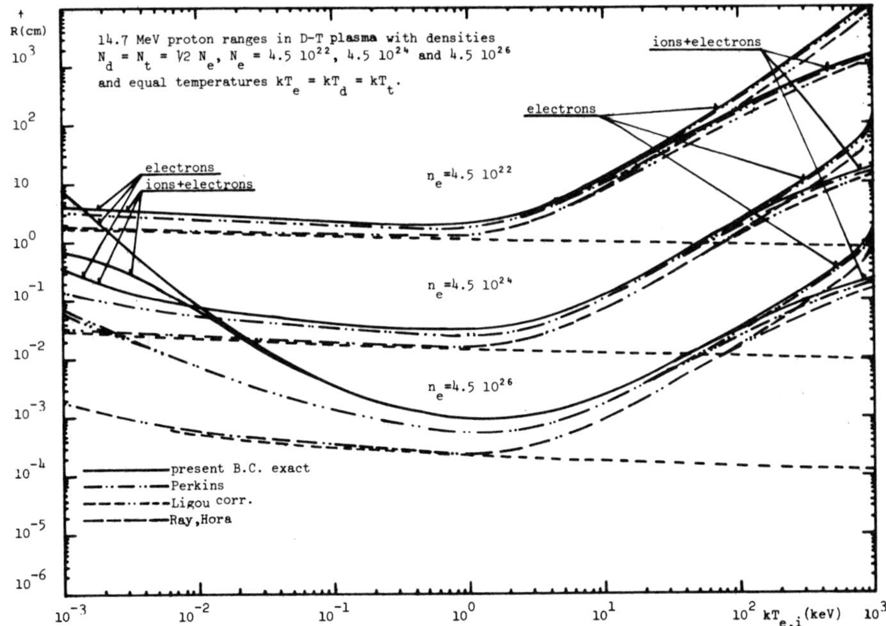

Fig. 7. 14.7 MeV proton ranges versus D-T plasma temperature.

Whereas Kammash's solution which was one of the best analytical
approximations for mirror and Tokamak-like plasma breaks down for
kT < 1 keV, the Ray, Hora's collective model gives satisfactory
results only for the lower densities, $n = 4.5\ 10^{22}$ and $4.5\ 10^{24}$ cm^{-3}
and low temperatures, kT < 100 eV.

The other approximations follow the exact solution of the binary
collision theory quite well. Only at very low temperatures and high
densities where the collective effects are large, are the deviations
also large. Especially for lower temperatures the Perkins approximation is closer to the exact solution than Ligou's results, because
Perkins considers the quantum approximation for electron and high
relative energies, whereas Ligou uses only the classical approximation.

Ray, Hora's "collective" approximation follows the exact solution for lower densities and lower temperatures, but breaks down
for high densities. For high and low temperature the energy deposition to the ions is greater than to the electrons. It would be
interesting to compare in range at low temperatures and high densities
the exact binary collision theory with the results of Sigmar, Joyce.
The minimum of the ranges is caused by higher energy loss rate in
temperature range of 0.1-1 eV, as shown in Fig. 5. Whereas the
maximal energy loss rate to electrons is at energy $\sim 1.6\ 10^{4}$ $kT_{e,i}$,

the maximal energy loss rate to ions is at about energy 3 $kT_{i,e}$.
Considering additionally that the ranges are calculated for energies
between $\frac{3}{2} kT_{i,e}$ and 3.5 MeV, the energy loss rates to ions becomes
much more important for higher temperatures. Therefore, the ranges
are shorter at higher temperatures if one considers electrons and
ions.

   3.5. In the last case the 14.7 MeV proton ranges were calculated for D-T plasma defined in the previous case. The range is
defined in the previous case.

   The proton ranges are much longer than the alpha ranges shown
in previous case. Also in this case, Perkin's approximation gives
better results than the Ligou's approximation.

ACKNOWLEDGEMENTS

   The author would like to express his thanks to Prof. W. Seifritz
for his valuable discussions during the work and to Mr. K. Herda
for preparing the manuscript and the figures.

REFERENCES

1. B. Trubnikov, "Reviews of Plasma Physics 1," Consultants
      Bureau, New York (1965).
2. L. P. Shkarofsky, T. W. Johnson and M. P. Bachynski, "The Parti-
      cle Kinetics of Plasmas," Addison-Wesley Publ. Co., Inc.,
      Reading, Mass. (1966).
3. K. Fowler, M. Rankin, Plasma Phys. 8 (1966).
4. D. J. Rose, Engineering Feasibility of Controlled Fusion, Nucl.
      Fusion, 9 (1969).
5. A. Husseiny, H. K. Forsen, Slowing Down of Fast Test Particles
      in Plasma, Phys. Rev., A1, 1483 (1970).
6. R. Mills, PPL-Report Matt-728 (1970).
7. G. Kelley et al., Plasma Phys., 13 (1971).
8. K. A. Bruckner and S. Jorma, Laser-Driven Fusion, Rev. of Modern
      Physics, Vol. 46, 2 (1974).
9. T. Kammash, "Fusion Reactor Physics," Ann Arbor Science Publ.,
      Inc., Ann Arbor, Mich. 48106, Chapt. 4 (1975).
10. P. S. Ray and H. Hora, Corrected Penetration Length of Alphas
      for Reheal Calculations, in: "Laser Interaction and Related
      Plasma Phenomena," Vol. 4 (1978).
11. J. R. McNally, Jr. and R. D. Sharp, Advanced Fuels for Inertial
      Confinement, Nucl. Fusion, 16, 868 (1976).
12. S. T. Perkins, Neutron-Induced Fission in a DT-Plutonium Plasma,
      Nucl. Sci. and Eng., 69, 137 (1979).

13. J. Ligou, Private communication (EPFL Lausanne).

14. D. J. Sigmar and G. Joyce, Plasma Heating by Energetic Particles, Nuclear Fusion, 11, 447 (1971).

15. D. Bohm, "Quantum Theory," Prentice-Hall, Englewood Cliffs, N.J., (1951).

16. S. Gasiorowicz, et al., Phys. Rev., 101 (1956).

17. E. Fermi and E. Teller, Phys. Rev., 72, 399 (1947).

CONCEPTUAL DESIGN OF 1200MWth LASER

FUSION REACTOR IN ILE OSAKA UNIVERSITY

C. Yamanaka, S. Nakai, S. Ido, T. Yabe, K. Imasaki,
M. Matoba, Y. Kitagawa, K. Yoshida, T. Mochizuki,
K. Mima, K. Nishihara, T. Sasaki, Y. Kato, Y. Izawa,
T. Yamanaka, M. Yokoyama

Institute of Laser Engineering
Osaka University
Suita, Osaka, Japan, 565

1. INTRODUCTION

The remarkable recent progress on pellet fusion experiments
and on high power driver technologies have urged us to the study
of the conceptual design of the Inertial Confinement Fusion (ICF)
reactor. The required technologies for the ICF reactor are far
different from that of the Magnetic Confinment Fusion (MCF)
reactor. This paper presents a design of a 1200MW(thermal)
commercial ICF reactor on the basis of our system study of the
ICF reactor which is proceeded to find out tecnical and physical
problems on a way to a commercial fusion reactor.

An original design of the reactor which is studied utilizes
thick Li flow inside a spherical cavity to moderate the first wall
loading and neutron fluence on solid structural materials. This
Li flow is also used as 'inner blanket' and it is guided by the
steady magnetic field (cusp field or mirror field) to form the
desired Li flow.

The items which are examined and designed are as follows.

(1) Pellet design —— design and fabrication of high gain pellet
     and energy spectrum of fusion burst.
(2) Design of 1 MJ energy driver.
(3) Conceptual design of the reactor system.
     (3)-1 Energy balance.
     (3)-2 Conceptual design of magnetically guided Li flow blanket.
     (3)-3 Thermal problems in the blanket.
     (3)-4 Mechanical problems in the reactor cavity.
     (3)-5 Pellet injection problem.
     (3)-6 Conceptual design of the cooling system.

(3)-7 Neutronics in the blanket.
(4)   Economics of the ICF reactor.
These designs are discussed in the following chapters; Pellet design in Chapter 2, Energy driver design in Capter 3, Reactor design in Chapter 4 and Economics in Chapter 5.

Designed parameters in our reactor system are as follows.

        (A) Laser Energy   1 MJ
                Peak Power density   200 TW
                Shot Cycle   1 Hz
                Pulse duration   100 nsec
                Efficiency   10 %
        (B) Pellet   Multi-shell
                                Radius $\sim$ 3 mm
        (C) Output Power   1240 MW (thermal)
                                             496 MW (gross electric)
                                             426 MW (net electric)
        (D) Reactor Vessel with magnetically guided inner Li
             flow
                SUS container   R = 5.64 m
                Cavity   R $\sim$ 5.00 m
                Inner Li flow   d $\sim$ 0.64 m

In this report, the design of the reactor system is examined to reveal the feasibility of the laser ICF reactor.

2. PELLET DESIGN

2-1 Scaling Law of Pellet Gain

The pellet gain Q (=fusion power output E_f/output energy of laser E_L) should be greater than 100 to achieve an total plant efficiency η_s of 30% for a power reactor as discussed in Chap.4-1. We here show that the pellet gain Q becomes of the order of 100$\sim$ 1000 for a laser of megajoule by means of an energy balance[1].

It is useful to write the pellet gain Q as

$$Q = \frac{E_f}{E_L}$$

$$= \left(\frac{E_a}{E_L}\right) \left(\frac{E_p}{E_a}\right) \left(\frac{E_f}{E_p}\right)$$

$$= \varepsilon_a \varepsilon_c g_f$$

$$= \varepsilon_{Lp} g_f \tag{2-1}$$

where ε_a is the absorption efficiency, ε_c is the fraction of absorbed energy converted to internal energy of the fuel at the instant of ignition, $\varepsilon_{Lp} = \varepsilon_a \varepsilon_c$ is the coupling efficiency between the laser output energy and D-T fuel and g_f is the fuel gain. The ignitor plasma of the fuel is required to be hot enough ($T_I > 5keV$) and large enough ($\rho_I R_I > 0.8gr/cm^2$) to ignite the thermonuclear reaction wave[27]. If the rest of the fuel is compressed isentropically, the internal energy of the fuel can be given by

$$E_p = 4\pi \frac{(\rho R)^3}{m_i \rho^2}[\alpha(\frac{\rho}{\rho_o})^{2/3} + (\frac{\rho_I R_I}{\rho R})^3 T_I], \quad (\alpha \geq 1) \qquad (2-2)$$

where ρ_o is an initial density of the fuel and α is the numerical facter which is unity if there is not preheating. Since the strong[3] reaction wave is an accelerated overcompressed detonation wave, its propagation velocity is so large that a homogeneous burning can be in good approximation to estimate energy balance. Fusion power output E_f is shown as follows.

$$E_f = \frac{4}{3}\pi R^3 \frac{\rho}{2m_i} YE_R$$

$$\simeq \frac{2}{3}\pi \frac{(\rho R)^3}{m_i \rho^2}(\frac{\rho R}{a+\rho R})E_R$$

where Y is the burningrate,

$$Y \simeq \frac{\rho R}{a+\rho R} \quad (a = \frac{2m_i v_s}{<\sigma v>}),$$

and E_R is the output energy of D-T reaction, 17.1MeV.

Thus the fuel gain $g_f = E_f/E_p$ can be written as

$$g_f = 2.9 \times 10^6 \frac{\rho R}{a+\rho R} \frac{1}{[\alpha(\frac{\rho}{\rho_o})^{2/3} + (\frac{\rho_I R_I}{\rho R})^3 T_I]} \qquad (2-3)$$

It is obvious from Eqs. (2-2), (2-3) that an optimum value for ρR exists to obtain the maximum gain for a given internal energy E_p. The maximum gain and the internal energy for the optimum value is given by

$$g_{fM} = 2.9 \times 10^6 \frac{\rho R}{a+\rho R} \frac{a+3\rho R}{13a+12\rho R} (\frac{\rho R}{\rho_I R_I})^3 \frac{1}{T_I}, \qquad (2-4)$$

$$E_p = 1.2 \times 10^7 \alpha^3 (\rho R)^3 \frac{(2a+3\rho R)^2 (13a+\rho R)}{(12a+9\rho R)^3}$$

$$\times \; (\frac{\rho R}{\rho_I R_I})^6 \; \frac{1}{T_I^2} \; (eV) \tag{2-5}$$

The dependence of the maximum gain Qmax on the laser energy is
shown in Fig. 2-1, while the parameters used are given in Table
2-1, when $E_L \sim$ 1MJ, therefore $\rho R \sim 3 \sim 4$ and $a \sim 2 \sim 6$. Table
2-2 shows the various parameters for the case II and IV and 1MJ
laser. The scaling law obtained from Fig. 2-1 can be written as

$$Q = 9.5 \times 10^5 \; \frac{E_L^{3/8} \varepsilon_{LP}^{11/8}}{\alpha^{9/8} (\rho_I R_I)^{3/4} T_I^{1/4} a^2} , \tag{2-6}$$

where E_L is in MJ.

2-2 High Gain Pellet

     The implosion of shell requires less peak optical power than
that of sphere[4-7]. On the other hand, the shell target is easier
to be destroyed by fluid instabilities, e.g. Rayleigh-Taylor
instability[8]. In order to overcome the demerit, the careful choice
of target structure is required.
     If we specify the aspect ratio and the initial radius of shell
ζ and R_o, respectively, the shell implodes optimally according[6,7]
to

$$R \sim R_o \sqrt{1-t/t_s} , \tag{2-7}$$

where t_s is the culmination time. Due to the acceleration

$$g \sim \frac{R_o}{4t_s^2} (1-t/t_s)^{-3/2}, \tag{2-8}$$

caused by this motion, surface perturbation ξ grows as

$$\xi \sim \xi_o (1-t/t_s)^{k/2} \sim \xi_o \eta_c^{k/3}, \tag{2-9}$$

where ξ_o, k are the amplitude and the mode number of initial
perturbations, respectively, η_c is the compression ratio. In
order that the amplitude of the most serious mode $(k \sim \zeta)$ may be

TABLE 2-1

THE PARAMETERS USED TO OBTAIN

THE PELLET GAIN IN Fig. 2-1

	T_I (KEV)	$\rho_J R_i$ (GR/CM^2)	ε_{LP}	$2M_I V_S / [\sigma v]$ (GR/CM^2)	α
I	10.0	0.5	0.07	2.0	1
II	5.0	0.8	0.07	2.0	1
III	10.0	0.5	0.05	2.0	2
IV	5.0	0.8	0.05	2.0	2
V	5.0	0.8	0.05	6.0	2
VI	5.0	0.8	0.03	2.0	2

TABLE 2-2

THE PARAMETERS FOR 1MJ LASER. M_F IS THE FUEL MASS, Y

IS THE BURNING RATE AND E_I IS THE INTERNAL ENERGY OF

THE IGNITOR

	CASE II	CASE IV
Q	860	250
ρR (GR/CM^2)	4.34	3.37
ρ (GR/CM^3)	302	368
R(μM)	70	92
ρ/ρ_0	1510	1840
M_F (MGR)	3.75	1.19
Y(%)	68	63
E_I/E_P	0.19	0.18

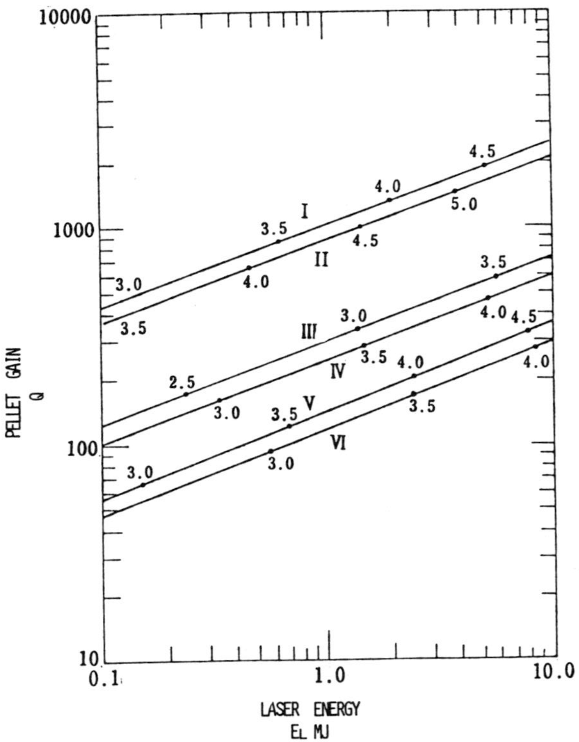

Fig. 2-1 The dependence of the pellet gain, Q, on the laser energy E_LMJ. The values in the figure correspond to the optimum ρR to obtain the maximum gain. The parameters used are shown in Table 2-1.

smaller than shell thickness, the limitation on the aspect ratio

$$\zeta < [3\frac{\log(\Delta R_o/\xi_o)}{\log\eta_c}]^2 , \tag{2-10}$$

must be fulfilled, where ΔR_o is the initial shell thickness.

The laser peak power of $\sim$100 TW may be conjectured in the last phase of implosion process, when the preheating of D-T fuel[9] by high energy electrons may be serious. The mean free path λ_M of momentum scattering by ions is

$$\lambda_M \simeq 0.22 \frac{(T_h/100\text{keV})^2}{<Z^2>(n/5\times10^{22}\text{cm}^{-3})} \text{ cm}, \tag{2-11}$$

and the penetrating distance λ_p may be estimated by

$$\lambda_p \simeq \sqrt{\frac{<Z^2>}{<Z>}}\ \lambda_M, \tag{2-12}$$

where T_h is the temperature of high energy electrons, n the ion density, Z the atomic number and <> denotes the average over ion species. To prevent the inner core from preheating, the pusher thickness of several times λ_p is needed. But the reduction of energy transport has an disadvantage that asymmetries of laser irradiation cannot be smoothed out. Accordingly the outer part of the pusher must be replaced by low Z ablator. The density of pusher materials must be close to that of the ablator to moderate the limiations due to instabilities between the pusher and the ablator.

Taking into account of requirements mentioned above, we choose a target shown in Fig. 2-2, by which the pellet gain of 500 $\sim$ 1000 may be expected.

2-3 Energy Spectrum

The fusion energy is distributed to the neutron and α-particle in the first step. The core plasma is heated up through the collision of energetic α-particles produced from the hot spot at the core center, i.e. formation of the thermonuclear reaction wave, so most of the energy of the α-particle is transported to the plasma[3]. Then the parameter of the plasma after thermonuclear burning is estimated as shown in Table 2-3.

Due to the large value of ρR of the core the neutron is scattered and delivers a part of its energy to the plasma[10]. From cross sections of elastic and inelastic collisions of neutron with D or T, the average number of scattering in the core, $n\sigma Rc$ is calculated to be 0.9 for 14 Mev neutrons, so that multiple

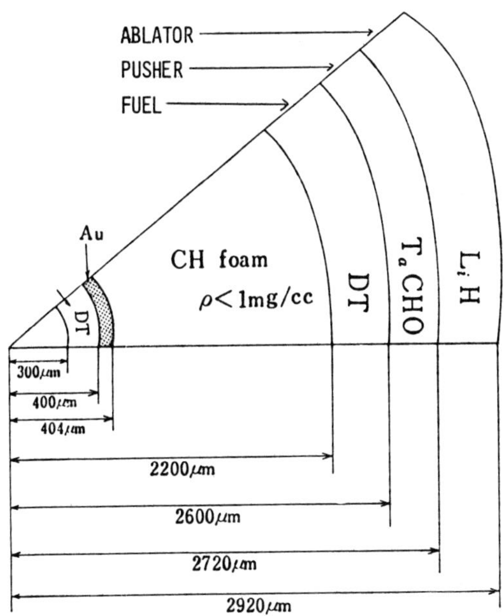

Fig. 2-2 The structure of high gain pellet. Outer radius of pellet in ~3mm, and the aspect ratio of pusher-ablator is 6~8. Inside the polyfoam, the small D-T ignitor is included.

TABLE 2-3

THE PARAMETER OF PELLET
FUEL PLASMA AFTER THE
THERMONUCLEAR BURN

Burn Rate	50%
Compressed Fuel Density	10^{26} cm^{-3}
Compressed Fuel Radius	130μm
Compressed Fuel Mass Density	400gcm^{-3}
Reaction Time	20ps
Electron Temperature	200keV
Ion Temperature	250keV

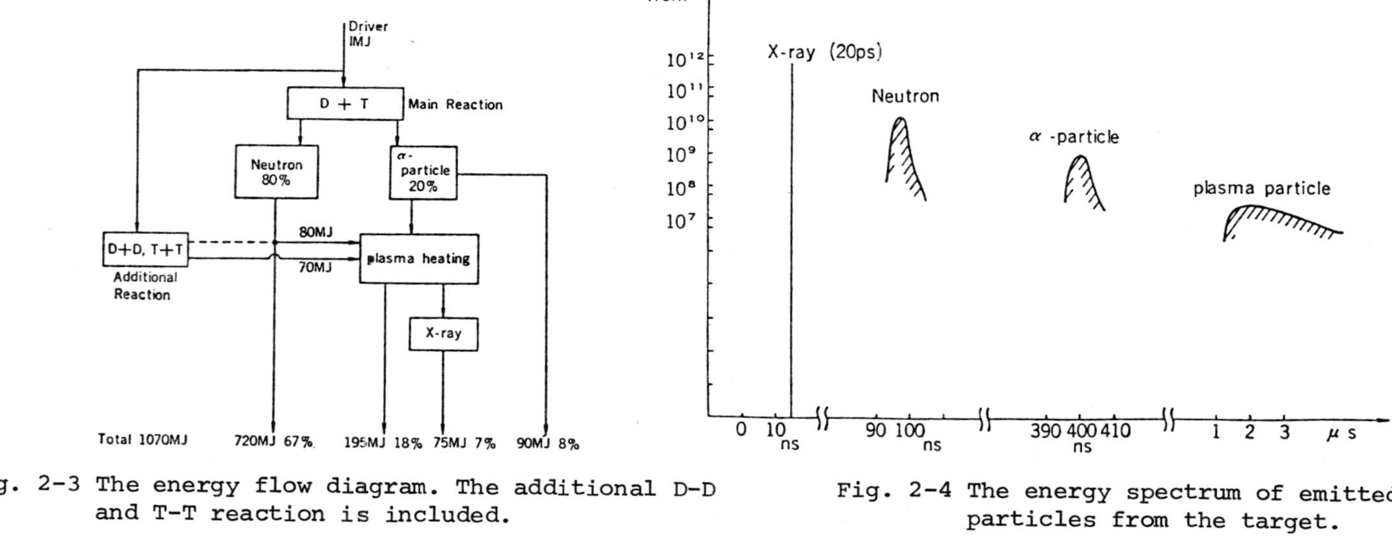

Fig. 2-3 The energy flow diagram. The additional D-D
and T-T reaction is included.

Fig. 2-4 The energy spectrum of emitted
particles from the target.

Fig. 2-5 Energy spectrum of neutron from
microexplosion.

scattering can be neglected.

Average energy deposition into plasma by single elastic collision is described by the following equation,

$$\Delta E_{el} = \frac{2A}{(1+A)^2} \cdot E_o ,$$

where A is the mass of the collision partner and E_o is the initial collision energy. Inelastic collision causes the following nuclear reaction,

$$n + D \rightarrow 2n - 3.8 \text{ MeV}.$$

Elastic collision cross section of such low energy products with plasma is so large that total kinetic energy is deposited into plasma. The deposited energy is 10.6 MeV. Taking into account D-D and T-T reactions, the number of neutrons which participate in the process has been estimated to be $1.4N_0$ where N_0 is the number of D-T neutrons. Thus the energy distribution of neutrons incident on a reactor wall has been obtained as shown in Fig. 2-5. In such length temperature plasma, the radiation through e-e and e-i bremsstrahlung is emitted. Fig. 2-3 shows a diagram of the energy flow produced by the fusion. Then the energy spectrum at the distance of 5 m from the pellet is estimated as shown in Fig. 2-4. It should be pointed out that the resultant neutrons have much lower and broader energy spectrum than the initially produced 14MeV neutrons from D-T reaction. Therefore the blanket should be designed for such energy spectrum of incident neutrons.

3. 1MJ ENERGY DRIVER

3-1 Requirements and Evaluations

The energy driver for a future inertial confinement power plant should have sufficient capabilities in 1: energy production, 2: energy propagation, and 3: energy deposition on a fusion target. Table 3-1 lists the required performances in these three characteristics.

These characteristics of lasers, relativistic electron beam (REB), light ion beam (LIB), and heavy ion beam (HIB) have been evaluated based on our present understanding, and the result is summarized in Table 3-2. Future developments of these energy drivers should be stressed on the following aspects.

Lasers: Efficiency; reliability, construction and maintenance; cost; radiation protection of beam propagation optics.

REB: repetitive operation; beam propagation; energy deposition on a target.

LIB: high power production; reliability; beam propagations

HIB: high power production, construction cost; beam propagation.

TABLE 3-1
REQUIREMENTS TO AN ENERGY DRIVER FOR ICF REACTOR

(1) ENERGY PRODUCTIN

OBJECTIVE		REQUIRED PERFORMANCES
1 To obtain suffi-cient (~1000) pellet gain	A Total output energy	~MJ
	B Peak power	~200TW
	C Pulse shape	Shaped pulse with total width of ~100ns
2 To improve total system efficiency	D Energy production efficiency	~3%[*][+]
	E Repetiton rate	$\gtrsim$1Hz[*]
3 High Reliability	F Cotinuous repe-titive operation	$>10^8$
	G Jitter	$\lesssim$1ns
	H Safty factor	$\gtrsim$2
	I Reproducibility	$\lesssim$3% for A~C
	J maintenance	Easiness for parts change and stock
	K Safty	Precausions against fire, explosion, toxity, and radiation
4 To reduce power production cost	I Initial construction cost	Within 10% of total construction cost
	M Maintenance cost	Within 10% of initial construction cost/year
	N System life	>30years

[*] These figures are based on a system having pellet gain of 1000 and repetition rate of 1Hz. If the drive efficiency is improved to over 10%, a reactor system with the pellet gain of 200 driven by a 500kJ energy driver at a rep. rate of a few Hz become possible.

[+] The drive efficiency could be reduced for a fusion-fission hybrid reactor.

(2) ENERGY PROPAGATION

1. To propagate energy from a driver to a target	A. Propagation distance	~ 100 m
	B. Divergence	$\lesssim 10^{-5}$ rad·cm
	C. Efficiency	$\gtrsim$ 90 %
	D. Directional control	Servo Controlled target tracking system
2. Radiation protection	E. Energy driver protection	Non straight Propagation
	F. Final focusing element protec-tion	Radiation protec-tion and easy maintenance

(3) ENERGY DEPOSITION ON A TARGET

1. Good deposition efficiency	A. Absorption	$\gtrsim$ 30 %
	B. Deposition property	Prevent decoupling of absorption layer from core
2. Good compression efficiency	C. Absorption layer	High density region
	D. Compression efficiency	$\gtrsim$ 15 %

TABLE 3-3

HIGH VOLTAGE POWER SUPPLY FOR 1 BEAM AMP

	Main discharge	E-beam
Power supply voltage	5.6 MV	5.6 MV
Output voltage	2.8 MV	2.8 MV
Output energy	1.07 MJ	100 kJ
Output pulsewidth	2 μs	2 μs
Output impedance	16.3 Ω	163 Ω

TABLE 3-4

PERFORMANCE CHARACTERISTICS OF THE 1MJ CO_2 LASER SYSTEM

Number of beam	8
Rep. rate	1 shot/s
Power for laser excitation	8800 kW
Power for laser switching	800 kW
Power for auxiliary systems	600 kW
Total operation power	10200 kW
Laser output (1 shot/s)	1440 kW
Overall efficiency	14 %

TABLE 3-2

COMPARISON OF ENERGY DRIVERS

		Laser	REB	LIB	HIB
1.	Energy producter				
a.	Total output energy	△	○	○	●
b.	Peak power	○	○	●	●
c.	Pulse shape	○	△	●	●
d.	Energy production efficiency	△	○	△	△
e.	Repetition rate	●	●	●	○
f.	Continuous repetitive operation	●	●	●	●
g.	Jitter	○	●	●	○
h.	Safety factor	△	●	●	△
i.	Reproducibility	△	●	●	△
j.	Maintenance	●	●	●	●
k.	Safety	○	△	△	△
l.	Initial construction cost	●	△	△	●
m.	Maintenance cost	△	△	△	△
n.	System life	●	●	●	△
2.	Energy propagation				
a.	Propagation distance	○	●	●	△
b.	Divergence	○	●	●	●
c.	Efficiency	○	●	●	●
d.	Directional control	○	●	●	△
e.	Energy driver protection	○	●	●	△
f.	Final focusing element protection	●	●	●	●
3.	Energy deposition on a target				
a.	Absorption efficiency	○	○	○	○
b.	Deposition property	△	△	△	△
c.	Absorption layer	△	●	○	○
d.	Compression efficiency	△	△	△	△

(NOTES) ○ Well established

△ Sufficient possibility

● Depends on future developments

Comparing these energy drivers, lasers have the most established understandings and technologies for scaling present systems to
a high energy driver for the ICF reactor. Among various high power,
high energy lasers, we have chosen a carbon dioxide laser as a model
system for the ICF reactor, beacause it has the highest efficiency,
good possibility for scaling up to $\sim$1MJ output energy and relatively low cost compared to other lasers.

3-2 Carbon Dioxide Laser

Carbon dioxide (CO_2) laser has complementary charcteristics to
glass laser. It has high efficiency and has capability for repetitive operation. On the other hand, technologies on CO_2 laser are
not so well developed as glass laser, and the effects of 10 μm wavelength on implosion prosesses are not well established. However,
we regard that CO_2 laser is one of the best candidate as a driver
for ICF reactor because of its potentiality shown below.
Main advantages of CO_2 laser as a reactor driver are
followings:
1. Very high overall efficiency of over 10 % is achievable.
2. High repetition rate operation of over 10 Hz is possible by fast
   cooling using rapid circulation of laser gas.
3. Accuracy and homogeneity requirements of optical components are
   less severe due to long wavelength (10.6 μm) of CO_2 laser.
4. Scaleup to 1 MJ system is fairly easy and construction cost will
   be modest.
Figure 3-1 shows the schematic presentation of a 1 MJ CO_2
laser system which we have designed. The laser is a regenerative
type amplifier having cavity length of 42 m and aperture area of
2m×2m. A laser pulse traverses and is amplified in the cavity 16
times in 2.5 μs. During this period, the laser gas remains active
due to electron-beam controlled discharge pumping (<2μs) and also
due to energy transfer from vibrationally excited nitrogen molecule
to CO_2 molecule. By this repeated extraction method, we can expect
overall efficiency of $\sim$14% when we can obtain extraction efficiency
of 20%.
Since the gain coefficient remains less than 3% cm^{-1} in this
design, the maximum g·L product is 8.8. Therefore parasitic
oscillation will not take place and saturable absorber will not be
required.
The length of the pumped region is $\sim$80 cm. High voltage pulse
forming networks are used for ionization and main discharge. Parameters of these power supply are shown in Table 3-3. These are 20
stage Marx circuits with output voltage of 5.6 MV. Optimum discharge field of 7 kV/cm·atm is applied in the He free laser gas
which has the pressure of 2 atm. The pumping energy density is
150 J/1·atm and the output energy is $\sim$180 kJ. The output energy
density in this case is 4.5J/cm^2 which is factor of 3 smaller than
anticipated damage threshold of gas dynamic window (Fig. 3-2) at
the pulse width of 100 ns.
The laser pulse in the cavity is switched-out by a large

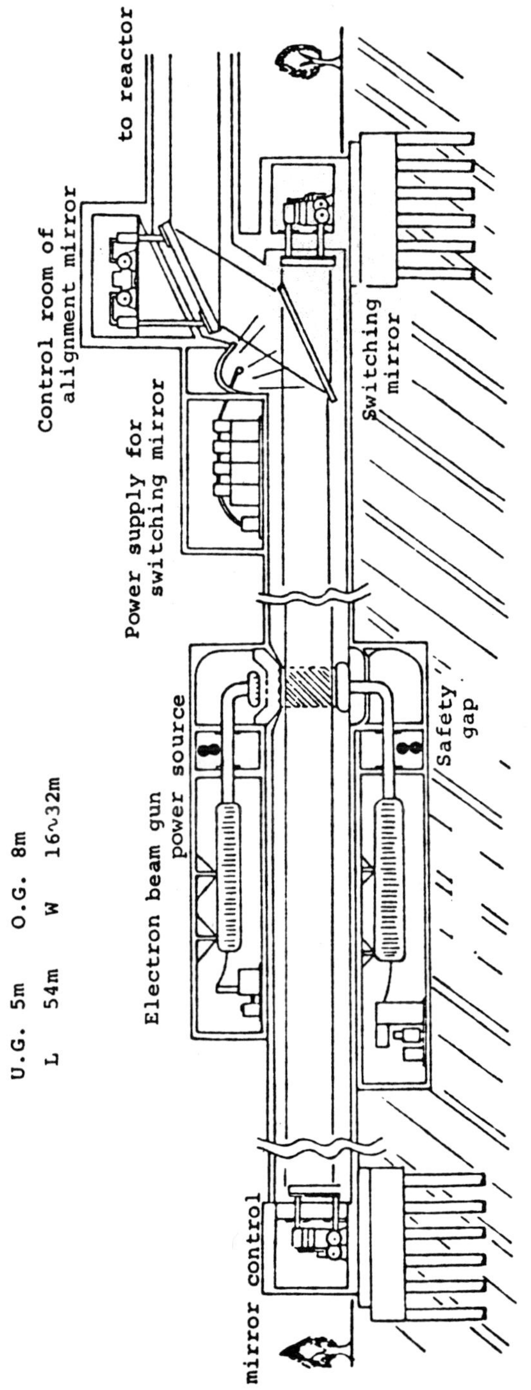

Fig. 3-1 Schematic of Laser Driver

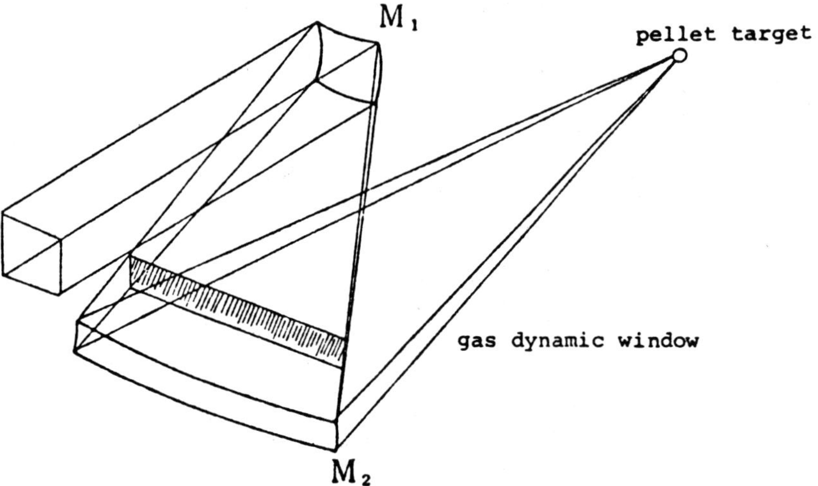

Fig. 3-2 Laser window using gas dynamic shielding window.

aperture switch refrector. Free carrier is generated in the
Brewster-angled Ge crystal plate by photo-excitation. In order to
obtain switching time of 100 ns, 64 kJ energy at 5000 A will be
required. Light emitting diode might be useful for this purpose.
Another approach by mechanical swithcing is also being investigated.

The switched out laser pulse is guided to the reactor chamber
by metal reflective mirrors. The laser chamber which is filled by
laser gas is isolated from the target chamber by gas dynamic window,
as shown in Fig. 3-2. By adopting free-vortex aerodynamic window,
we can sustain pressure difference of 2 atm by shock-free gas flow
without producing disturbance to laser beam propagation.

Operation lifetime of various components in this laser system
have been examined. Electrical components such as capacitor, high
voltage isolaion, and switch gaps can have maintenance-free life of
over 10^8 shots (>1 year at 1 Hz). Sufficient data are not available
on optical components, but probably overhaul once a year will be
necessary.

Performance parameters of this 8-beam laser system are
summarize in Table 3-4.

4. CONCEPTUAL DESIGN OF THE REACTOR SYSTEM

In this chapter, we discuss the conceptual design of 1200MW
(thermal) ICF reactor. The typical parameters of the reactor system
are summarized in Table 4-1. The concept of magnetically guided Li
flow is adopted as the 'internal blanket'. SUS-316 is used as the
structural materials because its material characteristics is well
known and it is reliable to weld and machine SUS. On the other
hand, temperature is supressed in the rather lower level (400∿500°C),
because compatibility between SUS and Li is not well known. In
spite of this low level of the limit of temperature, the well-
developed turbine engineering enables us to gain the sufficient net
electric power (426MW$_e$) in our system. The reliability of materials
is taken care of in our commercial design.

4-1 Energy Balance of the Total System

A total efficiency of the laser fusion reactor system is defined
to be a ratio of electric output energy to thermal output energy
from a thermonuclear microexplosion. The total efficiency is required to be larger than 30%. Since the wast energy may cause
various problems.

Let us introduce an energy flow diagram shown in Fig. 4-1.
Here, a pellet gain is defined by

$$Q = E_f/E_L , \qquad (4\text{-}1)$$

where E_f is the thermonuclear fusion energy produced by pellet
burning and E_L is an incident energy on the target from an energy
driver. Note here that the pellet gain Q is a function of the laser
energy E_L.

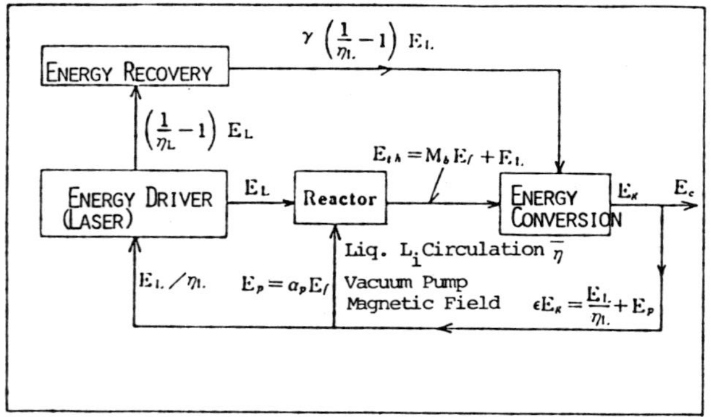

TABLE 4-1

REACTOR SYSTEM SUMMARY

LASER

OUT PUT(E_L)	1MJ
REPETITION	1H$_z$
EFFICIENCY(η_L)	10%
NUMBER OF BEAMS	8

BEAM TRANSFER EFFICIENCY   100%—(90%)

PELLET GAIN(Q)   1000

ENERGY FLOW

THERMAL OUT PUT E_{TH} 1000(0.8M + 0.20)

M: BLANKET GAIN = 1.3

M_B: 0.8M + 0.2 = 1.24

(NATURAL LITHIUM)

$$E_{TH} = 1240MJ$$

CONVERSATION EFFICIENCY	40%
GROSS ELECTRIC POWER	496MW
CIRCULATING POWER	70MW
NET ELECTRIC POWER	426MW

OVER ALL SYSTEM EFFICIENCY $\dfrac{E_e}{E_{th}}$ = 34.4%

REACTOR STRUCTURE

MAGNETICALLY GUIDED L_i FLOW

REACTOR CAVITY SUS 316

OPERATION TEMPERATURE	INLET	370°C
	OUTLET	500°C

E_e :   Net electric output energy

E_g :   Gross electric output energy

ϵ :   Fraction of recycled electric energy

E_L :   Laser output energy

E_{th}:   Thermal energy

E_f :   Thermonuclear fusion energy

M_b :   Blanket gain

$\bar{\eta}$:   Conversion efficiency

γ :   Recovery coefficient of thermal energy

η_L :   Laser efficiency

Fig. 4-1 Energy Flow diagram

For given Q, the gross electric output power E_g is estimated by

$$E_g = \bar{\eta} \{(M_b Q + 1) E_L + \gamma (\frac{1}{\eta_L} - 1) E_L\} , \qquad (4\text{-}2)$$

as shown in Fig. 4-1, where $\bar{\eta}$ is the conversion efficiency from thermal to electric, γ is the recovery coefficient of thermal energy in the driver system. The electrical power recycled to maintain the laser operation (E_L/η_L, η_L: laser efficiency) and to evacuate a reactor cavity ($E_p = \alpha_p E_f$) is assumed to be given by

$$\varepsilon E_g = \frac{E_L}{\eta_L} + \alpha_p (M_b Q + 1) E_L , \qquad (4\text{-}3)$$

where ε is the recirculating power fraction and $E_f = M_b Q E_L + E_L$. Therefore a net output energy is given by

$$E_e = E_g (1 - \varepsilon)$$

$$= \bar{\eta} [M_b Q + 1 + \gamma (\frac{1}{\eta_L} - 1)] E_L - E_L/\eta_L - \alpha_p (M_b Q + 1) E_L \qquad (4\text{-}4)$$

Finally the total plant efficiency η_s is obtained as follows,

$$\eta_s = \frac{E_e}{M_b E_f}$$

$$= \frac{\bar{\eta}}{M_b Q} \{M_b Q + 1 + \gamma (\frac{1}{\eta_L} - 1)\} - \frac{1}{M_b Q} \{\frac{1}{\eta_L} + \alpha_p (M_b Q + 1)\} \qquad (4\text{-}5)$$

We determine the pellet gain Q required to achieve a certain total efficiency η_s for a system with a laser efficiency η_L.

Here, we set $\bar{\eta} = 40\%$, $\gamma = 0$, $\alpha_p = 5\%$, $M_b = 1.2$ as standard values. The total efficiency η_s, is shown in Fig. 4-2. The figure shows that if the laser efeciency η_L is 10%, the practical power plant is available for the value of the pellet gain greater than 200.

In the next, we consider possibility of a fusion-fission hybrid reactor. In Fig. 4-3, the total efficiency η_s as a function of the pellet gain Q is shown for a hybrid system. The parameters indicated in this figure are the blanket gain, M_b. The solid and the dashed curve correspond to the laser efficiency of 5% and 10%, respectively. In D-T fusion reaction, blanket gain M_b is given by

$$M_b = 0.8M + 0.2 \qquad (4\text{-}6)$$

where M is an energy multiplication factor by neutron and is defined

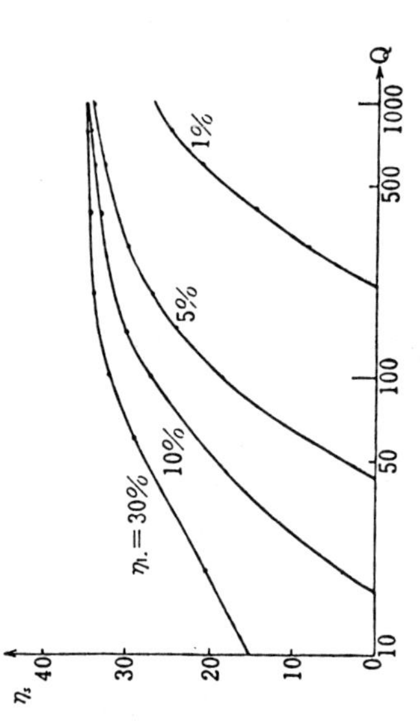

Fig. 4-2 Total plant efficiency η_s as a function of gain Q for various laser efficiency η_L.

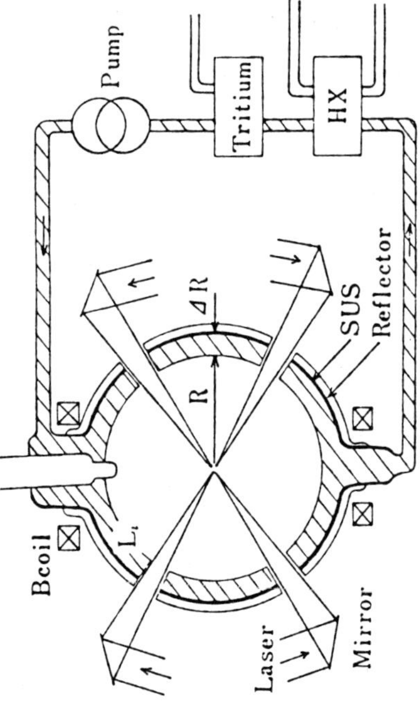

Fig. 4-4 Concept of magnetically guided L_i flow reactor.

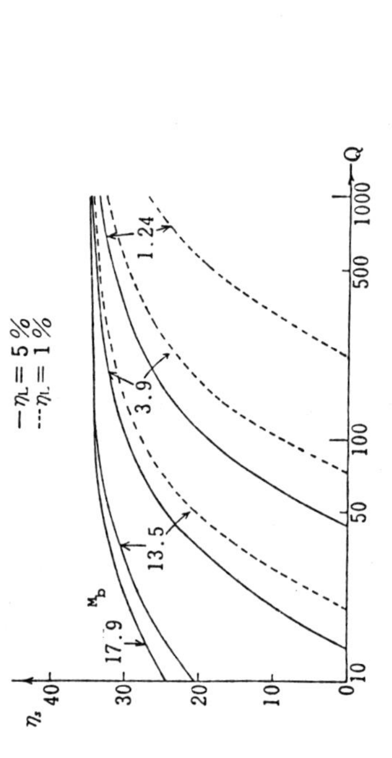

Fig. 4-3 Total efficiency η_s as a function of pellet gain Q with different blanket gain.

by the ratio of a total energy produced by neutron in the blanket
to the total neutron energy produced by the thermonuclear burning.
In a pure fusion reactor with natural Li blanket, M is about 1.3.
In the hybrid reactors, the values of M are 4.6, 16.6 and 22.1 for
^{232}Th, ^{238}U and natural uranium blanket respectively. When the
blanket gain M_b=17.9(multiplication factor M=22.1) and η_L=5%, only
the pellet gain Q of 20 is required for practical reactor.

Finally we determine the laser energy E_L required to achieve
the total efficiency η_s greater than 30% by using a scalling law of
$Q=250E_L^{3/8}$ (E_L in 1MJ), which is given in Eq. (2-6) and Case IV in
Fig. 2-1, Table 2-1 and Table 2-2.

Setting the laser efficiency η_L=10% and the coupling efficiency
between the laser output energy and the D-T fuel, ε_{Lp} = 3% (ε_{Lp}
defined in Eq. 2-1), we see the requirements for E_L to design a
practical reactor. For a pure fusion reactor, E_L>500kJ because
pellet gain Q must be greater than 200. On the other hand, for a
fission-fusion hybrid reactor, E_L will be less than 100kJ because
$Q\geq20$.

4-2 Li Flow Controlled by Magnetic Field

A concept of magnetically guided Li flow reactor is shown in
Fig. 4-4. The structural wall of SUS is covered by a thick Li layer
which flow down from top to bottom. To form thick flow layer along
the wall, it is guided by steady magnetic field. The thickness of
Li is designed to be 70cm to convert the neutron energy to thermal
energy in Li and to moderate the energetic neutrons. This ensures
a low level of neutron fluence on a structural wall and a long life
of the reactor cavity.

The energy deposited by X-ray, α- particles and plasma heats up
a very thin Li layer of order 10 μm above the boiling temperature to
evaporate. The fast temperature relaxation of the thin surface layer
by heat conduction ensures the cryopumping effect on Li surface by
recondensation during an interval of explosion of 1Hz.

The mechanical force or impulsive load are very much weakened
by using Li downflow with free inside boundary facing to vacuum
space. The established technique is available enough to construct
a reactor structure.

Natural Li can be used to obtain required tritium breeding.
The Li flow system provides a convenient method to remove the heat
and separate tritium from Li in the circularting path.

The necessary condition for the practical application of this
concept is the development of Li handling technique for circulation,
heat removal, tritium separation and material compativility.

Lithium 'water-fall' concept has many advantages for protecting
structural materials from high power X-ray, 14MeV neutrons and fast
charged particles which are generated in an imploded pellet. For
controlling 'lithium-fall', external magnetic fieldi is applied.
Without the external magnetic field, lithium flow velocity is accelerated and the flow thickness is then narrowed at the foot. Fur-

thermore when the lithium flow is free from the first wall, it is difficult to stabilize the flow. Those difficulties mentioned above can be overcome by using external magnetic field. The magnetic field decelerates the flow velocity through the Lorentz force $J \times B/c$ and the flow velocity and the thickness can be then controlled. We can also fix the flow on the first wall by the magnetic field pressure. The flow pattern is then shaped by the wall and the steady flow is obtained.

The overall configuration of the magnetic field line and the lithium flow line is shown in Fig. 4-5. Although cusp magnetic field is shown in Fig. 4-5, mirror configuration is also considered to be possible. Comparison of these two configurations has not been studied yet.

For simplicity, we analyze a flow along a straight wall with a straight external magnetic field as being shown in Fig. 4-6. Z axis is taken along the wall and x axis is perpendicular to the wall. Let us give the magnetic field beneath the wall by $B = B_0 (z \cos\theta - x \sin\theta)$. The stationary lithium flow is described by the following MHD equations,

$$0 = \frac{1}{c} \underset{\sim}{v} \times \underset{\sim}{B} - \rho g (z \cos\psi + x \sin\psi) - \nabla p \tag{4-7}$$

$$\underset{\sim}{j} = \frac{\sigma}{c} \underset{\sim}{v} \times \underset{\sim}{B}, \tag{4-8}$$

$$\nabla \times \underset{\sim}{B} = \frac{4\pi}{c} \underset{\sim}{j} \tag{4-9}$$

and the imcompressibility condition.

$$\nabla \cdot \underset{\sim}{v} = 0 \tag{4-10}$$

We take the following boundary conditions on the wall (x=0) and the liquid lithium surface $(x=-\Delta x)$. to solve Eqs. $(4-7) \sim (4-9)$.

$$v_x (x=0) = 0 \tag{4-11}$$

and
$$p(x=-\Delta x) = 0 \tag{4-12}$$

where we assume that the flow surface is parallel to the wall. Solving the equations (4-7) through (4-10) with the boundary conditions (4-11) and (4-12), the liquid pressure is given as follows,

$$p(x) = \rho g \, \cos\psi \, (x+\Delta x) \, [\cot\theta - \tan\psi - \frac{2\pi\rho g}{B_0^2} \times (x+\Delta x) \frac{\cos\psi}{\sin^2\theta} \tag{4-13}$$

The pressure on the wall p(x=0) should be positive for the flow to

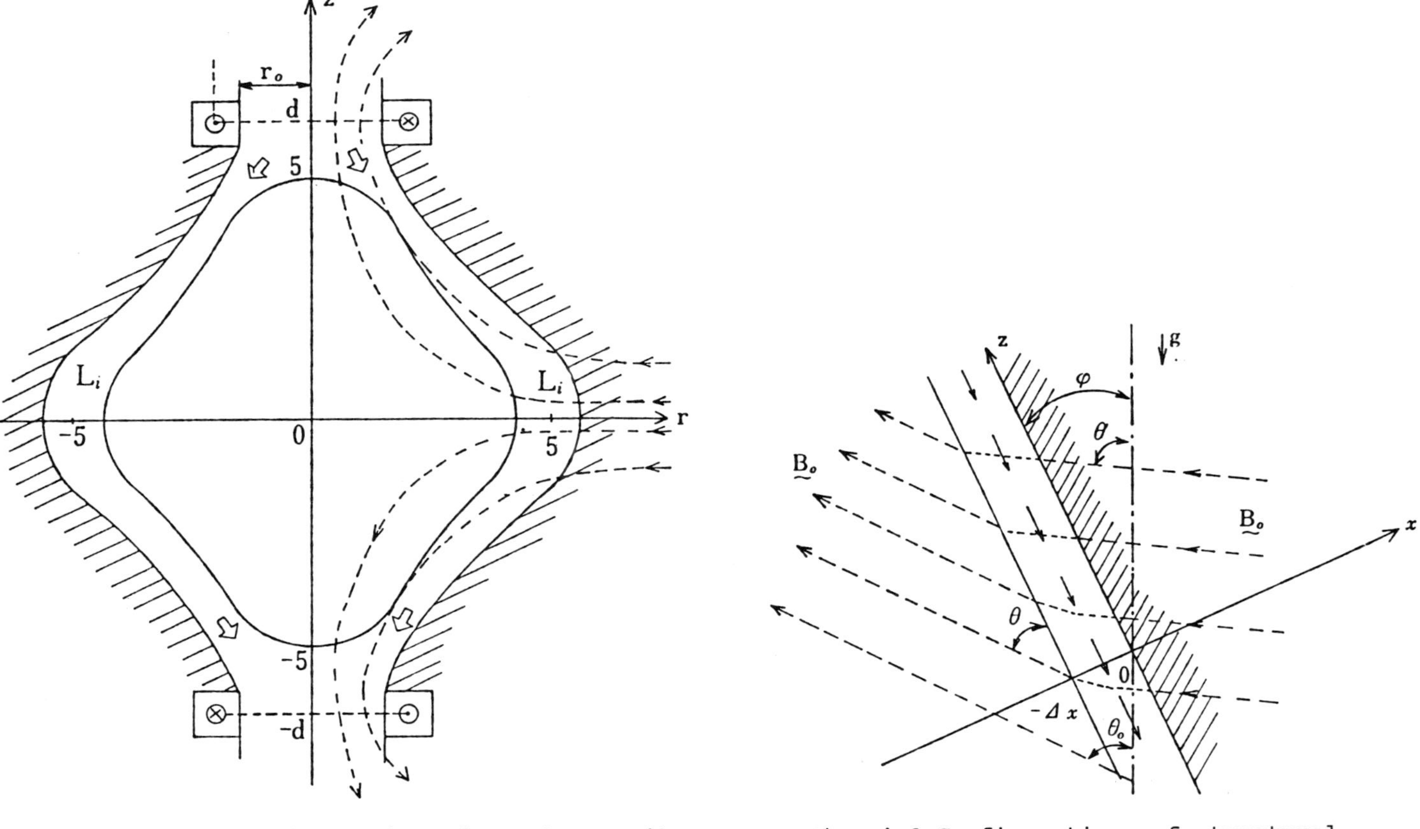

Fig. 4-5 Over-all configuration of reactor cavity.
Broken lines represent magnetic field lines.

Fig. 4-6 Configurations of structural
wall and lithium flow. Lithium
flow is indicated by solid arrows.

contact with the wall. This condition is met when

$$\cot\theta > \tan\psi + \frac{2\pi\rho\Delta x}{B_o^2}\,\frac{\cos\psi}{\sin^2\theta}, \qquad (4\text{-}14)$$

Therefore the angle between magnetic field and the flow has to satisfy

$$\frac{2\pi\rho g}{B_o^2}\,\Delta x\,\cos\psi \lesssim \theta \lesssim \frac{\pi}{2} - \psi \qquad (4\text{-}15)$$

On the other hand, self consistent magnetic field is as follows, inside the lithium fall,

$$B = B(z\,\cos\theta - x\sin\theta),$$

outside the lithium fall,

$$B = B_o\{z(\cos\theta - \frac{4\pi\rho g}{B_o^2}\,\Delta x\,\frac{\cos\psi}{\sin\theta}\,) - x\,\sin\theta\}, \qquad (4\text{-}16)$$

The difference of the magnetic field between the inside and the outside is due to eddy current in the lithium. Finally, the flow velocity of the stationary flow is given by

$$v_o = \frac{c^2}{\sigma}\,\frac{\rho g}{B_o^2}\,\frac{\cos\psi}{\sin^2\theta}. \qquad (4\text{-}17)$$

Employing material constants for lithium, $\sigma = 1.88\times10^{16}$ sec^{-1}, $\rho = 0.43$g/cm^3 at T=400 °C, the flow velocity is estimated to be $v_o = 0.2\,\cos\psi/\sin^2(\theta-\psi)$m/sec, where θ_o is an angle between B_o and the gravitational direction. For $\theta_o = \pi/6$, $\pi/4$ and $\pi/3$ and $B_o = 10^{-3}$ gauss, v_o is shown in Fig. 4-7.
    As an example, we set the shape of the wall to be r=5 sech(z/2.5) +0.7(m) and the magnetic field to be

$$B_r(r,z) \cong \frac{\pi I r_o^2}{c}\,[2(R^{-3}-R'^{-3})-3r(r+r_o)(R^{-5}-R'^{-5})] \qquad (4\text{-}18)$$

and

$$B_r(r,z) \cong 3\,\frac{\pi I r_o^2}{c}\,r[(z-d)R^{-5}-(z+d)R'^{-5}], \qquad (4\text{-}19)$$

where $R^2 = (r+r_o)^2+(z-d)^2$, $R'^2 = (r+r_o)^2+(r+d)^2$, d=6m, $r_o = 2$m and I is current intensity of a coil and r, z, d and r_o are shown in Fig. 4-5. For these magnetic field and wall configurations, z-dependence of a flow velocity and a flow thickness, $\Delta x(z)$ is shown in Fig. 4-8. Here, the flow thickness is determined by lithium flux conservation;

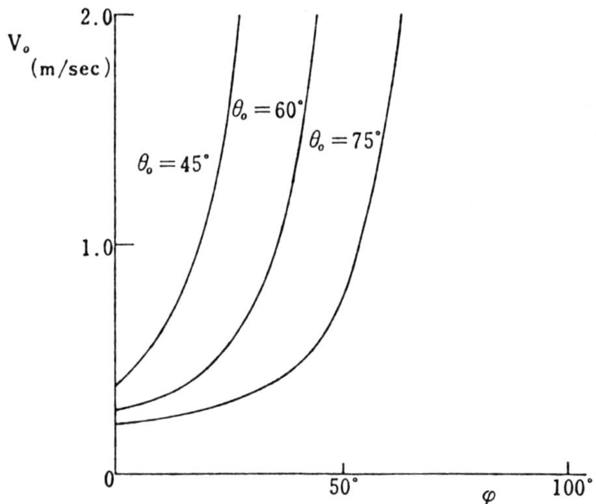

Fig. 4-7 Wall angle dependence of flow velocity. θ_0 is magnetic
field line angle.

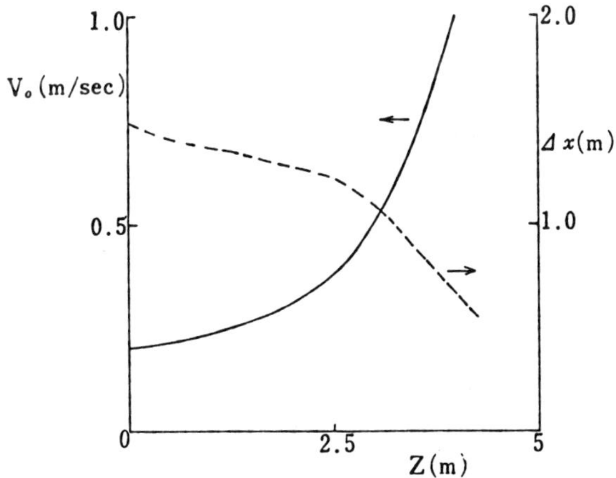

Fig. 4-8 Flow velocity and flow thickness. A solid line is for the
the flow velocity and a broken line is for the flow thickness.

$\pi\{r^2(z)-(r(z)-\Delta x(z))^2\}v_{z_0}$ =const. In Fig. 4-8, the flux is set to
be $9m^3$/sec. and B_0 =10^3 gauss. As a result, a stationary lithium
flow is formed along the reactor cavity wall. Furthermore, the
flow velocity is controlled to be appropriate by a reasonable
magnetic field strength.

4-3 Thermal Problems in The First Wall —— Surface Deposition
    and Thermal Conduction

     The energy of X-ray, α particles and plasmas is absorbed within
a thin layer on the Li surface. The deposition times T_{dep} are 30 ps
for X-ray, 350 ns for α-particle and 2 μs for plasma. If the
evaporation time T_{ev} or thermal conduction time T_{cond} is larger than
T_{dep}, the local temperature rise is given by

$$\Delta T(x) = \frac{q(x)}{\rho c_v} \qquad\qquad (4\text{-}20)$$

where $q(x)$ is specific absorbed energy, ρ is density and c_v is
specific heat capacity with constant volume. Assuming the energy
flux to be $I=I_0 \exp(-\sigma x)$ and 360 MJ in the total out-put of 1000 MJ
are carried out as X-ray, α-Particle and plasma, the distribution
of temperature rise on the Li surface of 5 m radius is given by

$$\Delta T(x)=56.76\sigma \exp(-\sigma x) \qquad\qquad (4\text{-}21)$$

where $\sigma(cm^{-1})$ is the average coefficient of energy deposition over
energy spectrum of different energy flux. The resultant temperature
rise for the values of $\sigma=10^3$ and $10^4 cm^{-1}$ are shown in Fig. 4-9.
Some part of heated region above 1317 °C is evaporated and heat
conduction occurred into the deeper Li flow layer. The surface
temperature of Li consequently lowers.
     To estimate the temperature decrease on the surface, heat
conduction process has been examined. The thermal conduction
equation for one dimensional case is

$$\frac{\partial T}{\partial t} = \frac{\partial}{\partial x} \chi \frac{\partial T}{\partial x} \qquad\qquad (4\text{-}22)$$

where χ is thermal diffusivity ($\equiv \kappa/\rho c_p$, κ:thermal conductivity,
ρ: density, c_p: specific heat). The exact solution of the equation
is given by Ya. B. Zel'dovich and Yu. P. Raizer

$$T = \frac{Q}{(4\pi\chi t)^{1/2}} \exp(-\frac{x^2}{4\chi t})$$

$$= \frac{41.3}{t^{1/2}} \exp(-1.35\frac{x^2}{t}) \qquad\qquad (4\text{-}23)$$

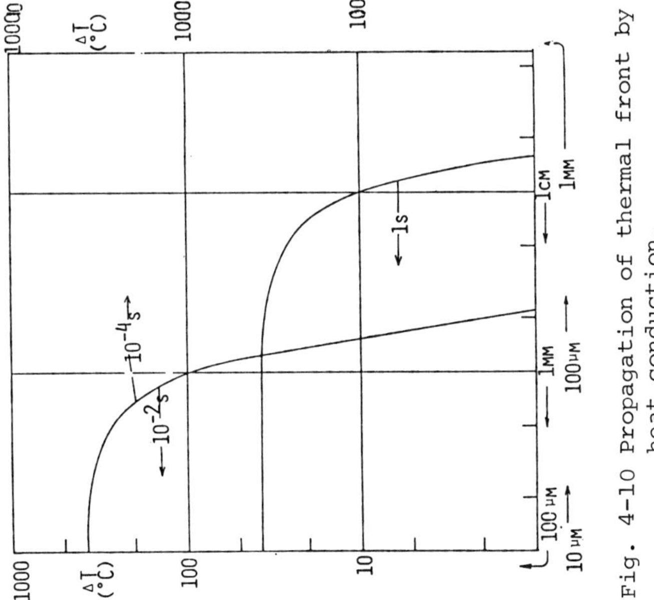

Fig. 4-10 Propagation of thermal front by heat conduction.

Fig. 4-9 Temperature distribution on L_i surface.

where $Q=\int_{-\infty}^{\infty} Tdx$ is given by $\varepsilon/\rho c_p$. ε is the total deposited energy on the surface which is used in the calculation of temperature rise. The propagation of heat front and the decrease of surface temperature are shown in Fig. 4-10 and Fig. 4-11.

In this estimation, the energy removal by evapolation is neglected. Therefore the obtained temperature is the highest possible value.

4-4 Structural Problems of the Reactor Cavity - Mechanical Force

A mechanical force onto the reactor cavity by the fusion explosion at the center may be devided into the following terms
(a) impulsive force by the bombardment of the pellet plasma
(b) local pressure rise by surface diposition of energy.
    (b-1) sharp temperature rise and consequent local pressure drives shock wave in Li
    (b-2) reaction of Li evapolation
    (b-3) impulsive force by surface energy diposition by recondensation
(c) reaction of the expansion of Li layer by deposition of neutron energy in volume.
(d) thermal stress of SUS container by the direct heat due to leakage neutron through Li layer
The evaluation of each force term is required to investigate the feasibility of our conceptual design.

(a) Impulsive force by plasma
The kinetic energy of plasma flow and the dynamic pressure are expressed by

$$E_{plasma} = 4\pi R^2 \frac{n_p Mv^2}{2}\Delta t \cdot v \qquad (4\text{-}24)$$

$$p = n_p Mv^2$$

Conbining the two equations, we get

$$p = \frac{E_{plasma}}{2\pi R^2 \Delta t v} \qquad (4\text{-}25)$$

where R, Δt and v are cavity radius (5 m), duration of plasma on the wall ($\sim 2\times10^{-6}$ s) and the flow velocity ($\sim 10^8$ cm/s). Using these values and $E_{plasma}=195$ MJ, the pressure acting on the Li surface becomes to be 6.9×10^6 dyne/cm^2. The pulse width of this force is very narrow, therefore the peak intensity is attenuated during the propagation in Li by viscosity and spreading of pulse. The impulsive force by plasma is well below the attainable strength of reactor cavity.

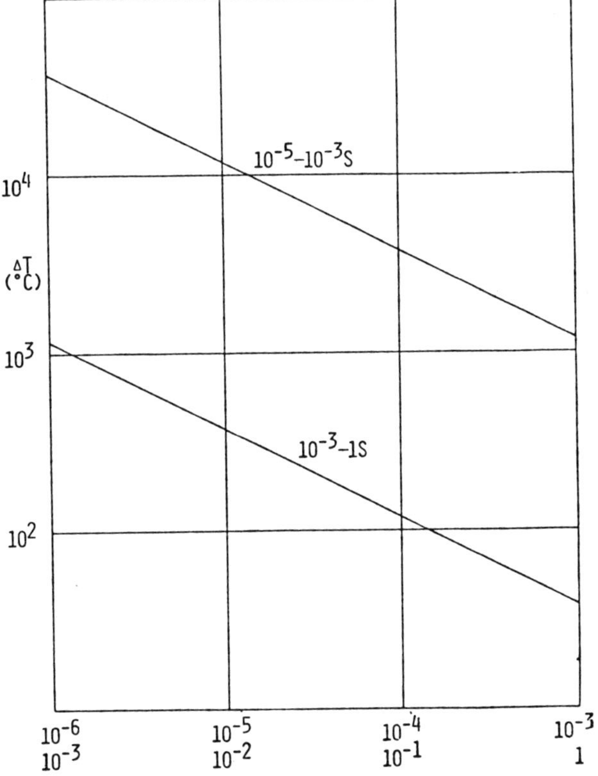

Fig. 4-11 Surface temperature relaxation by thermal
conduction.

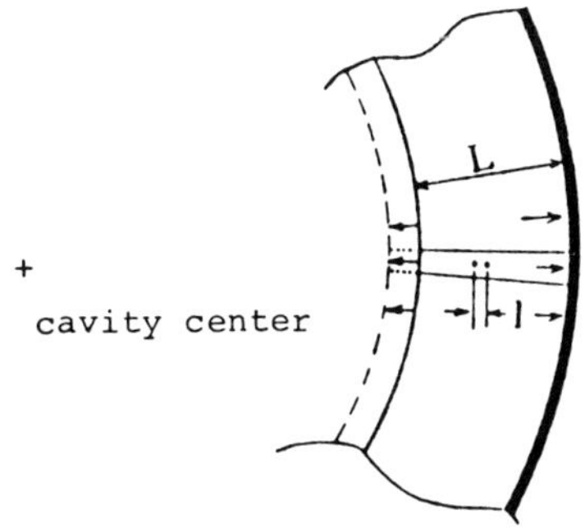

Fig. 4-12 Reactive force on cavity wall
by expansion of Li flow.

(b) Pressure rise by surface deposition of energy.
Let the deposition length of energy on the Li surface by X-ray, α-particle and plasma to be L_d, and sound velocity to be a_s. The response of surface on energy deposition depends on the relation of deposition time τ_d and sound transit time L_d/a_s. If $\tau_d > L_d/a_s$, the energy deposition is slow enough to be followed by the expansion of Li surface and pressure is kept constant. When $\tau_d < L_d/a_s$, a local temperature rise causes a pressure rise. The sound velocity $a_s = 4.3 \times 10^5$ cm/s at 800°k and $L_d = (30 \sim 100) \times 10^{-4}$ cm, then $L_d/a_s \approx (10 \sim 30) \times 10^{-9}$ s. The deposition of only X-ray energy causes the surface pressure rise. According to the results in the previous section, the pressure rise due to X-ray deposition is roughly estimated to be 6×10^{10} dyne/cm$^2 \sim 6 \times 10^4$ atm, which is very large but limited in very narrow region. The peak intensity decrease very rapidly with propagation by viscosity and spread of pulse. The expansion wave also catches up and attenuates the pressure wave when boundary is faced to vacuum as in this case. The attenuation of pressure peak by these effect is estimated to be more than 10^{-5} and the impulsive force to the first wall of SUS is smaller than 1 atm.

As for the recondentation of Li, the characteristic time is of the order of ms and $\tau_d > L_d/a_s$. Therefore this does not cause the local pressure rise.

(c) Reaction of the expansion of Li layer.
Thick Li layer expands rapidly when it is heated by neutron deposition in volume. The reactive force due to the expansion pushes the first wall, SUS container. The force is estimated as the reaction of the acceleration of the center of mass of the unit cross section as show in Fig. 4-12.

$$P = \frac{M \cdot \dfrac{\ell}{\Delta t}}{\Delta t} = \frac{\rho L \cdot \ell}{\Delta t^2} \tag{4-26}$$

ℓ is the lengh of expansion for one fusion explosion, L is the thickness of inner L_i flow and Δt should be estimated as the transit time of sound through the Li layer. For the 1000MJ fusion yield and 80% neutron fraction of total energy, the pressure is estimated to be 10 atm, which is well below the allowable cavity strength.

(d) Thermal stress of SUS container.
The neutron fluence on the Li surface for 1000MJ fusion explosion becomes to be 2.5MJ/m^2 with the reactor cavity of the radius R=5m. Most of the neutron energy is dissipated in the Li layer and the fluence on SUS container first wall is attenuated more than one order. Considering the simple geometry of sphere, the thermal stress on container due to neutron seems to be tolerable.

4-5 Pellet Injection

In a laser fusion reactor system, a pellet target is injected

into a reactor cavity. When the pellet reaches to the center of the
cavity, it is irradiated by the laser beams. In traveling to the
center, the pellet target receives radiations from the liquid Li
wall and also collides with the residual gases in the cavity. These
processes cause to heat and dissolve the solid D-T fuel. Here we
estimate the dissolving time of the solid D-T shell in the pellet
target.

Heat flux from the cavity wall, which the pellet receives, is
given by

$$P = \varepsilon_p \varepsilon_e s\sigma T^4 \tag{4-27}$$

where ε_p and ε_e is emissivity of the pellet and the cavity, s is the
surface area of the pellet, σ is the Stefan-Boltzman constant and T
is temperature of the cavity wall. We assume that the structural
materials of the pellet outer surface of solid D-T shell, such as
LiH, are transparent to the heat flux and radiations directly heat
the solid D-T shell. The dissolving time of the solid D-T shell is
given by

$$t = \frac{(\rho V/M)}{P} Q_F \tag{4-28}$$

where ρ is density of the solid D-T, V and M are volume and molecular
mass of the solid D-T shell and Q_F is the melting heat per mol of the
solid D-T. If we put ε_o=1, ε_p=0.3, and T=500 °C, we obtain t=0.5sec.
During the travelling of the pellet to the center of the reactor
cavity. collisions with the residual gas cause condensation of the Li
atoms on the pellet surface. Kinetic energies of the Li atoms is
converted to the thermal energy to heat the pellet. We assume that
the condensation coefficient is unity, the kinetic energy of colliding
atoms are deposited on the surface, and the solid D-T shell is heated
by heat conductions through the LiH layer. In one dimensional model
of heat conduction, the dissolving time was estimated to be $\sim$0.3sec.

From these estimated, a required injection velocity of the
pellet target to avoid heat influence is 50m/sec. Precise controls
of the initial velocity and direction of pellet injection are
required. Several methods are proposed to accelerate the pellet
to enough large velocity, such as mechanical, gas dynamical,
electrostatic, liquid jet on gas gun type techniques. But it cannot
be answered which method is best in laser fusion reactor at present.

4-6 The Cooling System of The Reactor

We have chosen liquid Li as coolant and breeder. We have also
chosen SUS - 316 as the structural material, because it is much used
in the atomic power plants, so its characteristics in neutron
fluence is well known and it is much more economical than other
materials such as Mo. On the other hand, compatibility between SUS

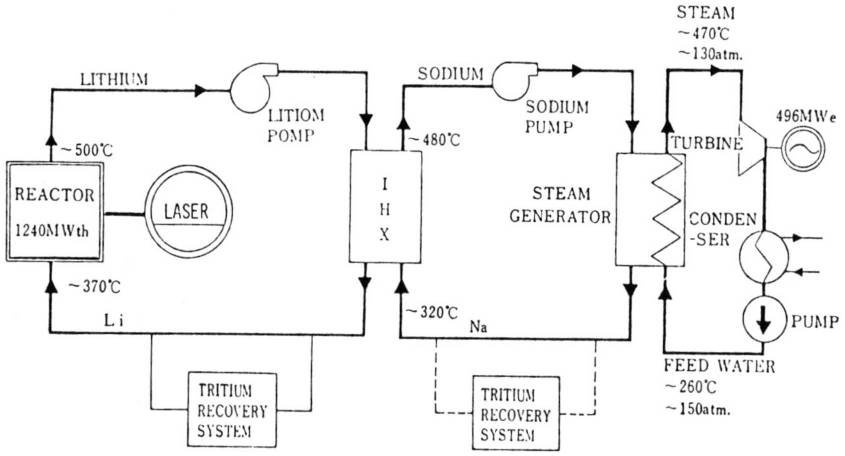

Fig. 4-13 Conceptual design of the cooling system of the reactor.

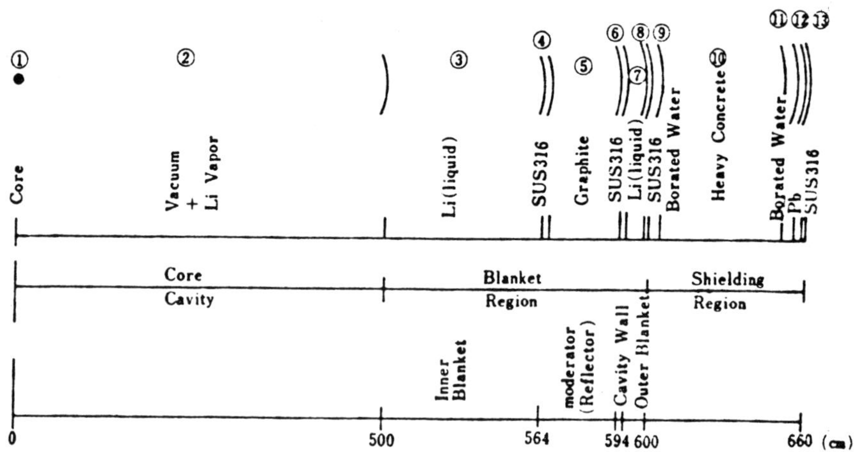

Fig. 4-14 Structure of Reactor Cavity and Blanket.

- 316 and liquid Li at high temperature is not known. Therefore
temperature is limited under 50ʋ °C in our conceptual design.
Reliability and economics of materials are important limiting
factors in our commercial reactor design.

The cooling system consists of 3 cycles through the heat
exchangers. In the first cycle, liquid Li flow cools the reactor
through in the inner blanket and the outer blanket. Li also breeds
tritium by the nuclear reactions Li (n,α)T and ^{6}Li(n,α)T. The
breeding ratio is discussed in the next section. As the intermediate cooling cycle (the secondary system), liquid Na flow is used.
Steam generators and it goes into the turbines. This 3 cycle system
is described in Fig. 4-13.

The Li temperature is 500 °C in the ontlet of the reactor and
370 °C in the inlet. The Na flow in the secondary cycle is heated
from 320 °C to 480 °C and the steam temperature is designed to be
470 °C, its pressure being 130 atm. 496MW gross electric power is
produced in the generator (turbines).

Tritium recovery systems are attached to liquid Li cycle and
liquid Na cycle.

4-7 Neutronics in the blanket

Neutron transport and tritium breeding ratio are calculated by
the 1-dimensional neutron transport code 'ANISN', in which nuclear
data is referenced to DLC-28, and energy spectrum are divided into
51 groups in neutron and 21 groups in γ-ray. In this calculation,
spherical symmetry is assumed, and both nuclear reactions

$$^6\text{Li}\ (n,\alpha)\ ^3\text{T}$$

$$^7\text{Li}\ (n,n'+\alpha)\ ^3\text{T}$$

are taken into account.

The structure of the calculated blanket in a spherical geometry
is shown in Table 4-2 and Fig. 4-14.

Neutron is produced only in the core (Fuel Pellet), the volume
is divided into three regions as described in Table 4-3, where 14
MeV neutrons are those that do not experience collisions, 6.7 MeV
neutrons are scattered ones and 1.5 MeV neutrons have lower energy
than the threshold level of ^{7}Li $(n,n'+\alpha)$ ^{3}T reaction (2.7 MeV).
Energy spectrum in each energy range is assumed to be flat.

The results of tritium breeding ratio calculated for each
energy range are summarized in Table 4-4. For each nuclear reaction
^{6}Li → ^{3}T, ^{7}Li → ^{3}T, the breeding ratio is under unity in all energy
ranges, but after summing the contribution of both nuclear reactions,
the breeding ratio is larger than unity for 14.2 MeV neutrons and
6.7 MeV neutrons. In the range of 1.5 MeV energy, ^{7}Li → ^{3}T reaction
does not take place, so the breeding ratio is smaller than unity.
Tritium is bred in both inner (64 cm thick) and outer (6 cm thick)
blankets. In the whole blanket, the breeding ratio is 1.704 for

TABLE 4-2
DESIGN OF BLANKET

Layer No.	Layer Name	Thickness(cm)	Inner and Outer Radius(cm)		Material
1	Pellet	1	0	~ 1	Solid DT
2	Vacuum	499	1	~ 500	Li Vapor $P=10^{-5}$ Torr
3	Inner Blanket	64	500	~ 564	Liquid Li $\rho=0.482 G/cm^3$
4	Support	1	564	~ 565	SUS-316
5	Reflector	28	565	~ 593	Graphite
6	Support	1	593	~ 594	SUS-316
7	Outer Blanket	6	594	~ 600	Liquid Li $\rho=0.482 G/cm^3$
8	Support	1	600	601	SUS-316

Li ; Natural Abandance

TABLE 4-3
ENERGY RANGE OF NEUTRON

Range No.	Averaged Energy (MeV)	Energy Range (MeV)
1	14.2	13.5 ~ 14.9
2	6.7	6.1 ~ 7.4
3	1.5	1.0 ~ 2.0

TABLE 4-4
TRITIUM BREEDING RATIO

Neutron Energy		Inner Blanket (64cm)	Outer Blanket (6cm)	Total Region (70cm)
14.2MeV	$^6Li \rightarrow T$	0.855	0.0639	0.919
	$^7Li \rightarrow T$	0.783	0.0015	0.785
	Total	1.638	0.0654	1.704
6.7MeV	$^6Li \rightarrow T$	0.853	0.0534	0.906
	$^7Li \rightarrow T$	0.369	0.0002	0.369
	Total	1.222	0.0546	1.275
1.5MeV	$^6Li \rightarrow T$	0.884	0.0458	0.930
	$^7Li \rightarrow T$	0	0	0
	Total	0.884	0.0458	0.930

TABLE 4-5
COMPARISON BETWEEN ILE AND WISCONSIN

		ILE	Wisconsin
Tritum Breeding Ratio	$^6Li(n,\alpha)^3T$	0.92	0.823
	$^7Li(n,\alpha)^3T$	0.41	0.413
	Total	1.33	1.236
Blanket	Material	Liquid Li	Li Oxide
	Thickness	70 cm	41 cm
Used Code		ANISN	ANISN

14.2 MeV neutrons, 1.275 for 6.7 MeV neutrons and 0.930 for 1.5 MeV
neutrons, we assume that neutron energy spectrum is same in the whole
energy ranges (15.2 MeV, 6.7 MeV and 1.5 MeV), the breeding ratio for
^{6}Li (n,α) T is 0.92 and one for ^{7}Li $(n,n'+\alpha)$ T is 0.41. Then the
total breeding ratio in the whole blanket (70 cm thick totally) is
1.33, larger than unity. Therefore it can be expected that tritium
breeding is processed successfully in our design. The results that
the breeding ratio is 1.33 are rather more encouraged than the
results by those in University of Wisconsin, 1.236, where Lithium
Oxide is used as the blanket material. This is summarized in Table
4-5. This improvement on the breeding ratio is probably dependent
on the fact that the thickness of the liquid Li blanket (70cm) is
larger than Wisconsin's one (41 cm Li Oxide).

From the neutron transport calculation by 1-dimensional code
'ANISN', we conclude that our blanket design is hopeful because
the breeding ratio of tritium is 1.33, larger than unity.

4-8 Power Plant Design

The design of the 1200 MW power plant is described in Fig. 4-15.
Arrangement of each equipment is temporarily decided so far. After
the detailed estimation, the scale size will become smaller.

5. COST EFFECTIVENESS OF ICF FUSION POWER PLANT

The total cost of power plant roughly consists of two parts,
capital cost for initial construction and operational cost. The
capital cost is estimated to be

$$M_{capitl} = AP_F + 2BP_L/f\eta_L + C \text{ (yen)} \qquad (5-1)$$

where, first term of right hand side is the cost of main plant
including a reactor cavity, a heat exchanger system, a generator
plant, a lithium cycle, a tritium cycle and pellet factory. This
term is propostional to the fusion plant output power P_F. A is a
constant of 200∿100 yen/watt estimated roughly from capital cost of
a present steam power plant. The second term is capital cost for an
energy driver (laser). P_L is averaged output power and P_L/f is the
energy of single shot, where f is operational frequency of driver.
Capital cost of energy driver is assumed to be proportional to
$P_L/f\eta_L$ the peak energy in the electrical system to pump the driver,
where η_L is the energy conversion efficiency of a energy driver.
Allowing the repair works of driver, two set of driver system are
necessary, so second term of right hand side is multiplied by 2.
The coefficient B is estimated to be 4×10^3 yen·efficiency/joule from
the capital cost of LEKKO VIII CO_2 laser system of ILE Osaka University (P_L/f=10kJ, η_L=4%, capital cost 1 billion yen). Finally, C is
the cost to protect the surroundings of fusion power plant. C is
not taken into account hereafter because of the difficulty of the
estimation.

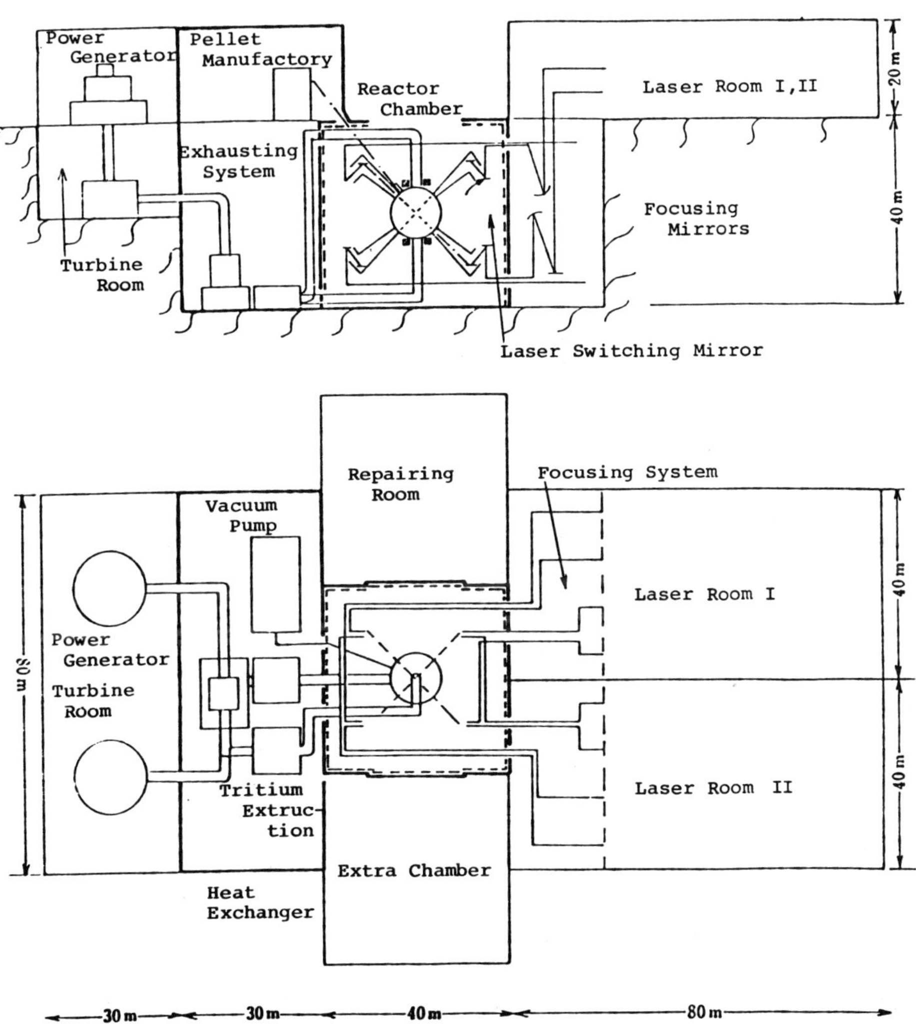

Fig. 4-15 Power plant design.

The operational cost is estimated to be

$$M_{operation} = [\alpha \frac{BP_L}{\eta_L} + D(Q)f]T \quad (yen) \tag{5-2}$$

The first term of right hand side shows the cost to repair the energy driver per a year, which is proportional to the capital cost of the driver. A coefficient α will be 0.1∿0.2 for a laser driver. The second term is the pellet production cost during one year with the operation of 1 Hz. $D(Q)=3.15\times10^7 g(Q)$, where $g(Q)$ is the cost per a pellet. As the high gain pellet has complex structure shown in Chap. 2, $g(Q)$ will increase depending upon pellet gain $Q(=P_F/P_L)$. Here we assume the following relation between g and Q,

$$g(Q)=g_o(\frac{Q}{1000})^n \quad\quad (n>0)$$

In our discussion, g is estimated to be 100 yen/piece for $Q=1000$. T is the lifetime of reactors.

The total cost is

$$M_{total}=AP_F+(2+\alpha T)\frac{BP_F}{\eta_L f}+3.15\times10^7 g_o(\frac{Q}{1000})^n fT.$$

As the effective electrical out put is $\eta_s QP_L(=\eta_s P_F)$ where η_s is total system efficiency shown in Chap. 4, depending on pellet gain Q, the cost per watt is

$$\frac{M_{total}}{\eta_s QP_L} = \frac{A}{\eta_s}+(2+\alpha T)\frac{B}{\eta_L \eta_s Q}+\frac{3.15\times10^4 Tg_o}{(P_L/f)\,\eta_s}(\frac{Q}{1000})^{n-1} \tag{5-3}$$

For the case of parameters, $\eta_s=10\%$, laser energy $E_L=P_L/f=10^6 J$, $A=100∿200$ yen/watt, $\alpha=10∿20\%$, $T=30$ year, $g_o=100$ yen, $B=4\times10^3$ yen· efficiency/Joule and $n=1$, the cost per watt are shown in Fig. 5-1 as a function of pellet gain Q. The parameter of 10% and 20% in Fig. 5-1 is α, which is the ratio of the cost for the repair work per year to the capital cost. It is shown in this figure that pellet gain $Q∿500$ in a commercial case (cost∿10 yen/KWH). Therefore it is concluded that our design, where $Q=1000$, is a much more effective system and that the realization of Inertial Confinement Fusion (ICF) power plant is sufficiently in our technology.

CONCLUSION

A new concept of 1240MWth (Net 426MWe) Inertial Cofinement Fusion Reactor design which utilizes thick Li flow guided by magnetic field inside a spherical cavity has been examined on the view point for commercialization of reactor system. From the analysis of laser-drived fusion reaction (fusion burst), we derive the scaling law of pellet gain Q(= fusion power output/output energy of laser)

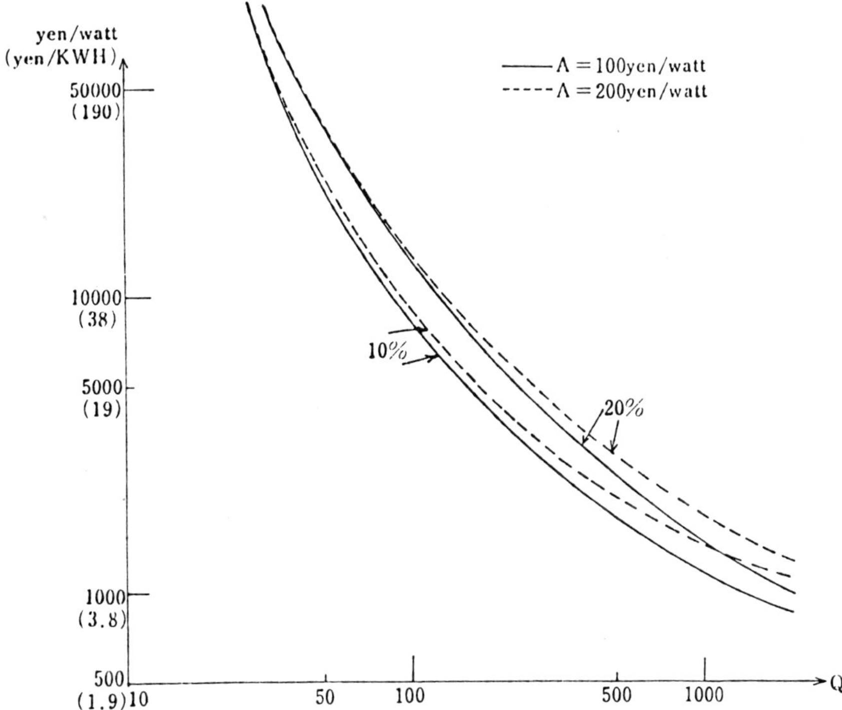

Fig. 5-1 Cost depending on pellet gain Q.

and obtain the conclusion that 1MJ Laser is sufficient for 1200MW
(thermal) Fusion Reactor, where required Q value is about 1000 for
pure fusion reactors and several ten times smaller for hybrid
systems. The CO_2 laser system is chosen as the 1MJ energy driver
after the evaluation and comparison of five types energy drived
(CO_2 Lasers, Glass Lasers, Relativistic Electron Beams, Light Ion
Beams and Heavy Ion Beams).

In a conceptual design, the idea of inner blanket of thick Li
flow is examined in detail, leading to the conclusion that it is
able to realize Li flow inner blanket in our technology. The
breeding ratio of tritium in Li blanket is shown to be greater than
unity by the computational calculations with the 1-dimensional
neutron transport code ANISN. Cost effectiveness of ICF reactors is
also estimated leading to the conclusion that high pellet gain
(Q=1000) laser fusion power system is much more effective than
today's steam power plant. By the above results, it is show that
difficulties on the laser ICF (Inertial Confinement Fusion) reactor
of this type are able to be overcome in our today's and near future's
technology. Detailed numerical analysis of the reactor cavity on
thermal, neutronic and mechanical responses to thermo-nuclear burning
are continued.

REFERENCES

1)   R. E. Kidder; Nucle. Fusion 16, 405(1976).
2)   N. Asano, K. Nishihara, K, Nozaki and T. Taniuchi, J. Phys, Soc.
     Japan 41, 1774(1976).
3)   K. Nozaki and K. Nishihara, J. Phys. Soc. Japan 43, 1393(1977)
4)   J. Nuckolls, L. Wood, A. Thiessen and G. Zimmerman: Nature 239
     (1972) 39.
5)   R. E. Kidder: Nuclear Fusion 14 (1974) 53, 797.
6)   J. S. Clarke, J. N. Fisher and R. J. Mason: Phys. Rev. Letters
     30 (1973) 89.
7)   R. E. Kidder: Nuclear Fusion 16-1 (1976) 3.
8)   J. D. Lindl and W. C. Mead: Phys. Rev. Letters 34 (1975) 1273.
9)   J. D. Lindl: Nuclear Fusion 14 (1974) 511.
10) S. Skupsky: Nuclear Fusion 18 843 (1978).
11) S. Maxson: Phys., Rev. A5 1630 (1972).
12) Ya. B. Zel'dovich and Yu. P. Raizer: "Physics of Shock Waves
     and High-Temperature Hydrodynamics Phenomena", Vol. II, p. 657,
     Academic Press, New York, London, 1967.

SYMMETRIC PLASMA COMPRESSION FOR FUSION BY LASERS BASED ON STUDIES OF NONLINEAR PHENOMENA*)

R. Castillo, H. Hora, E.L. Kane, G. Kentwell, P. Lalousis,
V.F. Lawrence, R. Mavaddat, M.M. Novak, P.S. Ray, and A. Schwartz

Department of Theoretical Physics, University of New South Wales,
Kensington-Sydney 2033, Australia

Analytic studies of the general nonlinear force (containing collision terms apart from the optical ponderomotive terms) arrive at new higher order terms for perpendicular incidence. For oblique incidence of radiation at plasma with collisions, a plasma motion parallel to the surface has been derived. A strong jet motion of plasma has been discovered as an alternative mechanism to explain the strong dissipation at resonance absorption. - The results of the transport of electromagnetic energy in plasma provided an access for a rediscussion of the Abraham-Minkowski-problem. - The interaction mechanism for a free-electron-laser using the nonlinear force were treated. - Computation of plasma compression by laser radiation are based on nonclassical nonlinear mechanism. The acceleration of fast blocks of plasma up to ion energies of some keV for subsequent compression and adiabatic heating for optimized fusion conditions, has been achieved. This ultrafast cold pusher by nonlinear forces is a unique alternative to the disadvantageous thermokinetic pusher and the ablation mode. For later times the density rippling is correlated to a behaviour given by the Korteweg-deVries equation in some regions and by the Benjamin-Ono equation in other parts. The soliton decay results in a fast corona heating necessary for the thermokinetic compression mode. This fast heating makes any asymmetric irradiation for resonance absorption obsolete. The shift of spectral lines of H-like high-Z atoms due to the electrostatic energy in the Debye sphere (polarization shift) is presented in a closed quantum mechanical solution. The calculations for fusion gains including anomalous reheat, fuel depletion and bremsstrahlung are adjusted to latest results of nuclear fusion cross sections and are extended to D^3He apart from the usual fuels (including $H^{11}B$). - In order to understand the problems of the interaction of cylindrical laser beams with plasma, the polarization independence of radial nonlinear forces in a beam has been derived from the quivering motion where the longitudinal field components are essential. The combination of the nonlinear forces with relativistic self-focusing has been studied in many numerical details. The axial generation of 10 MeV ions with pulses of present lasers within the first few psec (for Nd glass) have been confirmed.

*) Work supported by the Australian Research Grant Committee, ARGC No. 75/15538.

The studies on nonlinear, relativistic and quantum effects at laser interaction with plasmas performed during the preceding two years at the Department of Theoretical Physics of the University of New South Wales, were both analytically and numerically where a stimulating interaction with the experimental programmes at the Laser Physics Laboratory of the Australian National University and at the Laser Department of the University of Berne, have to be acknowledged. In a phase where the big laser systems are still under construction and where the preliminary interaction measurements at pellets are resulting in very slow increases of the neutron gains despite of a level of diagnostics and pellet technology of a post-apollo style, the contemplation of the basic strategies is important. The nonclassical age commenced with several unusual processes of parametric response, non-thermalized states and unusual electrodynamics at interaction of very intense laser radiation with high density plasmas. The processes reported in this contribution are directed mainly to the nonlinearities in the dynamics with applications and separate discussions of related problems for nuclear fusion.

1) HIGHER ORDER AND NON-PONDEROMOTIVE TERMS OF THE NONLINEAR FORCE

For the nonlinear plasma dynamics of laser irradiation [1], nonlinearities dominate in two of the equations of conservation: the equation of motion and that of energy while the equation of continuity if affected indirectly only. While the energy equation is affected by the intensity dependent (nonlinear) optical constants and the dynamical nonlinear absorption, the most straight forward nonlinearity is expressed in the equation of motion, given by the force density $\mathbf{f}$ in a plasma.

Historically, nonlinearities of forces are known for a long time, as ponderomotive forces, e.g. for static fields in electro striction or magneto striction, or for quasistatic fields as the forces driving an electro motor. The confinement of plasma by a magnetic field (see Eq. (5.14b) in Ref. [2]) is due to ponderomotive forces in their original definition as due to static (magnetic) fields. A generalization to high frequency fields of a frequency ω was done for microwaves by Gapunov and Miller [3] for low density plasmas to confine plasma in the modes of a standing wave [4], where the restriction to a WKB approximation was necessary [5] [6]. This generalization was called "Miller force" occasionally. The high frequency generalization for the laser-plasma interaction led to the essential property of the dielectric reaction in high density plasmas [7]. The hitherto most general formulation of these nonlinear forces including any density and being not restricted to the WKB approximation was published in 1969 by Hora [8] after separation of the "thermokinetic force" $\mathbf{f}_{th} = -\nabla p$ (due to the thermal pressure p)

$$\mathbf{f}_{NL} = \mathbf{f} - \mathbf{f}_{th} = \frac{1}{c}\mathbf{j}\times\mathbf{H} + \frac{1}{4\pi}\mathbf{E}\,\nabla\cdot\mathbf{E} + \frac{1}{4\pi}\nabla\cdot(\tilde{n}^2 - 1)\,\mathbf{E}\,\mathbf{E} \qquad (1.1)$$

where $\mathbf{j}$ is the current density, $\mathbf{E}$ and $\mathbf{H}$ the electric and magnetic field vectors of the laser radiation of frequency ω in the plasma and $\tilde{n}$ is the complex refractive index

$$\tilde{n}^2 = 1 - \omega_p^2/[\omega^2(1 - i\nu/\omega)] \qquad (1.2)$$

with the plasma frequency $\omega_p = 4\pi e^2 n_e/m$ (e ~ charge, n_e ~ a density, m ~ mass of electron) and the collision frequency ν. As will be shown here more generally than before, the nonlinear force contains terms which may be called optical ponderomotive terms and contains non-ponderomotive dissipative terms. This is the reason why the earlier use of "nonlinear force" [8 to 13] is more precise than the unspecific "poneromotive force".

Eq. (1.1) is formally identical [1,8] with the force density in a non-magnetic medium with **E** and **H** fields and a refractive index ñ given by Landau and Lifshitz [14]

$$\mathbf{f} = - \nabla p + \nabla \cdot \left(\underline{\mathbf{T}} + \frac{(\tilde{n}^2 - 1)}{4\pi} \mathbf{E}\,\mathbf{E}\right) - \frac{\partial}{\partial t} \frac{(\mathbf{E}\,\mathbf{H})}{4\pi c} \qquad (1.3)$$

where the Maxwellian stress tensor is

$$\underline{\mathbf{T}} = (\mathbf{E}\,\mathbf{E} + \mathbf{H}\,\mathbf{H} + (\mathbf{E}^2 + \mathbf{H}^2)\,\underline{\mathbf{1}}/2)/4\pi$$

and where $\underline{\mathbf{1}}$ is the unity tensor, however, for nondispersive and nondissipative media the generatization of Eq. (1.3) to the dispersive collisionless plasma was given by Pitajevski [15] based on the kinetic theory and by Hora [8] based on the macroscopic theory. Hora [8] also gave the generalization for dissipative (absorbing) plasmas arriving at the non-ponderomotive terms.

The predominance of the nonlinear force at high intensity laser interaction with plasmas has been confirmed numerically, where Shearer, Kidder and Zink [9] discovered the generation of density minima (cavitons) in agreement with numerous experiments (see [1] Chapter 10). While formula (1.1) or (1.3) remains its full generality for plasmas with collisions, its evaluation for special cases is not finally completed. Even for the simple case of plane electromagnetic waves perpendicularly incident on a stratified plasma, the non-ponderomotive collisional terms of the nonlinear force turned out to be of importance.

The remarkable result was that this collisional term (see last term in Eq. (25b) of Ref. [8]) of the nonlinear force was derived from Equation (1.3) for linearly polarized plane waves (E parallel to y, propagation direction +x)

$$\mathbf{f}_{NL} = - \mathbf{i}_x \frac{1}{8\pi} \frac{\partial}{\partial x} (\mathbf{E}^2 + \mathbf{H}^2) \qquad (1.4)$$

while Stamper [16], as well as the more general derivation of Miller et al [17], used the formulation of Eq. (1.1)

$$\mathbf{f}_{NL} = \frac{1}{c} \mathbf{j} \times \mathbf{H} \qquad (1.5)$$

While the phase shift between **j** and **H** was essential in Eq. (1.5) to reproduce the collisionless term, the derivation from Eq. (1.4) did not need the phase shift. To solve this discrepancy, we used the fact of the phase shifts as an access, as it will be reported now [18].

We consider plane electromagnetic waves perpendicularly incident on an inhomogeneous plasma of which the electron density is dependent on x only. Linear polarized radiation with $\mathbf{E}$ parallel to the y-direction results in

$$E_x = E_z = H_x = H_y = 0; \quad \frac{\partial}{\partial y} = \frac{\partial}{\partial z} = 0$$

From Eq. (1.3) the nonlinear force is

$$f_{NL} = -i_x \left[\frac{1}{8\pi} \frac{\partial}{\partial x} (E_x^2 + H_z^2) + \frac{\partial}{\partial t} \frac{E_y H_z}{4\pi c} \right] \tag{1.6}$$

In the following, average values of the force density during one period of the electromagnetic field will be considered. The Poynting term can be neglected as long as we assume that the switch-on process of the electromagnetic field is slow in comparison with the period of the laser light. Hence, we arrive at the time averaged force density

$$\overline{f}_{NL} = -i_x \frac{1}{8\pi} \frac{\partial}{\partial x} \overline{(E_y^2 + H_z^2)} \tag{1.7}$$

The expressions for E_y and H_z can be found by solving Maxwell's wave equations. We restrict ourselves to the WKB approximation within the following consideration.

The complex electric and magnetic field strenghts are then

$$E_y = \frac{E_v}{\tilde{n}^{1/2}} \exp i F_o = \frac{E_v}{\tilde{n}^{1/2}} \exp \left(\mp \frac{\tilde{k}(x)x}{2} \right) \exp i F \tag{1.8}$$

$$H_z = E_v \tilde{n}^{1/2} \exp\left(-\frac{\tilde{k}(x)x}{2}\right) \exp i F + i \frac{E_v c}{2\omega \tilde{n}^{3/2}} \exp i F_o \frac{d\tilde{n}}{dx} \tag{1.9}$$

where

$$F = \omega\left(t \mp \frac{1}{c} \int^x \operatorname{Re}(\tilde{n}(\xi))d\xi\right); \quad F_o = \mp ik(x)x/2 + F \tag{1.10}$$

and

$$\tilde{k}(x) = \frac{1}{x} \frac{2\omega}{c} \int^x \operatorname{Im}(\tilde{n}(\xi))d\xi \tag{1.11}$$

The upper sign determines a wave propagating in the positive x-direction while the lower sign describes propagation into the negative x-direction. Since our field strengths are complex we require the real components in our expression for the force density. Hence Eq. (1.7) is

$$\operatorname{Re} \overline{f}_{NL} = -i_x \frac{1}{4\pi} \left[\overline{\operatorname{Re}(E_y)\operatorname{Re}(\frac{\partial E_y}{\partial x})} + \overline{\operatorname{Re}(H_z)\operatorname{Re}(\frac{\partial H_z}{\partial x})} \right] \tag{1.12}$$

From Eq. (1.8) and Eq. (1.9)

$$\operatorname{Re}(E_y) = \frac{E_v}{\tilde{n}^{1/2}} \exp \frac{-\tilde{k}(x)x}{2} \left[\operatorname{Re}(\frac{1}{\tilde{n}^{1/2}}) \cos F - \operatorname{Im}(\frac{1}{\tilde{n}^{1/2}}) \sin F \right] \tag{1.13a}$$

and

$$\operatorname{Re}(\frac{\partial E_y}{\partial x}) = -\frac{E_v}{2} \exp \frac{-\tilde{k}(x)x}{2} \left[\operatorname{Re}(\frac{1}{\tilde{n}^{3/2}} \frac{d\tilde{n}}{dx}) \cos F - \operatorname{Im}(\frac{1}{\tilde{n}^{3/2}} \frac{d\tilde{n}}{dx}) \sin F \right]$$

$$- \frac{E_v \omega}{c} \exp \frac{-\tilde{k}(x)x}{2} \left[\operatorname{Re}(\tilde{n}^{1/2}) \sin F + \operatorname{Im}(\tilde{n}^{1/2}) \cos F \right] \tag{1.13b}$$

also

$$\mathrm{Re}(H_z) = E_v \exp(-\tilde{k}(x)x/2) \, [\mathrm{Re}(\tilde{n}^{1/2})\cos F - \mathrm{Im}(\tilde{n}^{1/2})\sin F]$$

$$-\frac{E_v \omega}{c} \exp(-\tilde{k}(x)x/2) \, [\mathrm{Im}\,(\frac{1}{\tilde{n}^{3/2}} \frac{d\tilde{n}}{dx})\cos F + \mathrm{Re}\,(\frac{1}{\tilde{n}^{3/2}} \frac{d\tilde{n}}{dx})\sin F \qquad (1.13c)$$

and

$$\mathrm{Re}\,(\frac{\partial H_z}{\partial x}) = -\frac{E_v \omega}{c} \exp(-\tilde{k}(x)x/2) \, [\mathrm{Re}\,(\tilde{n}^{3/2})\sin F + \mathrm{Im}\,(\tilde{n}^{3/2})\cos F]$$

$$-\frac{E_v c}{2\omega} \exp(-\tilde{k}(x)x/2) \; [\mathrm{Re}\frac{1}{\tilde{n}^{3/2}}\frac{d^2\tilde{n}}{dx^2} - \frac{3}{2}\frac{1}{\tilde{n}^{5/2}}(\frac{d\tilde{n}}{dx})^2]\sin F$$

$$+ \mathrm{Im}\,[\frac{1}{\tilde{n}^{3/2}}\frac{d^2\tilde{n}}{dx^2} - \frac{3}{2}\frac{1}{\tilde{n}^{5/2}}(\frac{d\tilde{n}}{dx})^2]\cos F \qquad (1.13d)$$

Hence Eq. (1.12) becomes

$$\mathrm{Re}\{\overline{f}_{NL}\} = \frac{E_v^2}{16\pi}\exp(-\tilde{k}(x)x) \, [\mathrm{Re}\,(\frac{1}{\tilde{n}^{1/2}})\,\mathrm{Re}\,(\frac{1}{\tilde{n}^{3/2}}\frac{d\tilde{n}}{dx}) + \mathrm{Im}\,(\frac{1}{\tilde{n}^{1/2}})\,\mathrm{Im}\,(\frac{1}{\tilde{n}^{3/2}}\frac{d\tilde{n}}{dx})]$$

$$-\frac{E_v^2}{16\pi}\exp(-\tilde{k}(x)x) \, [\,\mathrm{Re}\,(\frac{1}{\tilde{n}^{3/2}}\frac{d\tilde{n}}{dx})\,\mathrm{Re}\,(\tilde{n}^{3/2}) + \mathrm{Im}\,(\frac{1}{\tilde{n}^{3/2}}\frac{d\tilde{n}}{dx})\,\mathrm{Im}\,(\tilde{n}^{3/2})]$$

$$-\frac{E_v^2 \omega}{8\pi c}\exp(-\tilde{k}(x)x) \, [\,\mathrm{Im}\,(\frac{1}{\tilde{n}^{1/2}})\,\mathrm{Re}\,(\tilde{n}^{1/2}) - \mathrm{Im}\,(\tilde{n}^{1/2})\,\mathrm{Re}\,(\frac{1}{\tilde{n}^{1/2}})]$$

$$+\frac{E_v^2 \omega}{8\pi c}\exp(-\tilde{k}(x)x) \, [\,\mathrm{Re}\,(\tilde{n}^{1/2})\,\mathrm{Im}\,(\tilde{n}^{3/2}) - \mathrm{Im}\,(\tilde{n}^{1/2})\,\mathrm{Re}\,(\tilde{n}^{3/2})]$$

$$-\frac{E_v^2 c}{16\pi\omega}\exp(-\tilde{k}(x)x) \, \{\mathrm{Re}\,[\frac{1}{\tilde{n}^{3/2}}(\frac{d^2\tilde{n}}{dx^2}) - \frac{3}{2\tilde{n}^{5/2}}(\frac{d\tilde{n}}{dx})^2]\,\mathrm{Im}\,(\tilde{n}^{1/2})$$

$$- \mathrm{Im}\,[\frac{1}{\tilde{n}^{3/2}}(\frac{d^2\tilde{n}}{dx^2}) - \frac{3}{2\tilde{n}^{5/2}}(\frac{d\tilde{n}}{dx})^2]\,\mathrm{Re}\,(\tilde{n}^{1/2})\}$$

$$-\frac{E_v^2 c}{32\pi c}\exp(-\tilde{k}(x)x) \, \{\,\mathrm{Re}\,[\frac{3}{\tilde{n}^{3/2}}(\frac{d^2\tilde{n}}{dx^2}) - \frac{3}{2\tilde{n}^{5/2}}(\frac{d\tilde{n}}{dx})^2]\,\mathrm{Re}\,(\frac{1}{\tilde{n}^{3/2}}\frac{d\tilde{n}}{dx})$$

$$+ \mathrm{Im}\,[\frac{1}{\tilde{n}^{3/2}}(\frac{d^2\tilde{n}}{dx^2}) - \frac{3}{2\tilde{n}^{5/2}}(\frac{d\tilde{n}}{dx})^2]\,\mathrm{Im}\,(\frac{1}{\tilde{n}^{3/2}}\frac{d\tilde{n}}{dx})\} \qquad (1.14)$$

This is the hitherto most general evaluation of the general nonlinear force (Eq. (1.1) or (1.3) for plane waves perpendicularly incident on a stratified plasma for the WKB approximation. If we consider the case in which collisions are neglected then k = 0 and the imaginary terms vanish. Therefore the above expression yields then in

$$f_{NL} = \frac{E_v^2 \omega_p^2}{16\pi\omega^2 \tilde{n}^2}\frac{d\tilde{n}}{dx} - \frac{E_v^2 c^2}{32\pi\omega^2}[\frac{1}{\tilde{n}^3}\frac{d^2\tilde{n}}{dx^2}) (\frac{d\tilde{n}}{dx})] + \frac{3E_v^2 c^2}{64\pi\omega^2}\frac{1}{\tilde{n}^4}(\frac{d\tilde{n}}{dx})^3 \qquad (1.15)$$

The first term in Eq. (1.15) is the one obtained before [8] neglecting the phase term i.e.: the second term of Eq. (1.9) when using Eq. (7) as basis. It would be reasonable to assume that the higher order terms in Eq. (1.15) are due to the phase term. This is not so as can be seen by spatially differentiating our expressions (1.9) for the magnetic field strength **H**. When the phase term is differentiated one of its terms cancel with one of the derivatives of the first term of Eq. (1.9). Hence the nonlinear force arises from the phase term alone. This is in agreement with the model of quivering motion [19]. As described by Hora [8] the first term of Eq. (1.15) indicates a deconfining collisionless acceleration because the direction of the force density is towards decreasing plasma densities, which is clearly independent of the polarization of the incident laser light from symmetry considerations. The first of the higher order terms is a confining one whilst the other is a deconfining force term. These higher order terms may contribute to the momentum transferred to the homogeneous interior [8]. With collisions, the other terms can be interpreted as collision produced radiation pressure of the light, within the inhomogeneous plasma in analogy with the usual radiation pressure of homogeneous media, where, however, the very complex influence of the refractive index is included now.

2) OBLIQUE INCIDENCE WITH COLLISIONS AND JETS AT RESONANCE ABSORPTION

The nonlinear forces in an inhomogeneous plasma at oblique incidence of plane electromagnetic waves without collisions, were the main tool for deriving the general formulation of the force, Eq. (1.1) by observing the balance of the momentum of the photons [8]. For a standing wave at oblique incidence, a striated motion was concluded [20] and the case of plasma with collisions could be treated with some preliminary restrictions and analogies only. There was a strong need to derive the most general evaluation of the nonlinear force from the general expression (1.1) or (1.3) for oblique incidence for a plasma with collisions. We shall apply this for deriving the net (usual) radiation pressure at oblique incidence and for studying the resonance absorption.

We consider a collisional, inhomogeneous plasma in which a plane electromagnetic wave is obliquely incident at an angle α_0. In the arbitrary case the electromagnetic wave can be written in component form, with β the angle between the **E** vector and the plane of incidence. When $\beta = 0$ the electromagnetic wave is p-polarized ($\|$) and when $\beta = \frac{\pi}{2}$ the electromagnetic wave is s-polarized ($\perp$). Hence

$$\underline{E} = \underset{\sim}{i}_x (E_\|)_x \cos\beta + \underset{\sim}{i}_y (E_\|)_y \cos\beta + \underset{\sim}{i}_z (E_\perp)_z \sin\beta$$

and

$$\underline{H} = \underset{\sim}{i}_x (H_\perp)_x \sin\beta + \underset{\sim}{i}_y (H_\perp)_y \cos\beta + \underset{\sim}{i}_z (H_\|)_z \cos\beta$$

Using the above expressions, the Maxwellian stress tensor Eq. (14) is

$$4\pi \underset{\approx}{T} = \left\{ \begin{array}{ccc} \tfrac{1}{2}\left(E_x^2 - E_y^2 - E_z^2 + H_x^2 - H_y^2 - H_z^2\right) & E_x E_y + H_x H_y & E_x E_z + H_x H_z \\[2mm] E_y E_x + H_y H_x & \tfrac{1}{2}\left(-E_x^2 + E_y^2 - E_z^2 - H_x^2 + H_y^2 - H_z^2\right) & E_y E_z + H_y H_z \\[2mm] E_z E_x + H_z H_x & E_z E_y + H_z H_y & \tfrac{1}{2}\left(-E_x^2 - E_y^2 + E_z^2 - H_x^2 - H_y^2 + H_z^2\right) \end{array} \right\}$$

where now

$$E_x = (E_\parallel)_x \cos\beta \; ; \quad E_y = (E_\parallel)_y \cos\beta \; ; \quad E_z = (E_\perp)_z \sin\beta$$

and

$$H_x = (H_\perp)_x \sin\beta \; ; \quad H_y = (H_\perp)_y \cos\beta \; ; \quad H_z = (H_\parallel)_z \cos\beta$$

The x - y plane is the plane of incidence and hence

$$\frac{\partial}{\partial y}\,\overline{\underset{\approx}{E}\,\underset{\approx}{E}} = \frac{\partial}{\partial z}\,\overline{\underset{\sim}{E}\,\underset{\sim}{E}} = \frac{\partial}{\partial y}\,\overline{\underset{\sim}{H}\,\underset{\sim}{H}} = \frac{\partial}{\partial z}\,\overline{\underset{\sim}{H}\,\underset{\sim}{H}} = 0$$

since a non-zero time average force would violate the conservation of momentum.
Using the most general expression for the force density Eq. (1.3), we find

$$\begin{aligned}
\underset{\sim}{\overline{f}}_{NL} = &\; \frac{\underset{\sim}{i}_x}{4\pi}\frac{\partial}{\partial x}\left[(2\tilde{n}^2 - 1)\overline{(E_\parallel)_x^2}\cos^2\beta - \overline{(E_\parallel)_y^2}\cos^2\beta - \overline{(E_\perp)_z^2}\sin^2\beta \right. \\
&\; \left. + \left(\overline{(H_\perp)_x^2} - \overline{(H_\perp)_y^2}\right)\sin^2\beta - \overline{(H_\parallel)_z^2}\cos^2\beta \right] \\
&\; + \frac{\underset{\sim}{i}_y}{4\pi}\frac{\partial}{\partial x}\left[\tilde{n}^2\,\overline{(E_\parallel)_x (E_\parallel)_y}\cos^2\beta + \overline{(H_\perp)_x (H_\perp)_y}\sin^2\beta \right] \\
&\; + \frac{\underset{\sim}{i}_z}{4\pi}\frac{\partial}{\partial x}\left[\tilde{n}^2\,\overline{(E_\parallel)_x (E_\perp)_z}\cos\beta\sin\beta + \overline{(H_\perp)(H_\parallel)_z}\sin\beta\cos\beta \right]
\end{aligned} \tag{2.1}$$

This is equivalent to the expression derived by Hora [8] but with collisions

$$\tilde{n}^2 = 1 - \frac{\omega_p^2}{\omega^2 + \nu^2}\left(1 + i\frac{\nu}{\omega}\right)$$

Since we are now dealing with complex quantities we require the real components of
Eq. (2.1)

$$\begin{aligned}
\mathrm{Re}\{\overline{\underset{\sim}{f}}_{NL}\} = &\; \underset{\sim}{i}_x \frac{1}{4\pi}\left[\left(\mathrm{Re}^2\tilde{n} - \mathrm{Im}^2\tilde{n}\right)\mathrm{Re}(E_\parallel)_x \mathrm{Re}\,\overline{\frac{\partial (E_\parallel)_x}{\partial x}} + \mathrm{Re}(\tilde{n})\mathrm{Re}\left(\frac{\partial\tilde{n}}{\partial x}\right)\mathrm{Re}(E_\parallel)_x \right. \\
&\; \left. - \overline{\mathrm{Re}(E_\parallel)_y\,\mathrm{Re}\frac{\partial}{\partial x}(E_\parallel)_y} - \mathrm{Im}(\tilde{n})\,\mathrm{Im}\left(\frac{\partial\tilde{n}}{\partial x}\right)\mathrm{Re}^2(E_\parallel)_x - \overline{\mathrm{Re}(H_\parallel)_z\,\mathrm{Re}\frac{\partial (H_\parallel)_z}{\partial x}} \right]\cos^2\beta \\
&\; - \underset{\sim}{i}_x \frac{1}{4\pi}\left[\overline{\mathrm{Re}(E_\perp)_z\,\mathrm{Re}\frac{\partial (E_y)_z}{\partial x}} + \overline{\mathrm{Re}(H_\perp)_x\,\mathrm{Re}\frac{\partial (H_\perp)_x}{\partial x}} - \overline{\mathrm{Re}(H_\perp)\,\mathrm{Re}\frac{\partial (H_\parallel)_y}{\partial x}} \right]\sin^2\beta \\
&\; + \underset{\sim}{i}_y \frac{1}{4\pi}\left[\mathrm{Re}(E_\parallel)_x\left[\mathrm{Re}(\tilde{n}^2)\mathrm{Re}\overline{\frac{\partial (E_\parallel)_y}{\partial x}} + \mathrm{Re}\left(\frac{\partial\tilde{n}^2}{\partial x}\right)\mathrm{Re}(E_\parallel)_y \right] \right]
\end{aligned}$$

$$+ \operatorname{Re} \tilde{m}^2 \operatorname{Re}(E_\parallel)_y \, \overline{\operatorname{Re} \frac{\partial (E_\parallel)_x}{\partial x}} \bigg] \cos^2\beta$$

$$+ \underset{\sim}{i}_y \frac{1}{4\pi} \bigg[\overline{\operatorname{Re}(H_\perp)_x \operatorname{Re} \frac{\partial (H_\perp)_y}{\partial x}} + \operatorname{Re}(H_\perp)_y \operatorname{Re} \frac{\partial (H_\perp)_x}{\partial x} \bigg] \sin^2\beta$$

$$+ \underset{\sim}{i}_z \frac{1}{4\pi} \bigg[\operatorname{Re}(E_\parallel)_x \bigg[\operatorname{Re} \frac{\partial (E_\perp)_z}{\partial x} \operatorname{Re}(\tilde{m}^2) + \operatorname{Re}\left(\frac{\partial \tilde{m}^2}{\partial x}\right) \operatorname{Re}(E_\perp)_z \bigg]$$

$$+ \operatorname{Re}\tilde{m}^2 \operatorname{Re}(E_\perp)_z \, \overline{\operatorname{Re} \frac{\partial (E_\parallel)_x}{\partial x}}$$

$$+ \overline{\operatorname{Re}(H_\perp)_x \operatorname{Re} \frac{\partial (H_\parallel)_z}{\partial x}} + \operatorname{Re}(H_\parallel)_z \operatorname{Re} \frac{\partial (H_\perp)_x}{\partial x} \bigg] \sin\beta\cos\beta$$

$$\tag{2.2}$$

For the case of p-polarization ($\beta = 0$) then

$$\operatorname{Re}\{\overline{f_{NL}}\} = \frac{i_x}{4\pi}\bigg[(\operatorname{Re}^2\tilde{m} - \operatorname{Im}^2\tilde{m})\overline{\operatorname{Re}(E_\parallel)_x \operatorname{Re}\frac{\partial (E_\parallel)_x}{\partial x}} + \operatorname{Re}(\tilde{m})\operatorname{Re}\left(\frac{\partial \tilde{m}}{\partial x}\right)\operatorname{Re}^2(E_\parallel)_x$$

$$- \operatorname{Im}(\tilde{m})\operatorname{Im}\left(\frac{d\tilde{m}}{dx}\right)\operatorname{Re}^2(E_\parallel)_x - \overline{\operatorname{Re}(E_\parallel)_y \operatorname{Re}\frac{\partial (E_\parallel)_y}{\partial x}} - \overline{\operatorname{Re}(H_\parallel)_z \operatorname{Re}\frac{\partial (H_\parallel)_z}{\partial x}}$$

$$+ \frac{i_y}{4\pi}\bigg[\operatorname{Re}(E_\parallel)_x \bigg[\operatorname{Re}(\tilde{m}^2)\operatorname{Re}\frac{\partial (E_\parallel)_y}{\partial x} + \operatorname{Re}(E_\parallel)_y \operatorname{Re}\frac{\partial \tilde{m}^2}{\partial x} \bigg]$$

$$+ \operatorname{Re}\tilde{m}^2\operatorname{Re}(E_\parallel)_y \, \operatorname{Re}\frac{\partial (E_\parallel)_x}{\partial x} \bigg]$$

$$\tag{2.3}$$

and for s-polarization ($\beta = \frac{\pi}{2}$)

$$\operatorname{Re}\{\overline{f_{NL}}\} = -\frac{i_x}{4\pi}\bigg[\overline{\operatorname{Re}(E_\perp)_z \operatorname{Re}\frac{\partial (E_\perp)_z}{\partial x}} - \overline{\operatorname{Re}(H_\parallel)_x \operatorname{Re}\frac{\partial (H_\parallel)_x}{\partial x}} + \overline{\operatorname{Re}(H_\perp)_y \operatorname{Re}\frac{\partial (H_\perp)_x}{\partial x}} \bigg]$$

$$+ \underset{\sim}{i}_y \frac{1}{4\pi}\bigg[\overline{\operatorname{Re}(H_\parallel)_x \operatorname{Re}\frac{\partial (H_\perp)_y}{\partial x}} + \operatorname{Re}(H_\perp)_y \operatorname{Re}\frac{\partial (H_\parallel)_x}{\partial x} \bigg]$$

$$\tag{2.4}$$

Eq. (2.3) and (2.4) are the general expressions of the nonlinear force for obliquely incident plane waves in an inhomogeneous plasma with collisions [18].

We use the results of Eqs. (2.3) and (2.4) to demonstrate how the collisions will produce forces in the plane of the plasma surface. It was the earlier result [8] that for collisionless plasmas, only forces perpendicular to the plasma surface are being produced, where the magnitude of the forces was independent of the polarization up to third order and moderate angles of incidence. The forces parallel to the plasma surface in the case of collisions, will be seen by restricting the discussion to the WKB approximation only.

For the collisionless case at WKB approximation [8] the i_y components of f_{NL} vanish since

$$\overline{(H_\perp)_x (H_\perp)_y} = -\frac{1}{2} E_v^2 \cos\alpha_0 \sin\alpha_0 = \text{const}$$

and

$$\tilde{n}^2 \overline{(E_{\shortparallel})_x (E_{\shortparallel})_y} = \frac{E_v^2}{2} \cos\alpha_0 \sin\alpha_0 = \text{const}$$

so using Eq. (2.3), the i_y components vanish without collisions. Using the expressions for $(E_{\shortparallel})_x$ $(E_{\shortparallel})_y$ [8]

$$(E_{\shortparallel})_x = -\underset{\sim}{i}_x \frac{E_v(\cos\alpha_0)^{1/2}}{[\tilde{n}(x)\cos\alpha(x)]^{1/2}} \sin\alpha(x)\exp iF_0 = A\left[f(x) + i\,g(x)\right]\exp iF_0 \quad (2.5)$$

and

$$(E_{\shortparallel})_y = \underset{\sim}{i}_y \frac{E_v(\cos\alpha_c)^{1/2}\cos\alpha(x)\exp iF_0}{(\tilde{n}(x)\cos\alpha(x))^{1/2}} = -A\left[j(x) + i\,(k(x)\right]\exp iF_0 \quad (2.6)$$

where A is a constant, then for p-polarization the i_y component of f_{NL} is

$$\underset{\sim}{i}_y \frac{1}{4\pi}\frac{\partial}{\partial x}\tilde{n}^2(E_{\shortparallel})_x (E_{\shortparallel})_y = -\underset{\sim}{i}_y \frac{1}{4\pi}A^2\frac{\partial}{\partial x}\left[f(x)j(x) + g(x)k(x)\right]Re\,\tilde{n}^2\exp(-\tilde{k}(x,y)) \quad (2.7)$$

This is zero because $\tilde{n}^2 (E_{\shortparallel})_x$ $(E_{\shortparallel})_y$ is constant for the collisionless case (see Eq. (59)), in Ref. [8] but with collisions we find Eq. (2.5) and (2.6)

$$f(x) = Re\left[\frac{\sin\alpha(x)}{n(x)^{1/2}\cos\alpha(x)^{1/2}}\right] \quad ; \quad g(x) = Im\left[\frac{\sin\alpha(x)}{(\tilde{n}(x)\cos\alpha(x)^{1/2}}\right]$$

and

$$j(x) = -Re\left[\frac{\cos\alpha(x)}{(\tilde{n}(x)\cos\alpha(x))^{1/2}}\right] \quad ; \quad k(x) = -Im\left[\frac{\cos\alpha(x)}{(\tilde{n}(x)\cos\alpha(x))^{1/2}}\right]$$

hence

$$f(x)j(x) + g(x)k(x)$$

is depending on x and causes a y-component of the force f_{NLy}. For the s-polarization we similarly find a nonzero i_y component. The net momentum transferred to the plasma corresponds to the y-component of the momentum of the absorbed photons pushing the plasma in the plane of the surface (+ y-direction) [18].

An application of the general formula (2.3) for p-polarization is that for the resonance absorption. At oblique incidence of p-polarized plasma waves on an inhomogeneous plasma, the generation of striated jets in the outermost corona was derived leading to keV ions and a nonlinear absorption with a maximum of 25^0 angle of incidence [20] this process is competitive to the resonance absorption [1] where the generation of a local maximum of the E-field component directed to the vacuum at the critical density (within the evanescent field below the turning point) causes an absorption process. While three mechanisms are possible for this absorption [1] (acceleration of electrons of the fast part of the Maxwellian distributions; wavebreaking; quiver-drift in the field gradient), Kentwell [21] derived another mechanism. The nonlinear force in the direction of the plane of incidence and the plane of the plasma surface (y-direction) was found to be zero under simplified assumptions for a linear density profile. The careful rediscussion of the temporal averaging procedures arrived now [21] at a force

$$f_{NL,y} \approx \frac{1}{4\pi} Re\,\bar{n}^2 \frac{\partial}{\partial x} \overline{Re(E_x)Re(E_y)} + \frac{1}{4\pi} \overline{Re(E_x)Re(E_y)\frac{\partial}{\partial x}Re\,\bar{n}^2} = f_A + f_B \qquad (2.8)$$

near the resonance point $x = x_0$ (critical density $n_e(x_0) = n_{ec}$ where $\omega_p = \omega$). The first term f_A causes a striated jet, streaming to -y for $x < x_0$ and to +y for $x > x_0$. After reaching absolute maximum velocities, the forces vanish for x further away from x_0. The net motion is zero. The second term f_B causes a motion to +y with a maximum at $x = x_0$.

Examples for 10^{16} W/cm^2 Nd glass lasers at angle of incidence of 20° for p-polarization for a 500 eV plasma temperature and a linear profile of the refractive index of 2.5 μm (Denisov length L = 0.3 μm, 30μ laser beam diameter) resulted in f_A at $x = x_0$ of 1.87×10^{17} dyn/cm^3 producing 20 keV ions. The force f_B at $x - x_0 = \pm L/30$, where $| f_A | \ll | f_B |$, results in $\pm 4 \times 10^{17}$ dyn/cm^3 producing 45 keV ions.

3)   THE ABRAHAM-MINKOWSKI PROBLEM

Despite of the very special motivation of our studies of laser-plasma interaction for nuclear fusion, our results arrived in a new access for treating a very general and basic question in physics: In the relativistic description of the electromagnetic waves in media (plasmas) to be based on the Abraham or on the Minkowski tensor. The motivation was given initially by a remark by Basov [23] when he mentioned to the derivation of the nonlinear force, that Tamm had to use for his derivation of the Cerenkov radiation in some parts the Abraham and in other parts the Minkowski formulation.

While the momentum of photons in the Abraham picture is

$$p_A = \hbar\omega\tilde{n} \qquad (3.1)$$

and in the Minkowski picture

$$p_M = \hbar\omega/\tilde{n} \qquad (3.2)$$

(taking a real refractive index ñ for simplification) Hora derived [8] from the recoil of the c-w current of photons in an inhomogeneous plasma that the momentum of the electromagnetic energy is

$$p_\varphi = \frac{\hbar\omega}{2\tilde{n}}(1+\tilde{n}^2) = \frac{\hbar\omega}{\tilde{n}}(1-\frac{\omega_p^2}{2\omega^2}) = \frac{1}{2}(p_A + p_M) \qquad (3.3)$$

The same momentum was derived by Klima and Petrzilka [24] for a wave packet in a homogeneous plasma. The increase of momentum of the photons when entering a plasma discontinuously had to be compensated by reflection, where the Fresnel formula was achieved [25] as no further compensation of forces in a surface (by cohesive forces etc.) is possible in a plasma. The cohesive forces may be important for solids as Peierls mentioned [26] and as can be seen from the asymmetric damage of the surface of laser rods. For plasmas, after the numerous experimental agreement of the nonlinear force action (see Chapter 10 of Ref. [1]), we can base our considerations on the result (3.3) that half of the Abraham and half of the Minkowski momentum is valid.

This and the further need for a description of black body radiation in a plasma [1] by separating of the electromagnetic energy between that without the plasma electrons (bare photons = photoelectrically acting) and that including the energy exchanged with the electrons (dressed photons), led Novak to a basically new concept for the discussion of the Abraham-Minkowski problem [26].

The explanation of this result as well as the need for the correct description of the black body radiation inside plasma [1], was achieved [27] utilizing a basically new concept in the discussion of the Abraham-Minkowski dilemma.

We begin with the assumption (to be justified below) that the Minkowski momentum refers to a spinless photon whilst that of the Abraham describes a photon of spin unity. As is well known, the concept of spin occurs as a low frequency phenomenon only, as its contribution at the other end of the spectrum may be neglected.

Now, following the standard procedure, in investigating the black body radiation, the field energy density is

$$u(\nu)d\nu = \epsilon(\nu, T)\, g\,(\nu)d\nu \tag{3.4}$$

where the mean energy of oscillators is

$$\epsilon(\nu, T) = h\nu\,(\exp(h\nu/kT) - 1)^{-1} \tag{3.5}$$

and $g(\nu)d\nu$ is the density of states in terms of frequency ν. This density is usually defined in terms of the wave vector $\underline{k}$ as

$$g\,(\mathbf{k})\, d\,\mathbf{k} = \pi^{-2}\mathbf{k}^2 d\mathbf{k} \tag{3.6}$$

But the wave vector $\underline{k}$ in a medium is related to that in vacuo k_o by

$$\mathbf{k} = \widetilde{n}\,\mathbf{k_o} \tag{3.7}$$

where $\tilde{n}$ is the refractive index which for the collisionless plasma has the form

$$\widetilde{n}^2 = 1 - \omega_p^2/\omega^2 \tag{3.8}$$

Now, in order to express the density of states in terms of frequency, (3.6) is differentiated and we obtain using (3.7) and (3.8)

$$g(\nu)\, d\nu = \frac{8\pi}{c^3}\,\widetilde{n}\nu^2 d\nu \tag{3.9}$$

Insertion of the above and (3.5) into (3.4) leads for the expression for the Planck law in an isotropic, collisionless plasma to

$$u(\nu) = \tilde{n}\, u_c(\nu) \qquad (3.10)$$

where the subscript zero refers to vacuum.

As a further step, the equilibrium between the radiation and a medium is studied along the lines of Einstein [28]. Detailed calculation of this topic have been presented elsewhere [27]. It suffices to say that only the Minkowski form of the momentum density satisfies the equilibrium conditions. In order to obtain this result, one imposes a condition $h\nu \gg kT$, effectively, we may say that it holds in the geometrical optics approximation.

In order to study the whole range of frequencies, we intake the similarity between the dispersion relation

$$\omega^2 = \omega_p{}^2 + k^2 c^2 \qquad (3.11)$$

and the relativistic form of the energy E and momentum p of the particle

$$E^2 = m_o{}^2 c^4 + p^2 c^2 \qquad (3.12)$$

In view of the relations

$$E = \hbar\omega, \; \mathbf{p} = \hbar\mathbf{k} \qquad (3.13)$$

we assert that a photon, which before entering a plasma had a zero rest mass, acquired, whilst inside it, an effective rest mass of $h\omega_p/c^2$. As a consequence, the Maxwell field in a medium can now be replaced by the Proca field, describing vector mesons. We now utilize the methods of the Proca electrodynamics and obtain the canonical energy-momentum tensor

$$T^{\mu\nu} = \frac{\partial L}{\partial(\partial_\mu A^k)}\, \partial^\nu A^k - g^{\mu\nu} L \qquad (3.14)$$

where the proca Lagrangian L_p is

$$L_p = -\frac{1}{16\pi}\, F_{\alpha\beta} F^{\alpha\beta} - \frac{1}{c}\, j_\alpha A^\alpha + \frac{\mu^2}{8\pi}\, A_\alpha A^\alpha$$

with $\mu = m_o c/\hbar$ being the effective photon's rest mass in units of universe length. This tensor is not generally accepted. In order to satisfy the conservation laws, we require, following Belinfante [29], that

$$\theta^{\mu\nu} = T^{\mu\nu} + T_s^{\mu\nu} \qquad (3.15)$$

where $T_s^{\mu\nu}$ is interpreted as the spin energy-momentum tensor. It is then found that the total field tensor is symmetric and consists of an "orbital" part, which determines the energy and momentum, and a "spin" part, which does not contribute to either but is important in calculations involving the total angular momentum.

Hence, in accord with the promise of this work, the same conclusion must apply also to the tensor describing the electromagnetic field in a plasma. It is agreed that the Minkowski tensor readily adjusts itself to the canonical formalism. Then, in analogy to (3.15), the tensor

$$S^{\mu\nu}_{sp} = S^{\mu\nu}_A - S^{\mu\nu}_M \tag{3.16}$$

ought to be interpreted as the spin energy momentum tensor. In the component form, this tensor has the form

$$S^{oo}_{sp} = 0; \; S^{oi}_{sp} = 0; \; S^{io}_{sp} = \frac{\tilde{n}-1}{4\pi c} \; \mathbf{E} \times \mathbf{H} \tag{3.17}$$

Then, it follows from the above that the force corresponding to the spin is

$$f_{sp} = \frac{\tilde{n}-1}{4\pi c} \; \frac{\partial}{\partial t} \; \mathbf{E} \times \mathbf{H} \tag{3.18}$$

At high frequencies, the refractive index approaches unity and the force vanishes, whilst its value increases toward the low frequency end.

Thus, from the interchangeability between the radiation in a medium and a neutral vector meson in vacuo, we conclude that the correct form for the field energy-momentum tensor is given by Abraham. However, in the limit of high frequencies, where the spin contribution may be neglected, the asymmetric Minkowski tensor represents a valid approximation. This description neglects the interaction by photo - and Compton - effect. It has been shown that these quantum interactions are less than 5% [30] .

4) FREE ELECTRON LASER USING THE NONLINEAR FORCE

Free electron lasers have been built [31] where a multi-MeV electron beam crosses a rippled magnetic field. This laser concept was realized in 1952 by H. Motz with microwaves, as will be described in more details in the review by Schwarz in a historic way [32]. A basically different type of a free electron laser has been proposed by Schwarz [33], where the superposition of electron beams to produce an interference field with a quantum-modulated electron current is used. Modulation is possible by laser [34] or by the Aharonov-Bohm effect [33]. The long beating electron beam emits coherent radiation. This modulation type laser has a higher efficiency at higher frequency, while the efficiency of the first mentioned synchrotron radiation laser increases on the (3/2)th power of the wave length.

Another free electron laser uses the nonlinear force. The inverse process has been shown experimentally [35]: focussing a laser beam into helium of 10^{-5} Torrs at 10^{15} W/cm^2 intensity, electrons are ejected with energies up to 100 eV. This energy is equal to the half maximum quivering energy in the laser field which is transferred into kinetic energy. The laser beam loses energy by this complex non-linear process. The inverse is possible, if an electron beam of duration t is fired perpendicularly into the center of a laser beam, where the translative electron energy is changed into oscillation energy. If the laser beam has a duration t, its switching off process causes a finally resting electron and the laser pulse has been amplified by the translative electron energy. If the electron energy corresponds to the maximum laser intensity and if the laser beam radius corresponds to the energy spread $\Delta\epsilon_o$ of the electron beam, the amplification is [36]

$$A = \frac{2j\sqrt{m}}{e\pi m_{ec}\sqrt{\Delta\epsilon_o}} = \frac{j}{n_{ec}\sqrt{\Delta\epsilon_o}} \ 9.47 \ 10^{10}; \ [j] = Amp/cm^2; \ [\Delta\epsilon_o] = eV \ [n_{ec}] = cm^{-3} \quad (4.1)$$

where **j** is the electron current density. The amplification is small for $10\mu m$ wave length and needs a repetition by more than 10^4 times, if a usual 10^4 J CO_2 laser pulse should be amplified to 10^7 Joules [36]. The successive process has the advantage of a sufficient temporal spread of the electron energy to be stored. Space charge effects can be suppressed by using additional ion beams or by using neutral clusters.

5)   NUMERICAL RESULTS ON ULTRAFAST COLD PUSHERS BY NON– LINEAR FORCES

The usual way of compressing the plasma of pellets by laser irradiation starts from a heating of the plasma and follows the compression by the subsequent thermokinetic process. This way led to the ablative mode for relatively long pulses, where a high compression was achieved by the too low temperatures prevented interesting neutron gains from nuclear fusion reactions by hydrogen isotopes. Another way is to use shorter pulses which is called "fast pusher" compression, where very hot electrons are produced causing a preheat of the pellet core what prevents high densities at adiabatic compression. The neutron gains are also not very sufficient hitherto.

Our concept of driving the plasma by the nonlinear forces permits an acceleration of the plasma with nearly no heating [12, 36]. These "ultrafast cold pushers by nonlinear forces" have been studied in details, after the existence of the nonlinear force had been established analytically, numerically and in many experimental details (Chapter 10 in Ref. [1]). We used a one dimensional model of plane waves [37] where the nonlinearities were included in the equation of motion by the general expression of the nonlinear force, and in the energy equation by the nonlinear optical constants in the interaction part given from the complete solutions of the Maxwellian Equations. After a detailed analysis, the initial density profile of the plasma was of a "bi-Rayleigh-type", where two regions of a refractive index $n = 1/(1 + \alpha x)$ are connected. The plasma of 5 to 100 eV initial temperature had a thickness of 100 wave length for neodymium glass laser light and α was 10^3 to 10^5 cm^{-1}. Laser radiation growing as a $sin^2 at$ - profile to a constant value I (in W/cm^2) at 1 psec produce e.g. at 1.5 psec the $(E^2 - H^2)/8\pi$ - profiles given in Fig. 1a. The maxima are reached at $+30\mu m$, and not at 0 (as in a collisonless plasma), because of the absorption. The fully dynamic calculation arrives at velocity profiles at 1.5 psec, given in Fig. 1b. Thick blocks of plasma with nearly **no** change of the temperature are moving towards compression with velocities of 10^8 cm/sec and more. The dynamic absorption process was found to be within the analytically expected range [37]. The irreversible processes involved in the nonlinear absorption have been studied by Castillo and a formulation by Onsager coeffecents was found [38].

At later times, e.g. at 3 psec, the profiles of Fig. 1a are oscillating. The density of the outer block is getting rippled. At these times, the evaluation of density ripple profiles, of velocty v, and of the nonlinear force resulted in a relation of a Korteweg-deVries equation

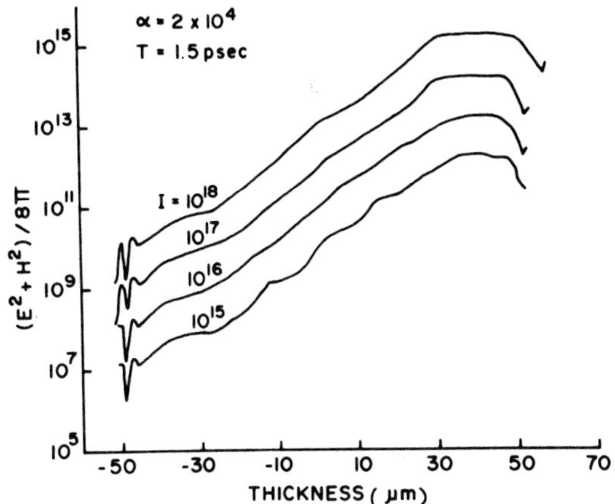

Figure 1a: For a $\alpha = 2 \times 10^4$ cm^{-3} bi-Rayleigh initial 100 eV electron density from -50 to $+50\mu$, irradiated by a Nd: glass laser of varying intensity I (in W/cm^2), the energy densities shown $(E^2 + H^2)/8\pi$ (in cgs) are produced at a time t = 1.5 ps.

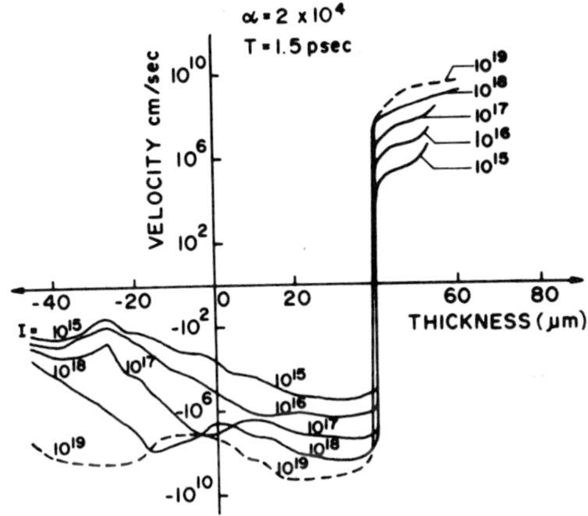

Figure 1b: The initially resting plasma has received a velocity profile at 1.5 ps in the case described in Fig. 1a.

$$\frac{\partial^3 v}{\partial x^3} = -\mu \frac{\partial}{\partial x}(\underset{\sim}{E}^2 + \underset{\sim}{H}^2)/8\pi \tag{5.1}$$

where the dispersion factor μ had the form of a Denisov type logarithmic derivative of the real part of the refractive index [1].

For these times, the block motion for nonlinear-force-compression was no more increased and a fast dynamic thermalization of the plasma corona happened. This is very useful for the gasdynamic compression scheme needing a fast thermalization by nonclassical effects. Therefore, after we have established the ultra-fast cold pusher at the earlier stage, the later fast ion heating by the soliton decay makes the use of thermalization by resonance absorption obsolete with its disadvantage of asymmetries of oblique incidence. Symmetric perpendicular irradiation of pellet results then possibly (in the earlier stage) in the nonlinear-force-compression, and later in thermokinetic ablation-compression. The Brillouin-ripple in the soliton phase quickly causes total reflection and zero coupling between laser field and plasma. This pulsating reflection and coupling within a time of 10 to 30 psec for $> 10^{16}$ W/cm^2 Nd glass plane wave irradiation heavily complicates the whole laser plasma compression. To avoid this, it is recommended to concentrate on very short pulses.

We have studied the soliton process for the times after the fast nonlinear-force pusher. It has been seen that the Korteweg-deVries relation is then fulfilled in the outermost part of the plasma, e.g. for a 15μ thick corona of a 50μ Rayleigh profile, Figure 2.

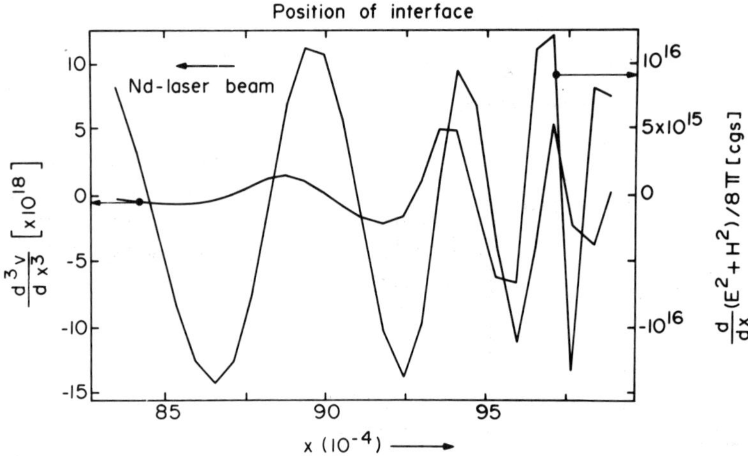

Figure 2: Output of a fully nonlinear dynamic calculation with plane waves (Nd-laser, 10^{16} W/cm^2) perpendicularily incident form r.h.s. on an initially Rayleigh type plasma after 2 psec. Correlation according to the Korteweg-deVries Eq. (5.1) can be seen from the fair correspondence of the extremata [39].

If we use a density permitting a deeper wave propagation, we find a correlation by the Benjamin-Ono euqation [40]

$$\frac{\partial}{\partial t} v + v \frac{\partial}{\partial x} v = -H \frac{\partial^2}{\partial x^2} v \qquad (5.2)$$

where the Hilbert transform H has to be unity. There is a direct understanding of this result from nonlinear force pushing of plasma to nodes of the (partially) standing wave.

6)   POLARIZATION SHIFT OF H-LIKE LINES IN PLASMAS

In order to receive a direct information about the density of the plasma in a laser compressed pellet, we have studied the shift of the H-like lines of ions with high charge Z in a plasma due to the energy

$$\varepsilon = e^2 / \lambda_D \qquad (6.1)$$

within a Debye sphere of radius given by the Debye length λ_D

$$\lambda_D = (kT/4\pi n_e e^2)^{1/2} \tag{6.2}$$

(k ~ Boltzmann constant; T ~ plasma temperature at local thermal equilibrium; n_e ~ electron density, e ~ electron charge). Each electron condensed in a plasma gains an energy

$$\varepsilon_B = \frac{e^2}{\lambda_D} \cdot \frac{1}{\frac{4\pi}{3}\lambda_D^3 n_e} = \frac{12\pi e^4 n_e}{(kT)^2} \tag{6.3}$$

which can be considered as a kind of "binding energy" [41]. The reality of this energy is seen in the fact of the "decrease of the ionisation energy" in plasmas [42].

This decrease of the ionisation energy has been included carefully in the numerical evaluation of Salzmann and Krumbein [43]. The question has been studied whether the electrostatic energy ϵ_B will not only affect the ionization energy but also the discrete energy levels of an electron in an atom (polarization shift) and experimental evidence was found [44]. Since low density approximations were achieved [44], we present here a quantum mechanical model for a closed solution derived by Hora [45].

A similar electrostatic method to describe the quantum state of an electron in a Coulomb potential is used as it was successful to describe the decrease of the ionization energy of atoms in a plasma [41]. The aim is to compare the energy ϵ_B with the energy of the electron when being bound to an ion or atom. If an electron within a Coulomb field **E** of Z protons $|E| = Ze/r^2$, is confined within a radius r_o. The electrostatic energy released by the electron when coming from $r = \infty$ is from the Maxwellian stress tensor

$$\varepsilon_s = \frac{Z}{8\pi} \int_{r_o}^{\infty} E^2 d^3\tau \, , \quad d^3\tau = 4\pi r^2 dr$$

$$= Z e^2/2r_o \tag{6.4}$$

The quantization requires that confinement of an electron to a radius r, corresponds to an increase of its momentum p (or energy $E = p^2/2m$) given by

$$r = n\hbar/\sqrt{2mE} \tag{6.5}$$

where n are integers n = 1, 2, 3 ... ∞. Apart from a factor of order 1, E can be considered as a Fermi-Dirac energy:

$$E = n^2 \hbar^2/(2mr^2) \tag{6.6}$$

The fact is that the increase of ϵ_s on decreasing r_o is slower than the increase of E on r. A stationary solution is obtained when both energies are equal, $E\,(r_o = r_s) = \epsilon_s(r_o = r_s)$ if

$$r_s = n^2\hbar^2/(m e^2 Z) = n^2 r_{Bohr}/Z \tag{6.7}$$

where the Bohr radius results for n = 1; Z = 1. It is well known from the solutions of the Schrödinger equation that the radius of an electron in the state n = 2 is $4r_{Bohr}$ as given in (6.7). Using r_s, the energy $E = E_n$ in Eq. (6.6)

$$E_n = \frac{Z^2 m e^4}{2\hbar^2 n^2} = \frac{Z^2 \, 13.6 \, eV}{n^2} \tag{6.8}$$

arrives in the ionization energy of hydrogen for n = 1, Z = 1 of 13.59 eV. In the case of an electron in the potential of a He^{2+} ion, Eq. (6.8) arrives in the ionization energy of He^+ at n = 1 of 54.38 eV and similar for higher Z in agreement with measurements. The terms

$$E_m - E_n = Z^2 R \left(\frac{1}{m^2} - \frac{1}{n^2} \right) \tag{6.9}$$

where R is the Rydberg constant $me^4/(2\hbar^2)$, define the energy level of transitions where e.g. for hydrogen (Z = 1) with m = 1 and n = 2, $E_1 - E_2$ = 10.24 eV from the Lyman series follows.

This model easily answers the question why an electron does not "fall into the proton" at electrostatic contraction of the electron radius r: the rise of the "quantum energy" E is faster and balanced at the Bohr radius. The electron has no orbital motion as in the usual quantum mechanical result (in difference to Bohr's model). This model has the advantage that the polarization energy ϵ_B per electron, Eq. (6.3) can be introduced immediately as correction to ϵ_s in a plasma, resulting then in the electrostatic energy

$$\epsilon_s^{plasma} = \epsilon_s - \epsilon_B$$

$$= \frac{Ze^2}{2r} - \frac{e^2}{\lambda_D} \frac{3}{4\pi \lambda_D^3 n_e} = \frac{Ze^2}{2r}\left(1 - \frac{3r}{2\pi \lambda_D^4 n_e}\right) \tag{6.10}$$

Equating this with the quantum energy E, Eq. (6.6), results in

$$\frac{n^2 \hbar^2}{2mr^2} = \frac{Ze^2}{2r} - \frac{3e^2}{\lambda_D^4 4\pi m e}$$

and hence in the algebraic solution for

$$r = \frac{\lambda_D^4 \pi n e}{3}\left[1 - \sqrt{1 - 6n^2 \hbar^2/(\lambda_D^4 n_e m)}\,\right] \tag{6.11}$$

The energy state E_n for the quantum number n is then from Eq. (6.6), using Eq. (6.11)

$$E_n = \frac{Z^2}{n^2} R \frac{A^2}{(1 - (1 - 2A)^{1/2})^{2}} ; \quad A = \frac{3n^2 \lambda_c^2 c^2}{\pi^2 Z^2 \lambda_D^2 v_e^2} \tag{6.12}$$

where the Compton wave length λ_c = b/mc and the mean electron velocity v_e was used. The polarization shift $E_n - E_m$ is of the same order as the low density approximation [43]. The advantage of the result (6.12) is an immediate inclusion of the plasma temperature.

The strongest shift of the levels occurs where the limiting case A = A^* = 1/2 in Eq. (6.12) is given. This means if 2A is close to one,

$$2A = \frac{6 n^2 \lambda_c^2 c^2}{\pi^2 \lambda_D^2 v_e^2} \leq 1 \tag{6.13}$$

The limiting case corresponds to the highest possible $n = n^*$ in a plasma similar to the Iglis-Teller continuum. For this case $E_{n^*} = \epsilon^2/(kT)$, using Eq. (6.1).

7) GAIN CALCULATIONS FOR NUCLEAR FUSION

In order to calculate the necessary conditions of energy input E_0 into an optimized (solid) pellet volume V_s before compression to a maximum density n_0 (in multiples of the solid state density n_s), the nuclear reaction gain G was computed as function of E_0, n_0, V_s for the reaction of deuterium tritium of hydrogen and boron (11), and deuterium and lithium (6) [46]. Fuel depletion, losses by bremsstrahlung and reheat based on a collective model were included. The cross sections for the last mentioned reaction have been re-evaluated [47]. The results have been published elsewhere [46] and were discussed by Miley [48].

The following subsequent developments have to be reported. For DT and HB(11), a volume ignition has been observed, while for DLi(6), the stronger losses by bremsstrahlung prevented the volume ignition. The ignition reduced the optimum temperature to 2 keV for DT or to 30 keV for HB(11). The HB(11) reaction is therefore the ideal goal, as no neutrons are produced, the net radioactivity per produced energy is ten times less than at burning coal, and energy is going into alphas only, permitting a direct conversion of the fusion energy into electricity with low heat pollution. Gains of 20 can be achieved for $E_0 = 1$ MJ, and compression to 10,000 times solid state.

There were extensive discussions, whether the reheat should be based on the collective model or on binary collisions. Both models agree for a range of low plasma temperatures only. The decision is: the model with the strongest reheat will dominate, what is the collective model (any stronger reheat mechanism has not yet been discovered). The results gained with the collective model agree for DT within 12% with those calculated by Nuckolls, where the reheat was taken from the experiments of large scale thermonuclear reactions [47].

Recalculation of our gains [47] using the latest cross sections [49] arrived at changes of the gains by less than 3%. In this connection it is interesting that the cross sections will be increased at high compression [50]. This has not yet been included in our calculations but can result in a more optimistic effect only.

It has to be mentioned that for our final zeal of HB(11), the present laser drivers are insufficient with respect to their efficiency, even if a 50% transfer of laser energy into the targets is possible by the ultrafast cold pusher by nonlinear forces. The efficiency of electron and ion beam driven pushers seems to be more promising at present.

8) THEORY OF FINITE LASER BEAMS IN PLASMAS

The step to nonclassical nonlinear effects teaches always the need to use the most exact solutions of the basic equations. One example is the description of the radial nonlinear forces due to a laser beam in plasma. Using the approximation of wave propagation in the +x-direction and describing E_y and H_z by a Gaussian decay of the amplitude in radial direction, the radial force is only in the y-direction given by the radial gradient of $(E_y^2 + H_z^2)/8\pi$ from Eq. (1.3), using the Maxwellian stress tensor of the third Eq. in Chapter 2. The forces are then zero in the y-direction because gradients are to be taken from $(E_y^2 - H_z^2)/8\pi$. We shall show, however, that the forces are the same as in the x-direction, saving all the preceding theories of non-linear-force self-focusing. The answer is given only [51], if the exact solution of the field of the laser beam is used which contains longitudinal x-components. Longitudinal components of an electromagnetic wave in vacuum (or low density plasma) look very strange, but they result in a solution of the problem.

The treatment will be given in terms of the quivering motion. For using a slab of laser radiation with a decay in the x-direction (parallel to the polarization direction of **E**) we reproduce the earlier result [52]. The E_y-field is then

$$E_y = E_0 (1 + \alpha_1 y + \alpha_2 y^2 + \cdots) \cos\left(\omega t - \frac{\omega}{c} x\right) \qquad (8.1)$$

with y being the displacement in the y-direction if the center of the beam is at $y = 0$. The acceleration of electrons has therefore the form

$$\frac{d^2 y}{dt^2} = -\frac{e}{m} E_0 (1 + \alpha_1 y + \alpha_2 y^2 + \cdots) \cos\left(\omega t - \frac{\omega}{c} x\right) \qquad (8.2)$$

Neglecting second and higher orders, a solution by small perturbation results in

$$\frac{d^2 y}{dt^2} = -\frac{e}{m} E_0 \cos\left(\omega t - \frac{\omega}{c} x\right) + \alpha_1 \frac{E_0^2 e^2}{m^2 \omega^2} \cos^2\left(\omega t - \frac{\omega}{c} x\right) \qquad (8.3)$$

Using $\alpha_1 = \frac{1}{E_0} \frac{\partial E_y}{\partial y}$ from a Taylor expansion at $x = x_0$ and averaging over a period we have [1]:

$$m \frac{\overline{d^2 y}}{dt^2} = -\frac{e^2}{4me^2} \frac{\partial}{\partial y} \overline{E_y^2} = -\frac{\partial}{\partial y} \overline{(E^2 + H^2)}/8\pi \qquad (8.4)$$

The generality of this solution for higher orders has been proved by reasons of energy conservation for the net acceleration of the electron to large values of x [52].

For the case of perpendicular polarization, we use a laser slab where the electric field polarization is again in y-direction but the field decay is in the perpendicular direction z. Similarly we can write:

$$E_y = E_0 (1 + \beta_1 z + \beta_2 z^2 + \cdots) \cos\left(\omega t - \frac{\omega}{c} x\right) \qquad (8.5)$$

with z being the displacement from the beam center at $z = 0$. The Maxwell equations arrive not only in the usual transversal component

$$H_z = E_0 (1 + \beta_1 z + \beta_2 z^2 + \cdots) \cos\left(\omega t - \frac{\omega}{c} x\right) \qquad (8.6)$$

but also in a longitudinal component. The longitudinal z-component of the magnetic field is given by

$$H_x = -\frac{E_0}{\omega}\left(\beta_1 + \beta_2 z + \ldots\right)\sin\left(\omega t - \frac{\omega}{c}x\right) \qquad (8.7)$$

which usually has been neglected. The acceleration of electrons, therefore, has the form

$$\underset{\sim y}{i}\frac{d^2 y}{dt^2} + \underset{\sim z}{i}\frac{d^2 z}{dt^2} = -\underset{\sim y}{i}\frac{e}{m}E_0\left(1+\beta_1 z + \beta_2 z + \ldots\right)\cos\left(\omega t - \frac{\omega}{c}x\right)$$

$$+ \underset{\sim z}{i}\frac{e}{m}\frac{dz}{dt}\frac{E_0}{\omega}\left(\beta_1 + 2\beta_2 z + \ldots\right)\sin\left(\omega t - \frac{\omega}{c}x\right) \qquad (8.8)$$

Considering again only the first order effect and substituting the zero order solution of (8.8) for y, z and $\frac{dz}{dt}$ in (8.8), we obtain

$$\underset{\sim y}{i}\frac{d^2 y}{dt^2} + \underset{\sim z}{i}\frac{d^2 z}{dt^2} = -\underset{\sim y}{i}\frac{e}{m}E_0\cos\left(\omega t - \frac{\omega}{c}x\right)$$

$$+ \underset{\sim z}{i}\frac{e}{m}\frac{E_0\beta_1}{\omega}\left(-\frac{e}{m}E_0\right)\sin^2\left(\omega t - \frac{\omega}{c}x\right) \qquad (8.9)$$

with $\beta_1 = \frac{1}{E_0}\frac{\partial E_x}{\partial z}$ at y = 0 and averaging over a time period results in a vanishing

y-component but an acceleration in the z-direction

$$\overline{m\frac{d^2 y}{dt^2}} = -\frac{e^2}{4m\omega^2}\frac{\partial}{\partial z}\overline{E_y^2} = -\frac{\partial}{\partial z}\overline{\left(E^2 + H^2\right)}/8\pi \qquad (8.10)$$

The general formulation of an optical beam in vacuum which satisfies the Maxwellian equations exactly, arrives always at a longitudinal component. The general formulation was given by Hermite functions [53] and the curvature of the wave field due to the longitudinal component corresponds to the direction of the first minimum of diffraction if the beam aperture is defined by the 1/e intensity of a Gaussian beam.

The main result is that the radial nonlinear force due to a laser beam in a plasma is given by the gradient of $(E^2 + H^2)/8\pi$ and all preceding calculations of self-focusing [1] are unchanged. An unexpected longitudinal component appeared in the exact description of laser beams in vacuum. We further experienced that only exact solutions lead to reasonable results at nonlinear effects.

9) SELF-FOCUSING AND MeV IONS

In order to understand the interaction of finite laser beams perpendicularly incident on a plasma and the subsequent self-focusing for very high intensities, estimations of the ion energies of Z-time ionized target ions arrived at MeV values. The beam was focused quickly within a short distance to nearly one wave length diameter by relativistic self-focusing [54] and the nonlinear forces were accelerating the ions to energies ϵ_i axially

$$\varepsilon_i = Z\,\varepsilon_{osc}/2 \tag{9.1}$$

where ϵ_{osc} is the maximum oscillation energy in the focus. This knowledge led to the first discovery of laser produced MeV ions [55] and was in agreement with numerous later experiments [56].

A detailed numerical study on the plasma dynamics at self-focusing based on nonlinear forces [1] and the instantaneous relativistic effect [54] was performed including very complete terms in the electro-dynamic equations which are neglected usually [57]. The generation of axial MeV ions has been confirmed. One example is shown in Figure 3.

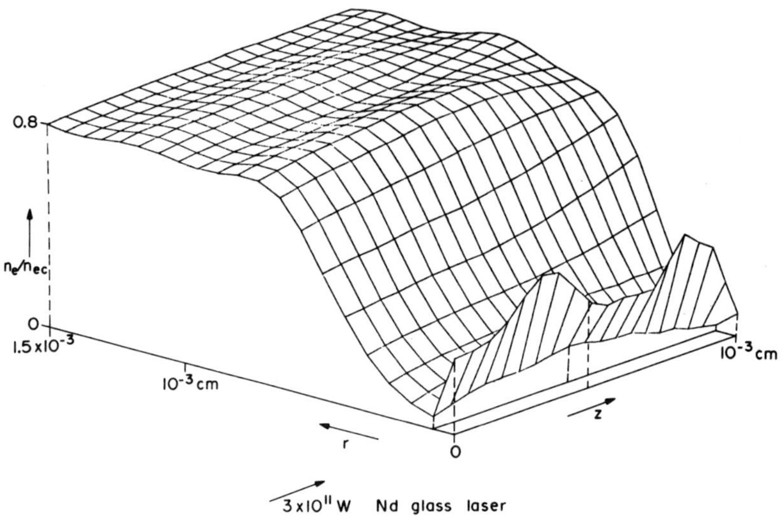

Figure 3: Density profile in an initially homogeneous H-plasma of 80% critical density fast interaction by a 10μ half-radial width and 30 psec duration after 9.2 psec. A hollow beam has been created with a residual plasma in the beam center [58].

A Nd glass laser beam of nearly Gaussian radial intensity profile with 10μ diameter of half intensity of power 3×10^{11} watts and a one pulse of a $\cos^2$ time dependence of 30 psec duration is incident on an initially homogeneous hydrogen plasma of 80% of the cut-off density. The relativistic effect causes an immediate shrinking of the beam while the nonlinear forces expel plasma from the interior of the beam. The radial velocities exceed 10^7 cm/sec. Because of the small radial field gradients in the center of the beam, a residual plasma remains and causes a hollow-beam after 9.2 psec operation, Fig. 3. This central plasma disappears e.g. at 18.4 psec due to fast axial motion.

REFERENCES

[1]   H. Hora, Nonlinear Plasma Dynamics at Laser Irradiation (Springer, Heidelberg, 1979), VII, 242 p.

[2]   H. Hora, Laser Plasmas and Nuclear Energy (Plenum, New York, 1975) p. 32.

[3]   V.A. Gapunov and M.A. Miller, Sov. Phys. JETP 7, 168 (1958).

[4]   S. Weibel, J. Electr. Cont. 5, 435 (1958).

[5]   H. Motz and C.J.H. Watson, Advances Electr. and El. Phys., L. Marton ed., (Acad. Press, New York, 1967) Vol. 23, p. 153.

[6]   H. Motz, Laser Produced Plasmas (Pergamon, Oxford, 1979).

[7]   H. Hora, D. Pfirsch, and A. Schlüter, Z. Naturforsch. 22A, 278 (1967).

[8]   H. Hora, Phys. Fluids 12, 182 (1969).

[9]   J.W. Shearer, R.E. Kidder, and J.W. Zink, Bull. Am. Phys. Soc. 15, 1483 (1970); J.W. Shearer, UC Livermore, LLL-Rept. UCID-15745, Dec 1970.

[10] A.J. Palmer, Phys. Fluids 14, 2714 (1971).

[11] R.B. White and F.F. Chen, Plasma Phys. 16, 565 (1974).

[12] J.W. Shearer and J.L. Eddleman, Phys. Fluids 16, 1753 (1974).

[13] M.E. Marhic, Phys. Fluids 18, 837 (1975).

[14] L.D. Landau and E.M. Lifshitz, Electrodynamic in Continuous Media (Pergamon, Oxford 1969) p. 242.

[15] L.P. Pitajevski, Sov. Phys. JETP 12, 1008 (1961).

[16] J.A. Stamper, Laser Interaction and Related Plasma Phenomena, H. Schwarz and H. Hora, eds. (Plenum, New York, 1977) Vol. 4B, p. 721.

[17] R.D.C. Miller and H. Hora, Plasma Phys. 21, 183 (1979).

[18] G.W. Kentwell and H. Hora, Plasma Phys. (submitted).

[19] H. Hora, Laser Interaction and Related Plasma Phenomena, H. Schwarz and H. Hora, eds. (Plenum, New York 1971) Vol. 1, p. 383.

[20] H. Hora, Phys. Fluids 17, 393 (1974).

[21] G.W. Kentwell, Honours Thesis, Univ. New South Wales (1979); (submitted for publication).

[22] J. Lindl and D. Kaw, Phys. Fluids 14, 371 (1971).

[23] N.G. Basov, 3rd European Conference on Laser Interaction with Matter, Paris 1969.

[24] R. Klima and V.A. Petrzilka, Cz. J. Phys. 1322, 896 (1972).

[25] H. Hora, Phys. Fluids 17, 1042 (1974).

[26] R. Peierls (private comm. 1975); Proc. Roy. Soc. A347, 475 (1976).

[27] M.M. Novak, PhD Thesis, Univ. New South Wales, 1979, Fortschr. Phys. July 1980.

[28] A. Einstein, Phys. Z. 18, 121 (1977).

[29] F.J. Belinfante, Physica 6, 887 (1939).

[30] H. Hora, Nuovo Cim. Lett. 22, 55 (1978); Phys. Stat. Sol. 86B, 685 (1978).

[31] D.A.G. Deacon, L.R. Flias, J.M.J. Madey, G.J. Ramian, H.A. Schwettman, and T.J. Smith, Phys. Rev. Lett. 18, 892 (1977).

[32] H.J. Schwarz, Laser Interaction and Related Plasma Phenomena, H. Schwarz et al. eds (Plenum, New York, 1980) Vol. 5, p.

[33] H.J. Schwarz, Phys. Rev. Lett. 42, 1141 (1979).

[34] H. Schwarz and H. Hora, Appl. Phys. Lett. 15 (1969).

[35] B.W. Boreham and H. Hora, Phys. Rev. Lett. 42, 776(1979); B.W. Boreham and B. Luther-Davies, J. Appl. Phys. 50, 2533 (1979).

[36] H. Hora, 2nd Int. Conf. Energy Storage, Venice Dec 1978; Plenum Press (in print); H. Hora, B.W. Boreham and J.L. Hughes, Sov. J. Quantum Electr. 9, 464 (1979).

[37] V.F. Lawrence, PhD Thesis, Univ. New South Wales, 1978; H. Hora, R. Castillo, R.G. Clark, E.L. Kane, V.F. Lawrence, R.D.C. Miller, M.F. Nicholson-Florcence, M.M. Novak, P.S. Ray,

J.R. Shepanski, A.I. Tsivinsky, Plasma Physics and Controlled Nuclear Fusion Research 1978, (IAEA Vienna, 1979) Vol. III, p. 237.

[38] R. Castillo, MSc Thesis, Univ. New South Wales, 1979; E. Schmutzer and B. Wilhelmi, Plasma Phys. 9 799 (1977).

[39] J.B. Lalousis (intern. report).

[40] H.H. Chen and Y.C. Lee, Phys. Rev. Lett. 43, 264 (1979).

[41] H. Hora and P.S. Ray, Atomkernenergie 31, 62 (1978).

[42] G. Ecker and W. Kröll, Phys. Fluids, 6, 62 (1963); H.R. Griem, Plasma Spectroscopy (McGraw Hill, 1964) p. 139.

[43] D. Salzmann and A. Krumbein, J. Appl. Phys. 49, 3229 (1978).

[44] B. Yaakobi and S. Goldsmith, Phys. Lett. A31, 408 (1970); M. Neiger and H. Griem, Phys. Rev. A14, 298 (1976).

[45] H. Hora, Rev. Laser Electr. 15, No. 1 (1980).

[46] see Figs. 1 and 2 of H. Hora et al Ref. [37].

[47] R.G. Clark, H. Hora, R.S. Ray, and E.W. Titterton, Phys. Rev. C18, 1127 (1978).

[48] G.H. Miley, Laser Interaction and Related Plasma Phenomena, H. Schwarz et al. eds. (PLenum, New York, 1980) Vol. 5, p.

[49] Ron McNally (ORNL - data 1979).

[50] C. Deutsch, Laser Interaction and Related Plasma Phenomena, H. Schwarz et al. eds. (Plenum, New York 1980) Vol. 5, p.

[51] R. Mavaddat and H. Hora (submitted).

[52] H. Hora, Laser Interaction and Related Plasma Phenomena, H. Schwarz et al, eds. (Plenum, New York 1977) Vol. 4B, p. 851.

[53] R. Castillo (internal report 1978).

[54] H. Hora, J. Opt. Soc. Amer. 65, 882 (1975).

[55] M. Siegrist, B. Luther-Davies, and J.L. Hughes, Opt. Com. 18, 605 (1976).

[56] H. Hora, E.L. Kane, and J.L. Hughes, J. Appl. Phys. 49, 923 (1978).

[57] E.L. Kane, PhD Thesis, Univ. New South Wales, 1979.

[58] E.L. Kane and B. Lalousis (unpublished).

COMPRESSION MEASUREMENT IN LASER DRIVEN IMPLOSION EXPERIMENTS[†]

D. T. Attwood, N. M. Ceglio, E. M. Campbell,
J. T. Larsen, D. M. Matthews and S. L. Lane

University of California,
Lawrence Livermore Laboratory
Livermore, California 94550 USA

CURRENT DIRECTIONS IN LASER FUSION EXPERIMENTS

The primary purpose of this paper is to discuss the measurement
of compression in the context of the Inertial Confinement Fusion
Program's transition from thin walled exploding pusher targets, to
thicker walled targets which are designed to lead the way towards
ablative type implosions which will result in higher fuel density and
ρR* at burn time. Before entering the primary discussion of compression diagnostics, we would first like to discuss, in qualitative
fashion, several general aspects relating to the state of
affairs in compression experiments. Such a discussion will permit us
to review some of the basic precepts [1-3] of inertial fusion and,
hopefully, provide some insight to the current preference for plastic
coated glass microballoons as "intermediate density" targets, as well
as an appreciation for their limited role as ablatively driven targets with currently available laser drivers.

The transition to targets which attain high fuel density is of
fundamental importance to the achievement of scientific feasibility

[†]This work was performed under the auspices of the U.S. Department
of Energy by the Lawrence Livermore Laboratory under contract number
W-7405-Eng-48.

*The abbreviation ρR will be used in this paper as a shorthand nota-
tion for $\int \rho\, dR$, generally with the intergral extending only over a
region of the target distinguished by one material, i.e., with separate integrals computed for the fuel region, the pusher, etc.

in the ICF program. Briefly, to ignite the thermonuclear fuel it is
necessary to simultaneously achieve both high temperature and high
density. Minimum ignition conditions are roughly 3-5 KeV and several
hundred g/cc for the central region of the DT fuel. To ensure an
efficient burning of the fuel it is further required that a spherical
burn wave, driven by alpha particle capture, propagate out superson-
ically from a centrally ignited core. This imposes the additional
constraint that the fuel be many α particle ranges thick, or equiva-
lently, that a DT ρR of approximately one g/cm^2 be achieved at these
temperatures. The fundamental role of high densities and high ρR's
is thus clear: alpha particle capture produces a propagating burn
in the fuel giving rise to high gain. This can be accomplished with
a minimal energy investment in the ignition of a small central core
of the fuel.

Unfortunately, present laser drivers do not permit us to attain
energy densities in the fuel required for simultaneously achieving
both high density and high temperature. Compromises in target design
and performance are thus required. As has been demonstrated for
several years with exploding pusher targets [4-8], the achievement of
high fuel temperatures is relatively easy. The far more difficult
task is that of compressing matter to high density. Achieving high
density encounters several formidable problems. The simplest to
understand is that heated matter requires more energy to compress
than cold matter -- the compressive forces or pressure that must be
brought to bear are proportional to the internal energy, or pressure,
of the fuel. The energy efficient procedure is to compress first and
then heat, (or to do so simultaneously as in the isentropic coalesc-
ing of weak shocks described in refs. [1-3]). Another, more difficult
aspect of the high density problem is that of maintaining spherical
uniformity during the implosion process. This is essential to the
achievement of the very high densities required. If the spherical
pusher-fuel interface deteriorates, mixing of the two materials
occurs, limiting the ultimate compression and thermonuclear yield.
The problem of maintaining a stable interface in the presence of an
adverse pressure gradient, even with a uniform accelerating field,
is well known in fluid stability studies, and is generally referred
to as the Rayleigh-Taylor problem. This situation is complicated in
the laser fusion case by non-uniformities in the laser drive and
target materials. In the former, for instance, local hotspots in
the laser may produce increased pressure at the ablation surface
which would cause localized filaments of the pusher to accelerate
inward faster than average values. Depending on pusher thickness,
the filament length may grow to a size comparable to the shell thickness and thus break through the pusher-fuel interface, causing mixing
thereof, and thus severely limiting the final fuel conditions which
might otherwise have been obtained. The study and control of unstable interfaces will eventually occupy a large part of our efforts.
This is, however, somewhat ahead of us, as we first attempt to demonstrate the achievement of ablatively driven implosions and perhaps

separately, the attainment of modest intermediate densities in the 2
to 20 g/cc range (10-100X liquid DT density) [9].

Although the achievement of intermediate densities through ablative implosions is of primary interest to the ICF program, our
efforts are currently limited by the unfavorable balance of energy
deposition into suprathermal electrons whose range, unfortunately, is
a considerable fraction of the target thickness. With current laser
intensities on target of 10^{15} to 10^{16} W/cm^2 we observe for 1.06 μm
lasers, and more so for 10.6 μm, a considerable fraction of energy
being deposited in very energetic, penetrating, suprathermal electrons [10]. For example, in a case to be discussed in Section IV
approximately 90% of the absorbed light is believed to be deposited
in suprathermals characterized by a temperature of 26 KeV. The mean
electron energy for such a distribution, 40 KeV, has a range of about
15 μm ($\sim$ 3 mg/cm^2) in normal density plastic. It is difficult to
ascribe an ablative nature to such a target, having comparable wall
thickness and electron deposition range. Rather, the implosion dynamics might be described by a milder form of the previous exploding
pusher target where now the pusher decompression is less violent,
resulting in a weaker shock propagating into the fuel, and thus producing less fuel preheat. These arguments are qualitative, but serve
notice that the term "ablatively driven" may not be appropriate for
experiments at many laboratories now aimed at attaining intermediate
densities.

To significantly improve this situation the energy of suprathermals must be reduced. Efforts toward this end must certainly include
the movement towards shorter laser wavelength and lower intensity,
each of which is known to reduce the number and energy of suprathermal electrons [11-13]. With sufficient uniformity this would permit
us to move on to fundamental work with ablative implosions, pusher
symmetry and stability, and thermal energy transport. Many of these
processes are currently obscured by the dominant role of suprathermals, and uncertainties in our understanding and characterization
thereof. With regard to our characterization of suprathermals, improved techniques are needed to more accurately determine the electron spectra, perhaps as suggested by the K-shell vacancy technique
of Hares and his colleagues [14].

EXPLODING PUSHER COMPRESSION MEASUREMENTS

In this section we briefly review compression measurements performed with exploding pusher targets. These experiments with thin
walled targets serve as an introduction to the subject in a situation
in which high temperatures provided sufficient amounts of kilovolt
x-rays and reaction products for diagnosing the implosions, and in
which the relatively low ρR of the surrounding glass ($\sim$ 5 x 10^{-4}
g/cm^2) permitted these emissions to escape for diagnostic purposes.
Figure 1 shows a schematic diagram in which a thin walled glass tar-

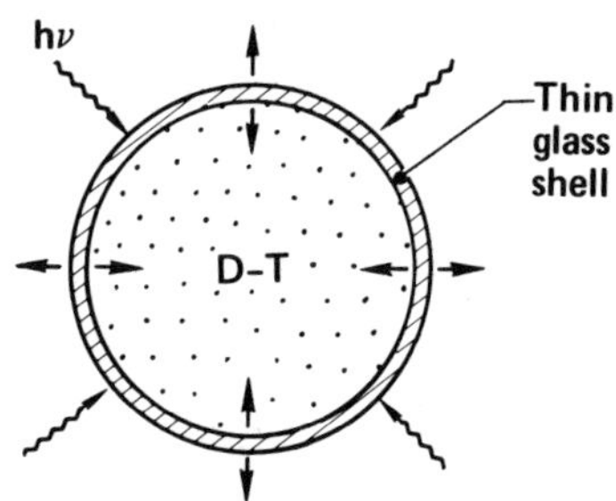

Fig. 1. Much of the early work in laser compressed targets was accomplished with thin walled "exploding pusher" targets. In these experiments short, intense laser pulses were used to produce fast electrons which then penetrated the entire shell, causing it to explode inwardly and outwardly, shock-heating and compressing the encapsulated fuel [4-6].

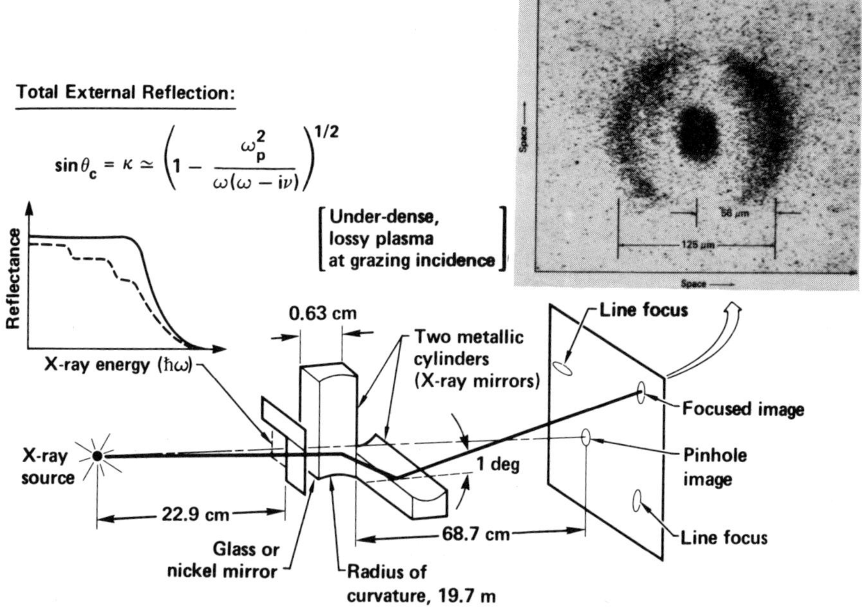

$$\sin\theta_c = \kappa \simeq \left(1 - \frac{\omega_p^2}{\omega(\omega - i\nu)}\right)^{1/2}$$

Fig. 2. Grazing incidence x-ray reflection from two cylindrical mirrors provides the basis for image formation in the Kirkpatrick-Baez x-ray microscope. A sample 2-3 keV x-ray micrograph of an early laser driven implosion is shown in the inset [15].

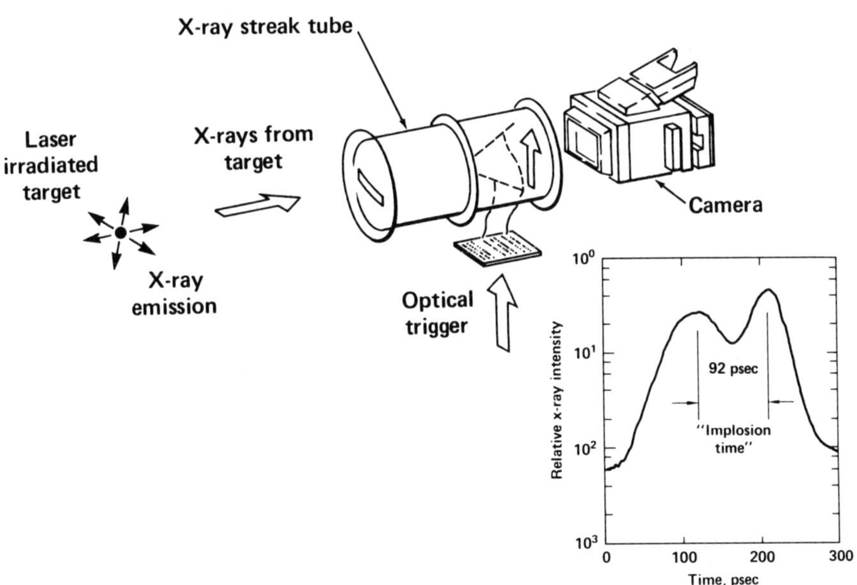

Fig. 3. Time resolution of picosecond x-ray pulses is obtained with a "streak camera". X-rays are converted to electrons in a thin gold photocathode. The electrons are then accelerated, imaged, and rapidly deflected in the vertical direction for temporal separation. The inset shows 2-3 keV x-ray data from a target similar to that in Fig. 2. Early time emission is synchronous with laser heating, while the later pulse is due to pusher stagnation at final compression. The time interval between the two gives a crude estimate of an "implosion time".

get, typically 70 μm in diameter, 0.5 μm thick, and filled with 10 atmospheres of DT ($2mg/cm^3$), is irradiated with a high intensity ($\sim 10^{16} W/cm^2$) laser pulse, typically 70 psec in duration and 0.5-2 TW peak power. Laser light is primarily absorbed in these short pulse high intensity experiments by resonance absorption, showering the target with fast electrons of sufficient energy to penetrate and heat the entire shell. The resulting explosive expansion of the glass shell causes a strong shock which pre-heats the fuel. Strong x-ray emission occurs from the initially heated target due to the high temperatures and densities in the glass. Upon decompression x-ray emission is reduced, reappearing near the target center as the inward propagating half of the pusher stagnates against the encapsulated fuel, compressing and further heating the fuel in the process. Figure 2 shows a 2-3 KeV time integrated x-ray micrograph [15] typical of exploding pusher experiments conducted during the period from mid-1974 when KMS first reported their results [16], to late 1976 when the above story was fairly well understood [4-6]. X-ray photographs such as this, whether by

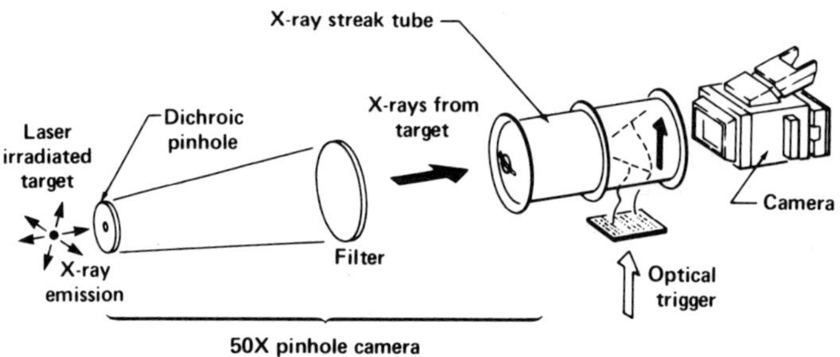

Fig. 4. Simultaneous space-time x-ray recording is obtained by combining 6 μm pinhole imaging with the 15 psec x-ray streaking capability [19].

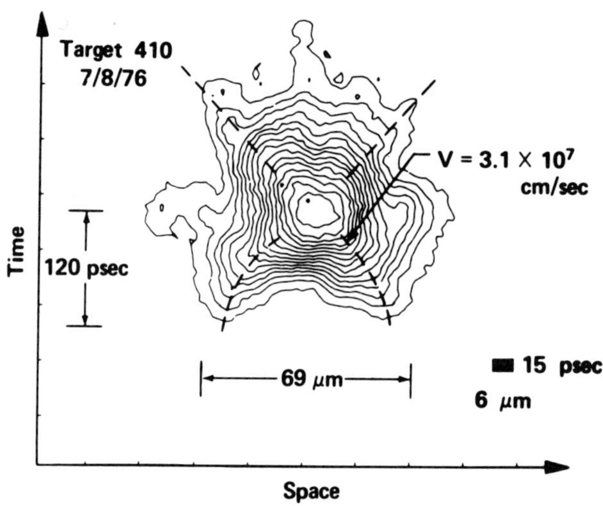

Fig. 5. Sample space-time data obtained with 2-3 keV x-rays, recorded as outlined in Fig. 4. Detailed velocity histories, and their implication for exploding pusher dynamics, is discussed in ref. [20,21].

pinhole camera or grazing incidence techniques, played an important role in the diagnostics of those early compression measurements. These diagnostics generally utilized x-ray emissions in the 1-3 KeV regime and provided resolutions of several to 10 μm. Figure 3 shows the complimentary use of an x-ray streak camera [17, 18] for the temporal measurement of implosion using x-ray emissions similar to

those discussed in the spatial image of Fig. 2. As shown in the
lower right inset, and resolved to about 15 psec, x-ray emission in
the 2-3 KeV band first rises due to laser heating of the shell in
its initial position. As the shell expands and density drops, emission from this region decreases. At a later time the inner portion
of the shell stagnates near the target core, again raising the glass
density and temperature leading to a second peak in x-ray emission.
Combining the "implosion time" obtained from these temporal peaks
with the spatial data of Fig. 2, one obtains an average implosion
velocity to be compared with numerical simulations. More detailed
information regarding implosion dynamics is obtained by combining
pinhole imaging and x-ray streak photography in a single recording,
as illustrated in Fig. 4 [19]. Typical data is shown in Fig. 5.
Detailed velocity histories, as obtained from this data, provided
significant constraints on numerical simulations of the experiment,
particularly with regard to the role of thermal transport inhibition
in limiting the value of final implosion velocities [20, 21].

The above diagnostics emphasize the dynamics and final state of
the glass pusher but do not provide data on the compressed fuel itself. By replacing the DT fuel with neon gas, Yaakobi and his colleagues at Rochester [22] were able to study compression by observing
stark broadened spectral lines of hydrogenic neon. An example of
their work is shown in Figs. 6 and 7, where the Lyman γ line of neon
is identified as being optically thin, and having a line profile
which is consistent with 300 eV electron and ion temperatures, and an

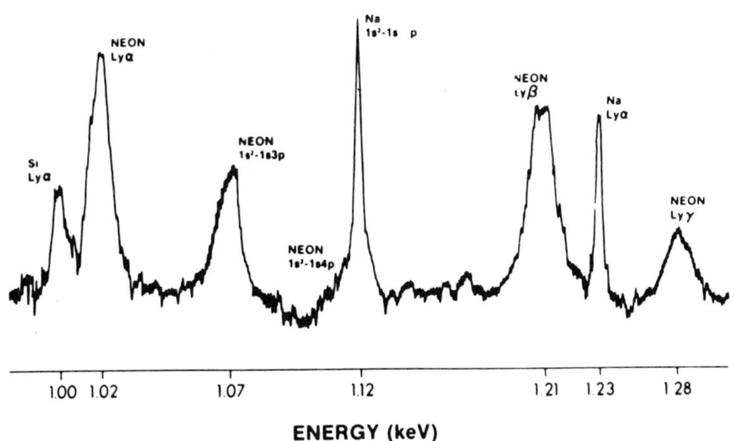

Fig. 6. X-ray spectra of hydrogenic neon are used to determine compression density in these experiments reported by Yaakobi et al. [22]
at the University of Rochester.

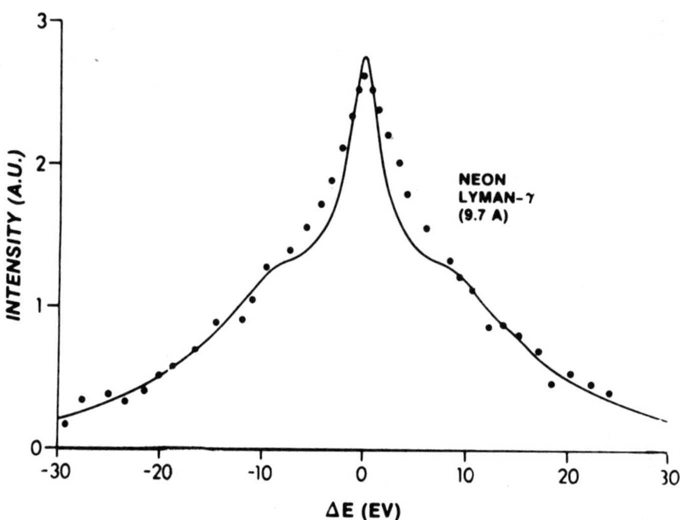

Fig. 7.   The optically thin Lyman-γ line of Fig. 6 is matched to a theoretical profile corresponding to a final electron density of 7×10^{22} e/cc, and equal electron-ion temperatures of 300 eV [22].

Spectrogram of plasma from DT, Ne-filled glass microsphere target

Compression of Ne Filled Glass Microsphere 10% Ne by Volume

Shot #	Target Fill DT + Ne	Laser Power (Watts)	Initial Diam. (μm)	Final Diam. (μm)	Neon Density Volume Measurement (cm^{-3})	Neon Density Stark (cm^{-3})	Volume Compression (Stark)
RR-11	8 ATM	3.24×10^{11}	189	62	$(3.2 \pm 1.3) \times 10^{20}$	4.1×10^{20}	28
RR-21	8 ATM	4.52×10^{11}	185	50	$(5.1 \pm 2) \times 10^{20}$	6.8×10^{20}	50
RR-26	7 ATM	8.22×10^{11}	181	44	$(5.5 \pm 2.4) \times 10^{20}$	5.6×10^{20}	70

Fig. 8.   Compression data obtained with a CO_2 laser by Mitchell et al. [24] from Los Alamos show that neon may be used as a seed gas to obtain compression data.

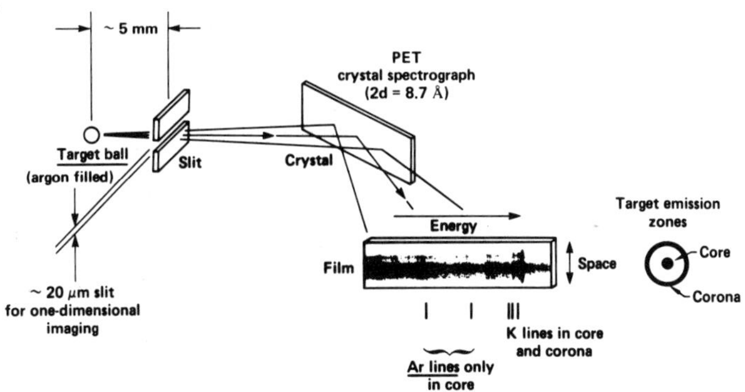

Fig. 9. Slit x-ray spectroscopy has been extended by Koppel et al
[25] to more energetic helium and hydrogen like lines of an argon
seed gas.

electron density of 7×10^{22} e/cc (equivalent fuel density $\sim$ 0.26 gm/cc
similar to results obtained elsewhere. Somewhat later Auerbach et al
[23] at Livermore, and Mitchell et al. at Los Alamos [24] extended
these techniques to a DT-neon mixture, in which the higher Z gas acts
only as a diagnostic seed, with little impact on target dynamics, and
allowing the achievement of thermonuclear reactions within the dominant DT gas. An example of that work is shown in Fig. 8. The agreement between stark and spatially determined densities appears fortui-
tious considering the complex spatial and temporal profiles expected,
and the size of slit employed. Figure 9 shows some later work at LLL
in which 3.1 KeV helium-like lines of an argon seed gas were observed
in the compressed core [25]. In those experiments with a slit spectrograph combination, the argon seed fill of 0.15 atmospheres was too
low and the slit too wide to allow detailed determination of the
density. Higher fill ratios will allow this technique to be extended
to somewhat higher ($\sim 5 \times 10^{-3}$ g/cm^2) ρR target compressions, as
will be discussed in a later section. It should be noted, however
that the geometry of compression is an important implosion issue
which is not addressed by diagnostic techniques providing 1-D images
or volume integrated data.

A significant advance in the diagnostics of laser compressed
targets occured with the development of zone plate coded techniques
in 1976 [26]. These techniques are particularly advantageous when
imaging small objects, working well for both particle and photon
emissions. Figure 10 describes the basis for this technique in which
incoherent zone plate shadow casting eventually permits the reconstruction of 3D images in a procedure somewhat reminiscent of holo-

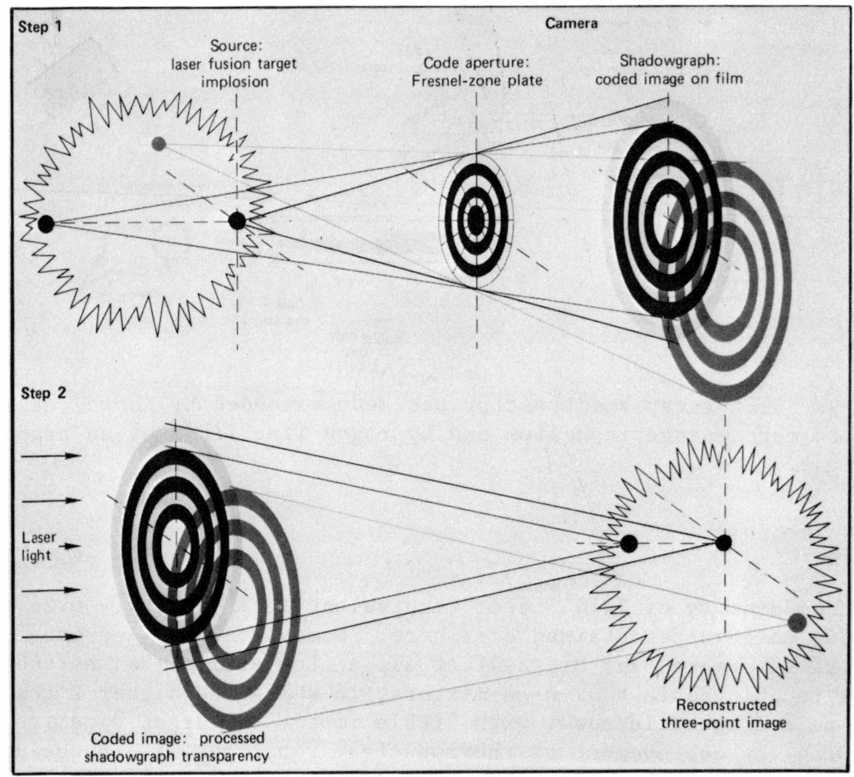

Fig. 10. Zone plate coded imaging is a technique particularly well
suited to small emitters, whereby three-dimensional images of incoherent photon or particle emissions can be reconstructed from simple
shadowgraphs. This technique has been used to record electron, proton, and alpha particle images, as well as x-ray images from 0.5 to
30 keV of laser driven targets [26, 27].

graphic techniques, but in which phase does not play a role in record
ing. As seen in Fig. 10, the location of three different emissions
is encoded by the size and location of the respective shadows. Emissions close to the coded aperture (zone plate) create larger shadows,
those off-axis to one side produce shadows to the other side, and
shadow intensity is proportional to emission intensity. The composite coded image is then reconstructed with coherent light in a second
process in which diffraction now plays a dominant role. Here the
zone plate provides its unique contribution to the encoding/decoding
process in that when illuminated with coherent radiation, a zone
plate produces a focused spot, with axial position (focal length)

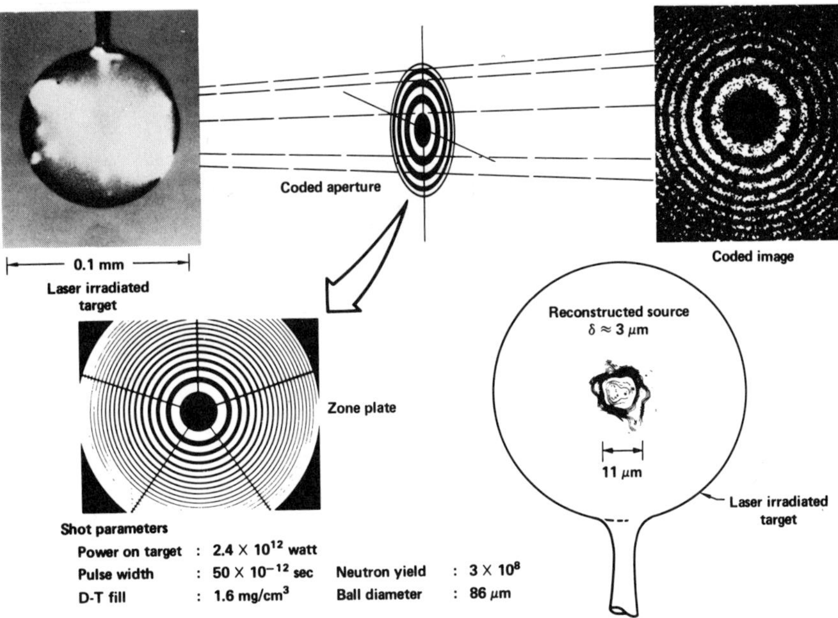

Fig. 11. The application of zone plate coded imaging to alpha particle imaging of thermonuclear burn is illustrated here. Third order
optical reconstruction of the coded image shows the emission region
resolved to 3 μm [27].

dependent on the size of the shadow. The composite zone plate shadows then faithfully reproduce the position and intensity of the original emission sources, as illustrated in Fig. 10. This technique
works well for both particle and photon shadow casting, within the
diffraction free approximation, presuming one uses appropriate filtering and detection techniques. Figure 11 shows a composite illustration of an experiment in which the coded aperture technique was used
with a cellulose nitrate track detector to image alpha particle
emission from the thermonuclear burn region of a laser imploded target [26]. This data, coupled with determinations of ion temperature
by alpha [7] and neutron [8] spectroscopy, provided our most detailed
measurement of final fuel conditions in an implosion experiment.

Figure 12 shows x-ray images obtained with this technique. Presented in isometric form to unambiguously display data over a wide
dynamic range, the illustrations show intensity as the vertical axis.
Both measured and computed images are shown, each with an asymmetric
outer ring, as seen previously in the microscope image of Fig. 2, and

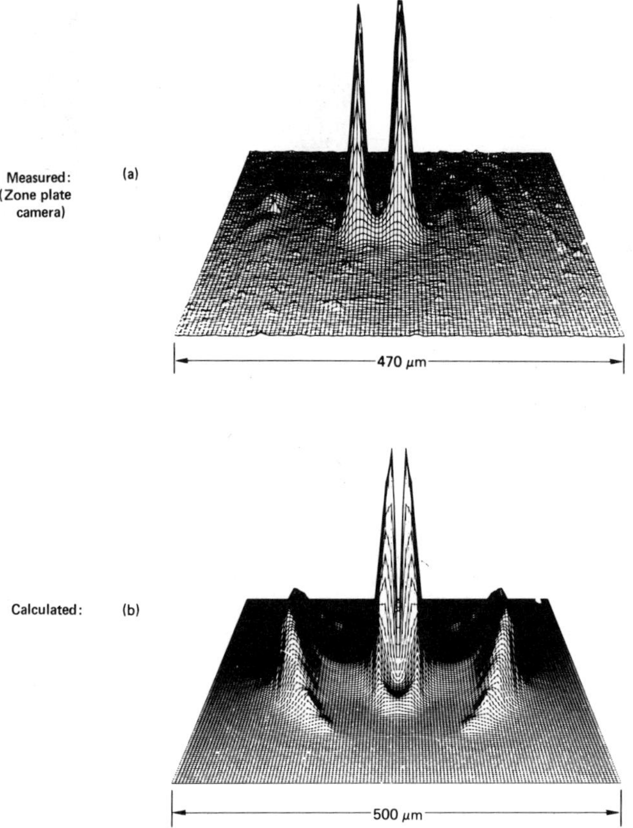

Fig. 12. Isometric images of 4-7 keV x-rays, obtained by zone plate coded imaging and numerical simulation, are compared for an exploding pusher target. Detailed comparisons are made in ref. [29].

strong emission peaks of the stagnated pusher surrounding the compressed fuel. The fuel does not show in these x-ray images because the atomic number, Z, of DT is small compared to the pusher glass, leading to a lower emission of 4-7 KeV x-rays. Figure 13 demonstrates how particle and x-ray images overlay when obtained on a single target, with a single camera recording both type emissions, a point we return to in a following section.

In summary a variety of newly developed techniques, providing data on the previously unexplored scale of microns and picoseconds,

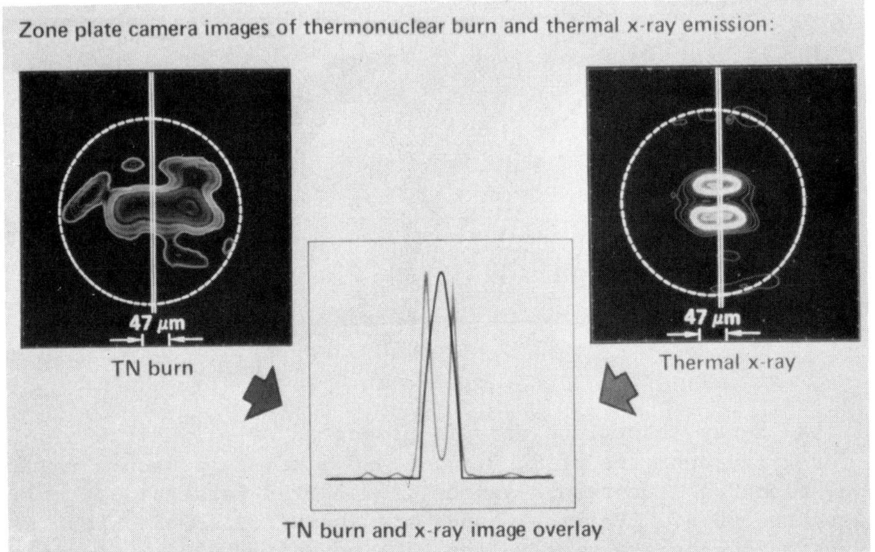

Fig. 13. Alpha particle and thermal x-ray images taken with a
single zone plate camera in which the cellulose nitrate track detector, x-ray filter and film, are placed behind a single zone plate.
Although the irradiation and subsequent implosion are particularly
non-spherical in this case, the data is nonetheless interesting in
that the alpha emission is pancaked between two discs of x-ray
emission from the stagnated pusher. It appears from the overlap
that the pusher 'disks' may have continued to move inward and emit
x-rays beyond the time of peak thermonuclear burn, an explanation
consistent with numerical simulation of the experiment.

were brought to bear on the diagnosis of explosively driven thermonuclear compressions. The techniques were successful in that they led
to and supported an adequate predictive capability for the performance of a rather complicated implosion process [4-6].

THE TRANSITION TO HIGHER DENSITY AND HIGHER ρR

The transition from exploding pusher targets to those achieving
intermediate densities [29] is accomplished by adding thick low Z
coatings such as plastic or teflon, which act primarily as preheat
protectors in that fewer electrons have sufficient range to deposit
their energy directly in the pusher, near the fuel-pusher interface.
Shock heating of the fuel is thus reduced, as are final pusher implosion velocities. Consequently, higher final densities and ρR are
achieved, but with lower temperatures. As a result, fuel and nearby

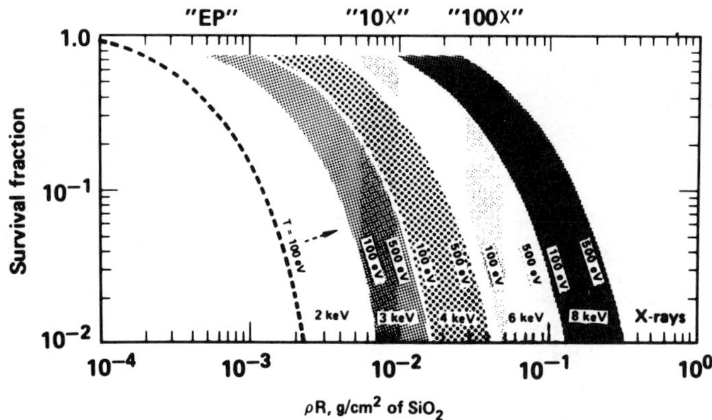

Fig. 14A. X-ray absorption data is presented as a function of glass
ρR, for x-ray energies of 2, 3, 4, 6 and 8 keV. In each x-ray band
shown, electron temperature varies from 100 eV (essentially cold
values) to 500 eV. Vertical bands are shown for some typical target
designs which achieve final glass ρR's as indicated. The vertical
bands correspond to exploding pushers ("EP") and relatively thick
walled glass shells with varying thicknesses of plastic coating
which are expected to achieve final fuel densities from "10X" to
"100X". The correlation of final fuel density and final glass ρR
will vary greatly depending on target design (e.g. initial glass
shell thickness), so that considerable horizontal shifting of the
vertical bands is to be expected as targets evolve.

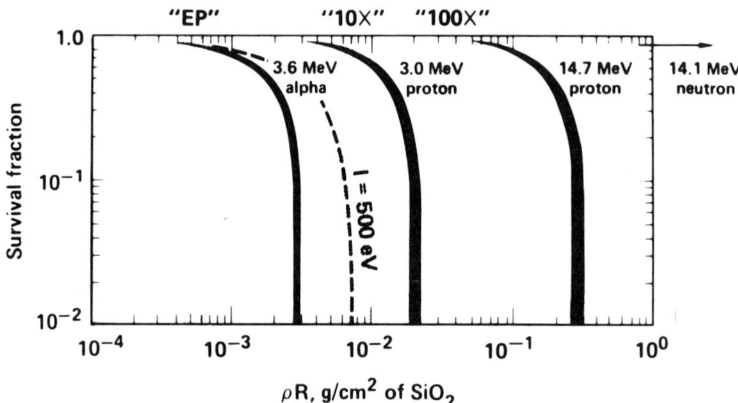

Fig. 14B. Energy survival fraction is plotted for particles emitted
from the target core in typical thermonuclear reactions (D-T, D-D,
D-He³). Again glass temperatures from cold to 500 eV are used. The
utility of this plot is that it can be overlaid with Fig. 14A to show
particle range data and x-ray extinction on a single graph. In this
manner one can readily appreciate the viability of various particles
and photons as potential sources of data regarding a compressed core
encapsulated by a surrounding glass shell.

pusher interface emissions are reduced, and transport difficulties are increased. Whereas with exploding pusher targets multi-kilovolt x-ray and thermonuclear reaction product emissions were sufficient for compression measurement with conveniently available diagnostics, they are significantly reduced with intermediate density targets. And whereas transport of one KeV x-rays and 3.6 MeV alpha particles was previously not a problem with thin pushers, the higher ρR glass and plastic coatings now prevent both of these values diagnostic observables from escaping the target.

Figures 14A and B address the problem of photon and particle transport through a compressed and heated glass shell. Figure 14A shows the fraction of 2, 3, 4, 6 and 8 KeV x-rays surviving after passage through a variable thickness of compressed glass, as given by the abscissa, for glass temperatures extending from 100 to 500 eV in each case. Also shown are vertical bands representing the measured or anticipated ρR values for compressed SiO_2 in exploding pushers (EP), and targets achieving densities of 10 and 100 times that of liquid DT ($\rho_{liq\ DT} = 0.2$ g/cc). The location of these vertical bands is somewhat dependent on initial target constraints and these may shift to some extent as target fabrication techniques advance, i.e., thinner mandrils (glass shells), thin high Z coatings, and thicker, lower Z ablators. It is worth noting that x-ray transmission problems will reduce with thinner mandrils, but the addition of high Z pre-heat shields will eventually more than offset those gains.

The significant point in Fig. 14A is that for exploding pushers all x-rays with energies above 2 KeV escape the target, whereas with the present 10X design even 3 KeV emissions (as, for instance, the helium-like argon line at 3.1 KeV) experience significant absorption, and for 100X type targets with ρR values approaching 100 mg/cm^2, 8 KeV x-rays are necessary for providing data concerning the compressed core. [28]. Figure 14B presents similar data for particles propagating through varying ρR thicknesses of glass, except that in this case the ordinate represents the surviving energy fraction, rather than number of particles. The important point here is that particle range data for reaction products are essentially presented in the same format as that of the x-rays, showing, for instance, that 3.6 MeV alphas from the DT reaction provides a useful diagnostic tool for exploding pushers, but do not escape the surrounding cold pusher associated with higher density targets. Although the 3.0 MeV protons of the DD reaction appear to be interesting for some targets, the presence of other proton emissions in this energy range may limit their use even in nominally hydrogen free targets. A more interesting option for imaging thermonuclear burn appears to be the 14.7 MeV D-He^3 protons, which if produced in sufficient quantity provide a viable diagnostic for a wide range of target conditions. It is worth noting that with extremely thick and complicated multishell high-Z targets, only the 14.1 MeV neutrons, with range values of order 10 g/cm^2, will be able to escape and provide information about the

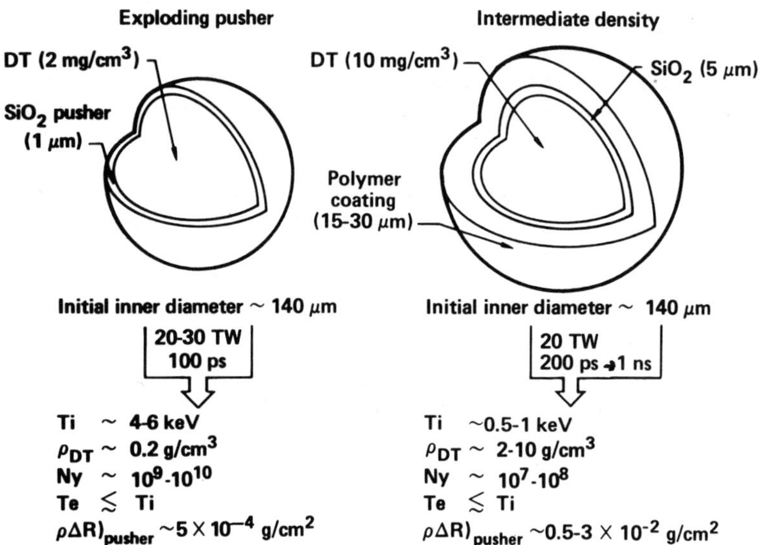

Fig. 15. Target parameters and final conditions are compared for exploding pusher and polymer coated intermediate density targets.

thermonuclear burn region. In a following section we will see how neutron activation is used in present experiments to diagnose pusher ρR conditions at burn time, and how these activation techniques may be extended to the more important question of fuel ρR with proper seed elements in near term experiments.

INTERMEDIATE DENSITY TARGETS

The first efforts to trade-off temperature for density involve the use of plastic coated glass microballoons, as described briefly in the previous section [29]. The use of low-Z ablators, rather than increasingly thicker glass, is to obtain as favorable as possible target to fuel mass ratio for our "spherical rocket". Figure 15 compares a target of this type with a more familiar exploding pusher, both of which have recently been shot at the Shiva laser facility at Livermore, the latter primarily for laser system or diagnostic systems check out. Note in particular the temperatures, densities, neutron yields, and glass ρR's achieved with the two target classes. The exploding pusher targets reach 4 - 6 KeV DT temperatures, and yields of 10^9 - 10^{10} neutrons, with final density and pusher ρR of only 0.2 g/cc (the usual "1X" performance for these targets) and

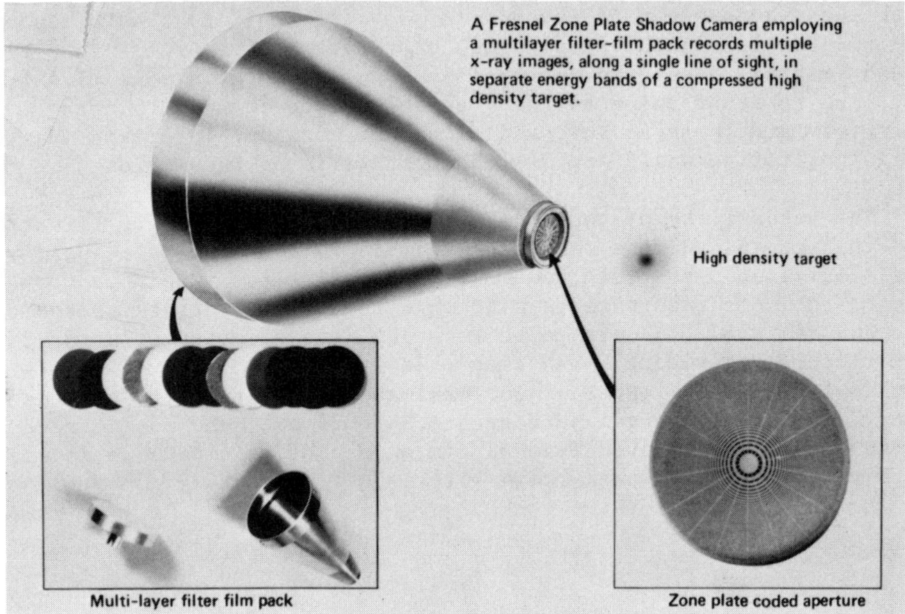

Fig. 16. Multi-spectral x-ray imaging of compressed intermediate density targets provides sufficient data to cover a wide range of final ρR conditions.

0.5 mg/cm^2, respectfully. By way of comparison, numerical simulations and recent experiments indicate that the thick walled targets achieve DT ion temperatures of only 0.5 to 1.5 KeV, while DT densities are pushed to 2 g/cc and glass ρR's of 6 mg/cm^2.

Referring back to Figures 14A and B, we observe that for these thicker targets the pusher ρR is beyond the range of 3.6 MeV alpha particles, so that alpha imaging of the burn region, as used previously with exploding pushers, is no longer possible. One also observes that at these higher ρR values in SiO_2, the choice of an x-ray energy for compression diagnostics is increased to the 4-8 KeV region depending on the type of target performance anticipated. Because of the low temperatures achieved with these targets, emission in these higher energy x-ray bands is significantly reduced with respect to exploding pusher targets. This general trend has a significant impact on the type of imaging diagnostic one uses. For instance, grazing incidence reflection optics are essentially limited to x-ray energies below about 4 KeV as presently configured, and high resolution x-ray pinholes present rather small collection solid angles, so that both techniques are hard pressed to provide significant data regarding compression in this new regime of target performance. Zone plate coded imaging techniques, using 10 - 40 µm thick gold zone plates, continue to be of value as they can be used to image with x-ray energies as high as 40 KeV, and of course continue to provide very large collection solid angle without sacrificing resolution.

The versatility of the zone plate imaging technique is illustrated in Fig. 16. Here a zone plate camera is shown with a capability for simultaneous particle and multi-spectral x-ray imaging. Located behind the coded aperture is a track detector appropriate to the target and reaction products to be recorded, and a series of x-ray films and filters, providing x-ray images in bands from 2 to 30 keV, with multiple films per band for maximum dynamic range to compensate for variations in target performance. It is not uncommon to record x-ray images which cover five or six orders of magnitude in intensity with the filter-film pack shown [31].

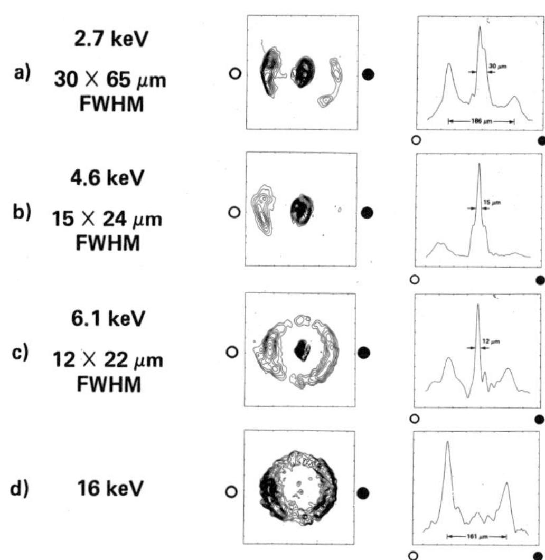

Fig. 17. Multi-spectral x-ray images from an intermediate density target irradiated from two sides. More symmetric irradiation is expected to yield clearer core emission, with a well defined fuel-pusher interface.

Figure 17 shows an example of the x-ray images obtained with a single camera of this type, upon irradiation of a 140 μm diameter, 5 μm thick glass target with a 15 μm CF (teflon) coating. Irradiation was accomplished with the 20 beam Shiva laser providing a total of 3.4 KJ in a 200 ps, 17 TW pulse. The initial target was filled to approximately 50 atmospheres pressure with equimolar DT. The broad band images shown are centered at 2.7, 4.6, 6.1 and 16 KeV. Several interesting features are observed. The lower energy thermal images display an asymmetry in both early time emission from the teflon and later from the ellipsoidal compressed core, each reflecting the asymmetry of the Shiva illumination scheme. Interesting aspects of the high energy symmeterization are discussed elsewhere [30]. Further analysis of core region emissions are shown, in each case, to the right side of Fig. 17. One observes a rather narrow emission spike emerging as one progresses towards the 6 KeV image. A detailed understanding of the compressed core geometry, and it's implication for compressed fuel volume are not yet in hand. Interpretation is particularly difficult in light of the illumination asymmetry on target. With a more symmetric irradiation pattern we would expect to see a more uniform outer ring, and a central emission region with a dip corresponding to the low-Z fuel. Analysis of such a simple image, with particular attention to sharp spatial and temporal profiles, would then give us a measure of final fuel conditions, to be compared with other techniques. Clearly more experience with these implosions is needed, with as many overlapping diagnostics as possible. This brings us to a major philosophical point of this paper, as outlined in Fig. 18.

- **Strong space and time dependence of thermodynamic quantities make simple interpretation of experimental observables difficult**

- **Sophisticated simulation codes are essential to properly interpret the experimental measurements**

- **Many independent measurements are necessary to constrain the simulation code calculations (sophisticated codes contain many approximations to complex physical phenomena)**

- **Symbiotic relationship between measurements and calculations is an essential part of a successful Laser Fusion Program**

- **Two-dimensional images are necessary to assess symmetry of implosion**

Fig. 18. A philosophical comment on the measurement of final fuel conditions in laser compressed targets.

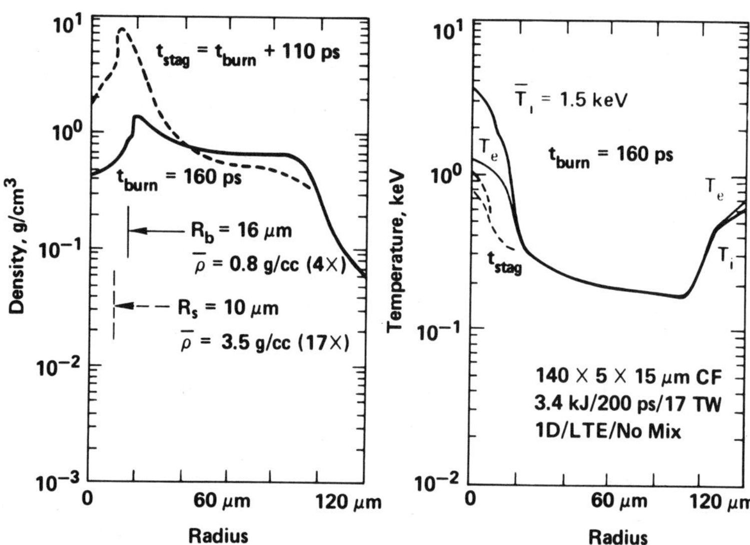

Fig. 19. Numerical simulations of target compression demonstrate the sharp radial and temporal dependencies of thermodynamic quantities, which strongly influence experimental observables such as x-ray images, spectral line shapes, and reaction product spatial profiles.

The accurate measurement of compression is a non-trivial exercise Thermodynamic quantities such as density and temperature, which have a strong affect on all our observables are varying widely in space and time on a scale of microns and tens of picoseconds. With these strong radial and temporal dependencies it is essential that several independent, direct,   quantitative measures of compression be obtained, and that the assumptions involved in the analysis of each be critically examined vis-a-vis the other observables. Because of the strong dynamical variations, and the differing dependencies of various emissions on local thermodynamic parameters, it is essential that a numerical simulation be used in an attempt to consistently explain all the observables.

Figure 19 is an example of the type of radial and temporal density and temperature profiles observed in numerical simulations of the experiment described above. Of particular interest are the sharp gradients of both ρ and T at peak burn time, and the significant differences in both 110 psec later (dashed lines), at the time of peak density. The magnitude of these variables and their gradients strongly affect the various experimental observables, such as x-ray images,

line profiles, etc., making comparisons non-trivial. We are presently studying these profiles with respect to the x-ray images of Fig.
17. Note in Fig. 19 that the fuel volume averaged quantities, corresponding to 80% of the fuel mass and denoted by superscript bars,
are also presented for comparison with experiments. Note also that
the fuel averaged densities, for instance, have increased four fold
in this calculation between the times of peak burn and peak density.
The point to be made is that overly simplified comparisons of the
results of one technique with another, without appropriate attendance
to details, are not likely to be of great value in guiding an experimental program. The emission sensitivities of each technique must be
carefully examined. Fuel region emissions, as seen for instance at
peak temperature with a high Z seed gas, are not likely to match
perfectly to emission images corresponding to the stagnated pusher,
which according to Fig. 19 might occur at a later time and smaller
radius in the case cited. Similar comments apply to comparisons
with reaction product images that one might obtain.

A significant diagnostic for exploding pushers, which carries
over to a certain extent to intermediate density targets, is the use
of high Z gases mixed in moderate amounts into the DT fuel. As point
ed out earlier, the 1.0-1.3 KeV Lyman series of hydrogenic neon provided significant data with lower ρR targets. At higher densities,
however, these lines are inappropriate because of the overly strong
stark broadening, self-opacity limitations, and difficult transport
problems in the temperature dependent high opacity pusher. Rather it
becomes important to use higher Z gases, such as argon. According to
Fig. 14, the 3.14 KeV helium-like line of argon is useful for glass
pusher ρR's up to about 5 mg/cm^2, for a glass pusher temperature of
about 500 eV. This corresponds approximately to the "10X" targets
presently used at Livermore, but could be pushed to higher density
implosions, as discussed earlier, with the use of thinner glass mandrils and appropriately thicker low-Z ablators. It is significant
to note in any event, that turning on the hydrogenic and heliogenic
lines of ever increasing Z seed gases is increasingly more difficult
as one trades off temperature for increased ablation and preheat
protection. As an example, ionization of lithium like argon to produce helium like argon requires an additional 918 eV. Production of
hydrogen like argon then requires an additional 4.1 KeV. With temperatures in the 0.5 to 1 keV range, helium like lines will be ever-
more difficult to form.

To obtain strong argon emission lines, the University of Rochester [32] has conducted a series of thick and thin target experiments
in which the entire fuel fill is argon. Although this prevents the
simultaneous study of thermonuclear reactions, it permits the development of a very valuable diagnostic technique. Figure 20 shows an
example in which the relatively weak, but fairly well understood,
Lyman β line of hydrogenic argon is compared for experiments with
thin (0.6 μm) and thick (2.9 μm) walled glass microballoons, with

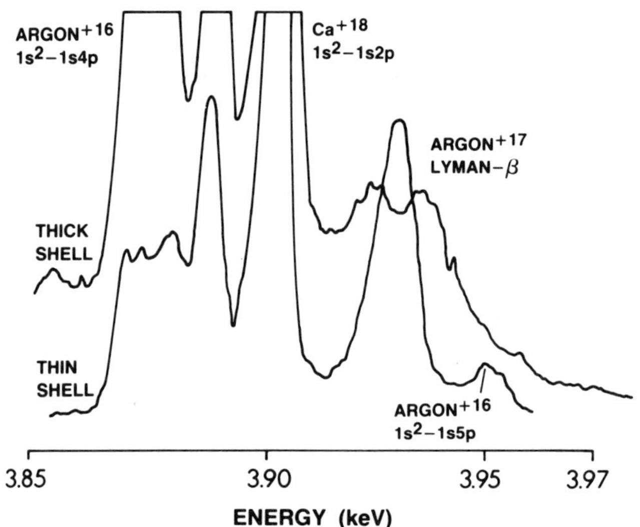

Fig. 20. Spectral broadening of hydrogenic argon lines is compared for thin and thick wall targets, demonstrating that the latter achieved higher final core density. ` Data obtained recently by Yaakobi [32] and his colleagues at Rochester is discussed in the text, and in the accompanying contribution by J. Soures.

diameters of 58 and 62 μm, and argon fills of 11 and 7 atmospheres, respectively. The laser parameters in the two cases were 75 J in 46 psec for the thin target, and 130 J in 65 psec for the thicker target. The higher achieved density in the thick shell case is illustrated by the broadening and splitting of the Lyman β line of Ar^{+17}, as predicted by the theory of stark broadening in dense plasmas [33, 34]. Note that the central dip of the Lyman β line is associated with the superposition of transitions from an odd (n = 3) numbered principal quantum state, and is not evidence of self-opacity effects. Comparison of the observed profile in the thick case with computed Stark profiles indicates a final argon density of approximately 4 g/cc. Other lines observed in Fig. 20, are the relatively stronger lines of helium like calcium and argon, which are off-scale in this illustration.

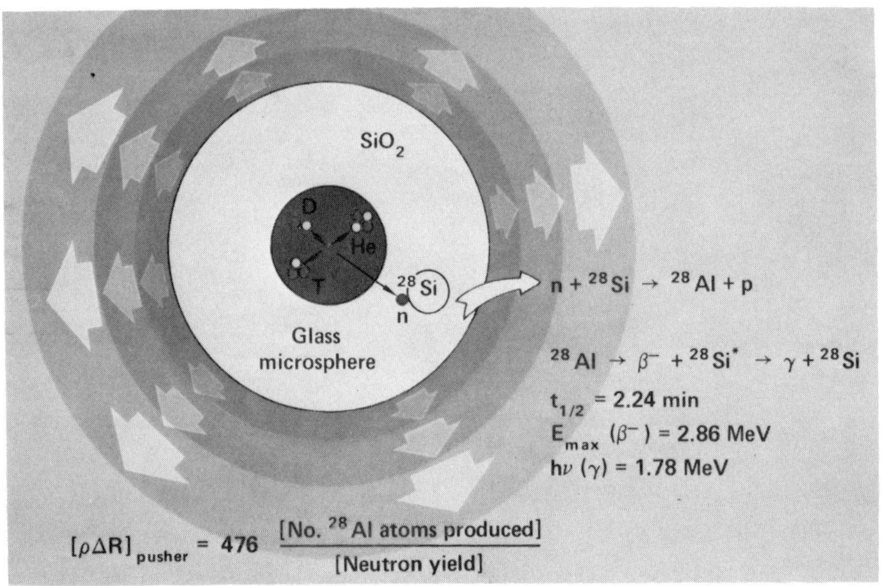

Fig. 21. The use of target generated neutrons to determine surrounding pusher ρR by neutron activation is illustrated, with appropriate
decay times.

A significant advance for the coming months would be to extend
these results to the case of an argon seeded DT fuel, and to obtain
multiple diagnostics on a single shot, such as line broadening, slit
spectroscopic imaging, pusher ρR via nuclear activation, and higher
energy two-dimensional x-ray imaging via zone plate coded techniques
in a symmetric implosion displaying a well defined dip at the fuel-
pusher interface. A consistent explanation of 10 - 30X type compression results, in the spirit described in Fig. 18, would permit us to
evaluate the ambiguities of interpretation in each case, and thus
allow us to continue to higher density implosion diagnostics with
greater confidence.

NEUTRON RELATED DIAGNOSTICS FOR HIGH ρR EXPERIMENTS

Diagnostic techniques which presently play an important role in
our experiments, and which will play an increasingly important role
as target ρR's increase, involve the direct or indirect use of neutrons [35]. A prime example of this, which has played a significant
role in our early efforts to achieve and diagnose implosions approach
ing 20 g/cc (100X) of DT, is that of neutron activation [36, 37]. In

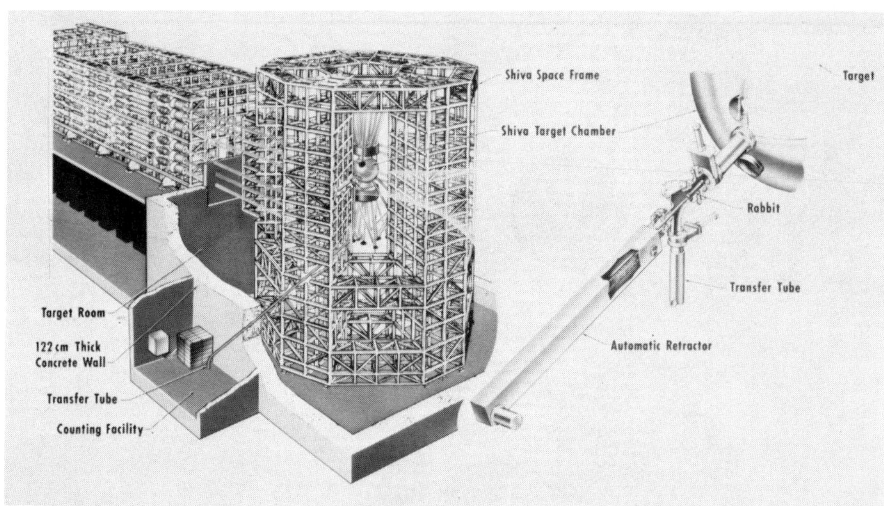

Fig. 22. The debris collecting "rabbit" used in neutron activation experiments at Shiva is shown. Experiments with short lived isotopes are simplified by this quick retrieval system.

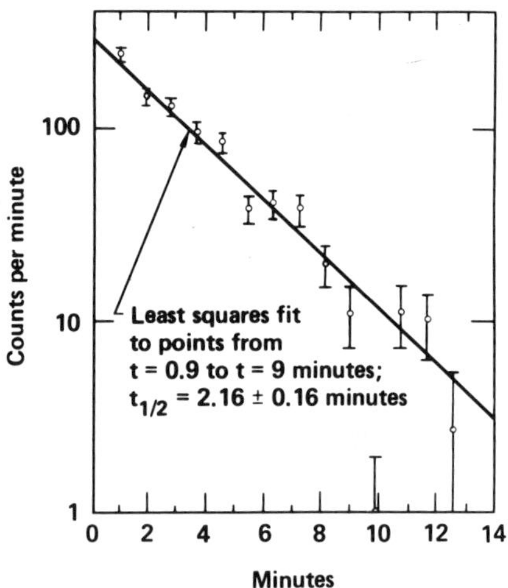

Fig. 23. Typical decay curve for activated ^{28}Al following implosion of a polymer coated intermediate density target at Shiva. Detected activations, measured neutron yields, and knowledge of collection and counting efficiencies provide data on final glass ρR.

a typical example of these measurements the 14.1 MeV DT neutrons
interact with atoms in the surrounding fuel and target layers to produce radioactive isotopes through various nuclear reactions such as
(n,p), (n,2n), etc., where the incoming neutron displaces a proton,
or two neutrons, respectively, producing a different element or isotope. Unstable isotopes are then captured and counted as they radio-
actively decay. The number of such activations is directly proportional to the neutron yield and the areal density (line integrated
density length product) of the material traversed by the neutrons.

In present experiments neutron yields, integrated ρR and target
fabrication difficulties have prevented us from directly measuring
the fuel region ρR, a quantity emphasized as being of primary importance to the ultimate success of ICF in Section I. Nonetheless, the
possibility of such measurements is not too far distant, a point we
shall return to later in this section. In the meanwhile, the activation technique has been used to measure spatially averaged pusher ρR
as indicated in Fig. 21. The 14.1 MeV neutrons liberated in the DT
fusion reaction interact with silicon atoms present in the surrounding glass pusher, producing radioactive aluminum in a ^{28}Si(n, p) ^{28}Al
reaction, as indicated in Fig. 21. After catching a measured fraction
of the target debris one observes, or counts, the decay of ^{28}Al back
to ^{28}Si by β^-, γ emission, with a 2.24 minute half life. Because the
total activation yield is extremely low (a few hundred atoms) for
present high density low yield target experiments, it is important
to catch as large a fraction of the activated debris as possible
and quickly transport the debris to an efficient counter.

The rather sophisticated collection system used at the Shiva
laser system is shown in Fig. 22. The system consists of a debris
collector, a transporter, and a high efficiency low background nuclear detector, capable of detecting total activation yields of as
little as 100 aluminum atoms. Depending on geometry employed [38, 39]
collection fractions as high as 70% are available. The transport
system delivers the debris to a counter within two minutes of activation, and the detector has an overall coincidence counting efficiency
of 40%. Typical experimental data illustrating the decay of activated ^{28}Al following an implosion experiment is shown in Fig. 23.
This data was obtained in the coated target experiment described
previously in Figs. 17 and 19. From the number of detected activations, the independently measured neutron yield, and the pre-determined collection and counting efficiencies, it was determined that
at neutron emission burn time the integrated ρR of the glass pusher
was approximately 6 mg/cm . To obtain fuel density and fuel ρR at
burn time it is then necessary, in present conditions, to rely on
simulation codes to model the implosion process, as was indicated
previously in Fig. 19.

Suggestions as to how this technique might be extended to the
direct measurement of fuel region ρR through the use of various seed

Fuel tracer candidates

— neutron yields required for 10 detected events

ρ fuel (g/cm^3)	^{14}N(n,2n)^{13}N	^{40}Ar(n,p)^{40}Cl	^{40}Ar(n,α)^{37}S	^{79}Br(n,2n)^{78}Br
2 ("10X")	1.3×10^{10}	2.7×10^{10}	1.3×10^{10}	1.3×10^{8}
20 ("100X")	2.7×10^{9}	6.7×10^{9}	2.7×10^{9}	4×10^{7}
200 (1000X")	6×10^{8}	1.3×10^{9}	6×10^{8}	6×10^{6}

Fig. 24. The issue of measuring final fuel region ρR is addressed by the suggestion of high activation cross-section seed gases, to be mixed with the initial DT. Neutron yields required for minimum detected counts are shown for several candidate seeds, as a function of final fuel density. In each case, 0.1 atmosphere seed is mixed with 50 atmospheres DT.

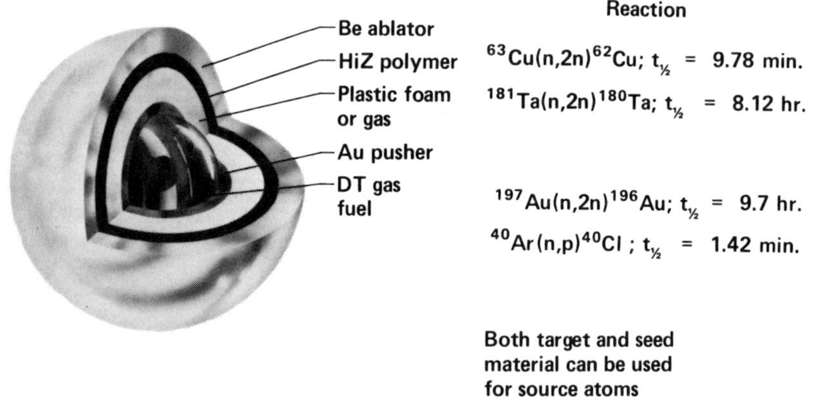

Fig. 25. The use of radiochemistry is suggested to separate diagnostic activations in the fuel and various layers of an anticipated high density target.

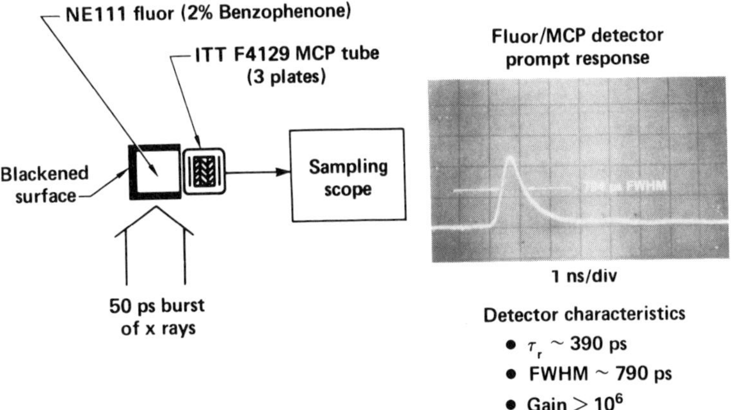

Detector characteristics
- $\tau_r \sim 390$ ps
- FWHM ~ 790 ps
- Gain $> 10^6$

Fig. 26. A quenched fluor for use as a fast neutron detector is described [40].

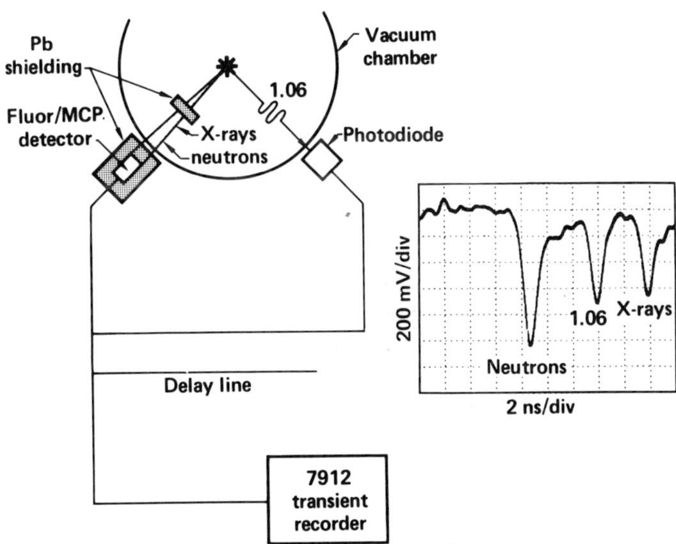

Fig. 27. The neutron time interval system described by Lerche and Ozawa [40] is shown with sample data. Fast detector responses, as in Fig. 26, are critical for studying implosion dynamics in complex multi-shell targets.

gases, are discussed in Fig. 24. Data is presented for three candidate seeds, ^{14}N, ^{40}Ar, and ^{29}Br. The bromine is particularly interesting because of its relatively high (n,2n) cross-section (862 millibarns) for 14.1 MeV neutrons. According to Fig. 24, direct fuel region ρR can be determined with a 0.1 atmosphere bromine seed, in a 50 atmosphere DT fuel driven to a final density of 20 g/cc and a yield of 4 x 10^7 neutrons, assuming collector and counting efficiencies typical of the Livermore activation experiments. Efforts to obtain results such as this are presently underway at Livermore. A further role of nuclear activation in which a number of activated species must be separated and counted, by nuclear chemistry techniques, is indicated in Fig. 25. In the conceptual multi-shell laser fusion target separate shells are shown for pre-heat protection, velocity multiplication, and symmeterization. Each shell including the seeded fuel, would be activated and separately counted. With targets of this complexity and high ρR, neutrons, or neutron related diagnostics, will provide the bulk of available data relating to the fuel and inner shell.

A diagnostic providing information on implosion dynamics which can be particularly valuable for double shell targets is that of "neutron interval timing", wherein an implosion time is deduced by recording both neutron emission and driving laser pulse with a single temporally resolved recorder. This is accomplished by use of a fast neutron fluor-fast photodiode combination, and a separate fast photodiode for laser light, both coupled to a single recording instrument. An example of the fast neutron detection system used at Livermore [40] is shown in Fig. 26, with sample data shown in Fig. 27. In these initial experiments the implosion times are not greatly larger than the overall system response time, limiting the preliminary impact of the technique. However, with larger double shell targets, this technique will rapidly become one of great value.

An additional concern is that of imaging thermonuclear burn in these complex multi-shell targets. With integral ρR's exceeding 1000 mg/cm^2, only neutrons will escape for diagnostic purposes, as was indicated in Fig. 14. As a consequence, there is considerable long term interest in neutron imaging. Unfortunately, the options currently available are limited. Neutron stopping power is so small, even in the best of materials, that very large thicknesses are required for minimal imaging contrast. For instance, the mean free path of neutrons is 35 g/cm^2 in copper. Considering that resolutions on the order of 10 μm are required, one readily appreciates the formidable nature of this task, particularly with regard to "optics" construction and alignment. Preliminary estimates of required neutron fluences for a neutron pinhole system are described by Ahlstrom et al. [35].

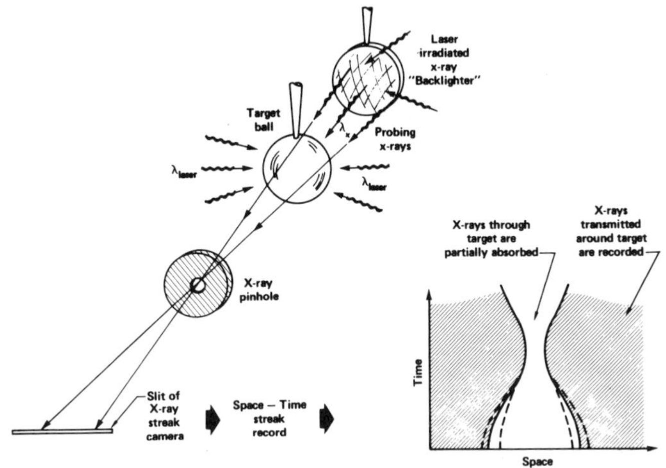

Fig. 28. Dynamics of a relatively cold, ablatively driven implosion can be studied with x-rays produced by an auxiliary source with independent laser irradition.

X-RAY PROBING OF DYNAMICS AND COMPRESSION IN ABLATIVELY DRIVEN TARGETS

In Section II on exploding pusher targets, we discussed the subject of pusher dynamics, showing that kilovolt x-ray self-emissions were sufficiently intense to permit streaked pinhole imaging of the implosion process. With ablatively driven targets such dynamical studies are more difficult because of the reduced self-emission and higher target opacity at several KeV, as we discussed previously in Section III. A solution to this problem, referred to alternately as x-ray probing or backlighting, is described in Fig. 28. A relatively cold, ablatively driven target is viewed with some appropriate x-ray optical system in the "light" of a background source of intense x-rays. The so-called "backlighter" is chosen to radiate with particular intensity in an x-ray line or band well suited to accentuate interface features of the imploding target. Both backlighter and target ball are driven by synchronously controlled laser pulses.

Two approaches appear feasible with current, or near term technology. With long pulse irradiation of the backlighter, streaked images of relatively cold pusher dynamics is possible. To study two-dimensional aspects of implosion symmetry and stability use would be made of short pulse irradiation of the backlighter, forming an x-ray flash of sufficiently short duration to avoid temporal smearing. The choice of techniques will depend on the emphasis of a given experiment. Figure 28 indicates how x-ray probing is used to study dynamics in the streaked mode. An x-ray image is placed on the slit of an x-ray streak camera. The space-time streak record of target opacity then permits a study of interface dynamics in a well designed experiment. This is a more difficult experiment than that for the self-emitting exploding pusher. Nonetheless, it is quite possible to accomplish. We will discuss a number of techniques in this section which could presently be brought to bear on this subject. It should be clear, however, that these techniques require a considerable investment of resource if useful results are to be obtained. For one, additional laser drive and synchronous techniques are required for the auxiliary source. Because high x-ray energies are required, and because x-ray flux levels are costly in terms of laser drive new high energy, large solid angle x-ray optics are required with several micron resolution. These stringent technical requirements, coupled with the knowledge that present target performances are dominated by deeply penetrating electrons in non-ablative fashion as discussed in Section I, have limited the pursuit of serious backlighting experiments to date. However, the role of ablative dynamics being an important one, we have continued to pursue the subject of x-ray probing in a generalized fashion which we hope to adapt to a variety of experimental situations when the use of shorter wavelengths and lower laser intensities combine to produce more interesting dynamical situations.

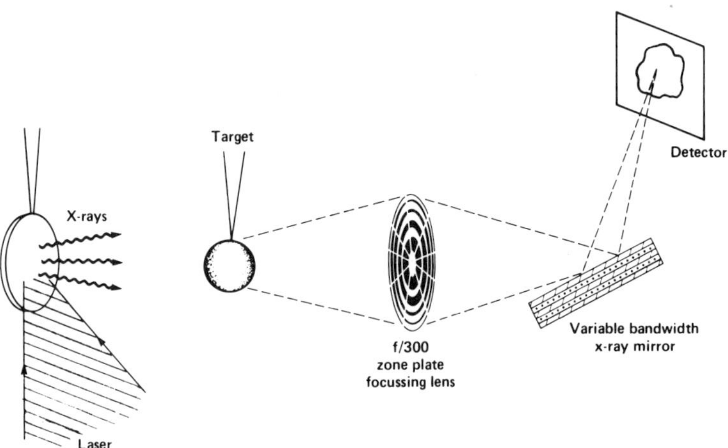

Fig. 29. The development of a flexible x-ray probing capability requires the production and characterization of x-ray line sources, with energy appropriate to targets of interest, as well as the development of x-ray imaging, reflection and detection capabilities.

Figure 29 shows a schematic outline of our efforts to develop a flexible x-ray probing capability. We have begun a program to engineer and characterize strong x-ray line sources by studying the spectra of K, L, and M shell radiations from various (target) elements in the kilovolt spectral regions of interest. K-shell transitions are particularly interesting because their narrow spectral width provides potential leverage for discriminating against broadband emissions from an imploding target driven by far greater laser energies. Interpretation of broadband opacity data is also avoided. An example of this work is shown in Fig. 30, where a titanium disk driven efficiently to a helium-like state radiates at a power level of 3×10^9 watts in the He α line at 4.7 KeV [41]. This corresponds to 1/2000 conversion from incident light at 1.06 μm to this single x-ray line (including its weaker lithium-like satellites). Strong line emission has obtained, at high laser intensities, with helium-like nickel and zinc lines at 7.8 and 9.0 KeV, respectively. X-ray streak records with broadband filters show that the temporal behavior of these lines to first order follows that of the 600 psec FWHM driving laser pulse [42]. Further studies are planned to clarify issues of intensity

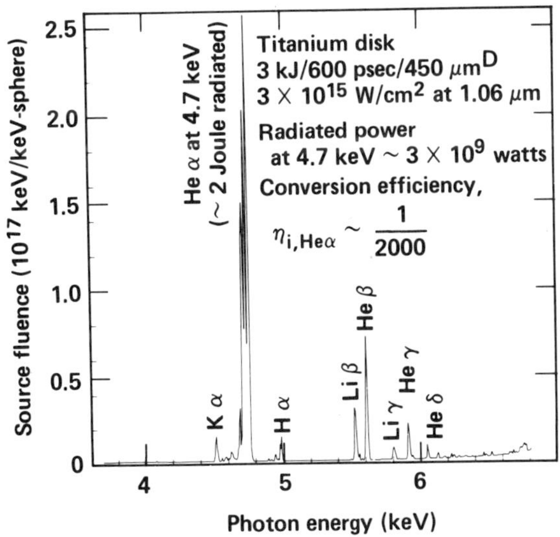

Fig. 30. X-ray emission from a laser heated titanium disk provides a relatively efficient source for probing studies. Here the 4.7 keV helium like line of titanium is shown to have an efficiency relative to incident laser light of 1/2000 for the conditions cited. An ionizational bottleneck at the helium-like level, for keV thermal temperatures, explains the unusually high conversion into line radiation.

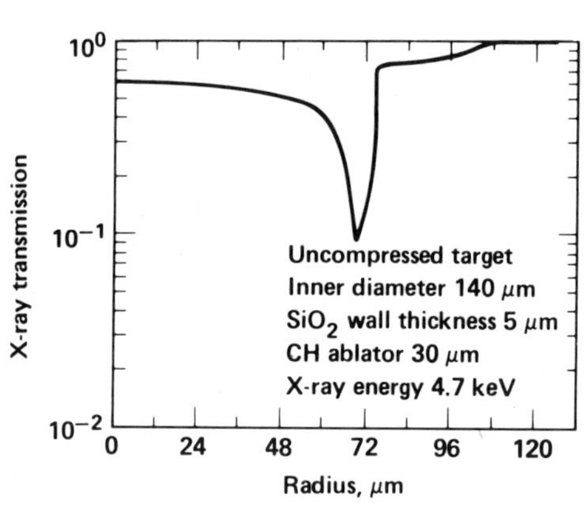
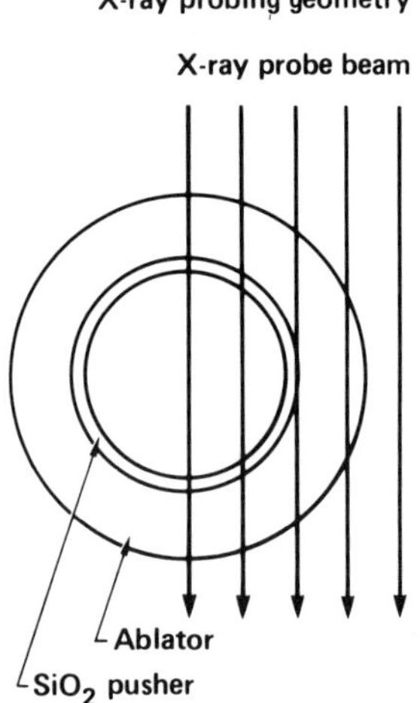

Fig. 31. X-ray transmission is shown for a typical "10X" target ball probed in its uncompressed state by 4.7 keV x-rays. During compression the fuel-pusher interface can be followed as a diagnostic of implosion dynamics. Final transmission profiles depend strongly on the driving electron distribution.

scaling for various energy lines, and to study temporal duration for short and long pulse irradiations.

Following the development suggested by Fig. 29, we discuss briefly the transmission profile of a target which might dictate the choice of a 4.7 KeV line source. Figure 31 shows the 4.7 KeV transmission profile, or radial opacity curve, for a target which initially consists of a 140 μm diameter, 5 μm thick glass mandril with a 30 μm thick CH coating. The glass-plastic interface is well identified, producing a transmission dip of a factor of ten, which should be easily identified in experiments. The glass-fuel interface is described by the gentle roll-off in the central region. Numerical simulations of this target (irradiation with laser conditions similar to those in the "10X" targets of Fig. 17 and 19, and achieving a final glass ρR of 6 mg/cm^2) show it to display a fairly well defined interface up to the time of peak compression, with x-ray transmission reduced at target center to about 2% of peak value. This factor of 50 is a convenient value for dynamic range in our instruments, and covers the entire implosion process fairly well. It is worth commenting, however, that final interfaces are not likely to

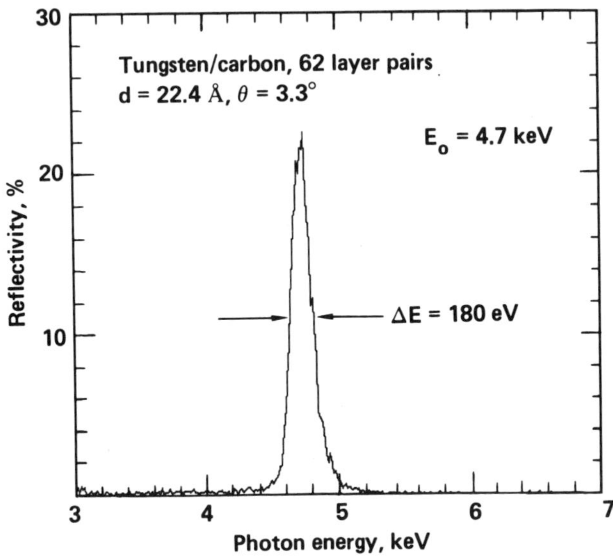

Fig. 32. Reflection properties of a synthetic x-ray mirror, produced by T. Barbee of Stanford [43, 44], are demonstrated for 4.7
keV x-rays. These multilayer interference mirrors are particularly
interesting because narrow bandwidth and high peak reflectance are
both achievable, depending on the number of layered pairs and materials chosen.

be sharply defined, unless pre-heat induced pusher expansion is significantly less than occurs during present implosion experiments.
Since x-ray probing is a line integral technique, spatially localized
perturbations will tend to be minimized in resultant images. The
combination of the above two effects, soft gradients and line integral smoothing, will limit to some extent, the impact of x-ray probing in studies of small scale pusher breakup. Nonetheless, the technique provides an important tool for studying pusher dynamics and
some larger scale aspects of compression symmetry in relatively cold
implosions.

In addition to strong x-ray sources, the development of a flexible x-ray probing capability requires the availability of x-ray

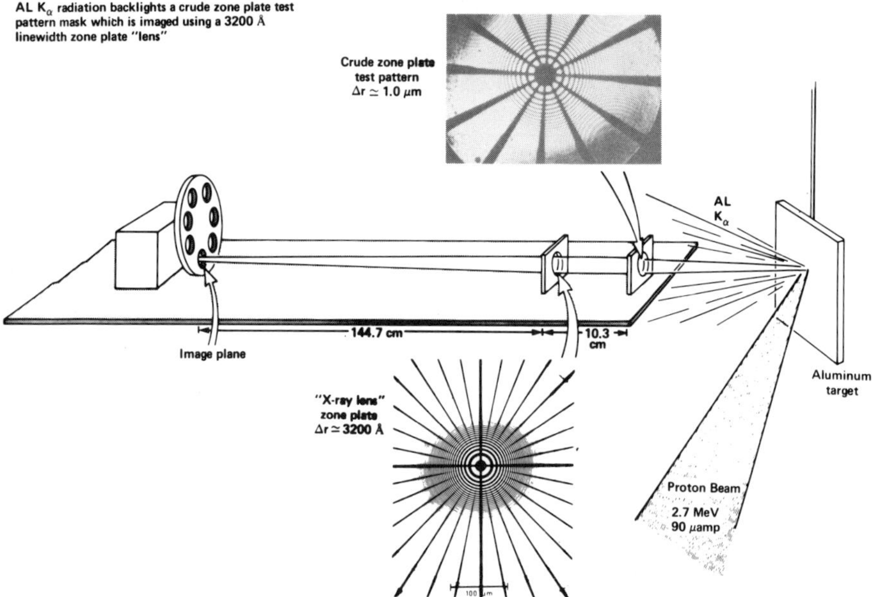

Fig. 33. Testing of a new 3200 Å outer zone x-ray lens [47] is accomplished with a proton pumped Al K_α source. A relatively crude 1.0 μm outer zone test pattern is used as a test pattern in this X14 magnification x-ray microscope. In these applications, the finer zone pattern makes full use of x-ray diffraction in producing images as opposed to the coded aperture case where undiffracted shadow casting is employed.

Microdensitometer analysis of x-ray image:

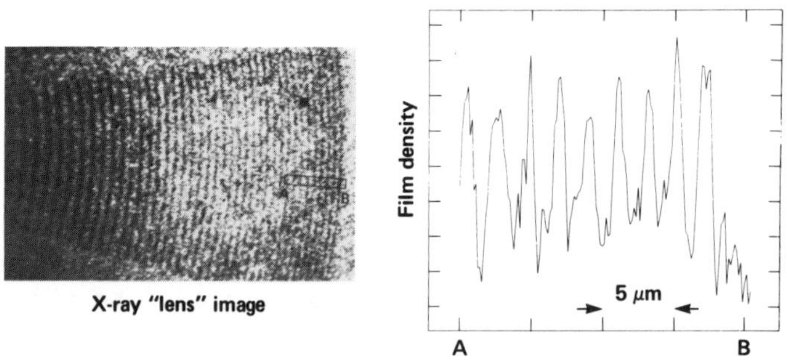

Conclude:

● 1.0 μm lines and spaces are clearly resolved

Fig. 34. An x-ray image demonstrating the sub-micron capability of the microscope shown in Fig. 33. The alternate 1.0 micron lines and spaces are clearly resolved [47].

lenses and mirrors. Shown in Fig. 31 is a multi-layer x-ray interference mirror produced by Barbee at Stanford University using well
controlled sputtering techniques [43]. The data in Fig. 32 shows
the mirror response to a broad band x-ray source at an angle of $3.3°$.
A peak reflectivity of 22% is obtained at 4.7 keV, with a measured
acceptance width of 180 eV. Accounting for detector response in
these measurements, the actual mirror acceptance is 130 eV, giving a
true quality factor $E_o/\Delta E \simeq 36$. A significant advantage of these
mirrors is their utility at both high and low x-ray energy, and the
ability to trade off peak reflectance and spectral acceptance width
by varying the number of dielectric layers. For instance, the layered mirror shown in Fig. 32 displays a peak reflectance of 40% at 9.6
KeV [44], falling to about 2% at 500 eV [45]. Better performance is
obtained near the carbon k-edge (283 eV) by use of magnesium in place
of carbon as the alternate low electron density material.

    A collaborative effort between LLL and MIT's Lincoln Laboratory
for the development of sub-micron resolution zone plate lensing elements is also underway [46]. Free standing micro-Fresnel zone plates
of diameter 0.63 mm, minimum linewidth 3200 Å, and made of 1.3 μm
thick gold have been fabricated by Shaver et al. [47] using scanning
electron beam lithography for pattern generation, and x-ray lithography for high aspect ratio pattern replication. A test experiment
to illustrate the resolution capability of these x-ray lensing elements is shown in Fig. 33 where the 3200 Å zone plate is used in a
14X microscope application. In this setup the zone plate lens (3200
Å minimum linewidth) is used to image a "crude" Fresnel zone plate
(1 μm minimum linewidth) used as a test pattern or resolution chart.
The test pattern is backlit with Al K_α radiation at 1.49 KeV produced
by proton bombardment of a water cooled Aluminum target. The x-ray
image thus produced is shown in Fig. 34, where the alternate 1 μm
wide lines and spaces are observed to be well resolved.

    In the introductory comments of this section we directed our
attention to the use of synchronous laser sources to drive target implosions and x-ray probing sources, with appropriate delay and jitter.
To produce the necessary assortment of short and long synchronous
laser pulses, the laser group at Livermore [48] has developed a very
flexible capability using actively mode locked oscillator concepts
and a synchronously driven regenerative amplifier, as shown in Fig.
35. This system is capable of providing well synchronized pulses
from 15 psec to 1 nsec, as indicated, assuring us of the necessary
technology to drive a wide variety of target probing experiments.

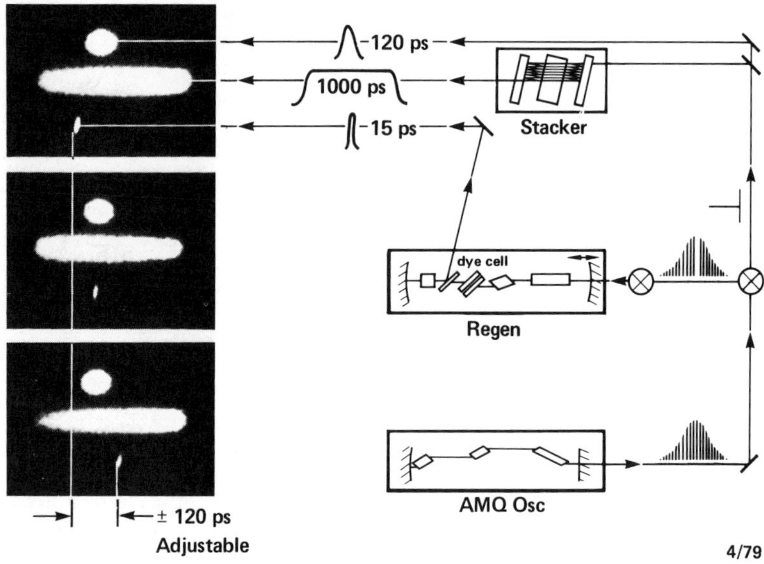

Fig. 35. A flexible array of synchronous short and long laser pulses
is achieved utilizing active mode locking techniques, as described by
Murray and Kuizenga [48] of Livermore. This capability will support
x-ray probing experiments in both short pulse "flash radiography" of
the final compression state, as well as long pulse probing of ablative implosion dynamics.

CONCLUSIONS

Within the confines of preliminary comments on important current
issues in laser driven implosion experiments, we have outlined diagnostic capabilities existing within the fusion community. In particular we have emphasized the transition from thin walled exploding
pusher targets to thicker wall targets which achieve higher compressed densities at the cost of reduced temperatures. These experiments
move us in a proper direction toward desirable reactor conditions,
but pose diagnostic problems because of reduced multi-kilovolt x-ray
and reaction products emissions, as well as increasingly more difficult transport problems for these emissions as they pass through the
thicker ρR pusher conditions.

We have identified solutions to these problems, pointing the way
toward higher energy two-dimensional x-ray images, new reaction product imaging ideas and the use of seed gases for both x-ray spectro-

scopic and nuclear activation techniques. We have also espoused the
attitude that not only are several independent measurement techniques
required to characterize the final implosion state, but that to do
so accurately and consistently it is important to recognize the sym-
bionic relationship between codes and experiments. With sharp gradients of density and temperature, varying significantly on the scale
of microns and picoseconds, it is extremely difficult to assess final
fuel conditions without recourse to numerical simulations. Likewise,
it is difficult to accurately model these complex implosion processes
without the variety of constraints provided by measurements of differing physical manifestations, i.e., two-dimensional x-ray and reaction
product images, stark broadened seed lines, nuclear activated elements, etc.

The development of a relationship such as this reached a rather
mature level several years ago for exploding pusher targets, but is
just beginning to develop with thick wall, high density implosions.
Hopefully this relation will evolve rapidly with the development and
refinement of new measurement techniques, as outlined in this paper.
The role of more flexible target fabrication capabilities will aid
in this development through the availability of lower Z ablators,
thinner mandrils and further options regarding seed gases. Lastly,
the development of truly ablative implosions through the availability
of short wavelength laser drivers cannot be overemphasized at this
critical junction in the development of high density, high ρR laser
driver implosions.

ACKNOWLEDGMENTS

This paper is an outgrowth of a series of recent review presentations by the authors of compression related measurements [49]. Nonetheless it is based to a great extent on a body of knowledge developed by and with their colleagues at the Lawrence Livermore Laboratory. The understanding of target physics and dynamics is a direct
outgrowth of implosion experiments designed, performed, and analyzed
by groups under the direction of John Nuckolls, Hal Ahlstrom, Bill
Kruer, Lamar Coleman, Erik Storm, Bill Slivinsky, Ken Manes, and
Chuck Hendricks.

Livermore colleagues deserving special acknowledgment for contributions to recent implosion experiments are Jerry Auerbach, Bill
Hatcher, John Hunt, Ralph Speck, Charles Swift, and the Shiva Operations Crew. Significant contributions to cited diagnostics were made
by Lou Koppel, Dick Lerche, Chuck Bennett, Jack Dellis, Dino Ciarlo,
Gary Howe, Claude Dittmore, Gary Stone, Ed Hee, Bill Herrmann, and
the LWPT reactor crew at Livermore. We also want to thank Barukh
Yaakobi of the University of Rochester for providing unpublished
data.

REFERENCES

1. J. Nuckolls, L. Wood, A. Thiessen and G. Zimmerman, Nature 239, 139 (1972).
2. J. Nuckolls, J. Emmett and L. Wood, Phys. Today, 26, 46 (August 1973).
3. J. Emmett, J. Nuckolls and L. Wood, Sci. Amer. 230, 24 (June 1974).
4. J. H. Nuckolls, "Exploding Pusher Targets", Laser Program Annual Report - 1976, Lawrence Livermore Laboratory, Rept. 50021-76 (1977), pp. 4-10 to 4-22, and references therein.
5. E. K. Storm, et al., "Laser Fusion Experiments at 4 TW", Phys. Rev. Lett. 40, 1570 (1978).
6. J. T. Larsen, "Exploding Pusher Targets", Laser Program Annual Report-1978, Lawrence Livermore Laboratory, UCRL-50021-78 (1978), pp. 3-2 to 3-10.
7 V. W. Slivinsky, H. G. Ahlstrom, K. G. Tirsell, J. Larsen, S. Glaros, G. Zimmerman, and H. Shay, "Measurement of the Ion Temperature in Laser Driven Fusion", Phys. Rev. Lett. 35, 1083 (1975).
8. R. A. Lerche, L. W. Coleman, J. W. Houghton, D. R. Speck and E. K. Storm, "Laser Fusion Ion Temperatures Determined by Neutron Time of Flight Techniques", Appl. Phys. Lett. 31, 645 (1977).
9. H. G. Ahlstrom, "Progress of Laser Fusion At Lawrence Livermore Laboratory", UCRL-82835 (1979) to be pushlished in J. Phys. (Orsay).
10. V. W. Slivinsky, H. N. Kornblum, and H. D. Shay, J. Appl. Phys. 46, 1973 (1975); updated for a variety of target conditions by V. W. Slivinsky in unpublished presentations and private communications.
11. K. G. Estabrook, E. J. Valeo, and W. L. Kruer, Phys. Fluids 18, 1151 (1975).
12. D. W. Forslund, J. M. Kindel, and K. Lee, Phys., Rev. Lett. 39, 284 (1977).
13. F. Amiranoff, R. Benattar, R. Fabbro, E. Fabre, C. Garban, C. Popovics, J. Virmont and M. Weinfeld, Bull. Amer. Phys. Soc. 24, 1054 (October 1979).
14. J. D. Hares, J. D. Kilkenny, M. H. Key and J. G. Lunney, Phys. Rev. Lett. 42, 1216 (1979).
15. F. Seward, J. Dent, M. Boyle, L. Koppel, T. Harper, P. Stoering, and A. Toor, Rev. Sci. Instrum. 47, 464 (1976).
16. P. M. Campbell, G. Charatis and G. R. Montry, Phys. Rev. Lett 34, 74 (1975).
17. D. T. Attwood, L. W. Coleman, J. T. Larsen, and E. K. Storm, Phys. Rev. Lett. 37, 499 (1976).
18. D. T. Attwood, L. W. Coleman, J. E. Swain, D. W. Phillion, K. R. Manes, D. S. Bailey, and Y. L. Pan, "The Correlation of X-ray Temporal Signatures, Neutron Yields, and Laser Performance," 10th European Conf. Laser Interaction With Matter, Palaiseau, France, Oct. 1976, UCRL-78739.

19. D. T. Attwood, B. W. Weinstein, and R. F. Wuerker, Appl. Optics 16, 1253 (1977).

20. D. T. Attwood, L. W. Coleman, M. J. Boyle, J. T. Larsen, D. W. Phillion, and K. R. Manes, Phys. Rev. Lett. 38, 282 (1977).

21. D. T. Attwood, IEEE J. Quant. Electr. QE-14, 909 (1978), and particularly references therein.

22. B. Yaakobi, D. Steel, E. Thoros, A. Hauer and B. Perry, Phys. Rev. Lett. 39, 1526 (1977).

23. J. M. Auerbach, D. S. Bailey, S. S. Glaros, L. N. Koppel, Y. L. Pan, L. M. Richards, V. W. Slivinsky and J. J. Thomson, J. Appl. Phys. 50, 5478 (1979).

24. K. B. Mitchell, D. B. Van Hulsteyn, G. H. McCall and P. Lee, Phys. Rev. Lett. 42, 232 (1979).

25. L. N. Koppel, D. L. Matthews, J. T. Larsen, S. M. Lane and N. M. Ceglio, "Diagnosis of Laser Produced Implosions Using Argon X-Ray Lines", unpublished, June 1979.

26. N. M. Ceglio and H. I. Smith "Micro-Fresnel Zone Plates for Coded Imaging Applications," Rev. Sci. Instrum., vol. 49, pp. 15-20, Jan. 1978; N. M. Ceglio, "Zone Plate Coded Imaging on a Microscopic Scale," J. Appl. Phys., vol. 48, pp. 1563-1565, Apr. 1977; N. M. Ceglio, D. T. Attwood, and E. V. George, "Zone Plate Coded Imaging of Laser Produced Plasmas," J. Appl. Phys., vol. 48, pp. 1566-1569, Apr. 1977; N. M. Ceglio and L. W. Coleman, "Spatially resolved emission from laser fusion targets," Phys. Rev. Lett. vol. 39, pp 20-24, July 4, 1977.

27. N. M. Ceglio, Energy and Technology Review, Lawrence Livermore Laboratory, pp. 1-8, UCRL-52000-78-1, January 1978.

28. Two keV x-rays present somewhat of a special case in glass of 500 eV temperature. In this case the outer shell electrons, including the L-shell, are removed leaving essentially helium-like silicon atoms with a K-edge elevated to a value somewhat in excess of 2 keV. As a result x-rays to 2 keV experience significantly reduced absorption in heated glass. However, the transport of these x-rays just below the K-edge is very temperature sensitive and thus not recommended for unambiguous diagnostic purposes. A better solution is to push towards the use of higher energy x-rays as suggested by Fig. 14A. Note that the use of plastic mandrils with beryillium ablators would significantly improve the transmission of x-rays for diagnostic purposes. On the other hand, these target changes would have little impact on our comments regarding particle transport, as energy losses there are dominated by the free electron densities which will be approximately the same for targets designed for a particular electron pre-heat environment.

29. W. C. Mead and Y. L. Pan, "Intermediate Density Targets: Design Constraints and Physics," Laser Program Annual Report - 1976, pp. 4-25 to 4-29; UCRL-50021 (1977).

30. N. M. Ceglio and J. T. Larsen, "Spatially Resolved Suprathermal X-Ray Emission From Laser Fusion Targets", submitted for publication, October 1979.

31. This wide dynamic range recording technique was developed by
    C. Dittmore of Livermore's Technical Photography Group.
32. B. Yaakobi, unpublished data.
33. R. J. Tighe and C. F. Hooper, Phys. Rev. A $\underline{17}$, 410 (1978).
34. H. R. Griem Phys. Rev. A. $\underline{17}$, 214 (1978); P. C. Kepple and H. R.
    Griem NRL Memorandum Report 3634 Sept. 1978.
35. H. G. Ahlstrom, L. W. Coleman, F. Rienecker, and V. W.
    Slivinsky, "Diagnostics of Shiva Nova High Yield Thermonuclear
    Events," J. Opt. Soc. Am. $\underline{68}$, 1731 (1979).
36. E. M. Campbell, S. M. Lane, Y. L. Pan, J. T. Larsen, R. J. Wahl
    and R. H. Price, "Determination of Fuel ρR of ICF Targets By
    Neutron Activation", submitted for publication, October (1979).
37. F. J. Mayer and W. J. Rensel, J. Appl. Phys. $\underline{47}$, 1491 (1976).
38. E. M. Campbell, et al, "Collection Fraction Determination Util-
    izing A Radioactive Tracer", submitted for publication, July
    (1979).
39. M. A. Yates, A. H. Williams, and J. T. Ganley "Radiochemical
    Diagnostics Development at Helios, Los Alamos Scientific
    Laboratory, private communication.
40. R. A. Lerche and J. T. Ozawa, "Neutron and X-Ray Emission Time
    Measurements", Bull. Amer. Phys. Soc. $\underline{24}$, 1099 (1979).
41. L. Koppel, unpublished data.
42. R. Kauffman, unpublished data.
43. T. W. Barbee and D. C. Keith, Stanford University, Center for
    Materials Research, unpublished.
44. L. N. Koppel and T. W. Barbee, Lawrence Livermore Laboratory and
    Stanford University, unpublished.
45. B. Henke, University of Hawaii, unpublished data.
46. N. M. Ceglio and H. I. Smith, "An Efficient Lensing Element for
    X-Rays", VIII Int'l Conf. X-Ray Optics and Microanalysis, Boston,
    August 1977.
47. D. C. Shaver, D. C. Flanders, N. M. Ceglio and H. I. Smith,
    "X-Ray Zone Plates Fabricated Using Electron-Beam and X-Ray Lith-
    ography", Lincoln Laboratory, MIT, unpublished, May 1979.
48. J. E. Murray and D. J. Kuizenga, "Regenerative Pulse Compression"
    Lawrence Livermore Laboratory, UCRL-83745 submitted for publica-
    tion, December 1979.
49. D. T. Attwood, Bull. APS $\underline{23}$, 814 (1978); N. M. Ceglio, Science
    $\underline{203}$, 1155 (1979); E. M. Campbell, Bull. APS. $\underline{24}$, 1031 (1979).

A REVIEW OF HIGH DENSITY, LASER DRIVEN, IMPLOSION

EXPERIMENTS AT THE LABORATORY FOR LASER ENERGETICS

J.M.Soures, T.C.Bristow, H.Deckman*, J.Delettrez,
A.Entenberg, W.Friedman, J.Forsyth, Y.Gazit**,
G.Halpern*, F..Kalk, S.Letzring, R.McCrory, D.Peiffer*,
J.Rizzo, W.Seka, S.Skupsky, E.Thorsos, B.Yaakobi,
T.Yamanaka***

Laboratory for Laser Energetics
University of Rochester
250 East River Road
Rochester, New York 14623

ABSTRACT

The six beam, 3×10^{12} Watt, ZETA laser facility is the
initial stage of the twenty four beam, 1.2×10^{13} Watt, National
Laser User's Facility, OMEGA. In this paper we review the results
of the first series of symmetrical illumination implosion experiments conducted on this facility. The program for these initial
experiments began with the implosion of thin-walled, DT filled
glass microballoons. Neutron yields in excess of 10^9 were generated with a laser pulse power less than 2×10^{12} Watts. In a
progression of experiments with increasingly thicker pushers,
higher density, lower temperature implosions were generated. In
some of these experiments DT densities in excess of 2 g/cm^3 were
achieved. Argon filled targets were also imploded, achieving
densities as high as 7 g/cm^3. Density measurements were made by
means of Stark broadening of x-ray lines emitted by high Z seed
gases in the fuel and x-ray imaging of the target using x-ray
pinhole cameras.

INTRODUCTION

The purpose of this paper is to review some of the laser
fusion experimental work that has been recently conducted at the
Laboratory for Laser Energetics of the University of Rochester.

We start by relating this work to some of the critical issues in current laser fusion research.

The basic processes involved in generating a laser driven micro-implosion are summarized in Fig. 1.   The laser beam energy is deposited close to the critical density region of the plasma. The energy must then be transported into the target's ablation region and converted to hydrodynamic motion driving the pellet implosion.

If the resulting fuel compression is high enough, thermo-nuclear ignition conditions and a propagating burn will result. The generally accepted ignition conditions are $\rho R > .3 g/cm^2$ (where ρ is the fuel density and R is the imploded fuel radius) and fuel temperature $T_f \geq 4$ keV[1].

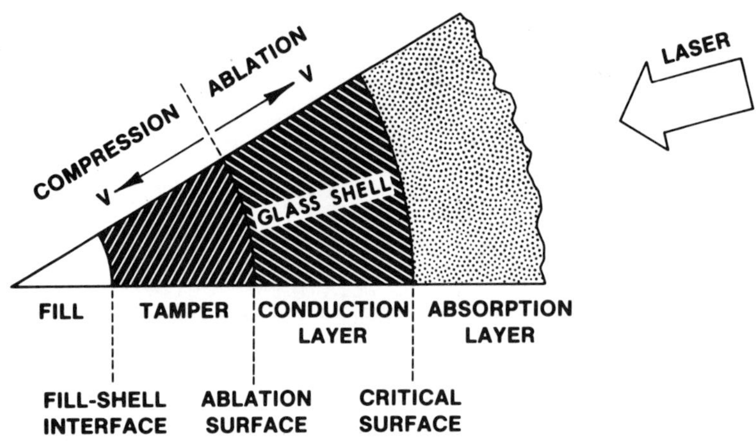

Fig. 1.   An illustration of the basic processes involved in generating a laser driven micro-implosion.

In the last few years, the most significant problems have arisen in the area of energy coupling. At a laser wavelength of 1 µm, stimulated Brillouin scattering has reduced the absorption for long pulse, high irradiance conditions ($\geq 10^{15}$ Watts/cm^2).[2,3] At these same flux levels, the production of suprathermal electrons has led to a considerable increase in the estimated energy requirements for targets driven by infrared lasers.[4] · For long laser wavelengths and high irradiance, the implosions are driven by high energy electrons. Preliminary data from shorter wavelength experiments is extremely promising.[5] However, until such time that a complete data base is developed for short wavelength irradiation experiments, the scientific feasibility of laser driven inertial confinement fusion will remain in doubt.

There are still other problem areas that need to be addressed independent of the energy coupling problem. These issues (which may be characterized as high energy density related problems) include hydro-stability, high density diagnostics, driver uniformity requirements for high compression implosions, an understanding of the ignition conditions pertinent to micro-core implosions and the understanding of the physics of matter under conditions never before attained in the laboratory.

At the Laboratory for Laser Energetics, we believe that one of our contributions is to address some of the high energy density issues with existing or soon to be available symmetric irradiation laser facilities.

In this paper some of the experimental work on high energy density implosions at the Laboratory for Laser Energetics will be described.

FACILITIES

The main laser facility at the University of Rochester's National Laser Users' Facility is OMEGA, a twenty-four beam Nd:glass laser with a peak power capability of .5TW/beam (12TW total at 50 psec (FWHM).[6] Fig. 2 is a schematic showing the placement of the beams and target area for this system.

Characteristics of this system include:

a. Extensive spatial filtering and imaging resulting in a beam capable of being focused to a 10 μm diameter spot.

b. Prepulse control resulting in a measured contrast $> 10^8$ at a time as close as 200 psec to the peak of the pulse.[7]

c. Symmetric placement of beams on target to maximize the irradiation uniformity.

d. A firing rate capability of one shot every 30 minutes.

The experiments to be reviewed here were conducted by a subset of this facility -- called ZETA -- a six beam symmetric irradiation facility making use of the first six beams of OMEGA.

The ZETA beams are directed into the target chamber shown schematically in Fig. 3. The beams are placed on target in a cubic geometry.

  J. M. SOURES ET AL.

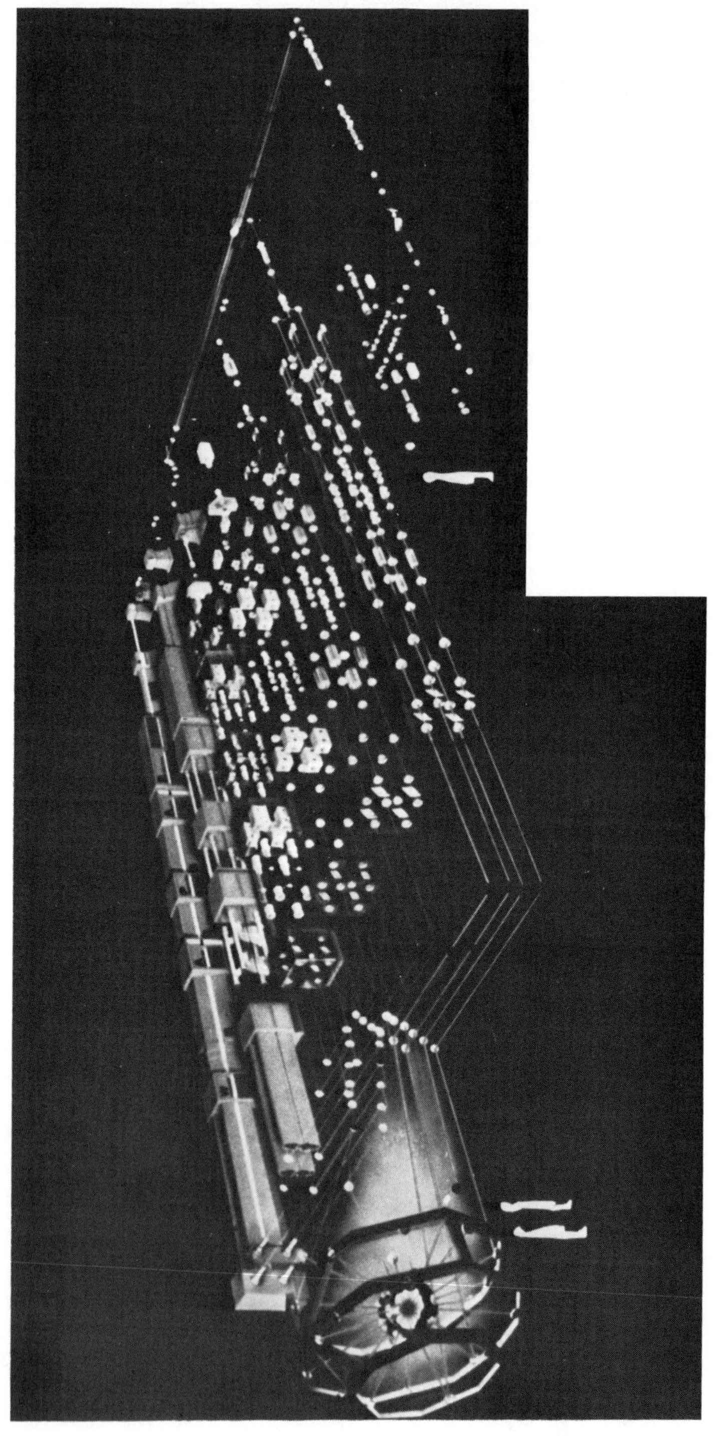

Fig. 2. The OMEGA laser facility consists of twenty-four beams arranged in six horizontally placed clusters of four beams. An open octagonal structure supports the final injection mirrors. Target illumination is symmetrical. For the six beam experiments a separate target chamber is used along with beams from the first two clusters shown. The total output of OMEGA is 12 TW in short pulses or 4.8 kJ in long pulses.

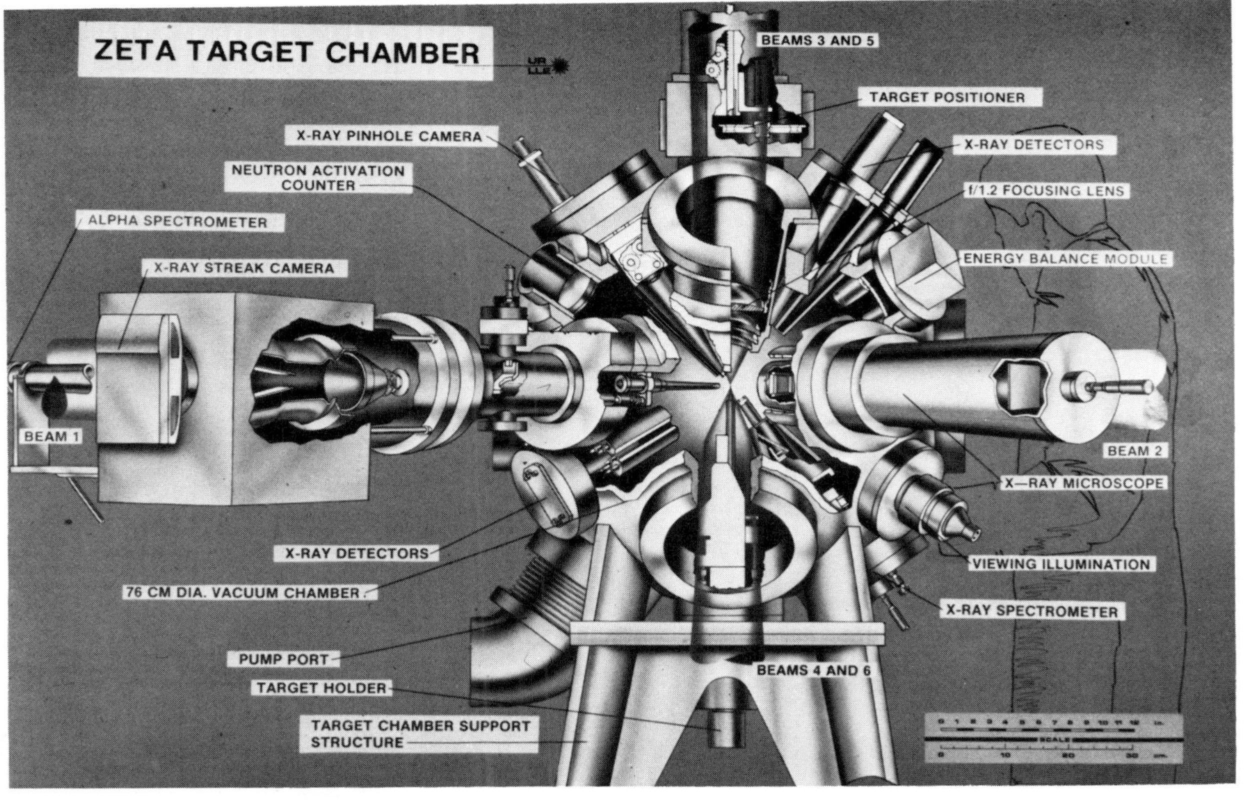

Fig. 3.   The six beam experiments are conducted in the ZETA target chamber. The general location of the beams and diagnostics are shown in this schematic. The focusing optics are f/1.2 four element lenses. The innermost element is a vacuum window. Focusing is controlled by moving the outermost element, which is in air.

Targets are stored in a carousel with a target capacity of 16 targets. These targets are automatically accessed and loaded into firing position. Diagnostics shown in Fig. 3 include:

a.   Plasma calorimeters.
b.   X-ray pinhole cameras.
c.   X-ray crystal spectrographs.
d.   X-ray, K-edge filter spectrometers.
e.   Alpha and proton time-of-flight, quadruple focusing, spectrometer.
f.   Copper activation and silver activation neutron counters.
g.   Neutron time-of-flight scintillator detectors.
h.   Alpha zone plate imaging camera.
i.   X-ray streak camera.

EXPERIMENTS

The target configurations used in the experiments to be described here are shown in Fig. 4. The target configurations included simple DT filled glass microballons, as well as plastic coated glass microballoons (GMB) filled with any of the following mixtures: D_2, DT, DT and Neon, Argon, Argon and Xenon. The highest fuel temperature (6 keV) was recorded for the exploding pusher targets (GMB filled with DT). The highest density was recorded with Argon filled targets (up to 7 g/cm^3).

In conducting these experiments one of the first questions to be answered was how well energy could be coupled into the target. In particular, the trade-off between uniformity of illumination and high absorption was investigated.

In a series of experiments (conducted on both uncoated and plastic coated glass microballoons), the absorbed energy was measured by means of plasma calorimeters for a variety of irradiation conditions (Fig. 5). The target irradiation pattern can be controlled by controlling the position of best focus of the laser beams with respect to the target position. Highest absorption ($\sim$ 50%) was measured when the six laser beams were each focused to the surface of the target. Surface focusing resulted in a beam spot $\sim$ 10 μm, with consequently very high intensities ($\sim 10^{18}$ Watts/cm^2). Highly asymmetric implosions were observed for tight focusing conditions. The best symmetry and optimum neutron yield were observed for an overlapping beam pattern attained by placing the beam focus at a position one target diameter behind the center of the target. Measurements of the supra-thermal x-ray spectrum accompanied these absorption measurements and are shown in Fig. 6. The peak hot temperature was found to coincide with surface focus positions - as might be expected. At the 1-2 TW (on target) power level, for 100 μm diameter targets and for overlapping illumination,

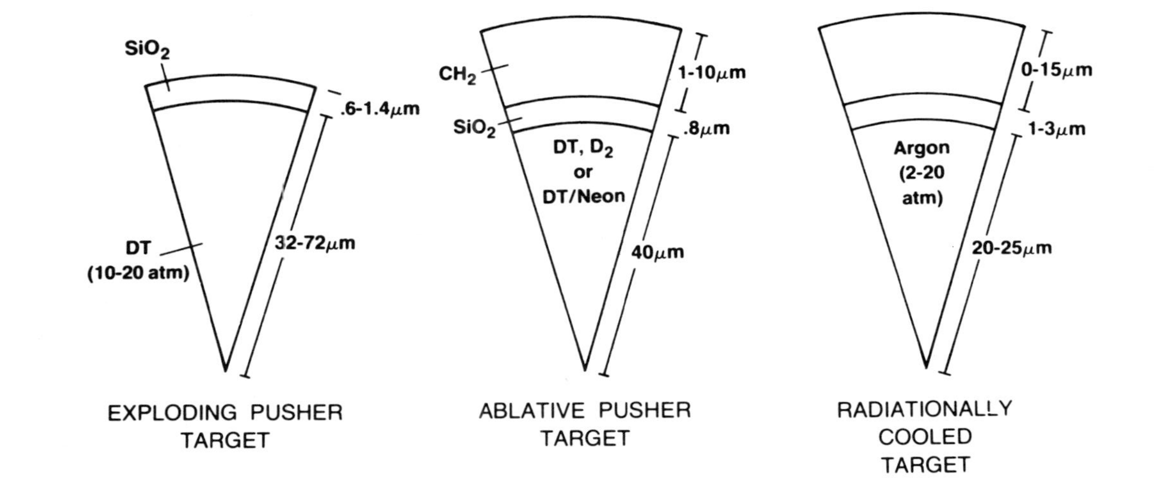

Fig. 4. Some of the target configurations used in the current series of implosion experiments at LLE are shown above. The Argon filled targets are created by laser drilling small ($\sim$ 1 μm) holes in the glass shell, filling to the required pressure and then sealing the hole with either a glass or plastic plug.

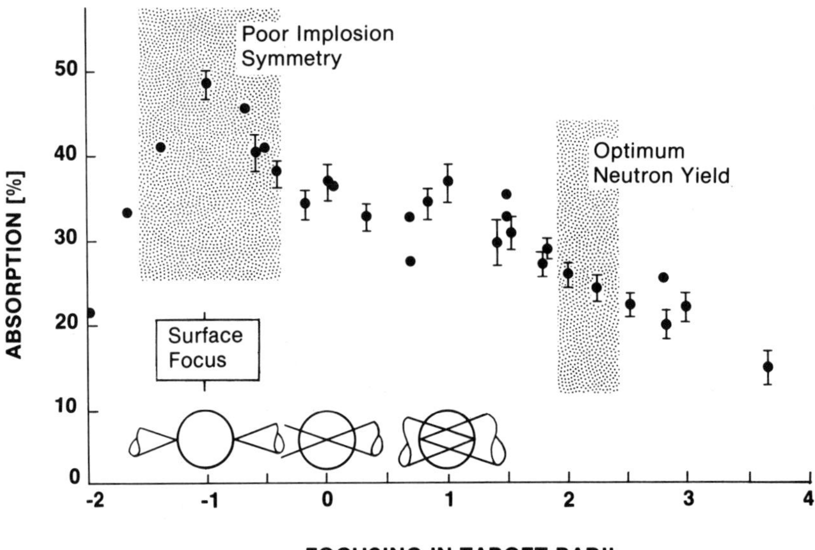

FOCUSING IN TARGET RADII

Fig. 5.   Absorption fraction of laser beam energy incident on
          spherical targets as a function of beam focus position
          for ZETA experiments. This data represents over 200
          shots taken with six beam irradiation of targets with
          diameters ranging from 50 to 100 μm and incident power
          ranging from 1.0 to 2.5 TW. The laser pulse width ranged
          from 50 to 130 psec (FWHM).

the suprathermal spectrum exhibited a hot temperature, $T_H \simeq 25$ keV.
This is consistent with previous measurements at an

$$I\lambda^2 \sim 10^{16} \text{ W/cm}^2\mu\text{m}^2 \qquad (8)$$

where I is the incident intensity and λ is the laser wavelength.

     Neutron yields from exploding pusher targets provide a good
test of the uniformity of the implosions and a useful calibration
of the irradiation facility. We have measured neutron yields up
to $(1.5 \pm .3) \times 10^9$ for laser powers of typically 2 TW.

     The scaling of neutron yields is best given in terms of the
specific absorbed energy, which is the absorbed energy divided by
the target mass. A simple scaling law which describes the data
is:

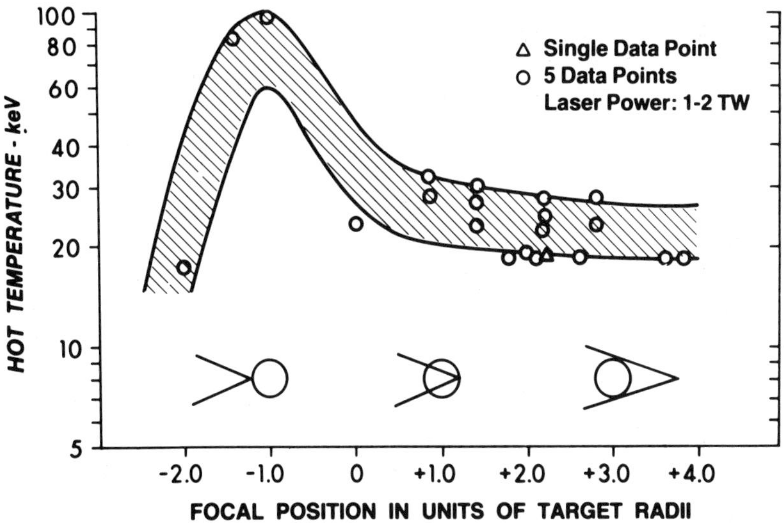

Fig.6.Characteristic suprathermal x-ray temperature plotted as a
       function of beam focus position for a number of six beam
       target shots. The general conditions were similar to those
       of Fig. 5.

$$Y \;=\; AR_0{}^4 \varepsilon^{-7/6} \exp\left(- \frac{B}{\varepsilon^{1/3}}\right)$$

where Y is the neutron yield, A and B are constants, R_0 is the
initial target radius, and ε is the specific absorbed energy.

This scaling law was used to examine the effect of beam
focusing on neutron yield as shown in Fig. 7. In this figure the
measured/calculated neutron yield is plotted versus the beam focal
position. The calculated yield is used to remove the dependence
on specific absorbed energy.

Best yield is obtained when the beam focus is placed two
radii behind the center of the target (2.0 R). For non-uniform
irradiation conditions, and especially in the region around surface
focus we observe a decrease in the neutron yield. This decrease
in yield is most likely due to reduced fuel trapping resulting
from a non-uniformly imploded pusher. For focusing beyond this
optimum yield region we also observe a decrease in the nuetron
yield. There is no current explanation of the later observation.

Evidence that substantiates the conclusions based on the

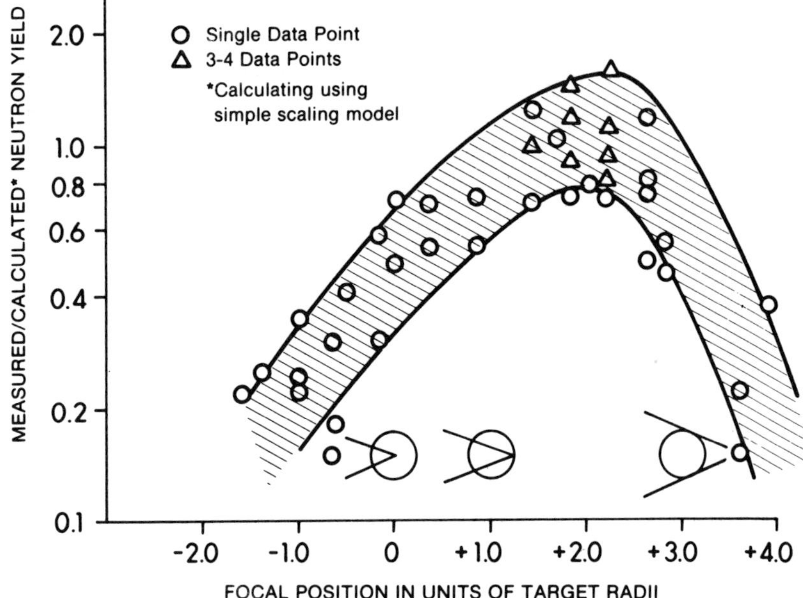

Fig. 7.   Comparison of measured and predicted neutron yield a
          function of beam focus position for six beam target
          experiments.   Conditions as in Fig. 5.

reduction in implosion symmetry, comes from a study of x-ray
images of the implosions.[9]   The implosion symmetry results were
obtained from spatially resolved x-ray images using pinhole
cameras.   The results show striking differences in the spatial
x-ray emission for different focusing conditions.   Fig. 8 shows
these results for three focusing conditions, center focus, 1.3
target radii behind center and 2.8 target radii behind center.
For center focus, Fig. 8 shows the fine structure of individual
beams in both the initial glass location (absorption region) and
the imploded glass.   We believe this observation results from each
beam imploding a section of the glass shell.   From this same
center focus result we observe that there is insufficient lateral
energy transport in the absorption region to symmetrize the shell
temperature.   As the foci of the beams are moved farther from
target center we find an improvement in the implosion symmetry.
For a focus at 2.8 target radii behind center the absorption
region and the imploded glass show no evidence of individual
beams.   Direct correlation of the symmetry as evidenced by x-ray
imaging and the core conditions has been accomplished by means of
alpha zone plate imaging of the burn region.

Shown in Fig. 9a and 9b are the x-ray pinhole image and alpha

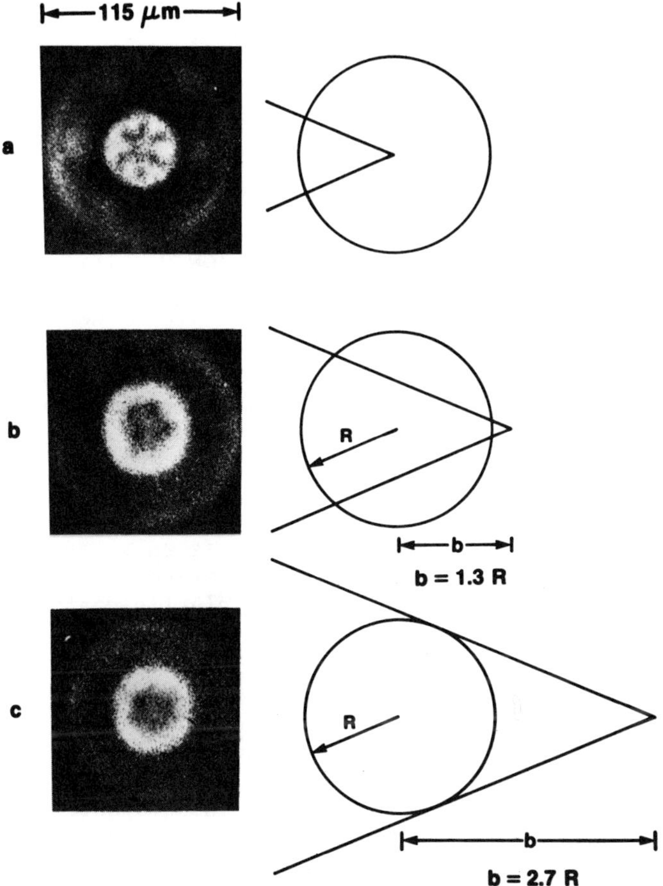

Fig. 8.   A sequence of x-ray images obtained with a pinhole camera
for six beam implosions. The laser beam focus for each
beam has been varied as indicated on the right. The
target diameters range from 115 to 120 μm. A 125 μm Be
filter has been employed in each case, yielding an
approximate x-ray window of 2-3 keV,
(a)   shot 1644, 2.0 TW, b = 0 (center focus),
(b)   shot 1690, 2.1 TW, b = 1.3R, and
(c)   shot 1787, 1.8 TW, b = 2.7R.

zone plate image, respectively, of a thin wall target irradiated with beams focused in the optimum 2R position. Notice that for this case, the α emitting region occupies the area enclosed by the ring of maximum xray emission from the core. The burn region is fairly symmetric and limited in size. If, however, non-overlapping beam focus is used, as shown in Fig. 10a, b, a considerably less symmetric implosion is evidence in both the x-ray and alpha images. The fuel region for this case is filamented and larger than the corresponding symmetric irradiation case (Fig. 9). The neutron yield is also a factor of two smaller for the non-symmetric case even though more energy was abosrbed by the target.

The exploding pusher targets discussed so far achieve a high temperature[10] (> 4 keV) but relatively low density; ρR is typically of order 10^{-4} g/cm^2. Even for these low compression experiments, however, the effects of non-uniform energy deposition can substantially reduce the target compression. Using the lessons learned from these low density experiments, a second series of experiments designed to achieve high compression ($\gtrsim 500$) were conducted.

In this second series of experiments, the density was increased by imploding targets with thicker pushers. An example of such an implosion is shown on Fig. 11b. Fig. 11a is an x-ray image from a typical exploding pusher target for comparison. Fig. 11b is the x-ray image of a target with a 7.4 μm of plastic coating. If we assume that the fuel is all within the x-ray emitting ring, we can infer a density of 2 g/cm^3 (10x liquid DT density). The fuel trapping assumption can be made if three conditions are met:

a. The focusing of the laser beams should be optimized for symmetric implosions;
b. The inner ring of emission from the stagnated pusher should be nearly spherically symmetric, as observed with more than one view of the implosion; and
c. The compression results should form a smooth and consistent trend to higher densities as the wall thickness is increased.

All of these conditions have been met in these experiments. Results for compressed DT density from x-ray imaging as a function of target wall thickness are shown in Fig. 12. The conditions for these experiments were, laser energy up to 140 J in 60 psec (FWHM) and a specific absorbed energy (total absorbed energy divided by target mass) of 8×10^7 J/g.

Note that for the range of pusher thickness covered, the density increases linearly with wall thickness. For the highest density measured (2 g/cm^3), the DT neutron yield was 2×10^7, the estimated $\rho R \sim 1.4 \times 10^{-3}$ g/cm^2 and the volume convergence is 470.

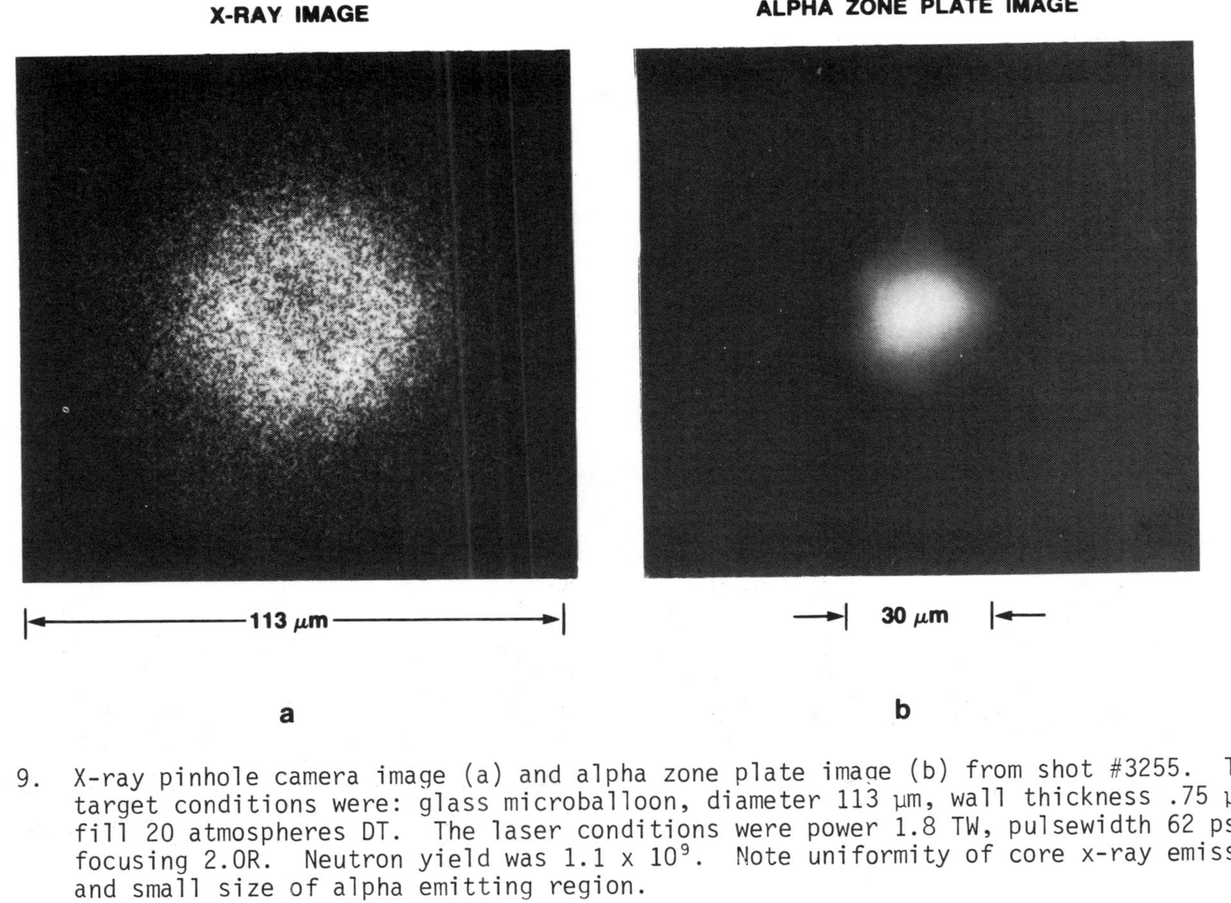

Fig. 9. X-ray pinhole camera image (a) and alpha zone plate image (b) from shot #3255. The target conditions were: glass microballoon, diameter 113 μm, wall thickness .75 μm, fill 20 atmospheres DT. The laser conditions were power 1.8 TW, pulsewidth 62 psec, focusing 2.0R. Neutron yield was 1.1×10^9. Note uniformity of core x-ray emission and small size of alpha emitting region.

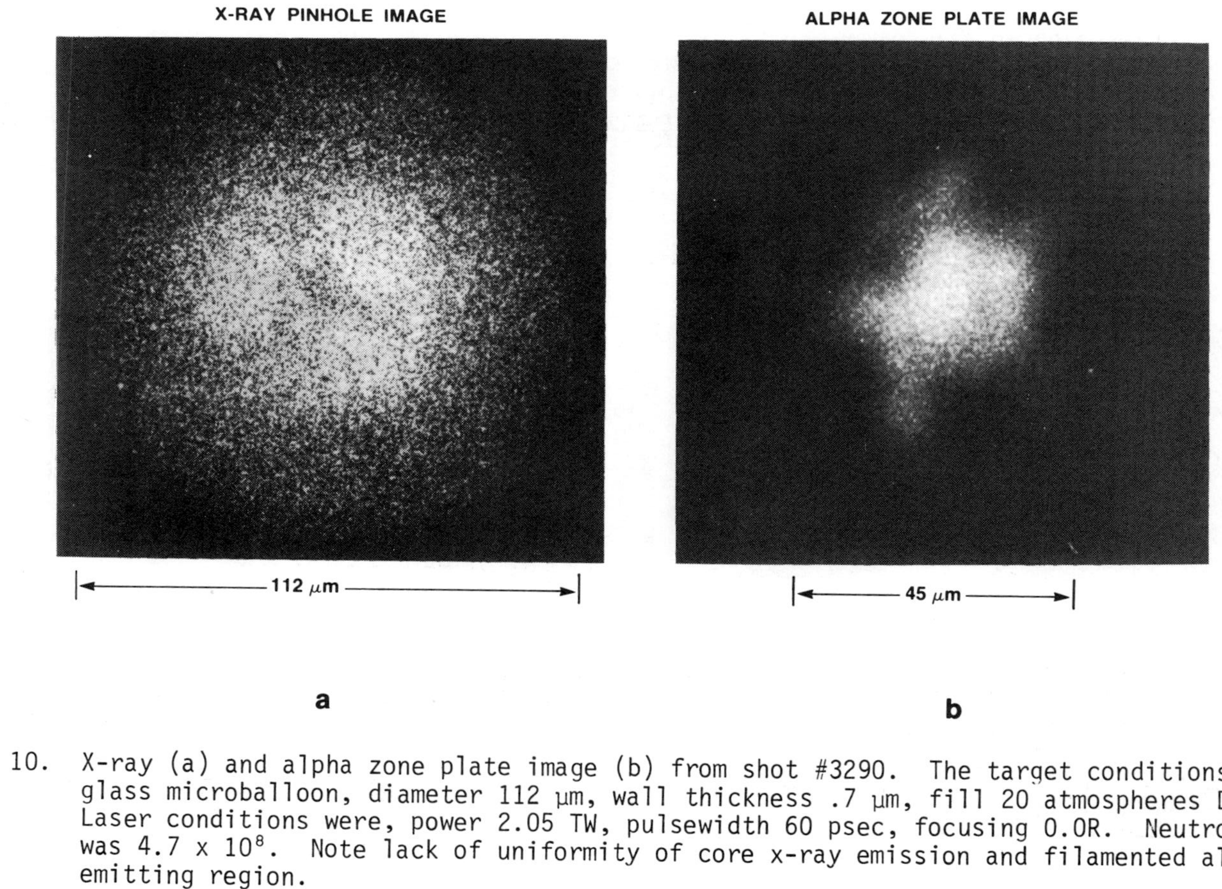

Fig. 10. X-ray (a) and alpha zone plate image (b) from shot #3290. The target conditions were: glass microballoon, diameter 112 μm, wall thickness .7 μm, fill 20 atmospheres DT. Laser conditions were, power 2.05 TW, pulsewidth 60 psec, focusing O.OR. Neutron yield was 4.7×10^8. Note lack of uniformity of core x-ray emission and filamented alpha emitting region.

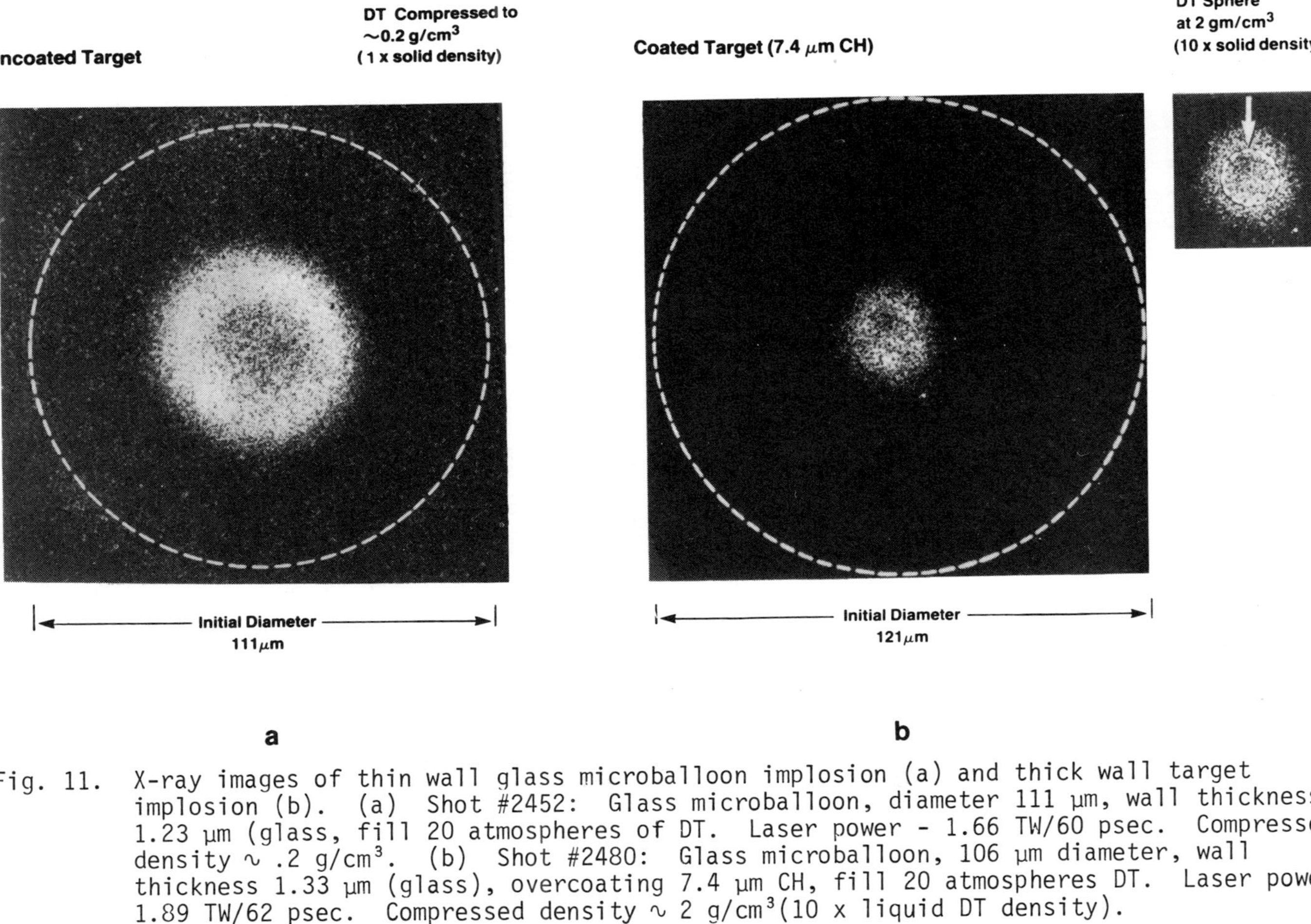

Fig. 11.   X-ray images of thin wall glass microballoon implosion (a) and thick wall target implosion (b).   (a) Shot #2452: Glass microballoon, diameter 111 μm, wall thickness 1.23 μm (glass, fill 20 atmospheres of DT. Laser power - 1.66 TW/60 psec. Compressed density ∿ .2 g/cm³.   (b) Shot #2480: Glass microballoon, 106 μm diameter, wall thickness 1.33 μm (glass), overcoating 7.4 μm CH, fill 20 atmospheres DT. Laser power 1.89 TW/62 psec. Compressed density ∿ 2 g/cm³(10 x liquid DT density).

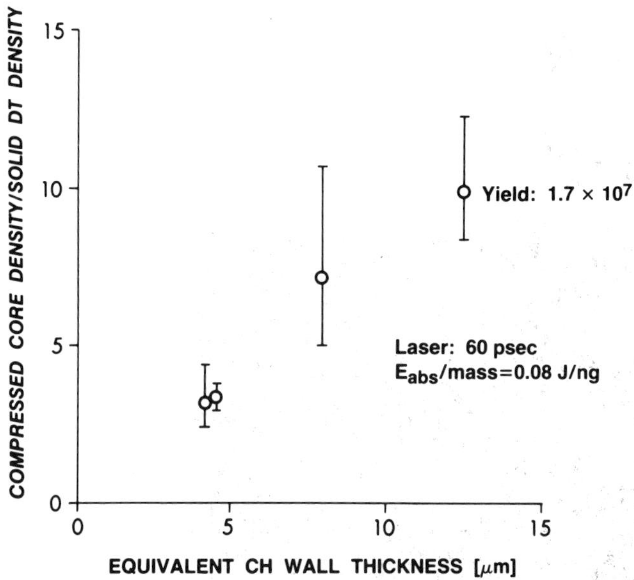

Fig. 12.   Measured compressed DT core density for thick pusher
           targets plotted as a function of equivalent CH wall
           thickness.

A second method of increasing the compressed density with
existing laser facilities is to use the concept of radiational
cooling. The idea behind radiational cooling is to reduce the
effect of the fuel preheat energy caused by the suprathermal
electrons and shock heating through radiation. This preheat energy
is dissipated by ionization and radiation of a high Z material
placed either in the fuel or preferably in the tamper. A test of
this concept has been carried out using Argon filled targets. An
advantage of using Argon in these initial experiments is that a
direct measurement of both the density and ρR may be made through
the x-ray spectrosopic techniques previously applied to Neon filled
targets.[12]

The Argon density is derived from the profiles of the 1s-3p
(Lyman-β) and the 1s-4p (Lyman-γ) transitions of Ar^{+17} as well as
the corresponding transitions in Ar^{+16}. The ρR product is derived
from the Lyman-α (1s-2p) profile.

The calculation of the Stark profiles used in the interpretation of these experiments is described in detail elsewhere.[13]
The method of fitting these profiles to the observed lines is

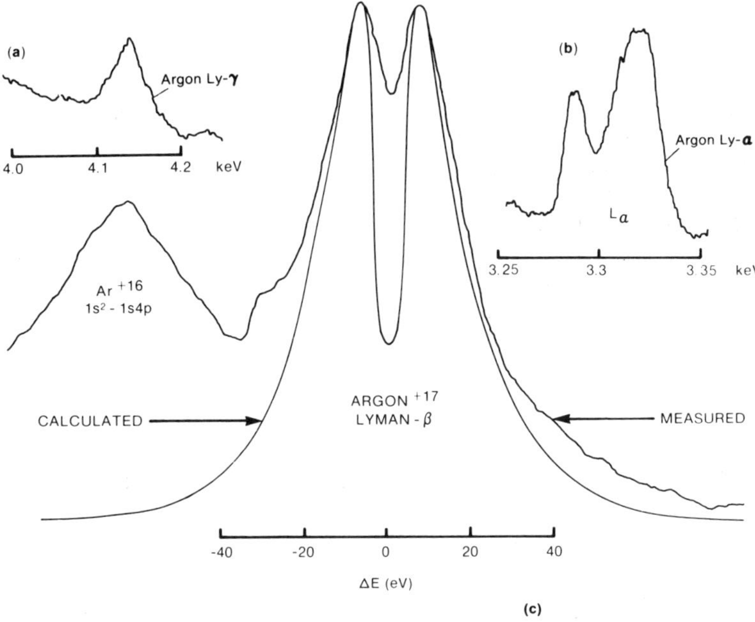

Fig. 13.   Argon lines from typical Argon filled target shot (the
           target is made of quartz so the impurities of K and Ca
           do not appear). The inserts show the profiles of the
           Lyman-α and Lyman-γ lines of Ar^{+17}. The density is
           measured by fitting the observed Lyman-β profile. The
           ρR is measured from the opacity broadening of the Lyman-α
           line.

similar to that used in neon-filled target experiments. The
various lines are predicted to have different widths and very
different shapes. A consistent agreement with this large number
of observables constitutes a highly reliable determination of the
compression.

     An examples of the results obtained is shown in Fig. 13.   The
conditions of the experiment illustrated by Fig. 13 were:

    a.   Laser Power - 1.8 TW
    b.   Target Diameter - 50 μm
    c.   Glass wall thickness - 1.9 μm
    d.   CH overcoating - 1.6 μm
    e.   Argon fill - 16 Atmospheres

Comparisons of the experimentally derived line profiles of Argon
Lyman-α and Lyman-β lines with the calculated profiles results in
a deduced Argon density of 6 ± 1.5 g/cm^3 and a

$$\rho R \sim (1.5 \pm 0.5) \times 10^{-3} \ g/cm^2 \quad .$$

Extensive simulations of radiationally cooled targets used in these experiments were done using the one-dimensional laser fusion computer code LILAC.[14] The simulations show Ar^{+17} lines to be emitted mostly during a $\sim$ 10 psec time interval when the average fill density is about 5-6 g/cm^3 and the temperature $\sim$ 0.8 keV.

In the near future these techniques will be extended to targets filled with Argon and DT mixtures, enabling simultaneous neutron yield and fuel density measurements.

SUMMARY

A series of symmetrical irradiation, laser driven implosion experiments have been initiated with the 3 TW, six beam Nd:Glass laser, ZETA. In the initial experiments, the effects of irradiation uniformity on exploding pusher targets were investigated. Under optimum irradiation conditions DT neutron yields of 1.5×10^9 were generated for as little as 1.7 TW incident on target. In experiments on targets with thick (10 mμ) pushers, DT densities of 2 g/cm^3 (10 x liquid density) were measured with a DT neutron yield of 2×10^7 with less than 2 TW on target. Finally, radiationally cooled, Argon filled targets were imploded with 2 TW pulses to densities as high as 6 g/cm^3 with volume compressions in excess of 500.

ACKNOWLEDGMENTS

This work was partially supported by the following sponsors: Exxon Research and Engineering Company, General Electric Company, Northeast Utilities, Empire State Electric Energy Research Corporation, New York State Energy Research and Development Administration.

*Permanent address: Exxon Research and Engineering Co., Linden NJ.

**Permanent address: Soreq Nuclear Research Center, Yavne 70600, Israel.

***Permanent address: Institute of Laser Engineering, Osaka University, Osaka, Japan.

REFERENCES

1. G. S. Fraley, E. J. Linnebur, R. J. Mason, R. L. Morse, Phys. Fluids 17:474 (1974).

2. B.H. Ripin, F. C. Young, J. A. Stamper, C. M. Armstrong, R. DeCosto, E. A. McLean, and S. E. Bodner, Phys. Rev. Lett. 39:611 (1977).

3. H. D. Shay, R. A. Haas, W. L. Kruer, M. J. Boyle, D. W. Phillion, V. C. Rupert, H. N. Kornblum, F. Rainer, V. W. Slivinsky, L. N. Koppel, L. Richards, and K. G. Tirsell, Phys. Fluids 21:1634 (1978).

4. R. E. Kidder, "Proceedings of Fifth Workshop on Laser Interaction, Laboratory for Laser Energetics, University of Rochester, November 5-9, 1979."

5. F. Amiranoff, R. Benattar, R. Fabbro, E. Fabre, C. Garban, C. Popovics, J. Virmont, M. Weinfeld, Bull. Amer. Phys. Soc. 24:1054 (1979).

6. J. M. Eastman, J. A. Boles, and J. M. Soures, Laser Focus 14: 48 (1978); J. Bunkenburg and W. Seka, paper 16.5 presented at 1979 CLEA IEEE/OSA Conference, Washington, D.C., 30 May - 1 June 1979 (unpublished).

7. G. Mourou, J. Bunkenburg, W. Seka (to be published).

8. K. Estabrook and W. L. Kruer, Phys. Rev. Lett. 40:42 (1978); D. W. Forslund, J. M. Kindel and K. Lee, Phys. Rev. Lett. 39:284 (1977).

9. E. I. Thorsos, T. C. Bristow, J. A. Delettrez, J. M. Soures, and J. E. Rizzo, Appl. Phys. Lett. 35:598 (1979).

10. Y. Gazit, J. Delettrez, T. C. Bristow, A. Entenberg, and J. Soures, Phys. Rev. Lett. 43:1943 (1979).

11. R. L. McCrory, S. Skupsky, Proceddings of 12th European Conference on Laser Interaction with Matter, 11-15 December, 1978, Moscow.

12. B. Yaakobi, D. Steel, E. Thorsos, A. Hauer, B. Perry, S. Skupsky, J. Geiger, C. M. Lee, S. Letzring, J. Rizzo, T. Mukaiyama, E. Lazarus, G. Halpern, H. Deckman, J. Delettrez, J. Soures, and R. McCrory, Phys. Rev. A 19: 1247 (1979).

13. C. F. Hooper, Jr., L. A. Woltz, Technical Report, University of Florida, Dept. of Physics and Astronomy, September, 1979 (unpublished).

14. Earlier version of LILAC is described in Report No. 16, 1973 and Report No. 36, 1976, Laboratory for Laser Energetics; S. Skupsky, ibid, Theory Group Report No. 11, 1979 (unpublished).

RECENT PROGRESS IN LASER INTERACTION AND ABLATIVE COMPRESSION

EXPERIMENT AT THE RUTHERFORD LABORATORY

J.D.Kilkenny, D.J.Bond, A.J.Cole, D.Everett,
T.J.Goldsack, J.D.Hares, R.W.Lee, J.Murdoch, S.J.Veats

Imperial College, London

L.Cooke, M.J.Lamb, C.L.S.Lewis, J.G.Lunney, A.Moore,
M.J.Ward

Queens' University, Belfast

R.G.Evans, M.H.Key, P.T.Rumsby, A.Raven, W.Toner

Rutherford Laboratory

T.A. Hall, T. Pugatschew, S.M.S. Sim

University of Essex

O.Willi

University of Oxford

ABSTRACT

Recent experimental results from the Rutherford Laboratory
Central Laser Facility pertinent to laser plasma interaction, energy
transport and compression are presented.

Interferometric measurements of electron density have shown
ponderomotive force steepening of the density profile with 100ps
1.06 μm laser pulses at an intensity I of 10^{16} W/cm^2. The density
jump across the critical density region is found to scale as $I^{0.17}$.
Time and spectrally resolved backscatter measurements from 1ns,
1.06 μm laser pulses at 10^{15} W cm^{-2} have shown that the backscattered
light from a wide range of targets is initially blue shifted and
later red shifted.

The production and transport of fast electrons produced by 100ps 1.06 μm laser pulses at $\sim 10^{15}$ W cm^{-2} have been studied by layered targets. A large fraction of the absorbed energy is shown to be converted to fast electrons causing heating at large depths ($\sim$ 2mgm/cm^2) within targets. The scaling of this fraction and the preheat range with intensity is presented. Target design to reduce this preheating effect by the use of low density gold is discussed and experimentally demonstrated. The properties of the thermal plasma produced by 100ps, 0.53 μm laser pulses at 5 10^{14} W cm^{-2}are spectroscopically examined. Comparison with computer modelling shows that the energy transport with this irradiation is essentially classical. With 1ns, 0.53 μm laser pulses, spectroscopic, and time resolved hard Xray emission indicated an ablation pressure scaling as I.

In thin wall high intensity compression experiments the core density has been diagnosed spectroscopically. The necessity of using full line profile calculations is illustrated. X ray and particle diagnostic results for two and six beam compression experiments are described.

Dense implosions created by irradiating large thin walled targets with 1.6ns laser pulses have been diagnosed by streaked X ray shadowgraphy. The results indicate that the compression is unstable. Preliminary results using an X ray microscope for streak shadowgraphy are presented. Target design to allow the diagnosis of core density by X ray shadowgraphy is discussed.

INTRODUCTION

This article reviews some aspects of the recent work of several University Groups and Rutherford Laboratory staff in the areas associated with laser plasma interaction, transport and compression. Some of this work is reported in more detail in the Rutherford Laboratory Annual Report to the Laser Facility Committee[1].

The neodymium laser at the Rutherford Laboratory Central Laser Facility is based on a Quantel rod laser, and two I.L.C. disk amplifiers. There are two main modes of operation, single beam, or double beam with each beam subsequently split into three. In short pulse operation the laser delivers about 40J in each beam, in a 90ps pulse. In long pulse operation a pulse as short as 600ps is switched from a Q spoilt oscillator and amplified to about 100J. Single beam experiments have also been performed with frequency doubled light at 0.53 μm. Focussing is by f/1 lenses; about half of the laser energy arrives on target.

This article concentrates on the plasma physics work at the Rutherford Laboratory C.L.F. associated with laser compression. Other areas of work at the C.L.F. are reported elsewhere[1]. The work is sectionalised here into laser plasma interactions, energy transport and compression, although the borderline is not always clear.

LASER INTERACTION PHYSICS

(a) Density profile measurements

Recent measurements[2] have confirmed theoretical predictions[3,4] that the radiation pressure P_r should modify the electron density profile, when P_r is comparable with P_c the plasma pressure at the critical density surface. Previous measurements[2] were with $P_r \sim 0.2 P_c$; measurements reported here[5] were at irradiances of 10^{16} W cm^{-2} where $P_r > P_c$. Initial steepening of the profile with a step-height scaling of $\Delta n/n_c \ \alpha I^{0.16 \pm .01}$, consistent with modelling[6] is observed.

The targets consisted of 40 μm diameter hollow glass micro-balloons, and were irradiated with 50ps f.w.h.m., 1.06μm laser pulses of energies between 0.5 and 2.5 J in a 15μm focal spot; irradiances on target were as high as 10^{16} W cm^{-2}. A probe beam of 150μJ in 25ps at a wavelength λ = 266nm was generated by frequency quadrupling a portion of the main beam. This beam was used with an intere-foremeter of the Nomarski type[7]. The imaging lens was an ultra violet X10 microscope objective, which was focussed with an accuracy of 10μm. Fig.(1) shows a typical interferogram. The fringes were enlarged, digitised, Abel inverted and used to obtain central electron density profiles. Fig.(2) is the density profile obtained at $I = 10^{16}$ W cm^{-2} and shows the upper and lower density shelves at $n_e = 1.8 \ n_c$ and $n_e = 0.4 n_c$ respectively. At $I = 2.5 \ 10^{15}$ W cm^{-2} the step height was reduced to 1.1 n_c implying an intensity scaling of the step height Δn of

$$\Delta n/n_c = 4 \times 10^{-3} I^{0.01 \pm 0.17} \text{; with I in W cm}^{-2}$$

A theoretical model based on isothermal flow through the critical density point predicts [6] $\Delta n/n_c \ \alpha \ (P_r/P_c)^{\frac{1}{2}}$. Some experiments(8,9) have shown that $T_e \ \alpha \ I^{2/3}$ and on this basis $\Delta n/n_c \ \alpha \ (I/nkT_e)^{\frac{1}{2}} \ \alpha \ I^{0.17}$, which is in suprisingly good agreement with the experimental measurements. However, more recent simple modelling in the absence of hydrodynamic flow, but with the inclusion of fast electrons[10] indicates

$$\frac{n_u}{n_c} = 3.17 \ I^* \ \lambda^{*2} /T_c^*$$

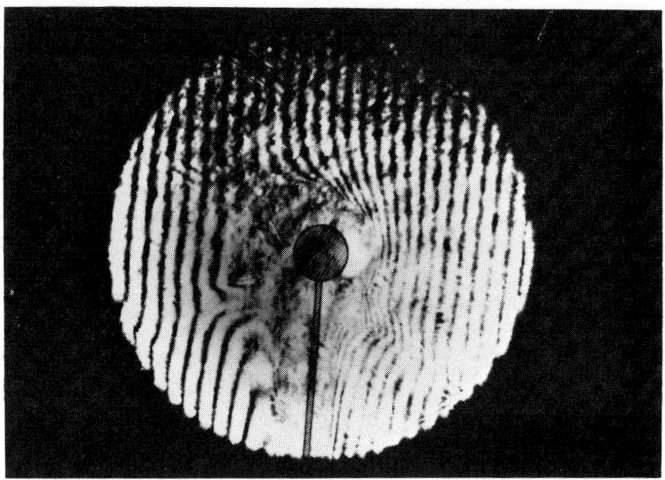

Fig. 1   Example of interferogram for a 45μm microballoon target
         irradiated with a 50ps pulse.

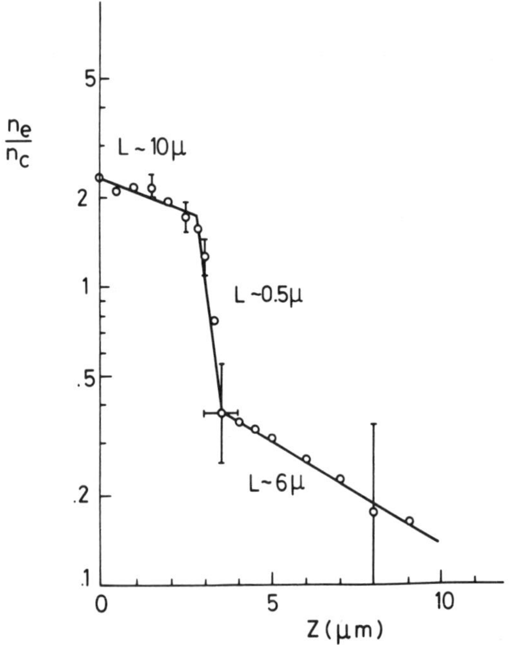

Fig. 2   Radiation pressure steepened profile during the main heating
         pulse at 10^{16} W cm^{-2}.   Solid lines are best fits to
         exponentials of the form exp(-Z/L)

with I* in units of 10^{16} W cm^{-2}, λ* in μm and T$^*_c$ in keV.   This model implies a much stronger intensity dependence than is observed.

Very recently the same intereferometer has been used with a 'stacked' main beam laser pulse.   After the probe beam had been split off, the main laser beam was split into five beams which were then added, with time delays between each of them, to give a 400ps long pulse with 50% intensity modulation on the beam.   The interaction of this long pulse with microballoons was measured with the same interferometer.   A typical interferogram, is shown in Fig.3, and illustrates a breakup of the interference fringes.   This may indicate a hydrodynamic instability or may be a manifestation of the temporal modulation of the main beam.   Further experiments with long pulses are planned to investigate this effect.

(b)   Time resolved measurements of stimulated Brillouin scattering
      and second harmonic generation.

Stimulated Brillouin scattering is a process which severely reduces the efficiency with which laser energy can be coupled to a target.   The process requires the presence of many wavelengths of

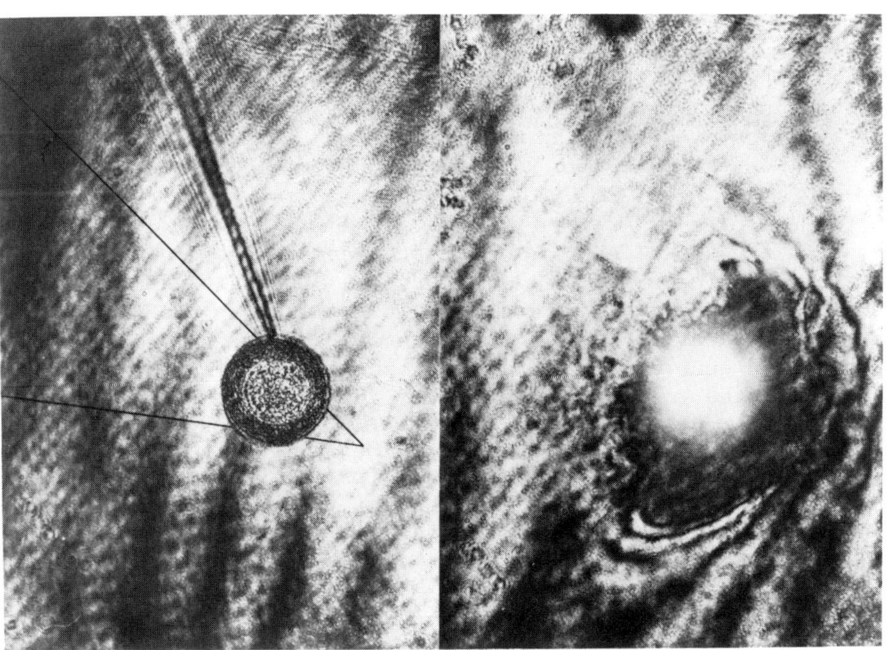

Fig. 3   Interferogram recorded for microballoon irradiation with
         the stacked laser pulse.   Notice the irregular modulation
         of the fringes with $\sim$ 10μm period.

underdense plasma and may be severe with laser pulse durations of more than $\sim$ 1ns such as used for laser compression studies.

The characteristic signature of Brillouin scattering is a frequency shift to the red of the scattered radiation from ω_0 the frequency of the incident wave. A convenient means of investigating Brillouin scattering is to study the spectral and temporal properties of the light backscattered through the focusing lens. Time-resolved spectra generally display an early red shifted component followed by an increasing blue shift [12] . It is indicated below that we may identify blue shifts with supersonic plasma flow and red shifts with subsonic flow. The late blue shift can then be understood as due to the build-up of a significant flow at large radii from the target.

The targets in this experiment were microspheres and micro-balloons in essentially two different sizes; 'small' of 70-90μm diameter and 'large' of 210-240μm diameter. A range of Z, atomic number is used from plastic polymer to U^{238} in the small targets and from plastic to manganese nickel alloy in the larger size.

The duration (1.6ns F.W.H.M) and shape of the laser pulses were nominally unchanged throughout these experiments except for one or two multi-peaked pulses caused by mode beating in the laser cavity.

For a fixed target type, the fraction of laser energy which was backscattered into the focusing lens was fairly constant over the limited range of irradiance that was studied although there was some evidence of saturation for the highest irradiances (5×10^{15} W cm^{-2}) on Aℓ coated microballoons.

Some typical time-resolved spectra of the backscattered ω_0 radiation are shown in Fig.4. The characteristic development of the spectra is as follows. A short burst of blue shifted ($\lesssim 5\text{\AA}$) light occurs very early in the incident pulse, followed by a long, red-shifted component. The red component is spectrally broader than the blue and generally lasts throughout the incident pulse. Most frequently, the two components appear separate in time on the streak photographs.

The red shift of the backscattered radiation during most of the laser pulse is consistent with Brillouin scattering but the frequency shift also contains a component $\Delta\lambda_D$ due to the Doppler shift of the ablating material.

We can write the Doppler shift $\Delta\lambda_D$ as

$$\Delta\lambda_D / \lambda_0 = -2v_a/c = -2Mv_{ia}/c$$

where v_a is the ablation velocity, v_{ia} is the ion acoustic phase

velocity $v_{ia}^2 = \dfrac{ZkT_e}{m_i}$, and M is the Mach number of the flow.

The red shift due to the ion acoustic wave generated by Brillouin scattering will be

$$\Delta\lambda_{B}/\lambda_0 = \frac{1}{\cdot}\frac{\omega_{ia}}{\omega_0} = \frac{2k_o v_{ia}}{\omega_0} = \frac{2v_{ia}}{c}$$

The net red shift is then

$$\Delta\lambda/\lambda_0 = \frac{2v_{ia}}{c}\,(1-M)$$

showing that the shift is to the red if the flow is subsonic and to the blue if the flow is supersonic.   Measurements with targets tilted at different angles[9] have discriminated between these two shifts and indicate that $M \sim 0.8$.   Therefore measurement of the reflected ω_0 spectrum is not a good indicator of the electron temperature in the under dense plasma because of the close cancellation of the two contributions to the wavelength shift.

Examples of the time-resolved $2\omega_0$ backscattered spectra are also shown in Fig.4.   At low irradiances ($\lesssim 10^{14}$ W cm^{-2}), the $2\omega_0$ emission is spectrally narrow and moves continuously in wavelength presumably due to a Doppler shift.   The spectral width of the emission (< 2Å), would suggest that this emission is due to a 'resonant' effect at the critical density without the involvement of any ion acoustic waves.   At slightly higher irradiances the $2\omega_0$ emission begins with a burst of spectrally broad emission, ~ 20Å, and then reverts to spectrally narrow emission as on the lower irradiance shots.   At irradiances of $> 10^{15}$ W cm^{-2} , the broad emission persists throughout much or all of the laser pulse and appears as a series of discrete temporal pulses.   Each pulse is close to the instrumental limit of time resolution and in many cases the pulses also break up into a set of fairly regular 'spots' in the spectral direction.

The time development of the backscattered fundamental frequency shows that the backscatter persists for the whole of the laser pulse but also exhibits very short duration modulations which are as yet unexplained.   The nature of the second harmonic emission suggests that the spectral broadening is associated with fast electron production through the ion turbulence set up by the cold electron return current.   Oscillations in the $2\omega_0$ emission occur only in

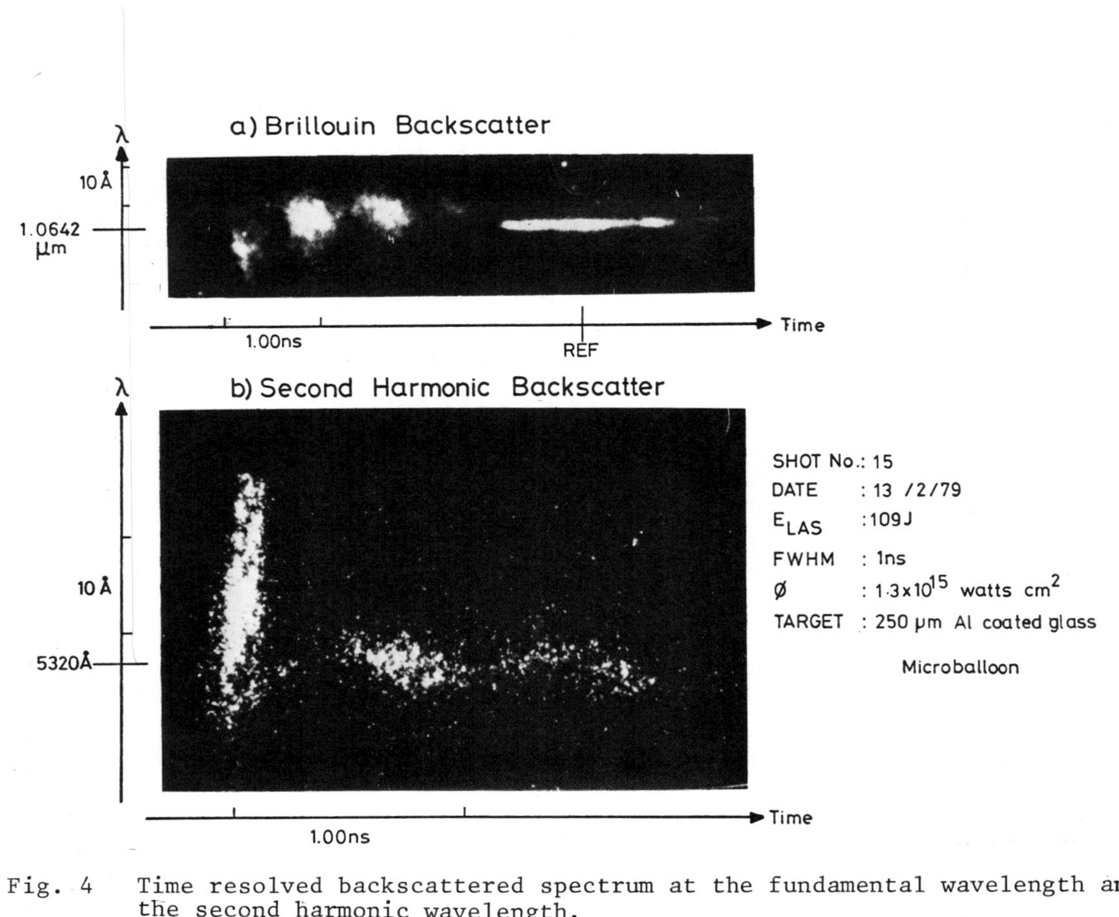

Fig. 4   Time resolved backscattered spectrum at the fundamental wavelength and the second harmonic wavelength.

the presence of steep density gradients and are suggestive of a
hydrodynamic flow instability.

EXPERIMENTS ON THE TRANSPORT OF ENERGY IN LASER PLASMAS

(a)   Experiments on energy deposition at depth from high intensity
      laser plasma interactions at 1.06μm.

In experiments with high intensity 1.06μm irradiation of solids
it is known from hard X ray continuum emission [10,13,14] and fast ion
emission[15] that a suprathermal tail to the electron velocity
distribution is formed.   Neither diagnostic localises the fast
electron, and neither diagnostic can easily measure the number of
fast electrons.   However, the K_α emission from targets can be used
to measure the degree of preheating at various depths within a target
and the number of fast electrons[16,17,18,19].   This work extends
earlier observations of K_α emission[17,18,19] which were probably
subject to interpetative difficulties from the use of low Z K_α
emitters which exhibit a saturation effect due to ionisation as well
as significant photoionisation induced K_α emission.

The targets for these experiments consisted of plane layers with
layer thicknesses in μm as follows

      0.1 Aℓ; 1.0 SiO;   3.0KCℓ;   2.5 Mylar; 2.5CaF$_2$
                                    12.5
                                    25
                                    50

The targets were irradiated normally by 20J of 1.06μm radiation in
100ps.   The intensity was varied by varying the focussing.   Two
absolutely calibrated X ray crystal spectrometers, viewed the targets
from the front and rear.   The spectral range of 7Å to 3Å showed
Cℓ, K, and Ca K_α lines and spectra from hydrogen and helium like Aℓ
and Si.   A typical spectrum recorded from the front of the target is
shown in Fig.5.

The penetration of the ablation front was limited to the Aℓ and
front of the Si and thus the K_α fluor layers (KCℓ and CaF$_2$) were only
affected by the fast electrons and the Xray flux.   Varying the mylar
thickness filtered the electron flux reaching the rear Ca fluor and
hence gave information on the energy spectrum.

Information on the fast electron energy spectrum can be obtained
from the K_α yields using Y(E), the K_α yield per incident electron of
energy E.   Y(E) was derived from earlier energy deposition
calculations and was also measured experimentally with a 0.1μA,
15-10 keV electron beam focussed onto duplicate targets and observed
with the same spectrometer as used in the laser experiments.   The
results are shown in Fig.6 for the K K_α and the Ca K_α lines with

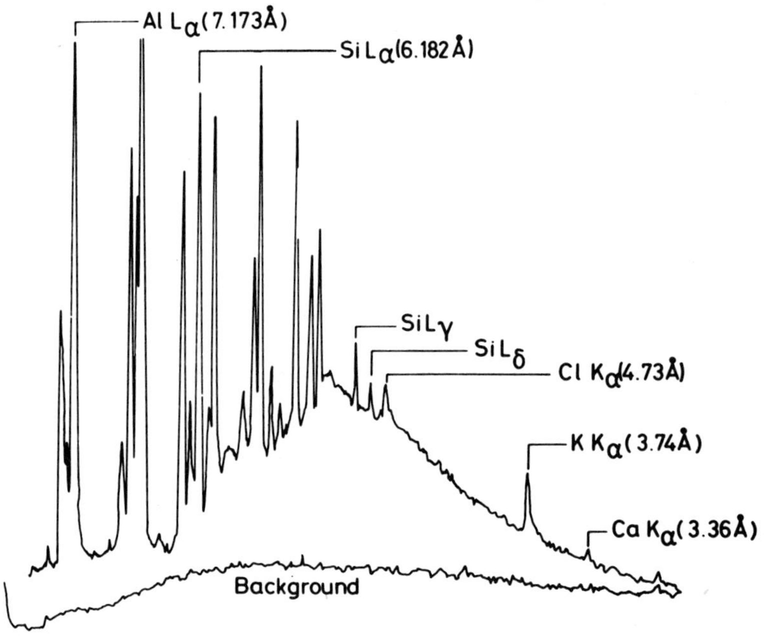

Fig.5   Typical X ray spectrum recorded from a layered target by the
        spectrometer at the front of the target.   The Ca K$_\alpha$ line is
        attenuated by the target self absorption but is recorded
        with little absorption by the rear spectrometer.

different thickness of mylar in the target.   If the distribution
function of electrons incident onto such a targer is n(E) the K$_\alpha$
yield would then be given by $\int_0^\infty$ Y(E) n(E)dE.   The distribution
function can then be found by fitting forms for n(E) to the observed
yields from fluors at different depths in the target.

There are however two effects which can complicate the above
procedure.   Firstly the Xray flux from the ablation plasma is very
efficient at pumping K$_\alpha$ lines for photon energies above the K
absorption energies.   As the X rays from the ablation plasma
$\alpha \exp(-k\nu/kT_e)$ with $T_e \sim 0.5$keV the choice of a high Z fluor

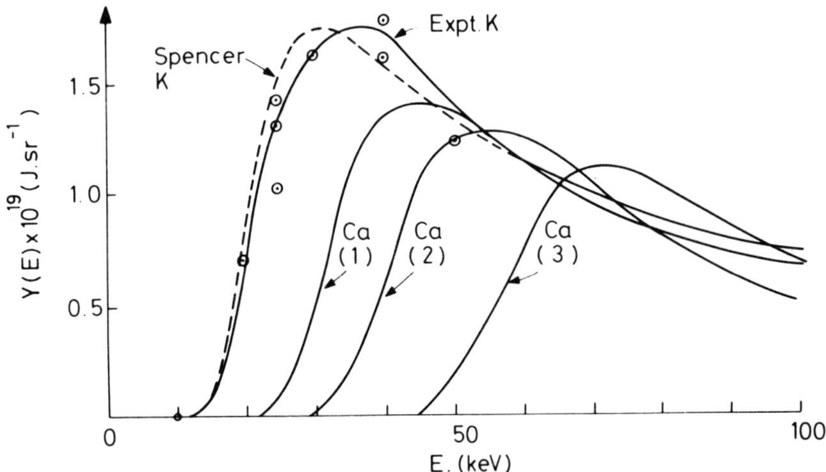

Fig. 6   The yield of K_α radiation per incident electron of energy
         E, from calculations and experimental measurements for
         targets with various mylar thicknesses. The mylar thickness
         1,2 and 3 are 2.5, 12.5 and 25μm respectively.

minimises this effect.   In fact the Cℓ in the KCℓ layer was mainly
radiatively pumped, whereas the K was mainly pumped by electrons.
Secondly there is an effective saturation effect due to large shifts
of the K_α line once the fluor become ionised into its L shell.   The
saturation energy for different Z fluors is shown in table 1, and
indicates the need for high Z fluors.

TABLE 1.   Saturation K_α yield for various elements

Fluor Element Z	Comment	Atomic density (cm^{-3})	Max Z for small shift	Te (eV)	Sat K_α yield per atom (eV)
Ne (10)	10B gas	2.7×10^{20}	1.5	8	0.017
Si (14)	Solid SiO	6×10^{22}	4.5	66	0.43
Cl (17)	Solid KCl	1.6×10^{22}	7.5	90	1.7
K (19)	Solid KCl	1.6×10^{22}	9.5	125	3.2

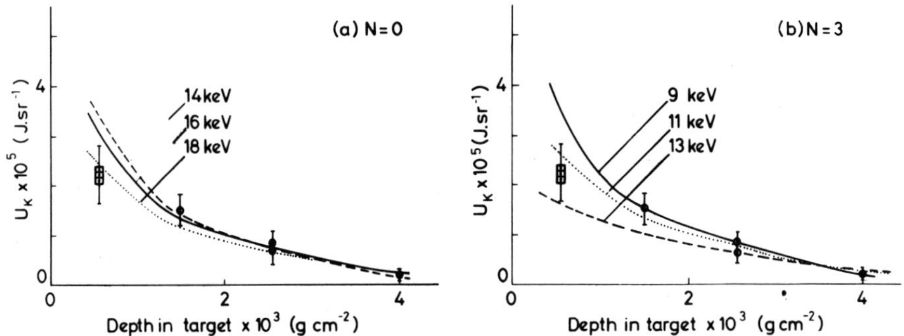

Fig.7   The K_α yields at different depths within the target derived
        from several shots with different mylar thicknesses.   Also
        shown are the predicted yields for various incident
        distribution functions of electrons.

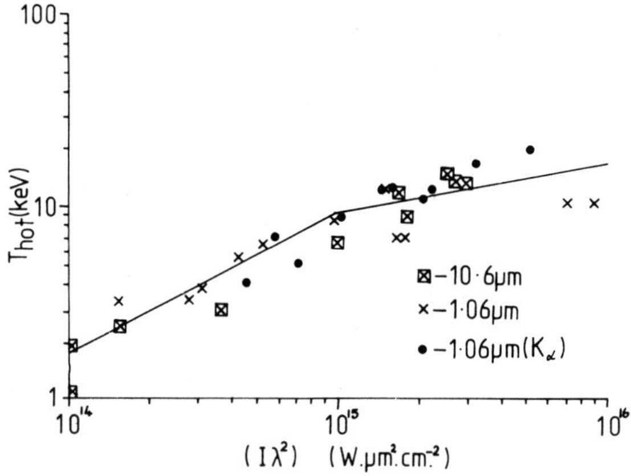

Fig.8   The fast electron temperature T_H as a function of intensity
        derived from the range measurements.   Also shown are the
        values of T_H from hard X ray continuum in ref.21.

Laser experiments were first done at an intensity of $2 \pm 0.5 \ 10^{15} \ W \ cm^{-2}$. The mylar thickness was varied from 2.5µm to 50µm and the measured K_α yields are shown in Fig.7 as a function of the depth within the target. The yields predicted for electron distribution functions of the form

$$n(E) = AE^{N/2} \exp(-E/kT_H)$$

are also shown. The best fit is obtained for N=3 T_H = 11 $\pm$ 2keV. This value of T_H is also consistent with hard X ray continuum emission measurements at the same intensity. (Fig.8). The total energy deposition in the target was 2.2 $\pm$ 0.4J for 20J incident, and presumably 6J absorbed. The energy deposition in the front fluor layer was 1.4 10^7J cm^{-3} which implies a pressure of 50 Mbar if an L.T.E. equation of state is used.

Experiments with varying focussing, and with the positions of the KCℓ and CaF_2 interchanged were also performed. Best fits to the relative yields gave values for T_H shown in Fig.8. The data from hard X ray continuum measurements[21] is also shown and illustrates the good agreement between the methods. The fraction of the incident laser energy in the distribution function is shown in Fig.9. At low temperatures the fluors sample only a small part of the distribution function and so the measured energies are sensitive to the model distribution function. For high intensities $>10^{15}$ W cm^{-2} the energy in the fast electrons is not sensitive to the distribution function and shows some dependence on intensity. However over the range of intensities observed there is very little dependence of the fraction of energy in the fast electrons on intensity.

(b) <u>An experimental demonstration of resistive inhibition of fast electrons.</u>

In the preceeding section it was shown that the fast electron temperature derived from their range agreed with the hard X ray continuum emission (Fig.8). An equivalent statement is that the range of fast electrons is given by the Bethe Bloch formula. However there is the possibility that the range of the fast electrons can be inhibited. Consider a laser beam incident on a semi-infinite plane target. A fraction of the absorbed laser power is converted into fast electrons flowing into the target having an associated current density j_H which is typically 10^{10} A cm^{-2} at an intensity of 10^{15} W cm^{-2}. The high energy deposition rapidly ionises the target which becomes a high electron density ($n_e \sim 10^{23}$ cm^{-3}) low electron temperature plasma ($T_e \sim 200eV$) of resistivity η. A return current of cold electrons must flow so that

$$j_c + j_H = 0$$

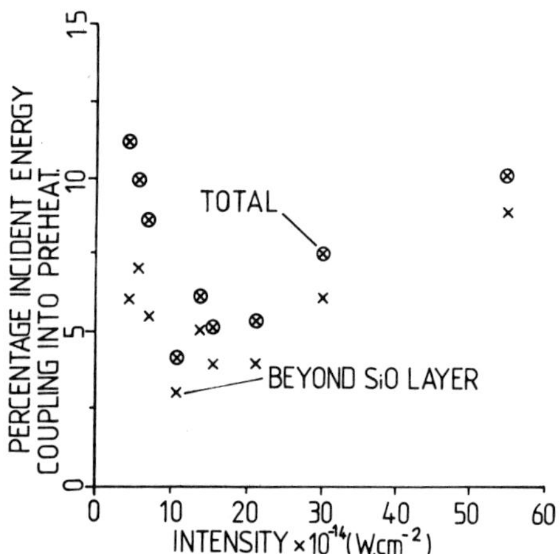

Fig. 9   Fraction of the incident laser energy deposited at depth in
the solid target as a function of intensity.

There will therefore be set up a resistive electric field

$$E = + \eta j_c = - \eta j_H$$

which decelerates the fast electrons.   If the collisional areal
range of the fast electrons is r_a(gm cm^{-2}) and the target density
is ρ then in the absence of electric field inhibition the fast
electrons will travel a distance r_a/ρ setting up a potential
$\sim \eta j_H r_a/\rho$.   However if $\eta j_H r_a/\rho > kT_H/e$ the resistive electric field
will appreciably impede the fast electrons.   For normal density
targets this potential is only $\sim$ 400V and the effect on fast

electrons would be negligible.   However for low density that is 1%
of solid density, gold the target potential can exceed 10kV and thus
the fast electrons will be inhibited to some extent.   The three
factors in low density gold which enhance this inhibition are (i)
the large stopping distance r_a/ρ at low density, (ii) the high state
state of ionisation (~ 30) and hence resistivity achieved for small
temperature rises in gold and (iii) the L.T.E. state of ionisation
being larger at lower electron density for a given T_e.

The effect of this resistive electric field has been quantitatively predicted by a Monte Carlo electron transport calculation.
The scattering is represented by randon Rutherford scattering as well
as collisional energy loss.   In between scattering events the
electrons follow parabolic trajectories representing the effect of
the resistive electric field.   The electric field is found
iteratively, consistent with the fast electron current density and
the energy deposition up to that time step.   Once a consistent
solution is found the local energy deposition, the state of ionisation
of the target and therefore the resistivity are then calculated for
the next timestep.

In the transport calculation the fast electron distribution
function used is of the form

$$N(E) = E^{3/2} \exp(-E/kT_H)$$

With kT_H = 14keV the calculated energy depositions for low and high
density gold, shown in Fig.10, illustrates the inhibition of the
fast electrons by the resistive electric field.

To confirm this effect an experiment similar to the layered
target experiments described in the previous section, except that
gold was used as an electron filter, was performed.   The gold layer
was deposited at various densities.   Solid gold layers were made by
vacuum evaporation and low density layers were made ($\sim 1\%$ solid
density) by slowly evaporating the gold in an atmosphere of 800mTorr
argon.   The low density gold target formed with voids in it on a
μm scale length.   Experiments were done in pairs with solid density
gold layers and then with a low density layer of the same areal
density.   The K_α yields from tracer layers either side of the gold
layer are shown in Fig.11.   It is clear that the K_α yield, and
therefore energy deposition is much less with low density gold than
it is with high density gold of the same area density.   The K_α
yields predicted by the transport model with a value of kT_H
appropriate to the experimental laser intensity is also shown.   The
predicted difference in the yields for low and high density gold is
considerably less than that observed.   This is probably due to
defects in the model, the most noticeable being the application of
the Spitzer formula for resistivity to a partially stripped ($Z \sim 30$)

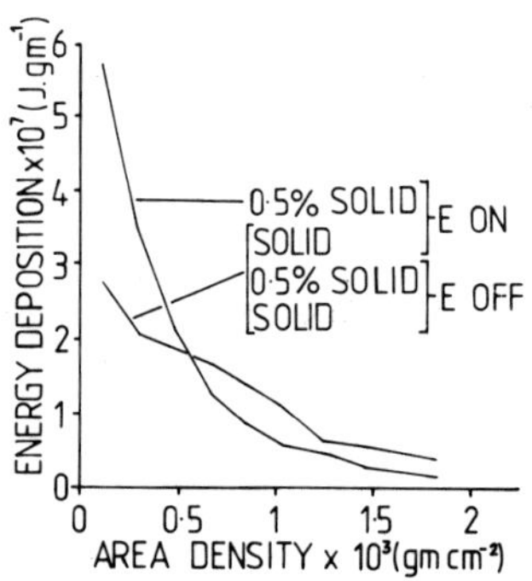

Fig.10   Predicted energy deposition as a function of depth in low
          and high density gold, showing the effect of resistive
          inhibition. Calculations are performed with and without
          the resistive electric field on. Notice that this resistive
          field only affects the low density gold.

plasma with a low logΛ where the minimum impact parameter is $\leqslant$ the
size of the atom.

        Thus a special target design has been demonstrated which can
inhibit the preheat of fast electrons. Use of this type of
inhibitor layer might extend the intensity at which ablative
acceleration, as opposed to exploding pusher behaviour of targets
can be achieved.

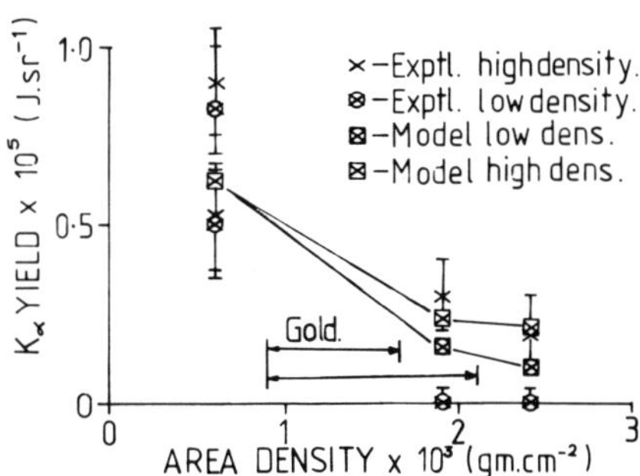

Fig.11   Experimental K_α yields for fluors either side of low and
high density gold layers for an incident laser intensity of
$2\ 10^{15}$ W cm^{-2}.   Also shown are the K_α yeilds predicted
by the Monte Carlo calculation for low and high density gold.

(c)   Experiments with 100ps, 0.53μm pulses at $5\ 10^{14}$ W cm^{-2}

Earlier experiment at the C.L.F   and elsewhere[22,23] with
1.06μm irradiation have shown that the penetration depth of the
ablation front is much less than expected, which implies that the
thermal conductivity is reduced from its classical value.
Experiments reported here measure the ablation depth for 0.53μm
irradiation and find that it is considerably larger than for 1.06μm
irradiation, although the spectroscopic density and electron
temperature are similar.

The glass laser was used in its 100ps mode, and was frequency
doubled in a KDP crystal.   The green beam was then transmitted to
the target chamber by four dielectric mirrors which gave a rejection
of 10^6 at 1.06μm, and was focussed on target with an f/1 ICOS
doublet lens.   The lens element spacing (1) was set to minimise
aberrations at 1.06μm.   As a result the focal spot for 0.53μm light

was dominated by spherical aberrations, and limited to 70μm diameter. The energies on target were 3-5J implying intensities of 10^{15} W cm^{-2}

The targets used in this experiment were either glass micro-balloons coated uniform with Al, or plane layered targets consisting of various thicknesses of Aℓ coated on 0.5μm of SiO. The X ray emission was viewed by two X ray crystal spectrometers. One spectrometer used a P.E.T. crystal and measured the Aℓ and Si hydrogen and helium like line and recombination continuum emission. X ray pin hole data, hard X ray emission and spectrally resolved backscattered 0.53μm light were also measured.

The burn depth of the ablation front was characterised by the ratio of the Si XIII $I^1S_0 - 3^1P_1$ line to the Aℓ XII $I^1S_0 - 4^1P_1$ line and is shown in Fig.12 together with 1.06μm data at $2\ 10^{15}$ W cm^{-2}. This observation confirms other measurements[24] which show that the burn depth is 3x larger for 0.53μm radiation,

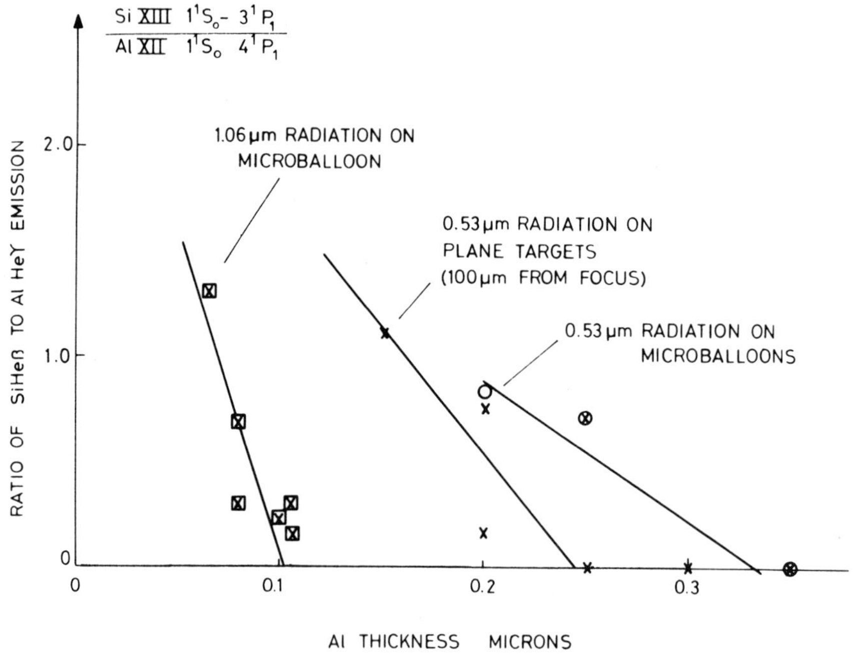

Fig. 12 Burn depths for 0.53μm and 1.06μm irradiation.

but uses a more direct method.   The cold electron temperature was
obtained from the slope of the recombination continuum in the range
$2.5\text{keV} < h\nu < 3.5$, and was $440 \pm 50\text{eV}$.   The harder X ray continuum
emission up to $h\nu = 12\text{keV}$ indicated a temperature of 5keV although
the reproducibility was poor.   The electron density of the X ray
emitting region was obtained by fitting the Stark broadened lines,
and was found to be $1 \pm 0.2 \ 10^{22} \ \text{cm}^{-3}$.

These results were modelled by a 1 dimensional simulation.
In this model the thermal conductivity and the absorption into the
electrons was varied: the runs were post processed to obtain values
of the two main experimental observables, the burn depth and the
X ray electron temperature.   Comparison with the experimental
results showed that fits could be obtained for classical thermal
conductivity and an absorption fraction of about 15%.   This is in
contrast to the results at 1.06μm for which fits could only be
obtained for severely flux limited thermal conductivity.   The
pertinent difference in the two cases appears to be in the ratio
of the electron mean free path to the scale length for temperature
change, which is $\sim 10^{-3}$ for the 0.53μm experiment and thus gives
rise to classical thermal conductivity.

(d)   Experiments with $\sim$1ns, 0.53μm pulses at $5 \ 10^{14}$ W cm^{-2}

The burn depth measured in the previous section is the combined
effect of a quasi steady penetration of the heat front ahead of the
critical density surface and the effect of material flowing out
through the critical density surface.   For long pulses the latter
effect dominates and thus the burn depth measures the rate of
ablation of material, $\dot{M}$.   If the asymptotic velocity of flow is v
then the rate of transfer of momentum to the ablated plasma is $v\dot{M}$,
which will be a measure of the ablation pressure.   Here very recent
experiments which directly measure $\dot{M}$ by a novel hard X ray pulsing
technique are described; preliminary analysis indicates pressures
of 45 MBar at intensities of $4 \ 10^{14}$ W cm^{-2} of 0.53μm radiation.

For this experiment the laser was used in its 'long' pulse
mode and gave 1.06μm pulses of minimum pulse length 0.6ns.   The
0.6ns, 1.06μm laser pulse was frequency doubled and gave a
0.53μm pulse which was 0.35ns long.   The power was limited by
damage to the doubling crystal, and was 25 GW onto target.
Experiments were also performed with longer 0.53μm pulses of length
1.3ns but of the same energy as the short pulses.   Focussing was
by the ICOS f/1 doublet, but for this experiment the spacing of the
components was set to minimise spherical aberrations at 0.53μm.
Most experiments were performed with the targets 125μm from focus,
which gave a focal spot of diameter 70μm, in agreement with the
size of X ray pin hole images.

Targets for these experiments consisted of plane layers of low
Z material in between two high Z layers.   The targets were
massive and were angled at 20° to the incident laser beam.   The
X ray emission was measured by an X ray streak camera with a CsI
photocathode.   It was hoped to observe motion and pulsing of the
soft ($h\nu$ = 2keV) X ray emission as the X ray emission should
decrease when the plastic is ablating.   To this end a slit was used
to image, with 30 X magnification, the laser plasma onto the photocathode, as shown in Fig.13.   In the X ray streak, Fig.13, the
intense emission is the soft X rays imaged through the slit.   The
ablation plume is evident, however there is no evidence of pulsing
of the soft X rays, as would be expected from the layered target.
However pulsing is evident in the emission either side of the soft
X ray emission.   This is hard X ray emission through the 30µm Cu
substrate of the slit.

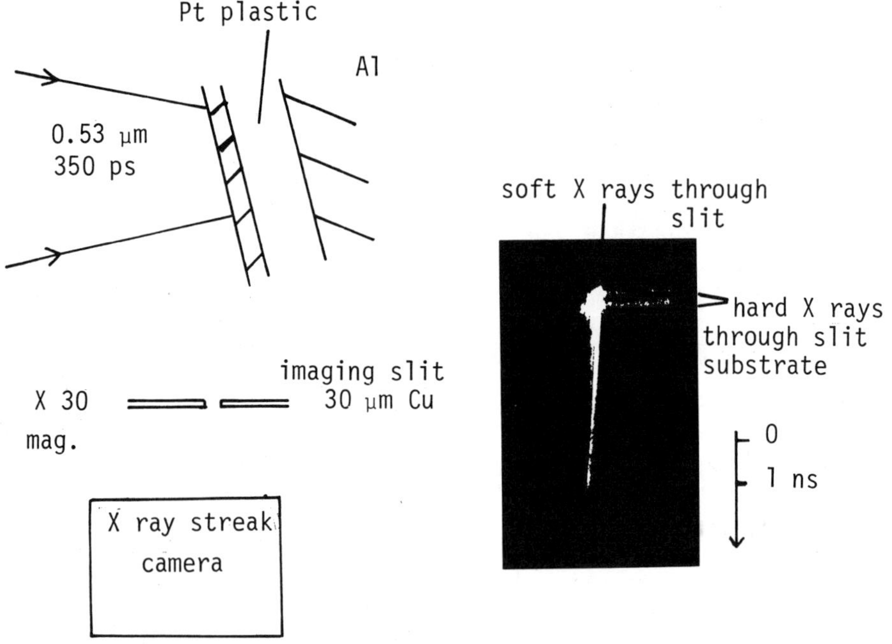

Fig. 13   Streaked X ray emission from layered targets.   The Pt
and plastic thicknesses were 20nm and 1.8µm respectively.

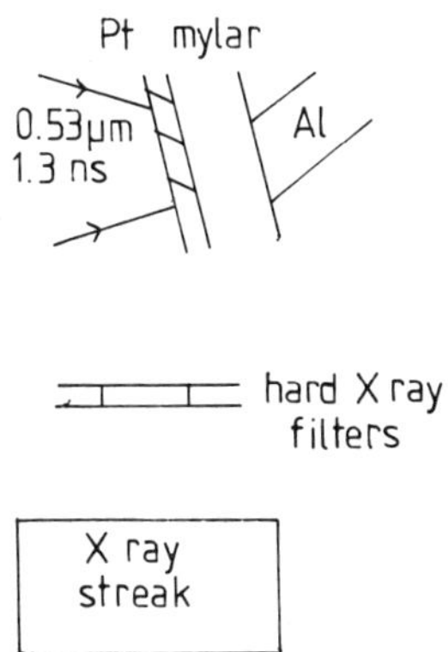

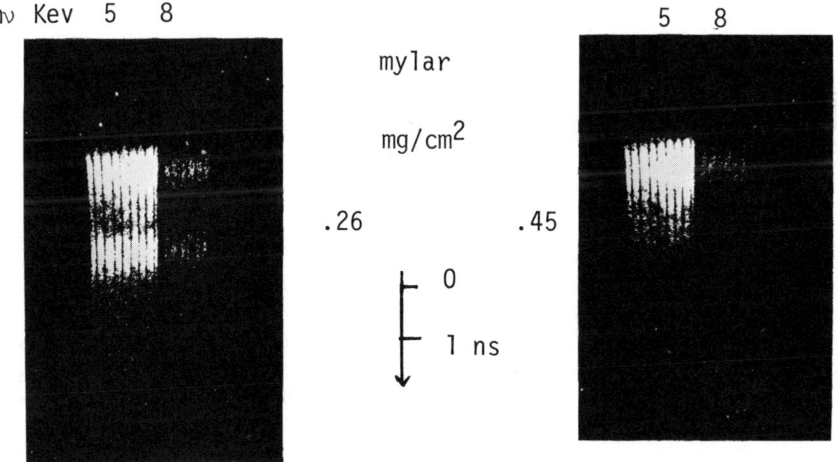

Fig. 14   Streaked hard X ray emission from layered targets.   The
Pt and mylar thicknesses were 20nm and 2µm respectively.

To confirm this pulsing effect, experiments were then performed with a hard X ray filter pack, and longer laser pulses.   Results are shown in Fig.14.   Again a pulsing effect in the hard X ray emission is evident.   It is postulated that the pulsing is associated with the absorption region being in a plasma of high Z.   It is known[25] that the hard X ray emission is greatly enhanced for high Z laser plasma targets.   This is partly due to X ray conversion from electrons being larger for high Z targets, but is also due to the source of fast electrons being greater from high Z plasmas.

The ion velocities were measured by charge collectors at 20^0 to the target normal.   For the shorter pulse experiments (Fig.13) at an intensity of $5\ 10^{14}$ W cm^{-2}, the ion velocity was $5\ 10^7$ cm s^{-1}, whereas for the longer pulse experiment (10^{14} W cm^{-2}, Fig.14) the ion velocity was $3\ 10^7$ cm s^{-1}.   The pressures, as derived from $\dot{M}v$ are 45 MBar and 9 MBar respectively.

The hard X ray emission was also monitored by an array of diodes.   The continuum slope in the region 5 - 10keV indicated a temperature of 3keV for the high intensity light focus experiments, and 1.3 keV for the experiment of Fig.14.

Even more recently the spectral emission from layered targets has been streaked.   These experiments measure directly the mass ablation rate, and confirm the pulsed hard X ray experiments.

COMPRESSION EXPERIMENTS

(a)   Compression experiments dominated by fast electrons

Compression experiments at the C.L.F. with thin wall glass microballoons have been performed with two beam irradiation[26] and recently with six beam irradiation.   Targets for these experiments were glass microballons of 70μm radius by about 1μm wall thickness. Various gas fills, D-T, Ne and A were used.   In the two beam compression experiments 15J per beam was usually delivered on target. In the six beam experiments up to 6J per beam was delivered on target in a 90ps pulse.   X ray pin hole photographs of some six beam implosions are shown in Fig.15.   They show improved symmetry in the core and on some shots structure in the core.

A space resolving crystal spectrometer was used to diagnose the properties of the cores.   The analysis of the spectra is discussed in detail elsewhere[27].   The electron temperature was deduced from

A space resolving crystal spectrometer was used to diagnose the properties of the cores.   The analysis of the spectra is discussed in detail elsewhere [27].   The electron temperature was deduced from

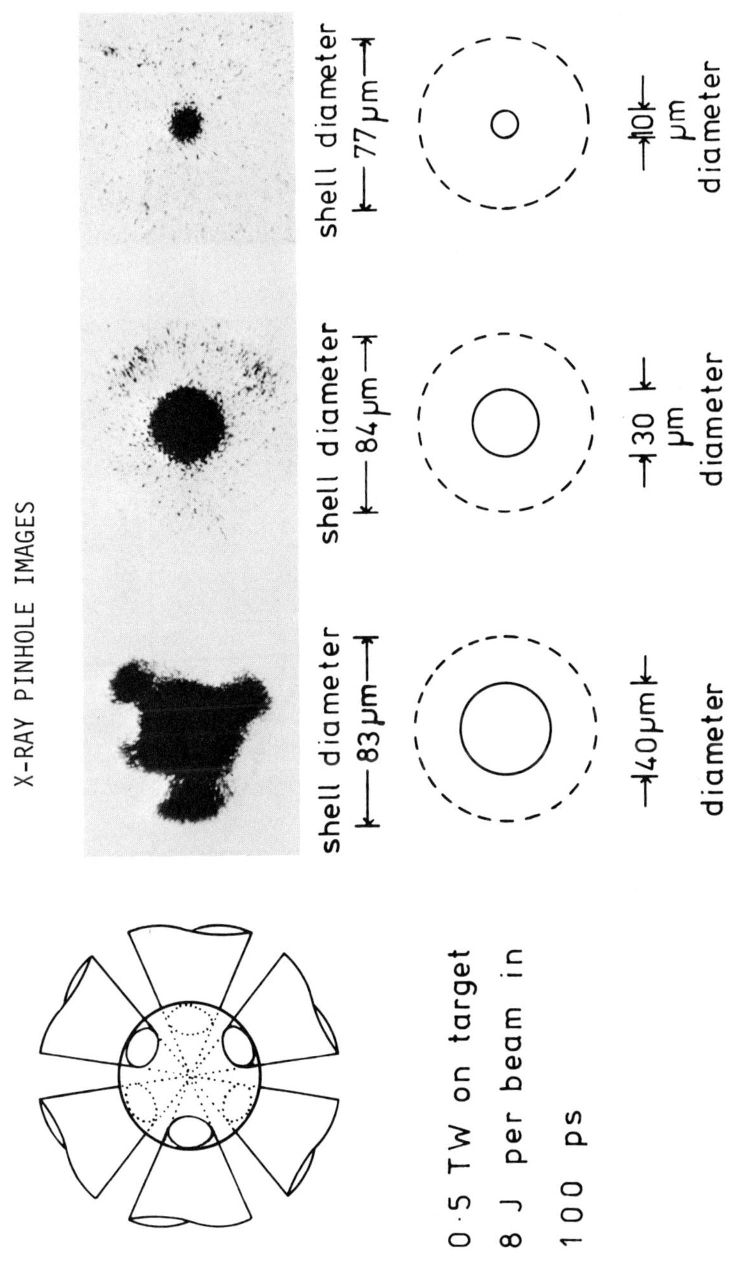

Fig. 15   Pin hole photographs of six beam implosions.

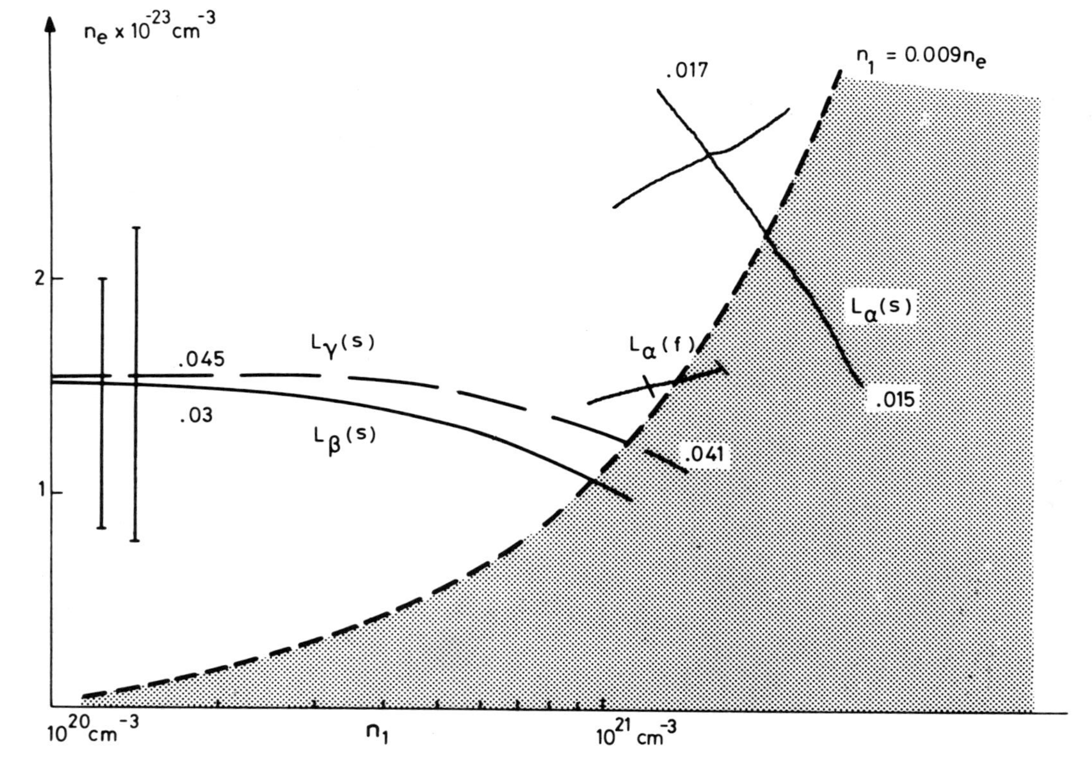

Fig. 16   Lines of best fit for Silicon Lyman lines from the core ($2r = 24\mu m$, shot 20a). The numbers on the lines of best fit indicate the quality of fits. F and s denote full and standard line broadening theory respectively.

the slope of the recombination continuum. The electron density was
deduced from the spectral broadening of the emission lines.   A
significant advance in our work in this area is due to new
theoretical work on line shapes by Lee[28].   This work has produced a
set of theoretical predictions which differ from the standard
theories in that high Z shifts and assymetries are included.   The
total effect of these improvements in the theory are most marked
for the α lines and for high Z plasmas.

Furthermore an improved method of using the line profiles as a
diagnostic has been developed.   Spectral profiles depend on the
electron density n_e and the opacity, which is proportional to $N_1 L$
(ground state density x line of sight depth).   Experimental He
like and H like line profiles are compared with theoretical line
profiles calculated for a homogeneous plasma slab.   Both n_e and
$N_1 L$ are varied to plot contours of the quality of line shape fit
in the n_e, $N_1 L$ parameter space.   The unique value of n_e and $N_1 L$
that gives a fit to <u>all</u> the lines is then sought.   An example of
such a parameter fit for a two beam implosion is shown in Fig.15.
The line marked s here is the line of best fit using the standard
line broadening theory.   The standard theory does not give a unique
fit for the L_α L_β and L_γ lines.   To get a unique fit the full theory
must be used and this is shown on Fig.15 for the L_α line.

From the analysis of many two beam implosions the value of n_e
and T_e are usually $1 \ 10^{2\ 3}$ cm^{-3} and 500eV.   Although in
approximate agreement with heuristic models of densities, there is
no clear dependence on the aspect ratio of the shell.

Argon core emission lines have been observed to be Stark
broadened.   In two beam implosions the A XVII $I^1 S_o - 3^1 P_1$ spectral
line has been observed spectrally broaden   In recent six beam
implosions AXVII and AXVIII   line have been observed.   Fig.17
shows a coarse microdensitometer scan of such an implosion.   It is
expected that an analysis of these line will indicate the need for
full theoretical profiles.

A Thomson parabola ion analyser was used to study the fast ions
generated in some of the six beam implosions[29]. Cellulose nitrate
detecting film was used and Fig.18 shows the tracks from a plastic
coated target and the tracks from a gold coated target.   It is
interesting to note that spectroscopic evidence on the latter shot
suggests that the state of ionisation of the gold is $\sim$ 50, whereas
the Thomson parabola only measures a state of ionisation of 30.
The difference is thought to be due to recombination in the
expansion phase.

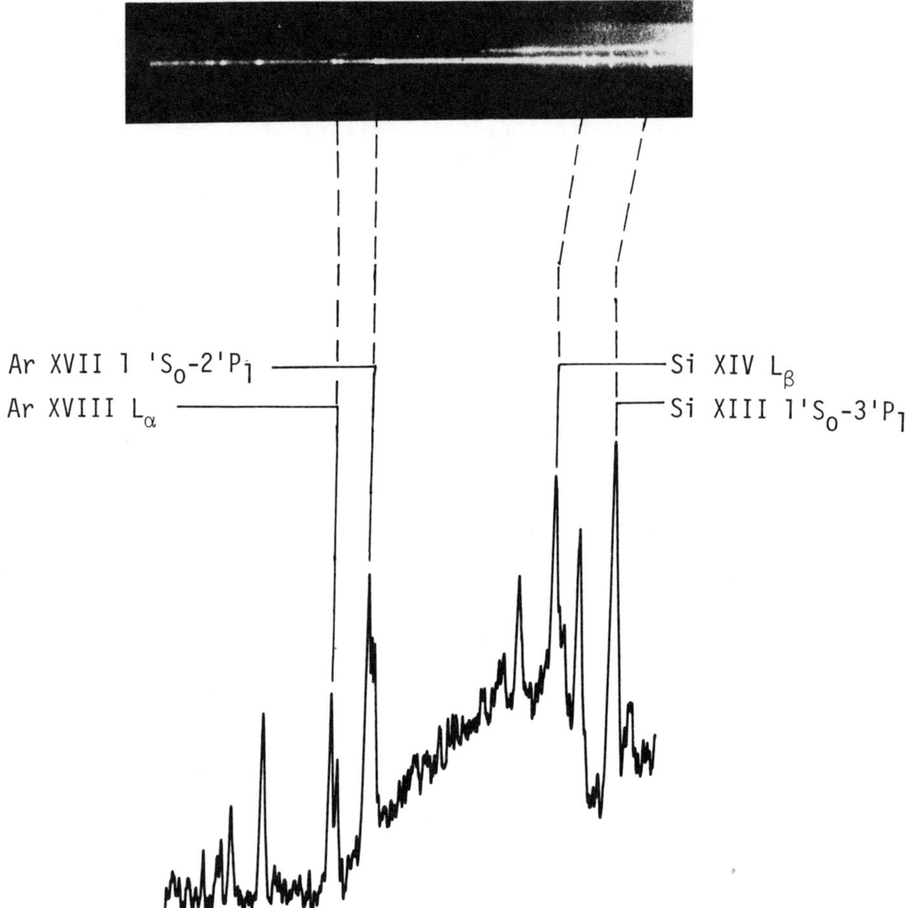

Fig. 17   Space resolved spectra of a six beam implosion of an argon
           filled microballoon.   Energy on target 35J in 90ps.
           Target diameter 69μm, 0.82μm wall thickness filled with 0.41B
           0.41B argon.   Argon XVII lines up to $1s^2 -$ 1s 5p and
           argon XVIII L_α and L_β are seen.

(b)   Time resolved X-ray shadowgraphy of laser driven ablative
      implosions.

      Realisation of higher density in laser driven implosions
requires minimum preheating of the compressed material and adiabatic
compression in order to maximise the density reached at a given final

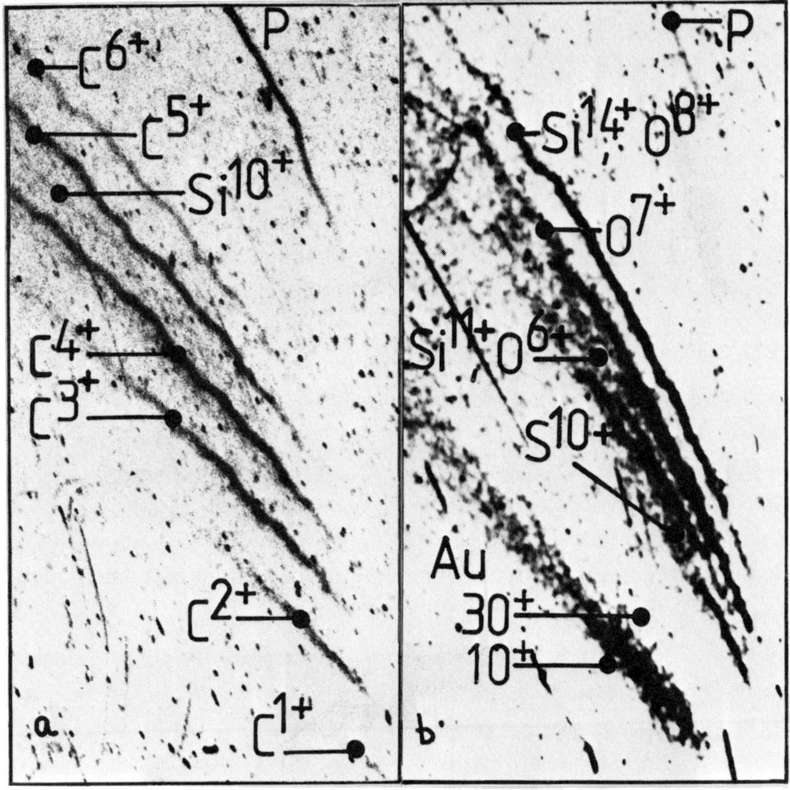

Fig. 18   Thomson parabola tracks from two six beam implosions.
          Fig.18a shows the parabolae obtained from a plastic
          coated microballoon.   Fig.18b shows the parabolae obtained
          from the implosion of a gold coated microballoon.

pressure.   In the ablative mode of compression laser irradiation
at lower irradiance ($I\lambda^2 \lesssim 10^{14}$ W cm^{-2} μm^2) gives neglibible fast
electron preheating but exerts pressure $\sim$ 10 Mbar through surface
heating and ablation[30,31].   Compression of gas filled microballoons
in this way should lead to higher density and lower temperature in
the gas than has been obtained in exploding pusher targets.   The
latter are readily diagnosed through their X-ray emission, but the
former require new methods of which one of the first to be
developed has been X-ray shadowgraphy.

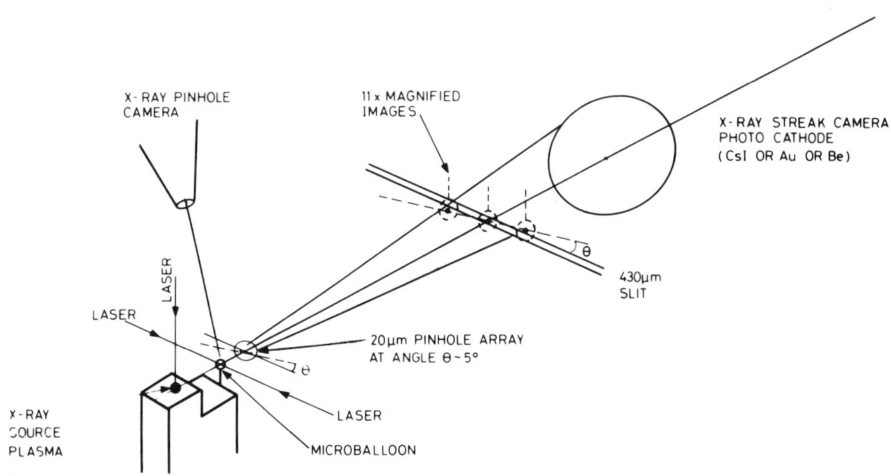

Fig. 19 Schematic system for streaked X ray shadowgraphy using 3 pin hole X ray imaging.

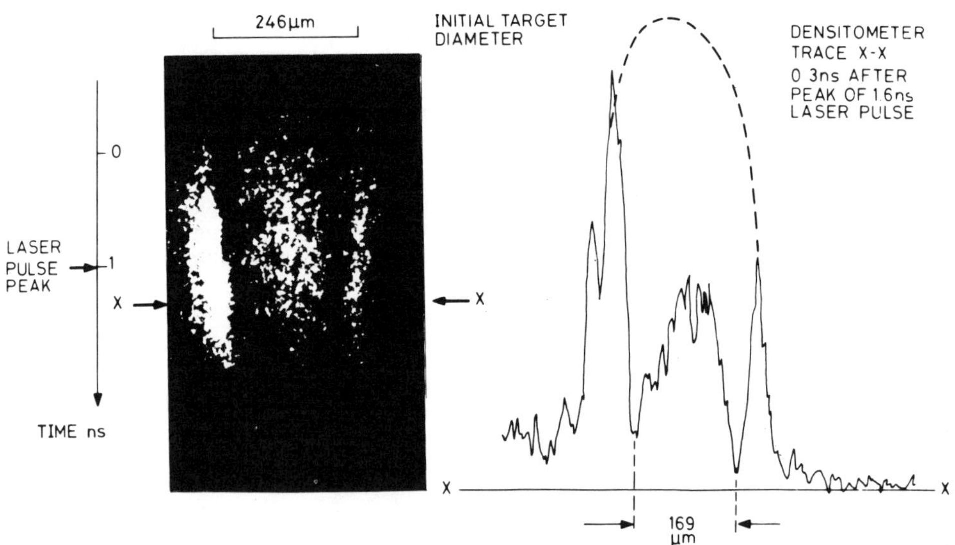

Fig. 20 A streaked X ray shadowgraph of the implosion of a glass microballoon coated with 6μm polymer. For this example there is negligible emission of the microballoon relative to the backlighting source.

Application of the shadowgraphy method to ablative implosions in which the irradiation continues throughout the implosion duration requires time resolution of the X-ray shadowgraph image. This is possible using an X-ray streak camera, to record the shadowgraph image as shown in Fig.19.

The first experiments of this kind[32] used 2 beam Nd glass laser irradiation of polymer coated glass microballoons wiht no gas fill. The targets were 240μm in diameter with 1.4μm glass and 1 to 6μm (variable) polymer wall thickness. The laser pulse was 60 J in 1.6ns giving an irradiance 4×10^{13} W cm^{-2}. A streaked X ray shadowgraph with spatial resolution $\sim$ 15μm and temporal resolution $\sim$ 250ps is shown in Fig.20. The filter selecting the backlighting X-ray energy had a cut off energy of 2 keV. There was usually a delay $\sim$ lns in the beam producing the X ray backlighting plasma so that the streak records show initially the relatively weak emission of X rays from the irradiated microballoon followed by intense back-lighting X ray emission which shows the formation of an absorbing implosion core of minimum diameter $\sim$ 100μm after about 1.8ns.

Fig.21 shows data from a similar implosion in which the radius of the region of opacity > 0.5 (opacity = ln 1/T where T is the transmission) is plotted as a function of time and compared with results from a computer simulation which showed density of 10 g cm^{-3} at peak compression in the imploded glass. The exponent of growth γt for the classical Rayleigh Taylor acceleration instability of wavelength equal to the shell thickness is also plotted and is seen to reach values large enough to suggest significant instability.

Some evidence for such instability was found in the experimental data, namely, (i) anomalously high intensity of X ray emission from the polymer coated targets suggesting mixing of glass and polymer, (ii) a scale length $\sim$ 30μm of the outer boundary of the opacity profile which is much larger than the computed value and the experimental resolution, (iii) anomalously high transmission of X rays through the shell in the early stages of the implosion suggesting breakup of the high Z glass layer.

Limitations of the experiment were that the 2 beam irradiation was of poor spherical symmetry, that the opacity of the glass shell was too high to measure except in the initial phase of the implosion dynamics were not optimised.

The limitation of too high an opacity of the shell to the backlighting X rays is overcome with low Z polymer shells. Fig.22 shows computed opacity profiles for 2.3 keV X rays at early, intermediate and final stages of compression for a polymer shell of 125μm

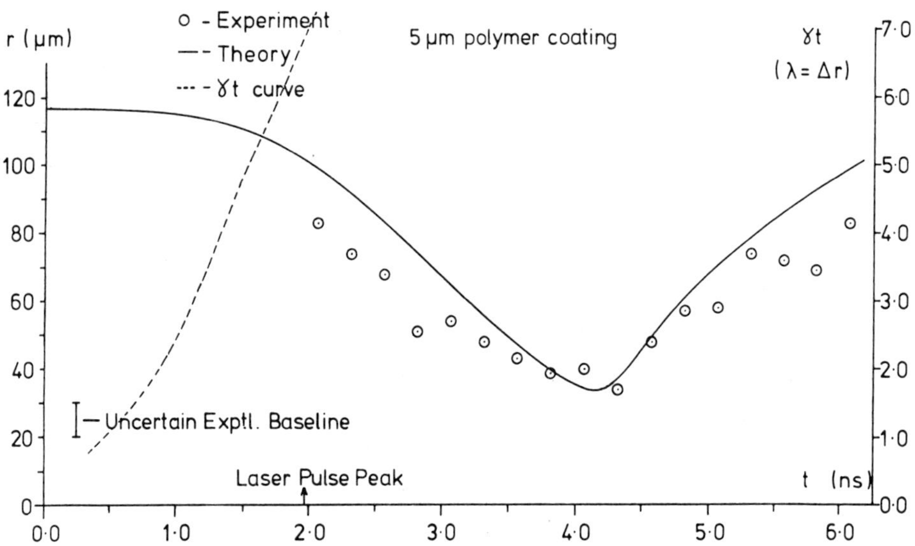

Fig. 21   Computer simulation (solid line) showing the radius as a
          function of time of the outer boundary of the region of
          opacity   0.5 to the experimental backlighting source.
          Experimental results are circles. The classical growth
          exponent γt for the Rayleigh Taylor instability is also shown

radius and 2µm wall thickness filled with 10^{-2} g cm^{-3} of gas and
irradiated at constant laser power of 230 giga watts giving an
irradiance of 1.2 x 10^{14} W cm^{-2}.   The contribution of the shell and
the fill gas to the opacity is shown separately.   Initially the
opacity profile falls well within the maximum measurable opacity of
about 2 (which is limited by the dynamic range of the imaging
system). This permits, in principle, study of the mass density
distribution in the imploding shell.   At peak compression the
compressed gas density reaches $\rho \sim$ 3 g cm^{-3} and the ρr value reaches
$\rho \sim$ 6 x 10^{-3} g cm^{-2}.   If the gas is chosen to be Argon or Neon (or
mixtures) a range of dominance of gas absorption over shell absorption is available as indicated in Fig.22.   Scaling down of the peak
opacity to measurable values < 2 requires increased X ray energy $h\nu$
(opacity $(h\nu)^{-3}$).   A direct measurement of the gas density from
the opacity profile is seen to be possible with practically feasible
X ray energies ($h\nu \sim$ 3 keV).

     The experimental irradiation symmetry has been greatly improved
with the six beam irradiation provided by newly constructed apparatus

at the SRC Central Laser Facility 1.   Streaked X ray shadowgraphy in
this system is accomplished with a seventh beam to provide an X ray
backlighting plasma and with a 15 x magnification Kirkpatrick Baez
microscope to relay the shadowgraph image to a slit at the streak
camera photo cathode.   Preliminary results obtained with this system
show a very narrow absorbing implosion core with dimensions $\sim$ 15μm.
The increased penetration of the X rays in the polymer target is
evident by comparison with Fig.20.

Work is continuing to measure the gas density directly and to
study the density profile and thus the stability of the imploding
shell.

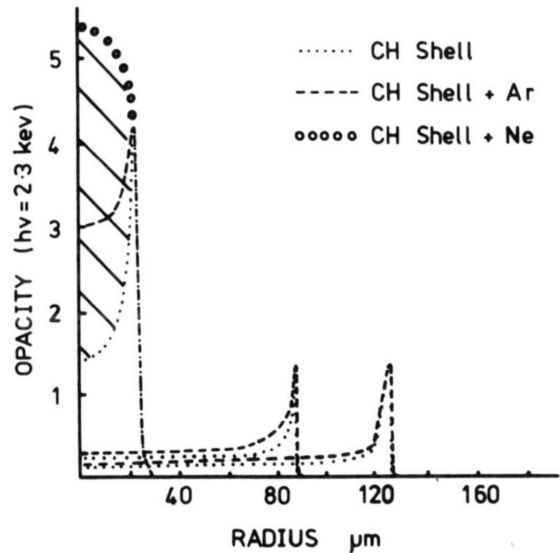

Fig. 22   Computer simulations showing opacity profiles for 2.3keV
          X rays during the implosion of a 250μm diameter 2μm wall
          thickness polymer microballoon filled with 10^{-2} g cm^{-3} of
          either A or Ne.

REFERENCES

1. Rutherford Laboratory Annual Report to the Laser Facility
    Committee 1979, RL-79-036.
2. D. Attwood et.al., Phys.Rev.Lett., 40, 184 (1978).
3. C.E. Max and C.F. McKee, Phys.Rev.Lett., 39, 1336 (1977).
4. J. Virmont, R. Pellat, and A. Mora, Phys.Fluids 21, 567 (1978);
    K. A.Brueckner and R.S. Janda, Nucl.Fusion 17, 451 (1977).
5. A. Raven and O. Willi, Phys.Rev.Lett., 43, 278 (1979).
6. R.E. Kidder, in Proceedings of Japan - U.S. Seminar on Laser
    Interaction with matter, Tokyo, 1975 (unpublished).
7. R. Benattar, C. Popovics, and R. Sigel, "A polarised light
    interferometer for laser fusion studies" (unpublished).
8. N. Minyanaga and K. Tanaka, Osaka University Report No.ILE-APR-
    77, 1977 (unpublished).
9. M.D. Rosen, D.W. Phillion, V.C. Ruperts, W.C. Mead, W.L. Kruer,
    J.J. Tomson, H.N. Kornblun, V.W. Slivinsky, G.J. Caporao ,
    M.J. Boyle, K.G. Tirsel, Phys.Fluids., 22, 2020 (1979).
10. D.W., Forslund, J.M. Kindel, K. Lee, Phys.Rev.Lett, 39, 284
    (1977).
11. W.L. Kruer, Lawrence Laboratory preprint U.C.R.L., 82701 (1979)
    (submitted to Phys.Fluids).
12. B.H. Ripin,et.al., Phys. Rev.Lett. 33, 634 (1974).
13. J.W. Shearer et.al., Phys.Rev., A6,764 (1972)
14. K.A. Brueckner, Nucl.Fusion 17, 1257 (1977)
15. L.M. Wickens, J.E. Allen and P.T. Rumsby, Phys.Rev.Lett., 41,
    243 (1978).
16. J.D. Hares, J.D. Kilkenny, M.H.Key, J.G.Lunney, Phys.Rev.Lett.,
    42, 1216 (1979).
17. B. Yaakobi, I.Pelah, J.Hoose, Phys.Rev.Lett., 37, 836 (1976)
18. A. Zigler, H. Zmora, and J.L. Schwob, Phys. Letts., 63A, 275
    (1977).
19. K.B. Mitchell, R.P. Godwin, J.Appl.Phys., 40, 3851 (1977).
20. L.V. Spencer, Energy Dissipation by Fast Electrons, National
    Bureau of Standards Mongraph No.1 (U.S., GPO, Washington D.C.
    1959).
21. S.J. Gittomer and D.B. Henderson, Phys.Fluids 22, 364 (1979).
22. B. Yaakobi and T.C. Bristow, Phys.Rev.Lett., 38, 350 (1977).
23. F.C. Young, R.R. Whitlock, R. Decoste, B.H. Ripin, D.J. Nagel,
    J.A. Stamper, J.M. McMahon, and S.E. Bodner, Appl.Phys.Lett.,
    30, 45 (1977).
24. F. Amiranoft et. al., 7th Int.Conf. on Plasma Physics and
    Controlled Thermonuclear Research, Innsbruck 1978, paper IAEA-
    CN-37-D4.
25. H.D. Shay, R.A. Haas, W.L. Kruer, M.J. Boyle, D.W. Phillion,
    V.C. Rupert, H.N. Kornblum, F. Rainer, V.W. Slivinsky,
    L.N. Koppel, L. Richards, and K.G. Tinsell, Phys. Fluids, 21,
    1634 (1978).

26. J.G. Lunney, C.L.S. Lewis, M.H. Key, J.D. Kilkenny, R.W. Lee,
    Proceedings of 9th European Conference on Controlled Fusion
    and Plasma Physics, Oxford 1979, p.119.
27. J.D. Kilkenny, R.W. Lee, M.H.Key, J.G. Lunney, submitted to
    Phys.Rev.
28. R.W. Lee, J. Phys.B 11, L167 (1978).
29. J.N. Olsen, G.W. Kuswa and E.D. Jones, J.Appl.Phys. 44,
    2275, (1973).
30. C.G.M. Van Kessel and R. Sigel, Phys.Rev.Lett., 33, 1020 (1974).
31. M.H. Key 'Lectures on the Physics of the Super Dense Region'
    to be published in the Proceedings of the Scottish Universities
    Summer School in Physics Laser Plasma Interactions (1979).
32. M.H. Key et.al., Rutherford Laboratory Report RL-79-014 (1979).

LASER INTERACTION WITH TARGETS

IN "COMPRESSED SHELL" REGIME

Yu.V.Afanasiev, N.G.Basov, N.N.Demchenko, A.A.Erokhin,
E.G.Gamaly, S.Yu.Gus'kov, A.I.Isakov, A.A.Kologrivov,
Yu.A.Merkuliev, V.B.Rozanov, A.A.Rupasov, A.A.Samarsky*,
A.S.Shikanov, G.V.Sklizkov, B.L.Vasin, P.P.Volosevich*,
Yu.A.Zakharenkov, N.N.Zorev.

P.N.Lebedev Physical Institute, USSR Acad.Sci.,
Leninsky prospect 53, Moscow, USSR
* Institute of Applied Mathematics, Moscow

INTRODUCTION

Most of the latest laser fusion investigations
are devoted to theoretical and experimental study
of heating and compression of spherical shell targets
/1/. Modern approaches in laser fusion may be subdi-
vided into two groups.

The first one includes experiments with laser
pulse duration less than the time of shell implosion
and a laser flux density of $\gtrsim 10^{15}$ W/cm^2 /2/ .In these
experiments a thin shell (wall thickness ~ 1 μm) is
heated by a thermal wave before the implosion collapse
time, and its density becomes lower than the initial
one (so-called "exploding pusher" regime). The final
shell and the enclosed gas densities are not large,
but due to high thermoconductivity there is a weak
influence of shell and irradiation assymmetry and
hydrodynamic instabilities on the processes in the
target under compression.

In the second group the laser pulses with time
duration near or more than the shell compression time
and the moderate flux densities $q \sim 10^{14}$ W/cm^2 have
been used /3,4/ . In this case the inner part of the
shell remains non-evaporated by the moment of maximal

gas compression, and its density may be much greater than the initial one (ablative, or "compressed shell" regime). The fact that the initial entropy in compressed thermonuclear fuel is introduced only by a shock wave, is an important feature of this regime, and that is why its value is much less than in the case of the "exploding pusher", when the entropy is added by hot electrons. As the result the greatest volume compressions ($\delta \sim 10^3$) along with relatively low gas temperatures ($T \sim 0.2$–0.5 keV) are realized. In "exploding pusher" regime the same parameters are: $\delta \sim 10^2$ and $T \sim 3$–7 keV.

As the theoretical study /5/ showed, the densities of the fusion fuel, required for fusion ignition, oan be reached in the "compressed shell" regime even at contemporary laser energy level ($\sim 10^2$–10^3 J) under certain conditions of irradiation and target symmetry. This regime seems to be dominant in the breakeven experiments, as a high value of the ρR parameter ($\rho R \gtrsim 0.1$ g/cm^2) will lead to development of fusion ignition wave /6/ . In the present paper the results of experimental investigations on "Kalmar" laser system and theoretical analysis of nanosecond laser interaction with shell targets and the processes in laser-produced plasma corona are discussed.

EXPERIMENTAL SET-UP

Results of the experiments carried out with "Kalmar" nine-beam laser system /7,8/ are presented. The output laser energy was up to $E_\ell \approx 150$ J, pulse duration on the base was $\tau_i \approx 2.5$ ns with rise time $\tau_4 \lesssim 0.5$ ns, full width on half maximum of radiation line was $\Delta\lambda \lesssim 10$ A. The laser radiation was focused onto the target by using nine two-lens systems with effective focal length F=20 cm. The focal plane of each system was located behind the target. One may satisfactorily use the fitting of the laser intensity distribution in the target plane with the following function:

$$I(\xi) = I_0 \exp\left\{-\left(\frac{\xi}{a}\right)^2\right\} + I_1, \quad \begin{array}{l} 0 \lesssim \xi \lesssim 350\,\mu m, \\ I_1/I_0 = 1.72\cdot10^{-2}, \end{array} \tag{1}$$

where I_0, is the maximum laser intensity; ξ , is the distance from an optical axis, a =70 um. The first term in (1) describes the part of laser radiation with low angular divergency ($\alpha < 2.10^{-4}$ rad), the second one corresponds to the "wings" intensity distribution, which are conditioned by the presence of large divergency component in the heating beam. With the energy on target ~ 100 J the maximum flux density of the laser pulse was $q \approx 2.10^{14}$ W/cm^2.

The targets were hollow and deuterium gas filled shells 70 - 300 μm in diameter ($2R_o$) and 0.5 - 4 μm wall thickness (ΔR). The methods of target manufacture, filling, selection and control were described elsewhere /9/. The shell control was performed by interferometric methods. The target was suspended on a rubber glue thread with the thickness of 0.5 μm. Thread was tightened on a V-shaped holder.

The main diagnostics methods used in the experiments are briefly listed below.

I. Energy balance measurement.
1. Calorimetric measurements of the passing the target, refracted and reflected laser radiation /10/;
2. Measurements of the absorbed energy with open-type caloremeters /10/;
3. Registration of a shock wave expansion in residual gas atmosphere with many-framed shadow and Schlieren photography /11/.

II. High speed interferometry of the plasma corona.
1. Slit scans of interferograms with the streak camera /12/;
2. Many-framed interferometry /13/.

III. Spatially and temporally resolved measurements of the evolution of the second harmonic emitting region /14/.

IV. The study of $2\omega_o$ and $3/2\omega_o$ spectral lines.
1. Spatially resolved measurements /15/;
2. Spectral slit scans with the streak camera /16/.

V. Many-channel soft X-ray pinhole plasma photography with high spatial resolution.

VI. Spectral measurements of X-ray continuum plasma emission.
1. X-ray films (1 keV $\lesssim h\nu \lesssim$ 10 keV) /17/;
2. Scintillators with photomultipliers ($h\nu > 5$ keV);
3. Thermoluminecsent detectors (TLD) (5keV $\lesssim h\nu \lesssim$ 15 keV) /18/;

4. Nuclear emulsions ($4\,\mathrm{keV} \lesssim h\nu \lesssim 17\,\mathrm{keV}$) /18/.

VII. Neutron yield measurements.
1. Time integrated neutron counters;
2. Time-of-flight neutron measurements with scintillators
   and photomultipliers.

VIII. Energy composition measurements of the plasma
      charged particles /19/.
1. Charged particle collectors;
2. Electrostatic charged particle analyzer.

LASER RADIATION ABSORPTION

Experimental investigation of the laser radiation
absorption at "Kalmar" set-up was carried out by dif -
ferent methods. They were: two calorimetric methods, ion
collector measurements and the method of a shock wave in
the residual gas. In the first calorimetric method the
absorbed energy E_{ab} was determined from the measurements
of a total laser energy on the target (E_i) and the energy
losses, caused by refraction (E_r) and by the laser radiation component with high angular divergency, which passed
the target (E_+). Calorimeter arrangement is schematically
presented in Fig. 1.

The second method is based on data comparison of the
adjacent open and filter covered caloremeters. Open detectors (6 and 7, Fig. 1) are sensible to the summary
energy ($E_{pl}+E_r$) of the expanding plasma particles, plasma
emission and reflected laser rays. The filter covered
ones measure only the refracted energy (E_r).

Dependences of different energetic balance terms,
normalized by incident laser energy, on the target diameter are presented in Figs. 2-4. Experimental and theore
tical ("Rapid" numerical simulation without anomalous
absorption mechanisms /20/) results are shown. Scattering
of the passed by energy data (Fig.2) increases with the
decrease of the target diameter.

This is, apparently, due to some deviation of the
heating beams' optical axes from the target center. So,
the relative contribution of this effect to the measurement error grows with the decrease in the target dimensions.

The energy of the refracted laser radiation is
about 30-40 % (Fig. 3).

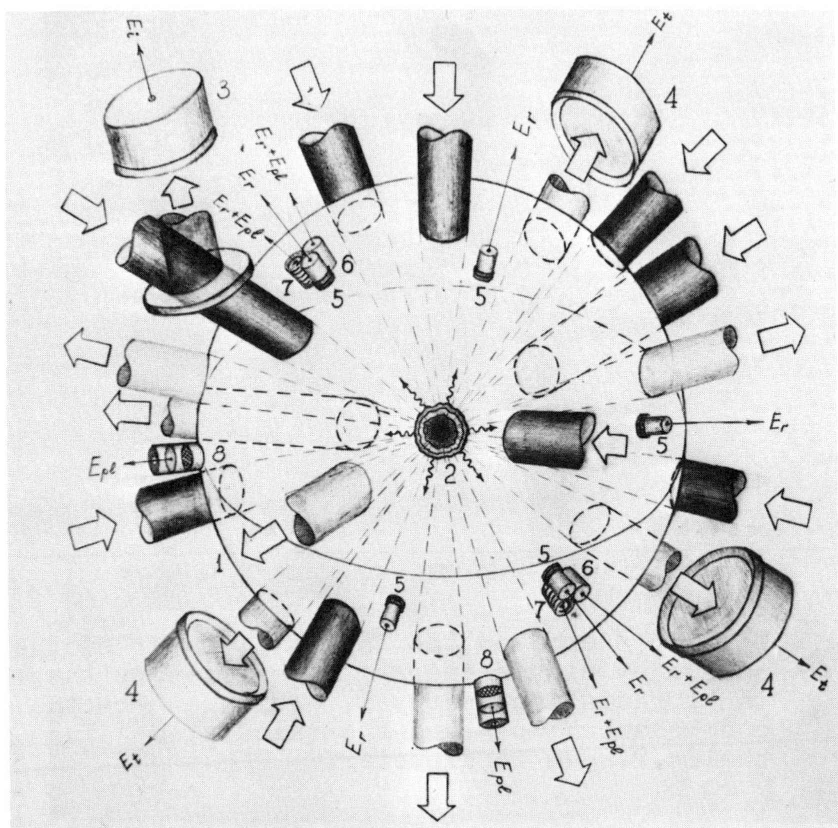

Fig.1. Scheme of disposition of the heating beams and
       calorimetric detectors: 1 - vacuum chamber;
       2 - target; 3,4 - KDS calorimeters; 5 - VChDM
       calorimeters covered by filters; 6 - opened
       VChDM calorimeters; 7 - opened DG calorimeters;
       8-electrostatic ion collectors; the arrows in-
       dicate the heating beams and the beams passing
       the target.

The fractional absorption, obtained with the first
calorimetric method, increases from 15% **up** to 50% with
increasing the shell diameter from 80 μm to 260 μm
(Fig. 4).

One of the possible causes for visible refracted
and absorbed energy data discrepancy between the theoretical and experimental results in case of a large (200-
-260 μm) shell diameter may be the destruction of the irradiation spherical symmetry, and, consequently, of the
corona hydrodynamic expansion. The shell dimensions in
this case considerably exceed the diameter of the heating
beams (see also /21/).

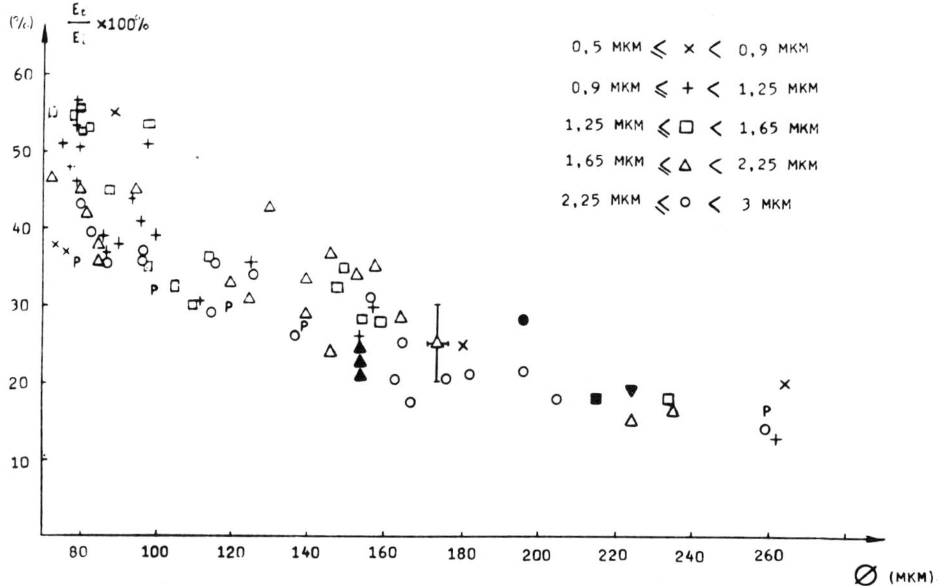

Fig. 2. The dependence on the target diameter of the passed by plasma energy, normalized by the laser radiation energy E_i, which enters the vacuum chamber; p - are the calculated values for shells 2.2 μm in diameter.

As numerical calculations for "Kalmar" experimental conditions have shown, the contribution of the resonance absorption mechanism is insignificant, about ∼ 1% of the absorbed energy value. This is conditioned firstly by the specificity of the focusing scheme used, and, secondly, by the fact that a relatively smooth density profile was formed in the plasma corona region at a sufficiently long pulse in the "compressed shell" regime. Thus, for the majority of the laser rays the distance from a turning point to the critical one considerably exceeds the laser wavelength. A typical corona density profile can be well fitted by the exponent function with the scale length about ≈ 30 μm.

One can assume the role of anomalous absorption mechanisms to be insignificant as well. At the present time there is no complete physical understanding of the anomalous absorption processes, which is needed for quantitative comparison with the experiment. For example, when

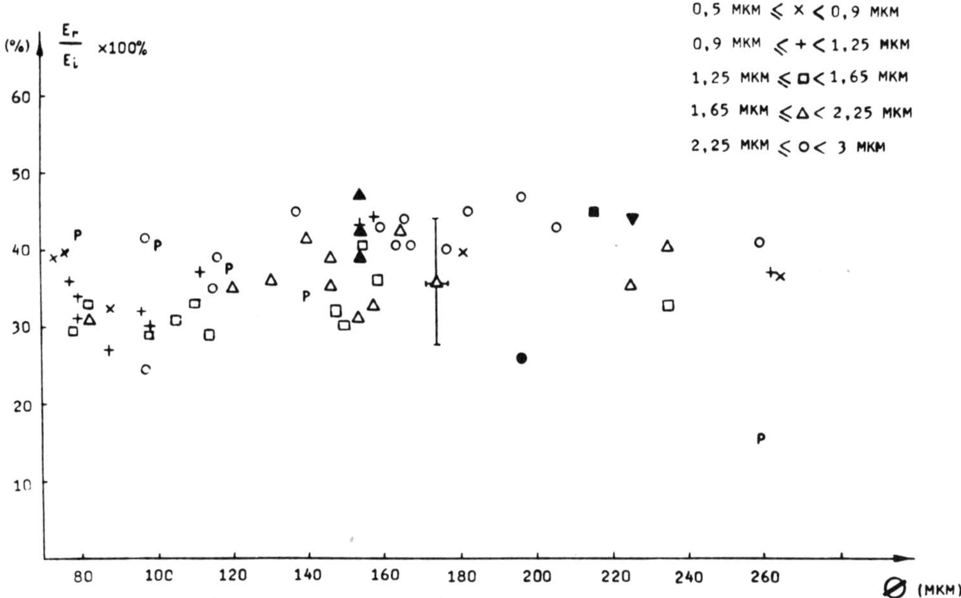

Fig. 3. The dependence of the refracted energy (normalized by E_i energy) on the target diameter; p – calculated values for shells 2.2 μm in dia - meter.

anomalous mechanisms were taken into account for uniform steady-state plasma /22/, then the absorbed energy value has increased by a factor of 1.5. Due to in-homogeneity and plasma expansion for "Kalmar" experiments with flux densities $q \sim 10^{14}$ W/cm^2 it is clear that the instability thresholds were only slightly exceeded.This gives a possibility to make an assumption that inverse bremsstrahlung mechanism is responsible for practically the whole value of the absorbed laser energy.

In plasma the absorbed energy transforms into the energy of the particle expansion and electromagnetic emis-sion, which has the maximum intensity in the X-ray spect-ral region. With many-channel film detectors the soft X-ray energy in the experiments was estimated to be ~ 0.5 J ($\sim 2\%$ of absorbed energy).

Charged particle collectors associated with an elec-trostatic charged particle analyzer provided the value of the absorbed energy with reasonable accuracy.Experimental

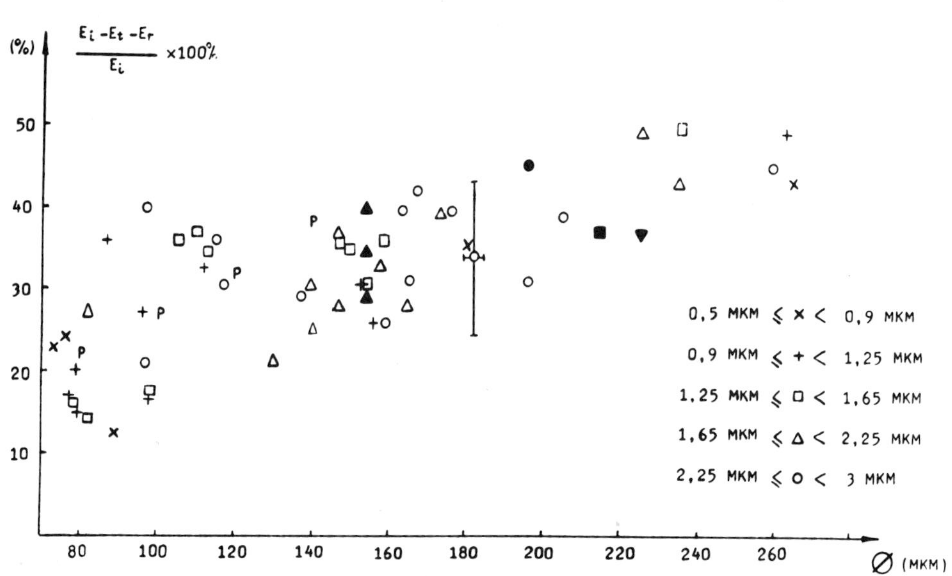

Fig. 4. The dependence of the absorbed by plasma
energy (normalized by E_i energy) on the
target diameter;
p – calculated values for shells 2.2 μm
in diameter.

results obtained by this method were in good agreement
with calorimetric measurements.

The last method for the definition of absorbed
energy consisted in the determination of the energy of
a spherical shock wave in the residual gas atmosphere
from R-t front expansion diagram, obtained with many-
-framed schlieren photography /11/. The laws of strong
ionization shock wave expansion in homogeneous atmos -
phere /23/ have been used for the purpose.

"COMPRESSED SHELL" PLASMA CORONA

Laser light absorption region is the source of the
waves of electron heat transport both into the dense
target and the underdense expanding plasma corona. Analysis of the experimental and theoretical results has
revealed that the plasma, being irradiated with nanosecond laser pulse with flux density $q \gtrsim 10^{14}$ W/cm^2, is

not characterized by model of anomalous flux-limited
electron energy transfer. The experiments are satisfactorily described in terms of the classical thermal transport theory. From numerical calculations it follows that
in the region in question $\ell_e \ll$ L(L=T_e/ ∇ T_e; ℓ_e is
the electron mean-free-path); hence, the classical Spitzer
thermal flux is considerably less than the upper bound
on the amount of energy which the electrons transport
without collisions, $n_e \upsilon_e \kappa T_e$, and is also less than
the flux corresponding to the ion velocity V_i :

$$q_{sp} = \varkappa_e \nabla T_e < n_e \upsilon_i \kappa T_e \ll n_e \upsilon_e \kappa T_e$$

Relaxation processes in the corona and dense target
matter have different character. In the corona, where
$\rho \gtrsim \rho_c$, the absorbed laser energy is transported in
general by electrons, and there is always $T_e > T_i$. In
the region of compression the energy transfers to the
hydrodynamic motion, and $T_e \gtrsim T_i$.

In the experiments time and space averaged value of
the corona electron temperature was determined with the
help of X-ray continuum spectral measurements (by the method of absorbers) and nuclear emulsions. The application
of the complex method of plasma X-ray registration by
many channels /18/ has allowed one to reconstruct the
spectrum of emission in the range of 3-25 keV. Different
reconstruction approaches /24/ gave practically the same
results.

Fig. 5 shows the spectral distribution of X-ray
emission from the plasma corona of glass shell $\sim 150\,\mu$m
in diameter and $\sim 1.3\,\mu$m wall thickness. It should be
noted that this spectrum has only an illustrative character and can not be used for the determination of an"ef -
fective" electron temperature within different spectral
regions. The reason of this fact lies in a strong dependence of the "effective" temperature on the slope of the
curve N(E). So, the determination of T_e requires higher
accuracy of the spectrum reconstruction. In our work T_e
was determined by the method of absorbers. It was $\sim$0.6 keV
in the spectral range 3-8 keV, and 2.7 keV in the range
15-25 keV. "Janus" and "Argus" "exploding pusher" spectra,
taken from paper /25/, are also drawn for comparison.

Generalization of the results obtained in a large
number of experiments shows that the plasma electron temperature values, time and space integrated, for different

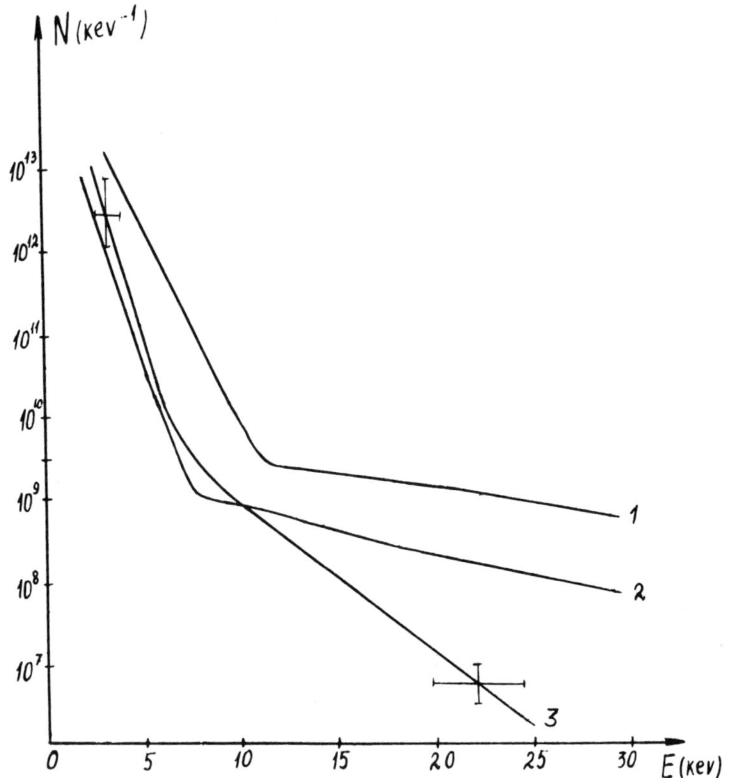

Fig. 5. Spectral distributions of X-ray quanta obtained with "Argus" (I), "Janus" (2) /25/ and "Kalmar" laser systems. Processing error in different spectral ranges is marked by a symbol ⊢╋⊣ .

shell targets lie within the range of 0.3-0.9 keV (3 - -8 keV range) and 2-3 keV (15 - 25 keV) range.

In order to provide time resolution in determining the plasma corona electron temperature we have studied time evolution of the spectral distribution in $3/2\,\omega_0$ harmonic radiation (generated in plasma region with the electron density of $n_c/4$. The experimental and theoretical study /26/ has shown that the spectrum of the $3/2\,\omega_0$ harmonic consists of two components, which are shifted in different sides from the nominal wavelength of $2/3\,\lambda_0$, and the spectral distance $\Delta\lambda_{3/2}$ between them is proportional to the plasma electron temperature:

$$T_e(t) \approx 1.5 \cdot 10^2\, \Delta\lambda_{3/2}(t)\, /\, \lambda_0$$

where λ_o, is the laser wavelength; T_e is presented
in keV. For example, for one of the shells /16/ the distance between the harmonic components was 38,44,32 $\mathring{A}$ for
0.8, 1.5, 2.3 nsec periods of time, and it corresponds to
the ele tron temperature of the $n_{cr}/4$ region 570, 660,
480 eV, respectively. Note, that the mean electron temperature, obtained by the abovementioned method is in a sufficiently good agreement with the X-ray measurements.

Fig.6 demonstrates a scheme of the laser plasma corona optical diagnostics used in the experiments.For the
determination of the plasma expansion spherical symmetry
many-framed interferometry was used /13/.

Simultaneously, we have performed slit scans of
shearing interferograms and the image of plasma region
where emissivity of the second harmonic of the heating
radiation occurs /14,27/ (slit images were separated due

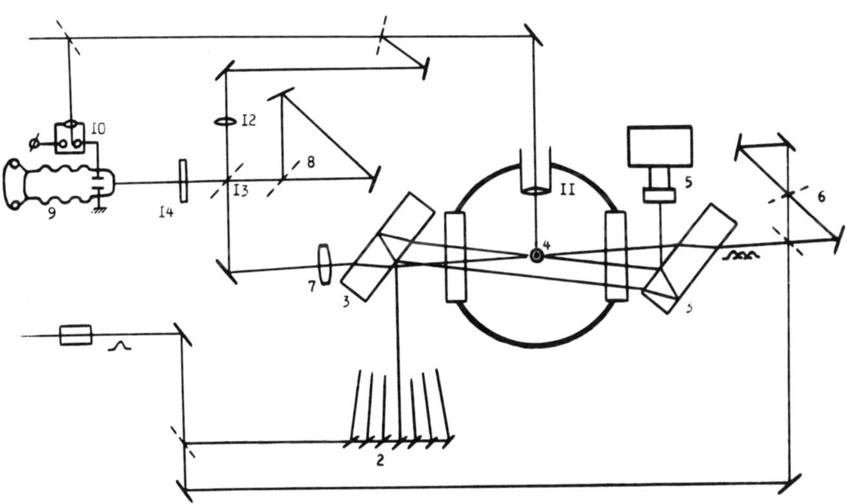

Fig. 6. The scheme of plasma optical diagnostics,
1 - KDP crystal; 2 - light delay system; 3-Jamin
interferometer; 4-target; 5-many-frame interferometric photocamera; 6- the system of formation
a combined light pulse; 7-objective; 8-shearing
interferometer; 9-photoelectron recorder; 10-
-discharger with laser ignition; 11-lens, focusing the heating radiation onto the target;
12- a lens; 13- semitransparent mirror; 14-interferometric light filter.

to optical delay). Measured electron density distributions $n_e(r)$ at different moments of time in two experiments are presented in Fig. 7, as well as the density
profiles, obtained by numerical simulations of the same
experiments according to "Luch" code /28/.

The analysis of these data disclosed not only general agreement in the dynamics of the calculated and
experimental distributions, but the proximity of their
absolute values as well. This enables one to assume that
numerical simulation describes well the state of the laser plasma corona, and such characteristics of the corona as the degree of ionization Z , and the evaporated
mass are close to the calculated ones. In the calcula -
tions Z was ~ 10, and the value of the evaporated
mass constituted 30 and 40 % for two experiments presented

One of the interesting features of the density
distribution is the presence of a perturbation, which is
formed by the beginning of the second nanosecond.Theore-
tical analysis has revealed that this perturbation was
formed in the corona by a shock wave, reflected from the
target centre and passing the shell compressed part.

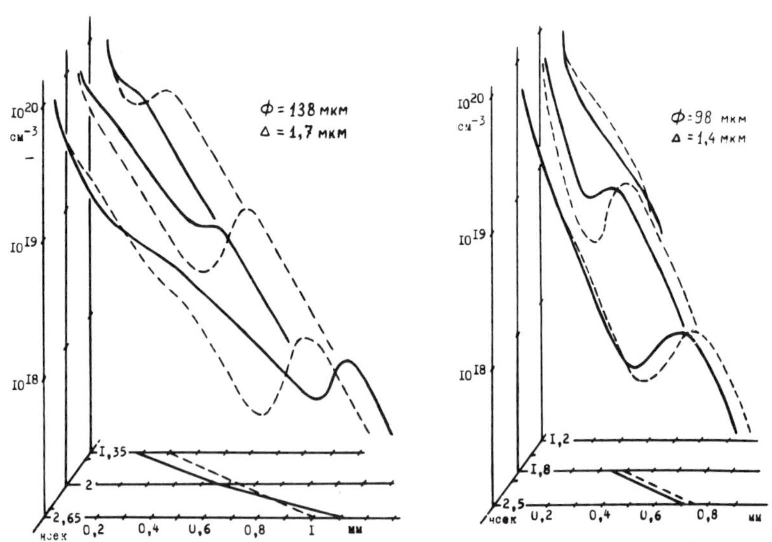

Fig. 7. Shock wave motion in the plasma corona.
        Dense lines illustrate interferometric
        measurements; dotted lines show the cal-
        culated electron density profiles.

The perturbation reaches the point R at the time t, which is combined of the shell collapse time t_* (one may approximately assume that just at that moment the shock wave from deuterium enters the compressed shell) and the time of propagation along the compressed shell Δt_s and the corona Δt_c. The numerical simulation for thin-wall shells gives $\Delta t_s \approx 0.1\text{-}0.2 t_*$; $\Delta t_c \approx R/\overline{u}$,where $\overline{u}$ is the mean velocity of the perturbation in the corona.

Thus, the study of the temporal evolution of the mentioned density perturbations makes possible to define such an important dynamic characteristic of the shell target compression as the collapse time:

$$t_* \approx t - R/\overline{u} \ .$$

According to the measurements of the perturbation position in the experiments discussed the $\overline{u}$ value was $6.5 \cdot 10^7$ cm/sec and $5 \cdot 10^7$ cm/sec, which gives $t_* \approx 1.0$ nsec and 1.1 nsec. It should be noted, that the problem of shock wave entering the laser plasma corona is close to the one of emerging of a shock wave at the star surface /29/. Analytical solution of the problem involves difficulties due to a complex character of the temperature,density and the matter velocity in plasma.

DYNAMICS OF THE SHELL MOTION UNDER COMPRESSION

The formation of the laser target corona presents the first stage of heating and compression of the targets irradiated by the laser. The second stage involves the shell motion under the action of the corona reactive impulse, the processes of the energy transfer to the compressed thermonuclear fuel, its heating and compression.

The equation of the motion of non-evaporated shell fraction has the form:

$$M\frac{du}{dt} = 4\pi R^2 \left[p + \rho(\upsilon+u)^2 - p_g \right], \ u(t=0)=0, \qquad (2)$$

where u , is the shell velocity; R, is the evaporation interface coordinate; p , ρ and υ , are the pressure, density and the velocity of the corona particles at the evaporation interface; p_g , is the compressed gas pressure. M mass value is determined by the evaporation regime:

$$\frac{dM}{dt} = -4\pi R^2 \rho_s^o D \ , \ M(t=0)=M_o \qquad (3)$$

where ρ_s^o , is the density of the non-evaporated shell; $\mathcal{D}$, is the velocity of the evaporation wave front in relation to non-evaporated shell.

At flux density of the Nd-laser $\gtrsim 10^{14}$ W/cm^2 the state of the corona is close to one of the limiting cases in the model considered in /30/. Under these conditions during the greater part of the period of the shell motion both the surface of the critical density ρ_c and the Jouguet surface are lying in the vicinity of the evaporation interface. Taking into account that $\mathcal{D} \ll u + \upsilon$, one may obtain:

$$\rho_s^o \, \mathcal{D} = \rho_c \, c \, , \tag{4}$$

$$p + \rho (\upsilon + u)^2 = 2 \, \rho_c \, c^2 \, , \tag{5}$$

where c is the sound velocity in the corona at the evaporation interface.

Eqs.(2) and (3) with account of the ratios (4) and (5) completely describe a self-consistent problem of the shell implosion under the pressure of the evaporated mass. At the stage of the shell acceleration ($p_g \ll p + \rho(u+\upsilon)$; $dE_{ab}/dt = const$) it follows from the solution of the equations:

$$M_* = M(t_*) = C_1 M_o \left[1 - \left(\frac{\tilde{M}}{M_o} \right)^{1/2} \right]^2 ; \quad \tilde{M} = \frac{5}{48} \left(\frac{R_o \, \rho_c}{\Delta_o \, \rho_s^o} \right) M_o , \tag{6}$$

where R_o and Δ_o , are the initial shell radius and wall thickness. As the shell deceleration time is signifi - cantly shorter th..n time of the acceleration, the collapse time t_* is also defined from Eq.(2) with a suf - ficient accuracy:

$$t_* = C_2 \left(\frac{5}{12} \right)^{1/2} \frac{\pi}{2} \left(\frac{M_*}{M_o} \right)^{1/2} \tilde{t} \left(\frac{R_o \, \rho_c}{\Delta_o \, \rho_s^o} \right)^{-1/2} , \tag{7}$$

$$\tilde{t} = R_o \left[\frac{2(\gamma - 1)}{3\gamma - 1} \frac{dE_{ab}/dt}{4\pi R_o^2 \, \rho_c} \right]^{-1/3}$$

Hydrodynamic transfer coefficient $\eta = E_g / E_{ab}(t_*)$ has the form:

$$\eta = C_3^2 \frac{\gamma - 1}{3\gamma - 1} \left[\left(\frac{12}{5} \right)^{1/2} \frac{1.43}{\pi} \right]^3 \left(\frac{R_o \, \rho_c}{\Delta_o \, \rho_s^o} \right)^{1/2} \left(\frac{M_o}{M_*} \right)^{1/2} \tag{8}$$

The solutions obtained have the only one dimensional parameter of the interface sound velocity and a dimensionless $R_0 \rho_c / \Delta_0 \rho_s^0$ parameter, which shows the ratio between the accelerating and accelerated surface masses. Thus, the dependences of the basic characteristics of the shell dynamics on the initial target parameters and the laser pulse are conditioned by similarity relations (6)-(8). The comparison with the numerical results obtained with "Luch" code /28/, modelling the "Kalmar" experiments, shows that formulas (6)-(8) give the values close to the calculated ones (the values of constants c_1, c_2, c_3 are 0.8, 1.4 and 1.6, respectively).

Experimental study of the target compression dyna - mics was performed by registering R-t diagrams of the plasma critical density surface. Under weak time dependence of the laser radiation flux, the plasma moves in accordance with the non-evaporated part of the shell. So, in the experiments we have studied the dynamics of the motion of the non-evaporated shell by the photographs of the plasma emission time scanned image at the second harmonic frequency of the heating radiation, and in parti - cular, we have defined the target implosion time (t_*).

Table 1 demonstrates the target implosion times obtained in the experiments /31/, and the t_* values obtained in the numerical simulations and from formula (7). It is seen that within the frames of the simple hydrodynamic model formula (7) shows correctly the dependence of t_* on the shell parameters and the energy absorbed.

Table 1.

SHELL FEATURES			ABSORBED ENERGY,	TIME OF IMPLOSION, NS		
DIAMETER, μM	THICKNESS, μM	PRESSURE, ATM	J	EXPERIMENTAL	NUMERICAL	ANALITICAL
125	2.25		15 ± 3	1.5 ± .15	1.5	1.02
125	2.1	15	9 ± 1.5	1.8 ± .15	1.7	1.1
100	1.2		13 ± 2.5	1.25 ± .15	1.1	.75
98	1.4		12 ± 2	1.4 ± .15	1.3	.86
90	1.1	22	11 ± 2	1.1 ± .1	.95	.66
75	.9	15	10 ± 2	.9 ± .1	.8	.53

Good agreement with theoretical results indicates the
fact that in the experiments the corona pressure and
the non-evaporated mass value were in the proximity with
the calculated ones. Numerical simulations give 10^5-10^6 atm
pressure near the surface with ρ_c ; $T_e(R_c) \approx 0.5$-0.7 keV,
the average degree of ionization ~ 10, $M_*^c/M_o \approx 0.6$-0.7,
hydrodynamic transfer coefficient $\eta \approx 5$-10 %.

   Shell dynamics has been studied by time resolved
measurements of the spectrum of $2\omega_o$ /16/ harmonic radiation. The $\Delta\lambda_2$ shift (in respect to a nominal value
$\lambda_o/2$) of the basic component in the harmonic spectrum,
which is caused by linear transformation of the heating
beam energy into longitudinal plasma waves, is conditi -
oned by Doppler effect in the moving generation region
/15/. Thus, measuring the time dependence of the $\Delta\lambda_2$
shift gives a possibility to reconstruct the time evolution of the velocity of the critical density region.

   In Fig. 8 is presented the V-t diagram, obtained
from measuring the time evolution of the second harmonic
spectrum in one of the experiments. Maximal velocity

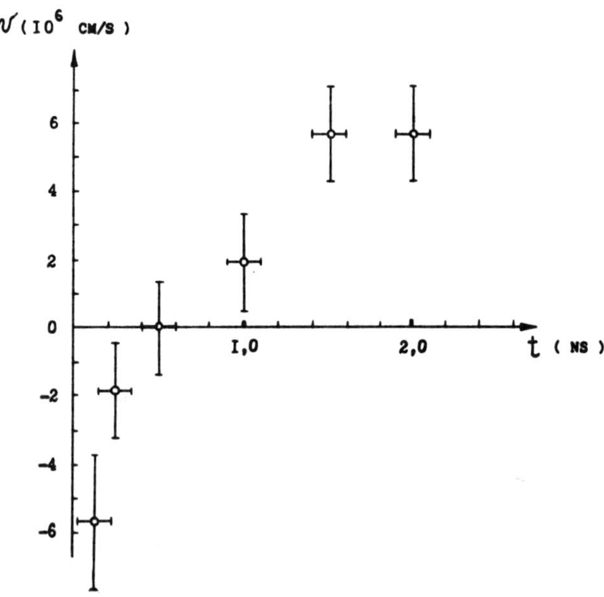

Fig.8. V-t diagram of the motion of the critical density
       region for target 163 μm in diameter, 4.6 μm
       wall thickness.

values of the shell motion towards the target center, measured in such a way, varied within the range (3.10^6-10^7) cm/sec, and the $t_* \approx R_o/V$ max value changed with respect to target parameter according to formula (7).

Important data on the shell motion is obtained by studying the dependence between the position of the plasma emitting region in the X-ray and the laser pulse and shell parameters. By using in the experiments a special laser pulse with two time-separated intensity maxima and 0.7-0.8 nsec time interval, two luminous circles were registered in the pinhole photographs. As the luminous circle regions corresponded to the occurrence of the laser radiation intensity maxima at the target, the usage of a special heating beam allowed one to define the position R of the corona region with electron density close to critical at fixed moments of time t. When the implosion time t_* exceeds the time t of the second heating maximum coming to the target one can find the mean velocity of implosion : $\bar{V} = (R_o - R(t))/t$. Usual values of $\bar{V}$ were in the range $(4-8).10^6$ cm/sec.

Fig. 9 shows the dependence on the shell thickness of the radius of the external circle region in the time integrated X-ray pinhole photographs, which were obtained in the experiments with a single laser pulse. This dependence illustrates the change in the dynamics of the shell motion and critical surface with the change of the shell thickness. With an increase in the shell wall thickness there takes place an increase in the following parameters: the distance covered by the critical density region outwards the target at the initial stage and the implo - sion time. It follows from the results obtained in the present paper and the conclusions from /30/. Up to the moments when the laser pulse intensity maximum reaches the target (0.5 nsec), the critical density region will be at different distances from the initial target radius, depending on the wall thickness. Here lies the explana - tion of the experimental data presented in Fig. 9.

HEATING AND COMPRESSION OF DEUTERIUM GAS

At shell deceleration, when the pressure in the co- rona becomes less than the compressed gas pressure, the whole kinetic energy of the shell is transformed into the internal energy of the compressed gas $\mathcal{E}_g(t_*)$ and the shell $\mathcal{E}_s(t_*)$:

$$\eta E_{ab}(t_*) = \mathcal{E}_g(t_*) + \mathcal{E}_s(t_*)$$

$$(9)$$

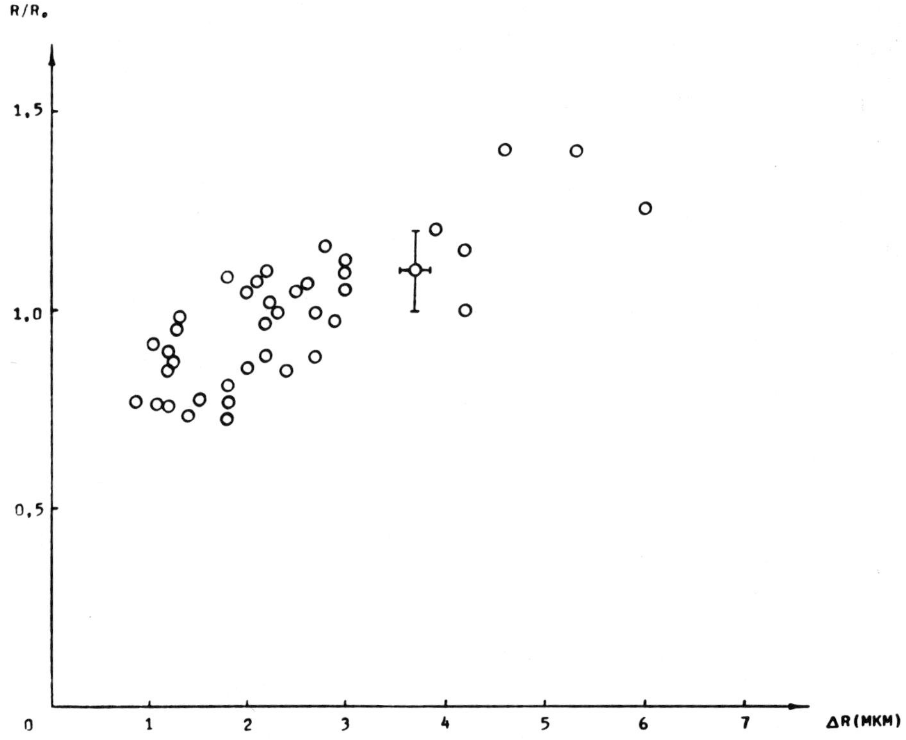

Fig. 9. The dependence on the shell thickness of
        the radius of the external circle region
        in the time integrated X-ray pinhole
        photographs.

In this case we may consider the gas pressure P_g (t_*)
and the shell pressure P_s (t_*) to be approximately equal.
For the mean gas temperature we have:

$$\overline{T}_g(t_*) = \frac{\eta\, E_{ab}(t_*)}{A_g\, m\left[1 + \frac{4}{3}\frac{\delta_g(t_*)}{\delta_s(t_*)}\frac{\rho_g^{\circ}}{\rho_s^{\circ}}\frac{M_*}{m}\right]} \, , \tag{10}$$

where m , is the gas mass; A_g , is its specific heat
capacity; δ_g (t_*), δ_s (t_*), are the degrees of the gas
and shell compression, which are determined from a sche-

matic consideration of the "compressed shell" regime
consisting of two stages: passing of a shock wave through
the shell and gas, and further adiabatic compression.

$$\delta_g(t_*) = \frac{\rho_g(t_*)}{\rho_g^0} \approx 20\,C_4 \left[\frac{M_*}{m\left(1 + \frac{4}{3}\frac{\delta_g(t_*)}{\delta_s(t_*)}\frac{\rho_g^0}{\rho_s^0}\frac{M_*}{m}\right)} \right]^{\frac{1}{\gamma-1}},$$

$$\frac{\delta_g(t_*)}{\delta_s(t_*)} \approx C_5\,(1+\sqrt{5})^{-2/\gamma}\left(\frac{\rho_s^0}{\rho_g^0}\right)^{1/\gamma} \tag{11}$$

Formulas (10) and (11) determine the similarity relations for the target temperature and density at the moment
of collapse, when the changing of the initial conditions
in the "compressed shell" regime takes place. These expressions give the values close to the ones obtained with
"Luch" code, and the coefficients C_4 and C_5 were 1.5
and 1.45, respectively.

In the experiments considered the compressed deuterium temperature lies within the range 0.1 - 1 keV and the
temperature dependence of the D+D reaction cross-section
is very strong. Therefore, almost all the neutrons are
produced during a short period of time close to the mo -
ment of maximum compression. The equation for the shell
motion at adiabatic compression and gas expansion has
the form (γ =5/3) :

$$M\frac{du}{dt} = \pm 4\pi R^2 p_g \;;\; p_g(t) = p_g(t_*)\left[\frac{R(t_*)}{R(t)}\right]^5 . \tag{12}$$

The solution for this equation gives average values of
the gas temperature and density:

$$\overline{T}_g(t) = \overline{T}_g(t_*)\left[1 + \left(\frac{t-t_*}{\tau}\right)^2\right]^{-2} \;;$$

$$\overline{\rho}_g(t) = \overline{\rho}_g(t_*)\left[1 + \left(\frac{t-t_*}{\tau}\right)^2\right]^{-3} \;; \tag{13}$$

where $\tau = \sqrt{M/3m}\,R(t_*)/C_g(t_*)$ is the typical time for the
inertial confinement; $C_g(t_*) = [2/3\,A_g\overline{T}_g(t_*)]^{1/2}$, is the
sound velocity in gas.

The gas temperature and space distribution may be
approximately determined from condition of a constant
electron heat conductivity flux $\varkappa\,T_g^{5/2}\,\partial T_g/\partial \varkappa \approx$ const,
and the equalizing of the gas pressure $\partial p_g/\partial \varkappa \approx 0.$

$$T_g(\tau,t) = T_g(0,t)\left(1-\frac{\tau}{R}\right)^{2/7}; \quad T_g(0,t) = 1.8\,\overline{T}_g(t) ,$$

$$\rho_g(\tau,t) = \rho_g(0,t)\left(1-\frac{\tau}{R}\right)^{-2/7}; \quad \rho_g(0,t) = \overline{\rho}_g(t)/1.8 \tag{14}$$

Using expressions (13) and (14) allows one to determine the number of thermonuclear neutrons:

$$N_{DD} \approx 3\cdot10^{46}\frac{\sqrt{\pi}}{4}\,\frac{257}{\left(K+\frac{3}{2}\right)\left(K+5\right)\left(K+\frac{17}{2}\right)}\,m\,\overline{\rho}_g(t_*)\,\tau<\sigma v>\Bigg|_{T_g=1.8\overline{T}_g(t_*)} \tag{15}$$

where the D+D-reaction cross-section is approximated by an exponential function of the temperature $<\sigma v> = A\,T_g^{K}$.

Registration of the neutron radiation in the experiments with "Kalmar" installation has been performed both by means of scintillation detectors with photomultipliers, located at different distances from the target, and the integral indium activation detector. Time-of-flight measurements performed with three scintillation detectors have revealed that the electron energy corresponded to D+D reaction.

Most of the experiments with gas-filled targets have been performed for the shells with small diameters $2R_0 \lesssim 100$ μm, and unstable neutron yield in this case was about 10^2-10^4 particles per shot. Maximal neutron yield $\sim3\cdot10^6$ has been obtained for a glass shell $2R_0 = 140$ μm in diameter, and the wall thickness $\Delta_0 \approx 2.2$ μm, and the initial deuterium pressure $P_g^0 = 35$ atm /27/.

Instability of the neutron yield in "Kalmar" experiments results from its very strong sensibility to the initial target parameters and experimental conditions at the gas temperature $T \sim 1$ keV ($N_{DD} \sim T^7$).

Moreover, the main contribution into the neutron yield is made by a small central region of deuterium 20-30% of the total mass) having the temperature close to the maximum of space distribution (14). Therefore, if the maximum in the temperature space distribution is not attained due to the compression non-symmetry, then the neutron yield may be considerably less than the expected one.

In our experiments the deuterium volume compression

was determined by plasma X-ray pinhole photographs with
space resolution $\gtrsim$ 10 μm. In Fig.10 is shown the plasma
image in X-ray radiation ($h\nu \sim$ 2 keV) for a glass shell
target 140 μm in diameter, 2.2 μm wall thickness, and
35 atm initial deuterium pressure. External emitting region corresponds to expanding plasma corona, and the internal one - to the glass layer adjacent to the compressed gas. By measuring dimensions of the internal region
one may estimate the gas volume compression.

   In a number of experiments with the following shell
parameters: $2R \approx$ 120-140 μm, $\Delta_o \approx$ 2-3 μm and deuterium
pressure of 30-35 atm, the value of the volume compres-
sion was $\sim 10^3$, which corresponds to compressed deuterium
density $\rho_{g2}(t_*) \approx$ 6-8 g/cm^3 and the parameter $\rho_g R \approx$
5.10^{-3} g/cm^2. These results, as well as the maximal compression velocity $V_{max} \approx 10^7$ cm/sec are in good agreement
with estimations of these parameters by formula (11) and
the calculation results obtained with "Luch" code, which
predict for the mentioned experimental conditions $\rho_g(t_*) \approx$
4-7 g/cm^3. Such an agreement gives one a possibility to
make a statement that the shell heating and compression
in "Kalmar" experiments is actually performed in the "compressed shell" regime. A c racteristic feature of this

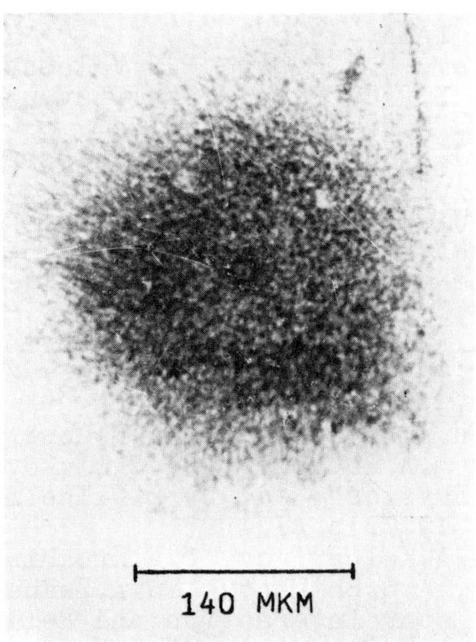

Fig. 10. X-ray pinhole photograph of shell target plasma.

regime lies in the fact that the initial entropy is introduced into the compressing matter only by a shock wave
without preliminary heating by fast electrons and heat
conductivity.

CONCLUSION

The mentioned results enable one to make a complete
model of the laser target heating and compression processes in "compressed shell" regime.

More than 20-30% of the energy is absorbed at laser
energy of ~ 100 J, flux density of $\sim 10^{14}$ W/cm^2 and shell
diameter 100-200 μm. The main contribution into absorp -
tion is due to inverse bremsstrahlung. Classical Spitzer
thermal conductivity gives a possibility to obtain good
agreement of the theoretical, experimental and numerical
results of the state and dynamics of the plasma corona.
The coincidence of the experimental and theoretical R-t
diagrams of the critical surface and the implosion time
at various laser and target parameters is an evidence of
good understanding of the implosion dynamics.

REFERENCES

1. Plasma Physics and Controlled Nuclear Fusion Research,
   v.2, Vienna, IAEA, 1975.
   Yu.V.Afanasiev, N.G.Basov, P.P.Volosevich, E.G.Gamaly,
   O.N.Krokhin, S.P.Kurdyumov, E.I.Levanov, V.B.Rozanov,
   A.A.Samarsky, A.N.Tikhonov; p.559.
   J.Nuckolls, J.Lindl, W.Mead, A.Thiessen, L.Wood,
   G.Zimmerman, p.535;
   G.S.Fraley, W.P.Gula, D.B.Henderson, R.L.McCrory,
   R.C.Malone, R.J.Mason, R.L.Morse, p.543.
2. Laser program annual report, LLL, Livermore, 1976,
   UCRL-50021-76.
3. N.G.Basov, A.A.Kologrivov, O.N.Krokhin, A.A.Rupasov,
   G.V.Sklizkov, A.S.Shikanov, Pis'ma v ZHETP, 23, 474
   (1976).
   N.G.Basov, Yu.A.Zakharenkov, N.N.Zorev, A.A.Kologrivov,
   O.N.Krokhin, A.A.Rupasov, G.V.Sklizkov, A.S.Shikanov,
   In: "Plasma Physics", ed. by H.Wilhelmson, Plenum
   Press, N.Y., 1977, p.47.
   N.G.Basov, A.A.Kologrivov, O.N.Krokhin, A.A.Rupasov,
   G.V.Sklizkov, A.S.Shikanov, Yu.A.Zakharenkov, N.N.
   Zorev. In :"Laser Interaction and Related Plasma Phe-
   nomena",ed. by H.Schwartz and H.Hora, Plenum Press ,
   N.Y., 1977, v.4A, p.479.

4. D.Billon, P.A.Holstein, J.Launspach, C.Patou, J.L.
    Rocchiccioli, D.Schirmann. Report on the 11th Euro -
    pean Conference on Laser Interaction with Matter,
    Oxford, 1977.
5. Yu.V.Afanasiev, N.G.Basov, P.P.Volosevich, E.G.Gamaly,
    O.N.Krokhin, S.P.Kurdyumov, E.I.Levanov, V.B.Rozanov,
    A.A.Samarsky, A.N.Tikhonov, Kvantovaya Elektronika,
    $\underline{2}$, 1816 (1975).
6. K.A.Brueckner, S.Jorna. Reviews of Modern Physics,
    v.$\underline{46}$, 325 (1974); S.Yu.Gus'kov, O.N.Krokhin, V.B.
    Rozanov. Nuclear Fusion, $\underline{16}$, 6 (1976).
7. N.G.Basov, O.N.Krokhin, G.V.Sklizkov, S.I.Fedotov,
    A.S.Shikanov, ZhETP, $\underline{62}$, 203 (1972).
8. N.G.Basov, Yu.A.Zakharenkov, N.N.Zorev, A.A.Kologrivov,
    O.N.Krokhin, A.A.Rupasov, G.V.Sklizkov, A.S.Shikanov,
    ZhETP, $\underline{71}$, 1788 (1976).
9. E.G.Gamaly, Yu.A.Merkuliev, A.I.Nikitenko, E.O.Rych-
    kova, G.V.Sklizkov, Trudy FIAN, $\underline{94}$, 29 (1977).
10. B.L.Vasin, N.N.Zorev, V.N.Radaev, A.A.Rupasov, G.V.
    Sklizkov, A.S.Shikanov, L.I.Shishkina. Preprint FIAN
    N 198 (1978).
11. A.A.Erokhin, Yu.A.Zakharenkov, N.N.Zorev, G.V.Skliz-
    kov, A.S.Shikanov, Fizika plazmy, $\underline{4}$, 648 (1978).
12. Yu.A.Zakharenkov, O.N.Krokhin, G.V.Sklizkov, A.S.
    Shikanov, Kvantovaya Elektronika, $\underline{3}$, N 5, 1068 (1976).
13. Yu.A.Zakharenkov, A.V.Rode, G.V.Sklizkov, S.I.Fedotov,
    A.S.Shikanov, Kvantovaya Elektronika, $\underline{4}$, N 4, 815
    (1977).
14. Yu.A.Zakharenkov, N.N.Zorev, O.N.Krokhin, Yu.A.Mikha-
    ilov, A.A.Rupasov, G.V.Sklizkov, A.S.Shikanov,
    Pis'ma v ZhETP, $\underline{21}$, N 9, 557 (1975).
15. N.G.Basov, V.Yu.Bychenkov, O.N.Krokhin, M.V.Osipov,
    A.A.Rupasov, V.P.Silin, G.V.Sklizkov, A.N.Starodub,
    V.T.Tikhonchyk, A.S.Shikanov. Kvantovaya Elektronika,
    6, 1829 (1979)
16. V.Yu.Bychenkov, Yu.A.Zakharenkov, O.N.Krokhin, A.A.
    Rupasov, V.P.Silin, G.V.Sklizkov, A.N.Starodub,V.T.
    Tikhonchuk, A.S.Shikanov. Pis'ma v ZhETP, $\underline{26}$, 500
    (1977).
17. A.A.Kologrivov, Yu.A.Mikhailov, G.V.Sklizkov, S.I.
    Fedotov, A.S.Shikanov, M.R.Shpol'sky. Kvantovaya
    Elektronika, $\underline{2}$, 2223 (1975).
18. A.A.Erokhin, S.A.Zverev, A.A.Kologrivov, V.V.Kushin,
    V.K.Lyapidevsky, A.A.Rupasov, G.V.Sklizkov, A.S.
    Shikanov, Kratkie soobshcheniya po fizike, N 9,(1979)
19. N.G.Basov, E.Wolowski, E.Voryna, S.Denus, Yu.A.
    Zakharenkov, S.Kaliski, G.V.Sklizkov, Yu.Farny,A.S.
    Shikanov, Preprint FIAN N 194 (1978).

20. Yu.V.Afanasiev, N.G.Basov, E.G.Gamaly, V.A.Gasilov,
    N.N.Demchenko, O.N.Krokhin, I.G.Lebo, V.B.Rozanov,
    A.A.Samarsky, V.F.Tishkin, A.P.Favorsky.
    Preprint FIAN (1977).
21. D.Billon, P.A.Holstein, J.Launspach, C.Patou, J.M.
    Reisse, D.Schirmann. In:"Laser Interaction and
    Related Plasma Phenomena", ed. by H.H.Schwartz and
    H.Hora, Plenum Press, N.Y., 1977, v.4A, p.503.
22. V.P.Silin. "Parametric influence of a powerful ra-
    diation on plasma", "Nauka", M., 1973 (in Russian).
23. N.N.Zorev, G.V.Sklizkov, A.S.Shikanov. "Investiga -
    tion of shock waves, which are formed under irradi-
    ation of spherical targets by laser radiation",
    report presented at XII Europ.Conf. on Laser Inter-
    action with Matter, 1978, Moscow (In Russian).
24. S.A.Zverev, V.K.Lyapidevsky. Proceedings of the 3rd
    All-Union Symposium on Luminescence detectors and
    X-ray image converters", Stavropol' , 1979, p.100.
    A.A.Alexandrov, T.V.Tishkina, S.A.Zverev, N.A.Klya-
    chin, V.V.Kushin, V.K.Lyapidevsky, O.V.Mikhailova.
    In sbornik: "Experimental techniques in nuclear
    physics", ed. by V.M.Kolobashkin, v.4, M., Atoizdat,
    1978, p.770.
    A.I.Abramov, Yu.A.Kazansky, E.S.Matusevich."Prin -
    ciples of experimental methods in nuclear physics",
    isd.2, M., Atoizdat, 1978, p.509.
25. E.K.Storm, H.G.Ahlstrom, M.J.Boyle, L.W.Coleman,
    N.N.Kornblum, R.A.Lerche, D.R.MacQuigg, D.W.Phillion,
    F.Rainer, V.C.Rupert, V.W.Slivinsky, D.R.Specl, K.G.
    Tirsell. Laser fusion experiments at 2TW, Preprint
    UCRL-78581, Rev.1 (1976).
26. A.I.Avrov, V.Yu.Bychenkov, O.N.Krokhin,V.V.Pustovalov,
    A.A.Rupasov, V.P.Silin, G.V.Sklizkov, V.T.Tikhonchuk,
    A.S.Shikanov. Pis'ma v ZhETP $\underline{24}$, 293 (1976);
    ZhETP $\underline{72}$, 970 (1977).
27. N.G.Basov, A.A.Erokhin, Yu.A.Zakharenkov, N.N.Zorev,
    A.A.Kologrivov, O.N.Krokhin, A.A.Rupasov, G.V.Skliz-
    kov, A.S.Shikanov. Pis'ma v ZhETP, $\underline{26}$, 581 (1977).
28. Yu.V.Afanasiev, P.P.Volosevich, E.G.Gamaly, O.N.
    Krokhin, S.P.Kurdyumov, E.I.Levanov, V.B.Rozanov.
    Pis'ma v ZhETP, $\underline{23}$, 470 (1976).
29. Ya.B.Zel'dovich, Yu.P.Raizer. Physics of shock waves
    and high-temperature gasdynamic phenomena", "Nauka",
    M., 1966 (in Russian).
30. Yu.V.Afanasiev, E.G.Gamaly, O.N.Krokhin, V.B.Rozanov,
    ZhETP $\underline{71}$, 594 (1976).
31. N.G.Basov,P.P.Volosevich,E.G.Gamaly,S.Yu.Gus'kov,
    Yu.A.Zakharenkov, O.N.Krokhin,V.B.Rozanov,G.V.Skliz-
    kov,A.S.Shikanov. Pis'ma v ZhETP $\underline{28}$, 135, (1978).

INERTIAL CONFINEMENT FUSION RESEARCH AT ILE OSAKA

C. Yamanaka, S. Nakai, Y. Kato, T. Sasaki, T. Mochizuki

Institute of Laser Engineering
Osaka University
Suita, Osaka, Japan 565

1. INTRODUCTION

The current objectives of laser fusion research are to obtain
understanding of the pellet implosion process in explosive and
ablative mode. The physical processes which contribute to inertial
fusion are the absorption of laser at cut-off density region, energy
transport into target core, compression dynamics and its stability,
and fusion burn. These processes are investigated to clarify the
effect of different types of compression mode.

The wave length scaling of the compression process is one of
the most interesting items for investigation. This is a reason why
we provide GEKKO systems (glass laser) and LEKKO systems (CO_2 laser)
in ILE Osaka University[1].

For implosion experiments, glass laser GEKKO IV and CO_2 laser
LEKKO II are used. For interaction and fundamental experiments,
GEKKO II, LEKKO II and long pulse CO_2 laser LEKKO I are used.
Simulation study is extended to cover the compression processes of
exploding mode and ablative mode. To aim the breakeven condition
"KONGOH Project" has been contemplated which will be extended for
ten years.

2. ENERGY DRIVER SYSTEMS

2-1 Overview of Developments

Energy driver development at ILE, Osaka University comprises
three areas; glass laser[2], carbon dioxide laser[3], and relativistic
electron beam. Development of reliable glass lasers, being the
main program at ILE, supports establishing physics of implosion
process and achieving high density ablative mode compression.

Carbon dioxide laser is developed as an efficient high energy laser
and to study wavelength scaling of implosion process. Relativistic
electron beam, being the most cost-effective driver, is studied to
control pulsed power technologies and to investigate electron beam
plasma interactions. Fig. 1 shows the energy drivers at ILE at
present and until 1985. The period of 1980-1984 is the first
phase of KONGOH Project and construction of 20kJ, 40TW GEKKO XII
glass laser is planned. During this period the feasibility study
will be taken. A 100kJ energy driver will be used for scientific
breakeven experiment in the second phase of KONGOH Project (1985-
1989).

2-2 Glass Laser

     Four beam glass laser system GEKKO IV has started target
irradiation experiments in July 1978. Peak focusable output power
is 4TW at 0.1ns. Target is irradiated from four tetrahedral
directions through F/1.2 aspheric lenses (Fig. 2). Until October
1979, 1050 target shots have been made with the pulse duration of
50ps $\sim$ 300ps for exploding as well as ablative mode compression
experiments by using various types of targets. Fig. 3 and Fig. 4
show the laser and target chamber of GEKKO IV System.
     Another two-beam glass laser GEKKO II is used for more basic
experiments using planar targets and to test new diagnostic techniques.
     During FY 1977-1979, GEKKO XII Module Program has been in
progress in order to develop various laser components and subsystems
which are necessary to construct a 20kJ GEKKO XII glass laser.
GEKKO XII Module consists of laser, alignment, laser beam diagnostics and target irradiation subsystems. Specifications of these
subsystems are shown in Table 1. Optical arrangement of the laser
and the one line of the modules are shown in Fig. 5 and Fig. 6,
respectively.
     A model of total GEKKO XII system is shown in Fig. 7. Final
detailed design will be based on the performance testings of GEKKO
XII Module.

2-3 CO_2 Laser

     There have been continued advances in CO_2 laser technology for
improvement of system efficiency, output power capability, and
reliability. Two beam LEKKO II laser system is used for testing
these properties as well as for plasma experiments. Design and
construction of the 10kJ LEKKO VIII laser system is in progress
based on these improvements.
     Present performance of the two-beam LEKKO II system (Fig. 8
and Fig. 9) having final aperture of 20cm are summarized in Table 2.
     A model of total LEKKO VIII system is shown in Fig. 10. Target

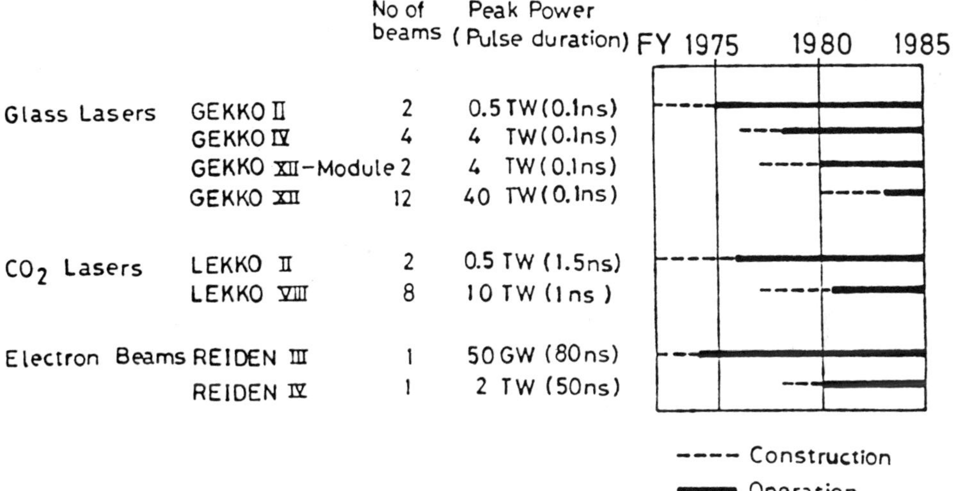

Fig. 1.   Energy drivers at ILE, Osaka University until 1985

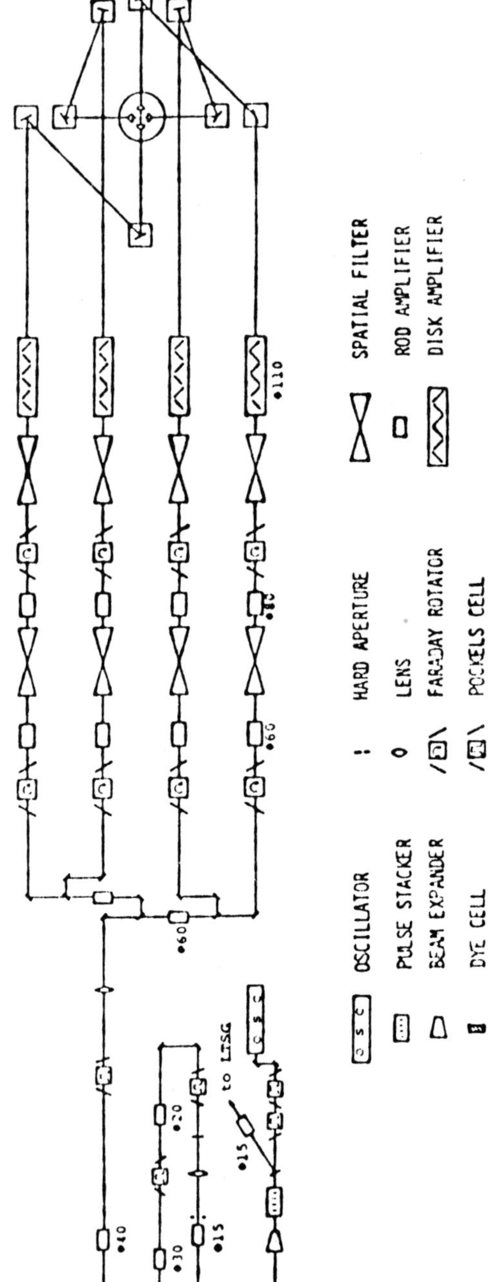

Fig. 2. Optical arrangement of GEKKO IV laser system

Fig. 3.   Glass laser GEKKO IV

Fig. 4.   Target chamber of GEKKO IV compression experiment

Table 1. Specifications of GEKKO XII Module

1. Laser System
      (1) Output power   2TW/beam (0.1ns)
      (2) Output energy   1 kJ/beam (1ns)
      (3) Pulse width   100 ps $\sim$ 2ns
      (4) Pulse shaping   pulse stacker
      (5) Stability of output energy $< \pm 10\%$
      (6) Background light $< 100\mu J/10^{-6}$ steradian/beam
      (7) Wavefront distortion   Linear $\lambda/2$, nonlinear λ
      (8) Protection from reflection Protected for 50% reflection
      (9) Repetition rate < 30min.

2. Alignment System
      (1) Laser alignment   centering $< \pm 1$mm
                              pointing $< \pm 5$sec
      (2) Focusing optics   centering $< \pm 1$mm
                                           $< \pm 5\mu$m(on target)
                              pointing $< \pm 1$sec
                              focusing $< \pm 10\mu$m
      (3) Beam timing $< \pm 10$ps

3. Laser Beam Diagnostics System
      (1) Oscillator diagnostics: mode locking, pulse selection
      (2) Amplification property: energy measurements at 11 points
      (3) Output beam diagnostics package: 8 items are measured

4. Target Irradiation System
      (1) Target Chamber   1200mm i.d. 10^{-6} Torr
      (2) Target positioning $\pm 5\mu$m, 16 targets selectable
      (3) Focusing lens   x, y movement precision $< \pm 5\mu$m
                                Z movement precision $< \pm 10\mu$m
      (4) Turning mirrors   angular movement precision < 0.5sec

5. System control
      (1) Laser controller
      (2) Alignment controller   Automatic control
      (3) Target controller   or remote manual
      (4) Laser beam diagnostic controller   control selectable
      (5) Main controller

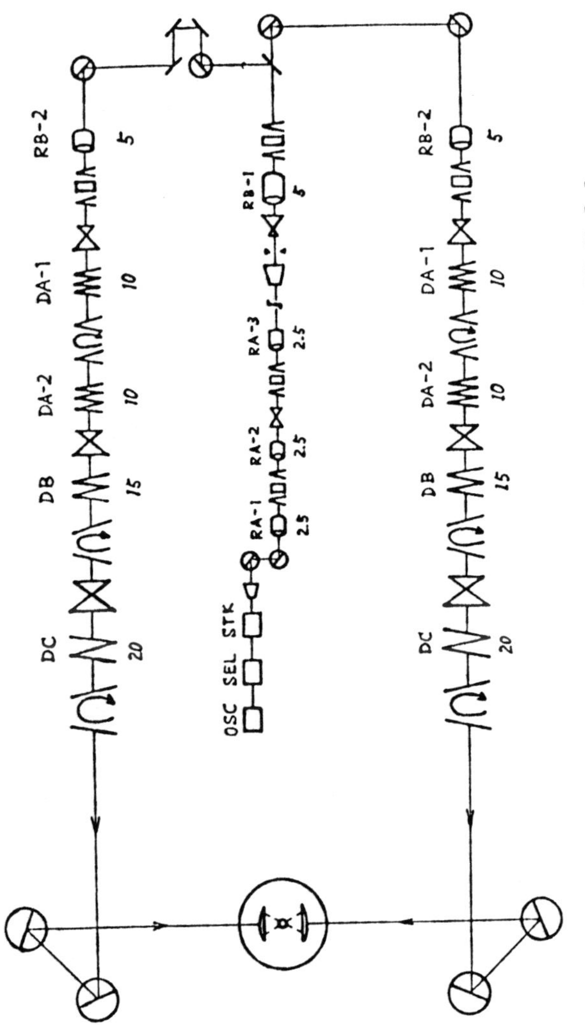

Fig. 5. Optical arrangement of GEKKO XII Module

Fig. 6. Glass laser GEKKO XII Module

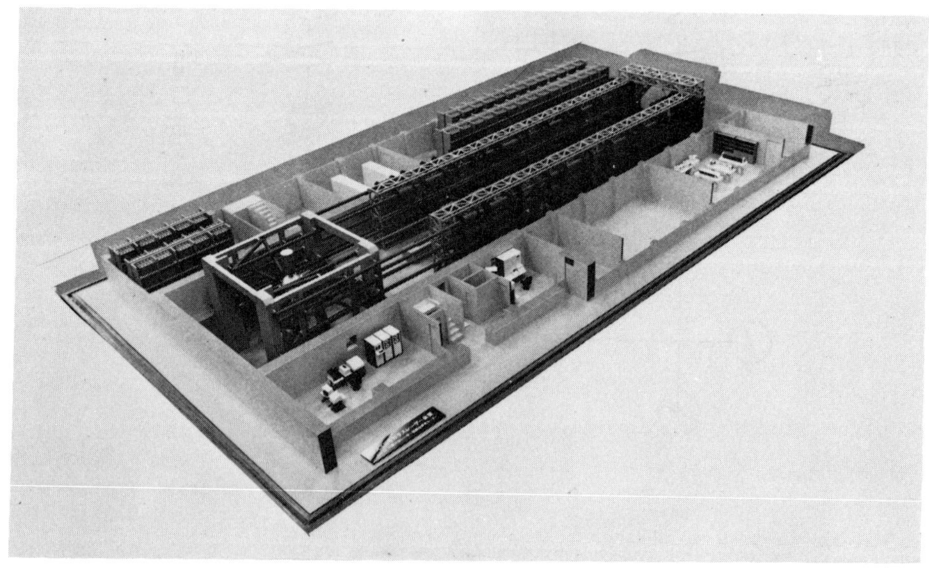

Fig. 7. A model of total GEKKO XII laser system

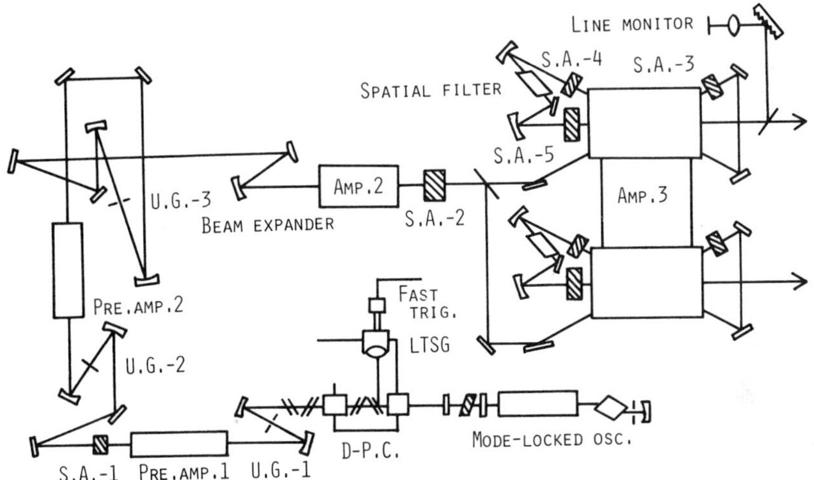

Fig. 8. Optical arrangement of LEKKO II laser system

Fig. 9. CO_2 laser LEKKO II

Table 2.   Present performance of LEKKO II laser system

<u>LEKKO II</u>

OUTPUT POWER	200 J/BEAM, SINGLE LINE
	400 J/BEAM, MULTI LINE
LASER PULSE DURATION	1.5 - 3 NS, POCKELS SHUTTER
	1.0 NS, MODELOCKED PULSE
OPERATION CYCLE	1 SHOT/3 MIN.
OUTPUT BEAM DIAMETER	20 CM
LASER GAS	$CO_2 : N_2 = 4 : 1$
TARGET EXPERIMENT	160 J/BEAM, SINGLE LINE
	$S/N > 10^5$, ON TARGET

Fig. 10.   Schematic layout of LEKKO VIII laser system

chamber is separated from the final amplifier by 60m in order to
reduce the prepulse energy due to parasitic oscillation. Completion
of this system is scheduled within FY 1980.

2-4 Relativistic Electron Beam

High energy electron beam machine REIDEN IV has been completed.
Design parameters are: maximum voltage of 1.4MV, peak current of
1.4MA with the output energy of 100kJ within 50ns. Fig. 11 is a
schematic view of REKDEN IV (Fig. 12) showing Marx generator, intermediate storage capacitor, pulse line, transmission line, and
the output diode. Characteristics of these parts are shown in
Table 3. Implosion of a target using this machine will start early
next year.

Smaller REB Machine REIDEN III has been used to study interaction between e-beam and plasma, and anomalous deposition of e-beam
energy has been found. Also small scale experiment of light ion
beam generation was tested. Beam current of approximately 5kA has
been observed.

3. COUPLING OF LASER ONTO TARGET

The fundamental process in coupling has been investigated on
plane and spherical pellet target using glass laser and CO_2 laser
to get clear understanding for wave length scaling.

3-1 Absorption

It has been well established that absorptivities on plane
target are good for both 1μm and 10μm (50~70%). At higher power
density, 10^{14} w/cm^2 for glass and 10^{12} w/cm^2 for CO_2 resonant
absorption becomes dominant process[4]. Density steepening was
observed directly by optical interferometry[5] and inferred from
optimum angle of resonant absorption[6]. Instability on critical
density surface was identified by anisotropy of scattered light and
interpreted theoretically[6]. Magnetic field generation was observed
by magnetic probe[7], magnetic recording tape or arroy, which was
related to resonant absorption.

3-2 Brillouin Scattering

Brillouin back scattering is suspected to be a possible loss in
long pulse regime for ablative compression. In CO_2 experiments,
back reflection was observed to be less than a few percents even
with long (~30ns) pulse or nsec pulse if the specular component was
excluded by oblique incident on plane target[7]. In glass laser, the
temporal behavior of back scattered light was investigated[1] in case
of 4π uniform irradiation on pellet target and suppression of back
scattering was observed as shown in Fig. 13 at the laser power

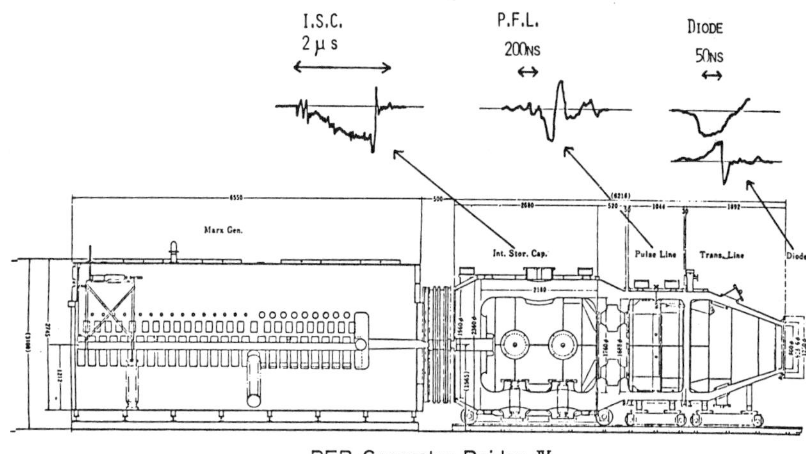

Fig. 11.   Schematic diagram of REIDEN IV electron beam generator

Fig. 12.   REB generator REIDEN IV

Table 3. Characteristics of REIDEN IV

CHARACTERISTICS OF REIDEN IV

DESIGNED VALUE	MEASURED VALUE
MARX GEN.	
43 NF	43 NF
20 μH	12 μH
1 OHM	1.5 OHM
I.S.C. (3 MV)	(2 MV<)
33 NF	41 NF
(87%)	(72%)
I.S.C.-P.F.L. SWITCH	
320 NH/CH	320 NH/CH
P.F.L.	
30 NF	30 NF
(65%)	
P.F.L. OUTPUT SWITCH	
100 NH/CH	120 NH/CH
(60%)	(50%)

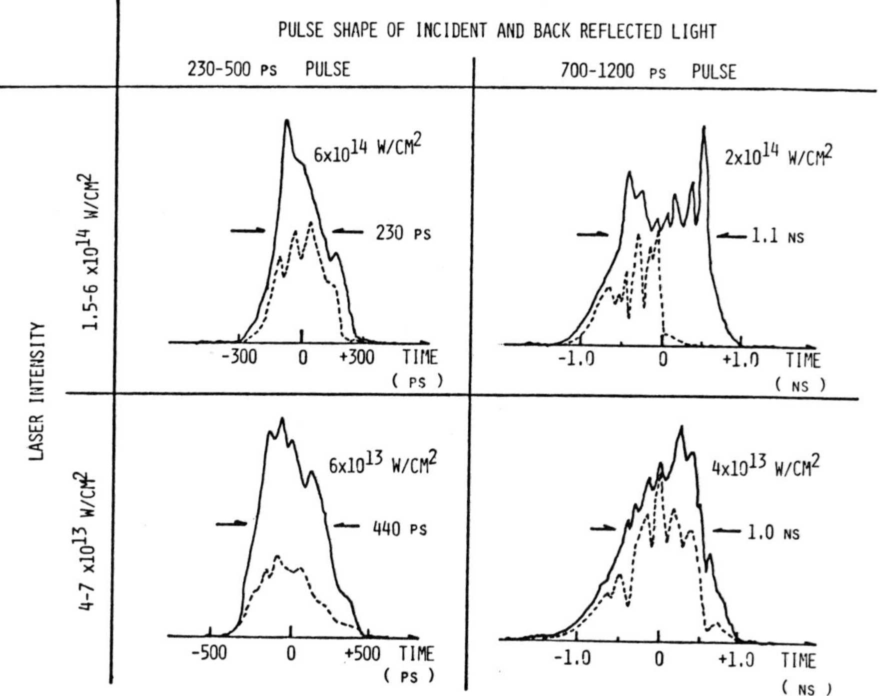

Fig. 13. Pulse shape of the incident (solid line) and the
          back-reflected (dashed line) light.

          At the high laser intensity ($>1.5\times10^{14}$ W/cm^2)
          and the long pulse duration (>700 ps), the
          reflected light is suppressed at the later
          half of the incident pulse.

density above 1.5×10^{14} w/cm^2. At this power region, density steepening was clearly observed and this might be the cause of the suppression.

The intensity dependent reflectivity was investigated taking into account the nonlinear ion Landau damping and the enhanced linear Landau damping due to ion heating by the ion acoustic wave itself. The results for the general reflectivity curves are shown in Fig. 14. In the low laser intensity, the reflectivity increases exponentially with the increase of intensity. Then, the nonlinear ion Landau damping becomes dominant to reduce the reflectivity. In the intermediate intensity, say $(Vos/Ve)^2 \gtrsim 0.01 - 0.1$, the ion heating sets on and reflectivity saturates and then starts to decrease. This is due to the fact that the linear ion Landau damping depends strongly on ZTe/Ti. In the higher laser intensity, the ions are strongly heated up to $ZTe/Ti \sim 1$ and the damping rate is saturated to be $Im\ \omega_{ac} \sim Re\ \omega_{ac}$. Therefore the reflectivity again starts to increase gradually. In Fig. 15. we show the comparison between our theoretical model and the CO_2 experimental results. The typical saturation phenomena due to the ion heating is observed.

3-3 Hot Electron and Fast Ion Generation

The dependence of T_{hot} and T_{cold} on the incident power and wave length are important properties in coupling and transport of energy. Fig. 16 shows the compiled data of glass laser and CO_2 laser. T_{cold} depends on incident intensity Φ with the power law of $\Phi^{1/4}$ and not depends on wave length. T_{hot} dependence seems to be $(\Phi\lambda^2)^{(1/3-1/4)}$ on intensity and wave length at the region $\Phi\lambda^2 > 10^{14}$.

The characteristics of fast ion generation were investigated by Thomson parabora ion analyzer, magnetic field deflection analyzer and charge collectors. A typical trace of Thomson parabora and derived velocity distribution are shown in Fig. 17 and Fig. 18 for polyethylene target. It should be noticed that there is high energy cut, the threshold energy of which is independent to the ionization state and the distribution profiles are similar to each other. It was also seen that the highest energy of carbon was three times larger than that of proton. These results infere the acceleration mechanism of ion in ambipolar field due to hot electrons. Lower charge state ion may be generated by recombination after acceleration and the acceleration time is same on proton and carbon. In CO_2 laser experiments, the electron temperatures which derived from the slope of velocity distribution assuming isothermal expansion are different each other with proton T_{hot}^{proton} and carbon T_{hot}^{carbon}. These temperatures are also different from T_{hot}^{X-ray} by X-ray measurements with the order $T_{hot}^{proton} > T_{hot}^{X-ray} \gtrsim T_{hot}^{carbon}$. This might reflect the dominant

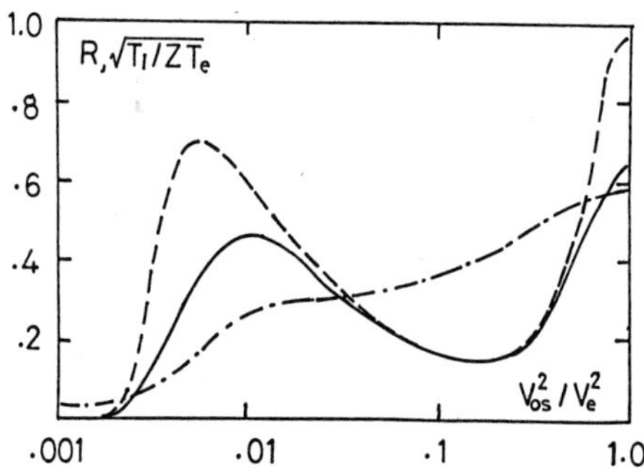

Fig. 14. A typical amplitude dependent reflectivity and the
          temperature ratio. $k_0 L=200$, $n/n_c=0.1$
          _______ ; reflectivity

          ------- ; reflectivity without nonlinear Landau damping

          —·—·—·— ; temperature ratio

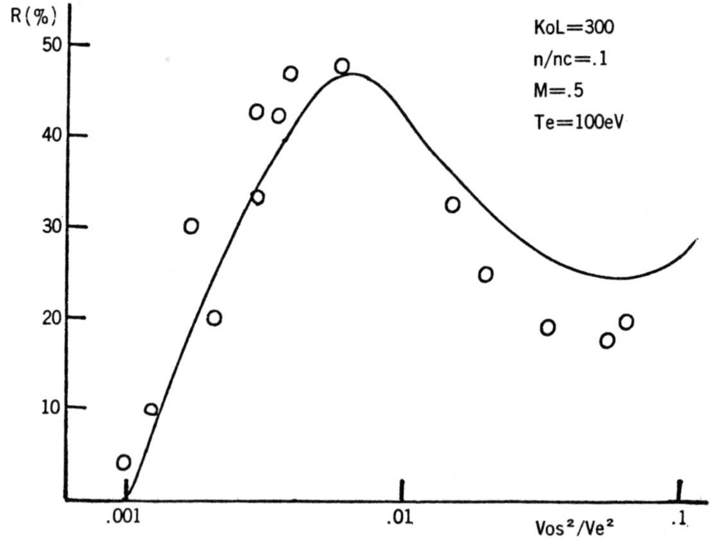

Fig. 15. Theoretical reflectivity (solid line) along with the
          experimental one (open circles).
          Parameters for a theoretical curve;
             $k_0 L=300$, $n/n_c=0.1$ and $T_e=100$ eV

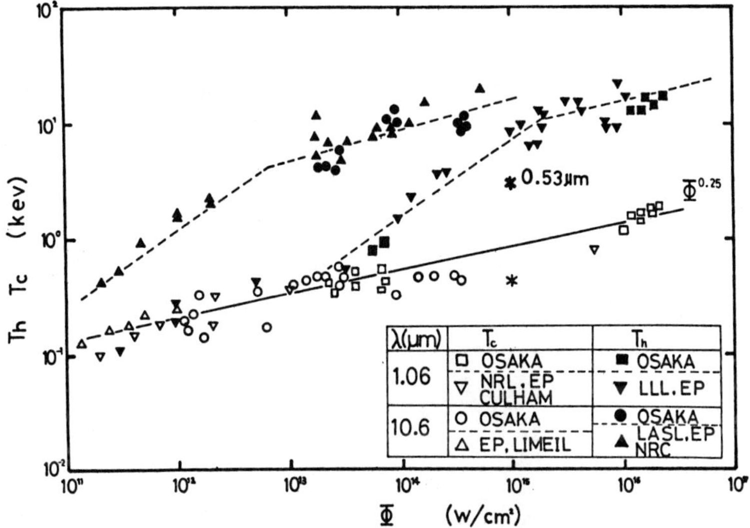

Fig. 16.   Temperature dependence of hot and cold electron
on incident laser power and wave length

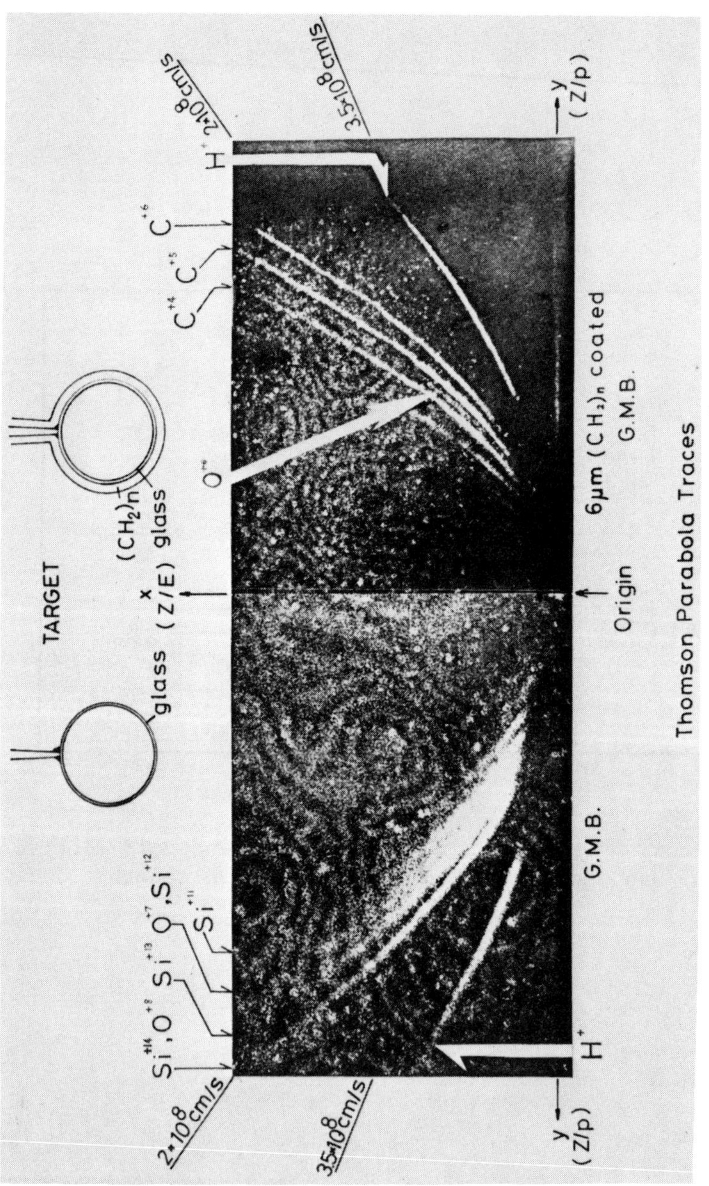

Fig. 17. Thomson parabola trace on nitrocellulose film for pellet irradiation by glass laser.

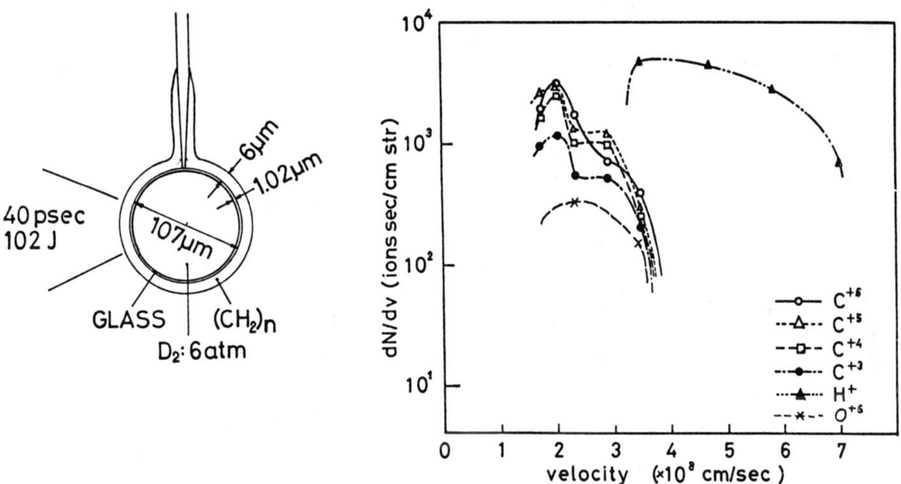

Fig. 18. Velocity distribution of fast ions by Thomson parabola.

source region of each species of proton, X-ray and carbon. Super fast
ions of MeV ranges are also observed even with short pulse irradiation
of 50 ps of glass laser.

3-4 Energy Transport

Energy transport was investigated by using multilayered target.
A typical result with CO_2 laser is shown in Fig. 19. An intensity
variation of $Al_{k\alpha}$ line gives the range of hot electrons, which is
consistent with the classical one if the electron energy is assumed
to be hot electron tmeperature by X-ray measurements. Al resonant
line may give us information on the depth of thermal conduction which
requires detail analysis with computer simulation, including induced
magnetic field and plasma instability. When the CH target thickness
is less than the depth of thermal conduction $(2000 \sim 3000A)$, the
expansion of plasma measured by charge collctor is symmetric to front
and rear side, which is explosive expansion. At thicker region where
only $Al_{k\alpha}$ line is observed, the plasma expansion becomes unsymmetric
and ablative mode.

A vacuum layer between $(CH_2)_n$ front and Al rear foil could
suppress the emission of k_α line when the thickness of $(CH_2)_n$ foil
is thicker than the thermal conduction range $(2000 \sim 3000A)$. This might
be due to the suppression of hot electron transport to rear foil by
electrostatic field. The energy transport by hot electrons and the
heat inhibition mechanism are extensively investigated experimentally,
theoretically and also computationally including the hot elctron
reflection effect by electrostatic field due to insufficient neutralization. Fig. 20 shows a simulation results which indicate trap
potential for hot electron at cut off region. It was also shown that
energy transport by hot electron has significant effects on the
ablation structure.

Figure 21 shows a typical energy flow in pellet irradiation with
glass and CO_2 laser. The energy balance and hydrodynamic efficiency
are under investigation as a function of focusing condition, laser
pulse shape and wavelength.

4. IMPLOSION EXPERIMENTS

A major interest of our current experiments on implosion is to
make comparison between exploding pusher and ablative mode compression processes.   This comparison was made by using uncoated thin
shell glass microballoons as exploding pusher type target and polyethylene coated glass microballoons as ablative mode compression
target. Glass microballoon (GMB) targets used in the experiments
have typical diameter of $50 \sim 200\mu m$ with wall thickness of $0.7 \sim 1.4\mu m$.
For the experiments of ablative compression, we have used GMB coated
by polyethylene up to the thickness of $10\mu m$. Polyethylene hollow
shell and solid ball targets have also been used to compare the
result with that of GMB.

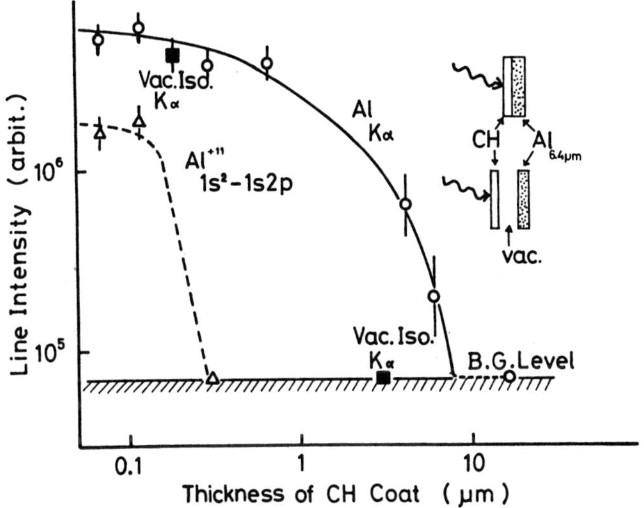

Fig. 19. X-ray line intensity decay as a function of foil
thickness, showing penetration of hot electron and
thermal conduction.

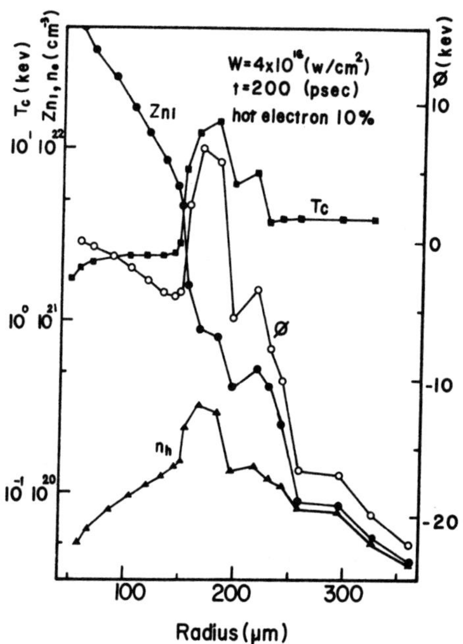

Fig. 20.   Simulation results for spatial distribution of potential ϕ, hot electron density n_h, cold electron temperature T_e and total density Zn_i.

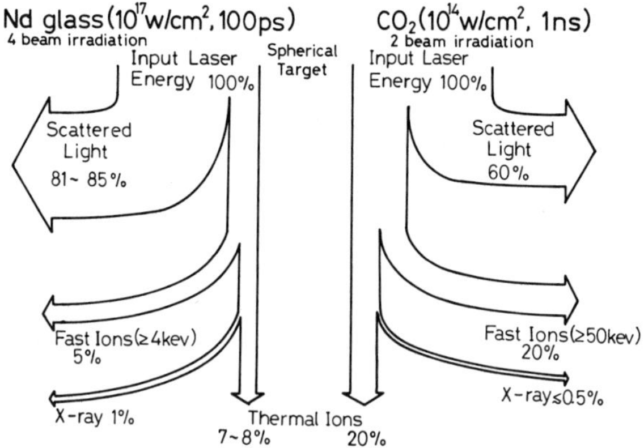

Fig. 21.   Energy flows in cases of pellet irradiation by glass laser and CO$_2$ laser.

GMB targets were filled with D_2, DT, neon and argon gas due to the purpose of experiments.

GEKKO IV glass laser system has been mainly used, which can deliver maximum power of 4TW in 100ps on a target from four tetra-hedral symmetric direction by f/1.1 aspheric lenses. Implosion experiments by LEKKO II CO_2 laser system has been started, which has maximum output of 800J and on target energy of 300J with a pulse of 1 ns FWHM.

4-1 X-ray Streak Measurements

Time and space-resolved X-ray image by an ultrafast X-ray streak camera is one of the important keys for understanding the implosion process. Fig. 22 shows the isodensity contours of a space resolved streak image. The implosion velocities V_1 and V_2 can be estimated from the dotted line in this figure to be 3.2×10^7 cm/s and 2.8×10^7 cm/s, respectively. Fig. 23 is the time evolution of spatial intensity distribution of X-ray emission obtained from Fig. 22. The implosion time can be determined by the time interval between the initiation of X-ray emission and the most intense time at the core region. Fig. 24 is the simulation results of plasma flow diagram for similar condition as the above experiment which shows good agreement.

Implosion time can also be determined by the temporal behavior of X-ray intensity as shown in Fig. 25. Implosion velocities are plotted in Fig. 26 as a function of specific energy to GMB with and without coating. The dependence of velocity on specific energy is $(v \propto (E/M))$ for ablation and $(v \propto (E/M)^{1/2})$ for explosion case.

Electron temperature variation with time can be measured by using various K-edge filters in front of streak slit. Fig. 27 shows the results. Temperature decreases with increasing thickness of coating.

4-2 X-ray Shadowgraphy

X-ray shadowgraphy method was applied to study implosion dynamics of a pellet.

The advantage of this diagnostics over an optical light probe or a conventional X-ray crystal spectrometer is that it enables us to measure spatial and temporal profiles of low temperature and high dense fusion plasmas which occur at ablative mode compression.

A schematic view of the experimental set-up is shown in Fig. 28. Material for X-ray source was Germanium which emitted about 1.4 keV photons. Laser beam for X-ray production had the same pulse duration ($\sim$100ps) as the implosion laser beam. Two pinhole cameras were used to obtain images of the imploded pellet. One was directed to record X-ray shadowgraphs of the pellet, and the other was mounted on a deflected axis in such a way not to see X-ray for shadowgraphy but only to record emission images of X-ray from the

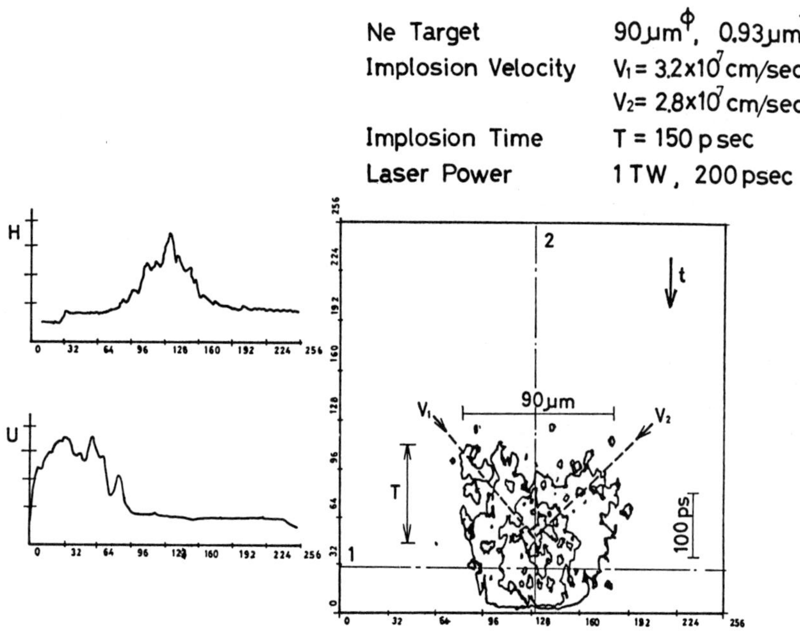

Fig. 22.   Isodensity trace of X-ray streak picture.

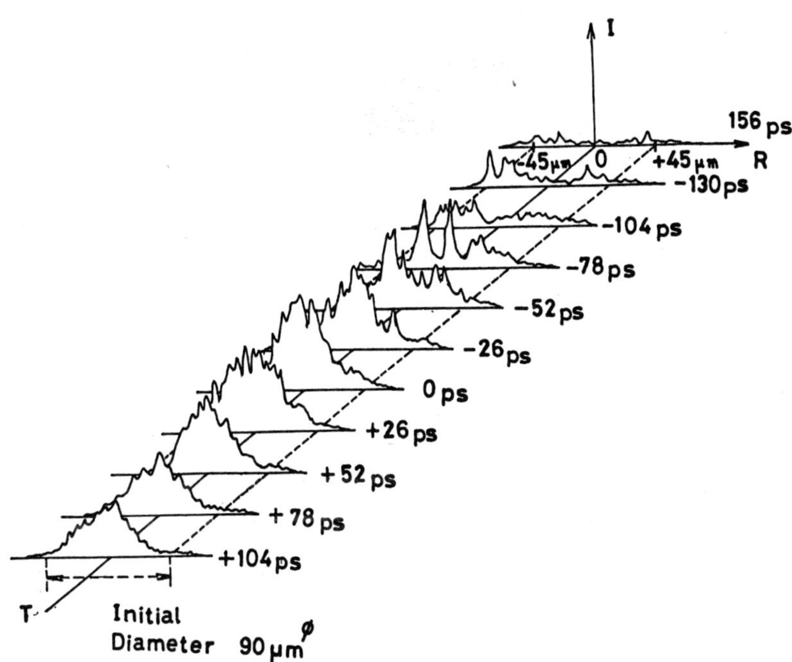

Fig. 23. Temporal behavior of spatial distribution of X-ray
          emission which was derived from X-ray streak picture.

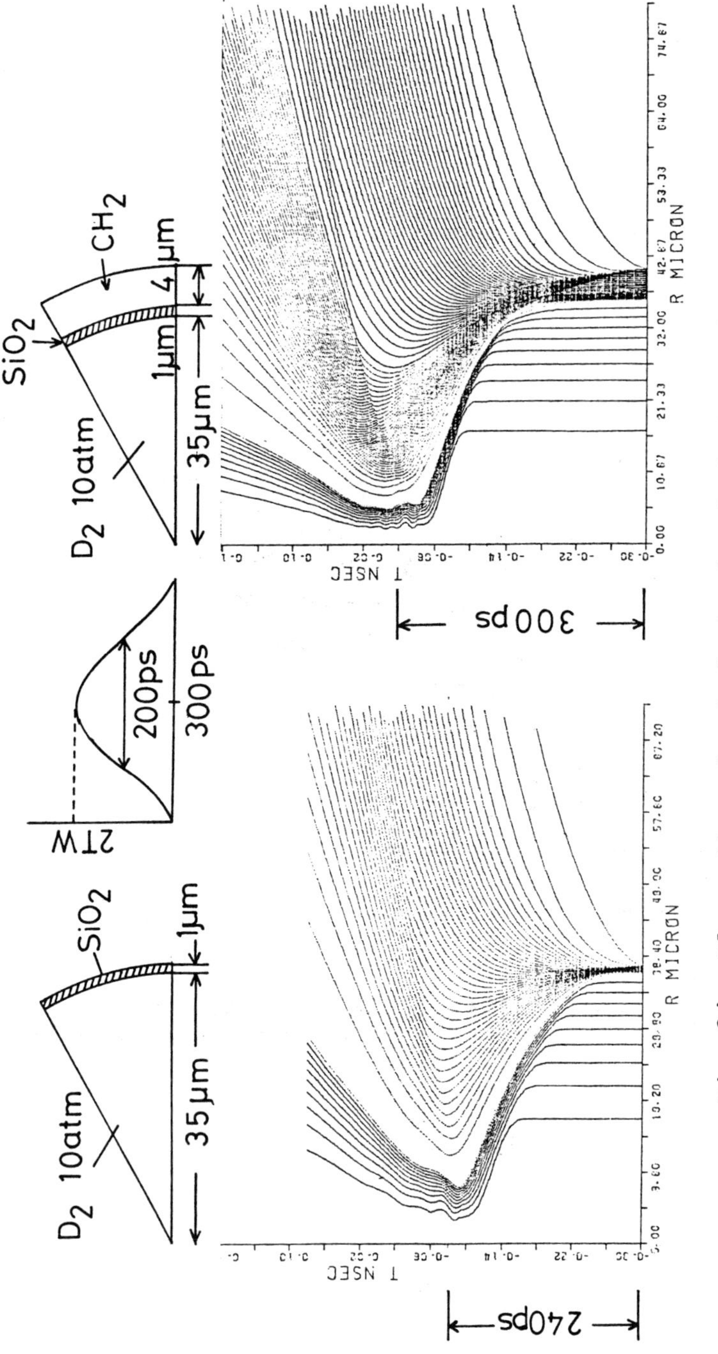

Fig. 24. Flow pattern for implosion by simulation to be compared with the results by X-ray streak picture.

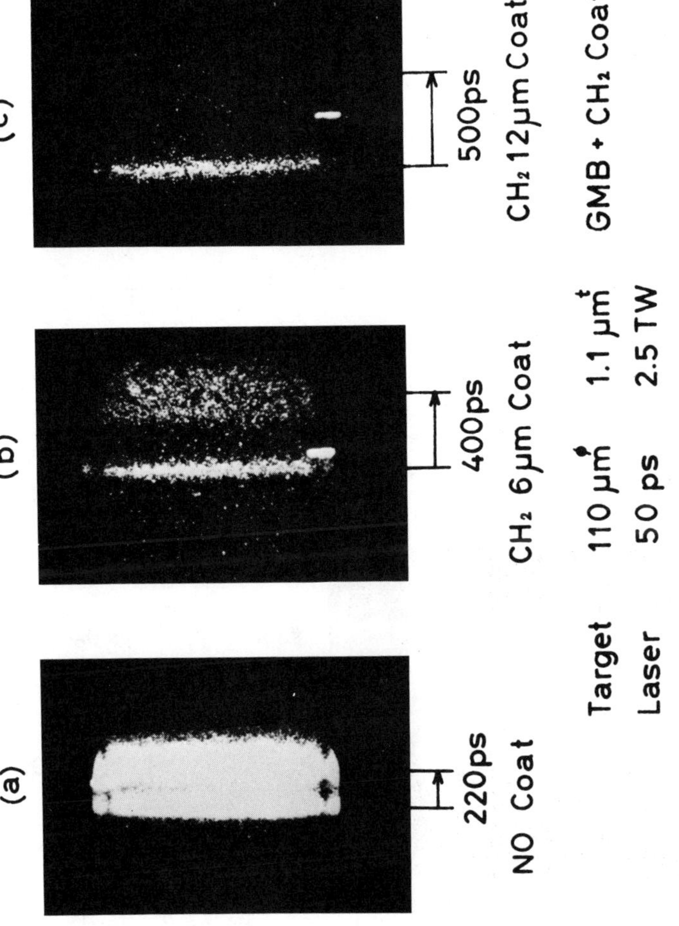

Fig. 25.   X-ray streak image without spatially resolving.

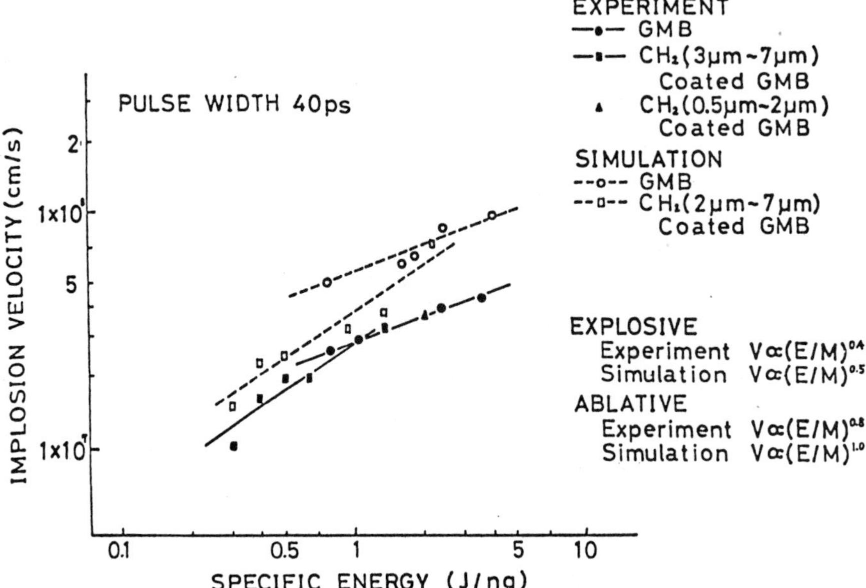

Fig. 26.   Implosion velocity as a function of specific energy
           (E/N) in comparison with simulation for simple GMB
           and low Z coated GMB.

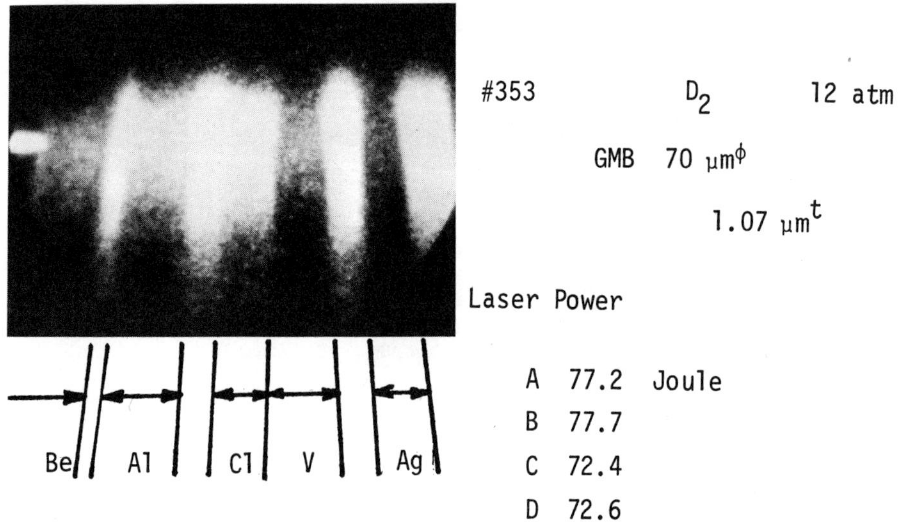

Fig. 27.   X-ray streak image for Be, Al, Cl, V and Ag K-edge
           filters.

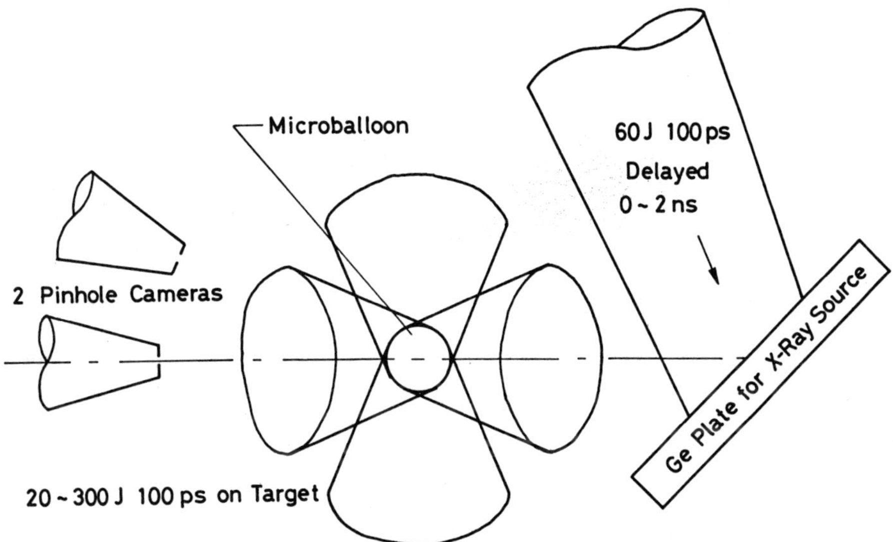

Fig. 28. Experimental set up for X-ray shadowgraphy.

imploded plasma. Filters of Be 25μm and Al 2μm thick were used in
the pinhole cameras so that photons of Germanium X-ray band were
selectively transmitted. Most of the shadow images were taken in
the condition that no emission image was observed in the second
pinhole camera.

4-2-1 Dependence of Implosion Symmetry on Focusing Condition

Implosion symmetry was observed by X-ray shadow images at
various laser focusing conditions. In Fig. 29(a) are shown images
at typical focusing conditions and in Fig. 29(b) calculated laser
absorption profiles on target surfaces. Calculation included the
ray traces of laser beam and the classical and resonance absorption.
One can see that implosion symmetry drastically depends on
laser focusing condition. Overlapped focusing gave smoother compressed surface of the pellet than the others. Exploding mode
compression is shown in Fig. 30. Characteristic behavior of this
mode is seen during compression. The ρR of pusher becomes so low in
exploding phase that the shadow image is transparent. Ablative mode
compression is shown in Fig. 31 and Fig. 32. Fig. 31(a) shows dy-
mic behavior of imploded 10μm CH_2 coated GMB of 80μm diameter and
Fig. 31(b) shows plots of pusher radius as a function of the delay
time to derive implosion velocity. The delay time is in reference
to the time of the laser peak. Fig. 32(a) and Fig. 32 (b) show the
case of CH_2 shell pellet of 150μm diameter and 5μm thickness.
In these images of ablative compression, there has been
observed whisker-like irregularity on the pusher surface in addition
to the anisotropy of compression due to localized heating of the
target surface by laser absorption. It might be resulted from
Rayleigh-Taylor instability produced in acceleration phase of
ablation.

4-2-2 Multi-frequency X-ray backlighting

Multi-frequency X-ray backlighting method was developed to
diagnose the temperature and ρR of the compressed core.
The experiment was done using GEKKO II laser system and in
exploding pusher mode compression.
3.5 atm Ne gas filled uncoated GMB targets were irradiated by
two 120 psec, 6J laser beams at overlapped focusing condition.
X-ray source was also Germanium. A schematic view of experimental
configuration is shown in Fig. 33.
A crystal spectrometer with a pinhole array as entrance
apertures enabled us to obtain separately the time resolved images
of hydrogen-like and helium-like neon line in the core as shown in
Fig. 34.
From the ratio of the opacities at hydrogen-like neon ion
absorption and helium-like neon ion absorption frequencies, we could
derive both T_e and ρR of the compressed core by assuming corona

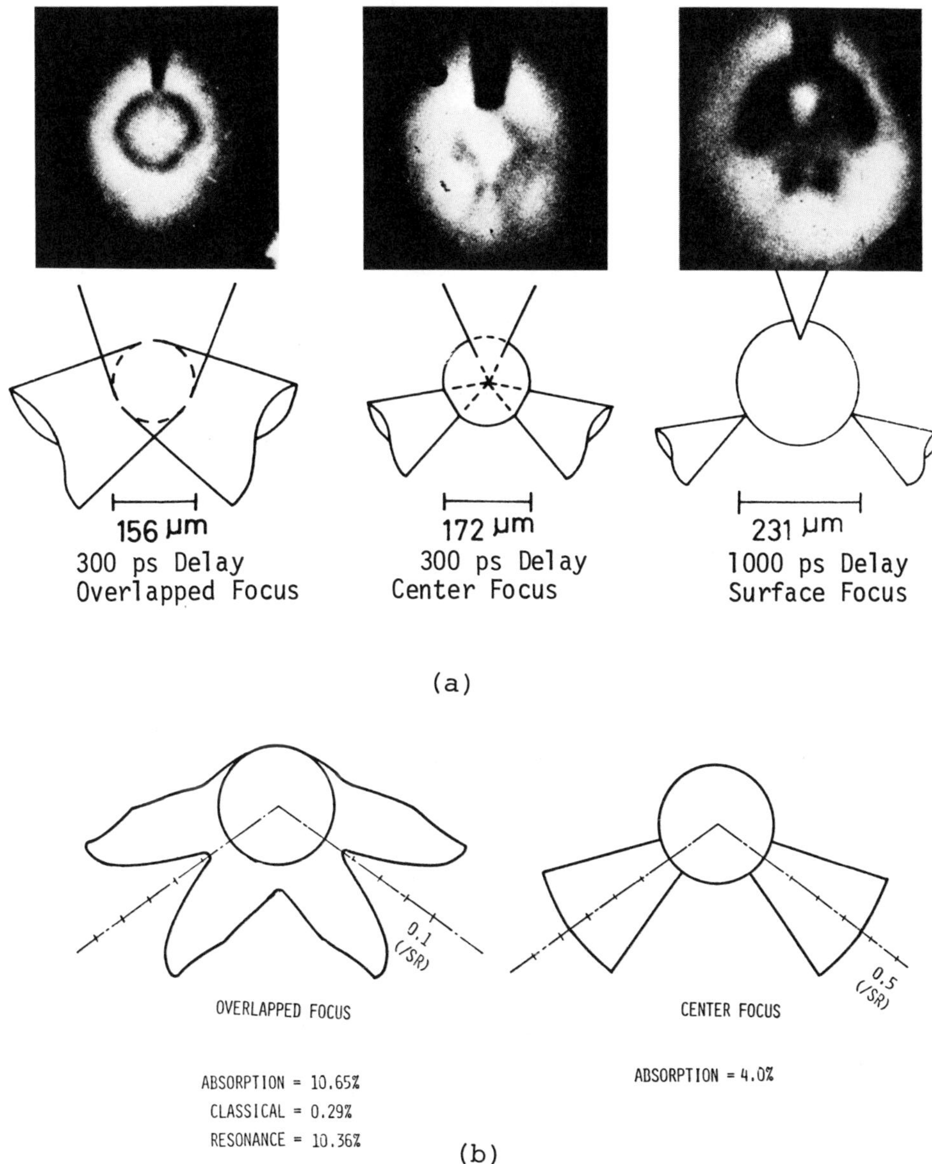

Fig. 29. Implosion symmetry dependence of focusing condition (a) and calculated distribution of energy deposition (b).

Target: UNCOATED G.M.B. $150^{\mu m \phi}$ $1^{\mu mt}$, EMPTY
LASER POWER: 0.70 TW/120 PSEC

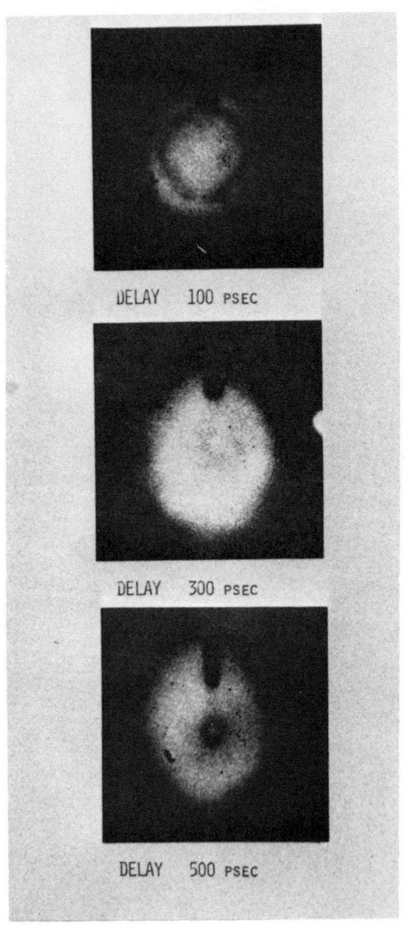

Fig. 30.   Image of X-ray shadow for simple GMB implosion.

model to the core plasma.   The result is shown in Fig. 35.   The absorption by oxygen in glass pusher was measured in a separate experiment which indicated that opacity of oxygen was comparable to experimental error bar in Fig. 35.

It should be emphasized that multi-frequency X-ray backlighting method yielded time resolved information on both temperature and ρR in the compressed core.

4-3 X-ray Spectrography

X-ray spectroscopic slit camera gives space resolved information of plasma parameters. Fig. 36 shows typical X-ray spectra obtained with different coating thickness. The initial pellet diameters are indicated at the bottom of the spectrographs. The broad spectrum at the left of Fig. 36(c) is a spatially integrated spectrum obtained with a 100μm slit whereas 10μm slit is used for space resolving.

From these spectra we found that the emission region of continuum and also Si line radiations became spatially narrower as coating thickness was increased. For the case of simple GMB (or GMB coated with thin polyethylne by 0.5μm) Si line were emitted not only from the inside but also from outside of the initial target surface. However, these emissions were only from inside when coating thickness was above 2.4μm. When the coating thickness was increased to 10μm, then dark line appeared in spectrum as shown in FIg. 36(c). These dark lines are due to resonant photo-excitation and photo-ionization originated from Si^{+11} and Si^{+12} ions, implying that the region of glass shell is relatively cold.

Electron temperature of the glass pusher is deduced from the ratio of Si^{+12} $1S^2$-1S3P and Si^{+13} 1S-3P (Lyβ) line intensities with assumption of corona equilibrium. Fig. 37 snows the spatial profiles of these line emissions and the electron temperature (solid line). As the coating thickness is increased, the intensity of these lines decreased and the peak position of Lyβ emission moved inward to the center of the pellet. These observations are in accordance with the expectation that higher compression is obtained with thicker coatings. The dashed line and dotted line in Fig. 37 show temperature profiles of neon and glass, respectively, obtained from computer simulations. These profiles are reasonably in good agreement with the experimental results.

The absorption line in Fig. 36(c) can be used to derive the parameters of glass pusher shell at the time of maximum compression. The densitometer trace of spectra are shown in Fig. 38(a) (CH_2:10μm). The resultant values are summarized in Table 4. The density of glass shell is higher than the initial density, and $\rho \Delta R$ has increased by a factor of about 20. This implied again that the glass plasma acted as a pusher.

The core density can be derived more reliably by the edge shift of radiative recombination spectrum. Fig. 39 shows the edge shift for the cases of simple and 10μm thick GMB implosion. The coating

TARGET: G.M.B., COAT $10^{\mu mt}$, N_e 27 atm

LASER POWER: 1.8 TW/100 PSEC

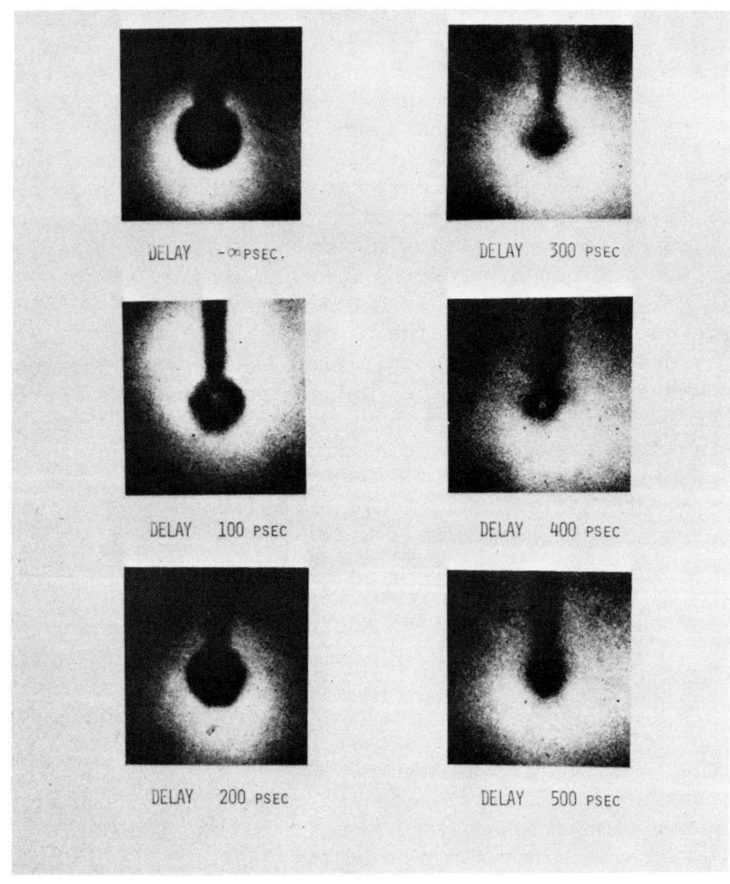

Fig. 31 (a)

Fig. 31.   X-ray shadow image for low Z coated GMB (a) and
its implosion dynamics (b).

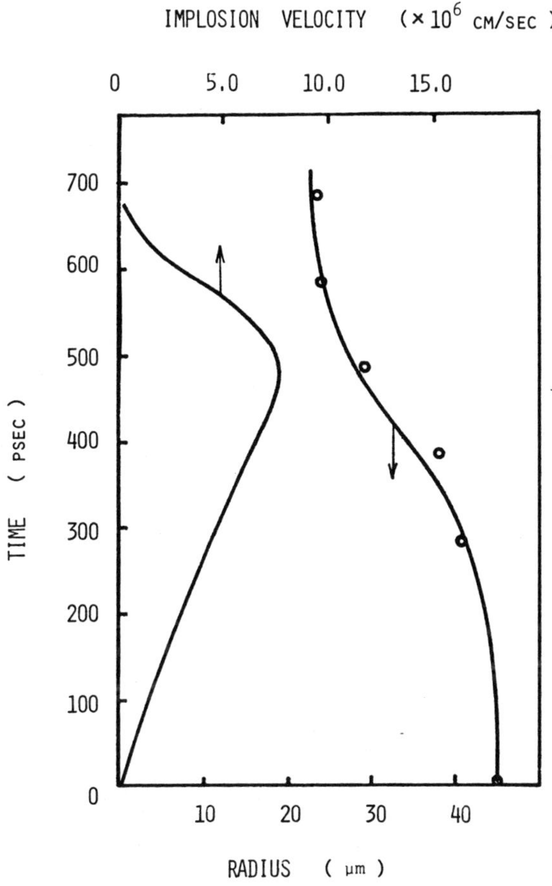

TARGET : G.M.B. $90^{\mu m \phi}$, CH_2 COAT $10^{\mu mt}$, N_e 27 atm

LASER POWER : 1.8 TW/ 100 PSEC

Fig. 31 (b)

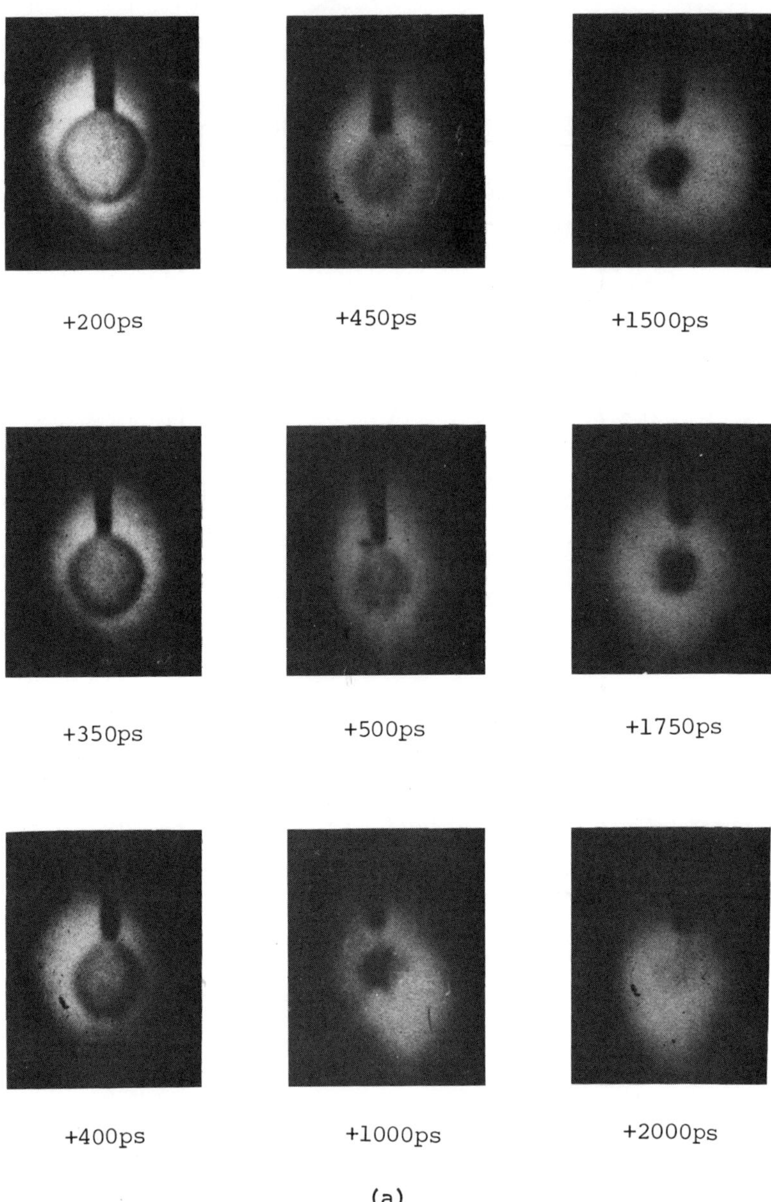

Fig. 32.   X-ray shadow image for $(CH_2)n$ shell (a) and its
          implosion dynamics (b).

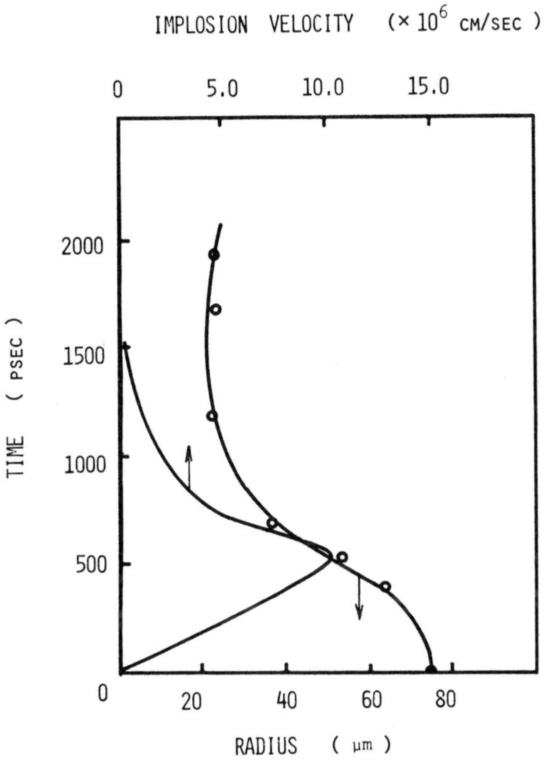

TARGET   : CH_2 SHELL   150 $^{\mu m \phi}$, 5 $^{\mu mt}$

LASER POWER : 0.67 TW/ 120 PSEC

Fig. 32 (b)

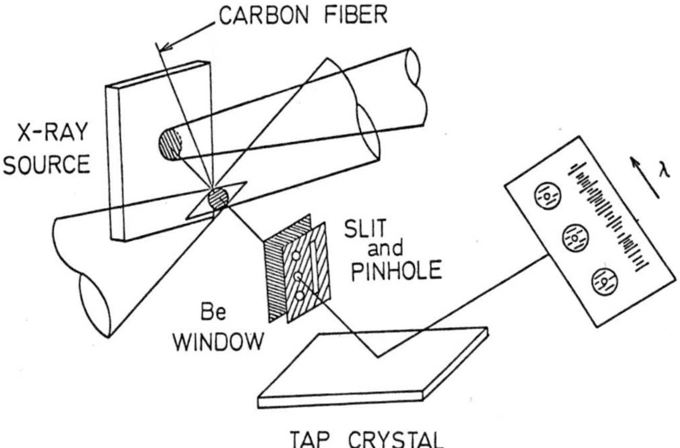

Fig. 33. Experimental set up for multi-frequency X-ray back lighting.

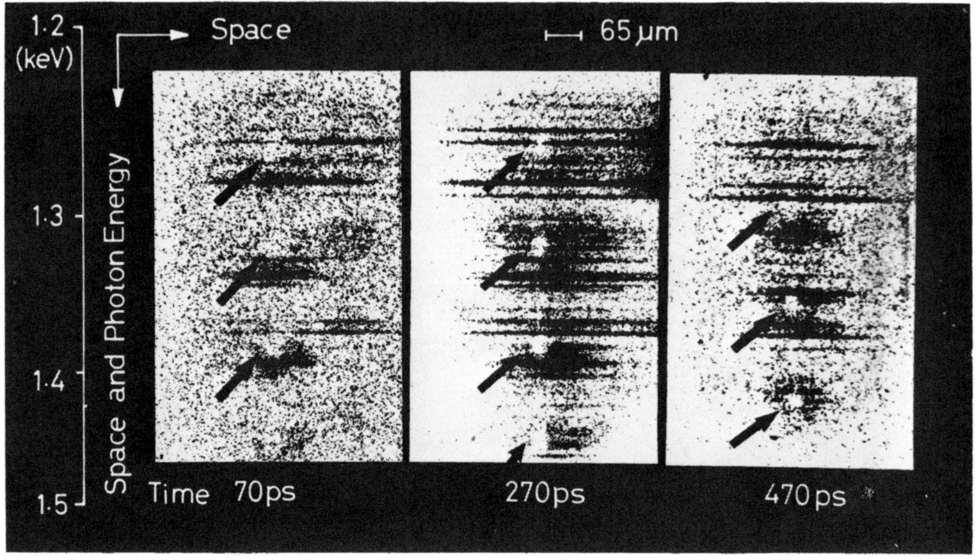

Fig. 34.   A series of the backlighting images for various delays from the laser pulse irradiating the microballoon. Shadows of the microballoon are indicated by arrows.

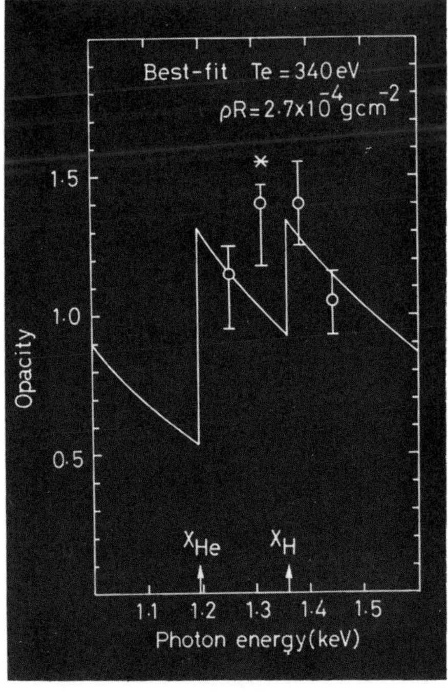

Fig. 35.   Spectral opacity plots for 270 ps delay along with the best-fit calculated curve. Absorption edges of He-like and H-like neon are indicated by X_{He} and X_H, respectively. Plot marked by an asterisk shows a shift of the absorption edge.

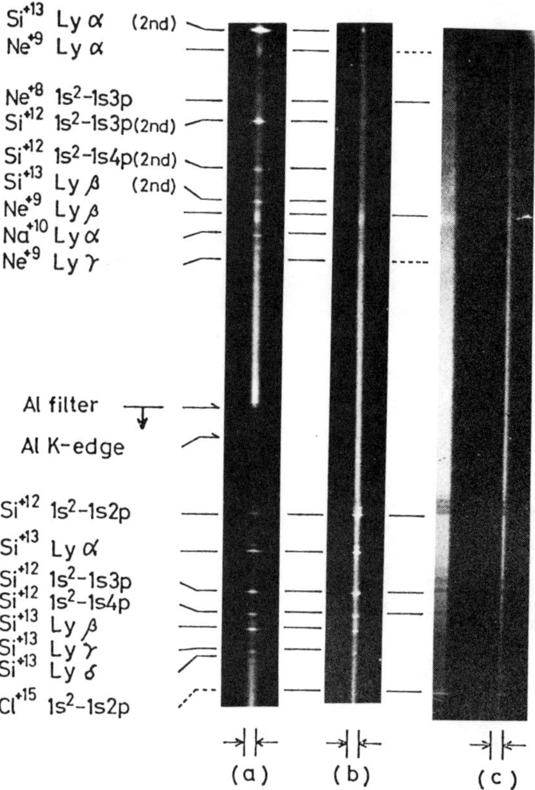

Fig. 36. Typical X-ray spectrographs in the spectrum range from 1 to 3 keV.

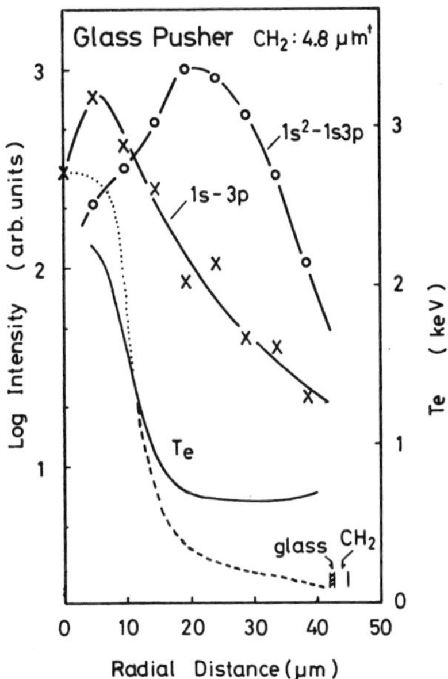

Fig. 37. Temperature determination of glass plasma. The spatial profiles of Si Ly β Si $1S^2$-1S 3P and electron temperature.

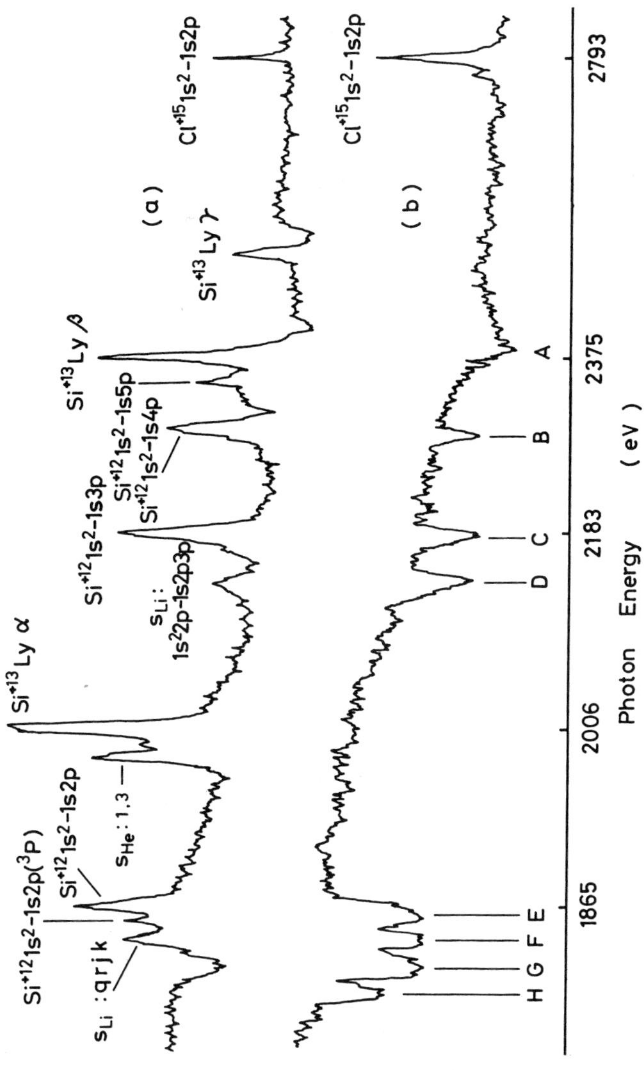

Fig. 38. Emission and absorption spectra of highly ionized
Si ions.

Table 4. Products of density and thickness of glass layers estimated from various dark lines.

A	$Si^{+12} + h\nu \; - Si^{+13} + e$ edge	$N^{+12} \; \Delta R = 6.0 \times 10^{17}$ cm^{-2}
B	$Si^{+12} \; 1s^2 \; -1s\,4p$	$N^{+12} \; \Delta R = 8.7 \times 10^{17}$ cm^{-2}
C	$Si^{+12} \; 1s^2 \; -1s\,3p$	$N^{+12} \; \Delta R = 7.1 \times 10^{17}$ cm^{-2}
D	$Si^{+11} 1s^2 2p - 1s\,2p\,3p$	$N^{+11}_{1s^2 2p} \Delta R = 1.2 \times 10^{18}$ cm^{-2}
E	$Si^{+11} 1s^2 2s - 1s\,2p\,2s$ (s,t)	
F	$Si^{+11} 1s^2 2p - 1s\,2p^2$ (a,b,c,d,j,k,l) $Si^{+11} 1s^2 2s - 1s\,2p\,2s$ (q,r)	
G	$Si^{+11} 1s^2 2p - 1s\,2p^2$ (e,f,g,h,i) $Si^{+11} 1s^2 2s - 1s\,2p\,2s$ (u,v)	
H	$Si^{+11} 1s^2 2p - 1s\,2s^2$ (o,p)	$N^{+11}_{1s^2 2p} \Delta R = 1.4 \times 10^{18}$ cm^{-2}

$$\tau_0 = \frac{\pi e^2}{m\,c} \cdot f_{abs} \cdot N^z \cdot \Delta R \cdot P_0$$

$T_{e,\text{glass pusher}}$ = 160 ± 30   eV   (from population ratio N^{+12}/N^{+11})

$\rho \Delta R_{\text{glass pusher}}$ = 6.6 × 10^{-3} g·cm^{-2} (from $N^z \Delta R$, T_e)

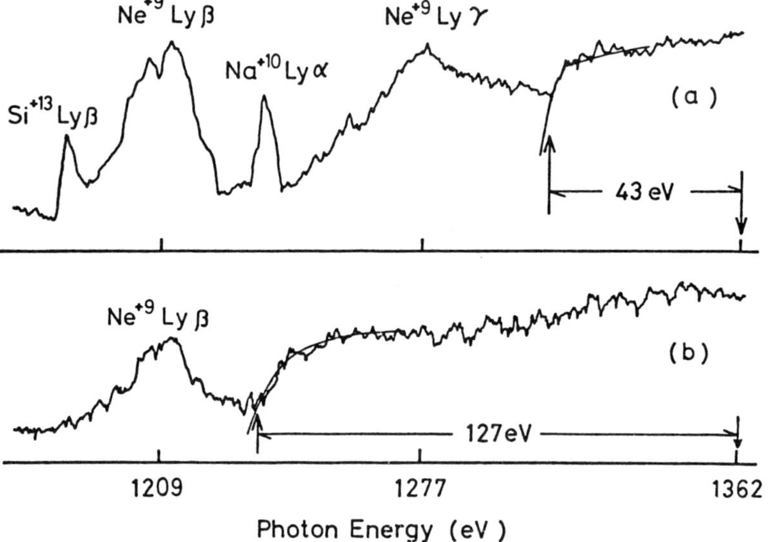

Fig. 39. Typical spectra from compressed Ne core; Ne^{+9} Ly β and free-bound continuum are shown. (a) simple GMB, (b) 10 μm thick coated GMB.

thickness dependence of the edge-shift is shown in Fig. 40. In
this figure, the dashed line shows the value of ΔE_s which was
calculated using T_e and ρ at the time of maximum compression from
computer simulation. Considering the reliability of the simulation
model which agrees with various experimental results, we can
conclude that we have obtained the core density of $5g/cm^3$. This
density is about 25 times higher than that obtained in exploding
pusher compression.

Exploding pusher type target filled with argon gas was used
to determine core parameters at high temperature. Electron
temperature of 2.1keV and ρR of 2.9×10^{-4} g/cm^2 were determined.
Also at another series of compression experiments, spherical targets
with high Z coating were used. Preliminary results indicate that
energy transport into the core is strongly inhibited by a layer
>0.3μm thickness.

Major results of these experiments are summarized in Table 5.

4-4 Fusion Particle Measurements

The neutron yield was measured by a silver activation counter
at different distance from target. The latter detectors were
calibrated using D-D neutrons generated from a plasma focus machine
against the silver activation counter which had been calibrated
carefully by continuous neutron source of ^{252}Cf and ^{241}Am-Be.

Fig. 41 shows the dependence of the neutron yield (Ny) on the
peak power of 100 ps laser pulse. From the experimental results, the
dependence was found to be Ny $\times$ $P^{1.6}$ for uncoated GMB with diameter
of 71μm and 117±4μm and 117±8μm, whicn was fairly in good agreement
with that of simulation. The neutron yield of low-Z coated GMB was
lower than that of uncoated GMB even with higher of compressed fuel.
This shows low temperature, high density compression with thicker
ablater on GMB.

With D-T fuel gas, neutron yield of 3×10^7/shot was obtained in
exploding pusher mode.

5. CONCLUSION

The understanding of fundamental process in laser plasma
interaction has been pursued. The results, such as good absorption
and coupling, low level of back reflection and suppression of that,
and weak dependence of hot electron on the incident power and wavelength indicate possibility of ablative compression by the presently
existing lasers to achieve high density.

The physical problems are energy transport in core. Energy
transport by radiation, high energy electrons and the inhibition
mechanism of thermal conduction have been investigated to clarify
the transition from explosive to ablative mode compression.

In compression experiments, detailed comparisons of explosive
and ablative driven implosion have been made by using glass micro-
ballon target with polyethylene coating of varying thickness.

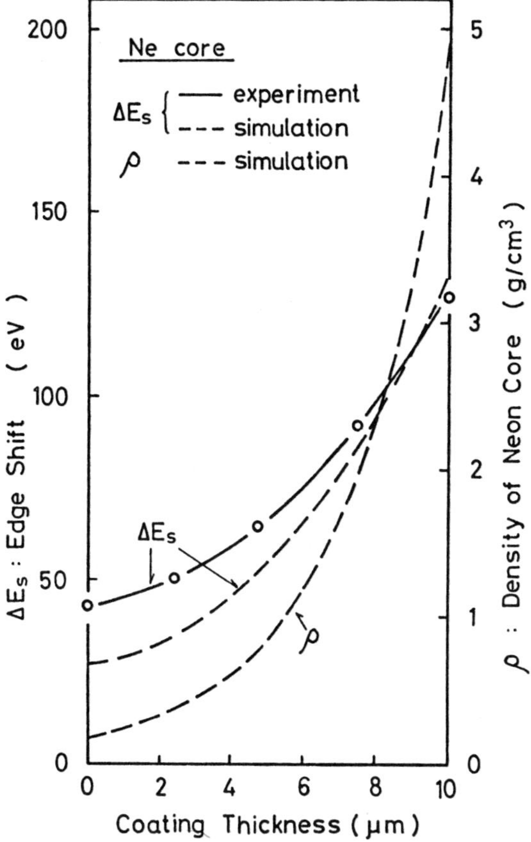

Fig. 40.   Coating thickness dependence of free bound edge shift and density of Ne core.

Table.5   Plasma parameters estimated from X-ray spectrography

1. Exploding pusher compression
   Ar core $2R = 24\,\mu m^{\phi}$ (initial $R_o=38\mu m$): compression ratio $\simeq 20$
   $T_e = 2150$ eV
   $\langle\rho\rangle = 0.15$ g/cm^3
   $\langle\rho R\rangle = 2.9\times10^{-4}$ g/cm^2 ($\langle n_i\rangle = 2.2\times10^{21}$/cm^3)

2. Ablative mode compression
   Thickness of CH$_2$ layer $\geq 5\mu m$
   Glass pusher $T_e = 160$ eV
   $\rho\Delta R = 6.6\times10^{-3}$ g/cm^2 (initial $(\rho\Delta R)_o = 2.9\times10^{-4}$ g/cm^2)
   Neon core $\rho = 5$ g/cm^3

3. High z coated pellets
   Implosion symmetry is not good.
   Energy transport by radiation is important.
   Thin gold layer acts as tamper.

4. Multi-layered pellets
   Hot electrons are stopped by thin gold layer. (Au $\geq 0.3\mu m^t$)
   Neon core $\rho_{Au-CHCl} \simeq (0.2-0.4)\rho_{CHCl}$

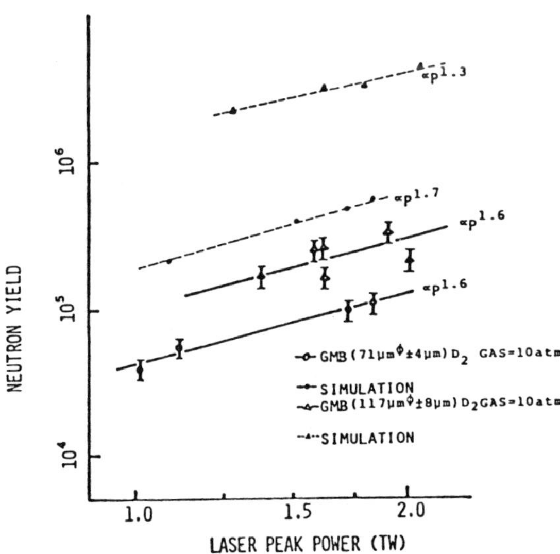

Fig. 41.   Neutron yield of GMB target as a function on laser peak power.

The major results are as follows:
(1) Compressed core density of $5g/cm^3$ was achieved in ablative
    implosion with thick low Z ablator, whereas it was $0.1\sim0.5$
    g/cm^3 in simple GMB compression.
(2) Implosion time increased and lower temperature pusher was
    achieved by increasing the coating thickness.
(3) The distinct feature of explosive and ablative compression was
    observed by X-ray shadowgraphic technique.
(4) Neutron yield depends on the compression mode which shows lower
    ion temperature in ablative compression than that in explosive
    compression. D-T neutron yield was 3×10^7/shot in exploding
    pusher mode.

Above results show the role of thick ablator which is in favor
for high density compression. To investigate the isentropic compression, larger output energy and some pulse tailering will be
required in the next step.

II. REB FUSION EXPERIMENTS

REB interaction with solid target has been investigated to
clarify the dissipation and transport mechanisms. The experiments
were carried out on Reiden III machine with parameters of 400kV,
100kA, 100ns. The experimental set up is shown in Fig. 42.

If the beam electron is stopped and deposits the energy mainly
on the front surface of the target[8], the target behavior is the
ablative mode. When the ablated thickness is negligibly small
compared to the initial foil thickness the velocity of the rear
side motion of the target is inverse proportional to the ρt where
ρ and t are the target mass density and thickness, respectively.

If the beam electrons can penetrate the target and deposit
some of their energy uniformly in the target, assuming the energy
deposition to be the classical collisional absorption, an explosive
behavior is considerable. In the explosive mode the target expansion occurs symmetrically on the front and the rear side of the
target and the velocity of rear side is almost constant and independent to ρt.

The dynamic behavior of the front (beam injection side) and
rear surface of the target was measured by shadowgraph method using
four N_2 lasers from the horizontal direction of the foil target.

The emitted X-ray image around the target was measured by a
filtered pinhole camera with a space resolution better than 150µm
from the perpendicular direction to the diode axis. Several sheets
of Kodak No-Screen film were mounted in a stacked array. Fig. 43
shows the ablative behavior of low Z target. The X-t diagram corresponding to Fig. 43(b) is shown in Fig. 44 which shows delayed
expansion of rear side. On the other hand, the behavior of thin Au
foil is different from that of low Z target as shown in Fig. 45(a).
For Au foil which is thicker than the classical range, the expansion
is similar to low Z ablative target as be seen in Fig. 43(b) and
Fig. 45(b).

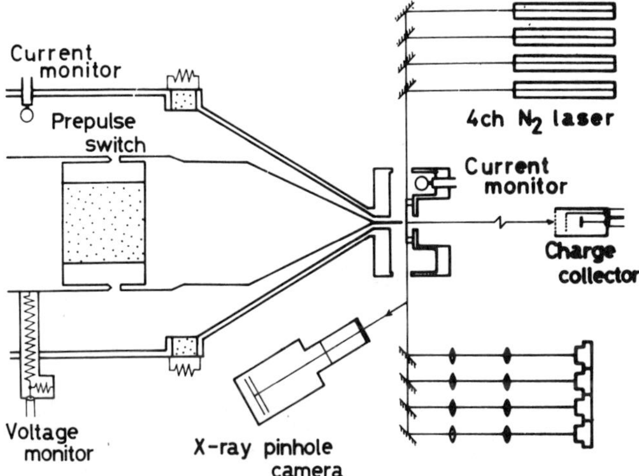

Fig. 42.   Experimental set up on REIDEN III.

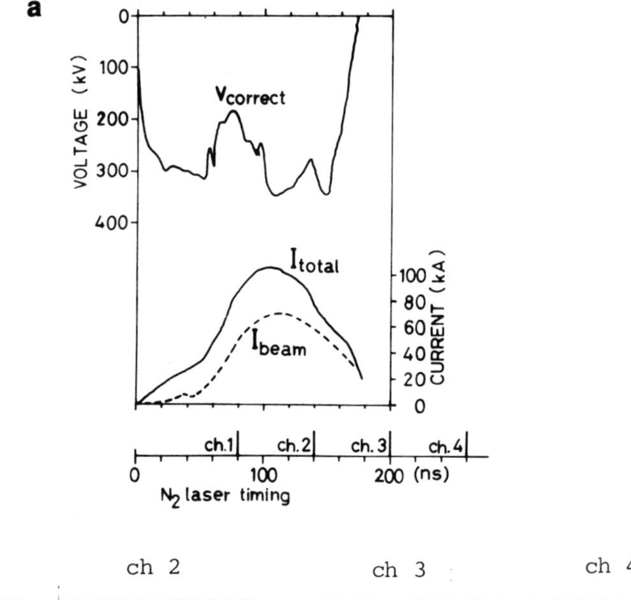

Fig. 43   (a) Diode characteristics and laser pulse timing.
(b) Shadowgraph of four channel N_2 laser.

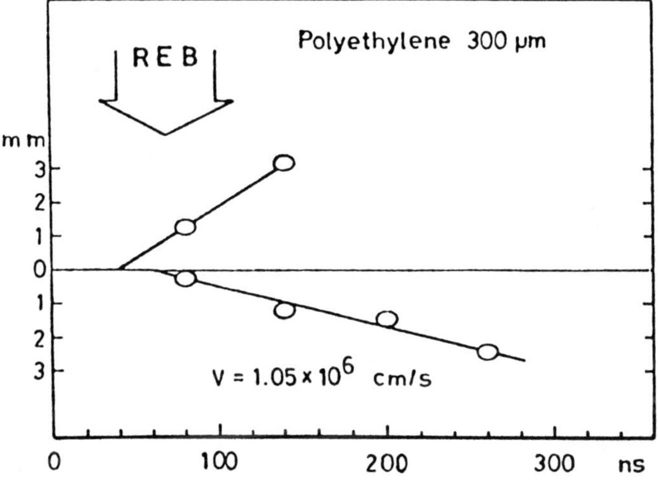

Fig. 44. Dynamic behavior of the target obtained from the shadowgraph.

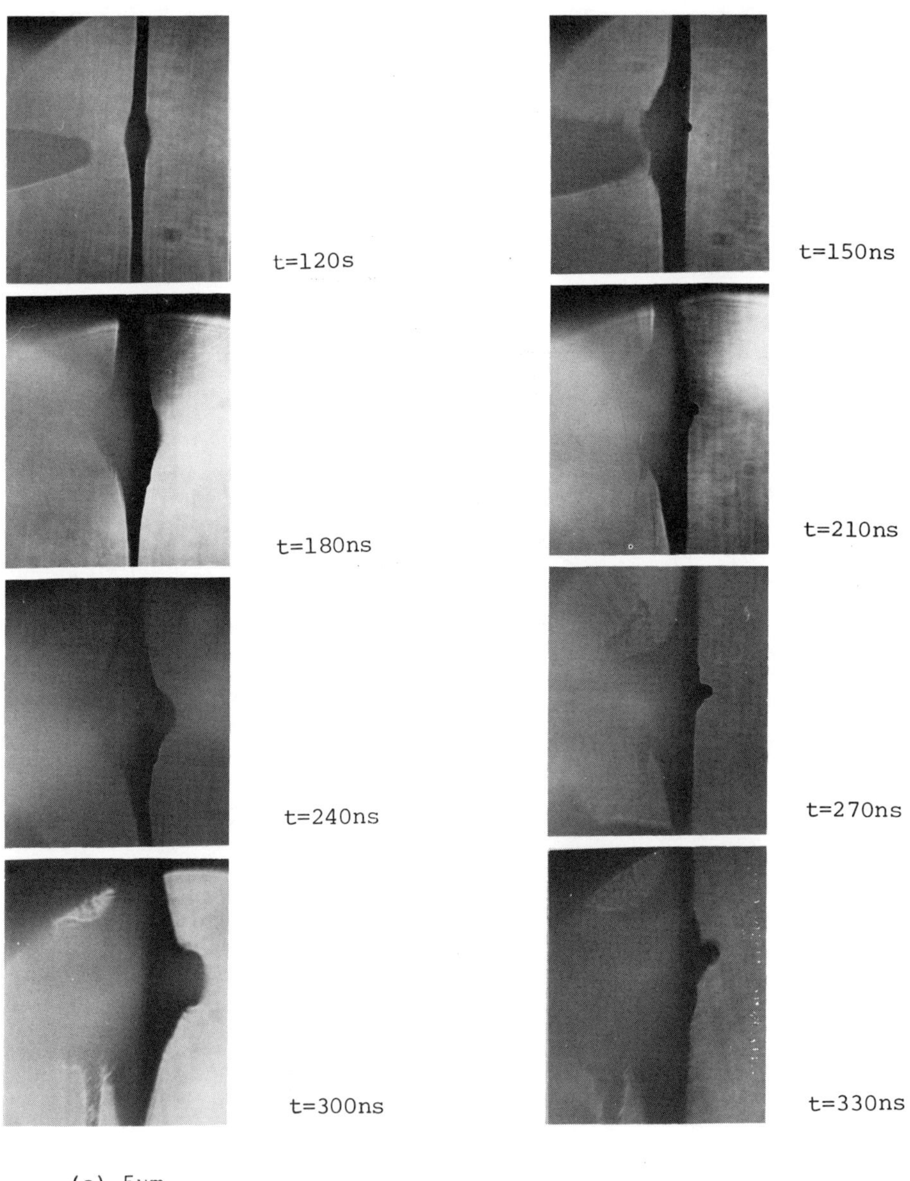

Fig. 45. Shadowgraphs of 5 μm and 50 μm gold foil target.

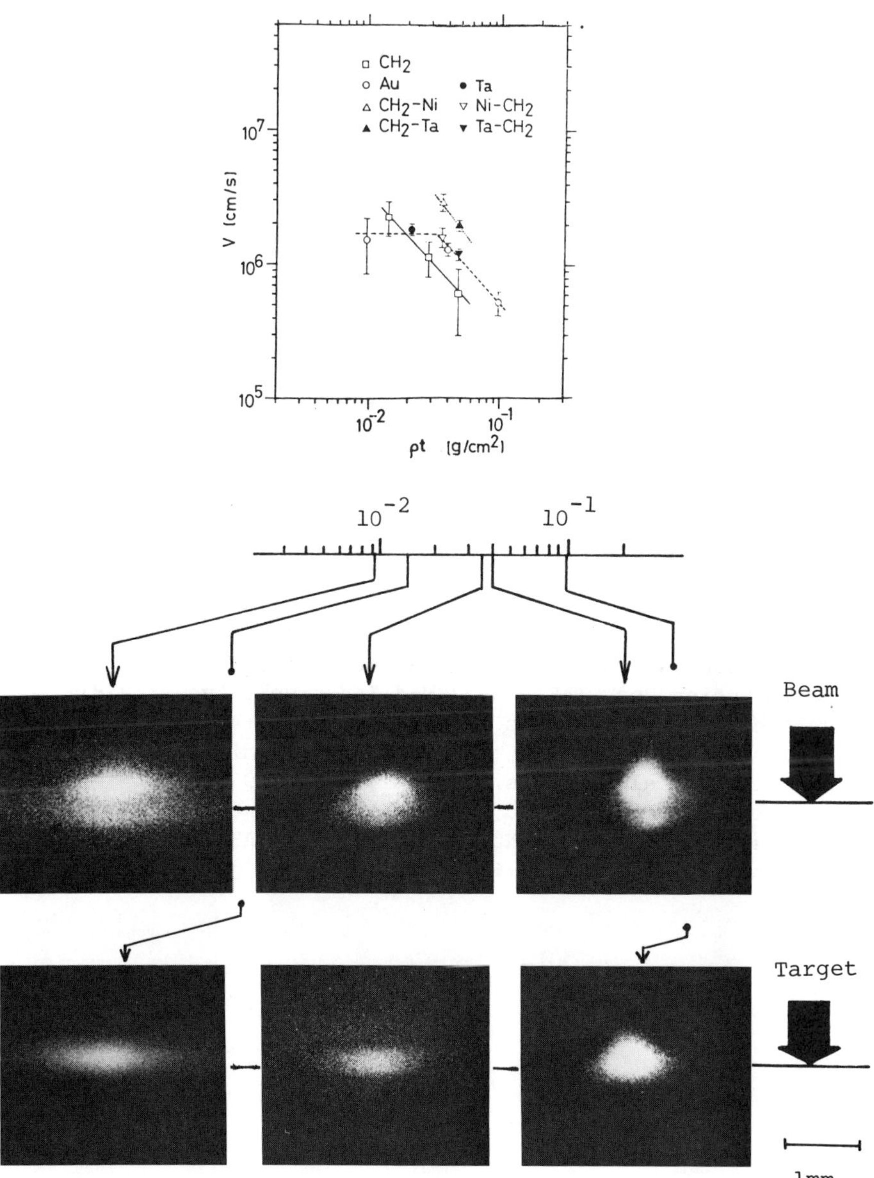

Fig. 46. The velocity of rear side of various targets.
With X-ray image of the targets.

The experimental results of the rear-sides velocities of various targets with corresponding X-ray images of targets are shown in Fig. 46.

In the case of Au and Ta foil targets the expansion velocity seems to be constant where the ρt is less than 3×10^{-2} gcm^{-2} (explosive mode) and is inverse proportional to the ρt of the target where the ρt is more than 3×10^{-2} gcm^{-2} (ablative mode). The X-ray emission was observed on the front and rear side of the 5µm thick Au target. In the case of 20µm thick Au foil, very weak emission was observed on the rear side of the target. When the target of a 50µm Au foil was used, the X-ray emission was observed only on the front side of the target. Data implied that the beam electron was completely stopped in 50µm thick Au foil and penetrated in 5µm thickness.

In the case of polyethylene target the dependence of the velocity of rear side motion is the ablative mode and the X-ray emission was observed only in the front side of the target. Data indicated the interaction between the beam and low Z target is due to an anomalous model.

References

1)   Annual Progress Report on Laser Fusion Program, ILE-APR-78:
     Institute of Laser Engineering, Osaka University (1978)
2)   K. Yamada, el al. : Tech. Rep. Osaka Univ. 29, 159 (1979)
3)   M. Matoba, et al. : Tech. Rep. Osaka Univ. 29, 147 (1979)
4)   C. Yamanaka, et al. : IAEA 7th Int. Conf. Plasma Phys. and Cont.
     Nucl. Fusion Res. IAEA-CN-37/M-4 Innsbruck, Aug. (1978)
5)   H. Azechi, et al. : Phys. Rev. Lett. 39, 1144 (1978)
6)   H. Nishimura, et al. : Tech. Rep. Osaka Univ. 28, 185 (1978)
7)   S. Nakai, et al. : Phys. Rev. A17 1133 (1978)
8)   S. Nakai, K. Imasaki and C. Yamanaka; IAEA 6th Int. Conf. Plasma
     Plasma Phys. and Cont. Nucl. Fusion Res. IAEA-CN-35/F8,
     Berchtesgaden, Oct. (1976)

FAST ELECTRON PREHEATING IN EXPLODING PUSHER EXPERIMENTS

C. Bayer, D. Billon, M. Decroisette, D. Juraszek,
D. Lambert, J. Launspach, M. Louis-Jacquet,
J.L. Rocchiccioli, D. Schirmann

Commissariat à l'Energie Atomique, Centre d'Etudes de
Limeil, B.P. N° 27, 94190 - Villeneuve-Saint-Georges-FRANCE

INTRODUCTION

Microballoon implosion experiments in the explosive pusher regime have been performed at CEL with the eight beams glass laser
system Octal.

Irradiations have been performed at laser irradiances of a few
10^{15} W.cm^{-2} in 50 ps. Numerous theoretical[1] and experimental works[2]
show that in such conditions, laser absorption occurs at critical
density in a high density gradient, mainly through resonant absorption, while energy transport processes imply fast electron occurrence and great thermal flux reduction. In a first part of this paper
we present an experimental study about the effect of a prepulse in
the contrast ratio range $10^{+4} - 10^{+6}$. Evolutions of specific absorbed energy, neutron yield, spatial profiles of Silicon resonance lines, ion distribution functions and X-ray pinhole pictures are described and tentatively connected to fast electron occurence.

In flat layered targets irradiation experiments[3] a strong reduction factor of the thermal conductivity of $(10^{-2}\ 10^{-3})$ and a fast
electrons preheating ratio of 10% have been deduced with respect to
the incident energy. Here suprathermal preheat has been diagnosed
by X-ray shadowgraphy. This diagnostic is of a particular value in ablatively driven laser implosions [4,5,6]. But for exploding pusher targets, X ray probing in the keV range provided quite useful informa-
tions about electron energy deposition in the shell. The backlighter
was a Cu-Zn target irradiated by a ninth chain (1 J ; 50 ps). The
timing of the probing pulse has been checked using an X-ray streak
camera. Time integrated transmission has been interpreted in terms
of glass shell heating versus laser intensity.

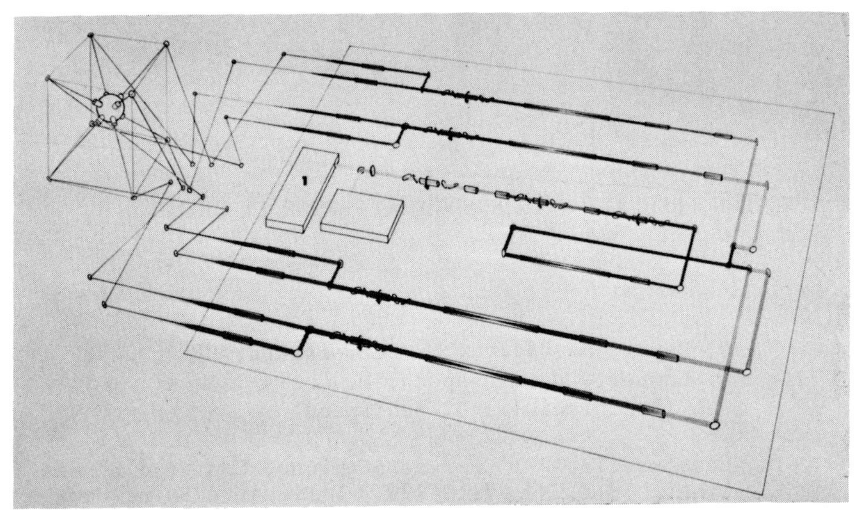

Fig 1 : Octal and Camelia implantation

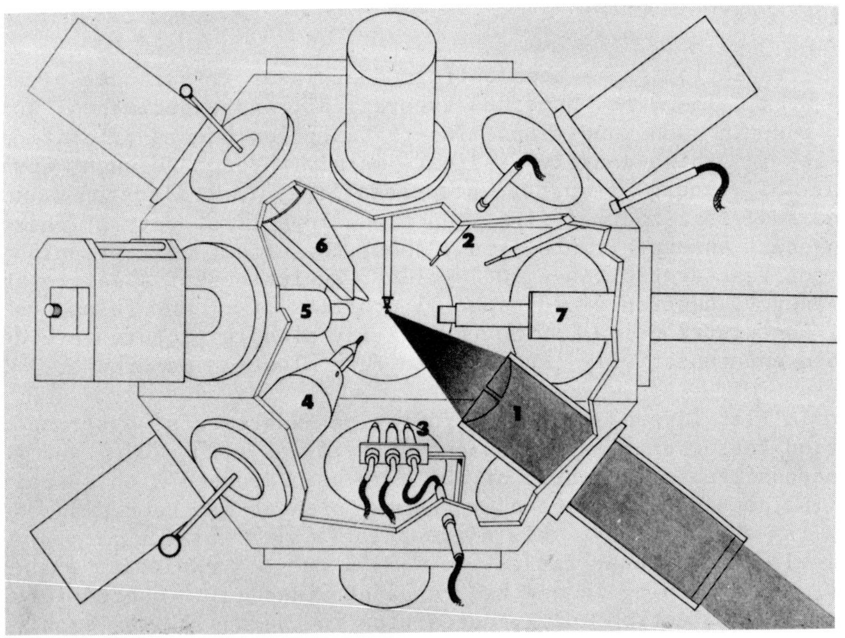

Fig 2 : Camelia chamber with diagnostics implantation

1. EXPERIMENTAL CONDITIONS

Experiments have been performed with the (8+1) beams Nd laser facility Octal, operated in short pulse regime (50 ps) ; the eight beams delivered typically 40J on target, the ninth one (1 J) being used for diagnostic purposes such as X-ray backlighting.

Target irradiation is performed in cubic geometry (fig. 1) in order to get isotropic energy deposition and beams are focused by f/2 aspherical lenses. More than sixty recording points are currently activated in the target chamber (fig.2) among which thirty two pyroelectric calorimeters (2) and twelve faraday cups (3) for energy balance and control of fast ions occurrence ; X ray pinhole cameras (4) for implosion symmetry and compression evaluation ; X-ray streak camera (5) for compression timing, X-ray plane crystal spectrographs (6) for electron temperature evaluation, T.O.F. magnetic-deflexion spectrometer (7) for alpha particules detection, and at last BF_3 detectors, activated silver counters and scintillator-photomultipliers devices for neutron measurements. Correct beam alignment on target was controlled after each laser shot by 2 ω_o time integrated pictures with a magnitude of 200 (fig. 3).

High energy contrast ratio on target was obtained by inserting a dye cell on each beam of Octal. Randomly occurring prepulses between 0.5 - 3 ns before the main pulse were measured in the power range 1-100 MW (fig. 4).

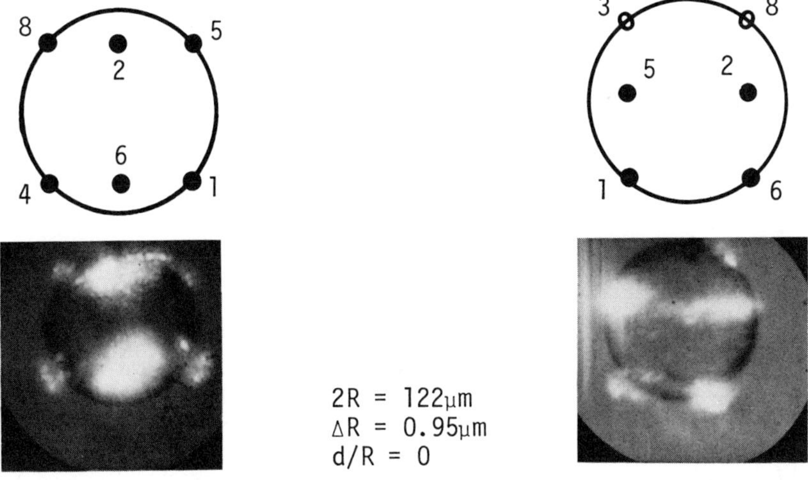

Fig. 3: 2 ω_0 time integrated pictures (without prepulse)

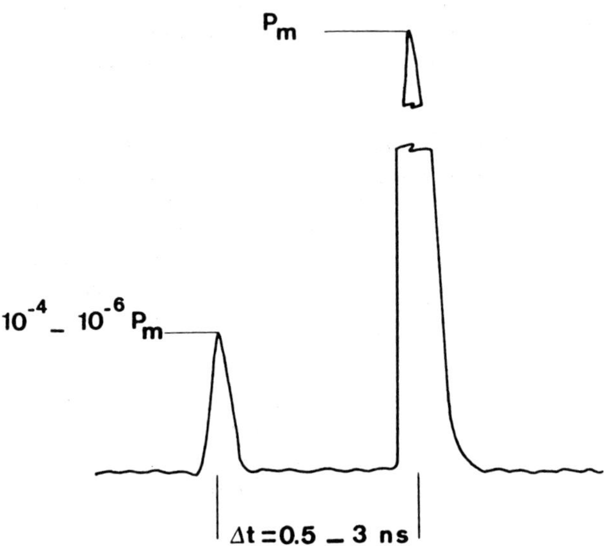

Fig. 4 : Schematic temporal evolution of laser prepulse and
         pulse powers.

2. TYPICAL RESULTS AND PREPULSE EFFECTS

Imploded targets were glass microballoons typically 80 μm in
diameter and 0.68 μm in wall thickness ; however for preheating
studies, variations on diameter and wall thickness (by p-xylene
coating)were performed.

Results will be presented with the mention "without" or "with
prepulse" respectively standing for $\simeq 10^{+6}$ or $< 10^{+4}$ power contrast
ratios.

2.1. Energy balance

Results on energy balance are summarized in table I and II,
and reported versus incident laser flux in fig. 5. For a high
contrast ratio ($\simeq 10^6$) in the incident laser flux range 2.8 10^{15} –
4.2 10^{15} W.cm^{-2}, the mean absorbed specific energy increases from
0.15 to 0.21 J/ng, variations being fitted by (E$_a$/M) J/ng $\simeq$
5.10^{-17} ϕ_1 (W.cm^{-2}). In table I, (R + T) stands for non absorbed
energy measured in the focusing lens cones, including refraction
and scattering (R) and transmission (T), T having been evaluated

in one beam irradiation, (r) stands for refracted and scattered
energy, measured in the target chamber outside of the lens cones
and (A) the overal absorption coefficient.(A) was evaluated within

Table I

P_{GW}		650	950
$\emptyset_i$	$W.cm^{-2}$	$2.8.10^{15}$	$4.2.10^{15}$
R + T	%	23.5	24
r	%	60.5	61.5
A	%	16	14.5
$\dfrac{E_a}{M}$	J/ng	0.15	0.21

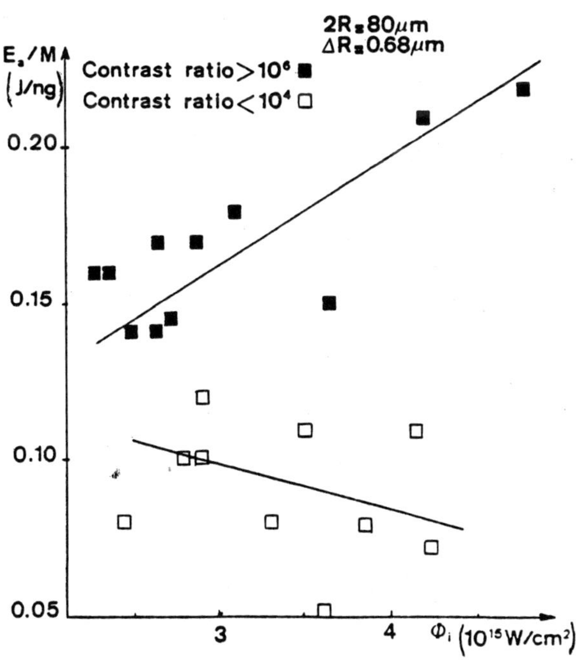

Fig. 5 : Mean absorbed specific energy versus incident laser
flux with and without prepulse

the assumptions of refraction and scattering isotropy which seem
valid since fluctuations of calorimeters signals

$$\Delta \cdot \left(\frac{dE_r}{d\Omega} \right) \; / \; < \left(\frac{dE_r}{d\Omega} \right) \; \simeq \; 0.24$$

were close to those of incident energy $\dfrac{\Delta E_i}{<E_i>} = 0.2$

In the case of a prepulse of relative intensity 10^{-4} (table II) the
absorption is about 50 % lower due to an increase of refraction and
scattering. This results in a mean absorbed specific energy slightly
decreasing around 0.1 J/ng.

Table II

P_{GW}		650	950
R + T	%	25	27
r	%	64	66
A	%	11	7
$\dfrac{E_a}{M}$	J/ng	0.1	0.08

These results suggest that the prepulse drives an early hydrodynamical flow carrying the critical density away from the initial
target surface, with a smooth density gradient. This leads first
to a lowering of the interacting flux for the main pulse and thus
a lowering of the resonant absorption contribution[7] ; moreover, a
larger subcritical plasma enhances non linear processes such as
stimulated Brillouin scattering[8].

2.2. Ion distribution functions analysis

Ion distribution functions are reported in fig. 6. They have
been deduced from time of flight measurements with faraday cups.
Isotropy of plasma expansion could be deduced from signals identity
within a few percent. But it appeared strong features modification
according to the prepulse level.

In the case of a high contrast ratio (10^6) the velocity
spectrum obtained with $\overline{Z} = 10$ and $\overline{M} = 20$ presents a high velocity
component for $v > 10^8 \mathrm{cm.s}^{-1}$. In fact, the distribution function
can be fitted by adding three maxwellian distributions (as indicated in the cartoon) characterized by a mean velocity respectively
$10^7 \mathrm{cm.s}^{-1}$, $2.5 \; 10^7 \mathrm{cm.s}^{-1}$ and $1.8 \; 10^8 \mathrm{cm.s}^{-1}$.

The second one, we shall call "thermal" seems to be attributed to corona ions directly expanding outwards from the target. The "cold" first one may be due to ions initially propelled inwards the target, and expanding off the core after stagnation. The third one is a hot distribution which may be interpreted as composed of fast ions driven by suprathermal electrons coming from resonant absorption.

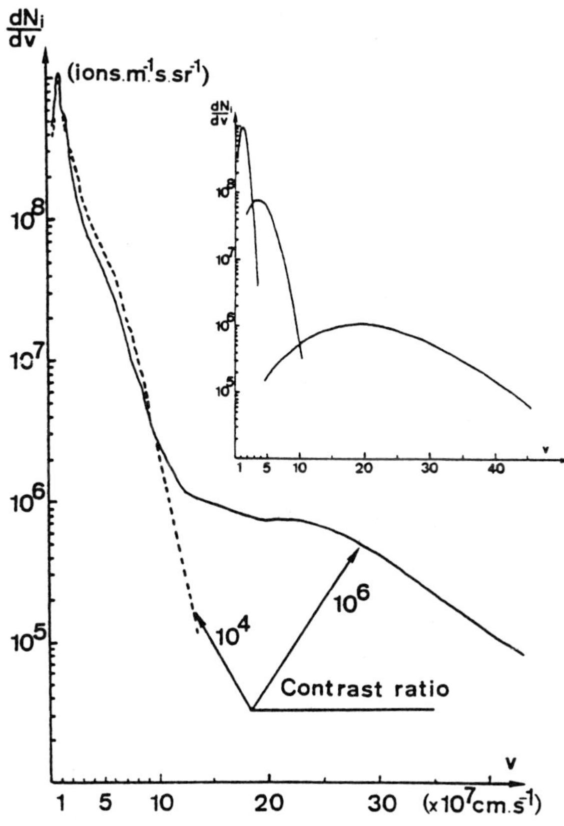

Fig. 6 : Ions velocity spectrum with and without prepulse ; the inset shows the three Maxwellian distribution fitting.

In this special case, the ion kinetic energy for velocities greater than $10^8 cm \cdot s^{-1}$ corresponds to 50% of the total absorbed energy, the "cold" and "thermal" distributions carrying respectively 12% and 28% of the total ion kinetic energy, and 78% and 20% of the total mass.

In the case of a prepulse (10^4 contrast) the "hot" distribution has quite disappeared (the proportion of energy carried away by ions velocities > 10^8 cm.s^{-1} being less than 5%) while the cold and thermal distribution account for respectively $\simeq$ 20% and $\simeq$ 80% of the kinetic energy. Desappearence of fast ions is difficult to connect unambiguously to fast electron disappearance, as the early plasma set up the prepulse may cause ions slowering down. However, masses carried out by the thermal distributions in shots with high and low contrast ratio correspond respectively to a glass ablated thickness of 0.14 μm and 0.15 μm, inferring that a thickness << 0.15 μm has been ablated by the prepulse. In the pessimistic assumption that 0.1 μm were ablated by the prepulse, calculation in spherical geometry of stopping power of Si^{14+} ions in SiO_2 absorbing material at different temperatures and densities, have been performed, taking into account electron collision (Bohr formulation) and nuclear collision (Bethe formulation). Table III presents the cut-off speed V_C such that ions whose velocities are greater than V_C, are not stopped by the ablated material. It can be seen that in every case V_C is lower than $10^8 cm.s^{-1}$

TABLE III

n_e \ absorbing material	Si O_2 Cold	Si O_2 1 keV
$7.5 \ 10^{23} \ (\rho = \rho_0)$	$1.2 \ 10^8$ cm/s	–
10^{21}	$9 \ 10^7$	$8.8 \ 10^7$
10^{20}	$6 \ 10^7$	$5.2 \ 10^7$
10^{19}	$3.5 \ 10^7$	$3.8 \ 10^7$

Thus fast ions disappearance seems to be related to a strong reduction in fast electron generation due to an important decrease in resonant absorption.

2.3. Neutron emission

Neutron yield for standard microballoons with 10-30 bar DT pressures was typically 10^6 - 10^7 in the absorbed specific energy range 0.1- 0.3 J/ng and a high contrast ratio. However it decreased of nearly one order of magnitude in the case of a prepulse. Similar results are described in ref. 9.

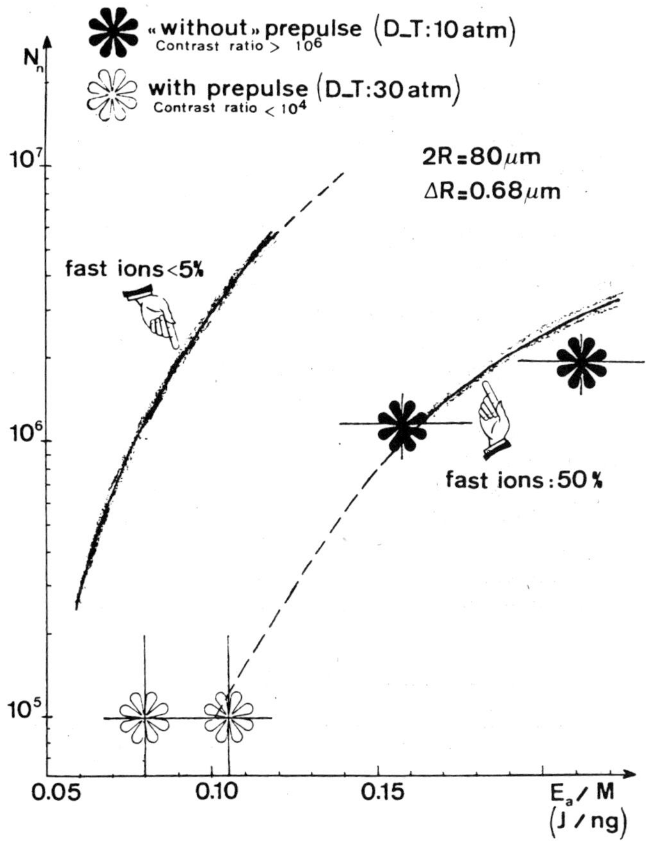

Fig. 7 : Experimental neutron yields versus mean absorbed specific energy, compared with theoritical model, with and without prepulse.

In fig. 7, experimental results have been compared to numerical evaluations performed using the analytical model of[2]. Two theoretical curves have been drawn for fast ion energy percentages 5 % and 50 % in view to fit with experimental observations with and without prepulse. High contrast ratio results appear in good agreement with the model, but low contrast ones are much below the related previsions. In fact, as all results correspond to the similar useful absorbed specific energy (as defined in[2]) this can signify that low contrast ratio results cannot be compared to the explosive pusher regime.

2.4. X-ray spectroscopy

In the followings, the word "core" will stand for the compressed zone, including both glass and D.T.

Prepulse consequenses were also seekeed for by X ray analysis in the 4 to 8 Å range, with spatial resolution about 10 μm. One dozen shots have been selected for this presentation, ported into two flux ranges series 800 - 850 GW and 950 - 1100 GW.

Emitted line power integration (table IV)

For helium-like Silicon α and β transitions, as for hydrogen-like Lyman α, no internal spectral resolution was possible, due to instrumental width exceeding Stark width. However, satellite-lines can be distinguished from resonance ones. Theoretical interest from introducing these satellites in line power ratios is low but it allows verification that their unwanted contribution would not qualitatively change any result. Satellites included figures are in brackets

Table IV

	lines energy ratios	
	with prepulse (A)	Without prepulse (S)
Low flux 800-850 GW	$\dfrac{H_\alpha}{He\,\alpha} = 0\cdot27$ (0.20)	$\dfrac{H_\alpha}{He\,\alpha} = 0.75$ (0.62)
	$\dfrac{He\,\beta}{He\,\alpha} = 0\cdot34$ (0·18)	$\dfrac{He\,\beta}{He\,\alpha} = 0\cdot58$ (0·35)
High flux 950-1100 GW	$\dfrac{H_\alpha}{He\,\alpha} = 0\cdot21$ (0.12)	$\dfrac{H_\alpha}{He\,\alpha} = 0.47$ (0.35)
	$\dfrac{He\,\beta}{He\,\alpha} = 0.11$ (0.10)	$\dfrac{He\,\beta}{He\,\alpha} = 0.65$ (0.39)

Core dimensions

Spatial analysis of the spectrograph films shows that for Helium like lines there is no distinction between core and corona emissions and the spatial line shape is a mere trapezium.

But for hydrogen-like Hα, the core emission may be figured as narrow trapezium superposed upon a wide one, representing corona emission.

The core diameter measured from trapezium basis is

With prepulse $<2 \ Ro \ >_A = 17 \ \mu$

$<2 \ Ro \ >_S = 23 \ \mu$

(A stands for with prepulse ; S for without prepulse)

This shows a 35 % decrease of diameter and it cannot be an artefact since, when omitting the extreme values into the averaging, the difference keeps higher than 25 %.

Interpretation of such facts can lie in :
- either the acceptance of better compression because of prepulse
- either the fact of a lower core-temperature with prepulse (that could agree with a lesser neutron production)
- either the fact that the prepulse partly breaks the microballoon letting it leak during compression (that could be demonstrated by a heavy gaz doping of D.T, Ne for example)

Corona dimensions

Since plasma-requirements for H-like and He-like emissions are not exactly the same, measurements of corona diameter from these two kinds of lines do not exactly coincide (apparent diameter from H-like analysis being narrower), but the ratio of figures with and without prepulse is quite the same in both cases.

We find by Heα analysis

With prepulse $< 2 \ R \ >_A = 230 \mu m$ dispersion less than 10 %

Without prepulse $< 2 \ R \ >_S = 190 \mu m$ dispersion less than 12 %

So the corona is 20 % larger after a prepulse.

If we admit that the difference between H-like and He-like emission borders is representative of the density gradient scale length, we find it 20 % larger with prepulse.

By Hα analysis, we see no diameter difference between "with" and "without" cases exceeding experimental resolution ; we measure a main diameter very similar to the original microballoon'one, as by pinholes cameras. Such a concordance between H-like spectrograph result and pinhole results but contradiction with He-like ones is not worrying since He and continous emissions require very similar conditions.

For He-like emission, density and temperature requirements are so different that, although we cannot distinguish temporaly different phenomena, we can state that the emissing matter elements are different, and confused only because of time integration.

So we can use different criteria for measuring H-like and He-like emissions diameters. The He-like figures listed above correspond to measurements of diameter between the intersections of zero-level with the two straight tangential lines to the spatial profile at half profile width and altitude

Spectral_lines_lumination

The values listed above integrate the total emission seen from a planar window

- materialy limited in one direction by the hedges of slit
- virtualy limited in the orthogonal direction by the cristal diffraction width and the dimension of plasma emitting zone itself.

Each position in the spatial analysis direction is then an image from a section of the plasma by a plane originated from the spectrograph slit.

Different values of these plane widths correspond to different values of λ, thus preventing us from determining a direct diagnostic from measured $\dfrac{H\alpha}{He\alpha}$ and $\dfrac{He\beta}{He\alpha}$ ratios, although their real knowledge would enable us to determine Te and Ne by the use of a suitable model.

So we must use only

$$\dfrac{\left(\dfrac{He\ \beta}{He\ \alpha}\right)_A}{\left(\dfrac{He\ \beta}{He\ \alpha}\right)_S} \quad \text{and} \quad \dfrac{\left(\dfrac{H\alpha}{He\alpha}\right)_A}{\left(\dfrac{H\ \alpha}{He\ \alpha}\right)_S}$$

Coronal temperatures

According to the model

$$\frac{\left(\dfrac{He\beta}{He\alpha}\right)_A}{\left(\dfrac{He\beta}{He\alpha}\right)_S} = \frac{e^{\frac{-1}{kT_A}}\,(h\nu_\beta - h\nu_\alpha)}{e^{\frac{-1}{kT_S}}\,(h\nu_\beta - h\nu_\alpha)}$$

The variation range of temperature in the prepulse occuring cases can be evaluated taking account of table IV, from temperatures of without prepulse cases.

We present the results for a fairly excessive range of T_S variation

For lower flux experiments

$T_S = 500$ eV $\quad T_A = 270\ (235)$ eV

$T_S = 2000$ eV $\quad T_A = 460\ (360)$ eV

For higher flux experiments

$T_S = 500$ eV $\quad T_A = 130\ (155)$ eV

$T_S = 2000$ eV $\quad T_A = 165\ (200)$ eV

Brackets figures are obtained from the ratios including satellites

So, for the coronal zone, prepulses induce a lowering of electron temperature, to be compared to the extending of characteristic density gradient lenght.

Compression evaluation

For the core investigation, the problem is there is no distinction possible about which part of helium-like emission is really due to the core-zone.

As the spatial shapes are the same kind of trapezia in both cases with and without prepulse, we may postulate that the ratios of emissions are similar

$$\left(\frac{He\alpha\ \ core}{He\alpha\ corona}\right)_S \sim \left(\frac{He\alpha\ \ core}{He\alpha\ \ corona}\right)_A$$

With the help of this hypothesis we can get an idea of core density. In ETL model, Saha equation implies

$$\frac{H\alpha}{He\alpha} = \frac{A_H}{A_{He}} \cdot \frac{N_H}{N_{He}} = \frac{c(kT)^{3/2}}{n_e} \frac{A_H}{A_{He}} e^{\frac{1}{kT}(h\nu_{He} - h\nu_H - Xi)}$$

where : Xi are ionization potentials for He like silicon and
(A_H, A_{He}) Einstein emission coefficients

So

$$\frac{n_{e_A}}{n_{e_S}} = \frac{\left(\frac{H\alpha}{He\alpha}\right)_S}{\left(\frac{H\alpha}{He\alpha}\right)_A} \cdot \left(\frac{T_S}{T_A}\right)^{3/2} e^{\left(\frac{1}{kT_A} - \frac{1}{kT_S}\right) \cdot \tilde{V}}$$

Where $\tilde{V} = h\nu_H - h\nu_{He} + Xi$

Experimentaly, we find

$$\frac{\left(\frac{H\alpha}{He\alpha}\right)_A}{\left(\frac{H\alpha}{He\alpha}\right)_S} \sim 0.4 - 0.3$$

As apparent core compression is higher with than without prepulse,
but neutron yield lower, the use of Saha equation is coherent with

$$T_{A_{core}} < T_S \text{ core} \qquad \left(\frac{T_A}{T_S}\right) \text{ core} \sim 0.3$$

$$n_{e_{A_{core}}} > n_{e_S} \text{ core} \qquad \left(\frac{n_{e_A}}{n_{e_S}}\right) \text{ core} \sim 10$$

$$T_{S_{core}} > 2 \text{ keV}$$

Lowering of glass compressed temperature seems to be
in agreement with a lowering of suprathermal shell preheat, modi-
fying the inner shell pressure profile.

Thus an increase of density may be interpreted as a
more ablative regime developped as a consequence of the prepulse
effect.

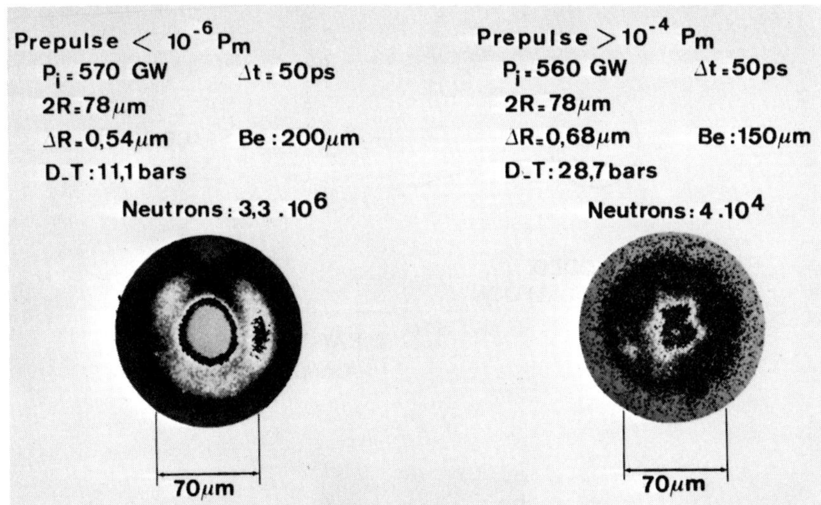

Fig. 8 - X-ray time integrated pin-hole pictures with and without prepulse.

2-5. <u>Volume compression</u>

Volume compression was inferred from X-ray pinhole photographs recording both X ray lines and continuum emission (fig. 8). In high contrast ratio experiments, the core was easily studied, as it looks spherical, with a high emittance compared to corona. Diameter was typically < 20 µm leading to volume compression of the order of 60. In the particular case of figure 8, film saturation prevents us from compressed DT analysis. With a prepulse, the core appears smaller in dimension, less spherical and rather perturbed, with a strongly reduced emittance.

Compression hydrodynamic was also studied in a simple way by coupling an X ray streak camera with a slit parallel to the temporal axis (figure 9). On the film, time delay between corona and core emissions as well as their durations are directly evaluated along the temporal axis, while corresponding dimensions are seen on the spatial axis. For low prepulse experiments, the collapse time can be identified as the time delay between corona and core emission, as both appear well separated, with fast rise time and sharp space gradients. In the case of important prepulse (fig.10) the core/corona contrast is reduced like in fig. 8, core and corona emissions present smoother rise times and space gradients and the collapse time appears smaller [9]. These results confirm spectroscopic ones as for the smoother density gradients, but the so called collaspe time cannot be compared to the previous, as obviously refering to a strongly different situation

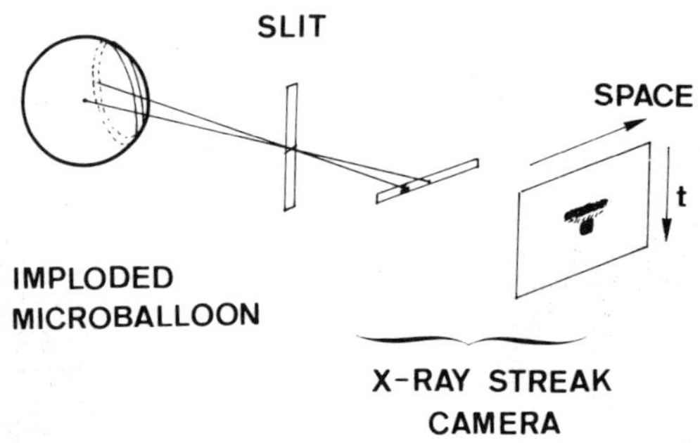

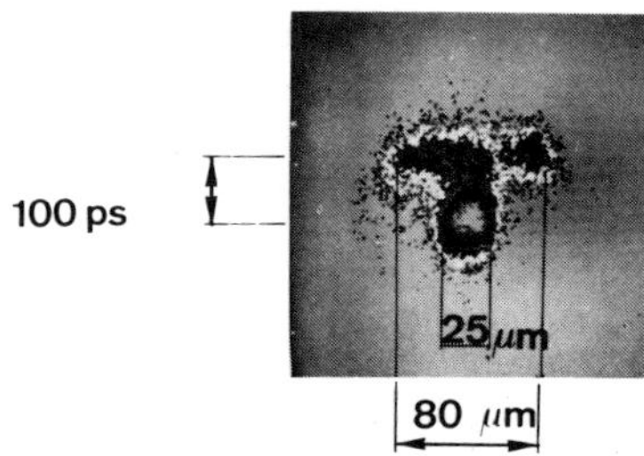

Fig. 9 - Temporally and spatially resolved (in planes) X-ray
streak camera recording, without prepulse.

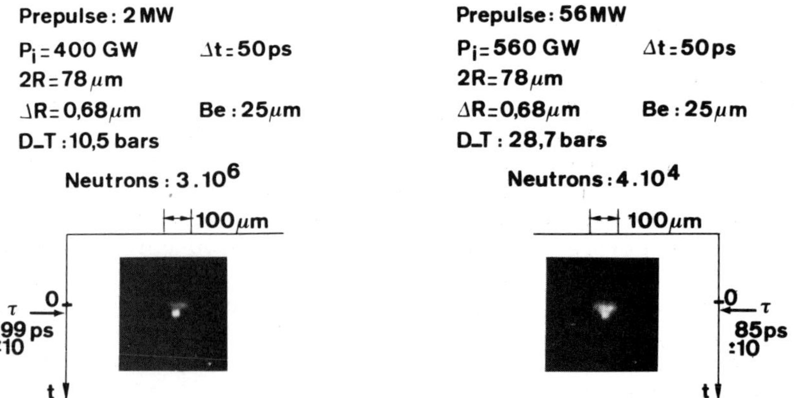

Fig. 10 - X-ray streak camera recordings with and without prepulse

2-6. Conclusion

According to these results, the following phenomenological description of the prepulse effect can be proposed :

The prepulse effect results in an increased critical radius and a smoothered density gradient. The first consequence is a lowering of the irradiation intensity $\emptyset_i$ for the main pulse and the development of scattering processes such as stimulated Brillouin scattering, lowering the absorption efficiency. The second effect is a decreased contribution of resonant absorption, both for photon absorption - as an increase of critical radius reduces the angle of incidence of P polarized elementary beams - and for fast electron emission, whose temperature varies as $\emptyset_i 0.4$ [10].

The last point appears as the most important, for the explosive pusher relies on fast electrons shell heating. Indeed fast ions disappearance seems to evidence a lesser suprathermal electron generation. At last, neutrons are no more explained by an explosive pusher analytical model, and arise from a colder and inhomogeneous but probably more compressed core.

All these results give evidence of a regime noticeably different from the explosive pusher. They seem to be related to recent plane targets experiments [3] which have shown in similar conditions (80 ps - 10^{15} W.cm^{-2}-1 μm wall thickness) that a slight flux reduction led to the transition from an explosive situation towards a more ablative one, with a rearwards motion of the whole target.

3 - X-RAY BACKLIGHTING

3-1. In order to test fast electron preheating, an X ray backlighting diagnostic was used. Experimental arrangement consisted of (fig. 11)

- An X-ray source obtained by focusing the ninth beam (1 J ; 50 ps) of the Octal facility, with a f/2 lens (f = 190 mm) on a brass plane target. A 50 to 100 μm diameter plasma was produced emitting a short pulse of soft X rays ($\simeq$ 50 ps duration) 1. to 1.5 keV spectral range.

- An X-ray pinhole type camera using two pinholes (15 μm diameter) and a kodirex film. For each shot, we analyse the following images; the X ray source alone providing with the incident X ray backlighting intensity ; the imploded microballoon alone, delivering its own emission which appears as a noise signal in X-ray probing ; at last the source observed through the target, providing us with data on X ray transmitted intensity added to microballoon emission. An example of the three pictures is also shown on the figure 11.

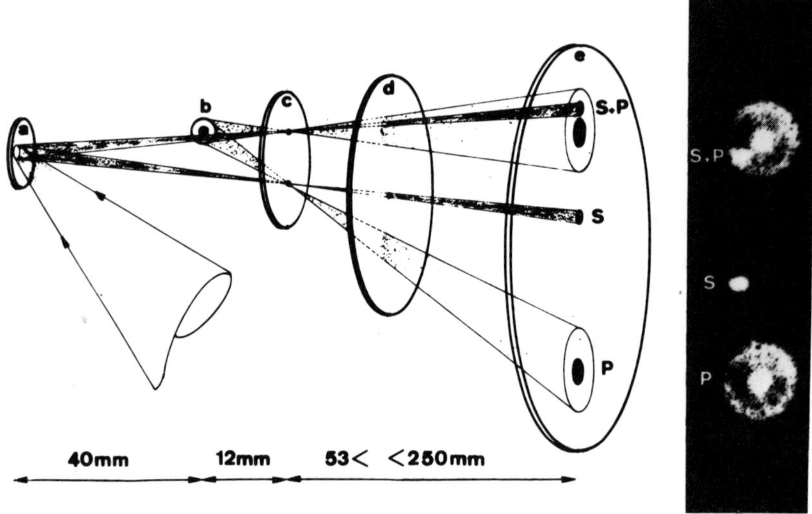

Fig. 11 - Experimental X-ray backlighting arrangement and recording
a) X-ray source ; b) imploded microballoon ; c) double pinhole-disc
(300 µm separation) ; d) aluminum foil ; e) recording film ;
S) image of X-ray source alone ; P) image of imploded microballoon
alone ; S+P) the source observed through the imploded microballoon

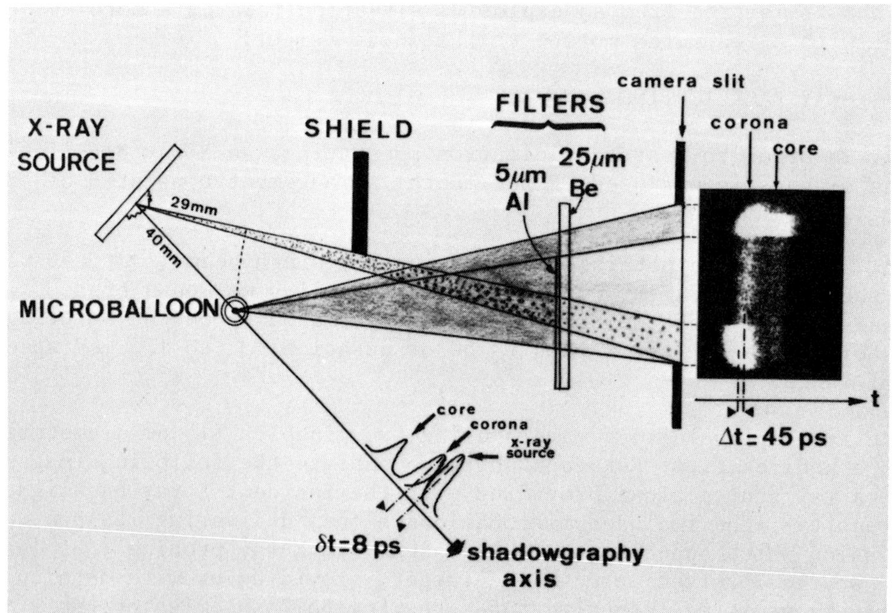

Fig. 12 - Experimental setting for temporal positioning of X-ray
source and microballoon emissions.

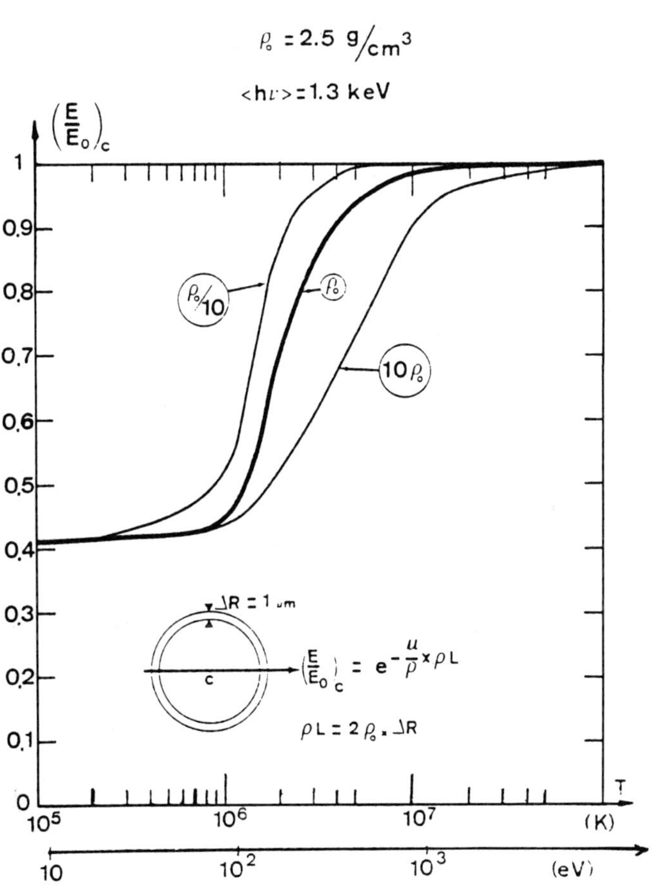

Fig. 13 - Glass microballoon transmission versus temperature.

In order to optimize the signal to noise ratio on the film,
X ray source and microballoon emissions were filtered by an aluminium foil (5 to 25 μm thickness). Thus, microballoon emission above
1,56 keV (Al K-edge) was strongly reduced. As an example, with 10 μm
Al thickness, transmission is 0.2 at 1.3 keV and 1.7 x 10^{-3} at 2 keV.

Moreover, the probing axis was slightly shifted from the target
center to reduce the influence of the core emission. But transmission
measurements E/Eo were normalized for the center (noticed(E/Eo)c)
using a simple geometrical assumption of a uniform spherical absorbing shell.

The aim of the experiment was to probe that the shell, although
heated by fast electrons, is not yet significantly exposed. The probing time had to be late enough to allow the fast electrons to be
generated, but it also had to keep previous to the disassembly
time occurring during the laser pulse. Practically in the experiment
we present here, it was adjusted to about - 10 ps with respect to
the maximum of laser pulse by means of an optical delay line inserted
on the diagnostic laser beam. It was measured by an X ray streak
camera[11] in another direction as shown on figure 12. A shield
parthy hid the X ray source emission on the camera slit in order
to distinguish both corona and core microballoon emissions and
X-ray source emission. Microballoon emissions were filtered by
two kinds of filters (25 μm Be alone or 25 μm Be with 5 μm Al)
to adjust intensities on the photocathode. Thus, it was verified
that duration of X-ray source emission was shorter than implosion
time. On the recording shown on figure 12, the delay (Δt = 45 ps)
measured between source and corona emissions corresponds to an
advance δt = 8 ps of the diagnostic pulse on shadowgraphy axes.

Pratically, a transmission measurement needed three shots :

- The first one with X ray source alone to compare pinholes
  transmission ;

- the second one with X-ray source through the microballoon not
  imploded to provide us with a reference of cold glass transmission ;

- the last one with the irradiated microballoon

Figure 13 shows theoretical variations of the transmission at
the center $(E/E_o)_c$ = exp - 2 $\mu\Delta$R, versus temperature, calculated
for different densities but with the product $\rho \cdot \Delta$R, kept constant
and equal to 2.5 $\times$ 10^{-4} g/cm^2 ; μ is the absorption coefficient
at $< h \nu > = 1.3$ keV (corresponding to X ray source) taking into
account bound-free, free-free and Compton effects and stimulated

emission. It is the large increase of the transmission above 100 eV
(relatively to the cold glass transmission) which made possible an
evidence of preheat in this temperature range

Moreover, for T $<$ 100 eV transmission is weakly dependant of
density. Consequently, position of probing time with respect to the
implosion beginning was not critical, and cold glass transmission
was a reference in itself for the study of the wall preheat.

Evolution of $\frac{(E)}{(Eo)C}$ versus the wall thickness is presented
on figure 14 fo 80 µm diameter microballoons, similar shots are
gathered in rectangles ; a, b, c correspond to about the same
incident laser flux (2 x 10^{15} W/cm^2), d corresponds to 5 x 10^{14}W/cm^2
All of them have been performed in the high contrast ratio situation.

Experimental values of cold glass transmission are also reported
in good agreement with the theoritical curve exp $-$ 2 µ.ΔR.

By looking at a, b c, cases, transmissions greater than those
in the cold glass case were only obtained with thinnest targets and
higher flux (a)

Moreover experiments (a) agree well with the model in ref 2
as for the neutron yield versus absorbed specific energy. Thus high
transmissions are related to an important fast electron preheat,
and are significant of typical explosive pusher regime.

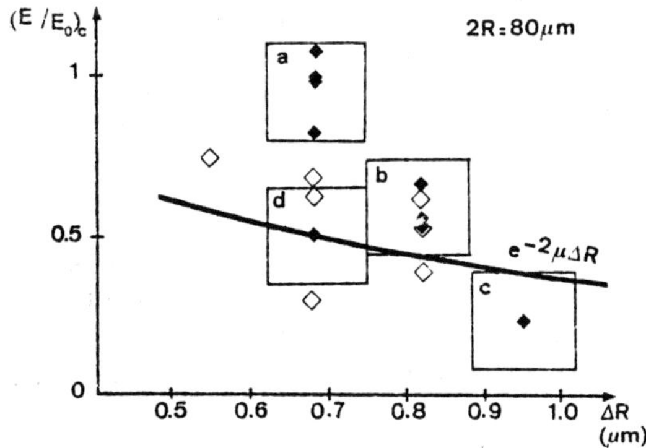

Fig. 14 – Measured transmissions versus wall thickness for 80 µm
diameter microballoons ; hollow squares correspond to cold glass
whose theoretical transmission is exp($-$ 2 µΔR) $-$ black squares
correspond to microballoons imploded at different fluxes :
(2 $\pm$ 0.5) 10^{15}W/cm^2 (a, b, c) ; 5.10^{14}W/cm^2(d). Vertical bars
correspond to the maximum experimental dispersion and horizontal
bars to the wall thickness uncertainty.

4.   Conclusion

Experimental studies concerning fast electron preheating in exploding pusher type implosion experiments have been performed at the level 1 TW with glass microballoons typically 80 µm in diameter and 0.8 µm in wall thickness.

First, effects of a prepulse on energy balance, neutron yield and X-ray emissions were interpreted as the consequence of a lack in fast electrons generation, transferring the implosion to a quite different regime.

Second, fast electron preheating was measured by X-ray backlighting. This diagnostic allows to determine the maximum wall thickness ensuring a microballoon to be imploded in explosive pusher regime.

The authors wish to thank J. C. Courteille and the laser team for OCTAL operation, the target fabrication and data acquisition teams for valuable contribution, and J. Bouard, H. Croso, G. Franzini, J. P. Godefroy, J. Kobus, D. Meynial, J. Turberville, for technical assistance.

REFERENCES

1.   E. B. Goldman, J. A. Delettrez and E. J. Thorsos, Nucl. Fus. 19, No. 5, 555 (1979).
2.   E. K. Storm et al., Phys. Rev. Lett. 40, 1570 (1978).
3.   F. C. Young et al., App. Phys. Lett. 30, 45 (1977);
      J. C. Couturaud et al., 9th Annual Conference on Anomalous Absorption of Electromagnetic Waves, Rochester (1979).
4.   J. Launspach et al., Annual Meeting of the American Physical Society, November 1975, St. Petersburg, Florida.
5.   M. H. Key et al., Phys. Rev. Lett. 41, 1467 (1978).
6.   D. T. Attwood, IEEE, Journal of Quantum Electronics Q.E. 14, 909 (1978).
7.   J. J. Thomson et al., Phys. Rev. Lett. 37, 1052 (1976);
      C. M. Armstrong et al., J. of App. Phys. 50, 5233 (1979);
8.   B. H. Ripin et al., Phys. Rev. Lett. 39, 611 (1979).
      D. W. Phillion, W. L. Kruer and V. C. Rupert, Phys. Rev. Lett. 39, 1529 (1977).
9.   D. T. Attwood, L. W. Coleman, J. E. Swain, D. W. Phillion, K. R. Manes, D. S. Bailey and Y. L. Pan, UCRL 78739 – 10th Conference on Laser Interaction with Matter, Palaiseau, France, October 18–22, 1976.
10.   UCRL 50021 – 76 LLL Laser Program Annual Report (1976).
11.   J. P. Gex, N. Fleurot and R. Sauneuf, Rev. de Phys. Appl. 12, 1049 (1977).

STUDIES OF LASER-PLASMA INTERACTIONS, EMPHASIZING ABLATIVE
ACCELERATION OF THIN FOILS[*]

J.A. Stamper, S.E. Bodner, D.G. Colombant, R. Decoste[a]
S.H. Gold[b], J. Grun[c], R.H. Lehmberg, W.M. Manheimer,
E.A. McLean, J.M. McMahon, D.J. Nagel, S.P. Obenschain,
B.H. Ripin, R.R. Whitlock, and F.C. Young

Naval Research Laboratory
Washington, D.C. 20375

ABSTRACT

Earlier studies at the Naval Research Laboratory (NRL) concentrated on interactions with short pulses (100 psec) focused at
high irradiance (10^{15} - 10^{16} W/cm^2) onto solid targets. In this
regime, such phenomena as thermal transport inhibition (including
magnetic fields) and stimulated Brillouin backscatter have shown
that there could be difficulties with the high irradiance pellet
designs. In addition, hydrodynamic instability may place a lower
limit on irradiance or ablation pressure. We are investigating an
intermediate irradiance regime which may be relatively free from
these problems.[1-3] Thus, recently we have been concentrating on
ablatively accelerating thin foil targets with longer pulses (3-4
nsec) and at a lower irradiance (10^{12} - 10^{14} W/cm^2). The results
have been quite encouraging. Targets have been ablatively accelerated to velocities (10^7 cm/sec), close to that required for fusion
applications, with a high (20%) hydrodynamic efficiency. An

---

[*] Work supported by DOE.

[a] Sachs-Freeman Associate; Present address: IREQ, Varennes,
    Quebec, Canada JOL 2PO
[b] NRC/NRL Resident Research Associate
[c] Physics Dept., University of Maryland, College Park, MD
    20742

introduction is given which discusses the irradiance constraints and some simple theoretical models. These simple models serve to introduce certain concepts and quantities needed to understand the experimental results. The NRL Pharos II Nd-glass laser system is outlined and a variety of diagnostics are described including: incident and scattered laser light; particle energy, momentum and velocity; x-ray imaging and spectra; visible spectroscopy; and laser probing, including Doppler studies, interferometry and dual-time shadowgraphy. Studies with these diagnostics are discussed and experimental results presented. Finally, a brief discussion is given of recent theoretical topics: optical ray retracing in stimulated backscatter and absorption and thermal flux inhibition due to ion acoustic turbulence.

INTRODUCTION

Irradiance Constraints

Pellet design for inertial confinement fusion (ICF) is evolving because of opposing considerations:[4,5] The higher irradiance designs are faced with difficulties due to enhanced backscatter, thermal transport inhibition and low hydrodynamic and driver power efficiencies. On the other hand, the lower irradiance designs require a high aspect ratio pellet for which the Rayleigh-Taylor instability poses a serious threat. Much of the prior target interaction studies at NRL and other laboratories have concentrated on the interesting interactions which occur at higher laser irradiance ($I \gtrsim 10^{14}$ W/cm^2). The presence of thermal transport inhibition due to magnetic fields[6-9] or ion-acoustic turbulence,[10,11] can result in lower hydrodynamic efficiency as well as increased target pre-heat due to superthermal electrons. In addition, a large fraction of the laser energy can be lost by stimulated Brillouin backscatter.[12,13] These high irradiance effects tend to lower pellet gain and increase the symmetry requirements.

The Rayleigh-Taylor instability has been recognized for some time as an important problem.[14] One can see from simple considerations why this instability may become important for high-aspect-ratio pellets and low laser irradiance. The number N of R-T e-foldings is the product of time t and the linear growth rate $\gamma = \sqrt{kg}$. The acceleration g (assumed constant for simplicity) is related to the pellet radius R by $R = (1/2)gt^2$ and the wavenumber k of the most dangerous mode corresponds to a wavelength $2\pi/k$ which is of the order of the shell thickness ΔR. Thus,

$$N \sim \sqrt{4\pi\ R/\Delta R}, \tag{1}$$

so that high aspect ratio $(R/\Delta R)$ pellets have a larger number of linear R-T e-foldings.

Now consider the dependence of N on ablation pressure and laser irradiance. For the high compression $(R_o >> R_f)$ of a single, thin $(\Delta R << R_o)$ shell one can express the ablation pressure P_a (now assumed constant in time) in terms of the aspect ratio. Thus, integrating $\rho d^2R/dt^2 = -\partial P/\partial R$ radially over the shell (after multiplying by $4\pi R^2$), multiplying by dR/dt, and integrating over time gives

$$P_a \sim (3/2)\rho v_f^2 \; \Delta R_o/R_o \tag{2}$$

where v_f is the final or maximum shell velocity. The coefficient would differ from 3/2 for other pressure pulse shapes. As shown later, P_a is given, in the steady-state planar limit as $2I_a/u$ where I_a is the absorbed laser irradiance. Thus, from Eqs. (1,2) and the dependence of P_a on irradiance, one would generally expect the number of R-T e-foldings to be greater at lower irradiances. However, quantitative results will depend on studies incorporating details of the laser pulse shape and interaction physics.

The earlier and lower-aspect-ratio pellet designs[14] required a highly shaped pulse with high peak irradiances. They would achieve high peak densities (10^4 x solid) and a moderate yield with low laser energy input. There are newer pellet designs at lower irradiance which require little pulse shaping but do not achieve the high peak densities. Laser energies in the megajoule range may be required for a reactor. A Soviet design[15], based on a single shell had an aspect ratio in the range 60-100 and would require an irradiance of about 10^{13} W/cm^2. Double shell designs have been considered at Los Alamos and Livermore[16]. These have an aspect ratio in the range 8-10 and a peak irradiance of about 10^{14} W/cm^2. At present, we can only hypothesize that there is an irradiance window within which the pellet designs are relatively free from the high irradiance plasma effects and the R-T shell break-up at lower irradiance.

Background Discussion

In order to achieve the high energy densities or pressures required to ignite a compressed pellet it is first necessary to efficiently accelerate the shell radially inward. A background discussion is given here which simply describes the physical processes involved in ablatively accelerating the shell and achieving these pressures. The discussion also serves to introduce certain concepts and quantities needed to understand the experimental results - presented later.

The maximum pellet yield occurs when most of the fuel is un-heated, and kept on a low isentrope[14]. The fuel pressure is then only a few times the Fermi-degenerate pressure. But the required pressure is still huge: 10^5 MB. For a laser driver, one starts with a radiation pressure $P_r = I/2c$ where I is irradiance and c is the velocity of light. For an irradiance of 10^{14} W/cm^2, the radiation pressure is .02 MB. The required pressure multiplica-tion $P_c/P_r \sim 10^7$ is achieved, as noted by Nuckolls et. al.[14], through the physical processes of ablation and compression.

We now consider, in more detail, how the ablation process results in the required pressure multiplication. Firstly, note that ablation pressure is itself a factor of $4/\beta$ (where $\beta = u/c$) greater than the radiation pressure. Secondly, with a target volume compression K, there is a still larger core pressure: Assuming all of the kinetic energy $(1/2)$ $M_s v_f^2$ of the imploding shell (M_s is shell mass) goes into the energy of the core, then P_c is $(1/2)$ $M_s v_f^2/V_c$. However, from Eq. (2) one can see that the ablation pressure is $(1/2)$ $M_s v_f^2/V_o$ where $V_o = (4/3)\pi R_o^3$ is initial target volume. These results show that $K = V_o/V_c$ can be expressed as P_a/P_c. Thus, the required pressure multiplication to go from a radiation pressure P_r to a core pressure P_c can be expressed approximately as

$$P_c = \frac{4K}{\beta} P_r \qquad (3)$$

Since u is around 3×10^7 cm/sec, β is around 10^{-3}. For a thousand-fold compression of the material in the shell and an initial shell volume approximately of one-tenth the target volume, then K is about 10^4. Thus, a radiation pressure of the order of .01 MB will result in the required core pressure of several times 10^5 MB.

In order to efficiently concentrate the energy into the core the ablation process must efficiently accelerate the shell to its final kinetic energy $(1/2)Mv_f^2$ while keeping it on a low isentrope. The rocket action of the ablating material can be described by a simple model.[2] This leads to a simple expression for the hydrodynamic efficiency η_h or fraction of the absorbed laser energy E_a which ends up in directed target motion.

The external force (which vanishes) can be expressed,[17] through conservation of momentum, in terms of the (laboratory frame) target v and exhaust or ablation u' velocities as $d(mv)/dt - u'\,dm/dt$. Defining $u = u' - v$ as the (constant) ablation velocity relative to the target or rocket gives,

$$m \frac{dv}{dt} = -u \frac{dm}{dt} \qquad (4)$$

where we assume a planar rocket with m being the mass per unit area.

The absorbed irradiance I_a is the sum of the energy flux $d((1/2)mv^2)/dt$ delivered to the target and the energy flux $(1/2)(-dm/dt)(u-v)^2$ carried by the exhaust or ablated material. Using Using Eq. (4), this gives,

$$I_a = \frac{u^2}{2} \frac{dm}{dt}$$ (5)

One can integrate Eqs. (4) and (5) over time to show, respectively, that v/u is $\ln (m_o/m)$ and that the absorbed energy per unit area E_a is $(1/2) u^2(m_o-m)$.

The hydrodynamic efficiency can thus be expressed as,[2]

$$\eta_h \equiv \frac{(1/2)mv^2}{E_a} = \frac{(v/u)^2}{e^{v/u}-1}$$ (6)

The hydrodynamic efficiency is a function only of the ratio of the target velocity v to the ablation velocity u. <u>For highest efficiency one strives to match the ablation velocity to the desired target velocity</u>. From Eq. (6), one finds a maximum hydrodynamic efficiency of 0.65 when v/u is 1.6 and 80% of the mass has been ablated. In this case, some of the ablated material is moving in the direction of the target in the laboratory frame. For small ablated mass ΔM, η_h approaches $\Delta M/M_o$ or v/u and the ablated mass moves oppositely to the target in the laboratory frame.

The effects of an angular distribution in the target and ablation blow-off can be included in the planar expression (Eq. (6)) by defining an effective blow-off cone angle θ for the target and the component of ablation velocity normal to the target $u_\perp$. The modified expression is[1,2]

$$\eta_h = \frac{(v \cos\theta/u_\perp)^2}{e^{v/u_\perp}-1}$$ (7)

Deviation from the idealized planar picture will also result from target edge effects such as lateral heat flow and finite target width. These effects are discussed in Refs.(1-3).

We have, in this simplified treatment, ignored radiation losses (which are indeed small) and have also implicitly assumed a steady-state. This is approximately achieved in the asymptotic region where the ablation and target velocities are measured and where most of the thermal energy has already been converted (via

pressure gradients closer to the deposition region) into directed
energy. The thermal energy flux is small compared to the directed
plasma energy flux in the asymptotic region. However, the
temperature in the asymptotic region is not necessarily small [18-20]
compared to that in the deposition region. Computer simulations
confirm that a steady-state and asymptotic conditions are reached
in about the first nanosecond of a multi-nanosecond pulse. Also,
the experimentally determined velocities have an ablative signature[21] (single, narrow velocity peak) on the charge collectors.
Finally, despite any restriction, the rocket model gives good
agreement with experimental results.[1,2]

The rocket model can be used to derive expressions for the
ablation pressure and ablation depth. These are important quantities in the experimental studies - to be discussed next. The
ablation pressure P_a is defined as the thrust $m\,dv/dt$ responsible
for the acceleration of the target. Using Eqs. (4,5), the ablation
pressure can be expressed as $2I_a/u$. For $I_a = 10^{13}$ W/cm^2 and
$u = 3.5 \times 10^7$ cm/sec, this is several megabars. From Eq. (5) one
can write $I_a = (1/2)\,\rho u^2 V_f$ where $V_f = x_a/\tau_L$ is the velocity of the
ablation front, τ_L is an effective laser pulse width and x_a is
the ablation depth. The ablation depth can thus be expressed as,

$$ x_a = \frac{2 I_a \tau_L}{\rho u^2} \qquad (8) $$

For a typical NRL experiment using a plastic target ($\rho = 1$ g/cm^3),
our laser pulse width (3 nsec), the experimental irradiance
(10^{13} W/cm^2) and ablation velocity (3.5 x 10^7 cm/sec), Eq. (8)
predicts an ablation depth of 5 microns which is approximately
that which is observed.

EXPERIMENTAL STUDIES

<u>Experimental Conditions</u>

In order to study ablative acceleration it is necessary for
the laser and target to be within certain parameter ranges.
First, in order to set up a steady-state it is necessary to have
a laser pulse duration greater than about one or two nanoseconds.
We have already discussed how (to avoid plasma effects and shell
break-up) the irradiance must fall within a certain range, assumed
here to be 10^{12} to 10^{14} W/cm^2. The focal spot size should be
greater than about 500 μm in order to minimize edge effects.[1-3,22]
Taken together, these conditions imply that the laser energy must
also be in a certain range. For example, a few hundred joules of
laser energy is required to produce an irradiance of 10^{13} W/cm^2
over a 1 mm focal spot for a pulse width of a few nanoseconds.

An important reason for choosing thin foil targets was to
allow viewing the cold inner (rear) side of the target. One can
thus study the entire problem: target (rear side) and ablation
(front side) variables as well as transport through the target.
The foil targets can be thought of as a small section of an
imploding spherical shell. The target width must be large enough
that edge effects are minimized and understood. Lateral variations,
of a lesser degree, must also be understood for a spherical target
since there is always some lateral variation in the laser irradiance.
Targets are chosen to be either wide compared to the focal spot
or comparable in size (limited mass disc targets). Large targets
are limited by the effects of lateral heat flow whereas limited
mass targets are limited by energy flowing around the edge and
directly heating the rear surface. Neither problem is as severe
if the focal spot is sufficiently large (e.g., 500 micron). The
target should be sufficiently thick that it is not grossly preheated by particle or x-ray energy deposition. At the same time
the target must be sufficiently thin that the ablation depth (Eq.
(8)) is an appreciable fraction of its original thickness. Target
thicknesses in the range 2 to 15 micron have been found useful
for studies of target motion.

NRL, Pharos II Laser System

The NRL, Pharos II laser system[23] was used to irradiate
the targets in the experiments discussed here. It is a 2-beam
system with Nd-doped phosphate glass ($\lambda = 1.054 \mu$m) rod and
disc amplifiers. The phosphate glass, with its high gain and low
non-linear index allows one to maximize the laser efficiency and
thereby minimize the laser cost. The beams are relayed with a
series of telescopes so that the good beam uniformity at the
amplifier outputs (before edge diffraction fringes have developed)
is maintained to the target chamber lens.

The earlier studies, discussed in Refs. (1-3) were done
with 5-30 J in a 3 nsec (FWHM) pulse. In 1979, the Pharos II
system was up-graded both in beam energy and beam uniformity.
Both beam relays and, eventually, soft apertures are used to
improve the beam uniformity. The output disc amplifiers are 10.5
cm in diameter and are shown in Fig. 1. One beam has delivered
650 joules in 3.5 nsec.

A passively Q-switched, Nd-doped YLF (Yttrium-Lithium
Fluoride) Quantel oscillator is used to produce a single-mode
pulse of duration 3 nsec (FWHM). The temporal shape of the
output pulse is shown in Fig. 2.

The laser beam is focused onto the target with an f/6,
1.2 m focal length lens. The target is usually rotated 6 degrees

Fig. 1. Pharos II laser system, showing 10.5 cm disc amplifiers.

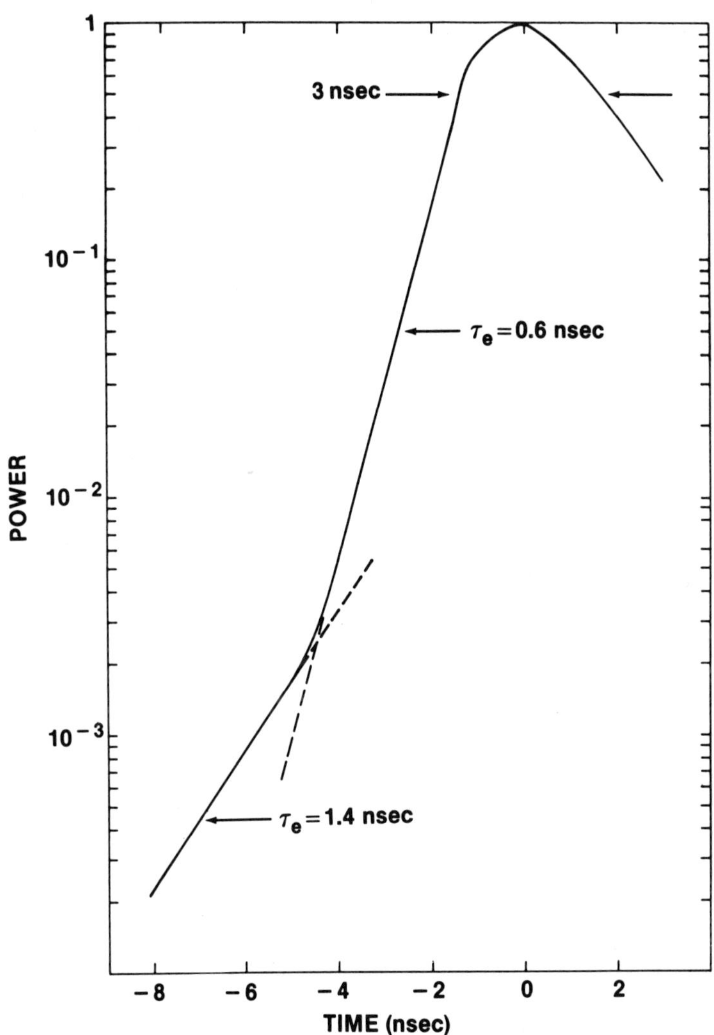

Fig. 2. Temporal shape of the incident laser pulse.

about a vertical axis so that the target normal is displaced
6 degrees horizontally from the laser axis. The specular and
back-reflected laser light can then be distinguished and the
plasma blow-off along the target normal is more accessible to
diagnostics.

Experimental results utilizing lower energy, as reported
previously,[1-3] and more recent experiments with higher energy,
and larger focal diameters (1 mm) follow.

Diagnostics and Energy Balance

A variety of calorimeters are used to check the energy balance.
The incident and back-reflected (through the focusing lens) laser
light are monitored by large calorimeters. The distribution of
scattered laser light energy and energy in plasma blow-off are both
monitored by an arrangement of minicalorimeters.[1] These are small
tantalum disks attached to thermocouples and are arranged in
pairs about the target. One calorimeter of each pair is open to
receive both light and plasma energy while the other is covered
with a glass filter which allows only the laser light through.
The total scattered laser light is also monitored, on some shots,
with a box calorimeter which receives light scattered into all
solid angles except the entrance and exit lens cone angles ($\pm 3^{\circ}$).
Energy balance at the 15-20 J level is achieved, over a range of
irradiances, to within about 10% - as shown in Table I.

Plasma velocity, on both the target front (laser) and back
sides, is measured with a variety of diagnostics. Charged
particle, time-of-flight detectors yield target material velocity
on both the front and back. Several diagnostics depend on a
short-pulse probing laser beam. Part of the main laser beam is
split-off after the oscillator, frequency doubled to 5270 Å
(green), and shortened to 300-500 psec by either a (in earlier
studies) Pockels cell or a (in later studies) regenerative amplifier system. Side-on interferometry and shadowgraphy with the
laser probe beam allows both front and rear surface velocity measurements of plasma density structure. Velocity of the cold, dense
rear surface is also measured with the probe beam by reflecting
it at near-normal-incidence and measuring the Doppler shift.

The plasma energy (minicalorimeters) and velocity (charge
collectors) measurements are complemented by ballistic pendula
which measure the momentum directly. The front and rear surface
momenta balance to within about 30%.

A variety of x-ray diagnostics yields further information.
The temperature of the bulk of the electrons as well as information
about any superthermal distributions is obtained from absolute

Table I.   Energy Balance (% of incident laser energy)

W/cm^2		$3x10^{12}$	10^{13}	$7x10^{14}$
Scattered light, 4π				
	mini-cal's	9 ± 5	20 ± 5	44 ± 10
	box-cal (±10)	–	15	47
Particles, UV & X-Rays, 4π				
	mini-cal's (±10)	90	78	52
TOTALS				
	mini-cal's (±10)	99	98	96
	box-cal (±10)	⊢—— 105 ——⊣		

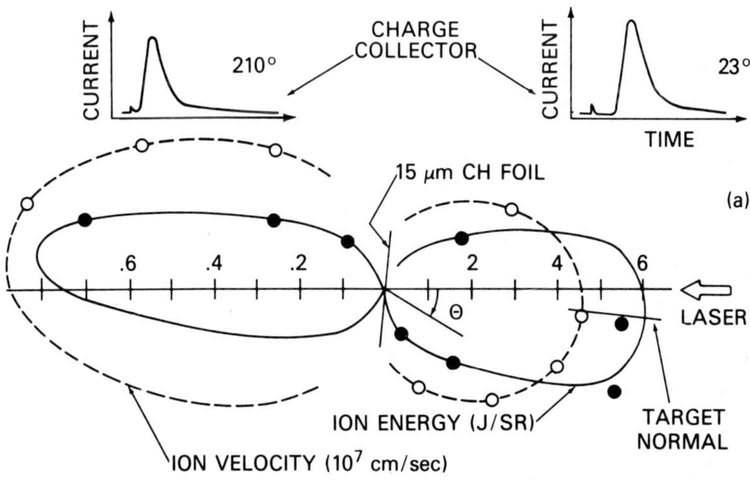

Fig. 3. Measurement of ablation parameters – analysis of a
typical shot.

x-ray continuum measurements. X-ray pin-hole photographs yield information about the spatial distribution of thermal energy. Subnanosecond response x-ray detectors when used with layered targets, yield information about the velocity of the ablation front.

Visible-UV spectroscopy when used side-on (line spectra) yields information about the velocity of particular ion species. The continuum spectrum observed from the rear side yields information about the rear side temperature.

<u>Studies of Ablative Acceleration</u>

Some of the particular studies, carried out by the diagnostics just described, are now discussed with regard to the ablative acceleration concepts. Fig. 3 illustrates these ideas. The data was taken at an irradiance of 10^{13} W/cm^2 for a 3 nsec laser pulse on a 15 micron CH (plastic) foil. The angular distribution of ion energy (minicalorimeters) and ion velocity (charge collectors) is shown for the front (ablated material) and back (target material). Note the expanded scale on the back and that the plasma blow-off is centered along the target normal. Ion charge collector traces (with an ablative signature) are also shown for the front and back. The average ablation velocity u (3.3×10^7 cm/sec) and target velocity v (5.1×10^6 cm/sec) are obtained by integrating over the front and back-side distributions, respectively. The directed target energy, and thus the hydrodynamic efficiency η_h (6.6%), is obtained by integrating the energy over the rear-side distribution. The ablated mass fraction ($\Delta M/M_0 = .25$) is obtained by integrating twice the energy divided by the velocity squared over the front-side angular distribution.

Velocities of the accelerating target or ablation blow-off (near the laser focus) can be measured by optical methods. The velocity of the cold, dense rear-side target plasma is of direct interest for ablative acceleration studies. Fig. 4 shows the arrangement for observing the Doppler shift of a probing second harmonic (5270 A) laser beam scattered at near-normal-incidence, from the back-side of a reflective target to infer its velocity. The scattered light can be observed up until the time (typically just after the peak of the main laser pulse) that the rear surface becomes hot enough to strongly absorb the probe beam. Since the image is focused onto the entrance of a stigmatic spectrograph, Doppler shifts (and thus velocities) can be measured across a diameter of the focal region. Such a Doppler inferred velocity profile is shown as the vertical dashes in Fig. 5. Here, a 9-micron thick carbon foil was irradiated at 5×10^{12} W/cm^2 and the probe pulse arrived at the peak of the main laser pulse. For comparison, the spatial profile (on target) of the main laser beam is shown in dots. The velocity profile is somewhat broader

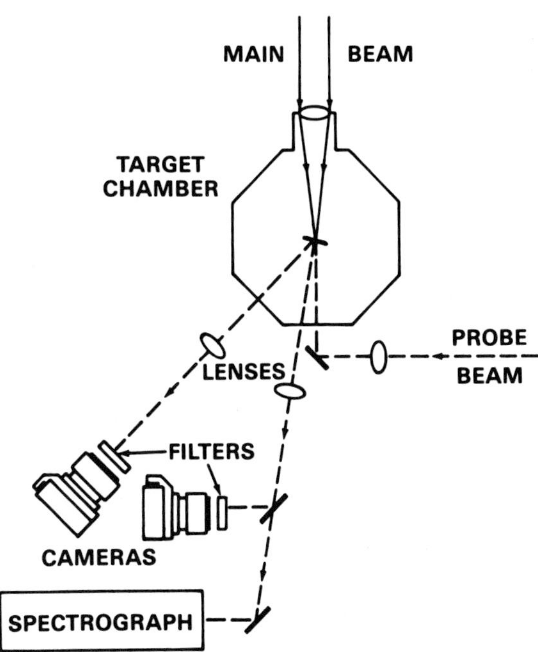

Fig. 4.   Arrangement for measuring rear-side velocity by the
Döppler shift of a probing laser beam.

than the laser profile - as might be expected due to lateral
thermal conduction.

Side-on shadowgraphy and interferometry are used to obtain
the two-dimensional velocity profiles of density structure during
the acceleration phase. Refraction by the rather large plasma
limits the maximum density sampled to a few times 10^{19} cm^{-3} for
the f/2 to f/6 collection apertures used. The interferograms in
Fig. 6 show the early time plasma development on the front side
of a 7 micron thick aluminum foil. Image shifters were used in
these studies to modify a Jamin interferometer into a folding
wave-front interferometer. A 400 psec, second harmonic probe
was used and the main laser beam produced an irradiance of
3×10^{13} W/cm^2 in a 3 nsec pulse. A detectable plasma can be
seen as early as 8 nsec before the peak of the main laser pulse.
A front-side axial velocity variation is obtained by following
the axial position of the 10^{18} cm^{-3} fringe in several shots. Such

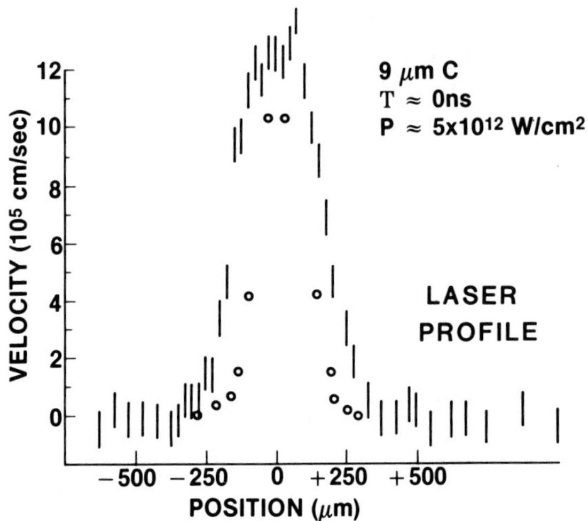

Fig. 5.   Velocity profile of target (vertical lines) from
          Doppler shifts compared with the laser energy profile
          at the target location (dots).

a velocity variation is shown in Fig. 7. At early times the
velocity is about 7×10^6 cm/sec. By around 2 nsec before the
peak of the main laser pulse the plasma has accelerated to its
final velocity of 3.5×10^7 cm/sec. This is near the asymptotic
ablation velocity observed for these shots on the charge collectors.

By changing the target thickness for different shots one can
vary the ablated mass fraction and thus the asymptotic target
velocity. Figure 8 shows the variation of hydrodynamic efficiency
and ablated mass fraction with the observed ratio of (asymptotic)
ablation to target velocities. For comparison, the predictions
of the planar rocket model (Eq. 6) and modified (for angular
distributions) rocket model (Eq. 7) are shown as solid and dashed
lines, respectively. The hydrodynamic efficiency agrees rather
well with the rocket model - particularly the modified form. The
observed ablated mass fraction is a bit high but may be corrected

by accounting for the effects of peripheral target heating.[22]
Hydrodynamic efficiencies as high as 20% are seen when the targets
are thin enough for the target velocity to become a large fraction
(50%) of the ablation velocity.

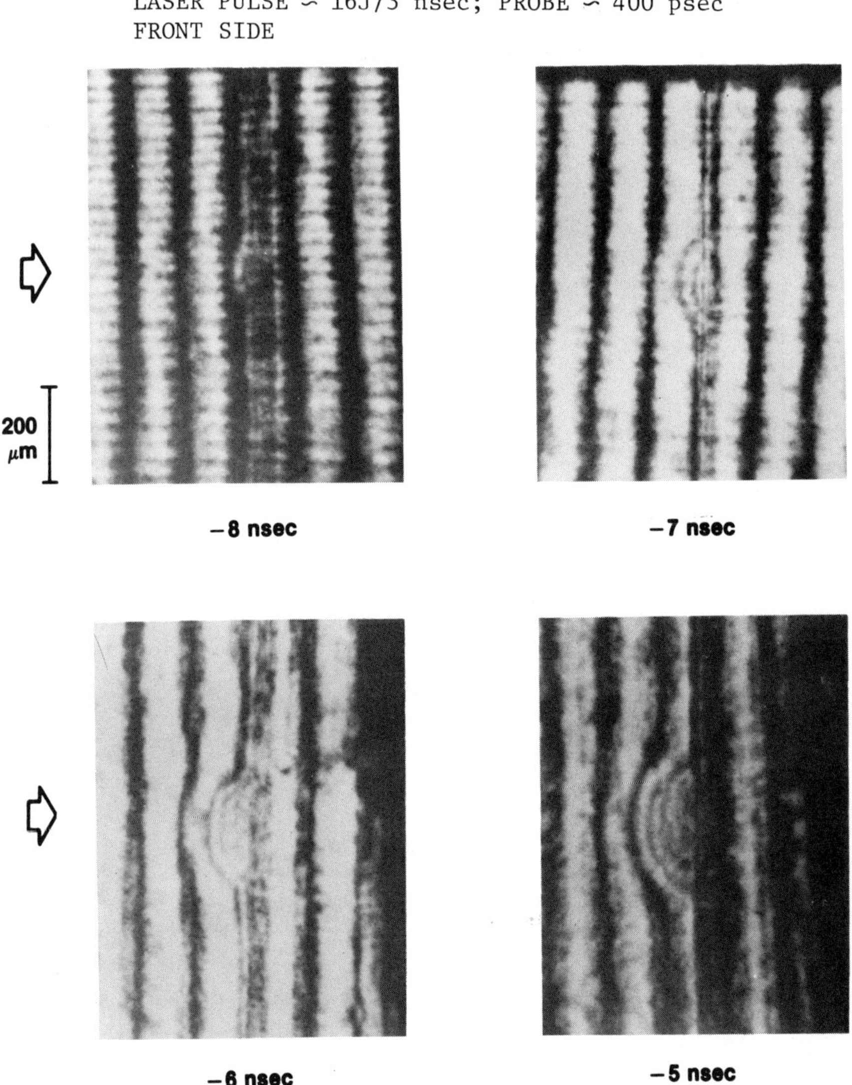

Fig. 6.   Interferograms showing the early-time plasma development
on the front side of a 7-micron aluminum foil.

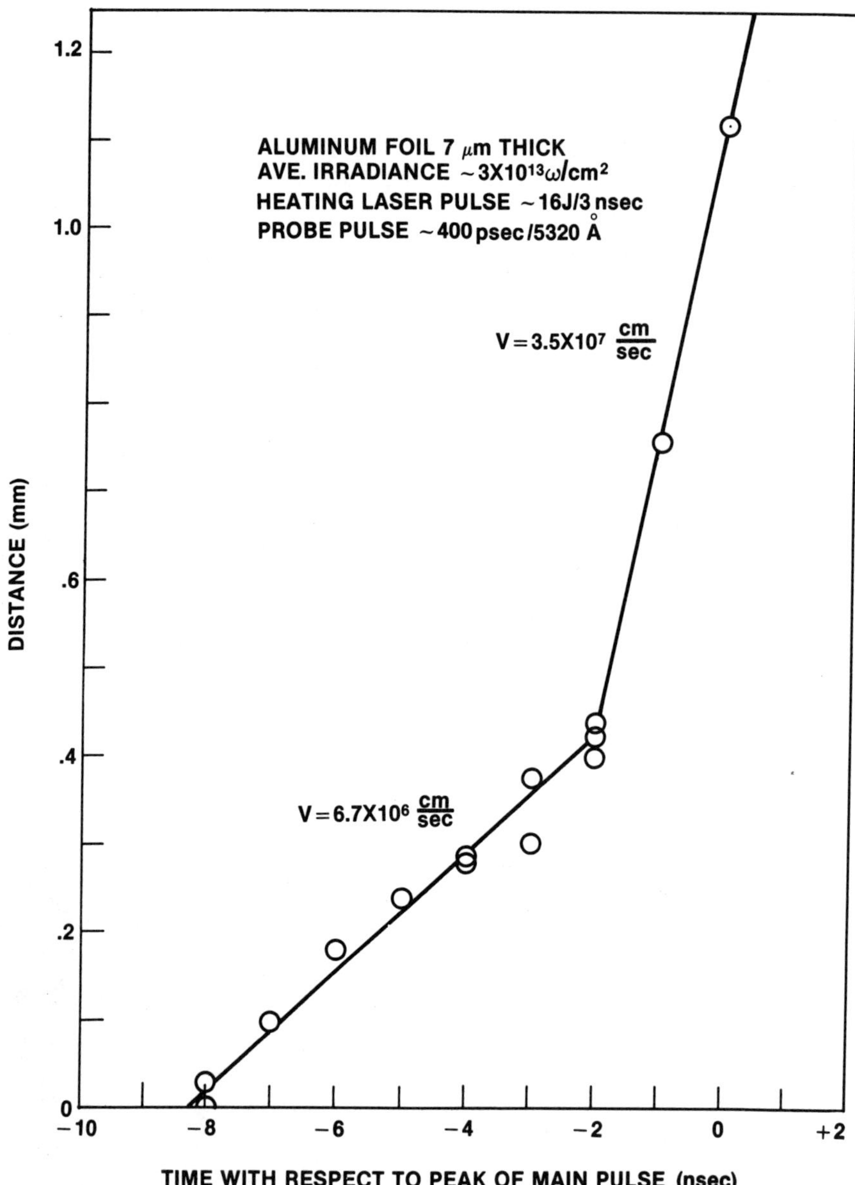

Fig. 7.   Axial velocity variation on the front side of a
7-micron aluminum foil.

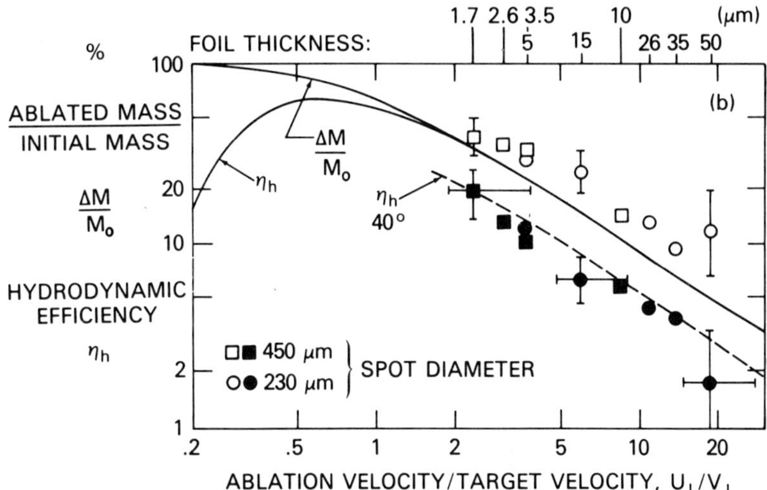

Fig. 8.   Comparison of experimental hydrodynamic efficiency and
ablated mass variations with the rocket model.

Axial Energy Transport in the Target

A more detailed study of the physics behind ablative
acceleration must involve an understanding of energy transport
in the target. The problem of axial and lateral heat conduction
is discussed in Refs. (1,3). Here, two experiments are discussed
which give information about energy transport into the target-
along the target normal or axis. These experiments utilize
x rays and visible light to determine energy transport, respectively,
to a layer of a structured target and to the rear of a homogeneous
target. First, let us consider the electron energy distribution
within a homogeneous target - as implied by x-ray continuum spectra.
Such spectra are shown in Fig. 9 for a CH target. The spectra
were obtained by an array of filtered PIN diodes.[1,24] These show
that, at all irradiances, most of the laser energy is deposited
into thermal electrons with an energy around 250 eV. No
evidence is seen for significantly energetic electrons until the
irradiance is above 10^{14} W/cm^2.

Layered targets were used to study axial heat transport and
to estimate the ablation depth. The targets consisted of thin
layers of CH plastic deposited on an aluminum substrate. The
laser was incident onto the CH side. Since aluminum is a much
stronger emitter of xrays than CH, there is a large increase in

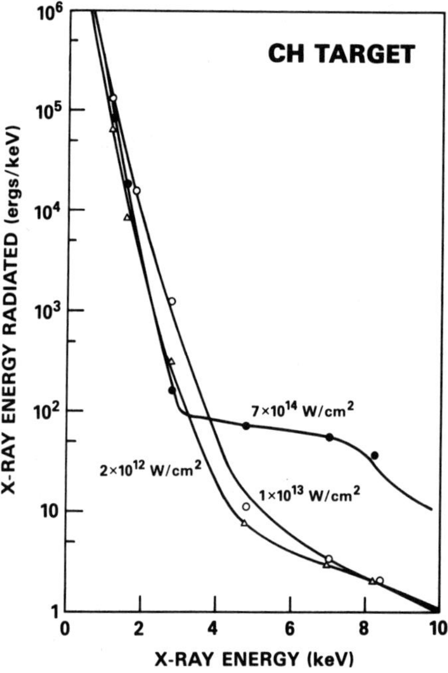

Fig. 9.　Absolute x-ray continuum spectra for a CH target at several irradiances.

x-ray yield when the plastic is thin enough that it is ablated down
to the aluminum during the laser pulse. The results are shown in
Fig. 10 for an irradiance of 1 x 10^{13} W/cm^2 and a 3 nsec (FWHM)
laser pulse width. The ratio (left side) of x-ray intensity for
CH layers on aluminum to the intensity for pure CH is seen, as
expected, to decrease with increasing CH thickness. The most
sensitive x-ray energy range to the aluminum heating (because of
line emission) is the 1 to 3 keV range. Here the intensity ratio
drops rapidly as the CH thickness is increased beyond 1 micron,
with no substrate heating seen for a 5 micron CH layer. From
these results one estimates that the ablation depth is 2 to 3.5
microns. This compares favorably with the ablation depth (3.5
microns) estimated by Eq. (8) for the 3 x 10^7 cm/sec ablation
velocity observed.

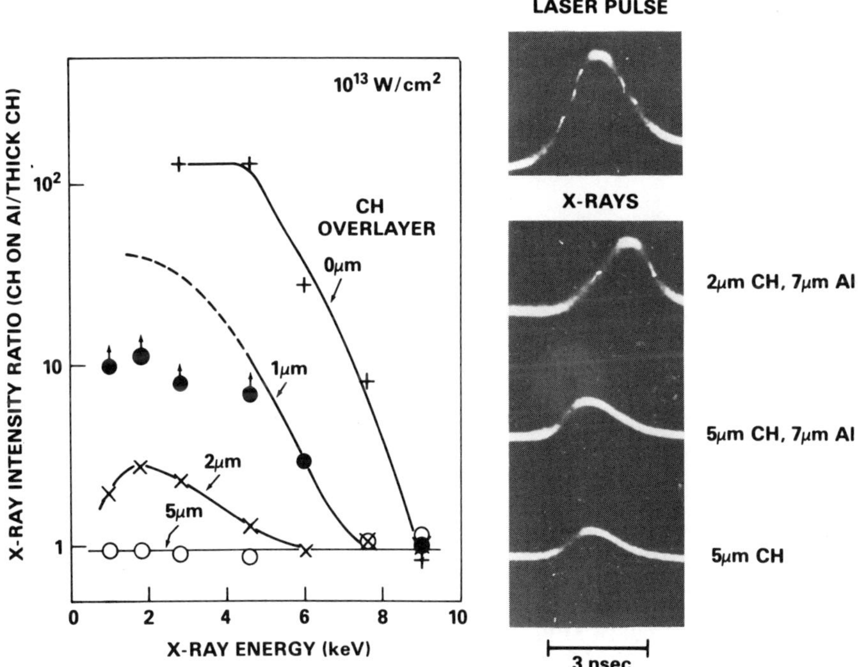

Fig. 10.   (Left) Ratio of x-ray intensity for CH on aluminum
           to the intensity for pure CH.
           (Right) Laser and x-ray time histories for pure CH
           and CH layers on aluminum.

Temporal behavior of the x-ray emission from layered targets
was monitored with a Be-filtered sub-nanosecond response PIN diode.
The results are shown on the right in Fig. 10 with the laser pulse
time variation shown at the top. A pure CH target (bottom) and
targets with 2 and 5 microns of CH on a 7-micron aluminum substrate
were used. For the pure CH and 5-micron CH overlayer targets the
x-ray time history is about the same as for the laser pulse.
However, when the CH overlayer thickness is decreased to 2
microns the x-ray peak is delayed with respect to the laser peak
and the emission is more intense. The delayed x-ray peak is due
to the finite time (2 nsec) required for energy to reach and heat the
aluminum substrate sufficiently to emit xrays. This gives an
ablation front velocity (2 microns in 2 nsec) which is in agreement
with that estimated from the previously estimated ablation depth
(3.5 microns) divided by the laser pulse width (3 nsec).

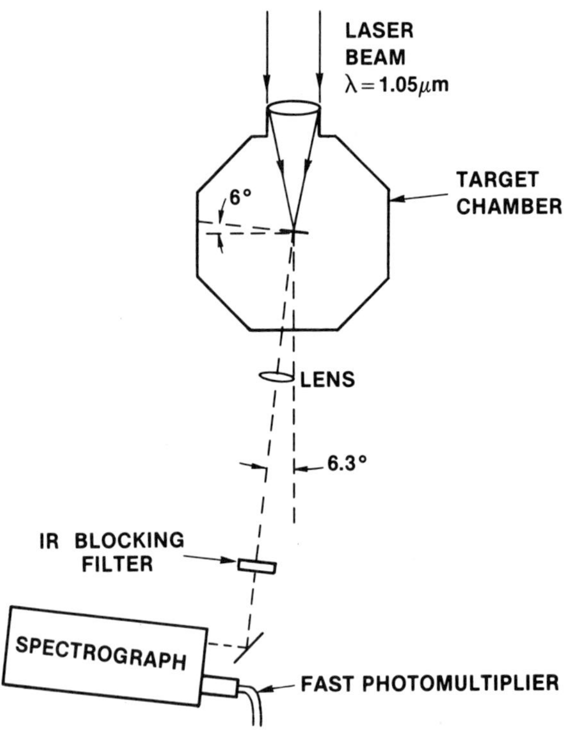

Fig. 11. Arrangement for measuring rear-side temperature by
measuring spectral variation of visible continuum.

Heating of the rear side of the target is being studied by
visible spectroscopy.[25]   This is an important study since preheat
of the pellet fuel can cause a pronounced decrease in pellet yield.
The setup for the study is shown in Fig. 11.   The normal of the
target (a 3 mm wide strip of foil) is tilted at 6 degrees to the
laser axis with the spectroscopic observations being made along
the target normal.   The rear side of the target is imaged onto
the entrance slit of the spectrograph with a single lens and a
one-nanosecond-response photomultiplier is located at the exit

DUAL-TIME PROBING VIA POLARIZATION – ENTRANCE OPTICS

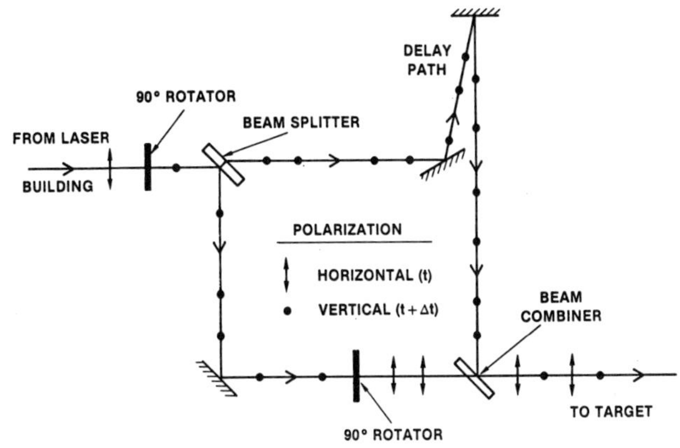

DUAL-TIME PROBING VIA POLARIZATION – COLLECTION OPTICS

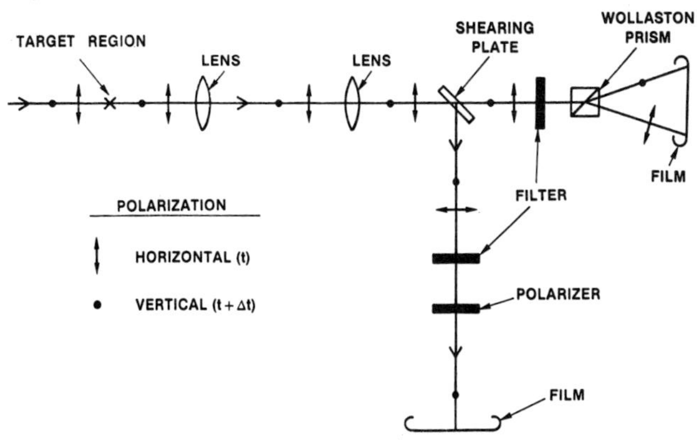

Fig. 12.   Arrangement for utilizing the linear polarization of
          a probe laser beam to separately record events at two
          different times.

slit. The time history of a narrow band (33 Å) of the rear side
continuum emission is recorded by a fast (1 GHz) oscilloscope.
The entire optical train, including the spectrograph and the photomultiplier, is calibrated _in situ_ on an absolute scale with a
calibrated tungsten lamp.

The spectral variation of the continuum emission can be
compared, at a given time, with the blackbody curve to get a
color temperature. Alternatively, the absolute emission at a
single wavelength yields a brightness temperature. For 5×10^{12} W/cm^2
and a 1 mm spot size, rear surface temperatures have been measured
over the range of 2.5-18 micron thick CH foils and 4-23 micron thick
Al foils with peak temperatures ranging from sub-eV for the thickest
targets to tens of eV for the thinnest. Calculations indicate that
preheating due to xrays and energetic electrons is not sufficient
to account for the heating. Further studies are necessary to
evaluate the relative contributions of xray and energetic electron
deposition, shock waves and thermal transport.

Density Structure Observed with Dual-Time Probing

Plasma temperatures measured by visible and x-ray spectroscopy
correspond to a high density region inaccessible to side-on laser
probing. However, side-on shadowgraphy and interferometry can
monitor the motion of the lower density plasma which is being
accelerated (through pressure gradients) by the hot-dense plasma.
The velocity of the rear-side plasma observed with late-time
(+4 nsec) shadowgraphy will be seen to correspond to the asymptotic
target velocity deduced from charge collector data. These side-on
studies also show some interesting small and large-scale density
structure which is not yet well understood.

The arrangement to accomplish the dual-time probing is
shown in Fig. 12. Due to the linear polarization of the probing
laser beam, one can separately utilize two time-separated pulses
traveling along the same optical path. The horizontally polarized
incident probe beam (300 psec, 5270 Å) is first rotated 90° so
that it is vertically polarized as it enters the beam-splitter.
There is a variable optical delay in one leg and a 90-degree
polarization rotator in the other leg. The two beams are then
re-combined in a polarization dependent beam combiner. The
combined beam consists of two, time-separated probe pulses which
are orthogonally polarized (vertical beam is delayed) and
traveling along the same optical path. As shown at the bottom
of Fig. 12, the combined beam is directed into the target region
parallel to the target surface. Since both pulses travel along
the same path, a single, fast collector can be used. A diffraction-
limited f/2 lens was used here as the collector. A second lens,
positioned at the sum of the focal lengths of the two lenses after

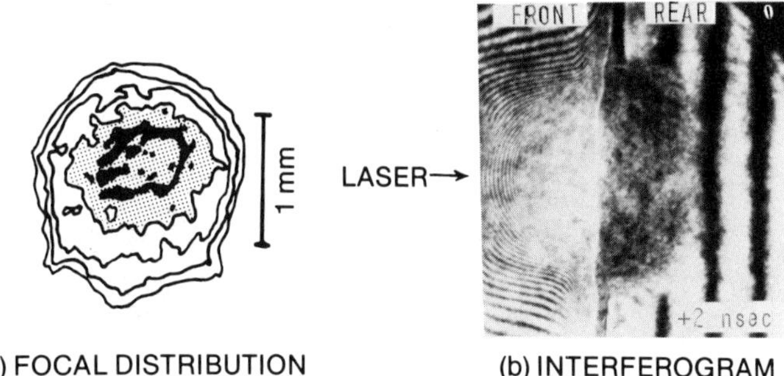

(a) FOCAL DISTRIBUTION   (b) INTERFEROGRAM

Fig. 13.   (a) Laser focal distribution and (b) shearing interferogram showing front-side vs rear-side density structure
for a 4.5 micron thick CH target.

the first, is used to focus the initially scattered (or refracted)
light (which was collimated by the first lens) and re-collimates
the un-scattered light (focused by the first lens).

We first consider light which is reflected at 45 degrees from
the two surfaces of a shearing plate. This is a 5 cm diameter
by 1 cm thick flat plate of fused quartz, with a 2 minute wedge.
The front and back reflections interfere and the wedge produces a
background fringe pattern with a spacing typically around 500
microns. Due to the 1 cm plate thickness, the phase-disturbed
(by the plasma) region of the beam for one surface reflection falls
on an undisturbed region of the beam from the other surface which
serves as a reference beam. The shearing plate thus serves as
a folding-wave-front shearing interferometer. The magnified
disturbed region was larger than the aperture of available
polarizing prisms (which would allow both polarizations or times
to be recorded) and so one or the other polarizations (times)
was selected with a large plastic sheet polarizer.

The plasma development near the target surface can be seen in
the shearing interferogram shown in Fig. 13b, along with the laser
focal distribution (13a). The target (a 4.5 micron thick strip
of CH) was irradiated with 154 J in a 1 mm focal spot. In the
focal distribution shown in Fig . 13a, this corresponds to an
average irradiance of 7×10^{12} W/cm^2. Note that the dense target
plasma (black region) occurs only on the rear side and is well
localized, i.e., there is practically no fringe bending beyond
the dark region in the back. (Localized fringe bending just at

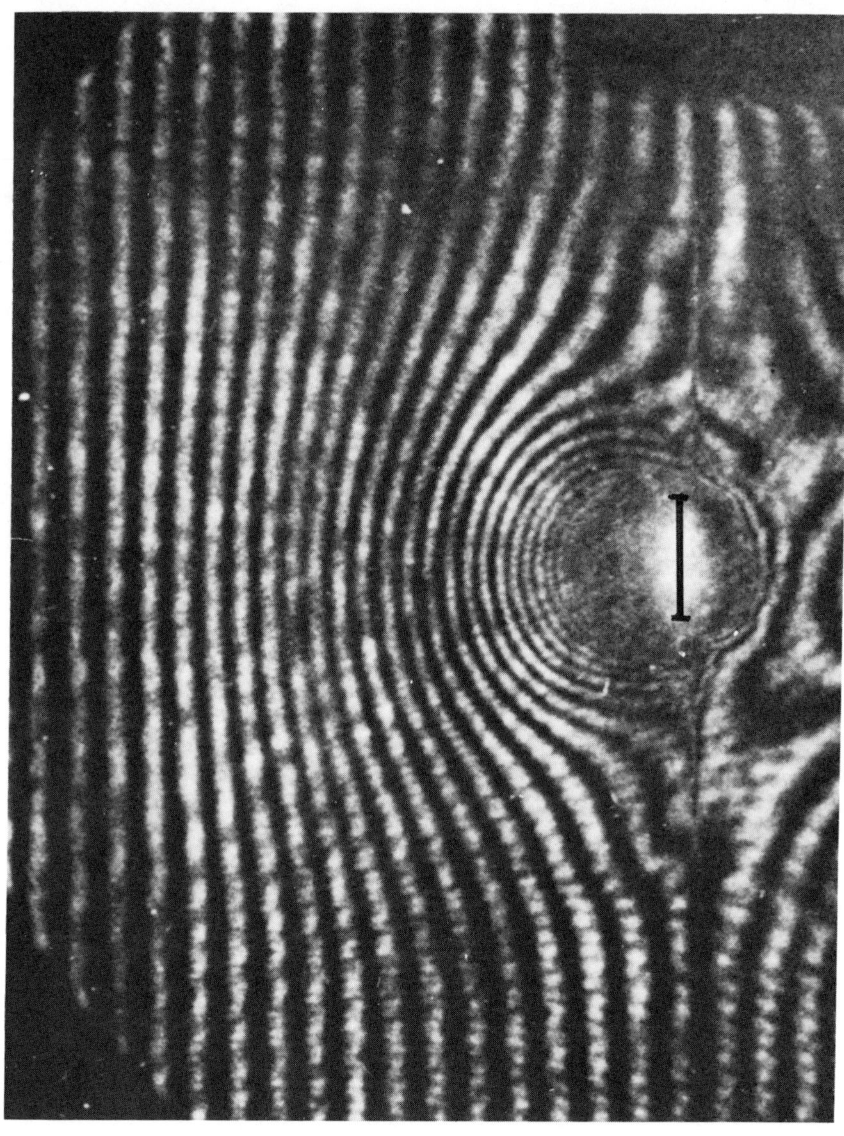

Fig. 14.   Shearing interferogram for a 600-micron diameter
           by 4.5 micron thick, limited mass disc target of CH.

the edge of the dark region is more noticeable on most of the
interferograms.) However, the fringe bending on the front side
shows an extensive and low density plasma. The full extent of
the front-side plasma can be seen in Fig. 14 which is a shearing
interferogram for a 600 micron diameter by 4.5 micron thick CH
limited mass target mounted on 5 micron fibers. Here the
irradiance was 2×10^{12} W/cm^2 with 50J of laser energy on
target and the time was +2nsec. A 500 micron distance marker
is placed near the initial target location. Note that plasma can
be seen on the front (left) side as far out as 2 mm. Abel inversion
shows, near the maximum density sampled (4×10^{19} cm^{-3}), a density
scale length of 400 microns on axis and 700 microns at 45 degrees
to the axis. These long density scale lengths are expected for
material that is smoothly ablated to provide acceleration of the
dense target material to the rear. Indeed, the simultaneously
recorded dual-time shadowgrams (which depend on large density gradients) show no effect on the front side but do show the dense plasma
expanding to the rear. These are discussed next.

We now consider (in Fig. 12) light that has been transmitted
through the shearing plate. Here, a somewhat smaller region with
large gradients is responsible for the shadowgrams. A polarizing
prism· (7-degree Wollaston prism) was used to angularly separate
the polarizations so that both polarizations (times) could be
recorded side-by-side to give a dual-time shadowgram.

A dual-time shadowgram for a wide (3mm), 6.5 micron thick CH
target is shown in Fig. 15. The irradiance was 5×10^{12} W/cm^2
over a 1 mm focal diameter. Times noted are relative to the peak
of the main laser pulse. A smoothed curve was drawn along the
transition region to the solid dark area. An on-axis velocity
of 1.1×10^7 cm/sec was obtained from the displacement (220
microns) of the on-axis positions of the curve between +2 and
+4 nsec. By varying the foil thickness, and thus the ablated
mass ratio, one is able to see how such velocities compare with
the velocities expected (via momentum conservation) from the
front-side, asymptotic ablation velocity. Integrating Eq. (4)
over time shows that the ablated mass ratio $\Delta M/M_o$ or ratio of
ablation depth x_a to initial target thickness z_o is given by $1 -
\exp(-v/u)$ where u and v are ablation and target velocities.
The target velocity calculated (via momentum conservation) from
this relation with u (4.7×10^7 cm/sec) and x_a (1 micron) from
charge collector data is denoted by V_{calc} in Table II.
The rear-surface velocity observed in the dual-time shadowgram is
denoted by V_{obs}.

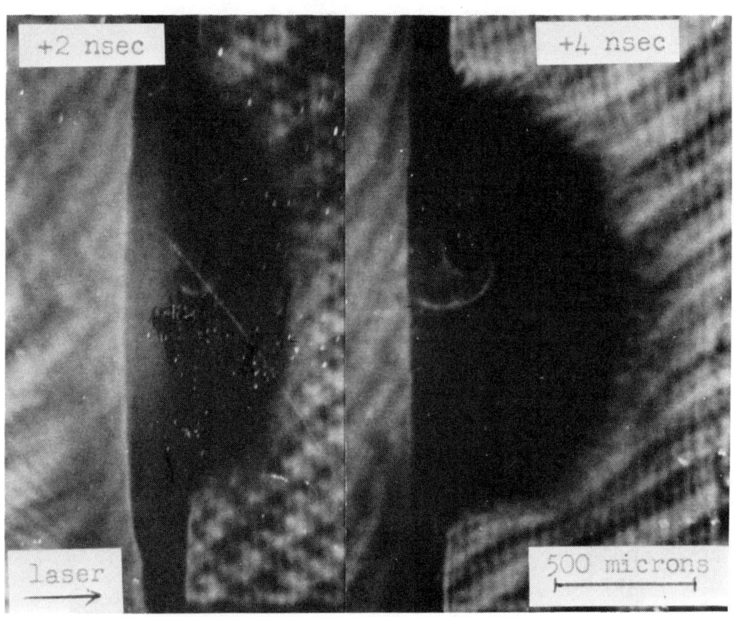

Fig. 15.   Dual-time shadowgram for a wide, 6.5 micron thick CH
           target showing development of the dense, rear-side
           plasma over the interval +2 to +4 nsec.

Table II.   Comparison of target velocities observed in dual-time
shadowgrams with that calculated (via momentum conservation) from
ablation velocity and depth obtained with charge collectors.

Shot No.	Z_o (microns)	V_{obs} (cm/sec)	V_{calc} (cm/sec)
8532	12	0.5×10^7	0.4×10^7
8535	6.5	1.1×10^7	0.8×10^7
8539	3.5	1.5×10^7	1.6×10^7

     The agreement seen in Table II indicates that the velocities
seen in late-time (+4 nsec) dual-time shadowgrams do correspond
to the asymptotic target velocities. Since, with the f/2
collector lens used, densities of only a few times 10^{19} cm^{-3} could
be seen, it is concluded that the (so far unobservable) dense
target material is moving inside the dark region at about the same
velocity as that of the observed edge of this region.

The results of various studies described here quantitatively
confirm the simple picture of the foils being accelerated by the
rocket-like action of the ablating material. However, there is
also evidence for more complicated phenomena - not yet well
understood. Thus, in Fig. 15 one can see at +4 nsec small-scale
structure at the leading edge of the expanding dense plasma.
An earlier dual-time shadowgram (Fig. 16) taken on a 6-micron
CH foil at a higher irradiance (1×10^{13} W/cm^2) and with a less
uniform main laser beam showed large-scale plasma structure. This
could be due to partial foil burn-through or, perhaps, a hydrodynamic instability.

Discussion of Experimental Results

The experimental studies presented here have confirmed the
general picture of an efficient ablative acceleration of foil
targets to velocities in the range required for fusion applications.
Hydrodynamic efficiencies as high as 20% have been seen and
target velocities in the range 2×10^7 cm/sec were observed.
The target and ablation velocities are simply related by momentum
conservation (Table II) while hydrodynamic efficiency and

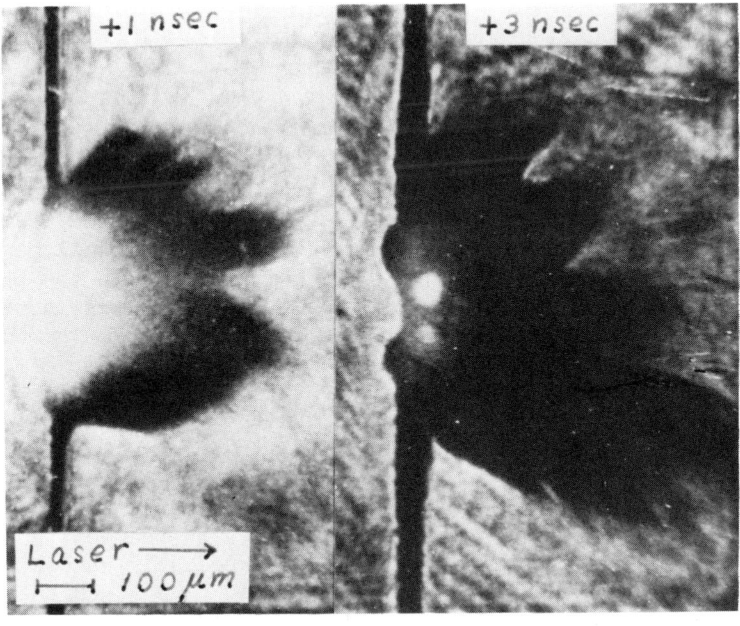

Fig. 16. Dual-time shadowgram at higher irradiance
          (1×10^{13} W/cm^2) and less beam uniformity, showing
          gross target break-up.

ablated mass ratio are well-described in terms of the ablation-to-target velocity ratio by a simple rocket model (Fig. 8) which depends only on momentum and energy conservation. These studies were carried out in an irradiance regime (10^{12} - 10^{14} W/cm^2) which is relatively free of deleterious high-irradiance plasma effects and shell break-up at lower irradiance. In this regime there is efficient (60-80% of laser energy absorbed) and thermal coupling to the target. Further studies are necessary to extend the original results over a broad range of target and laser pulse conditions, to explore the limits of the tolerable irradiances, and to better understand various aspects of the relevant physics - particularly hydrodynamic instabilities.

THEORETICAL STUDIES

<u>Numerical Study of Optical Ray Retracing in Laser-Plasma Backscatter</u>

Numerical studies[26] are currently being performed at NRL to investigate optical ray retracing in stimulated Brillouin back-scatter from laser-produced plasmas.[27,28] These studies use a two dimensional electromagnetic propagation code BOUNCE, which treats steady-state behavior in the strong damping limit. At the present time, the code includes filamentation due to the ponder-motive force, but ignores ion wave saturation and pump depletion.

Ray retracing phenomena can be grouped into two limiting cases. In the "whole beam" limit, the pump radiation at the lens has a broad spatial profile that can be focused to a long narrow waist within the plasma. This geometry selectively amplifies only those initial noise components that are propagating back along the axis, where the net gain is highest. The following results were found: (i) In simulations where a half-blocked beam is focused near the critical surface of an inhomogeneous plasma, only a small portion (5%) of the backscatter returns to the lens in the shadow region. (Fig. 17, left). This demonstrates that the reflection is non-specular, and is in good qualitative agreement with a recent NRL experiment.[28] (ii) Because of spatial gain narrowing effects within the focal waist, the spatial profile of the backscatter at the lens tends to have broader wings than that of the pump beam. (iii) The backscatter profile at the lens is relatively insensitive to filamentation provided that the Brillouin gain is sufficiently high (e.g., $G \sim 100$). At lower gain, filamentation can cause the retracing to exhibit a pronounced threshold with respect to variations of density or pump irradiance, especially when the plasma is placed just beyond the focal plane. (iv) When the target is moved far out of focus, so that the beam width becomes com-parable to the length of the gain region, the retrace effect disappears, and the angular distriubtion of the backscatter becomes essentially that of amplified noise.

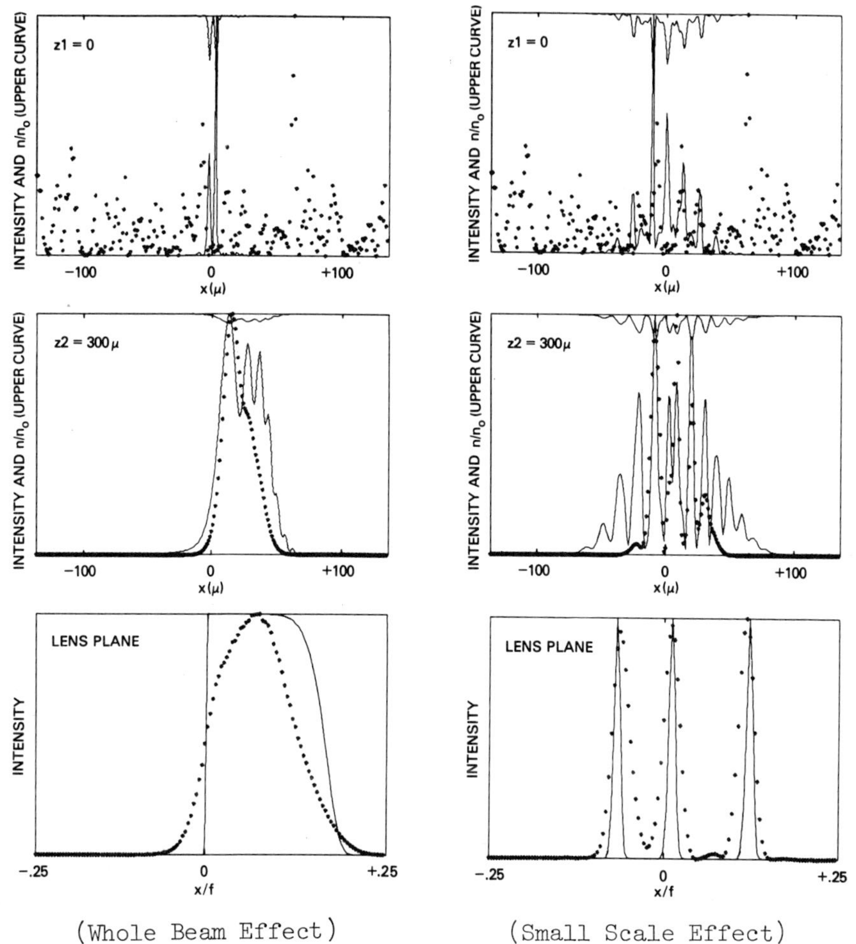

(Whole Beam Effect)   (Small Scale Effect)

Fig. 17.   (Left) Transverse profiles of 1.06-micron pump radiation
           and backscatter from a 300-micron plasma of density
           .2n$_c$ exp (-z/100),showing partial ray retracing of a
           half blocked pump beam at the lens. The stochastic noise
           appears as random dots at the focal plane z1 = 0. Normal-
           ized pump intensity and pondermotive density variations
           n/n$_c$ are denoted by solid curves. Backscatter intensity
           is denoted by dotted curves.
           (Right) Transverse profiles for a 300 micron plasma of
           density .2 n$_c$ exp (-z/100), showing nearly complete re-
           trace of the three pump beams at the lens.

In the opposite limit, where the pump field is dominated by small scale transverse structure, the backscatter can reproduce this structure in detail (i.e., exhibit wavefront reversal)[29] under certain conditions. This is illustrated in Fig. 17 (**right**) in which the pump distribution at the lens consists of three isolated hot spots. The effect occurs because the beams from these hot spots overlap near focus to create an interference pattern in the plasma. This pattern behaves as an active volume hologram; i.e., it selectively amplifies only those initial noise components propagating oppositely to any one of the three beams, and simultaneously Bragg-diffracts energy back along the other two.[30] According to this model, effective wavefront reversal requires that the gain occur slowly within typical diffraction lengths ℓ_D. Excellent agreement was found between a simple coupled wave analysis of the model[30] and the numerical solutions of an idealized test problem where two truncated plane waves formed an interference pattern within a uniform plasma of 100 micron thickness.

In more realistic simulations, the pump profile consisted of three isolated hot spots at the lens (e.g., Fig. 17, right), giving a 100 micron wide interference pattern near focus. These simulations produced the following results: (i) wavefront reversal is enhanced if the plasma has a monotonically decreasing gain coefficient away from the target. For example, a plasma density variation $\exp(-z/\ell)$, where $\ell \gg \ell_D$, produces significantly better wavefront reversal than a uniform slab of thickness ℓ and comparable net gain. This enhancement arises from the better spatial filtering action that occurs in the lower gain region $z > \ell$. (ii) Spatial gain narrowing effects, which were present in all cases, restrict the number of periods within the backscatter interference pattern. At the lens plane, this tends to eliminate higher spatial frequencies and broaden the hot spots in the backscatter. (iii) Filamentation appears to have a minimal effect on the wavefront reversal, provided that $\ell \gg \ell_D$ and the width of the backscatter interference pattern is large in comparison to its period. If this width decreases below a critical value, however, then whole beam filamentation takes over and destroys the wavefront reversal. This is especially likely in cases where gain narrowing effects are prominent, or where the pump beams are focused a few hundred microns in front of the target. (iv) If the target is moved into the quasi near field of the lens, so that the hot spots no longer interfere, the wavefront reversal disappears, and the backscatter profile assumes a random noise character. The formation of a volume hologram is therefore essential for wavefront reversal.

In future studies, an attempt will be made to incorporate phase mismatch, pump depletion, and ion wave saturation into the code.

Anomalous Flux Limitation and Absorption Due to Return Current
Driven Ion Acoustic Turbulence

There is now a large body of experimental evidence which
indicates that electron thermal energy flux can be substantially[31,32]
reduced in a laser produced plasma.   One possible mechanism
is ion acoustic turbulence driven by the return current. That
is, if there is an energy flux of energetic electrons to the right,
there must be a return current of slow electrons to the left in
order for there to be no net electron current. If the energy
flux is denoted by Q, and the electron number density and thermal
velocity are n and V_e, the return current velocity is Q/nmV_e^2.
This return current velocity can drive an ion acoustic instability
in very much the same way as an actual current.[11]

If this instability is excited, quasi-linear theory shows
that the following anomalous effects may become important.[11]
First, there is a reduction in electron thermal conduction. If
the fluctuating potential is ϕ, then the electron thermal conduction is given roughly by $K \sim V_e \lambda_D (e\phi/T_e)^{-2}$. Thus, thermal conduction
is reduced as the instability is more strongly excited. Second,
the ion density fluctuations produced by the instability can give
rise to anomalous absorption of the laser light in the underdense
plasma. Third, there is an anomalous electron cooling and an equal
ion heating. Fourth, there is an electron current generation in
the direction of the energy flux (or depending on boundary conditions, the generation of an electric field which decelerates electrons in the direction of the energy flux). Also, it is important
to note that there have recently been experiments in dense pinches
which demonstrate electron thermal energy flux limitation by return
current generated ion acoustic turbulence.[33]

One can make use of these transport effects to calculate the
anomalous absorption and flux limitation in a steady state[34] or
dynamic[35] plasma. The flux limitation factor $f = Q/nmV_e^3$ is not
assumed, but rather is calculated as a function of time and/or
space. Typically it maximizes around the critical density at
a value of roughly 0.1 or 0.2. For short pulse experiments,
anomalous absorption by ion acoustic turbulence can be important,
but is generally less important than resonant absorption. However
for longer pulse experiments, it can be the dominant absorption
mechanism.

Another possible physical effect which can lead to inhibition
of electron thermal transport is spontaneous magnetic field
generation.[8,9] In this case return current generated ion acoustic
instability leads to enhanced electron thermal conduction.[36] In
fact one can recover the formula for K for either a magnetized or

unmagnetized plasma by assuming an amomalous collision frequency $\nu \sim \omega_{pe}$ x $(e\phi/T_e)^2$. In addition to enhancements in K due to instability, it can also be enhanced by perturbations and irregularities in the field.[37] Finally we note that steady state calculations of absorption and flux limitation, analogous to those in Ref. 34 have been done for a magnetized plasma.[38]

REFERENCES

1. A detailed discussion of the experimental studies is given by B.H. Ripin, R. Decoste, S.P. Obenschain, S.E. Bodner, E.A. McLean, F.C. Young, R.R. Whitlock, C.M. Armstrong, J. Grun, J.A. Stamper, S.H. Gold, D.J. Nagel, R.H. Lehmberg, and J.M. McMahon in Phys. Fluids, to be published (1980), Also, see NRL Memo Report No. 3890 (1978), B.H. Ripin, Ed.

2. The ablative acceleration experiments and the rocket model are discussed by R. Decoste, S.E. Bodner, B.H. Ripin, E.A. McLean, S.P. Obenschain, and C.M. Armstrong in Phys Rev. Lett. 42, 1673 (1979).

3. The associated laser-plasma interactions (absorption, thermal transport, etc.) are discussed by B.H. Ripin, R.R. Whitlock, F.C. Young, S.P. Obenschain, E.A. McLean, and R. Decoste in Phys. Rev. Lett. 43, 350 (1979).

4. S.E. Bodner, B.H. Ripin, and J. McMahon, Bull. Am. Phys. Soc. 24, 1074 (1979).

5. R.E. Kidder, in Laser Interaction and Related Plasma Phenomena, edited by H. Hora, H. Schwarz, M.J. Lubin, and B. Yaakobi (Plenum Press, New York 1980).

6. B.H. Ripin, P.G. Burkhalter, F.C. Young, J.M. McMahon, D.G. Colombant, S.E. Bodner, R.R. Whitlock, D.J. Nagel, D.J. Johnson, N.K. Winsor, C.M. Dozier, R.D. Bleach, J.A. Stamper, and E.A. McLean, Phys. Rev. Lett. 34, 1313 (1975).

7. W.C. Mead, R.A. Haas, W.L. Kruer, D.W. Phillion, H.N. Kornblum, J.D. Lindl, D.R. MacQuigg, and V.C. Rupert, Phys. Rev. Lett. 37, 489 (1976).

8. J.A. Stamper, E.A McLean, and B.H. Ripin Phys. Rev. Lett. 40, 1177 (1978).

9. A. Raven, O. Willi, and P.J. Rumsby Phys. Rev. Lett. 41, 554 (1978).

10. R.C. Malone, R.L. McCrory, and R.L. Morse, Phys. Rev. Lett. 34, 721 (1975).

11. W.M. Manheimer, Phys. Fluids, 20, 265 (1977).

12. B.H. Ripin, F.C. Young, J.A. Stamper, C.M. Armstrong, R. Decoste, E.A. McLean, and S.E. Bodner, Phys. Rev. Lett. 39, 611 (1977).

13. D.W. Phillion, W.L. Kruer, and V.C. Rupert, Phys. Rev. Lett. 39, 1529 (1977).

14. J. Nuckolls, L. Wood, A. Thiessen, and G. Zimmerman, Nature 239, 139 (1972).

15. Yu, V. Afanasev, N.G. Basov, P.P. Volosevich, E.G. Gamalii, O.N. Krobkin, S.P. Kurdyumor, E.J. Levanov, V.B. Rozanov, A.A. Samarskii, and A.N. Tikhonov, J.E.T.P. Lett. 21, 68 (1975).

16. See J.H. Nuckolls and J.D. Lindl in the digest of technical papers presented at the Topical Meeting on Inertial Confinement Fusion, February 7-9, 1978, San Diego, CA. Published by the Optical Society of America.

17. R. Resnick and D. Halliday, "Physics" (Wiley, New York, 1978), Vol. I, p. 178.

18. M.K. Matzen and R.K. Morse, Phys. Fluids 22, 654 (1979).

19. P.J. Moffa, J.H. Orens, and J.P. Boris (to be published).

20. S. Gitomer, private communication.

21. J.P. Anthes, M.A. Gusinow, and M.K. Matzen, Phys. Rev. Lett. 41, 1300 (1978).

22. R. Decoste, B.H. Ripin, J. Grun, J.A. Stamper, S.E. Bodner, S.P. Obenschain, and F.C. Young, Bull. Am. Phys. Soc. 24, 1074 (1979).

23. J.M. McMahon, Bull. Am. Phys. Soc. 24, 1074 (1979).

24. F.C. Young, in NRL Memo Report No. 3591 (1977) (ed. S.E. Bodner) p. 41-49.

25. E.A. McLean et al., to be published.

26. R.H. Lehmberg, Lasers '78, First International Conference on Lasers, ed. V.J. Cororan (STS Press, McLean, VA. 1979); Submitted to Phys. Rev.

27. K. Eidman and R. Sigel, in Laser Interation and Related Plasma Phenomena, Vol. 3B, edited by H.J. Schwarz and H. Hora (Plenum Press, New York, 1974); J.A. Stamper, et al., ibid., p. 713.

28. B.H. Ripin, F.C. Young, J.A. Stamper, C.M. Armstrong, R. Decoste, E.A. McLean, and S.E. Bodner, Phys. Rev. Lett. 39, 611 (1977); B.H. Ripin, NRL Memo Report 3864 (1977), unpublished.

29. B. Ya Zel'dovich, V.I. Popovichev, V.V. Ragul'skii, and F.S. Faizullov, ZHETF Pic. Red. 15, 160 (1972). (Sov. Phys. JETP Lett. 15, 109 (1972)).

30. R.H. Lehmberg, Phys. Rev. Lett. 41, 863 (1978).

31. F.C. Young, R.R. Whitlock, R. Decoste, B.H. Ripin, D.J. Nagel, J.A. Stamper, J.M. McMahon, and S.E. Bodner, Appl. Phys. Lett. 30, 45 (1977).

32. W.L. Kruer, Comm. Plasma Phys., to be published.

33. D.G. Gray, et al., private communication.

34. W.M. Manheimer, D.G. Colombant and B.H. Ripin, Phys. Rev. Lett. 38, 1135 (1977).

35. D.G. Colombant and W.M. Manheimer, NRL Memo Report 4083, Oct. 1979 (Submitted for publication).

36. W.M. Manheimer, C.E. Max, and J.J. Thomson, Phys. Fluids
    $\underline{21}$, 2009 (1978).
37. C.E. Max, W.M. Manheimer, and J.J. Thomson, Phys. Fluids
    $\underline{21}$, 128 (1978).
38. W.M. Manheimer and D.G. Colombant, Phys. Fluids $\underline{21}$, 1818
    (1978).

TARGET FABRICATION FOR EXPERIMENTS AT THE UNIVERSITY OF ROCHESTER'S
LABORATORY FOR LASER ENERGETICS

H. W. Deckman and G. M. Halpern

Exxon Research and Engineering Company

Linden, New Jersey 07036

INTRODUCTION

Current experimental attempts to demonstrate the scientific
feasibility of laser fusion employ targets designed to elucidate
fundamental physical processes governing laser-plasma interaction
phenomena. Key issues involve the mechanisms by which the laser
radiation is absorbed by the target and the subsequent conversion
of this absorbed energy into the compression and heating of the
target core to the densities and temperatures required for significant thermonuclear yield. To address these fundamental issues the
experimental program at Rochester[1-2] requires a variety of well-
characterized targets for establishing a data base upon which a
consistent understanding of laser fusion can be built. Consequently,
today's targets are hand-crafted research-type objects which probably bear little resemblance to the high gain fuel pellets that
might someday be used for commercial reactor applications[3].

This paper will review the fabrication and characterization of
several types of spherical targets which have been used in recent
experiments at the University of Rochester's Laboratory for Laser
Energetics (LLE). The specific design of the targets was constrained
by the operating characteristics of the laser[4] as well as the nature
of the physical processes being investigated. Targets at LLE are
being irradiated symmetrically with a six beam laser system, ZETA,
which operates at a wavelength of 1.054 μm and delivers a maximum
focusable power of $\sim$3 TW in a 50 psec pulse. In these experiments,
the gas within the target has been compressed to more than 10^3 times
its initial density[5] with a few hundred joules of energy output from
the laser. In this regime the symmetrical illumination capability
of ZETA is a key factor in attaining these compressions.

Moderate constraints on target uniformity are also imposed; in particular, spherical symmetry of a few percent is required and surface finishes of $\sim$.5µm are sufficient. Future experiments with OMEGA,[6] a 24 beam system designed to produce 12-15 TW of focusable power, will produce substantially higher compressions with concomitantly more stringent requirements on target uniformity and surface smoothness. Ultimately, for a fuel pellet to produce at least as much energy as it absorbs from the laser, it will be necessary to compress the fuel to densities of 1000 or more times the density of liquid deuterium-tritium (0.2 g/cm^3) in order to initiate a propagating burn[7]. Theoretical estimates indicate that such future targets will require a surface finish of $\sim$100 Å to ensure a stable compression[8].

The simplest targets used in the experiments[9] at LLE are hollow glass shells (microballoons) having diameters in the range 50-150µm with wall thicknesses varying from .5-1.5µm. Equimolar mixtures of deuterium and tritium (DT) are permeated into the interior of the shells at pressures ranging from 2-50 atm. To hold the target at the focal point of the laser, each shell is glued to the tip of a hollow glass stalk. The diameter of the tip is generally 5-10% of the diameter of the glass shell to minimize the perturbing effect of the stalk.

Bare glass microballoons filed with DT are primarily used in "exploding pusher" - type experiments. In this mode[10] a short, high intensity laser pulse is absorbed at the target surface, creating superthermal electrons that uniformly heat and "explode" the glass shell. The exploded shell then shock heats the DT fuel to high enough temperatures to produce thermonuclear neutrons. However, the resultant compressions are relatively low and it is generally agreed that the yield from an exploding pusher target cannot be scaled with increasing laser power (or energy) to reach a breakeven level. Nevertheless, exploding pusher targets are extremely useful in order to: (i) compare thermonuclear yields with theoretical predictions[10], (ii) determine the effect of laser illumination uniformity on the implosion symmetry[9] and the fraction of incident light absorbed by the target, and (iii) study the physics of fast electron and ion production[11] which may influence the design of more advanced, higher yield targets.

In order to interpret these experiments it is necessary to characterize[12-15] the uniformity, thickness, diameter, composition and density of these targets as accurately as possible. Significant shell nonuniformities can deleteriously affect target performance. The most common imperfections are small particulate-like defects on the exterior, pits associated with compositional inhomogeneities on the inner surface and the decentration of the inner surface of the shell relative to its exterior surface. Decentrations can be quantified by optical interferometry,[16-17] which is also capable of resolving isolated defects larger than one micron. Scanning electron microscopy (SEM) is also useful for discerning sub-micron features on the target surface[18]. It should be noted that all experimental evidence to date indicates that the performance of exploding pusher targets is insensitive to the presence of a few micron-sized particulates or pits.

Gas fills must be characterized with respect to composition and pressure[19]. The fuel mass is typically .1-10 ng which is nondestructively assayed for each target to an accuracy of $\pm 15\%$. This mass should be compared with the mass of the glass shell (10-100 ng) which can be determined from measurements of the shell wall thickness, density and diameter to better than 10%. Presently, wall thickness, diameter and fill pressure are measured for each target, whereas glass composition and density are characterized for representative batches of targets. It is important to develop accurate characterization techniques for glass microballoon targets since they are so extensively used in fundamental laser fusion experiments and will probably serve as elementary components in the more advanced targets of the future. Table 1 summarizes the accuracy with which these parameters are measured for typical exploding pusher targets.

TABLE 1

CHARACTERIZATION OF EXPLODING PUSHER TARGETS

Target Parameter	Accuracy of Measurement
Glass Shell:	
Diameter (50-150μm)	$\pm$ 1μm
Wall thickness (.5-1.5μm)	$\pm$ .05μm
Decentration of inner bubble with respect to exterior surface of shell	10% of wall thickness
Glass density (2.5 g/cm^3)	$\pm$ .2 g/cm^3
Index of refraction (1.5)	$\pm$ .05
Gas Fill:	
Nondestructive measurements	
Fill pressure (5-50 atm)	$\pm$ 15%
Destructive	
Fill pressure	$\pm$ 15%
D/T ratio (1:1)	$\pm$ 5%

With the availability of increased laser power it becomes possible to drive the implosion of thicker (i.e more massive) shells. The advantage of using thicker shells is that the more energetic electrons which are generated in the outer absorption region are stopped before reaching the interior fuel. Thus, in contrast to the exploding pusher mode, the inner part of the shell

does not disintegrate but remains intact to compress the shielded
target core in a nearly adiabatic and hence more efficient manner.

Experiments employing thicker shells are generally designed
to produce high compressions of the target core. Using argon as
a fill gas the compressions can be measured directly by spectroscopic
methods[20-21] involving the Stark broadening of argon x-ray emissions.
Glass shells thicker than those used in exploding pusher targets
are usually adequate for these experiments, since the argon emissions
can penetrate through the glass.

Other experiments, designed to produce more moderate final
densities,[21] employ DT or DT/neon fills. The principal diagnostic
method uses pin hole cameras to image the compressed portion of the
target. The use of thick glass shells in these experiments is not
appropriate because the geometric imaging of the core will be
distorted due to the absorption by the glass of the softer x-ray
emissions. To circumvent this difficulty these experiments utilize
a glass inner shell overcoated with a hydrocarbon plastic which
provides adequate stopping power for the fast electrons while
permitting the transmission of x-rays from the target interior for
accurate imaging diagnostics. Thus, depending upon the purpose of
the experiment, the composition of the shell material can be tailored
to the chosen diagnostic, providing that the areal density (g/cm^2) of
the shell is sufficient to stop the energetic electrons.

Imaging experiments[21] employ glass microballoons, filled with
up to 50 atm. DT or D_2 and, sometimes, 5 atomic percent neon. The
filled microballoons are coated with up to 15μm of plastic. To
date, we require that the microballoons be mounted on stalks during
the coating operations. Hence the completed target has a thickly
coated stalk which must be removed, since it represents an unacceptably large perturbation to the required spherical symmetry. Micromachining in a miniature lathe is feasible, but tedious; alternate
methods such as pulsed laser machining[22-23] are also effective.
Techniques for coating unmounted precharacterized targets are currently
under development and should be applicable, eventually, to both
plastic and inorganic coating materials.

Experiments to be diagnosed by spectroscopic (line-broadening)
techniques may also employ plastic coatings which are considerably
thinner. For example, these targets may involve glass shells in
the diameter range 35 to 70μm and plastic coatings up to 8μm thick.
The perturbative effects of the thinner coated stalks are less severe
but clearly non-negligible.

The remainder of this article gives a detailed accounting of
the fabrication and characterization procedures we have developed
and employed to provide the Laser Fusion Feasibility Project with

the aforementioned targets. Exploding pusher targets are first
discussed, including glass microballoon selection and measurement,
gas permeation filling procedures, fill gas characterization methods
and mounting operations. A novel technique for filling glass shells
with several atmospheres of argon is also presented. This "drill,
plug and fill technique"[24] is also applicable to other fill gases
which are being comtemplated for future experiments. Three independent plastic coating approaches are also outlined. Next, a section
on advanced configurations describes some recent work on multi-shell
fabrication using hemispheres[25] produced by microelectronics techniques.
This section also includes a description of a versatile physical
vapor deposition apparatus used in the production of these hemispheres,
as well as in the coating of individual levitated[26] glass microballoons.
Finally a section on advanced characterization deals with a unique
x-ray microradiograph generation and readout technique employing a
commercial scanning electron microscope[27].

EXPLODING PUSHER TARGET FABRICATION

Exploding pusher targets consist of thin-walled glass shells
filled with equimolar mixtures of deuterium and tritium (DT).
These shells, commonly referred to as microballoons, have found
extensive application in laser fusion programs throughout the world.
The method for filling targets by permeating DT at elevated temperatures through the glass walls of commercially available shells was
first reported in the open literature by Lewkowicz[28]. Subsequently,
a number of laboratories have reported on several variations[29-30]
of the technique.

Commercially available hollow glass spheres are used as fillers
in plastics and are produced by the 3M Co., Emerson and Cuming Inc.
and Philadelphia Quartz Inc. The 3M glass shells have been used
extensively in experiments at LLE. These shells are produced[31]
by blending a fine glass powder with a gas-releasing blowing agent
and passing the mixture through a hot zone. The elevated temperature causes the blowing agent to decompose and release gas which then
expands through the molten glass particle to form a bubble. Analysis[19]
of the residual gas shows that a typical balloon will contain
1/8-1/3 atm. of SO_2 with significantly smaller amounts of N_2 and O_2.
The composition of the shell material is approximately that of soda
lime glass and a representative balloon would be comprised[32] of (mole
percent): 80% SiO_2, 10% Na_2O, 6.5% CaO, 2% B_2O_3 with other oxides
making up the remaining 1.5%. Emerson and Cuming have used variations
of a process licensed from the Standard Oil Company of Ohio[33] to
produce glass bubble products from a solution-derived dried gel[34].
Glass shells are formed in a heated furnace where a blowing agent
which was incorporated in the original gel material expands to create
a bubble in the molten glass. Emerson and Cuming also produces
shells with a composition similar to quartz. These have been

employed in many of the high density experiments at LLE. Since commercially available microballoons are produced in bulk quantities, they are covered with glass shards and dirt. Once cleaned, almost all of these balloons have an exterior surface finish which is better than 2,500 Å; however, their inner surface is characterized by submicron to micron-sized bumps associated with compositional inhomogeneities[35]. The major geometrical defect in these shells is a decentration of the inner bubble with respect to the exterior surface of the glass microballoon. The percentage of cleaned microballoons which could be selected as targets for exploding pusher experiments is tabulated in Table 2. Inspection of the table quickly reveals that at least 50-1000 shells must be inspected before a suitable target can be found for exploding pusher experiments.

TABLE 2

QUALITY FACTORS* FOR SELECTED GLASS MICROBALLOON STOCKS

Supplier and Product	Average Wall Thickness for Target Quality Balloons (μm)	Quality Factors* for Balloons with $100<\text{diameter}<125\mu$m (%)	Quality Factors* for Balloons with $75<\text{diameter}<100\mu$m (%)
Emerson Cuming			
VT	1.1	.4	1
FTD-202	1.0	.5	.6
SI	1.2	.6	2
IG-101	1.1	.2	.2
3M Co.			
B12AX	.6	.1	2
B15BX	.6	1	1
B18A	.7	.1	2
B22A	1.0	.6	2
B23/500	.9	.4	.1

*Quality factors are the % of balloons interferometrically inspected which could be used as targets in exploding pusher experiments. At least 400 balloons were viewed for each entry. All balloons were cleaned and sieved before examination.

In order to obtain high quality targets with surface finish approaching 100 Å, Lawrence Livermore Laboratory[36-38] (LLL) and KMS Fusion Inc.[39-40] have developed new processes to fabricate glass microballoons. LLL has sucessfully implemented a process based upon a liquid droplet technique. In this approach, microballoons are created by spraying aqueous solutions containing water soluble glass forming compounds downward through a vertical hot zone. As the droplet is heated, the glass forming compounds transform to glass and some of the resulting water vapor is trapped within the sphere. The water vapor acts as an internal blowing agent creating a hollow glass shell. By appropriate selection of droplet size and temperature profile in the hot zone, hollow glass microspheres of uniform diameter, with surface finishes approaching 100 Å have been produced[41]. Some of the high quality spheres produced using this technique are subject to surface weathering which can be prevented by either surface treatment or storage in ethanol[42].

KMS has produced glass shells with low permeabilities by drying a gel containing glass forming compounds and passing the resulting frit material (crushed and sieved pieces of the dried gel) through a vertical hot zone. The frit material used in the KMS process has a distribution of sizes and shapes so that the time-temperature history for each balloon produced will be slightly different. As such, no two shells have identical physical and chemical properties; however, surface finish and gas permeability can be controlled by judiciously choosing the chemical composition of the starting material. Because there are several distinct methods with which to prepare the starting gel, the technique is amenable to the production of glass shells with a wide variety of chemical compositions. Several of the high quality KMS shells have been used in exploding pusher experiments at LLE.

The composition of the shell material is important because it determines the permeation rate for gas to diffuse into the hollow shell interior as well as the useful shelf life of the filled target. Diffusion through the shell wall is approximately governed by Fick's first law[43-45] and thus the pressure in the target varies exponentially with time. It is easily shown that, for a given gas, the pressure retention half life of a thin shell of inner diameter D and thickness t, is given by

$$t_{1/2} = \frac{tD}{3RTK} \ln 2$$

where R is the gas constant, T is the temperature and K is the permeability of the shell material to the particular gas. Permeability depends sensitively upon the temperature as well as the relative percentage of non-network formers in the glass[46-47] such as Na_2O, K_2O, CaO; (network formers are SiO_2, B_2O_3, P_2O_5). For instance, when filled with hydrogen, a shell composed of soda lime glass (25% molar non-network formers) has a shelf life 10^5 times greater than an SiO_2 (0% non-network formers) shell of the same dimension.

Exploding pusher targets are filled by diffusing equimolar mixtures of deuterium and tritium through the glass wall at temperatures ranging from 350-425°C. At these temperatures, the 3M and KMS shells of interest can be filled to virtual equilibrium in a period of less than 20 hr. Balloons have been packaged for permeation either by loading them in bulk into a quartz vial or placing them individually into dimples of an "eggcrate" made from aluminum, or

quartz. The eggcrate packaging technique is preferred because: (i)
typically 10-25% of the balloons rupture during permeation thus
contaminating the local environment with glass shards, (ii) each
target can be characterized prior to permeation and only 100-200
balloons need be selected for each run, and (iii) the total radiation
inventory of tritium in the laboratory is minimized by limiting the
number of filled balloons - each of which may contain up to $\sim$100 µCi
of tritium.

Occasionally during permeation, microballoon surfaces become
contaminated with micron-sized crystalline and/or non-crystalline
growths. Removal of these growths can usually be effected by
chemical cleaning; however, preliminary experiments indicate that
either ozone or plasma cleaning methods used in the microelectronics
industry[48-49] provide a superior surface finish. Presumably, the
growths are formed by vapor phase deposition of contaminants volatil-
ized when the permeation cell is heated.

Before they are packaged into eggcrates and filled, the microballoons are individually inspected to ascertain the thickness and
uniformity of the glass shell. At LLE, characterization is
accomplished[17] using a Leitz Mach-Zender interference microscope.
In a transmitted light interferometer such as the Mach-Zender, a
fringe is seen whenever the optical path difference (O.P.D.) between
a ray traveling through the reference arm and the corresponding ray
traveling through the object is an integral number of optical wavelengths. Thus when the interferometer is adjusted so that the object
and reference rays are collinear, a target quality microballoon will
exhibit a series of circularly symmetric fringes. An example of
such a fringe pattern is shown in Figure 1 for a shell of outer
diameter 112µm and wall thickness 1.47µm.

After locating such a target, a white light source is used and
the sample and reference beams are adjusted so that a series of
parallel fringes appears in the background of the field of view. The
darkest fringe is easily identified and corresponds to the zeroth-
order destructive interference between the sample and reference
beams. The test balloon is positioned against this fringe background
until a portion of the darkest fringe appears at the center of the
balloon image, as seen in Figure 2. It is easily shown[50] that the
distance between the background reference fringe and its location
on the center of the balloon (measured in units of the background
fringe spacing and denoted by x) is related to the wall thickness
according to the formula $t = x\lambda/2(n-1) = x\lambda$ for glass having an
index of refraction $n = 1.5$. In practice, x is measured using monochromatic light ($\lambda = 0.5461$ µm) so that the fringes appear as sharp
as possible. The value of t yielded by this method is, in reality,
the average glass thickness traversed by the central rays of the
sample beam. The accuracy of this measurement has been estimated
to be better than ± 0.10µm.

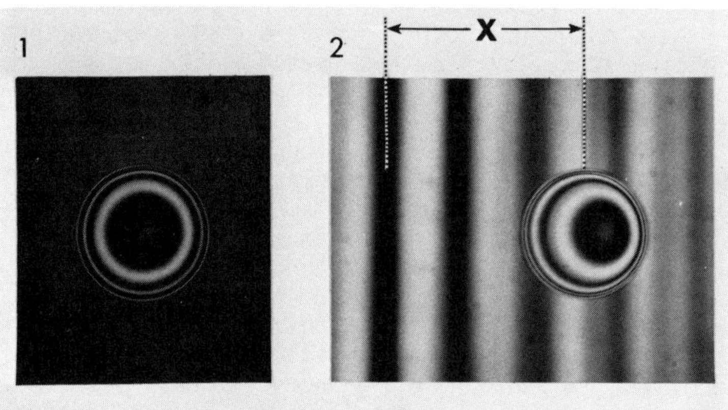

Fig. 1 (LEFT)   Interferometric fringe pattern of target quality
                   microballoon.
Fig. 2 (RIGHT)   Microballoon of Fig. 1 showing shift of reference
                   fringe due to optical pathlength through balloon.

     Using the results from a series of ray tracing calculations[51]
it is possible to estimate the minimum decentration which can be
detected when the balloon has been rotated and viewed in two
orthogonal directions. For a microballoon having an average wall
thickness, t, and an outer diameter, OD, we find $\Delta t/t \simeq 12.5/OD$
for this interferometer.

     To quantify and detect smaller scale length variations in the
shell wall, a phase-lock interference microscope[52-54] (PLIM) has
been developed. The PLIM measures optical phase in real time with
analog output voltages proportional to phase and position which can
be digitized to form an accurate picture of the shell. The instrument samples a $9\mu m^2$ area of the microballoon surface and is capable
of detecting variations in shell thickness as small as $.015\mu m$.
In order to utilize the information properly the balloon should be
continuously rotated during inspection so that a complete map of
the surface is obtained. Lawrence Livermore Laboratories has
developed a micromanipulator[55-56] to examine the entire balloon
surface during inspection. Apart from performing a complete
inspection, defects can be isolated by digitally processing images
taken in two orthogonal views[57].

     The tritium content of each target can be accurately analyzed
in a nondestructive fashion by exploiting the radioactivity of the
gas[58-60]. For example, energetic beta particles emitted from the
decay of tritium will excite x-ray emissions from the glass walls,
which can be quantitatively correlated with the activity (hence
tritium pressure). At LLE the beta particles which escape from the

target are counted directly[61-62] by placing the target in a flow
gas proportional counter. To provide an example of representative
counting rates associated with this method, note that a microballoon
with an inner diameter of 100μm and filled with an equimolar DT
mixture at a total pressure of 20 atm. contains 1.3×10^{-9} g of
tritium. The rate of beta particle production within this target
is 458,000/sec. If the target has a glass wall thickness of 1μm,
approximately 6% of these beta particles will escape. The resulting
count rate of 27,200 particles/sec is more than adequate to provide
good counting statistics with a 10 second sampling interval.

In order to relate the measured beta particle flux to the tritium
content it is necessary to know the wall thickness of the microballoon-
a parameter which is measured for each target before it is filled
with DT gas. Given the beta particle count rate and the wall thickness, the remaining step in completing the tritium assay requires
a knowledge of the transmission fraction for each target. The quantity
f is defined as the fraction of beta particles generated within the
target which are able to penetrate through the glass shell. Originally f was calculated as a function of wall thickness using known data
for the transmission of electrons through thin films. More recently,[63]
a Monte Carlo simulation code was employed to provide an improved
estimate of the transmission fraction, which also accounts for the
moderation and absorption of the beta particles by the fill gas in
the target. For targets of current interest to LLE (D $\leq$ 200μm,
t $\leq$ 2μm, total pressure p $\leq$ 50 atm), an accurate representation is
given by $f = \exp[-(2.717 \times 10^{-5} \, pD + 2.769 \, t)]$. Using this counting
technique it is possible to perform a rapid, nondestructive assay
of the tritium content of glass microballoon targets with an accuracy
of better than ±15%.

It should be noted that a measurement of the gaseous tritium
content suffices to infer the deuterium content of the target as
well. This follows from our observation[19] that glass microballoon
targets always contain the same deuterium:tritium ratio as the fill
gas mixture. The importance of this finding is extremely significant
because commercially available glass microballoons, in general, exhibit a wide range of pressure loss rates due to leakage. Figure 3 shows
the pressure retention characteristics of a group of 3M B-18A microballoons which were stored at room temperature. When stored at
liquid nitrogen temperature, the useful lifetime of the targets has
been extended to more than a year. It should be pointed out that
the pressure in the targets can also be determined interferometrically
because the optical path difference is usually measured prior to
filling. However, variations in target wall thickness limit the
accuracy of this measurement to ±7 atm, and the beta-counting
technique provides a superior measurement of the DT pressure in
exploding pusher targets.

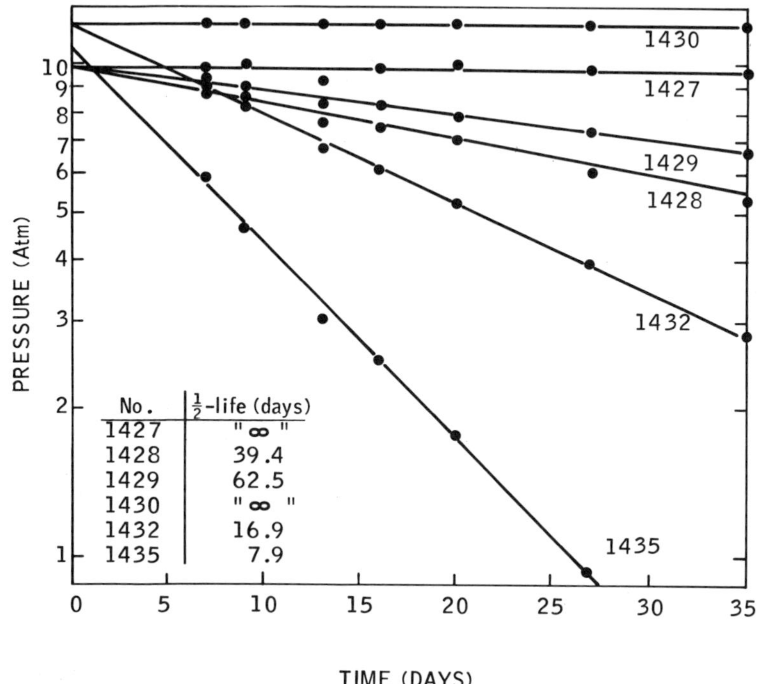

Fig. 3 Pressure (atm) versus time (days) for microballoons stored
        at room temperature.

Representative samples of a fill batch are sometimes chosen
for a gas analysis of the microballoon contents. Total gas
pressure can be determined to an accuracy of $\pm$15% by breaking
the balloon under glycerin[64]. Destructive measurement of the gas
composition and pressure has been accomplished using gas chromato-
graphy[65] and mass spectrometry[19]. A mass spectroscopic examination
involves recording the signal of interest immediately (<100 msec)
after the balloon is broken, and calibrating the response of the
quadrupole mass spectrometer. For analysis of balloons containing
D_2 and T_2, the D/T ratio can be determined to an accuracy of $\pm$5%.

ARGON FILLED MICROBALLOONS

Incorporating argon into the interior of microballoons[20-21]
serves a twofold purpose: 1) the high atomic number gas partially
mitigates the effect of preheat through radiational cooling, thereby
enabling higher compressions, 2) spectrally resolved x-ray emission
from argon serves as a direct diagnostic of the compressed core
density (ρ) and the quality of confinement (ρR).

Argon has been permeated[66-67] through a limited number of
glass compositions, and shells have been filled to pressures of
less than one atmosphere during the microballoon manufacturing
process[68]. These techniques have a limited range of applicability
and thus we have developed a new technique[24] for filling microballoons
with argon and other gases which do not permeate readily through
glass. Each microballoon is filled through a laser-drilled micron
sized hole which is sealed using a plastic or glass plug of comparable
dimension. The plug is melted over the hole to form a seal, and
the resulting perturbation to the surface finish is .3-1.5μm with
a width of 2-6μm. Key elements of the method used to drill and
fill microballoons are: 1) the microballoon is affixed to a substrate
during the entire process and removed only after filling is completed,
2) a micron or submicron hole is laser drilled through the top
surface of the microballoon, 3) a single glass or plastic plug is
transferred onto the hole 4) the plugged balloon is transferred into a
pressure vessel where the plug is melted under conditions required for
sealing and 5) after filling, the integrity of the seal is certified
by measuring interferometrically the gas pressure in the balloon.

By affixing the microballoon to a substrate, the hole can be
accurately located in the top of the balloon, facilitating transfer
of a comparably sized plug onto the hole. Salt crystals are used
to glue the balloon onto a substrate and they are dissolved away
after filling is completed. Figure 4 shows an electron micrograph
of a microballoon anchored with salt crystals onto the surface of a
gold coated glass cover slip. The reflective gold coating is
deposited on the substrate so that an interferometric measurement
of fill pressure can be obtained.

Holes are drilled using 40-80 psec laser pulses of wavelength
1.05μm which are switched out of a mode locked oscillator[69]. The
beam is routed through a microscope objective and brought to a focus
on the top surface of the microballoon. Due to the short pulse
length, high aspect ratio holes are created with diameters smaller
than the focused laser spot size. Figure 5 shows an electron micrograph of a hole drilled through a 3.5μm thick glass shell. Approximately 90% of the laser energy was concentrated within a disk of
diameter 3.2μm whereas the characteristic dimension of the hole
is ∼1μm. The small hole size is due to the fact that only the
central portion of the focal spot, where the beam is sufficiently
intense, is responsible for the drilling phenomena. Holes have
been drilled with aspect ratios as high as 3.5:1, although typically
the energy in the laser beam is adjusted to produce .7-2.0μm holes
with aspect ratios of 1:1.

Plug materials which have been used to seal argon inside microballoons include 2μm polystyrene spheres, 2-5μm saran particles
and irregularly shaped pieces of low melting point glass. Since
the plugs are chosen from materials which can be melted to form the

Fig. 4 Electron micrograph of microballoon adhered with salt
crystals to a gold coated glass cover slip.

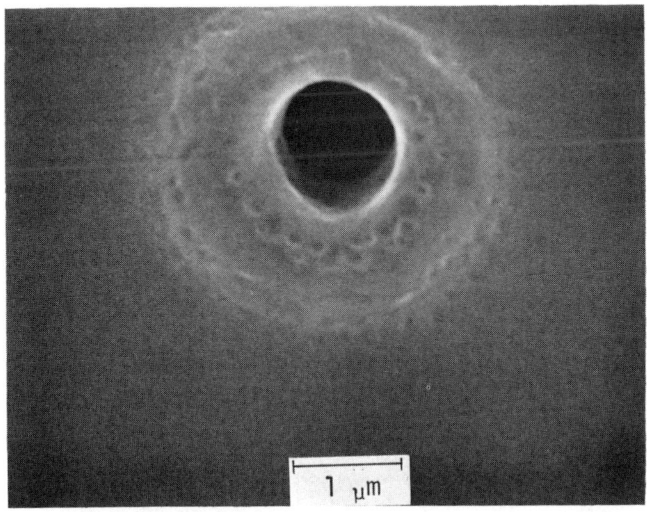

Fig. 5 Electron micrograph of a micron sized hole which was laser
drilled through a 3.5μm glass microballoon wall.

seal, they are transferred onto the hole under ambient conditions.
The small hole size aids in the transfer which is accomplished by
contacting the hole with a plug lying on the undersurface of a flat
glass slide. Alignment of hole and plug is done by micromanipulating

them to the focal point of an incident bright field microscope with
a magnification of ∿400. Although the plug cannot be dislodged from
the hole, a gas-tight seal is not effected until the microballoon is
placed in a heated cell which contains the desired fill gas (e.g.
argon). After the cell has been pressurized, the system is heated
in order to cause the plug to melt and form a seal over the hole.
For saran, polystyrene and glass plugs, temperatures of 180, 220
and 400°C, respectively, are adequate to produce effective sealing.
Figure 6 shows a 2μm polystyrene sphere which has been melted to
seal ∿20 atm. of argon inside a microballoon.

To determine the argon content, the optical path length is
measured before and after filling, with a reflected light interferometer. The change in optical path length is a direct measure of the
gas pressure inside the balloon. Since the microballoon orientation
is fixed during the entire process, the interferometric measurement
can be made quite accurately and argon pressures as small as 3 atm.
can be detected inside a 50μm diameter balloon. When stored at
room temperature, the pressure inside the balloon varies exponentially
with time and the half life is directly proportional to the interior
volume and inversely proportional to the permeability of the plug
material. It is found that the pressure retention half life of
argon sealed with a polystyrene plug in a 50μm diameter microballoon is 15-30 hr. while for glass or saran it is in excess of
3 months.

It should be pointed out that LASL has also developed a similar
filling technique. Instead of using a material which can be melted
to form the seal, epoxy plugs[70] have been used to trap Ar, H_2S and
Cl_2 within glass microballoons.

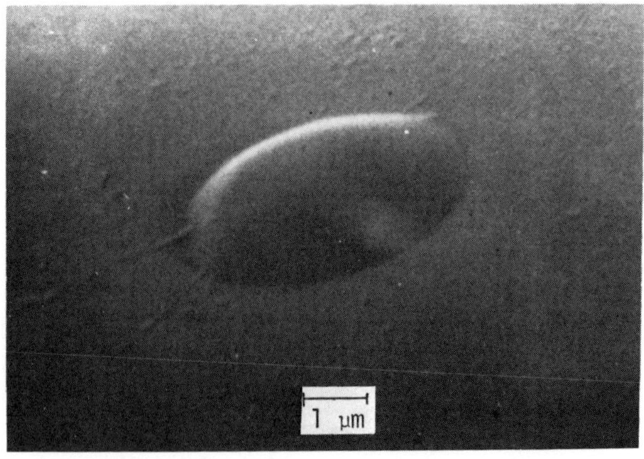

Fig. 6   Electron micrograph of a 2μm polystyrene plug which was
melted to trap 20 atm. of argon.

The drill, fill and plug technique provides a method for filling targets with a wide variety of gases which do not diffuse readily through the shell wall. At LLE it has proven extremely useful in current experiments which are diagnosed using x-ray spectroscopy. Future applications of the technique may involve the introduction of other gases, such as HBr, which will allow for the radiochemical diagnosis of target compressions.

PLASTIC COATINGS

The literature is rich with organic coating procedures[71-76]; however, most of these approaches were developed for applications which do not share the unique and stringent requirements encountered in laser fusion target fabrication. Existing coating techniques provide a basic resource for target fabrication, but must be altered and refined for the present application. In particular, the constraints on coating uniformity and surface finish define those aspects of coating technology in need of further development.

In the Laser Fusion Feasibility Project at the University of Rochester, three separate approaches have successfully provided plastic coatings on glass microballoon targets. Microwave discharge polymerization has produced hydrocarbon coatings up to $10\mu m$ thick with surface finishes of $\sim.3\mu m$. The vacuum pyrolysis of modified p-xylylene dimers has also yielded plastic coated targets with similar properties. Finally, a colloidal coating approach unique to this Laboratory has been employed to apply coatings in excess of $1\mu m$ composed of multiple layers of sub-micron latices of acrylate or styrene.

Microwave Discharge Polymerization

Glow discharge (plasma) polymerization is an established technique[77-78] for producing films from organic vapor monomers. Workers at Los Alamos Scientific Laboratories[79-80] have employed a parallel plate discharge system operating at ~ 1 kHz with p-xylene to obtain highly cross-linked hydrocarbon coatings up to $100\mu m$ thick. Because flaking is a potential contamination problem, the electrodes are generally replaced after each $15\mu m$ of deposition. Furthermore, it has been found that ultra-smooth surface finishes ($<0.1\mu m$) can be produced by applying periodic pulses of air[81] to the gas inlet, contributing $\sim 5.5\%$ oxygen to the as-deposited film. Unmounted balloons can be coated in this system by using mechanical agitation to bounce a batch of targets in a random fashion during the deposition run.

A different approach has been taken at Lawrence Livermore Laboratories. Here, a helical resonator cavity[82-83] operating at 10.5 MHz creates a plasma medium for polymerization which is decoupled from the targets to be coated. Batches of glass microballoons have been coated with fluorocarbon and hydrocarbon coatings with excellent surface finishes (better than .1μm). A voltage spike is applied periodically to the pan holding the targets in order to keep the microballoons freely rolling. Special care must be taken throughout to maintain surface cleanliness, since it has been clearly established that deleterious film defects will grow from localized areas of contamination. Recently this apparatus has been successfully coupled to a molecular beam levitation system, providing uniform coatings on unmounted, pre-characterized microballoons.

At LLE the approach to plasma polymerization[84] employs a flowing monomer tubular geometry in which the plasma is excited by an Evenson cavity (frequency = 2.456 GHz) external to the tube. An objective of this design has been to minimize the problem of contamination from electrode flaking by isolating the electrodes from the vacuum medium and using a flowing monomer; in this case, p-xylene or ethylene. A sketch of the apparatus is provided in Figure 7. The periodic addition of oxygen to the discharge has been suggested as a means of limiting polymer chain length and thus reducing nodular growth on the film surface. In this laboratory, a steady admixture of air has been found to produce similar benefits.

With plasma polymerization it is well known that an attempt to achieve a high deposition rate will result in gas phase polymerization and powder formation[85]. It is preferable, however, to operate in a regime where the monomer polymerizes as a film; since the small particles of powder will "rain" on the targets giving rise to unacceptably rough coatings. It has been shown[86] that film formation is promoted by operating in a regime of high flow rates and high monomer pressure. In the present apparatus it is also believed that the flowing gas provides the further benefits of protecting the target from powder accumulation and cooling the target surface, hence increasing the deposition rate. Excellent coatings up to 12μm thick, with surface finish better than .3μm have been produced at very high deposition rates (on the order of 40μm/hr). The system is simple to construct and operate, and especially amenable to molecular levitation for coating unmounted targets. Initial attempts to coat levitated microballoons have been moderately successful; however, the results are preliminary and further development of this approach is required.

POLYMERIZATION OF DI-PARAXYLYLENE (DPX) AND ITS DERIVATIVES

The vacuum pyrolysis of di-paraxylylene (DPX), a commercially available material[87-88], to form a coherent film of poly-paraxylylene (PPX) has been studied extensively[89-90]. The reaction chain for this polymerization mechanism is illustrated in Figure 8. The sequence consists of: i) sublimating the solid DPX at 60-80°C, ii) thermally cleaving the vapor molecules in a furnace heated to 650°C to form

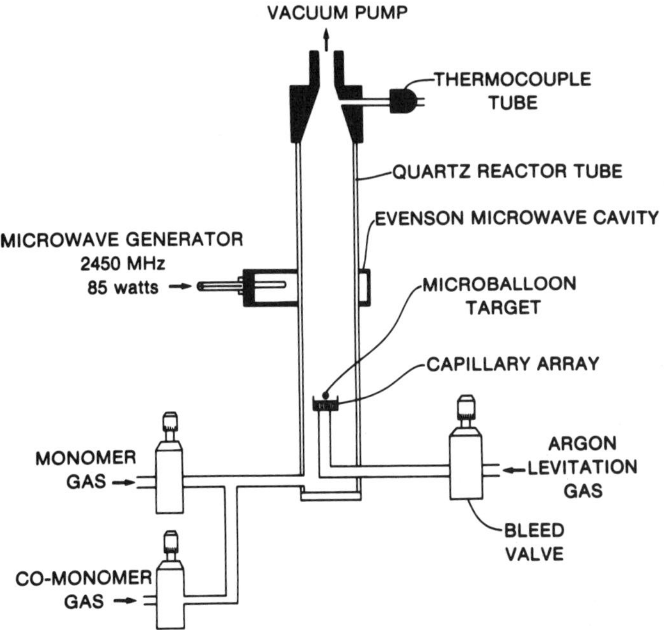

Fig. 7 Microwave plasma polymerization apparatus

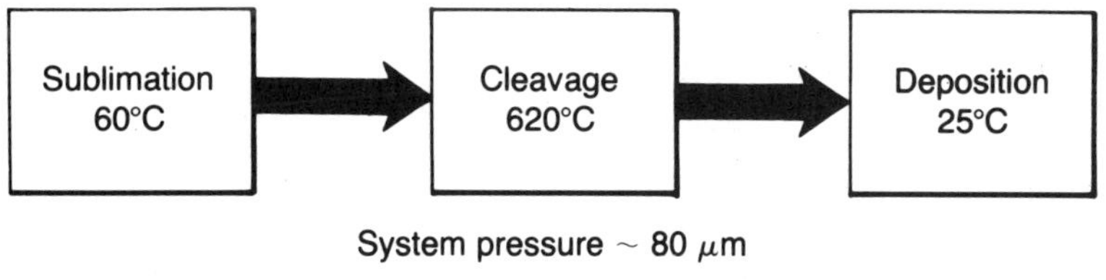

Dimer (DPX)   Fragment (PX)   Polymer (PPX)

Fig. 8 Reaction chain for polymerization of di-paraxylylene by vacuum
pyrolysis.

the diradical monomer paraxylylene (PX), and iii) allowing the PX
to condense and polymerize to PPX on the cooled substrate (i.e.
target) desired to be coated.

The spontaneous polymerization of PX offers certain advantages
over other coating techniques. Because the process operates at
a relatively high pressure (typically 80μm) the deposition is
omnidirectional and no special means are required to rotate the
targets to assure uniformity of coating. The morphology of the
polymer film is related to its degree of crystallinity--namely,
highly crystalline PPX material is susceptible to nodular growth
producing unacceptable coatings for laser fusion targets. This
problem can be mitigated, somewhat, by reducing the deposition rate
through a lowering of the vapor pressure of the dimer. To achieve
an acceptable coating surface, this rate must be kept below 1μm/hr;
hence, it becomes highly impractical to achieve coating thicknesses
in excess of a few microns.

Considerable work has been performed by this Laboratory to
modify the basic DPX method to yield a technique for producing
polymer coatings of improved surface quality at increased deposition
rates. The approach, which has proven highly successful, consists
of altering the chemical composition of the starting material to
produce a dimer with greatly improved deposition characteristics.
The basic idea is to add functional groups to the DPX molecule,
producing a substituted paraxylylene dimer (SPX) less prone to
crystallization. Two important benefits have been realized by
this approach. First, the new species (a derivative of DPX) is
limited in its ability to form crystallite domains due to steric
hinderances in the bulkier molecule. Reduction of crystallinity
correlates directly with a reduction in nodular growth on the
film's surface. Second, due to the increased molecular weight of
the SPX dimer, the kinetics of condensation and polymerization
are altered, giving rise to a dramatic increase in the rate of
deposition.

The choice of particular additive groups is constrained by the
desired chemical composition of the final film and the ability of
the substituted dimer to form a polymeric film at all. A series
of useful SPX compounds has been synthesized via standard Friedel
Crafts alkylation methods[91]. The reaction consists of replacing
a hydrogen atom on one of the phenylene groups in the DPX molecule
by an alkyl group, such as ethyl (CH_2CH_3), propyl ($CH_3(CH_2)_2$), or
butyl ($CH_3(CH_2)_3$). As discussed above, the net effect of using
an alkyl-substituted dimer is that the resultant film is produced
at an enhanced rate and is less crystalline in nature. Due to
this rate benefit, it is unnecessary to deliberately cool the
targets to provide an acceptable coating rate, as must be done
with ordinary DPX. A comparison of the room temperature coating rate
of DPX and three of its alkyl-substituted derivatives is shown in
Table 3.

TABLE 3

EFFECT OF PENDENT GROUPS ON THE DEPOSITION
RATE OF P-XYLYLENE AND ITS DERIVATIVES

Designation	Pendent Group	Deposition Rate (μm/hr)*
Normal DPX	H	1
Ethyl	CH_2CH_3	2.5
Propyl	$CH_3(CH_2)_2$	3.2
Butyl	$CH_3(CH_2)_3$	4.0

*The rate of polymer deposition was measured at room temperature.
These rates are only meaningful in terms of the specific geometry
of the coating apparatus and the location of the sample.

The improvement in surface smoothness is depicted in Figure 9
which shows a glass microballoon surface coated with poly(paraxylylene)
and poly (iso-propyl-paraxylylene). Note that the surface finish
of the latter coating is better than .25μm even though the deposition
was performed at more than three times the rate of the former system.

Target coatings employing DPX and mixtures with its commercially
available chlorine-substituted variant have been reported by workers
at LASL[92]. They have also begun molecular beam levitation experiments
using these materials and have had some success in coating unmounted
glass microballoons[93]. Similar experiments are also in progress at
LLE. Since molecular beam levitation is a delicate and difficult
operation, the advantages of providing a significantly enhanced
deposition rate are self-evident. Furthermore, it should be noted
that the concept of chemical modifications of the dimer can be
generalized to enhance the utility of this coating approach. For
example, it should be possible to produce films with significant
concentrations of fluorine or even metal atoms.

Colloidal Methods

A third approach to the problem of applying plastic coatings
to laser fusion targets involves the sequential deposition of
multiple layers of colloidal particles. Alternate layers are
oppositely charged and are held in place by electrostatic forces.
Because the colloidal particles which compose each layer are generally
much smaller than .5μm, many layers are required to achieve the
coating thicknesses which are of interest for target fabrication.
This coating technique is similar, in principle, to the work of
Langmuir and Blodgett[94-95] who produced films of fatty acids by
multiple dipping procedures. Whereas the Langmuir-Blodgett method
dealt exclusively with the transfer of single molecular layers from

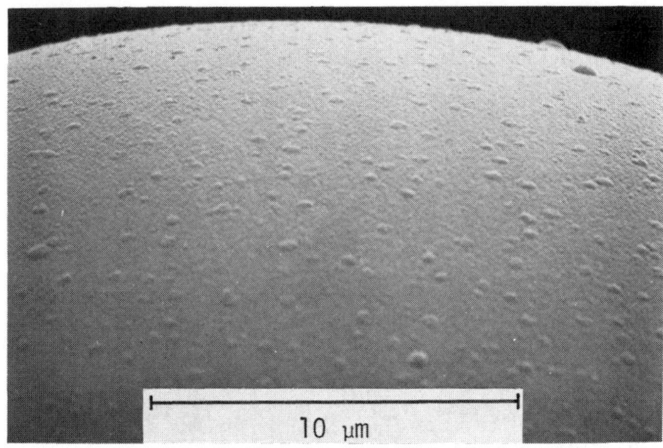

Fig. 9a   Surface of glass microballoon with a 5µm coating of
          poly(paraxylylene).

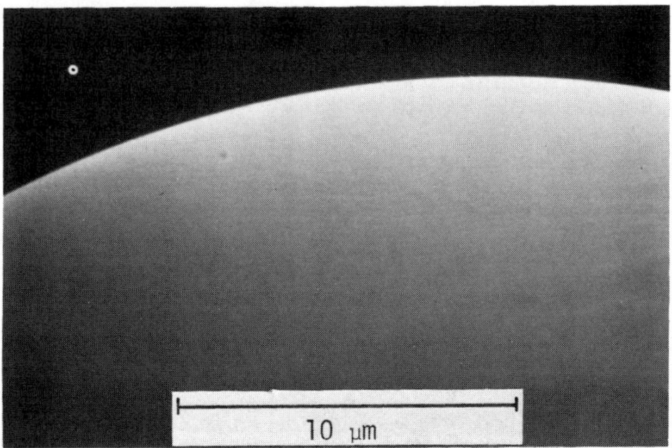

Fig. 9b   Surface of glass microballoon with a 5µm coating of
          poly (iso-propyl-paraxylylene).

liquid surfaces, the colloidal method involves the deposition of
discrete particles[96] directly from solution.

At LLE coatings a few microns thick have been produced[97] using
an automated, computer-controlled multiple dipping apparatus. The
technique has been successfully applied to the coating of mounted
and unmounted glass microballoons with organic and inorganic material
Principle advantages are the ability to control the coating thickness
and to choose from a wide variety of materials which are readily
available in colloidal form.

This method exploits the fact that most glass surfaces, including
silicate glass microballoons, acquire a negative charge when placed in
an aqueous solution with pH adjusted to 3.0. Other materials, such
as alumina, are known to acquire a positive charge when dispersed
in a similar solution. Thus a glass microballoon immersed in an aqueous
solution containing 50-100 Å sized alumina fibrils will become coated
with a monolayer of these particles and will exhibit, subsequently, a
net positive charge. The thinly-coated, charge-reversed microballoon
is then transferred to a second aqueous solution which contains the
negatively-charged colloidal particles which are to comprise the
target coating. These particles may be polymer latices of styrene
or acrylate, for example, and are in the form of uniformly sized
spheres in the diameter range 500 Å - 1μm. After the deposition
of a monolayer of these larger colloidal particles, the surface
reverts to its original state of negative charge.

The process can be repeated, in principle, without limit.
Each cycle causes the additional deposition of a single layer of
polymer particles. Since the alumina fibril thickness is negligible
compared with the polymer latex diameter, the incremental coating
thickness per cycle is controlled entirely by the polymer sphere
size. Furthermore, it is clear that the composition of the coating
is essentially that of the polymer, since the alumina contributes a
negligible mass and serves only as an "electrostatic glue" to bond
the layers of polymer particles.

Inorganic coatings are also amenable to this method; in fact,
colloidal coatings have been produced by depositing alternate layers
of silica and alumina particles onto glass microballoons. These
particles are comparably sized (typically 50-150 Å) and thus ∿100
layers must be deposited for each micron of coating thickness. To
do this an automated programmed dipping apparatus has been constructed to execute a pre-determined number of coating cycles- including
rinsing and drying steps. Unmounted microballoons have been
successfully transferred through up to 500 cycles, producing inorganic coatings up to 5μm thick. Solution cleanliness, immersion rates
and drying procedures are critical factors in the production of
target-quality coatings.

For inorganic coatings the surface smoothness is controlled
by adjusting the size of the colloidal particles. For some organic
coatings a smoothing effect accompanies the use of polymers such as
acrylates, whose glass transition temperature lies below room
temperature. In these cases the final layer dries to produce a
smooth film in which the discrete nature of its constituent particles
is virtually undetectable. The colloidal coating method, due to
its versatility and control is a potentially valuable technique for
laser fusion target fabrication.

ADVANCED FABRICATION AND CHARACTERIZATION METHODS

The preceding discussion has dealt with the production of
targets which have been utilized in the six beam ZETA campaign.
Forthcoming experiments with the twenty-four beam laser system,
OMEGA, will require new types of targets which, in general, will be
more complex and difficult to build. Advanced fabrication and
characterization techniques for OMEGA targets are currently being
developed, and highlights of some of these procedures will be
presented in the remainder of this article.

Microfabrication of Hemispheres for Structured Targets

Experiments using targets incorporating the velocity multiplier
concept[98] become feasible with the availability of large laser systems
such as OMEGA (power output = 12-15 TW). Prototypical target designs
of this type incorporate a multi-shell configuration in which an inner,
possibly metal coated DT-filled glass microballoon is positioned
concentrically within an outer spherical shell. The space between
the spheres may be void, or filled with a gas or a low density foam.
The outer spheres may be glass, plastic or metal.

A proposed sequence for building such a target is summarized
as follows. A hemisphere with the parameters of the outer shell is
positioned concave side upward, and a thin (100-200 A) formvar film
is stretched across the exposed rim. The inner sphere is a glass
microballoon which has been selected, filled and characterized by
previously described methods, and is carefully positioned at the
center of the film. A second formvar film is stretched over the
balloon, thus sandwiching it in place at the center of the hemisphere's
base plane. Finally, a second, identical hemisphere is mated with
the first; resulting in a concentric, two-shell configuration. Issues
such as the precise centering of the inner shell, and the sealing of
the hemispheres have not, as yet, been seriously addressed. Our
belief is that these issues are difficult, but resolvable in the light
of past and recent accomplishments in the field of target fabrication.
Thus, the main technological challenge in this approach is the development of a method for the production of identical hemispheres with
suitable compositions and dimensions.

Other laser fusion laboratories have recognized the utility of
hemispheres for the fabrication of structured targets. Diverse
methods have been proposed and implemented for their production
including single point diamond turning,[99] micromachining of spheres[100]
and the application of semiconductor device technology[101,25]. Of
these approaches, the third is by far the least labor-intensive and
is the only one which offers the potential for "mass production"
capabilities.

At LLE we are developing a microfabrication molding technique
which will be capable of mass-producing identical hemispherical
shells of a wide variety of materials. The initial step involves
the careful selection of a spherical article (e.g. a glass microballoon whose sphericity and diameter define the properties of the final
hemispherical form. Using steps more fully described below, a
hemispherical section is formed from which a master mold is cast.
From the master, many disposable identical molds are replicated.
The replicated molds are then coated; for example by sputtering,
with the desired target material. Finally, the coated molds give
rise to free standing hemispherical shells when the mold is either
dissolved or etched away.

Before describing these steps in greater detail, it is important
to note that the key feature which distinguishes this approach from
others relates to the method of hemispherical cavity formation. For
example, the KMS approach involves the photolithographic production
of circular openings in a metallic mask which has been deposited on
a silicon wafer. Hemispherical cavities in the silicon are produced
by performing an isotropic acid etch through these openings.

A problematic aspect of this approach is the use of acid to
effect a well-controlled, sufficiently isotropic etch in order to
produce a series of identical cavities of the required sphericity.
This difficulty is circumvented in the LLE method in which the
characteristics of an already formed object of known diameter and
sphericity are propagated throughout the entire sequence. Since
this master object can, in principle, yield an unlimited number of
hemispherical shells, a great deal of labor can be invested in its
selection.

Glass microballoons were selected as models to create the
hemispherical section because they are available with a high degree
of sphericity over a broad diameter range. Several techniques have
been employed successfully for producing the hemispherical section
from the chosen microballoon. To date, the best results have involved
the imbedding of the balloon in epoxy (degassed to remove all bubbles)
and subsequently oxygen plasma etching[102] the epoxy until exactly
one half of the microballoon becomes exposed above the remaining
epoxy surface. By operating the oxygen plasma discharge in the
appropriate regime, the etching can be made totally vertical. The
benefit of a vertical etch is that the microballoon acts as its own
mask and some overetching can be tolerated due to the almost total
absence of undercutting with this technique.

Following the plasma etching of the epoxy, the glass microballoon is totally dissolved away with hydrofluoric acid, leaving
a hemispherical depression in the epoxy. Using RTV silicon rubber
molding compound, a (positive) replica of the cavity is cast. This

RTV hemispherical dome is the master mold for all future operations.
For example, a series of secondary disposable molds can be cast from
the RTV master by spinning a layer of photoresist over the RTV.
These secondary molds are easily released from the master to provide
any desired number of hemispherical cavities used in the final
fabrication steps. To complete the sequence, the photoresist cavities
are coated with a thin film of suitable thickness and material
commensurate with the specifications of the outer target shell. The
step of removing excess film material from around the cavity can be
accomplished by mechanical lapping or etching. Finally the photoresist mold is dissolved away in acetone (for example), leaving a free-
standing hemispherical shell. An outline of this fabrication sequence
is given in Figure 10.

Results thus far have been quite encouraging and no major
obstacles are forseen in the application of this technique to the
mass production of hemispherical shells for multishell laser fusion
targets. Figure 11 shows a copper hemisphere $1\mu m$ thick and $210\mu m$
in diameter which has been produced by sputtering onto a final mold.
In addition, plastic hemi-shells of di-chloroparaxylylene have also
been fabricated using the thermal polymerization unit described
in the previous section on plastic coatings.

The positioning of a glass microballoon, sandwiched between
two formvar films, at the center of the baseplane of a single hemisphere has been accomplished. The complete assembly, involving the
mating and sealing of a second hemisphere to the above configuration
will be attempted in the near future.

PHYSICAL VAPOR DEPOSITION AND LEVITATION

The fabrication of multishell and layered targets requires
facilities for the deposition of smooth, uniform metal and other
inorganic coatings on spherical target substrates. These coatings
have been proposed as pushers and tampers in ablatively-driven
targets[/], and as components of multishelled targets based upon
the concept of radiation cooling[103]. Physical vapor deposition
techniques, especially evaporation and sputtering, are well suited
to these applications. Sputtering is a particularly valuable
technique because of its versatility, and its ability to provide
uniform films with high quality surface finishes.

Two separate approaches to physical vapor deposition have
been implemented at LLE; incorporating evaporation and sputtering.
Of the two methods, evaporation is less useful because: 1) it is
very difficult to evaporate downward onto a levitated target,
2) evaporation is a highly directional coating method and not well
suited to spherical substrates, 3) evaporation generally gives
rise to higher substrate temperatures, resulting in poorer surface

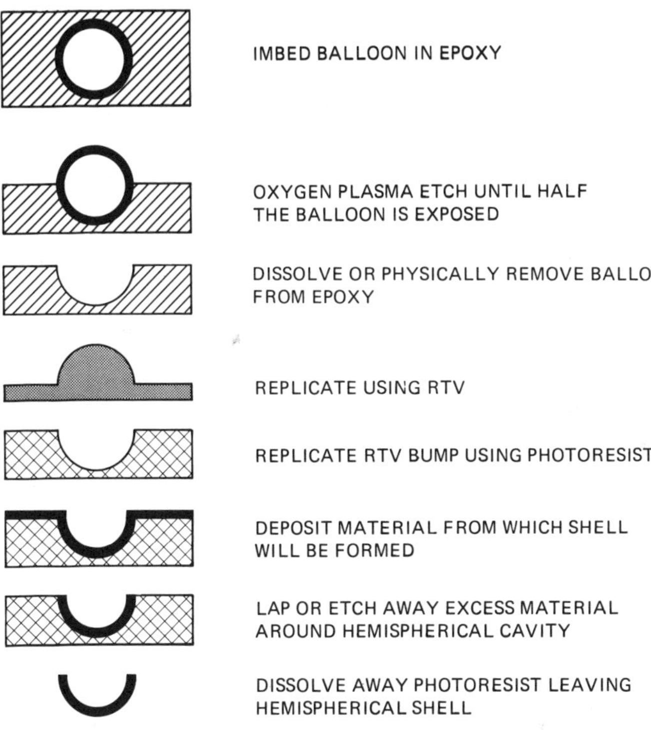

Fig. 10 Fabrication sequence for hemispherical shell production.

Fig. 11 Electron micrograph of freestanding copper hemishell
         approximately 1 μm thick.

finishes, and 4) the variety of materials which can be evaporated readily is somewhat limited.

In spite of the above limitations, evaporation can be a useful technique for certain target fabrication applications. For example, a resistively heated evaporation unit was used in the preparation of thin (1000-2000 Å) aluminum coatings on glass microballoon targets[104]. Sufficient uniformity was attained by using an extended source and rotating the stalk mounted targets in a motorized fixture inside the vacuum chamber. Because of the thinness of these coatings, no significant additional perturbation resulted from performing the coating on supported balloons.

In a separate evaporation system heated by an electron beam source, planar targets[105] of aluminum have been prepared by deposition through a mask. Used in single beam experiments designed to elucidate electron transport phenomena, these targets are in the form of free standing circular discs 100μm in diameter, with thicknesses ranging from 1 to 10μm.

When high quality coatings in excess of 1μm are required on spherical substrates, sputtering is a preferred technique. These requirements are encountered in the fabrication of hemispheres according to the methods of the previous section, as well as in the coating of spherical target substrates such as the surfaces of glass microballoons. In order to realize the full potential of sputtering, a versatile system has been designed and built[26] in which three separate sources are available for coating applications. Also incorporated, is a gas levitation system whereby an individual target can be supported and rotated to produce a uniform coating.

Gas (or molecular beam) levitation is accomplished by supporting the target on a gentle stream of gas, such as argon. Other laboratories have reported the use of this technique in the production of metal[106-107] and plastic[83] coatings on glass microballoons. The approach at LLE utilizes a focused glass collimated hole structure in which the 50μm diameter channels are angled to form a focus at a point 25 mm above the surface. The top surface of the structure is concave with a radius of 25 mm and a beveled ring is used to constrict the flow. A major problem in the art of gas levitation of small particles relates to the difficulty of obtaining a stable configuration at the low pressures encountered in the coating chamber. At higher pressures ($\lesssim$1 atmosphere) the damping effect of the background gas helps considerably in achieving and maintaining a stabilized levitation. Using cleaned microballoons, the careful design of the hole structure and argon flow system has enabled us to perform levitation experiments routinely at pressures of 1μm. Such pressures are required for the operation of the sputtering sources in a regime in which acceptably smooth finishes are produced on the target surface.

Three sputtering sources are incorporated in the vacuum system: a Sloan 3 inch sputter gun, a Sputtered Films sputter gun and an Ion Tech ion gun deposition source. Each source provides certain unique advantages, depending upon the particular application. For example, the Sputtered Films gun, modified to our specifications, produces a relatively high deposition rate with the use of a small sputtering target. The ion gun source involves an active plasma region which is well shielded from the target to be coated, thus minimizing the substrate temperature. Low substrate heating by the source is essential when ultrahigh quality surface finishes are required.

Using the Sputtered Films d.c. source we have produced copper coatings thicker than 1μm on levitated glass microballoons with a local surface finish of ∿250 Å. The coating rate for these experiments was in excess of 200 Å/minute. Further optimization of the surface finish is being carried out for copper and other metals. In addition, more experiments are in progress using r.f. power supplies and the ion beam source.

MICRORADIOGRAPHY

High resolution x-ray microradiographic analysis of laser fusion targets is a useful technique for detecting and quantifying defects[108-110]. A contact radiograph is produced by recording with a magnification of unity, the shadow cast by the target when it is illuminated with an x-ray beam. A 2-D computer analysis of the recorded image is required to locate and quantify defects in the target. Thus a microradiographic analysis system is comprised of an x-ray source which exposes the film, a readout technique to interrogate the recorded image and a 2-D computer image analysis program to interpret the data.

The x-ray source must be intense enough to provide reasonable exposure times and must satisfy several geometric and spectral requirements[111]. The contrast in the recorded image is determined by the wavelength and spectral purity of the x-ray beam. For targets of current interest, optimum contrast is obtained using soft (.3-5 kev) monochromatic x-rays. The finite x-ray source size produces a smearing of the recorded image due to the penumbral shadow cast by the target. The width of penumbra can be reduced by increasing the source to film distance with a concomitant loss in x-ray intensity at the film plane. In the interest of achieving the shortest possible exposure times, there is no point to decreasing the width of the penumbral shadow much below the resolution limit of the film. Thus in practical radiographic applications, resolution and intensity considerations define an optimized set of experimental parameters.

At LLE, exposures are made using a point, soft x-ray source generated with the tightly focused electron beam of a scanning electron microscope. Figure 12 shows the schematic arrangement of the x-ray radiography system as configured to produce contact radiographs of plastic coated glass microballoons. The distance between the x-ray source and film plane was chosen so that the width of the penumbral shadow equals the resolution limit of the film ($\sim$.1-.2μm). The x-ray source is formed using a 3 μA electron beam accelerated to 5 kv and focused to a 2μm spot on the surface of an AL anode. Heat dissipation from the 2μm spot permits the use of current densities as high as 100 A/cm^2 with no detectable anode damage or contamination. This unusually large current loading on the anode yields a high emittance x-ray source ($\sim$100 W/cm^2) which in turn permits short exposure times. Under such conditions Kodak type 120-02 plates can be exposed to optical density 3 in a period of 5 min. More than 75% of the x-ray photons reaching the film correspond to Al(Kα) emission. With such a high degree of monochromaticity, adequate contrast is achieved for analyzing hydrocarbon or plastic coatings on glass microballoons. A radiograph of a plastic coated microballoon is shown in Figure 13.

Electron microscopes have been used previously to provide sources for x-ray projection microscopy[112] and low resolution contact radiography[113]; however, to our knowledge this is the first application of an SEM soft x-ray source in a system capable of submicron resolution. The SEM x-ray source can be used with a variety of anode materials (including insulators) and the beam voltage can be varied continuously to optimize excitation of characteristic radiation. Due to the short source to film distance (1-2cm) the lateral magnification is anisotropic and can change by as much as 5% across the field of view. This distortion in the recorded image is readily correctable.

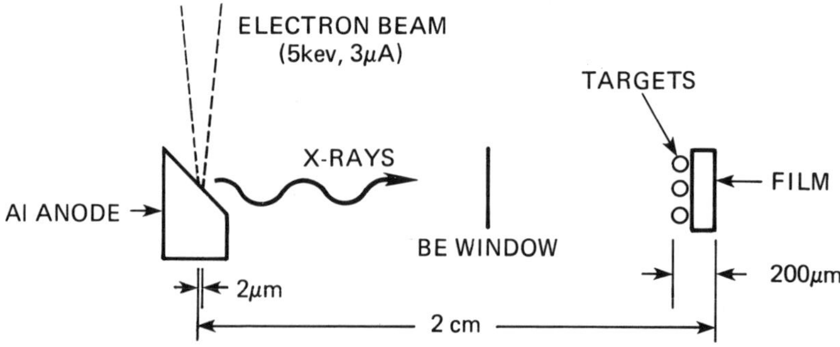

Fig. 12   Schematic diagram of x-ray radiography system. The electron beam of a scanning electron microscope is focused to a 2μm spot to produce a soft monochromatic x-ray source.

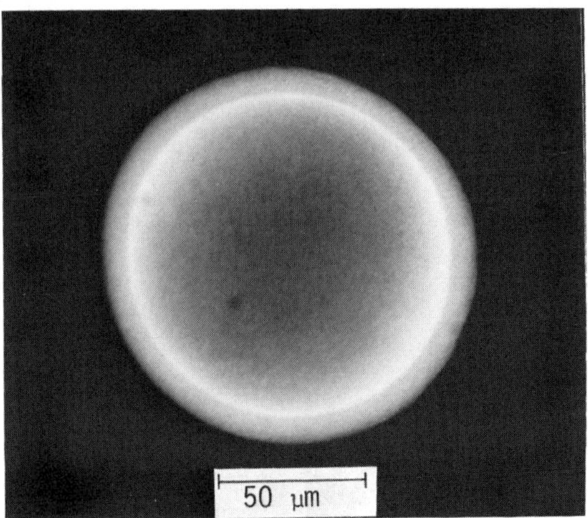

Fig. 13   X-ray radiograph of microballoon coated with 10µm of
          poly(chloro-paraxylylene).

     Usually a developed radiograph is examined optically[108-111] at
high magnification to extract measurements of the target parameters.
In this analysis the optical density of the film is directly related
to the x-ray attenuation through the target. At LLE, the optical
density of the developed film is measured in the SEM by scanning the
electron beam over the image and detecting electrons which are
backscattered from silver in the emulsion layer. For silver-halide
films exposed with soft x-rays, a significant fraction of the
developed grains are located in the top 1/2-1µm of the emulsion.
These grains, which embody the information content of the radiograph
can be probed with electrons backscattered from a 20-100 kev electron
beam. The number of electrons backscattered from the exposed silver
grains is 2-10 times greater than that from the surrounding gelatin.
Thus, the backscattered electron current is a measure of the areal
density of silver-bearing grains which can be directly related to
the optical density of the film. Figures 14 and 15 show an optical
and backscattered electron image of the same radiograph. It is seen
immediately that the backscattered and optical images are reversals
of one another. In the optical image, the regions with lowest silver
density scatter the least light and thus appear brightest. In the
backscattered electron image, the areas with lowest silver concentration backscatter the fewest electrons and thus appear darkest. An
advantage of the backscattered electron imaging technique is that it
is free from diffraction effects which limit the resolution capabilities of optical systems. Figure 16 shows a portion of the edge of a
microballoon image and demonstrates that ∿1/4µm spatial resolution is
achievable in this radiography system.

     To determine defects in targets, the backscattered electron
readout technique is being combined with a 2-D computer image analysis

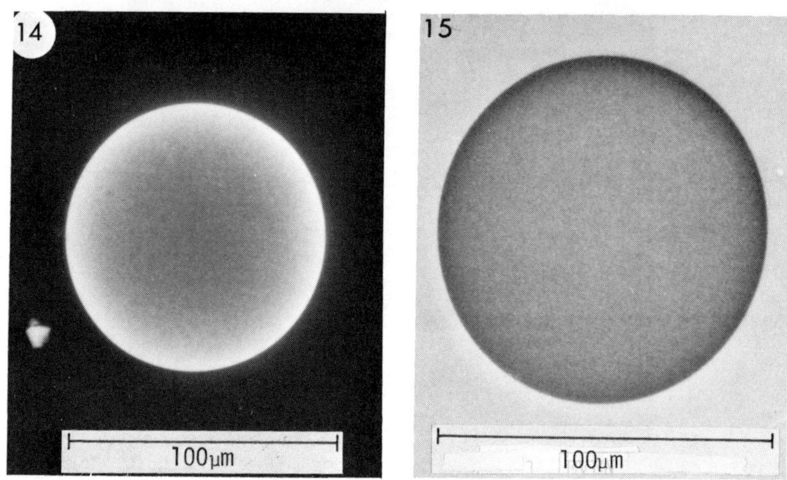

Fig. 14 (LEFT)   Optical image of radiograph of glass microballoon.
Fig. 15 (RIGHT)   Backscattered electron image of radiograph shown
                  in Fig. 14.

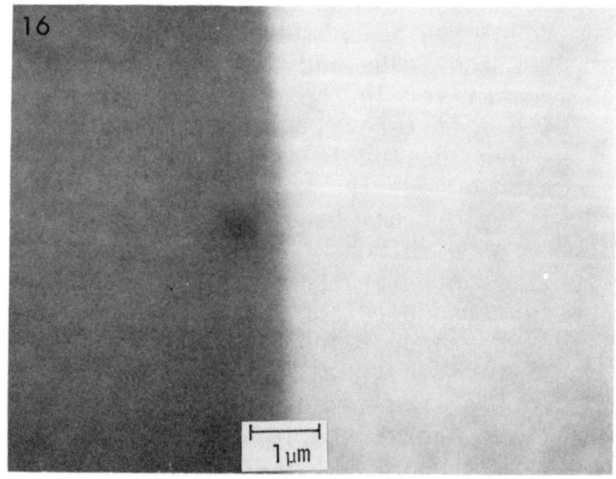

Fig. 16   Highly magnified backscattered electron micrograph of the
          edge of microballoon image shown in Fig. 15. Change in
          optical density at edge of microballoon demonstrates $\sim.25\mu m$
          spatial resolution in radiography system.

program. Ultimately, the system should be able to measure decentrations
in coated and uncoated targets as well as detecting localized defects
within the target which are not accessible by optical means.

ACKNOWLEDGEMENT

       This work was funded in part by the Laser Fusion Feasibility
Project, Laboratory for Laser Energetics, University of Rochester
and was partially supported by the following sponsors: Exxon
Research and Engineering Company, General Electric Company,
Northeast Utilities, Empire State Electric Energy Research Corpora-
tion, New York State Energy Research and Development Administration.

REFERENCES

1. B. Yaakobi, D. Steel, E. Thorsos, A. Hauer, and B. Perry,
     Phys. Rev. Lett. 39:1526 (1977).
2. B. Yaakobi, D. Steel, E. Thorsos, A. Hauer, B. Perry, S. Skupsky,
     J. Geiger, C. M. Lee, S. Letzring, J. Rizzo, T. Mukaiyama,
     E. Lazarus, G. M. Halpern, H. W. Deckman, J. Delettrez,
     J. Soures, and R. McCrory, Phys. Rev. A19:1247 (1979).
3. G. M. Halpern and H. W. Deckman, AIChE Symposium Series No. 191,
     75:201 (1979).
4. J. Hoose, SPIE 103:22 (1977).
5. B. Yaakobi, R. McCrory, S. Skupsky, H.W. Deckman, C.F. Hooper, Jr.,
     P. Bourke, T.C. Bristow, W. Seka, J. Delettrez, and J. Soures,
     "Digest of Topical Meeting on Inertial Confinement Fusion",
     Feb. 26-28, 1980, San Diego, CA.
6. J. Eastman and J. Soures, "Digest of Topical Meeting on Inertial
     Confinement Fusion", Feb. 7-9, 1978, San Diego, CA.
7. J. Nuckolls, L. Wood, A. Thiessen, and G. Zimmerman, Nature
     239:139 (1972).
8. J. R. Freeman, M. J. Clauser, and S. L. Thompson, Nuclear Fusion
     17:223 (1977).
9. E. Thorsos, T. C. Bristow, J. Delettrez, J. Soures, and J. Rizzo,
     Appl. Phys. Lett. 35:15 (1979).
10. E. Goldman, J. Delettrez, and E. Thorsos, Nuclear Fusion 19:555
     (1979).
11. Y. Gazit, J. Delettrez, T. C. Bristow, A. Entenberg, and J. Soures,
     Phys. Rev. Lett. 43:1943 (1979).
12. "Lawrence Livermore Laboratory Laser Program Annual Report":
     UCRL-50021-77.
13. "Progress Report of Los Alamos Scientific Laboratory Laser
     Fusion Program": LA-6982-PR.
14. "KMS Fusion Inc. Annual Report on Laser Fusion Research" (1977).
15. "Laboratory for Laser Energetics Annual Report"(1977).
16. R. R. Stone, D. W. Gregg, and P. C. Souers, J. Appl. Phys.
     46:682 (1975).
17. G. M. Halpern, J. Varon, D. C. Leiner, and D. T. Moore, J. Appl.
     Phys. 48:1223 (1978).
18. C. Ward, C. Hendricks, and B. Weinstein, in "Scanning Electron
     Microscopy/1978", Chicago Press, Chicago (1978).

19. H. W. Deckman and G. M. Halpern, "Digest of Topical Meeting on Inertial Confinement Fusion", Feb. 7-9, 1978, San Diego, CA.

20. B. Yaakobi, S. Skupsky, R. McCrory, C. Hooper, Jr., H. W. Deckman, P. Bourke, and J. Soures, (to be published).

21. J. Soures, T.C. Bristow, H.W. Deckman, J. Delettrez, A. Entenberg W. Friedman, J. Forsyth, Y. Gazit, G. M. Halpern, F. Kalk, S. Letzring, R. McCrory, D. G. Peiffer, J. Rizzo, W. Seka, S. Skupsky, E. Thorsos, B. Yaakobi, T. Yamanaka, Proceedings of the Fifth Workshop on Laser Interaction with Matter, Rochester, New York, (1980).

22. E.H. Farnum and J. Feuerherd, Bull. Am. Phys. Soc. 24:1071 (1979).

23. H. W. Deckman, J. Drumheller, and J. Rizzo, J. Opt. Soc. Am. 69:1468 (1979).

24. H. W. Deckman, J. Dunsmuir, G. M. Halpern, and J. Drumheller, "Digest of Topical Meeting on Inertial Confinement Fusion", Feb. 26-28, 1980, San Diego, CA.

25. I. S. Goldstein, J. Trovato, and F. Kalk, ibid.

26. I. S. Goldstein and J. Varon, ibid.

27. H. W. Deckman and J. H. Dunsmuir, ibid.

28. I. Lewkowicz, J. Phys. D: Appl. Phys. 7:L61 (1974).

29. D. E. Solomon and T. M. Henderson, J. Phys. D: Appl. Phys. 8:L85 (1975).

30. P. C. Souers, R. T. Tsugawa, and R. R. Stone, Rev. Sci. Instr. 46:682 (1975).

31. W. R. Beck and D. L. O'Brien, U. S. Patent 3,365,315 (1968).

32. R. T. Tsugawa, I. Moen, P. E. Roberts, and P. C. Souers, J. Appl. Phys. 47:1987 (1976).

33. F. Veatch, U.S. Patent 2,797,201 (1957).

34. C. Henderson, U.S. Patent 3,699,050 (1972).

35. R. L. Nolen, Jr. and D. E. Solomon, J. Non-Cryst. Solids 19:337 (1975).

36. A. Rosencwaig, J. L. Dressler, J. C. Koo, and C. Hendricks, Lawrence Livermore Lab. Report: UCRL-81421 (1978).

37. C. Hendricks, A. Rosencwaig, R. L. Woerner, J. C. Koo, J. L. Dressler, J. W. Sherohman, S. L. Weinland, and M. Jeffries, Lawrence Livermore Lab. Report: UCRL-82530 (1979).

38. J. C. Koo, Lawrence Livermore Lab. Report: UCRL-83397 (1979).

39. R. G. Budrick, F. T. King, R. L. Nolen, Jr., and D. E. Solomon, U. S. Patent 4,021,253 (1977).

40. R. L. Nolen, Jr., R. L. Downs, W. J. Miller, M. A. Ebner, N. E. Doletzky, and D. E. Solomon, "Digest of Topical Meeting on Inertial Confinement Fusion", Feb. 7-9, 1978, San Diego, CA.

41. R. L. Woerner, V. F. Draper, J. C. Koo, and C. Hendricks, "Digest of Topical Meeting on Inertial Confinement Fusion", Feb. 26-28, 1980, San Diego, CA.

42. A. Rosencwaig, S. L. Weinland, J. C. Koo, and J. L. Dressler, Lawrence Livermore Lab. Report: UCRL-81436 (1978).

43. R. T. Tsugawa, I. Moen, and P. E. Roberts, J. Appl. Phys. 47:1987 (1976).

44. P. C. Souers, I. Moen, and R. T. Tsugawa, J. Phys. D 10:1079 (1977).
45. P. C. Souers, I. Moen, R. O. Lindahl, and R. T. Tsugawa, J. Am. Ceram. Soc. 61:42 (1978).
46. J. E. Shelby, J. Appl. Phys. 45:2146 (1974).
47. J. E. Shelby, Sandia Labs Rep. SLL-73-0259 (1973).
48. K. L. Mittal (Ed.), "Surface Contamination: Genesis, Detection and Control", Vols. 1 & 2, Plenum, New York (1979).
49. D. T. Hawkins, J. Vac. Sci. Tech. 16:1051 (1979).
50. B. Weinstein, J. Appl. Phys. 46:5305 (1975).
51. T. Powers (private communication).
52. D. C. Leiner and D. T. Moore, Rev. Sci. Instr. 49:1702 (1978).
53. G. W. Johnson, D. C. Leiner, and D. T. Moore, Opt. Eng. 18:46 (1979).
54. P. O. McLaughlin and D. T. Moore, J. Opt. Soc. Am. 67:A1386 (1977).
55. B. Weinstein, C. Hendricks, C. Ward, and D. L. Willenborg, Rev. Sci. Instr. 49:870 (1978).
56. J. A. Monjes, B. Weinstein, D. L. Willenborg, and A. L. Richmond, "Digest of Topical Meeting on Inertial Confinement Fusion", Feb. 26-28, 1980, San Diego, CA.
57. T. Powers and G. M. Halpern, ibid.
58. R. J. Fries and E. H. Farnum, Nuc. Instr. and Methods 126:285 (1975).
59. E. H. Farnum and R. J. Fries, U. S. Patent 3,940,617 (1976).
60. I. Moen, C. Hendricks, and B. Weinstein, Bull. Am. Phys. Soc., 21:1137 (1976).
61. H. W. Deckman and G. M. Halpern, J. Appl. Phys. 50:132 (1979).
62. H. W. Deckman and G. M. Halpern, "Digest of Topical Meeting on Inertial Confinement Fusion", Feb. 7-9, 1978, San Diego, CA.
63. G. M. Halpern and R. Sampath, "Digest of Topical Meeting on Inertial Confinement Fusion," Feb. 26-28, 1980, San Diego, CA.
64. W. B. Rensel, T. M. Henderson, and D. E. Solomon, Rev. Sci. Instr. 46:787 (1975).
65. R. G. Schneggenburger, W. S. Updegrove and R. L. Nolen, Jr., Rev. Sci. Instr. 49:1543 (1978).
66. P. T. Rumsby (private communication).
67. R. L. Nolen, Jr. (private communication).
68. J. C. Koo, J. L. Dressler, and C. Hendricks, Lawrence Livermore Lab. Report: UCRL-51788 (1979).
69. W. Seka and J. Bunkenburg, J. Appl. Phys. 49:2277 (1978).
70. S. Butler and B. Cranfill, "Digest of Topical Meeting on Inertial Confinement Fusion", Feb. 26-28, 1980, San Diego, CA.
71. P. H. Carnell, J. Appl. Polym. Sci. 9:1863 (1965).
72. M. Xanthos and R.T. Woodhams, J. Appl. Polym. Sci. 16:381 (1972).
73. G. Mengoli, S. Daolio, U. Giulio, and C. Folonari, J. Appl. Polym. Sci. 23:2117 (1979).
74. G. Mengoli and B. M. Tidswell, Polymer, 16:881 (1975).
75. H. L. Gerhart and R. E. Smith, ACS Organic Coating and Plastic Chem. 37:405 (1977).

76. E. B. Mano and L. A. Durao, J. Chem. Ed. 50:228 (1973).
77. D. Peric, A.T. Bell, and M. Shen, J. App. Poly. Sci.21:2661(1977).
78. H. Yasuda, in "Plasma Chemistry of Polymers", (M. Shen ed.),
       Marcel Dekker, New York (1976).
79. A. T. Lowe and R. J. Fries, Surface Science 76:242 (1978).
80. A. T. Lowe and R. J. Fries, "Digest of Topical Meeting on
       Inertial Confinement Fusion", Feb. 7-9, 1978, San Diego, CA.
81. R. Liepins, M. Campbell and R. J. Fries, in "Progress in
       Polymer Science Vol. 7", Pergamon Press, New York, (1980).
82. W. L. Johnson, C. W. Hatcher, C. Hendricks, S. A. Letts,
       L.E. Lorensen, Lawrence Livermore Lab.Report:UCRL-80162(1979).
83. S. A. Letts and D. W. Myers, "Digest of Topical Meeting on Iner-
       tial Confinement Fusion," Feb. 26-28, 1980, San Diego, CA.
84. D. G. Peiffer, T. J. Corley, G. M. Halpern, and B. Brinker,
       (to be published).
85. H. Kobayashi, A. T. Bell, and M. Shen, J. Appl. Polym. Sci.,
       17:885 (1973).
86. J. R. Hollahan and A. T. Bell, "Techniques and Applications of
       Plasma Chemistry", J. Wiley, New York (1974).
87. W. F. Gorham, U. S. Patent 3,288,728 (1966).
88. W. F. Gorham, U. S. Patent 3,342,754 (1967).
89. L. A. Errede and M. Szwarc, Q. Rev. Chem. Soc. 12:301 (1958).
90. W. F. Beech, Macromolecules 11:72 (1978).
91. G. A. Olah, "Friedel-Crafts and Related Reactions", John Wiley
       New York (1963).
92. R. Liepins, M. Campbell, and R. J. Fries, ACS Organic and Plastic
       Chem. 40:175 (1979).
93. A. T. Lowe and C. D. Hosford, J. Vac. Sci. Tech. 16:197 (1979).
94. K. B. Blodgett and I. Langmuir, Phys. Rev. 51:964 (1937).
95. G. L. Gaines, Jr., Insoluble Monolayers at Liquid Gas Inter-
       faces", Interscience, New York (1966).
96. R. K. Iler, J. Colloidal and Interface Sci. 21:569 (1966).
97. D. G. Peiffer, H. W. Deckman, T. J. Corley, and J. H. Dunsmuir,
       "Digest of Topical Meeting on Inertial Confinement Fusion,"
       Feb. 26-28, 1980, San Diego, CA.
98. G.S. Fraley, W.P. Gula, R.C. Malone, R.J. Mason, R.L. McCrory,
       and R. L. Morse, in "Plasma Physics and Controlled Nuclear
       Fusion Research", IAEA, Vienna, (1974).
99. C. Hendricks, "Digest of Topical Meeting on Inertial Confinement
       Fusion", Feb. 26-28, 1980, San Diego, CA.
100. E. H. Farnum and J. K. Feuerherd, ibid.
101. K. D. Wise, T. N. Jackson, N. A. Masnari, M. G. Robinson, D. E.
       Solomon, J. Vac. Sci. Tech. 16:936 (1979).
102. J. M. Moran and D. Maydan, Bell System Tech J. 58:1027 (1979).
103. S. Skupsky, R. L. McCrory, and B. Yaakobi, Bull. Am. Phys. Soc.
       24:945 (1979).
104. K. Tanaka and E. Thorsos, Appl. Phys. Lett. 35:853 (1979).
105. J. Varon, Rev. Sci. Instr. 48:941 (1977).
106. J. K. Crane, "Digest of Topical Meeting on Inertial Confinement
       Fusion", Feb. 26-28, 1980, San Diego, CA.

107. W. E. Anderson, J. M. Bunch, R. Liepins, and A. T. Lowe, ibid.
108. T. M. Henderson, D. E. Cielaszyk, and R. J. Simms, <u>Rev. Sci.</u>
      <u>Instr.</u> 48:835 (1977).
109. R. L. Whitman, R. H. Day, R. P. Kruger, and D. M. Stupin, <u>Appl.</u>
      <u>Opt.</u> 18:1266 (1979).
110. R. M. Singleton, B. W. Weinstein, and C. Hendricks, <u>Appl. Opt.</u>
      18:4166 (1979).
111. R. H. Day, T. L. Elsberry, R. P. Kruger, D. M. Stupin, and
      R. L. Whitman, in "VIIIth Int. Conf. on X-ray Optics and
      Microanalysis", Science Press, Princeton, NJ (1979).
112. H. R. F. Horn and H. G. Waltinger, <u>Scanning</u> 1:100 (1978).
113. H. Malissa and H. Haelbig, <u>Microchim Acta</u> 2:376 (1967).

SOLITON THEORY AND NON LINEAR LASER PLASMA INTERACTION

J.L. BOBIN

C.E.A. - Centre d'Etudes de Limeil

B.P. n° 27 - 94190 Villeneuve Saint Georges - FRANCE

ABSTRACT

In order to describe non linear interaction of high intensity electromagnetic (laser beams) with plasmas, one uses equations which are known to have soliton solutions. Some relevant results of enveloppe soliton theory are reviewed first. They are applied to self focusing, profile steepening, the resonance peak, underdense shelves, cavitons.

I - INTRODUCTION

Solitary waves appear as solution of non linear partial differential equations. Since the discovery of stricto sensu "solitons" i.e. waves whose shape and velocity are not altered by interactions with similar objects /1/, the theory has been developped extensively in large number of directions and seems to apply to almost all subfields of physical sciences /2/.

Even in the case of a comparatively weak coupling, wave plasma interaction often takes place non linearly . Accordingly, as pointed out by N.J. Zabusky:
"...PLASMA PHYSICS WITHOUT NON LINEARTIES IS LIKE A ZOO WITHOUT ELEPHANTS".

It is the aim of this paper to review some aspects of soliton theory which are connected to laser plasma interaction.

Incidentally note that although the word was not coined yet, solitons were associated with the laser almost from the begining
when self focusing was first investigated /3/.

II - A LAGRANGIAN APPROACH TO SOLITON THEORY : /4/-/5/-/6/-

Let Φ be the field relevant to the problem at hand. The
equation of motion having soliton solutions derives from a general variational principle. One starts from the Lagrangian density

$$(2-1) \qquad L (\Phi, \Phi_t, \Phi_x)$$

after which the action S is defined as the functional

$$(2-2) \qquad S = \int \underline{dx} \; dt \; L \; .$$

Setting the first variation δS equal to zero and using the
usual procedures of the calculus of variations, one gets the
Euler-Lagrange equation*

$$(2-3) \qquad \frac{\delta S}{\delta \Phi} = \frac{\partial L}{\partial \Phi} - \frac{\partial}{\partial t} \frac{\partial L}{\partial \Phi_t} - \frac{\partial}{\partial x} \frac{\partial L}{\partial \Phi_x} = 0$$

which is the equation of motion of interest.

From the physics, one usually obtains directly the equation
of motion both dispersive and non linear, which serves as a
starting point. However the knowledge of the corresponding Lagrangian density provides a simple way to demonstrate usefull properties as it is to be shown after the examples displayed on TABLE I.

Since Φ is a complex quantity, Φ and its complex or hermi-
tian conjugate Φ^* should be treated as independant variables.
They obey similar equations of motion. Now, the Lagrangian densities are invariant under the transformation :

$$(2-4) \qquad \Phi \longrightarrow e^{i\theta} \; \Phi$$

(gauge invariance) from which, using the variational principle

$$(2-5) \qquad \frac{\delta S}{\delta \theta} = 0,$$

-----

* A formalism onedimensional in space is used throughout this
  and the following section. However, the results apply to mul-
  tidimensional cases as well.

TABLE 1

	LAGRANGIAN	NON LINEAR EQUATION
KLEIN GORDON	$- \partial_\mu \Phi^* \partial_\mu \Phi - m^2 |\Phi|^2 + g^2 |\Phi|^4$	$(\dfrac{\partial^2}{\partial t^2} - \nabla^2 + m^2)\, \Phi - 2\, g^2\, |\Phi|^2\, \Phi = 0$
CUBIC SCHRODINGER	$\dfrac{i}{2} (\Phi\, \Phi^*_t - \Phi^*\, \Phi_t) - |\Phi_x|^2 \pm g^2 |\Phi|^4$	$i\, \Phi_t + \Phi_{xx} \pm g^2 |\Phi|^2\, \Phi = 0$
SCHRODINGER COUPLED TO WAVE EQUATION	$\dfrac{i}{2} (\Phi^*\, \Phi_t - \Phi^*_t\, \Phi) - |\Phi_x|^2 - u_t |\Phi|^2$ $+ \dfrac{1}{2} |\Phi|^4 - \dfrac{u_t^2 - u_x^2}{2}$ $\rho_t = u_{xx}\ , \qquad u_t = \rho + |\Phi|^2$	$i\, \Phi_t + \Phi_{xx} - g^2 \rho\, \Phi = 0$ $(\dfrac{\partial^2}{\partial t^2} - \nabla^2)\, \rho = - (|\Phi|^2)_{xx}$

it follows the current conservation law ($\mu = x$, ict) :

(2-6) $\partial_\mu \, J_\mu = 0$

where $J_\mu = -i \, (\Phi^* \, \partial_\mu \Phi - \Phi \partial_\mu \Phi^*)$. Accordingly

(2-7) $Q = \int i \, (\Phi^* \, \Phi_t - \Phi \, \Phi^*_t) \, \underline{dx}$

is a constant of the motion (conservation of the number of quanta).

     Another general result of importance comes out from the way
the coupling constant g enters the Lagrangian densities of TABLE I.
When g is zero, the equations reduce to the linear wave propagation equation whose solution is a dispersive plane wave packet.
However since one has only terms of the form :

(2-8) $g^2 \, |\Phi|^4$ or $g^2 \, \rho |\Phi|^2$,

under the transformation

(2-9) $\Phi = \dfrac{\sigma}{g}$, $\rho = \dfrac{\tau^2}{g^2}$

the Lagrangian density becomes

(2-10) $L = \dfrac{1}{g^2} \, L_{\sigma\tau}$

where $L_{\sigma\tau}$ is g independant. It still contains the non linear
terms which lead to soliton solutions. Consequently, these cannot
be neglected when g is small (but non zero). Soliton solutions
are thus singular when g ($\neq$ o) $\longrightarrow$ o (a simple pole in the case
of classical fields. The corresponding quantum mechanical behavior
is obtained by setting $g^2 \simeq \hbar$).

III - STATIONARY SOLUTIONS IN PHASE SPACE

Obviously, the Lagrangian densities displayed in table I may be
cast in the general form (with summation over repeated indices)

$$(3\text{-}1) \qquad L = - \frac{\partial \Phi^*}{\partial x_\mu} \frac{\partial \Phi}{\partial x_\mu} - U(\Phi^* \Phi).$$

The equation of motion, derived from the variational principle (3-3) is

$$(3\text{-}2) \qquad \frac{\partial^2 \Phi}{\partial x_\mu^2} - \Phi \frac{d}{d(\Phi^* \Phi)} U(\Phi^* \Phi) = 0.$$

Such an equation has solutions of the form

$$(3\text{-}3) \qquad \Phi = E(x) e^{-i\omega t} \qquad \text{with real } E(x),$$

if $E(x)$ satisfies

$$(3\text{-}4) \qquad \frac{d^2 E}{dx^2} + \omega^2 E - E \frac{d}{dE^2} U(E^2) = \frac{d^2 E}{dx^2} + \omega^2 E - \frac{dU(E)}{2dE} = 0.$$

This is the equation of motion of a pointlike particle with "position" E moving in "time" x across a potential

$$(3\text{-}5) \qquad P(E) = - \frac{1}{2}(U - \omega^2 E^2).$$

Assume $P(E)$ has the shape shown on fig.1 . Then one may construct after (3-4) the integral curves in phase space E_x, E, shown on fig.2 . The singular points defined by

$$(3\text{-}6) \qquad E_x = \frac{d}{dE} P(E) = 0 ,$$

are either centers or saddle points. Only the singular curve S passing through the origin is such that $E_x = E = 0$ at both infinities now multiplying all terms of (3-4) by E_x one gets the first integral

$$(3\text{-}7) \qquad E_x^2 = U - \omega^2 E^2.$$

Hence the solution :

$$(3\text{-}8) \qquad X = \int_S \pm \frac{dE}{\sqrt{U - \omega^2 E^2}} ,$$

Where the integration is taken along S. The solution has the form shown on fig.3 It is confined in space at all times. It is called an <u>enveloppe soliton</u>.

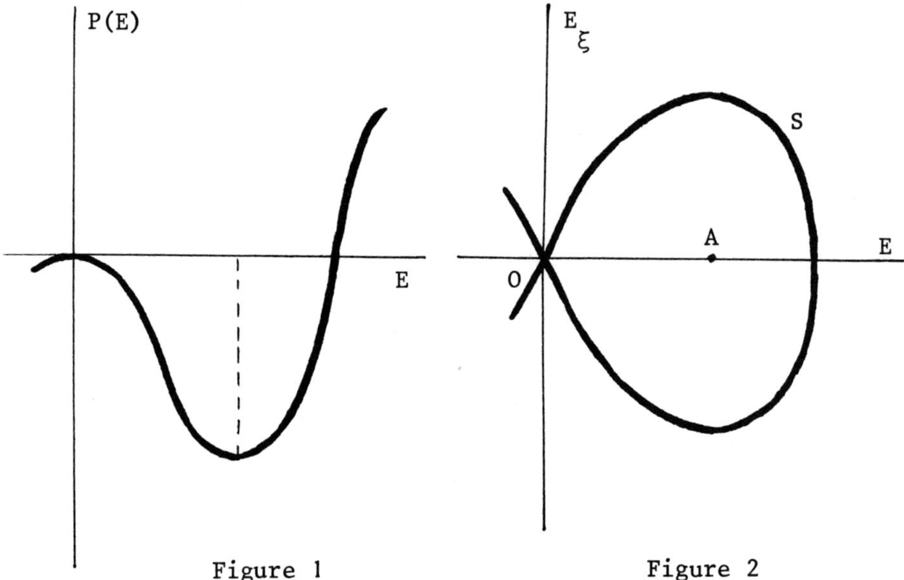

Figure 1   Figure 2

Potential well for eq(3-4)   Solution of (3-4) in phase space

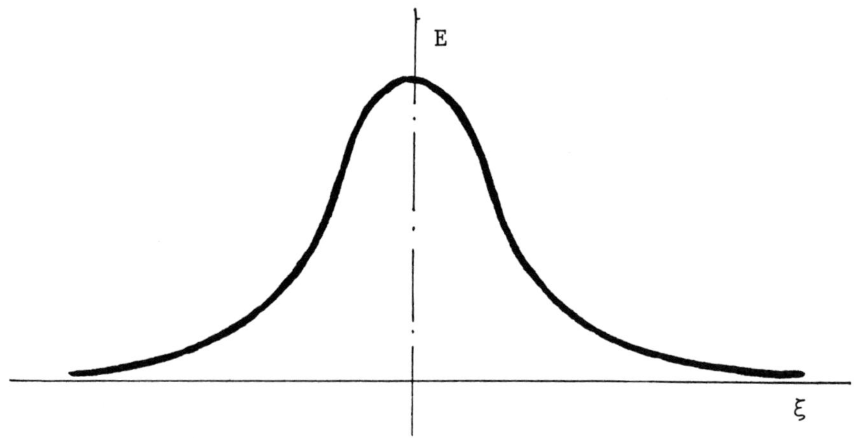

Figure 3 - Bell shaped enveloppe Soliton

As an example, one may choose the cubic non linear Schrödinger equation. In order to put it in a time independent form, set

$$(3\text{-}9) \qquad \Phi = E\,(x,t)\,e^{ikx}$$

so that

$$(3\text{-}10) \qquad \Phi_{xx} = E_{xx}\,e^{ikx} + 2\,ik\,E_x\,e^{ikx} - k^2\,E e^{ikx} \ .$$

After substitution, the imaginary terms cancel out provided one performs the change of variable

$$(3\text{-}11) \qquad \xi = x - V\,t \qquad \text{with} \quad V = -\,2k \ .$$

The equation reduces then to

$$(3\text{-}12) \qquad E_{\xi\xi} - k^2\,E + g^2\,E^3 = 0$$

with the first integral

$$(3\text{-}13) \qquad (E_\xi)^2 - k^2\,E^2 + \frac{g^2}{2}\,E^4 = 0 \quad ,$$

and the final solution

$$(3\text{-}14) \qquad \xi = \int \frac{\pm\,dE}{E\sqrt{k^2 - g^2\dfrac{E^2}{2}}} = \frac{g}{k\sqrt{2}}\ \text{arg sech}\ \frac{gE}{k\sqrt{2}}$$

or alternatively

$$(3\text{-}15) \qquad E = \frac{k\sqrt{2}}{g}\ \frac{1}{\text{ch}\ \dfrac{k\sqrt{2}}{g}\,\xi}$$

which has the shape shown on Fig. 3 as expected.

IV – ENVELOPPE SOLITONS IN PLASMAS

In order to keep a uniform pressure, a necessary requirement for a steady state in some reference frame, a local high intensity field is accompanied by a density dip. One then has to deal with the coupled equations in the third row of table I instead of a mere cubic non linear Schrödinger equation. Performing the transformations (3-9) and (3-11) one gets the following system

$$(4\text{-}1) \qquad E_{\xi\xi} - k^2 E - g^2 \rho E = 0$$

$$(4\text{-}2) \qquad (v^2 - 1)\, \rho_{\xi\xi} = E^2_{\xi\xi} \quad ,$$

with the boundary condition $\rho = \rho_o$, $E = 0$, $\rho_\xi = E_\xi = 0$, at both infinities. A trivial integration of (4-2) thus yields

$$(4\text{-}3) \qquad \rho = \rho_o + \frac{E^2}{v^2 - 1} \qquad \text{with } v^2 < 1 \; .$$

$\rho(E)$ is represented on fig. 4. This is compatible with the results of the previous section : single valued potential $P(E)$, a single arch of E in the profile (fig. 5).

In another case of interest, the density profile results from an isothermal flow, E^2 acting as an external potential. Its gradient is the so-called ponderomotive force which denoting by ρ_c the critical density, reads (see section VIII) :

$$(4\text{-}4) \qquad F_p = - \frac{\rho}{\rho_c} \, |E|^2_\xi \; .$$

In steady state, one has to couple (4-1) with the gas dynamical system

$$(4\text{-}5) \qquad (\rho u)_\xi = 0$$

$$(4\text{-}6) \qquad \rho\, u u_\xi = - (Z + 1) \frac{T}{m} \rho_\xi - \frac{\rho}{\rho_c} \, |E|^2_\xi$$

where u is the particle velocity, T the temperature (in energy units), Z and m are the ion charge and mass respectively. Obvious first integrals are :

$$(4\text{-}7) \qquad \rho u = J \qquad \text{(constant mass flow)}$$

$$(4\text{-}8) \qquad \frac{E^2}{\rho_c} = - \frac{J^2}{2\rho^2} - (Z + 1) \frac{T}{m} \ln\rho + \frac{J^2}{2\rho_o^2} + (Z + 1)\frac{T}{m}\ln\rho_o \; .$$

As shown on fig. 6, ρ is now bi valued in E^2 and so is $P(E)$. The corresponding potential wells, phase space plot and shape of the solution are displayed on the figure 7. The cavity in the fluid encompasses two archs of the oscillating field.

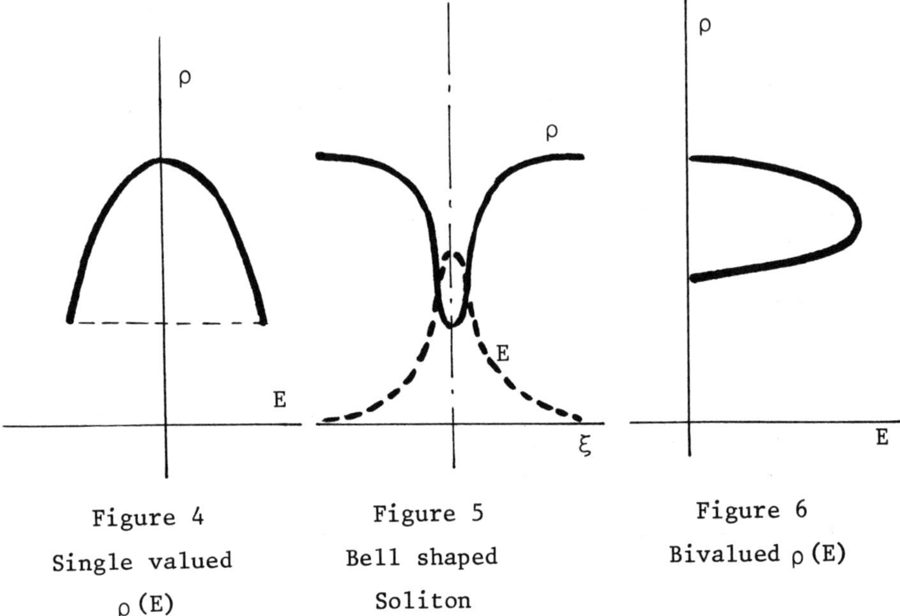

Figure 4
Single valued
ρ (E)

Figure 5
Bell shaped
Soliton

Figure 6
Bivalued ρ (E)

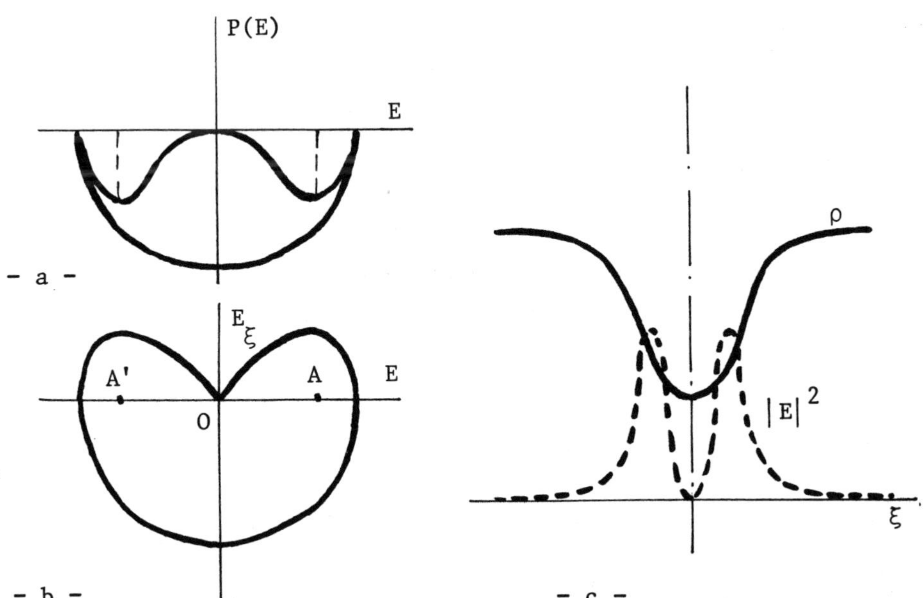

Figure 7 — Solution for a bivalued ρ (E)

-a- bivalued potential well, -b- Solution in phase space
-c- Camel shaped soliton

V - STABILITY

As noted in the introduction, one remarkable properties of the solutions of a non linear dispersive partial differential equation is their indestructibility. Solitons with the shape (3-15) are conserved in head on collisions. Not all solitary waves have this property. We still call then solitons provided they keep their amplitude width and velocity between collisions. A minimal requirement is then : they should be stable against small perturbations.

Numerical simulations show that, for the example dealt with in the previous section, this is not always the case. A Eulerian gas dynamical code was modified in order to include $\dfrac{\rho \, \nabla |\varepsilon|^2}{8\pi \rho_c}$ as an external force, the complex field ε being described by a non linear Schrödinger equation.[*] An exact analytical solution of the type shown on fig. 7 was used an initial condition. It turns out that such a profile is unstable and rapidly evolves to a cavity with a single arch of the field, immersed in the fluid and moving with the flow velocity (Fig. 8).

An analytical perturbative treatment can also be made setting

$$(5\text{-}1) \qquad\qquad E = E + E^{(1)} \quad , \quad \rho = \rho + \rho^{(1)}$$

and linearizing the equations. Thus the system (4-1) (4-2) yields

$$(5\text{-}2) \qquad\qquad E^{(1)}_{\xi\xi} - k^2 \, E^{(1)} - g^2 \, \rho^{(1)} \, E - g^2 \, \rho E^{(1)} = 0$$

$$(5\text{-}3) \qquad\qquad (v^2 - 1) \, \rho^{(1)}_{\xi\xi} = (2EE^{(1)})_{\xi\xi} \quad .$$

The latter is immediately integrated to give

$$(5\text{-}4) \qquad\qquad (v^2 - 1) \, \rho^{(1)} = 2 \, EE^{(1)}$$

$$(5\text{-}5) \qquad\qquad E^{(1)}_{\xi\xi} - \left[k^2 + g^2 \left(\frac{3 E^2}{v^2 - 1} + \rho_o \right) \right] E^{(1)} = 0 \quad .$$

The stability condition is then

$$(5\text{-}6) \qquad\qquad k^2 + g^2 \left(\frac{3 E^2}{v^2 - 1} - \rho_o \right) \langle 0 \quad .$$

---

[*] the computational work was made by J.P. MANGIN and Mrs J. PORNET whose contribution is gratefully aknowledged.

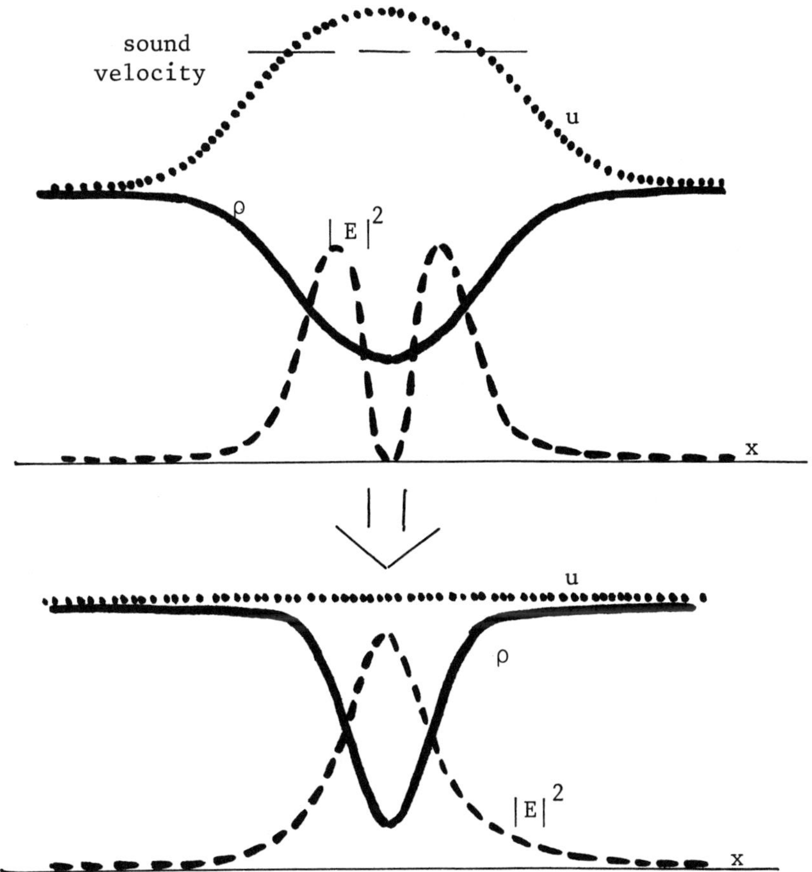

Figure 8 - Instability of the camel shaped soliton

-a- Initial profile , -b- Resulting bell shaped soliton

similarly, one gets from (4-6)

$$(5\text{-}7) \qquad \left| u\,u^{(1)} \right|_\xi = -\frac{(Z+1)}{m}T\,\frac{\rho^{(1)}_\xi}{\rho} - \frac{2(EE^{(1)})_\xi}{\rho_c}$$

which can be integrated <u>for a given</u> ρ

$$(5\text{-}8) \qquad u\,u^{(1)} = -\frac{(Z+1)T}{m}\,\frac{\rho^{(1)}}{\rho} - \frac{2\,EE^{(1)}}{\rho_c} \quad,$$

or since $\rho u^{(1)} + \rho^{(1)} u = 0$,

$$(5\text{-}9) \qquad \left[\frac{J^2}{\rho^2} - \frac{(Z+1)}{m}T\right]\frac{\rho^{(1)}}{\rho} = \frac{2\,EE^{(1)}}{\rho_c} \quad.$$

The equation for $E^{(1)}$ is accordingly

$$(5\text{-}10) \qquad E^{(1)}_{\xi\xi} - \left\{k^2 + g^2\left[\frac{2\rho}{\rho_c}\,\frac{E^2}{\dfrac{J^2}{\rho^2} - \dfrac{(Z+1)T}{m}} + \rho\right]\right\}E^{(1)} = 0 \quad.$$

Obviously when

$$(5\text{-}11) \qquad \frac{J^2}{\rho^2} > \frac{(Z+1)T}{m}$$

i.e. The flow is locally supersonic, the perturbation grows exponentially. In the above computational example, there are two sonic points in the initial flow which is supersonic in the vicinity of the bottom of the cavity. The state is then unstable for any unavoidable numerical perturbation. The resulting profile has no flow in its own reference frame and is actualy stable.

VI - THE EFFECT OF DAMPING

The soliton solution of a non linear partial differential equation results from a balance between the non linearity and dispersive effects. Actual physical systems are also dissipative. So far the problem has been investigated following two lines of thought. First one can look for steady state solutions in the reference frame moving with the soliton (as e.g. Karpman's treatment of the K.D.V. Burgers equation /7/). In an alternative approach one applies the inverse scattering transform to the

equation written in the laboratory frame. When dealing with the
same equation the two methods yield different results.

Let us discuss the consequences of adding a damping term to
(3-12) or (4-1). Let Γ be a phenomenological coefficient. The
damped cubic non linear Schrödinger equation is then

$$(6-1) \qquad E_{\xi\xi} + \Gamma\, E_{\xi} - k^2\, E + g^2\, E^3 = 0 \quad .$$

The solution in phase space $(y = E_{\xi}, E)$ is an integral curve of
the system

$$(6-2) \qquad y_{\xi} = -\Gamma\, y + k^2\, E - g^2\, E^3 \quad .$$

$$(6-3) \qquad E_{\xi} = y$$

The singular points are the origin and

$$A\left\{ y = 0, \quad E = \frac{k}{g} \right\} \qquad A'\left\{ y = 0, \quad E = -\frac{k}{g} \right\}$$

Furthermore y_{ξ} is zero along a cubic parabola Π , and E_{ξ} is zero
along the E axis. The nature of the singular points is determined
after the characteristic equation :

$$(6-4) \qquad \lambda^2 + \Gamma\, \lambda - k^2 + 2\, g^2\, E^2 = 0 \quad .$$

The origin is always a saddle point. A and A' are attractive nodes
provided $\Gamma > 2\,k$ (strong damping) and attractive foci if $\Gamma < 2k$,
(weak damping). In both cases the solution is a singular curve
which in the latter situation spirals around A. The figure 9
summarizes the results and also displays $E(\xi)$.

For a bi-valued potential, one has to deal with

$$(6-5) \qquad y_{\xi} = -\Gamma\, y - k^2\, E - g^2\, \rho\, E$$

instead of (6-2), $\rho(E)$ having the shape shown on fig. 6. y_{ξ} is
then zero along a figure of eight curve whose archs with the
larger $|E|$ belong to the potential well centered at E = 0. The
origin is an attractive node or focus for this potential well
whereas it is still a saddle point for the other one. The

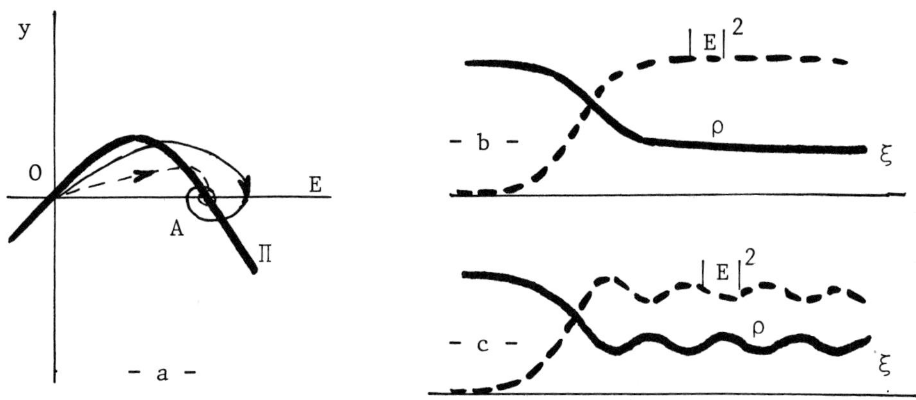

Figure 9 - Effect of damping with single valued $\rho(E)$

-a- phase space plot, -b- profiles for strong damping, -c- profiles for weak damping

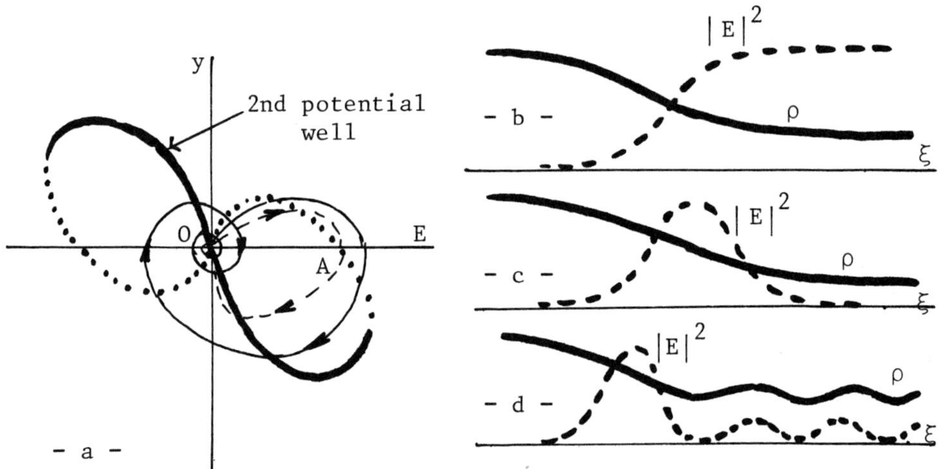

Figure 10 - Effect of damping with bi-valued $\rho(E)$

-a- phase space plot, -b- very strong damping, -c- strong damping
-d- weak damping

resulting cases of interest :

- very strong damping   :   A is the end point (node)

- strong damping   :   O is the end point (node), A is a
                                                         saddle
- weak damping   :   O is the end point (focus)

are presented on fig. 10. In all cases the densities are different
for the end points. The introduction of damping in the non linear
equation changes the solution. It is no longer a soliton but a
shock-like front with modulations in the case of a weak damping.

So far, the influence of damping has been considered only in
connexion with stationnary solutions. This is not the whole story.
Recent investigations /8/ - /11/, most of them using the inverse
scattering technique have evidenced a different behavior : a
solitary wave is generated with a time dependant amplitude and
shape. Of course, the conservation laws of the type (2-6) (2-7)
no longer apply. A final problem directly relevant to laser
plasma interaction was also solved /12/ : a source with given
amplitude E_o and frequency ω_o balances the damping so that one
starts from the equation

$$(6-6) \qquad \Phi_t - i\,\Phi_{xx} - 2\,i\,\Phi^2\,\Phi^* = -\Gamma\,\Phi - E_o\,e^{i\omega_o t} \quad .$$

The conservation law (2-7) is now replaced by

$$(6-7) \qquad \frac{\partial}{\partial t} \int_{-\infty}^{\infty} \Phi\Phi^*\,dx = -2\,\Gamma \int_{-\infty}^{\infty} \Phi\Phi^*\,dx - \left[E_o\,e^{i\omega_o t} \int_{-\infty}^{\infty} \Phi^*\,dx + E_o\,e^{-i\omega_o t} \int_{-\infty}^{\infty} \Phi\,dx \right]$$

and by analogy with the solution of the Schrödinger equation
(3-12) let

$$(6-8) \qquad \Phi \sim 2\,\eta\,\operatorname{sech} 2\,\eta\,(x - x_o)\,e^{-2i(\sigma + \frac{\pi}{4})} \quad .$$

where η and σ are time dependant. Then (6-7) yields

$$(6-9) \qquad \eta_t = -2\Gamma\eta + \frac{\pi E_o}{2}\,\sin\chi \quad , \quad \chi = \omega_o t + 2\,\sigma \quad .$$

Since, on the other hand, for the non linear Schrödinger .equation.
$\sigma_t = -2\,\eta^2$, one gets

$$(6\text{-}10) \qquad \chi_t = \omega_o - 4\,\eta^2$$

so that it is possible to perform a phase space (χ, η) analysis.

Singular points exist if

$$(6\text{-}11) \qquad E_o^2 \;>\; \frac{4\,\Gamma^2 \omega_o}{\pi^2}$$

i.e. a large enough amplitude of the pump field. They are attractive foci when $\eta \cos \chi > 0$, saddle points in the opposite case. Attractive foci correspond to phase locked solitons whose amplitude depends upon the frequency of the pump and not on its amplitude.

VII – NON LINEARITIES IN LASER PLASMA INTERACTION

A high frequency oscillating fieds obeys a propagation equation which is different for electromagnetic waves

$$(7\text{-}1) \qquad \frac{\partial^2 E}{\partial t^2} - c^2 \nabla \times \nabla \times E = -\,4\pi e\,\frac{\partial}{\partial t}(nu)$$

and for Langmuir waves

$$(7\text{-}2) \qquad \frac{\partial^2 E}{\partial t^2} = -\,4\pi e\,\frac{\partial}{\partial t}(nu)$$

where n is the density (particle number) of the electrons with velocity u. The term in the right hand side is a source of non linearities. It may be expanded as

$$(7\text{-}3) \qquad -\,4\pi e\left[\, n_o\,\frac{\partial u_1}{\partial t} + u_1\,\frac{\partial n_1}{\partial t} + n_1\,\frac{\partial u_1}{\partial t} + n_o\,\frac{\partial u_2}{\partial t}\,\right]$$

where the subscripts 0, 1 and 2 denote zeroth first and second order quantities respectively.

Since n_o is also representative of the ion density, it can either oscillate at the (low) ion-acoustic frequency Ω or vary according to fluid-dynamical equations with a slow velocity, the inertia being carried by the ions. Thus $n_o\,\frac{\partial u_1}{\partial t}$ introduce any one of the following non linear couplings :

- transverse (ω) to $\Big\}$ $\Big\{$ transverse ($\omega-\Omega$) field (stimulated Brillouin scattering)
 longituninal (Ω) fields or
 longitudinal ($\omega-\Omega$) field (parametric decay)

- flow to transverse field $\longrightarrow$ density profile modifications.

The first order variables u_1 and n_1 have frequencies close either to ω or to the plasma frequency $\omega_p = (4\pi e^2 n_o/m)^{1/2}$ whereas u_2 has twice that frequency. Resonances are expected and actually found for $\omega = \omega_p$. The three terms $u_1 \frac{\partial n_1}{\partial t} + n_1 \frac{\partial u_1}{\partial t} + n_o \frac{\partial u_2}{\partial t}$ introduce more non linear couplings viz :

- transverse (ω) to $\Big\}$ $\Big\{$ longitudinal (ω_p) field (parametric resonance at $\omega = 2\omega_p$)
 longitudinal (ω_p) fields transverse ($\omega \pm \omega_p$) (antistokes (+) or Stokes (−) Raman scattering)

- transverse (ω) to transverse (ω) field $\longrightarrow$ second harmonic generation

Still more can be found by going to higher orders (e.g. three plasmon recombination /13/ after $n_1 \partial u_2/\partial t$).

Now setting

(7-4) $$E = E(x, t)\, e^{-i\omega t}$$

where E is a slowly varying function of t, one has

(7-5) $$\frac{\partial^2 E}{\partial t^2} = -\left[2i\omega\, \frac{\partial E(x, t)}{\partial t} + \omega^2\, E(x, t) \right] e^{-i\omega t} \quad .$$

The electron equation of motion is

(7-6) $$\frac{\partial u}{\partial t} + \nu u = \begin{cases} -\dfrac{eE}{m} & \text{transverse (E.M.) field} \\[2ex] -\dfrac{eE}{m} - \dfrac{3T}{mn_o}\nabla n & \text{longitudinal field} \end{cases}$$

where ν ($\ll \omega$) is an effective collision frequency, and the temperature T is in energy units. Keeping in (7-3) the term $n_o \partial u_1/\partial t$ only, one then gets after some algebra, for transverse waves

$$(7\text{-}7) \qquad 2i\omega \frac{\partial E}{\partial t} + c^2 \frac{\partial^2 E}{\partial x^2} + \left(\omega^2 - i\frac{\nu}{\omega}\omega_p^2 - \omega_p^2\right) E = 0$$

and for longitudinal ones

$$(7\text{-}8) \qquad 2i\omega \frac{\partial E}{\partial t} + \frac{3T}{m}\frac{\partial^2 E}{\partial x^2} + \left(\omega^2 - i\frac{\nu}{\omega}\omega_p^2 - \omega_p^2\right) E = 0$$

which since ω_p is proportional to the density ρ are non linear Schrödinger equations of the type (4-1). Using the transformation (3-9) (3-10) in which now k has an imaginary part whose value is choosen in order that imaginary terms in E cancel out, transforming the variables by (3-1) with $V = Re k\, c^2/\omega$ or $Re k\, 3T/m\omega$, and neglecting higher order terms in ν/ω, one gets the steady state equations :

$$(7\text{-}9) \qquad c^2 \frac{d^2 E}{d\xi^2} - \frac{\nu}{\omega}\frac{\omega_p^2}{Re k}\frac{dE}{d\xi} + \left(\omega^2 - \omega_p^2\right) E = 0$$

$$(7\text{-}10) \qquad \frac{3T}{m}\frac{d^2 E}{d\xi^2} - \frac{\nu}{\omega}\frac{\omega_p^2}{Re k}\frac{dE}{d\xi} + \left(\omega^2 - \omega_p^2\right) E = 0$$

Since one has $\omega_p^2 = h\rho$, defining the critical density $\rho_c = \omega^2/h$, equations (7-9) and (7-10) may be rewritten in a form similar to (4-1) :

$$(7\text{-}11) \qquad \frac{c^2}{\omega^2} E_{\xi\xi} - \frac{\nu}{\omega Re k}\frac{\rho}{\rho_c} E_\xi + \left(1 - \frac{\rho}{\rho_c}\right) E = 0$$

$$(7\text{-}12) \qquad \frac{3T}{m\omega^2} E_{\xi\xi} - \frac{\nu}{\omega Re k}\frac{\rho}{\rho_c} E_\xi + \left(1 - \frac{\rho}{\rho_c}\right) E = 0$$

Alternatively one may choose k in such a way that the real part in E in (7-7) or (7-8) cancels out. One then recovers the linear dispersion relation for the kind of wave of interest. Then assuming E(x, t) is slowly varying with x, E_{xx} is negligible in the expansion (3-10) so that considering any term in (7-3) one ends up with three wave couplings of the form

$$(7\text{-}13) \qquad \frac{\partial E_1}{\partial t} + \nu g_1 E_1 = g_1 E_2^* E_3^*$$

$$(7\text{-}14) \qquad \frac{\partial E_2}{\partial t} + \nu g_2 E_2 = g_2 E_3^* E_1^*$$

$$(7\text{-}15) \qquad \frac{\partial E_3}{\partial t} + \nu g_3 E_3 = g_3 E_1^* E_2^*$$

Such a system has been extensively studied and is known to have soliton-like solution /14/ /15/

VIII - THE PONDEROMOTIVE FORCE

Equations such as (7-11) (7-12) have to be coupled to fluid dynamical equations. Let us first take up the case of transverse e.m. waves. They act directly on the electrons whose equation of motion including non linear terms is

$$(8\text{-}1) \qquad \frac{\partial u}{\partial t} + (u.\nabla)\, u = - \frac{e}{m}\left(E + \frac{u}{c} \times B\right) - \nu u \quad .$$

Separating slowly varying from oscillating parts one gets for the former

$$(8\text{-}2) \quad \frac{\partial \overline{u}}{\partial t} + (\overline{u}.\nabla)\,\overline{u} = \frac{e}{m}\left(\overline{E} + \frac{\overline{u}}{c} \times \overline{B}\right) - \left\langle (\tilde{u}.\nabla)\,\tilde{u} \right\rangle - \frac{e}{c}\left\langle \tilde{u} \times \tilde{B} \right\rangle$$

where the averages $\langle\ \rangle$ are taken over one period of the pump. Linearized equations for the oscillating parts are

$$(8\text{-}3) \qquad \frac{\partial \tilde{u}}{\partial t} = - \frac{e}{m}\tilde{E} - \nu \tilde{u} \quad , \qquad \tilde{B} = - \frac{mc}{e}\nabla \times \left[\tilde{u} + \nu\,(\delta r)\right]$$

with $\delta r = \displaystyle\int_0^t u\, dt'$. The averages $\langle\ \rangle$ in (8-2) are then readily

transformed into

$$(8\text{-}4) \qquad \frac{f_p}{m} = - \frac{\nabla \langle \tilde{u}^2 \rangle}{2} - \left\langle \tilde{u} \times \nabla \times \nu\,(\delta r)\right\rangle$$

from which one gets the global force

$$(8\text{-}5) \qquad F_p = n f_p = - \frac{\omega_p^2}{\omega^2}\left(1 - \frac{\nu}{\omega}\right)\frac{\nabla \langle E^2 \rangle}{8\pi} \quad .$$

For the case of longitudinal waves, B is zero and the electron equations of motion for the slow and the oscillating parts are

$$(8\text{-}6) \quad \frac{\partial \overline{u}}{\partial t} + (\overline{u}.\nabla)\,\overline{u} = -\frac{e\overline{E}}{m} - \frac{3T\nabla n}{m\overline{n}} - \left\langle (\tilde{u}.\nabla)\,\tilde{u} \right\rangle + \frac{3T}{m\overline{n}^2}\left\langle \tilde{n}\,\nabla\,\tilde{n} \right\rangle$$

$$(8\text{-}7) \quad \frac{\partial \tilde{u}}{\partial t} = -\frac{e\tilde{E}}{m} - \nu\tilde{u} - \frac{3T}{m\overline{n}}\nabla\,\tilde{n} \; ,$$

hence in first order:

$$(8\text{-}8) \quad \frac{3T}{m\overline{n}}\nabla\,\tilde{n} = \frac{\nu}{\omega}\,\frac{e\tilde{E}}{m} \quad .$$

On the other hand, one has the Poisson equation

$$(8\text{-}9) \quad \nabla.\tilde{E} = -4\pi e\,\tilde{n}$$

so that the averaged terms in (8-6) eventually take the form

$$(8\text{-}10) \quad \frac{f_p}{m} = -\left\langle (\tilde{u}.\nabla)\,\tilde{u} \right\rangle + \frac{\nu}{\omega}\,\frac{\left\langle (\nabla.\tilde{E})\,\tilde{E} \right\rangle}{4\pi\overline{n}} \quad .$$

The global force is then in one dimension

$$(8\text{-}11) \quad F_p = nf_p = -\left(\frac{\omega_p^2}{\omega^2} - \frac{\nu}{\omega}\right)\frac{\partial\langle E^2\rangle}{8\pi\,\partial x} \quad .$$

Equations (8-5) and (8-11) define the so called ponderomotive force* which is introduced in the fluid dynamical equation as an external force and provides the coupling with the field propagation equations.

IX - SOLITONS IN LASER PLASMA FLOWS

In a laser irradiated plasma, electrons are submitted to two external forces, the ponderomotive force derived in the previous section plus the electrostatic force deriving from the self consistent potential ϕ . Ignoring damping terms, the equation of motion reads

$$(9\text{-}1) \quad \frac{\partial u}{\partial t} + (u.\nabla)\,u = -\frac{\nabla Pe}{mn} + \frac{e}{m}\nabla\phi - \frac{1}{mn_c}\nabla\,\frac{\langle E^2\rangle}{8\pi} \sim 0$$

since for the cases of interest, the flow velocity u is much

---
* more detailed and accurate derivations can be found in /16/

smaller than the electron thermal velocity $(T_e/m)^{1/2}$. Only the self consistent potential acts on the ion with charge Z whose equation of motion is thus

$$(9\text{-}2) \qquad \frac{\partial u}{\partial t} + (u.\nabla)\, u = -\frac{\nabla p_i}{\rho} - \frac{Z e n_i}{\rho} \nabla \phi$$

Solving (9-1) for $\nabla \phi$ substituting in (9-2) and assuming the plasma is electrically neutral, one gets

$$(9\text{-}3) \qquad \frac{\partial u}{\partial t} + (u.\nabla)\, u = -\frac{\nabla (p_i + p_e)}{\rho} - \frac{\nabla \langle E^2 \rangle}{8\pi \rho_c} \quad .$$

This equation can be coupled to the continuity equation in two different ways so that one obtains either flow equations of the type (4-5) (4-6) or after linearization, a wave equation of the type (4-2). Furthermore, if the neutrality condition is relaxed, the electrons are Boltzmanian

$$(9\text{-}4) \qquad n = n_o\, e^{\left(e\phi - \langle E^2 \rangle / 8\pi\right)/T} \quad .$$

Eliminating ϕ with the help of the Poisson equation, and expanding the exponential up to the second order in ϕ yields the Boussinesq type equation

$$(9\text{-}5) \quad \lambda_D^2 \frac{\partial^4 \rho}{\partial t^2 \partial x^2} = \frac{\partial^2 \rho}{\partial t^2} - a^2 \frac{\partial^2 \rho}{\partial x^2} - \frac{1}{8\pi} \frac{\rho_o}{\rho_c} \frac{\partial^2 \langle E^2 \rangle}{\partial x^2} - \frac{a^2}{2\rho_o} \frac{\partial^2 (\rho^2)}{\partial x^2} \quad ,$$

where $\lambda_D = (T/m\omega_p^2)^{1/2}$ and $a = (ZT/M_i)^{1/2}$ are the Debye length and the ion sound velocity respectively.

Now, the flow influences the propagation of the waves by means of the term $g^2 \rho E$. Assume for instance that ρ has the form (4-3) and look for a steady solution $\partial E/\partial t = 0$. Transvere e.m. waves,' $E(x)\, e^{i(kx - \omega t)}$, where $E(x)$ is a slowly varying function propagating the x direction according to the usual dispersion relation $(\omega^2 = \omega_p^2 + k^2 c^2)$. $V.E$ is zero and $\nabla \times \nabla \times E = \nabla^2 E$. The relevant equation is then

$$(9\text{-}6) \qquad 2i\, k\, \frac{\partial E}{\partial x} + \nabla_T^2 E + \frac{E^2}{V^2 - 1} E = 0$$

where ∇_T^2 denotes the Laplacian operator in the transverse direction. This is a non linear Schrödinger equation in which the coordinate x plays the role of the time. The soliton solution

found in section 4 do not apply to the problem of self focusing
which has to be solved numerically /17/ /18/. Indeed the transformation (3-9) does not work since the phase factor is no longer
a linear function of the transverse coordinate only. However setting

$$(9\text{-}7) \qquad\qquad E = A\, e^{i\theta} \quad , \quad u = \frac{\nabla_T \theta}{k}$$

where θ is a function of x and of the transverse coordinate, one
gets the coupled equations

$$(9\text{-}8) \qquad \frac{\partial A}{\partial x} + (u.\ \nabla_T)\, A + A\, (\nabla.u) = 0$$

$$(9\text{-}9) \qquad \frac{\partial u}{\partial x} + (u.\ \nabla_T)\, u = \frac{\nabla_T}{2k^2} \left[\frac{A.\nabla_T^2 A}{A^2} + g^2 A^2 \right].$$

Let A_0 be an unperturbed (x independant) solution. A first order
equation resulting from the above system is

$$(9\text{-}10) \qquad \frac{\partial^2 u}{\partial x^2} = - \frac{g^2 A_0}{2k^2}\ \nabla_T.A_o\ \nabla_T.u$$

with a solution

$$(9\text{-}11) \qquad u = ax^2 + b(y + z) + c \quad , \text{ with } \frac{a}{b} = - \frac{g^2 A_0 \nabla_T.A_o}{2k^2} \quad .$$

The singularity u = 0 on the x axis is thus evidenced. Depending
on the sign of u, it corresponds either to self focusing into a
plasma cavity or beam break up, provided the intensity is above
thrreshold.

An alternative way of dealing with this problem uses the
Kaup-Newell theory /12/. The equation of interest is now

$$(9\text{-}12) \qquad \frac{\partial E}{\partial x} - i\, \frac{\partial^2 E}{\partial y^2} - 2i\, E^2\, E^{\textstyle *} = -\Gamma E - E_o e^{-ik_o x} \quad ,$$

with a solution of the form (6-8) i.e. a soliton in the transverse direction y. Now η and σ are x dependant and satisfy the first
order system

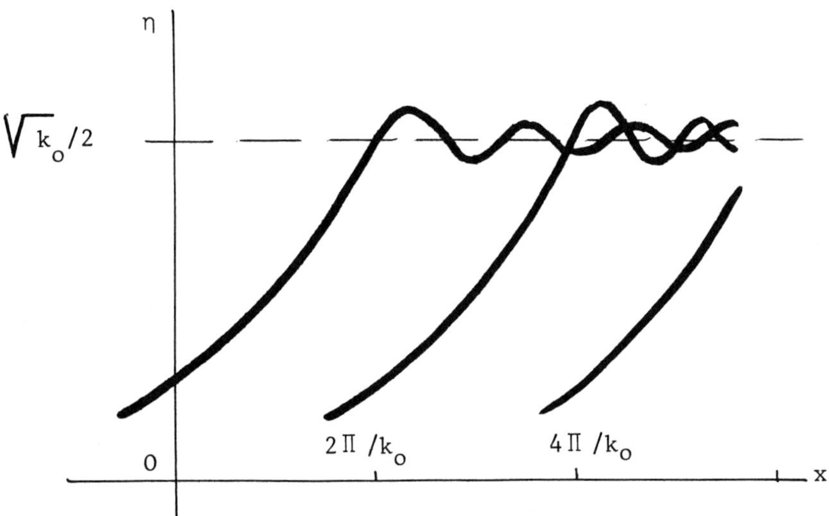

Figure 11 - Axial variation of the field amplitude
for self focusing

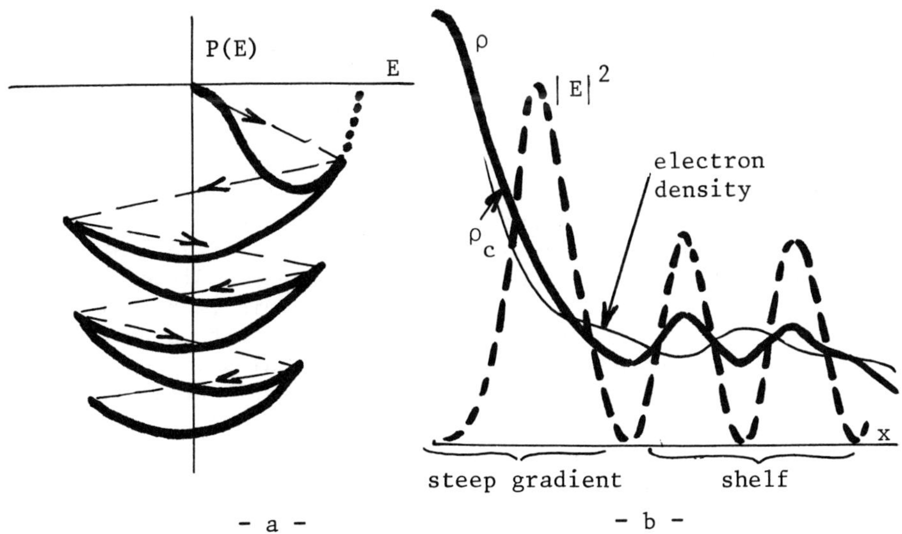

Figure 12 - Ponderomotive force dominated flame
-a- multi valued potential well, -b- resulting density and field
profiles

$$(9\text{-}13) \qquad\qquad \eta_x = -\,2\,\Gamma\,\eta + \frac{\Pi E_o}{2}\ \sin\,\chi$$

$$(9\text{-}14) \qquad\qquad \chi_x = k_o - 4\,\eta^2 \quad .$$

The threshold condition for phase locking is

$$(9\text{-}15) \qquad\qquad E_o^2 \ \rangle\ \frac{4\,\Gamma^2 k_o}{\Pi 2} \quad .$$

Then attractors exist. η increases with x and reackes an asymptotic value either monotonously after an overshoot when the damping is strong, or in an oscillatory way when the damping is weak. (Fig. 11). Due to the sine function in (9-13), η is periodic along the x axis. To the best of the author's knowledge, the Kaup-Newell theory is the only one which accounts for both the periodicity and the long range behavior of self-focusing. Also note that the threshold is proportional to the spatial frequency k_o. In a laser driven plasma, light is reflected at the critical density for which locally k_o is zero. The threshold is thus lowered and has to be calculated after the density gradient.

Other cases of interest in laser driven plasma flows can be handled following the line of thought of sections 4 to 6. As an example, it was found in /19/ that ρ (E) is multivalued in a laser driven flame structure. So is the potential P(E) with the results displayed on fig.12. It corresponds to normal incidence with weak absorption. For oblique incidence with p polarization, a resonance peak is formed with fractional absorption Φ (τ)/2 where Φ (τ) is the angle dependant Ginsburg function /20/. One then is in the case of strong damping shown on fig. 10. The case of very strong damping may apply to what happens in the transverse direction for a circularly polarized axially symetric beam. This is actually the situation dealt with at the end of section 6 with a pump compensating for absorption. Although the analytical solution may not be used for cylindrical coordinates, the result still holds. A cavity is formed when the intensity is greater than a critical value.

X - CONCLUSION

Soliton theory may be used to revisit laser plasma interaction. CO_2 experiments seem to evidence underdense shelves /21/ and cavities /22/. A surprising feature is the occurence of simultaneous peaks of ion density and of field squared, in the case of multiple potential wells fig. 10 and 12. However it should be

noted that the electrons are still forced to bunch in the vicinity
of the nodes of the fields (fig. 12) as it can easily be shown by
considering the corresponding B.G.K. equilibria.

REFERENCES

/1/ - N.J. ZABUSKY, M.D. KRUSKAL - Phys. Rev. Lett. 15 (1965) 240.

/2/ - Special Issue of Physica Scripta - 20 (1979) n° 3/4.

/3/ - R.Y. CHIAO, E. GARMIRE, C.H. TOWNES - Phys. Rev. Lett. 13
      (1964) 479.

/4/ - V.G. MAKHANKOV - Physics Reports 35 (1978) 1.

/5/ - T.D. LEE in Problèmes théoriques liés aux nouvelles particu-
      les. F.M. RENARD, P. SABATIER - ed. C.N.R.S. Paris (1978)

/6/ - D. Ter HAAR in Plasma Physics - H. WILHELMSSON - ed. Plenum
      (1977)

/7/ - V.I. KARPMAN - Non linear Waves in dispersive Media -
      Pergamon (1975)

/8/ - A.C. NEWELL in Solitons and Condensed Matter Physics -
      Springer (1978)

/9/ - Y.H. ICHIKAWA in réf /2/ p. 296

/10/ - V.I. KARPMAN in réf /2/ p. 462

/11/ - J.C. FERNANDEZ, C. FROESCHLE, G. REINISH in réf /2/ p. 545

/12/ - D.J. KAUP, A.C. NEWELL - Proc. Roy. Soc. Lond. A 361 (1978)
       413

/13/ - A.I. AVROV, V.Yu. BUCHENOV, O.N. KROKHIN, V.V. PUSTOVALOV,
       A.A. RUPASOV, V.P. SILIN, G.V. SKLIZKOV, V.T. TIKHONTCHUK,
       A.S. SHIKANOV - Soviet Phys. J.E.T.P. 72 (1977) 970.

/14/ - J.WEILAND, H. WILHELMSSON - Coherent non linear interaction
       of waves in plasmas - Pergamon (1977)

/15/ - D.J. KAUP - Stud. Appl. Math. 55 (1976) 9.

/16/ - R.D.C. MILLER, H. HORA - Plasma Physics 21 (1979) 183.

/17/ - V.N. LUGOVOI, A.M. PROKHOROV in Progress in Lasers and
       Lasers fusion. B. KURSUNOGLU, A. PERLMUTTER, S. WIDMAYER ed.
       Plenum (1975)
/18/ - K. KONNO, H. SUZUKI in réf /2/ p. 382

/19/ - J.L. BOBIN, WEE WOO, J.S. De GROOT - J. de Phys. 38 (1977)
       769

/20/ - V.L. GINSBURG - The propagation of Electromagnetic Waves
       in Plasmas - Pergamon (1964)

/21/ - R. FEDOSEJEVS, W. TOMOV, N.H. BURNETT, G.D. ENRIGHT,
       M.C. RICHARDSON - Phys. Rev. Lett. 39 (1977) 932.

/22/ - J.L. BOCHER, J.P. ELIE, J. MARTINEAU, M. RABEAU, C. PATOU
       in Laser Interaction and Related Plasma Phenomena.
       H.J. SCHWARZ, H. HORA ed. 4B Plenum (1977)

TAYLOR INSTABILITY IN FUSION TARGETS

R. L. McCrory
University of Rochester, Laboratory for Laser Energetics
Rochester, New York 14623

L. Montierth
R. L. Morse
C. P. Verdon*

University of Arizona, Department of Nuclear Engineering
Tucson, Arizona 85721

ABSTRACT

Optimum performance in laser driven fusion targets is enhanced
by the use of high aspect ratio shells in commonly employed spherically symmetric designs. Taylor instability occurs in these systems
at (1) the ablation surface when the acceleration is in the same
direction as the local density gradient, and (2) later in the implosion process when the fuel decelerates the pusher. Results for
a linear stability analysis of the ablation-driven Taylor instability are obtained from a perturbation analysis of the two parameter
stationary ablative flow model. The linear growth rates are shown
to be in agreement with full two-dimensional numerical simulations.
From the linear analysis, the potentially most damaging unstable
mode is identified, and full two-dimensional numerical simulations
are performed. The two-dimensional calculations determine the
nature and saturation behavior associated with the unstable mode.
Our results indicate that saturation of the ablatively driven
Taylor instability does occur. This saturation occurs at an amplitude which is sufficiently large to be a possible cause of difficulty in using large aspect ratio shells in fusion targets, but is
seen to prevent these shells from breaking up and becoming turbulent. It appears plausible that such distorted but laminar, i.e.
non-turbulent, shells could be successfully employed in fusion
targets.

*Present address: University of Rochester, Laboratory for .Laser
 Energetics, 250 E. River Rd., Rochester, New York 14623

I.   INTRODUCTION

Taylor instability is the instability which occurs at the
interface of a heavier fluid superposed over a lighter fluid.[1]
Gravity causes ripples at the interface to grow until the heavier
fluid falls through the lighter, and in the end the heavier fluid
comes to rest on the bottom of the container. A modified version
of this instability occurs in spherical implosion systems in
inertial confinement fusion (ICF) targets, with radial accelera-
tions playing the role of gravity. The most effective implosion
systems are particularly vulnerable to this type of instability
because they consist of one or more relatively thin spherical
shells which accelerate through many times their initial thickness
during implosion.[2-7] Figure 1, adapted from Ref. 2, shows density
profiles of the standard types of systems. Type IV with "levitated
fuel", which is a more elaborate version of Type II, in general
gives higher compressions for given constraints on the driver
pulse, i.e., laser or particle beam pulse, than do Types I and III.

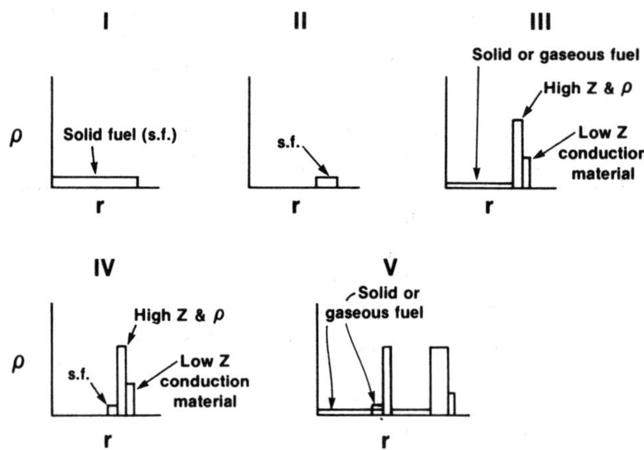

FIG. 1   Five common types of spherically symmetric inertial
confinement fusion pellets. (Adapted from Ref. 2).

In general, the larger the initial aspect ratio, $A_0 \equiv r_0/\Delta r_0$ (where r_0 and Δr_0 are initial shell radius and thickness), of the shells, the better the performance of the system. Typical desirable values of A_0 are in the range $10 \lesssim A_0 \lesssim 100$. The currently popular use of multiple high density shells, of which Type V is a typical example, makes possible various implosion timing improvements at some cost in efficiency, but does not introduce any instability phenomena which are qualitatively different from those found in Types III and IV.

Taylor instability occurs in these systems when and where acceleration is in the same direction as the local density gradient, so that in effect a heavier fluid is above a lighter one in the local gravitational field. This occurs first at the outside surface of the outer shell as illustrated in Fig. 2. The role of the lower density fluid is played by the material which has been heated directly by thermal conduction from the outer high-temperature corona region and is being ablated outwardly from the shell.

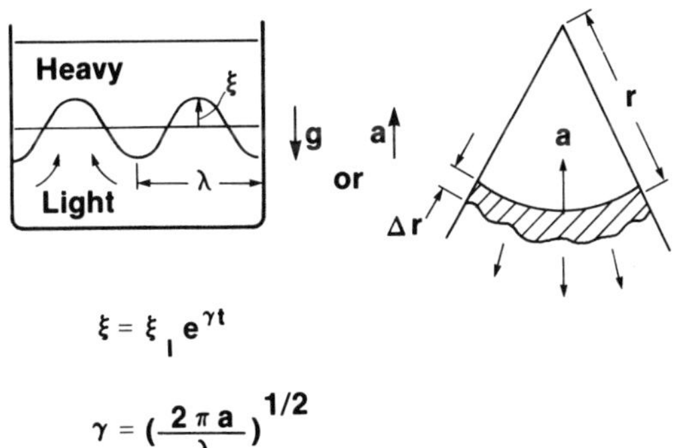

FIG. 2 Classical Taylor instability conditions are illustrated by the container of two fluids in a gravitational field, as illustrated in the sketch on the left. The analogous situation for the ablation driven Taylor instability is illustrated on the right.

The inward acceleration causes ripples in the ablation front region
to grow unstably. Later, when the fuel has reached the center, the
interface between the fuel and the high density pusher shell in
Type III or IV systems also becomes unstable, because the pusher is
decelerated by fuel pressure.[2]

When successful implosion systems are designed for optimum performance they will probably come as close as possible to failure
from Taylor instability without failing. In the manner of failure
that would be approached, outside instability could cause growth of
shell distortions but not to the point of complete loss of shell
integrity at the inner surface. The inside instability would then
cause further growth until the interface between the fuel and the
pusher shell became so distorted that fuel and pusher material would
become intermixed; further fuel compression would cease, and ignition precluded or any thermonuclear burn in progress would be quenched.
Because both the inside and outside instability contribute to this
failure mode, it will eventually be necessary to calculate the behavior of both in detail to predict and to avert failure. However,
at present the outside instability is in a sense more important,
because much less is known about it, and because it appears possible
that it may not be as troublesome in all cases as has previously been
predicted from linear analysis.[2]

The inside instability is usually classical, i.e., ablative
phenomena usually do not play a significant role. Classical analytic estimates of growth rates[1] are, therefore, approximately correct,
and the nonlinear development can be anticipated from previous numerical simulations of the classical Taylor instability which show the
bubble and spike structure.[8-10] In section II below we give a brief
description of the analytic theory of these classical growth rates
and their application to the implosion of shells. In contrast with
the state of understanding of the approximately classical inside
instability, there has been no complete systematic treatment of
either linear growth rates or nonlinear development of the outside
instability, which is more complicated because the heat flow and
convection associated with ablation play an important role. Research
done on linear growth rates[2,11-14] shows that in cases of interest,
ablation phenomena reduce the linear growth rates significantly at
shorter wavelengths, but that the growth rates are still large enough
to cause serious difficulty.[2,11] There has also been an indication
from numerical simulations that nonlinear saturation mechanisms may
in some cases limit shell distortion enough to prevent failure.[14]
This effect could be very important if it occurs over a sufficient
range of parameters to make usable large aspect ratio systems which
would be expected to fail from a simple extrapolation of linear
theory.

Preliminary results are presented below from two separate sets of calculations of the behavior of ablation-driven Taylor instability. The first (Section III) is a parameter study of linear growth rates which is based on zero-order stationary ablative flow solutions. The second (Section IV) is a numerical simulation study of nonlinear development done with a two-dimensional, triangular zone, Lagrangian hydrodynamic and heat flow code which was developed for this purpose. From the linear growth rates, it will be possible to determine, for a given implosion system, the range of modes which grow enough that failure should be expected from extrapolation of linear theory. The two-dimensional simulations should then show which of these modes would be prevented by nonlinear effects from growing to sufficiently large amplitude to cause failure and, therefore, which systems should be expected to implode successfully.

II. CLASSICAL ESTIMATES OF SHELL INSTABILITY

Figure 2 shows the classical Taylor instability conditions in the container of the two fluids in a gravity, g, on the left and the analogous situation on the outside of the shell on the right, which is being given an inward acceleration, a, by ablation pressure at the outside surface. Suppose the amplitude of the ripple of the interface between the two fluids in the container has the form,

$$\xi = \xi_0 \, \exp(ikx + \gamma t), \tag{1}$$

where x is the coordinate parallel to the interface,

$k \equiv 2\pi/\lambda$, and

ξ_0 is the amplitude at t = 0.

If the fluids are incompressible and inviscid, the thicknesses of the fluid layers are great compared to the ripple wavelength, λ, and the heavy fluid is much more dense than the light fluid; then the growth rate, γ, is

$$\gamma = (gk)^{1/2} . \tag{2}$$

To the extent that classical theory is at all applicable to the ablative outside surface, the assumption that the heavy fluid is much more dense than the light is a good approximation. This is so, because typical density ratios across the ablation front from the shell, which is the heavy fluid, to the ablated material near the isothermal sonic point of the flow (which is a reasonable place to define the lower density) are of the order of 10^2.

<u>Failure</u>

From this classical theory, estimates can be made of the conditions under which shell failure would occur. The subject of
failure is not well understood, but it is expected that failure
would occur if the surface ripple amplitude became of the order of
or larger than the shell thickness at the time of failure. If we
accept this criterion for failure, then from the definition of
aspect ratio, $A \equiv r/\Delta r$, the failure amplitude, ξ_f, is

$$\xi_f = \Delta r = r/A \ . \tag{3}$$

Here all quantities are taken not at initial time but at the time
of t_f, when failure occurs. That is, t_f is the time when the
failure criterion is first met, even though shell distortions may
increase after this time. To obtain numerical estimates that indicate the magnitude of the problem, we must choose the values of
k that will be most troublesome, and this we do according to the
conventional wisdom, to which the authors at least partly subscribe.
That is, the most troublesome mode is the fastest-growing mode that
can cause the shell to break up. From Eq. (2) the growth rate
increases with k. On the other hand, the e-folding depth of
penetration of the k mode into the fluid from the interface is k^{-1},
and nonlinear calculations indicate that growth slows from exponential to linear, or in some cases, saturates turbulently and stops
when the amplitude reaches this penetration length. Modes with
values of k^{-1} less than the shell thickness, the conventional wisdom dictates will not cause the shell to break up and are not as
troublesome as the slightly slower modes that can cause break up.

This worst value of k is then $k_w \simeq \Delta r^{-1}$, or since for large ℓ
$k \simeq \ell/r$, we also have a corresponding worst ℓ, and, from Eq. (3)

$$\xi_f = k_w^{-1} = \frac{r}{\ell_w} = \frac{r}{A} \ . \tag{4}$$

Substituting Eq.(4) into Eq.(2) gives the growth rate for the
"worst" mode:

$$\gamma_w = \left(\frac{Ag}{r} \right)^{\frac{1}{2}} \ . \tag{5}$$

It should be noted that A here is the in-flight aspect ratio, as
discussed below, not the initial aspect ratio A_0. Substituting
Eq.(5) into Eq.(1) gives the amplification of the "worst" mode,

$$\xi = \xi_0 \exp \left[\left(\frac{Ag}{r} \right)^{\frac{1}{2}} t \right] \ . \tag{6}$$

The larger t can be, that is, the longer the instability has to
grow, the smaller ξ_0 must be to prevent failure, and practical

limits on surface finish put lower limits on ξ_0. The procedure to
follow is then to estimate the minimum value of ξ_0 and the maximum
value of t and to ask if the resulting ξ exceeds ξ_f from Eq.(3).
It is tempting to assume that failure can occur as late as the
full implosion time, τ_I, defined as

$$\tau_I \equiv \left(\frac{2r_0}{g} \right)^{1/2} . \tag{7}$$

This is simply the time required for a shell to reach the origin
if it implodes with a constant acceleration, g. This, however, is
probably a bit pessimistic in the sense that it allows more time
for the instability to develop than is actually available, for
several reasons. First, the outer shell surface does not travel
all the way to the origin, because the compression is finite. In
high compression ICF systems, however, the outer surface does
travel well over half way. Spherical convergence makes a further
contribution to limiting growth time. In most cases of interest,
as the shell moves toward the origin, its aspect ratio first increases rapidly because the shell is made thinner by shock and
acceleration-induced compression. Typically A increases from A_0
by a factor of 4 or a bit more. A then changes slowly for an
intermediate period during which the effects of convergence and
ablative mass removal roughly cancel. In cases of interest, which
do not burn through, the convergence then finally wins; the shell
thickens while the radius decreases, and A decreases again. It is
during the intermediate period when A is largest and the shell is
near its initial radius, r_0, that the shell is most vulnerable to
break up by the larger ℓ (or k) faster growing modes. Thereafter,
as A decreases, it is expected that further growth of the higher
ℓ modes will have relatively less disruptive effects on the shell.
For these reasons in applying Eq.(6), we choose $t_f = \tau_I/2$,
$A = 4A_0$, and $r = r_0$. These choices of "round numbers" all appear
to err a bit in the direction of underestimating growth of
unstable modes. Substituting into Eq.(6) gives the initial perturbation amplitude, ξ_0, which would cause failure:

$$\xi_0 = \frac{r_0}{4A_0} \exp \left[-\left(\frac{4A_0 g}{r_0} \right)^{1/2} \frac{\tau_I}{2}\right]$$

$$= \frac{\Delta r_0}{4} \exp \left[-(2A_0)^{1/2}\right] . \tag{8}$$

Notice that the acceleration, g, has cancelled out. To see what
this means, consider a typical case of interest: $A_0 = 25$ and
$r_0 = 100$ μm. From Eq.(8) the requirement to avert failure is
$\xi_0 < 8.5 \times 10^{-8}$ cm. Without going into the particulars of the
possible spectra of shell surface imperfections responsible for
ξ_0, it can be seen that this requirement places severe

restrictions on the surface finish. If instead A_0 = 10, then the
requirement is $\xi_0 \leq 2.8 \times 10^{-6}$, and if A_0 = 5, then $\xi_0 \leq 2.1 \times 10^{-5}$.
Only the requirement for A = 5 represents a surface finish which
seems clearly attainable.

Linearized calculations presented in Ref. (2), which include
ablative effects, obtain amplifications which are a bit smaller
than, but within an order of magnitude of those obtained from
Eq.(6) above. From these more realistic calculations and slightly
more conservative estimates of attainable surface finish, Ref. 2
also concluded that $A_0 \leq 5$ is required to avert shell breakup.
Clearly if anything can be done to relax this restriction, it
would be very desirable, because the performance expected from
shells with $A_0 < 5$ is rather disappointing.

Compressibility Correction

There is a further correction to the classical linear growth
rates that is caused by compressibility, which may be important,
and should be pointed out before the ablative calculations are
presented. In the classical calculations[1] it is assumed that the
fluids are incompressible. For present purposes this is essentially equivalent to assuming that the linear growth time of a
mode, $\tau \equiv \gamma^{-1}$, is greater than the acoustic transit time, τ_s,
across the characteristic length of the mode structure. That is,
since the characteristic length of a Taylor mode is k^{-1}, it is
implicitly assumed that

$$\gamma \leq k\, C_s \, , \tag{9}$$

where C_s is the sound speed in the shell. A Taylor mode cannot in
fact grow much faster than the upper limit given by the equality
in Eq.(9), because the pressure differences that cause the flow
cannot be transmitted across the mode structure on a shorter time
scale. In ideal gases, the sound speed is

$$C_s = \left(\frac{5P}{3\rho} \right)^{1/2} \, , \tag{10}$$

where P and ρ are pressure and density, and the pressure in a
shell at the outside, or ablation, surface is approximately given
by the effective weight per unit area of the shell on the ablation
surface,

$$P = (\rho a \Delta r_f) \, . \tag{11}$$

Here for consistency with the estimates of growth rates above, the
final shell thickness, Δr_f, is used as an estimate of the in-flight
thickness, Combining Eqs. (9), (10), and (11) gives

$$\gamma \leq k\left(\frac{5a}{3}\,\Delta r_f\right)^{1/2} = (ak)^{1/2}\left(\frac{5}{3}k\,\Delta r_f\right)^{1/2} = k\left(\frac{5}{3}\,\frac{ar_f}{A}\right)^{1/2} \,. \quad (12)$$

Comparison of Eq. (12) with (2) above, with a = g, shows that in the "worst case" discussed above, for which $k_w\,\Delta r_f = 1$, Eq. (12) places an upper limit on γ which is larger than the classical growth rate but only by a factor of $(5/3)^{1/2}$. For smaller values of k, it is seen that Eq. (12) limits γ to values which are smaller than the classical incompressible result. It will in fact be seen in section III that for a range of smaller values of k below k_w, Eq. (12) becomes approximately the correct expression for γ, instead of Eq. (2). If nonlinear saturation prevents shell breakup from modes with $k \leq k_w$, this reduction in linear growth rates for smaller k may then be important.

III. CALCULATION OF LINEAR GROWTH RATES USING THE STEADY FLOW MODEL

Exact calculation of amplification of linearly unstable Taylor modes during an implosion requires calculation of the development in time of first-order perturbations of a time-dependent zero order flow. The separation into modes occurs naturally through an expansion of the angle dependence in the spherical harmonics, $Y_\ell^m(\theta,\psi)$. The various first-order quantities then appear in the form $f_{1\ell}(r,t)\cdot Y_\ell^m(\theta,\psi)$, where $f_{1\ell}$ is one of the first-order quantities, $\rho_{1\ell}$, $T_{1\ell}$, the radial component of the perturbed velocity, $V_{1r\ell}$ or ξ_r, the perturbed displacement, or the divergence of the corresponding angular components of these two quantities, $\nabla\cdot\bar{V}_{1\Omega\ell}$ and $\nabla\cdot\xi_\Omega$. The system of coupled differential equations for these $f_{1\ell}$'s is degenerate with respect to m and decouples with respect to ℓ.[12-14] The different modes are, therefore, identified by their ℓ number. Solution consists of numerical integration in time of the system of equations for the $f_{1\ell}$'s for as many different ℓ's as are desired. The zero-order quantities, the $f_0(r, t)$'s, which describe the spherically symmetric implosion, appear as coefficients in the system of equations for the $f_{1\ell}$'s and can be computed simultaneously or computed previously and stored.

This exact method of solution has been applied successfully to implosions of shells[2,13,14] and will probably be necessary for detailed examination of specific fusion target designs in the future. However, the method in a sense gives too much information for making qualitative, i.e., order of magnitude, judgments about broad classes of implosion systems. Total amplifications of perturbations are obtained for a specific implosion system driven by a specific incident driver pulse. In practice, this solution space has so many parameters that extracting patterns of behavior and insight from it can be a bit tedious.

What is needed to complement these exact calculations of total linear amplifications is the instantaneous linear growth rates of the Taylor modes as a function of ℓ for the minimum number of fundamental parameters needed to describe the instantaneous state of an unstable region. The classical theory, of which Eq. (2) is a simplified form, fills this need for non-ablative unstable regions when, as discussed below, the compressibility corrections are added. The numerical calculations of growth rates of ablation-driven Taylor instability presented here are intended to fill the similar need for ablative regions where heat flow and convection are important.

An appropriate family of zero-order ablative flow solutions from which to derive instantaneous growth rates is obtained from the steady flow model.[15,16] In cases of interest, the instability growth time is short compared to the implosion time, i.e., the time for shell density, temperature, and velocity profiles to change significantly. However, the growth time is not necessarily shorter, and can be longer, than the characteristic time for a fluid element to flow through the ablation front region. By being stationary in time while including the effect of convection, the steady flow model meets the requirements of this ordering of times, but is simple enough that the family of solutions may be described by a two parameter family. A brief description of the steady flow model and solutions will be given next, before the stability calculations are presented.

Steady Flow Model

In this model, the flow is spherically symmetric, radially outward, and stationary. Stationary flow is a reasonable approximation in most cases of interest, because the time required to establish the flow between the pellet surface and the region where the absorbed energy is assumed to be deposited is less than the characteristic implosion time and, because the pressure gradient forces in the ablation region are larger than the forces associated with the average inward acceleration.

For our purposes, very little generality is lost and considerable simplicity is gained, by assuming a single temperature instead of separate electron and ion temperatures. The steady-state hydrodynamic equations are then:

Continuity:

$$\frac{d}{dr}\left(\rho v r^2\right) = 0 ,\tag{13}$$

Momentum:

$$\rho v \frac{dv}{dr} = - \frac{dP}{dr} - \rho \frac{d\psi}{dr} , \quad \text{and} \tag{14}$$

Energy:

$$\frac{1}{2} v^2 + \frac{\gamma_c}{\gamma_c - 1} \frac{P}{\rho} + \psi - \Omega = \frac{1}{2} v_b^2 , \tag{15}$$

where ρ is the fluid mass density, v is the radial fluid velocity, r is the radius, P is the pressure, ψ is the gravitational potential, γ_c is the ratio of specific heats, and v_b is the Bernoulli energy flow constant. The gravitational potential, ψ, which is introduced to represent inward acceleration must be included to yield shell-like density profiles. The quantity Ω is defined to be $\Omega = (\sigma\kappa/S)(dT/dr)$, where $S = 4\pi r^2 \rho v$ is the mass flux, a constant of the motion by Eq. (13). κ is the thermal conductivity coefficient given by

$$\kappa = G(Z) \, T^\alpha \, \rho^{-\beta} . \tag{16}$$

Following Spitzer[17] we take $\alpha = 5/2$, $\beta = 0$, and $G(Z) = \kappa_0/Z$, where κ_0 is a constant, for most of what follows. We have ignored the weak temperature and density dependence in the $\ln\Lambda$ term which appears in the conductivity,[17] T is the temperature, and we have defined $\sigma = 4\pi r^2$. P is related to T through the fully ionized ideal gas equation of state $P = \rho RT/\mu$, where R is the gas constant, $\mu = \mu_0/(1 + Z)$, μ_0 is the atomic mass of the fluid, and Z its atomic number. After some manipulation, Eqs. (13), (14), and (15) become:

$$\rho v \sigma = S , \tag{17}$$

$$v\left(1 - \frac{RT}{\mu v^2}\right) \frac{dv}{dr} = \frac{RT}{\mu} \left(\frac{2}{r} - \frac{1}{T} \frac{dT}{dr}\right) - \frac{d\psi}{dr} , \tag{18}$$

$$\frac{1}{2} v^2 + \frac{\gamma}{\gamma - 1} \frac{RT}{\mu} + \psi - \frac{\sigma}{S} \kappa \frac{dT}{dr} = \frac{1}{2} v_b^2 . \tag{19}$$

Because transonic solutions are of particular interest to us, it is useful to scale the equations in the following way. Define $\tilde{\rho} = \rho/\rho_s$, $\tilde{T} = T/T_s$, $\tilde{v} = v/v_s$, $\tilde{\psi} = \psi/v_s^2$, and $\tilde{r} = r/r_s$, where ρ_s, T_s, and v_s are the values of the flow variables at the *isothermal* sonic point r_s; that is, at r_s, $v_s = (RT_s/\mu)^{1/2}$. Using this scaling, Eqs. (18) and (19) become:

$$\frac{d\tilde{v}}{d\tilde{r}} = \frac{1}{\tilde{v}} \; \frac{\tilde{T}\left(\frac{2}{\tilde{r}} - \frac{1}{\tilde{T}} \frac{d\tilde{T}}{d\tilde{r}}\right) - \frac{d\tilde{\psi}}{d\tilde{r}}}{(1 - \tilde{T}/\tilde{v}^2)} \quad , \text{ and} \tag{20}$$

$$\frac{d\tilde{T}}{d\tilde{r}} = \frac{M\,[\tfrac{1}{2}(\tilde{v}^2 - \tilde{v}_b^{\,2}) + \frac{1}{\gamma - 1}\tilde{T} + \tilde{\psi}]}{\tilde{r}^2(1 + \beta)\tilde{T}^\alpha\tilde{v}^{+\beta}} \quad . \tag{21}$$

The dimensionless quantity M is defined as

$$M = \frac{r_s\rho_s^{\,1 + \beta} v_s^{\,3}}{G(Z)T_s^{\,1+\alpha}} = \frac{r_s\rho_s\, v_s^{\,3}}{k_s T_s} \quad , \tag{22}$$

and is the only place in the equations where dimensional parameters appear explicitly.

For use in Eq. (20), we define $\tilde{g} = -\,\partial\tilde{\psi}/\partial\tilde{r}$. The form of ψ that has been used is $\psi = -g(r - r_s)$, or $\tilde{\psi} \equiv -\tilde{g}(\tilde{r} - 1)$, where the dimensionless form of the gravity constant then becomes $\tilde{g} \equiv gr_s/v_s^2$.

The numerical solution of Eqs. (20) and (21) is accomplished by integrating inward ($\tilde{r} < 1$) and outward ($\tilde{r} > 1$) from the singularity of Eq. (20) at $\tilde{r} = 1$. We further restrict our attention to solutions which are subsonic for $\tilde{r} < 1$ and supersonic for $\tilde{r} > 1$. We require that $d\tilde{v}/d\tilde{r}$ be finite at $\tilde{r} = 1$. Thus, since the denominator of the right-hand side of Eq. (20) becomes zero at $\tilde{r} = 1$, the numerator must also be zero there. Values of the temperature and velocity derivatives at this singularity are determined as a function of M and $\tilde{g}$ by requiring that the solutions be continuous there. Limits on the range of values of M are also determined, as a function of $\tilde{g}$, by conditions at $\tilde{r} = 1$, i.e., $r = r_s$. These constraints, as well as the procedure for treating the laser energy deposition at a critical surface, and some asymptotic analytic solutions for $r \rightarrow \infty$ are derived in Ref. 15.

It is also shown that values of M near the lower limit of the allowed range correspond to larger values of the ratio $\tilde{\rho}_p \equiv \rho_p/\rho_s$ of the peak density at the outer shell surface ρ_p, to the sonic point density, ρ_s.

For present purposes, the driver energy is assumed to be deposited at larger radii than are directly involved in the stability calculations, i.e., at $r \rightarrow +\infty$. In other applications of the steady flow model, the driver energy is explicitly introduced at finite radii[15], such as the critical surface of laser light, and one or more additional parameters are introduced, which are not necessary for the present requirements here.

The solutions of Eqs. (21) and (22) form a two-parameter family. The two parameters are M and $\tilde{g}$, but M, whose physical meaning is not transparent, is equivalent to the ratio of peak shell density to the sonic density. That is, M is found to be equivalent to $\tilde{\rho}_p \equiv \rho_p/\rho_s$ where ρ_p is the peak density of the shell or pellet. Cases of interest for laser fusion applications have values of $\tilde{\rho}_p$ of the order of 10^2. All other things being equal, larger $\tilde{\rho}_p$ corresponds to smaller incident power. Values of $\tilde{\rho}_p$ of the order of 10 or less approach the rapid thermal burn through limit where exploding pusher behavior accurs. It may be interesting to the reader to note from Eq. (20) that $\tilde{g}$ (defined as $-\partial\tilde{\psi}/\partial\tilde{r}$) is effectively a measure of the relative importance of spherical divergence and acceleration in determining the flow in the sonic region. The limit $\tilde{g} \ll 1$ gives solutions in which acceleration is unimportant, and the shells are so thick that $A \lesssim 1$. As $\tilde{g} \to \infty$, $A \to \infty$, spherical curvature is not important, and flat slab solutions are obtained for this limit. For ICF applications, solutions with $\tilde{g}$ of order 1 are often most relevant, as we discuss below. Figure 3 shows the steady flow solution profiles for the particular case $\tilde{\rho}_p = 50$ $\tilde{g} = 1.0$, and represents one of the particular cases for which instability growth rates and mode forms are obtained by the procedures described in the following section.

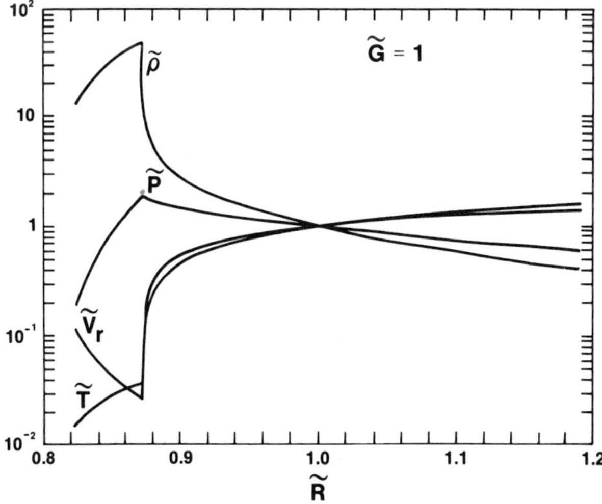

FIG. 3 Stationary ablative flow profiles for the normalized
         density ($\tilde{\rho}$), pressure ($\tilde{P}$), velocity ($\tilde{V}_r$), and tem-
         perature ($\tilde{T}$) obtained for the case $\tilde{G} = 1$. The flow
         variables are normalized at the isothermal sonic point
         ($\tilde{R} = r/r_s = 1$).

A further transformation of parameters is useful for applications of the steady flow solutions and their stability properties to implosion systems. The aspect ratio, A, defined in Section II above, is defined from each computed solution as the width of the shell density profile at half height. This is the first transformed parameter. The second parameter is the fraction of the shell mass that is removed by ablation during the implosion time τ_I, defined in Section II. If M and ΔM are the initial mass and the ablated mass increment respectively, this burn through parameter is then

$$B \equiv \frac{\Delta M}{M} = \frac{S\tau_I}{M} = \frac{(\rho_s r_s^2 v_s)\tau_I}{M} = \left(\frac{r_p}{r_s}\right) \frac{\tau_I}{\tilde{g}} \left(\frac{r_s^3 \rho_s}{M}\right). \tag{23}$$

Figure 4 illustrates the dependence of A on $\tilde{g}$ and $\tilde{\rho}_p$. Note the linear dependence of A on $\tilde{\rho}_p$ and on $\tilde{g}$ for large $\tilde{g}$ and $\tilde{\rho}_p$. Figure 5 shows the full mapping from the solution parameters $(\tilde{\rho}_p,\tilde{g})$ to the parameters (A,B). Maximum energy transfer is achieved from the energy source driving the ablation to part of the imploding shell that is not ablated if $\Delta M/M$ is about 2/3, just as one would expect in analogy with a "rocket model"[18] to which this system is approximately equivalent. Taking into account the approximations involved, one then expects the optimum solutions to be in the range $0.5 < B < 1.0$. Other implosion system design constraints could, however, lead to the use of solutions outside of this range. From Fig. 5 it is seen that this range of B values and aspect ratios in the popular range $10 < A < 100$ requires solutions in the neighborhood of $\tilde{\rho}_p = 50$ to 100 and $\tilde{g} = 1.0$.

Method of Calculating Perturbations

The method used here is similar to that used in Refs. 12-14, and described briefly above, except that the equations for the time dependence of the perturbed quantities are written in Eulerian form and solved on a regular fixed grid instead of being written in Lagrangian form. This procedure was adopted because the zeroth order steady flow solutions are computed on a regular fixed Eulerian grid, as opposed to the procedure used when the exact zeroth order solutions of Refs. 12-14 are computed which employ a Lagrangian grid moving with the fluid. In either case, it is desirable, for simplicity and accuracy, to have the first and zeroth order computations done on the same grid. The first order equations are obtained by expansion of the angle dependence in spherical harmonics, $Y_\ell^m(\theta,\psi)$, and make use of the fact that the resulting system of equations decouple with respect to ℓ and are degenerate with respect to m. These equations are, therefore, the same as those derived in Ref. 12, except for the differences required by the change from Lagrangian to Eulerian coordinates and will not be written out here.

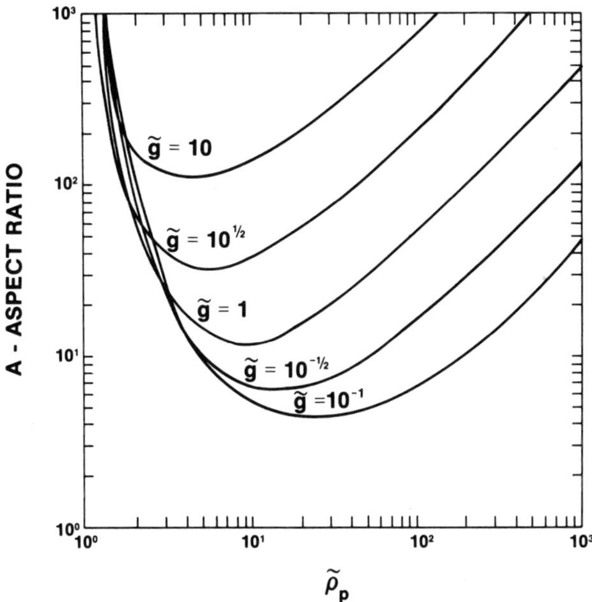

FIG. 4   Dependence of the aspect ratio, A, as a function of $\tilde{\rho}_p$ for various values of the normalized acceleration, $\tilde{g}$ obtained from the two parameter stationary ablative flow model. Note the linear dependence of aspect ratio on $\tilde{\rho}_p$ for large $\tilde{\rho}_p$, $\tilde{g}$.

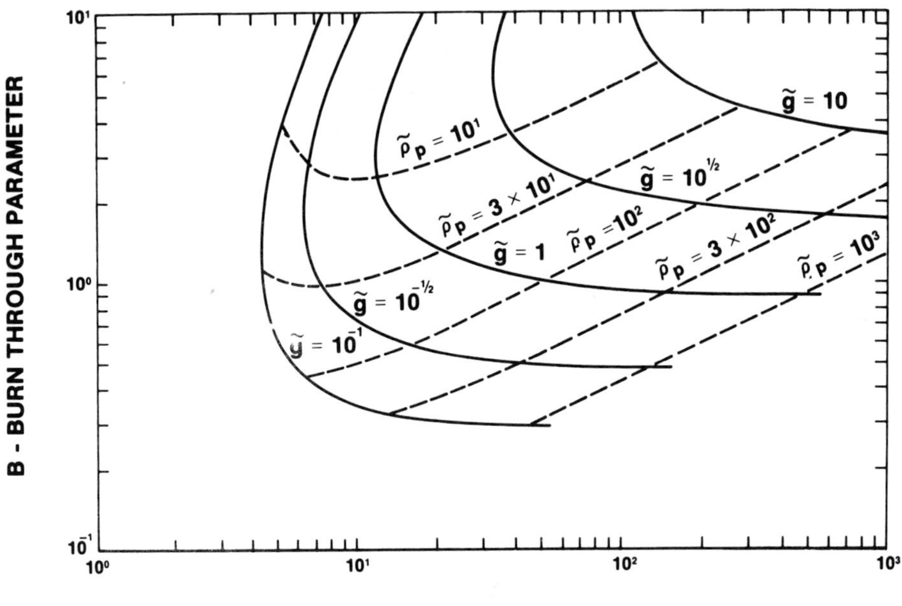

FIG. 5 Full mapping from the solution parameters $(\tilde{\rho}_p, \tilde{g})$ to the parameters (A,B) from the stationary ablative flow model. A is the shell aspect ratio, and B is the burn through parameter, defined in the text, (Eq. 23), B = ΔM/M.

The first order variables are T_1, ρ_1, $\bar{v}_1$, and $\bar{\xi}$, the perturbed temperature, density, velocity, and displacement. All scalar quantities have the form $T_1 = T_1(r,t)Y_\ell^m(\theta,\psi)$ as described near the beginning of this section. The zeroth order quantities obtained from the steady flow model appear as coefficients in the equations of motion for the first order variables. The displacement, $\bar{\xi}$, is the distance between the perturbed and unperturbed positions of a fluid element and is introduced primarily as a diagnostic variable. $\bar{\xi}$ is a measure of shell distortion and is obtained by integrating

$$\frac{D\bar{\xi}}{Dt} \equiv \frac{\partial\bar{\xi}}{\partial t} + (\bar{v}_0\cdot\bar{\nabla})\bar{\xi} = \bar{v}_1 \; . \tag{24}$$

The mode form and growth rates for a given ℓ are obtained from the equations of motion for the perturbations by introducing an arbitrary initial perturbation (in practice a Gaussian perturbation of $\rho_1(r)$ localized near the ablation surface of width $2\pi r/\ell$ was used) and integrating the equations in time until the form of the mode and the growth rate of its amplitude become approximately constant. In this way, the fastest growing mode with the given ℓ and its growth rate are obtained.

Results

Figure 6 shows the growth rates for three zero-order steady flow solutions with $\tilde{\rho}_p(\equiv \rho_p/\rho_s) = 50$. The zeroth order solution with $\tilde{g} = 1$ is the case shown in Fig. 3. These three zeroth order solutions can be identified in Figs. 4 and 5. Recall that the dimensionless growth rate is $\tilde{\gamma} \equiv \gamma\, r_s/v_s$. The scale time r_s/v_s, is a characteristic convection time for the ablative flow (although perhaps this time is a bit longer than is typical of flow through the ablation front because r_s is the full radius of the system). Consequently, solutions with $\tilde{\gamma} \gg 1$ should not be greatly influenced by convection, while those with $\tilde{\gamma} \ll 1$ should be influenced by convection, in whatever way this influence expresses itself. It is interesting that there appears to be no qualitative change in the growth rate curves (see Fig. 6) from one side of $\tilde{\gamma} = 1$ to the other.

For larger ℓ, i.e., larger k, it is seen that ablative effects cause $\tilde{\gamma}$ to reach a maximum and then decrease to zero, instead of increasing indefinitely as classical, non-ablative inviscid, incompressible theory would predict. For smaller values of ℓ, it is seen from Fig. 6 that $\tilde{\gamma}$ becomes independent of $\tilde{g}$ and A and approximately proportional to ℓ. That $\tilde{\gamma}$ is approximately independent of A and $\tilde{g}$ for smaller ℓ is a consequence of the approximately linear dependence of A on $\tilde{g}$ for larger $\tilde{\rho}_p$ and $\tilde{g}$ (see Fig. 4). The

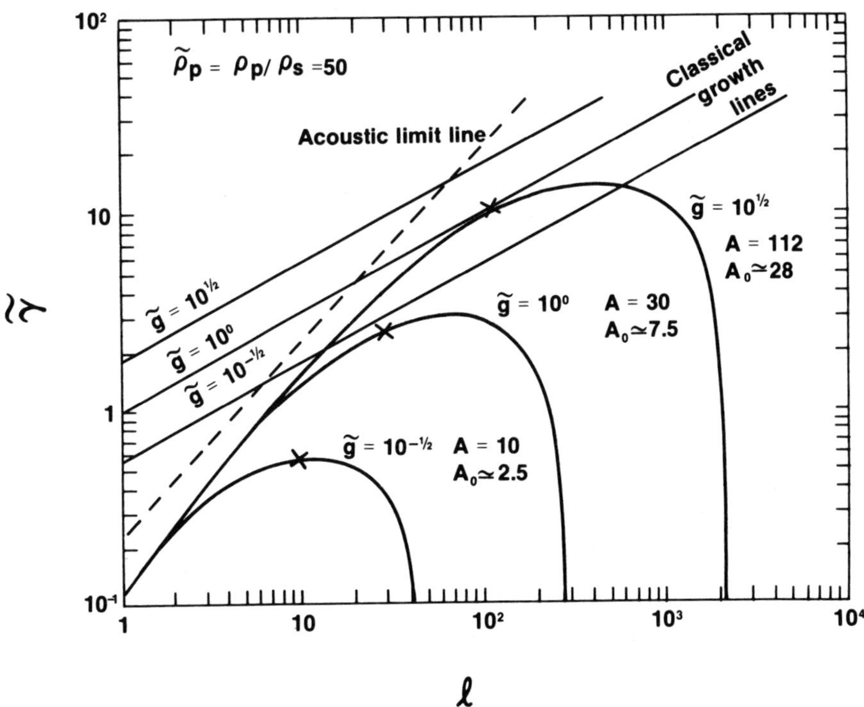

FIG. 6   Growth rates obtained for perturbation calculations performed for three stationary flow solutions with $\tilde{\rho}_p = 50$.
The steady flow solution profiles for the $\tilde{g} = 1$ case are
those of Fig. 3. The classic incompressible inviscid growth
lines are given for the three cases; $\tilde{\gamma}(\text{classic}) = \sqrt{\ell\tilde{g}}$. The
acoustic limit line, $\tilde{\gamma}(\text{limit}) = \ell\sqrt{.6(\tilde{g}/A)}$, is also shown.
This line illustrates the effects of compressibility,
important at low ℓ values, which is responsible for the large
departure of the growth rates for these ℓ's from the
classical value. The "worst case" ℓ values given by the
approximate formula, $\ell = A$, are marked by crosses.

approximately linear relationship between $\tilde{\gamma}$ and ℓ seen in Fig. 6
is in approximate agreement with the equality in Eq. (12) and the
limitation on $\tilde{\gamma}$ imposed by the acoustic transit time. Recall that
for $\ell \gg 1$, $k \simeq \ell/r$. When numerical values from Fig. 4 are substituted into Eq. (12), the acoustic limit line of Fig. 6 is obtained. It is seen that $\tilde{\gamma}$ nowhere exceeds this limit but lies
rather close to it for small ℓ, where it also appears that
ablative effects have not significantly reduced $\tilde{\gamma}$. On the other
hand, for these smaller values of ℓ, and therefore k, $\tilde{\gamma}$ lies far
below the incompressible value given by Eq. (2), which further

indicates that compressibility is determining the growth rates
for small values of ℓ. An additional effect, not treated in the
above argument, occurs for small values of ℓ: $k \sim \ell/r$ is not applicable, because the wavelength of the mode is no longer small compared
to the radius of the shell, and curvature effects must be considered.
The lines with slope 1/2 on Fig. 6 are those given by Eq. (2) with
$k = \ell/r_s$, for the different values of $\tilde{g}$. The "worst case" values
of ℓ, at which $\ell = A$ are indicated in Figure 6. It is interesting
that these classical "worst case" values of ℓ are quite close
(within a factor of 2) to the values of ℓ at which the maxima of $\tilde{\gamma}$
occur. The values of these maxima $\tilde{\gamma}$ at these "worst case" values of
ℓ are seen in all cases to be significantly less than the $\tilde{\gamma}$ one
obtains from the classical estimate associated with that ℓ (see
Fig. 6). From this result, we see it is possible to increase significantly the value of the lowest aspect ratio, A at which shell break-
up should be expected to occur.

Figure 7 shows the form of the perturbed radial displacement
ξ_r, for the case $\tilde{\rho}_p = 50$, $\tilde{g} = 1$, and $\ell = 50$. This case is seen from

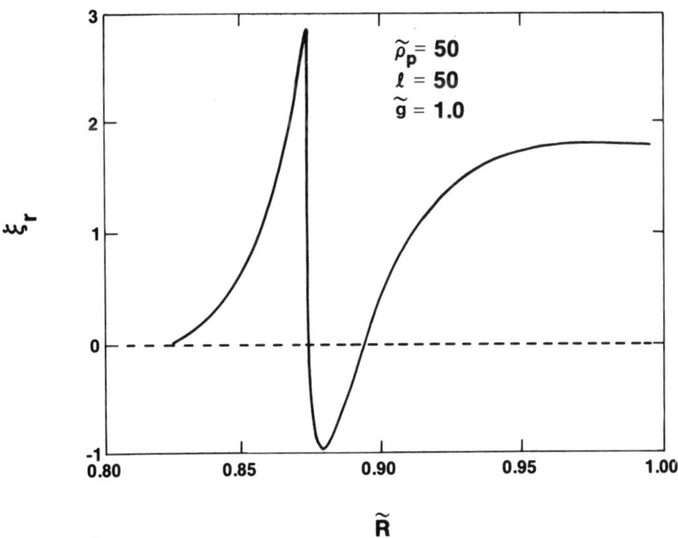

FIG. 7 Perturbed radial displacement ξ_r for the case $\tilde{\rho}_p = 50$, $\tilde{g} = 1$,
and $\ell = 50$. This case has nearly the maximum growth rate
(see Fig. 6). Note that ξ_r penetrates the shell to a
depth $\sim \ell^{-1}$, and is extended by convection far into the
exhaust region to the right.

Fig. 6 to have approximately the maximum growth rate for $\tilde{g} = 1.0$ $\tilde{\rho}_p =$ 50. It is seen that, as would be expected, the flow distortion, ξ_r, penetrates the shell to a depth of about k^{-1}, or ℓ^{-1}, and is extended by convection far into the exhaust or blow-off region to the right.

IV. NUMERICAL SIMULATION

In order to learn about nonlinear development of the ablation-driven Taylor instability, two-dimensional numerical simulations have been done. These have been done with a Langrangian hydrodynamic and heat-flow code named DAISY[19] which was written with triangular zones in order better to treat the highly distorted flows that occur during the large amplitude development of the instability.

It may be interesting to the reader to know that developing the numerical method for treating the heat flow was a much more challenging task than that presented by the hydrodynamics. The problem is that most of the schemes that come to mind give an inaccurate approximation to $\bar{\nabla}T$ when the shape of the zones are far from being equilateral triangles, and can even cause heat to flow "up hill" for some cases of highly irregular meshes. These problems also occur in codes employing quadrilateral zones, but only after the usually approximately rectilinear initial zoning has been distorted by the flow. In some ways, these problems are then more insidious in a quadrilateral code than they are in a triangular code in which many of the difficulties manifest themselves near the initialization time and therefore are more obvious. These additional difficulties encountered in a triangular code force one to devise a numerical scheme which, from the outset, gives an accurate approximation of ∇T, for highly irregular mesh configurations. The resulting method turns out to be more elaborate than those usually formulated for quadrilateral two-dimensional codes. The method that was developed to calculate heat flow on triangular grids has been found to be satisfactory for a variety of difficult (nonlinear) heat flow problems and was described in Ref. 19.

Initialization

The class of problems we have chosen for studying nonlinear development are flat slab targets which represent the limit of $A \to \infty$. In the context of the steady flow model discussed above this also corresponds to the limit $\tilde{g} \to \infty$. This limit was chosen for simplicity, because no particular finite aspect ratio seemed singularly interesting, and it is expected that almost all of the nonlinear ablative phenomena of interest should also be found to occur in the flat targets.

To initialize the two-dimensional code DAISY, an equivalent one-dimensional code is run until approximately steady accelerating ablative flow develops. The resulting one-dimensional temperature and density profiles are transferred to the two-dimensional code in a planar configuration; a perturbation is then applied that is localized near the ablation surface, and the two-dimensional simulation is begun. This procedure not only reduces the amount of computing required, but also minimizes certain types of transients[9,20] which can obscure interpretation when the problem is begun with a perturbation, and facilitates comparisons with the steady flow model.

The family of cases presented here was begun with a carbon slab 3 μm thick. Dimensionless, rather than physical variables, could have been used in this presentation of the two-dimensional results, but it was felt that representative physical values would make it easier for the reader to relate the cases presented to other experience with laser driven ablation. The absorbed laser irradiation has a constant value of 10^{15} w/cm^2 absorbed at a critical density of $ne_c = 10^{21}$ cm^{-3}. That is, we have assumed a step function pulse of 1.06 μm light. Simulations with 2.5 and 5.0 x 10^{14} w/cm^2 have also been done and will be discussed elsewhere. Figure 8 shows a sequence of temperature and density profiles generated by the planar one-dimensional calculation. Transfer of the results to initialize the two-dimensional code was performed at 60 psec.

Results

In discussing these flat targets it is useful to use a burn through time, defined as

$$\tau_B \equiv \left(\frac{1}{M} \frac{dM}{dt} \right)^{-1} , \qquad (25)$$

because we no longer have an implosion time, τ_I from which to obtain the burn through parameter, B, discussed above. In practice, in order to maximize transfer of energy to kinetic energy of an imploding shell, one would choose $\tau_B \simeq \tau_I$, which is equivalent to B $\simeq$ 1. In thinking about the relevance of flat target cases to spherical shells, it is therefore useful to think of τ_B as being equivalent in this sense to τ_I. A flat target is then more unstable (according to the classical theory discussed above), in the sense that the amplification of perturbations is larger, if in the time τ_B it accelerates through distances which are correspondingly greater multiples of its thickness.

From Fig. 8, it is seen that, if the slab remained planar, burn through would occur around 130 psec. when the peak density is beginning to decrease and the slab has moved about 25 μm from its

position at t = 0 or about 20 μm from its position at 60 psec.
when initialization of the perturbation was performed. This latter
distance of 20 μm is about 20 times the shell thickness at
t = 60 psec. (about 1 μm). If we consider t = 60 psec. to be
the beginning of the instability calculation then $\tau_B \simeq$ 70 psec.

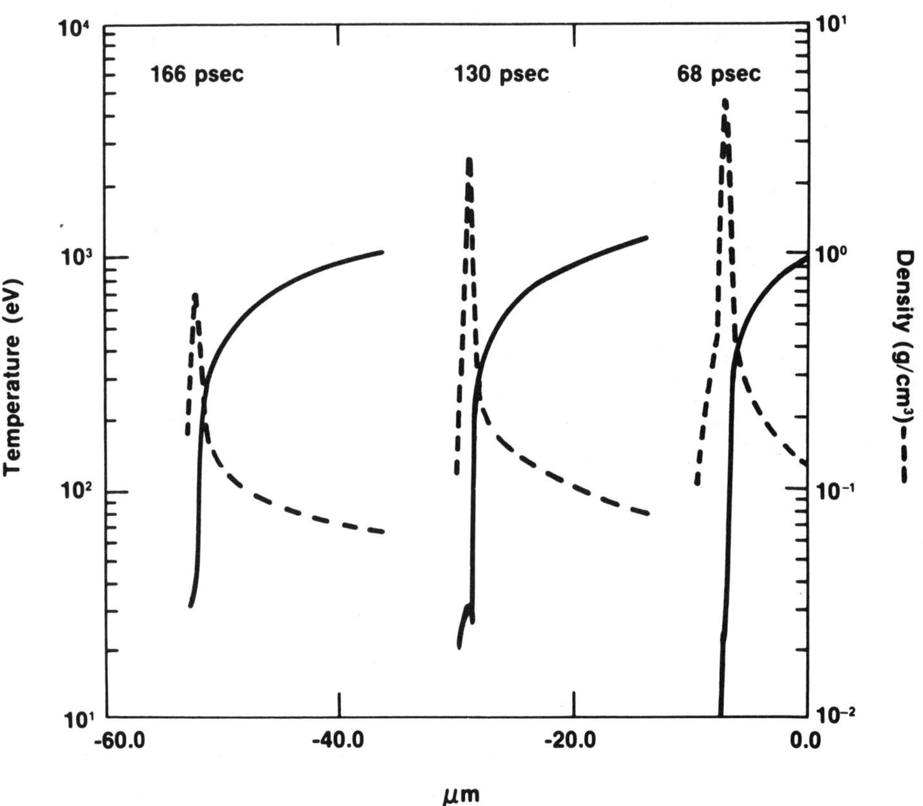

FIG. 8   A time sequence of temperature and density profiles generated
        by a one-dimensional planar simulation of the laser driven
        ablation of a 3 μm carbon slab. The laser irradiance,
        incident from the right, was 10^{15}W/cm^2.

The applied perturbation $\overline{\xi}$, is made to be incompressible by
choosing it to have the form:

$$\overline{\xi} = \nabla \times \vec{A} \text{ , where } \vec{A} \text{ is a vector potential.} \tag{26}$$

The problem is done in Cartesian geometry with the unperturbed
target surface planes of y = constant, the laser is incident in the
x-direction from the right, and the acceleration is in the negative
x-direction. The form of $\vec{A}$ chosen is

$$A_x = 0 \, , \quad A_y = 0$$

$$A_z(x,y) = (\xi_0/k_y)\cos(k_y y)\mathrm{sech}[k_y(x-x_m)] \, , \qquad (27)$$

where k_y is the wave number of the mode in y, the coordinate parallel
to the surface and X_m is the point of maximum density, which is
approximately the left edge of the ablation front and where the
density is almost discontinuous. For $|k_y(x-x_m)| \gg 1$ Eq. (27) has
the standard form of the classic Taylor mode at a density dis-
continuity[1] and is a smooth function of x at smaller $|k_y(x-x_m)|$.
Of the forms of A_z that have been used, Eq. (27) appears to be the
most physical in the sense that for a given value of the initial
amplitude number, $k_y\xi_0$ this form causes the least transient disturbance immediately after initialization.

Figure 9 shows the triangular grid at t = 60 psec. after initialization of a perturbation of the form given by Eq. (27) with
$k_y\xi_0 = \pi/10$ and $k_y = 2\pi/(2.5 \times 10^{-4}$ cm). The two-dimensional
computation is done in a region which is infinite in x and bounded by
rigid slip surfaces in y. These surfaces are separated in y by
one half wavelength of the perturbation, i.e. by a distance
$\Delta y = \pi/k_y$.

Both in Fig. 9 and Fig. 10 discussed below we illustrate the
results by showing one and one half wavelengths which are obtained
by folding over (and thereby periodically reproducing) the computational grid two more times in the y direction to help visualize
the results. That the mode being considered is a typical mode of
interest can be seen from Figs. 9 and 10. The characteristic structure length of the mode is approximately equal to the shell thickness.
This mode might, therefore, be expected to be capable of causing shell
breakup, and moreover, at least to be in the range of the fastest
growing mode capable of doing so. In fact the growth rates obtained
from this simulation and others with different k_y's are in agreement
with planar steady flow stability calculations and show this case
to have a growth rate near the maximum for all modes, i.e. for all k_y.
We illustrate this point below in Fig. 12. This mode, being both
in the range of the fastest growing mode and of sufficiently long
wavelength to cause breakup is, therefore, one of the most interesting cases to consider.

Figure 10 shows a sequence in time of grid plots from the simulation done with both the initial conditions and boundary conditions
described above. In contrast with Fig. 9, only those grid lines
("i" lines in the DAISY calculation) are shown in Fig. 10 which

were initially in the plane of the unperturbed slab target and
therefore, initially vertical. These lines show the distortion of
the slab and are made easier to follow, at the expense of omitting
some information, by omitting the other two families of grid lines.

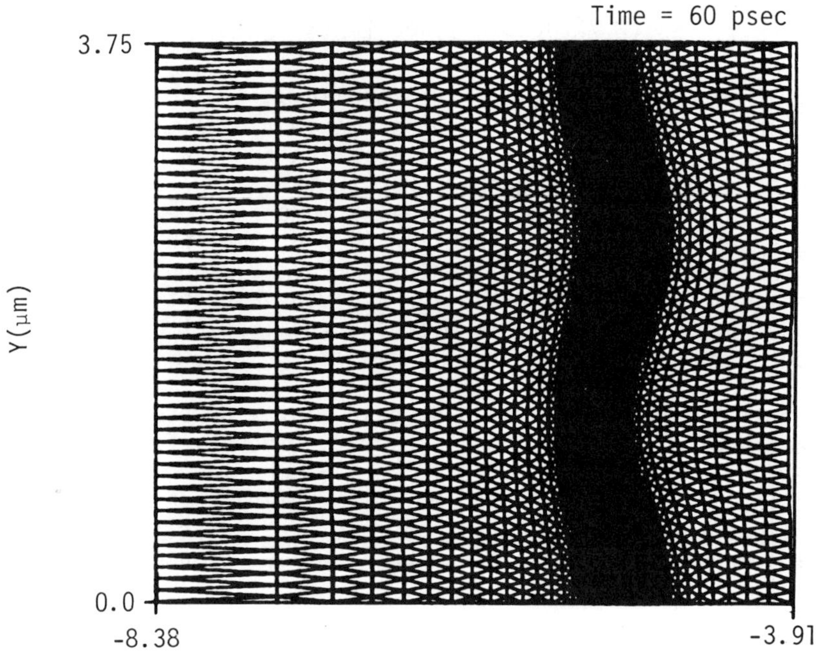

FIG. 9   Mesh plot from the two-dimensional Lagrangian hydrodynamic/
         heat-flow code DAISY. The mesh distortion illustrates the
         $k_y \xi_0 = \pi/10$ mode initialized at the shell surface for a
         2.5 μm wavelength disturbance at t = 60 psec. The simula-
         tion involves approximately 4000 zones with a y-resolution
         of 72 points per wavelength.

     As the amplitude of the distortion increases, its form is seen
from Fig. 10 to resemble the bubble and spike structure seen in
two-dimensional incompressible simulations of the classic Taylor
instability.[8,9] The bubbles are relatively smoothly varying protrusions of the distorted shell to the left. The spikes are the
pointed protrusions to the right, toward the incident laser beam
and in the direction of the effective gravity and the ablative
flow. In the nonlinear simulations of the classic instability, the
spikes are seen to extend to the right until the fluid in the
spike is essentially in free fall, and then continue to extend

indefinitely. Here we see a different behavior. The amplitude
of the shell distortion, and in particular, the length of the
spikes, does become of the order of k_y^{-1} or larger, certainly large
enough to cause concern about loss of symmetry of the final compression of fuel in an implosion system. However, the spikes do
not grow indefinitely. Instead, it appears that when the spikes
have become sufficiently long and thin, heat flow ablates material
from the tips sufficiently rapidly to prevent them from any larger
growth. At the same time, as the amplitude increases, the rate of
growth of the bubble-like structure also appears to decrease.

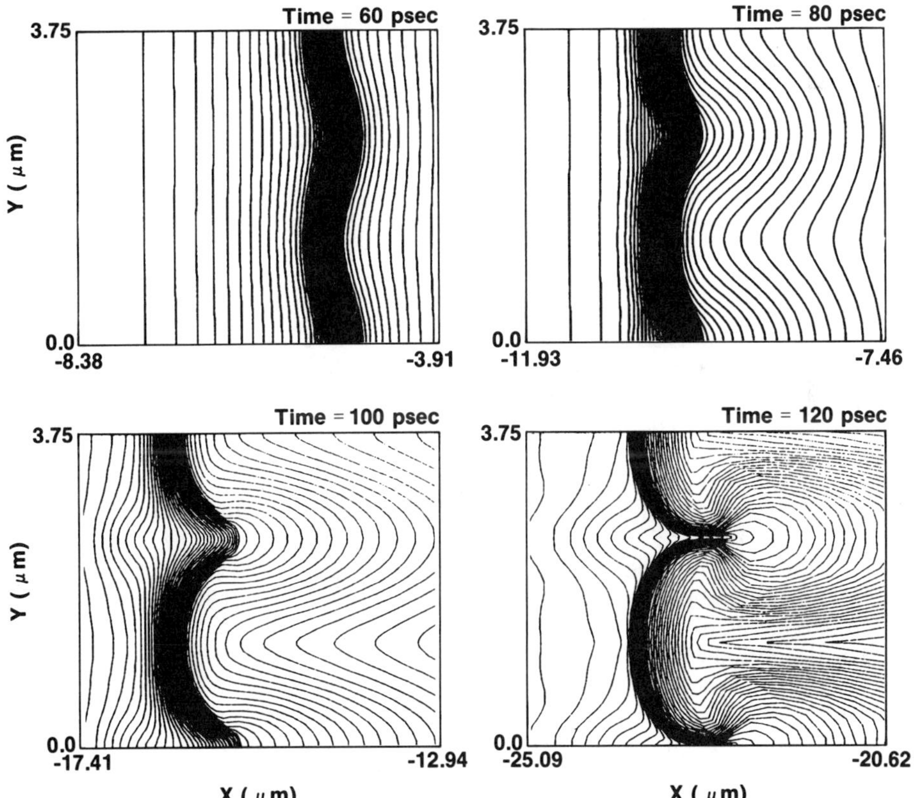

FIG. 10   Time evolution of the unstable 2.5 μm wavelength case
           (see Fig. 9) until burn through (t ≃ 130 psec.) from
           the DAISY simulation. The amount of mass between each
           line (i-line) is a constant throughout the calculation.

       Some insight into the nonlinear development can be acquired from
the RMS (root mean square) deviation of (initially vertical) grid
lines (i-lines) shown in Fig. 10. This measure of distortion is

defined as

$$<\Delta x>_i \; \equiv \; [\; (x_i - <x_i>)^2 \;]^{\frac{1}{2}}\;, \tag{28}$$

where i is the index of a grid line and the averages are taken over the position of the mass points in the ith line. Figure 11 shows the time histories of these $<\Delta x>_i$'s. Only a group of lines are shown which were near the ablation surface during the time of interest. At $t = 60$ psec. the largest value of $<\Delta x>$ is seen to be $10^{-1}\mu m$ which is the RMS amplitude of the maximum of the initial perturbation described above. The line on Fig. 11 passing through $10^{-1}\mu m$ and at $t = 60$ psec. is, therefore, the time history of the grid line nearest the point of maximum density at this time. (Recall that the point of maximum perturbation, x_m, was chosen to coincide with the maximum density of the one-dimensional profile at initialization time.) All other grid lines have smaller $<\Delta x>_i$ at this time. A dip is seen in the maximum $<\Delta x>$ between 60 and 70 psec. which is caused by a transient response to imperfections in our assumed form of perturbation, Eq. (27). Introducing initial perturbations of velocity as well as density would probably be more nearly correct and reduce this transient effect, and will be tried in future simulations. The behavior in which we are interested begins at about 70 psec.

In Fig. 11 can be seen a locus of inflection points of the time history curves, which has been shaded to call attention to it. Each curve passes through this inflection as the grid line to which it corresponds passes through the ablation surface. This inflection, or change in slope, occurs because the rate of change of $<\Delta x>$ is greater on the downstream side of the ablation surface. This can be understood from Fig. 10 where it is seen that the largest distortions occur on the right of the ablation surface, particularly near the tip of the spikes where the ablation rate is greatest. The large flow velocity of the low density ablated material convects it rapidly through this region of large and changing distortion and, therefore, causes a larger rate of change of $<\Delta x>$ for a given i grid line than that same i-line experiences on the higher density upstream side of the ablation surface. All points on an i-line do not pass through the ablation surface at the same time, but the times of passage, particularly in the tip region, are near enough to being simultaneous to give the clear inflections that are seen. This locus of inflections in Fig. 11 is, therefore, the history of the amplitude of distortion of the ablation surface. In addition, the slope of the locus of inflection points on this log-linear graph gives, therefore, the growth rate of the instability. The curves below the ablation surface locus at anytime after 70 psec. correspond to i-lines that are interior to the dense shell material that has not yet been ablated. The slopes of these curves are slightly greater than the slope of the ablation surface locus, because the corresponding i-lines are being convected toward the ablation surface where distortion is greater, as well as experiencing the local rate of growth of the mode. That is, there is a convective as well as a local growth contribution to the growth rate of $\langle\Delta x\rangle_i$ in the frame of a given i-line.

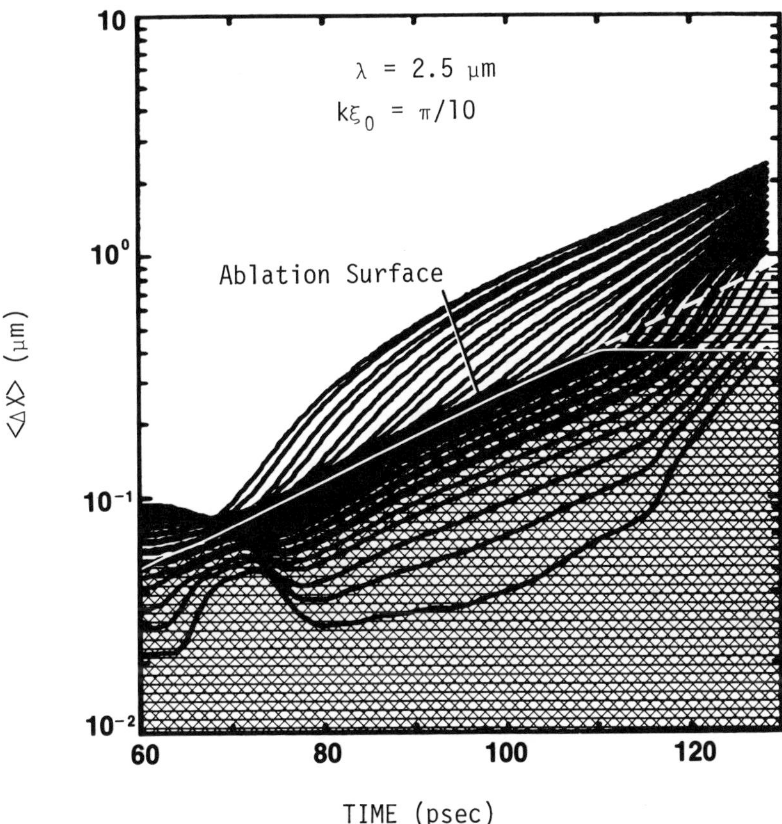

FIG. 11.   Time history of the RMS distortion ($\langle\Delta x\rangle$) for various DAISY Langrangian markers (i-lines). Each curve shows a change of slope (inflection) as the grid i-line to which it corresponds passes through the ablation surface. The locus of these inflection points corresponds to the ablation surface time history. The decrease in the growth rate due to saturation accounts for the change in slope of the ablation surface line near $t \simeq 110$ psec.

Figure 12 shows growth rates for different values of k_y obtained from the slope of the ablation front locus from a set of simulation runs. These runs were done with a small initial amplitude, $k_y\xi_0 = \pi/50$, so that the mode growth would be in the linear regime. To identify the case shown in Figs. 9 and 10 on this graph, recall

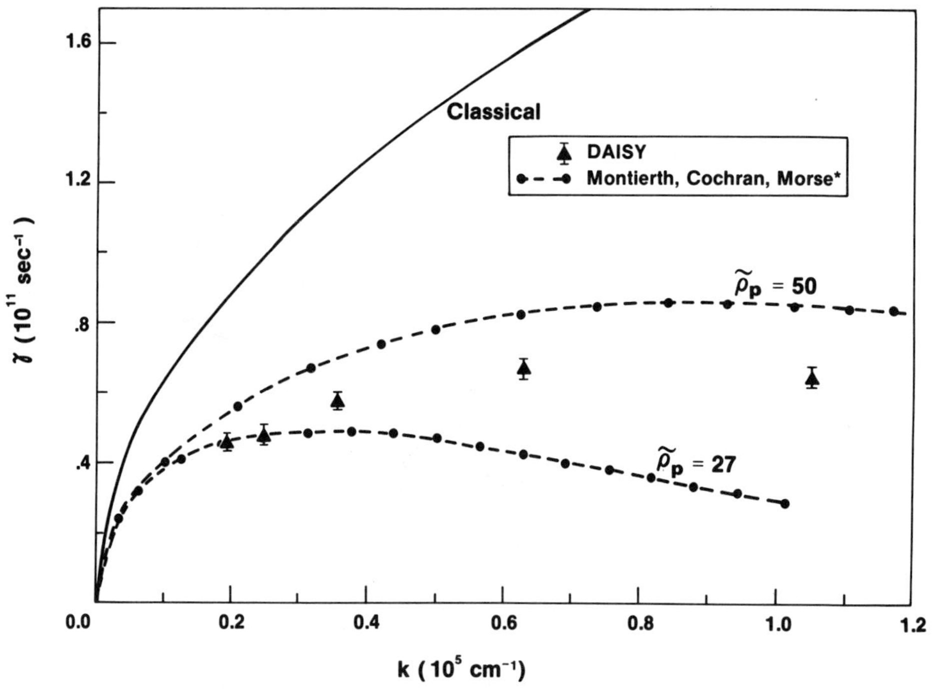

FIG. 12 Growth rate as a function of wave number for small initial
amplitude ($k_y \xi_0 = \pi/50$) modes where linear theory is expected to hold. The DAISY simulations used an absorbed
laser irradiance of 10^{15}W/cm^2 on a carbon slab. Observed
simulation density profiles were approximately stationary
in time with an observed ratio of ablation surface density
to density at the isothermal sonic point of approximately
35. Note the agreement between the simulations and the
perturbation results from the two parameter stationary
ablative flow calculations which bracket our simulation
results in this linear regime. The "classical" growth
rate, $\gamma = \sqrt{ka}$, where the acceleration is determined from
the simulation, is given for reference.

that k_y of the initial perturbation in that case is $k_y = 2\pi/(2.5 \times 10^{-4} \text{cm})$ $\simeq .25 \times 10^5$ cm^{-1}. For comparison with the simulation growth rates, linear growth rates calculated with the planar stationary ablative flow model are also shown for two values of $\tilde{\rho}_p$ ($\equiv \rho_0/\rho_s$, see section III); viz. $\tilde{\rho}_p = 27$ and $\tilde{\rho}_p = 50$. These values of the density ratio $\tilde{\rho}_p$ approximately bracket the values seen in the one-dimensional simulation at the initialization time, $t = 60$ psec. The agreement between the growth rates obtained these two different ways has given us some confidence in both techniques. It is seen from Fig. 12 that, as stated above, the linear growth rate of the mode $k_y \sim .25 \times 10^5$ cm^{-1}, from which the run shown in Fig. 10 was initialized, is near the maximum. From Fig. 12 and the one-dimensional burn through time (Eq. (25)), $\tau_B \simeq 70$ psec, obtained above, it can also be seen that $\gamma \tau_B \simeq 3$ for this mode.

At approximately 100 psec. in Fig. 11 it can be seen that the slope of the ablation surface begins to decrease and is almost horizontal by 120 psec.: that is, by 120 psec. the RMS distortion of the shell near the ablation surface is no longer growing. This behavior corresponds to the indications of saturation seen above in Fig. 10. Of course, the mode cannot become completely stationary because of ablative removal of mass from the shell. Similar saturation behavior has been seen for other values of k_y and $\gamma \tau_B$, including larger values of both.

Conclusions

The quasi stationary form that the Taylor mode is seen to have acquired in Fig. 10 discussed above is quite distorted and, therefore a potential source of disruption of the final fuel compression. However, at 120 psec. in Fig. 10, the side of the shell away from the laser, which would correspond to the inside of a spherical shell, is seen to be much less distorted than the outside. This only moderately distorted inner surface, while not as desirable as a perfectly flat (or spherical) surface, would probably not be mixed with lower density fuel inside it and might give somewhat better final fuel compression than would the inner surface of a shell that had clearly lost its integrity due to turbulent mixing.[3] It appears plausible that future successful high aspect ratio ICF implosion systems may, therefore, be designed to operate with shell distortions of the kind (and amplitude) like that illustrated in Fig. 10.

Acknowledgement

This work was partially supported by the following sponsors: Exxon Research and Engineering Company, General Electric Company, Northeast Utilities Service Company, New York State Energy Research and Development Authority, The Standard Oil Company (Ohio), The University of Rochester, and Empire State Electric Energy Research Corporation.

In addition, this work was also partially supported by the United States Department of Energy (USDOE).

Such support does not imply endorsement of the content by any of the above parties.

REFERENCES

1. S. Chandrasekhar, Hydrodynamic and Hydromagnetic Stability, Chapter X, 428 (Oxford: 1961).
2. G. Fraley, W. Gula, D. Henderson, R. McCrory, R. Malone, R. Mason, and R. Morse in Plasma Physics and Controlled Nuclear Fusion Research, 543, International Atomic Energy Agency (Vienna: 1974).
3. R. L. McCrory and R. L. Morse, Phys. Fluids 19, 175 (1976).
4. L. Montierth, F. L. Cochran, and R. L. Morse, Bulletin of the American Physical Society, 24, 945 (1979).
5. J. R. Freeman, M. J. Clauser, and S. L. Thompson, Nuc. Fusion 17, 223 (1977).
6. R. J. Mason, Nuc. Fusion 15, 1031 (1975).
7. J. D. Lindl, Lawrence Livermore Laboratories Report, UCRL-79735 (1979) unpublished.
8. B. J. Daly, Phys. Fluids, 10, 297 (1967).
9. F. H. Harlow and J. E. Welch, Phys. Fluids 9, 842 (1966).
10. G. Birkhoff, Los Alamos Scientific Laboratory Report LA-1862 (1955) and LA-1927 (1956) unpublished.
11. S. E. Bodner, Phys. Rev. Lett., 33, 761 (1974).
12. D. B. Henderson and R. L. Morse, Phys. Rev. Lett., 32, 355 (1973); see also J. N. Shiau, E. B. Goldman, and C. I. Wong, Phys. Rev. Lett., 32, 352 (1973).
13. D. B. Henderson, R. L. McCrory, and R. L. Morse, Phys. Rev. Lett. 33, 205 (1974).
14. R. L. McCrory, R. L. Morse, and K. A. Taggart, Nuc. Sci. Eng. 64, 163 (1977).
15. S. J. Gitomer, R. L. Morse, and B. S. Newberger, Phys. Fluids 12, 234 (1977).
16. L. Montierth, F. L. Cochran and R. L. Morse, Bulletin of the American Physical Society, 24, 945 (1979).
17. L. Spitzer, Physics of Fully Ionized Gases, Interscience (Princeton: 1961).
18. K. A. Brueckner, Siebe Jorna, Rev. Mod. Phys. 46, 325 (1974).
19. C. P. Verdon, R. L. McCrory, and R. L. Morse, Bulletin of the American Physical Society, 24, 945 (1979) and R. L. McCrory, R. L. Morse, and C. P. Verdon, Bulletin of the American Physical Society, 24, 945 (1979).
20. L. A. Elliott, Proc. Roy. Soc. A, 284, 397 (1965).

ELECTRON TRANSPORT EFFECTS

IN LASER PRODUCED PLASMAS[+]

R. J. Mason

Los Alamos Scientific Laboratory

Los Alamos, NM 87545

ABSTRACT

A Monte Carlo hybrid scheme for the study of transport inhibition is outlined. Its use indicates that return-current thermal
convection thru density troughs may be a significant inhibition
mechanism. Inhibition by thermal electron trapping is also suggested.

I.   INTRODUCTION

Experiment indicates that while suprathermal electrons appear
to transport classically in laser produced plasmas the thermal
electrons are inhibited.[1-3] The Evidence has come from Nd 1.06 μ
experiments where in certain instances there has been excessive
energy loss to fast ions, late collapse times (in comparison with
simulation) for microballons, and/or reduced line emissions from
the depth of layered targets. All these have been attributed to
severe inhibition of the thermal transport, typically at 1/20th of
classical expectations.

____

[+]Work performed under the auspices of the USDOE.

In this paper we outline important features of a Monte Carlo hybrid simulation scheme that has been developed to help us acquire a fundamental understanding of the causes of this anomalous inhibition. Application is made to the transport of electrons in foil-like geometries.

II. THE MONTE CARLO HYBRID MODEL

The transport scheme treats the hot electrons as weighted PIC particles. The weights allow the emission of a constant number of particles/cycle, even when the energy deposited in hot electrons is changing with time. The cold electrons are treated as a fluid when the scheme is used in its "hybrid mode", and as a separate set of Monte Carlo particles when the scheme is in its "full-particle mode". The electric field is calculated either (i) from the current form of Poisson's equation by <u>dilating</u> the effective local <u>plasma period</u> up to the order of the calculational time step Δt, or (ii) by a moment method which sets the predicted total hot and cold current sum to zero. The model applies in 1-d only, although extensions to 2-d are obvious and would follow from the use of $\partial B/\partial t = -c\nabla \times E$.

The restriction to one-dimensional studies means that we have no opportunity to explore the inhibiting effects of B fields. Previously[4], we have shown that B fields can effect even the supra-thermal transport, when the illumination is sufficiently aniso-tropic. But the experimental inhibition described in the intro-duction has all been seen in essentially one-dimensional configura-tions, for which our model should suffice. Also, there are some indications[5] that in actual applications the self-consistant B fields may only arise outside the critical surface, say a 1/4 critical, where their effects on the bulk thermal transport would be minimal.

In the scheme particle hot electrons undergo scattering and Coulomb drag. In a scattering event, they are given a gaussian distribution of angular deflections θ about their initial direction. The mean square deflection angle is

$$\langle\theta\rangle^2 = 8\pi \; e^4 m^2 c^{-3} Z \; (Z + 1) \; n_i \Delta t \; \log \lambda. \tag{1}$$

The particle velocities are then rotated thru a random azimuthal angle ϕ in an isotropic distribution. The drag reduces the parti-cle speeds c in accordance with

$$\frac{dc}{dt} = 4\pi^4 m^{-2} n_c c^{-2} \; \log \lambda \tag{2}$$

where $c^2 = u^2 + v^2$, with u the longitudinal velocity component and v in the transverse direction. The energy lost in drag is deposited

in the thermal background. See Ref. 6 for details. In the full
particle mode the drag is neglected--although this limitation will
be eliminated in a future version of the scheme. The particles are
accelerated and moved by standard PIC procedure including area
weighting. An exception is that across large density gradient
regions, where the large confining E might substantially slow a
particle, we introduce subcycling so that the longitudinal velocity
is never changed by more than 10% in a subcycle.

In the hybtid mode, the cold electrons are treated as a fluid[6]
using donor-cell[7] hydrodynamics. This has the advantage that as
the E field in the corona draws down the cold density, it can never
drag this density below zero--a problem that has plagued various
alternate transport schemes[8] and one that was previously solved in
Double-Diffusion by <u>change limitation</u>.[4] Thus, we retain the inertia
term in the thermal electron momentum equation which frees us from
the need for a flux limiter on the cold currents. The cold drift
speed is thereby determined uniquely, as is the degree of thermal
Joule heating. [Under flux limitations this heating is <u>arbitrarily</u>
set as we fix j in j^2/σ by our choice of the limitation parameter
f--classically equal to 0.6]. For the cold fluid we use an extension[6] of the full set of Braginskii equations. The extension is
due to C. Cranfill, and initially included an equation for the temporal advancement of the heat flex vector q. It also provides
analytic forms[10] for the Z dependent Branginskii cofficients.

For the thermal scattering frequency we use Braginskii's
classical rate

$$\nu_{cl} = \frac{4}{3} \left(\frac{2\pi}{m} \right)^{1/2} e^4 T_c^{-3/2} Z^2 n_i \log \lambda \tag{3}$$

$$(T \text{ in keV})$$

where $\log \lambda$ ($\simeq 10$) is given in Ref. 10, and we add to this the
anomalous ion-acoustic collision frequency determined by Lindman et
al. from PIC simulation. This is

$$\nu_{ia} = 3 \times 10^{-5} \omega_{pc} \left(\frac{m}{m_i} \right)^{1/4} \frac{T_c}{T_i} \frac{u_c}{(T_c/m)} 1/2, \qquad T_c \simeq 300 eV. \tag{4}$$

The Lindman calcualtion was made with the lowest level of numerical
noise to date, and for the lowest value of $u_c/(T_c/m)^{1/2} \simeq 0.3$ [u_c
is the thermal drift speed]. It predicts frequencies nearly an
order of magnitude lower then the earlier Biskamp rate, and roughly
four orders of magnitude lower then the rates obtained with a
Kadomtsev spectrum for the turbulence.

The E field is calculated by one of two alternate schemes. In the first, the local plasma period is dilated to the order of the time step. Thus, in

$$\frac{\partial E}{\partial t} = 4\pi\varepsilon (n_h u_h + n_c u_c)$$

we locally decrease ε from its physical value e to a generally smaller value

$$\varepsilon = C\ e,\quad C \equiv G\left[\frac{4\pi e^2}{m}\ \Delta t\left(\frac{n_h}{\nu_h'} + \frac{n_c}{\nu_c'}\right)\right]^{-1} \le 1.0, \qquad (6)$$

in which $\nu_{h,c}' = \nu_{h,c} + 1/\Delta t$, and, typically G = 0.01. By using this smaller charge we find, for example, that in the corona, when $n_c \to 0$ and $\nu_h \to 1/\Delta t$, the effective plasma frequency becomes

$$\left.\omega_p^2\right|_{eff} = \frac{4\pi e\varepsilon}{m}\ n_h \to G\Big/(\Delta t)^2 \qquad (7)$$

or that $\left.\omega_p\right|_{eff} \Delta t \to G^{1/2} \to 0.1$, permitting stable calculations of E at Δt. Our Δt is now set by a Courant condition for hot electrons crossing a cell. Without the dilation a much smaller Δt would be required in dense regions of a target. The $\nu_{h,c}$ values use Eq. (3) with $T \to T_{h,c} + u^2_{h,c}/3$, respectively, with the addition of an ion-acoustic contribution to ν_c given by Eq. (4). The detailed form for C derives from the use of mean momentum equations for $n_{h,c}\ u_{h,c}$ in conjunction with Eq. (5). Stability and accuracy requirements on the resultant explicit equations for E fix C. Details will be given elsewhere.[6] Our experience with Eq. (5) has shown that in a few Δt it establishes a steady state $n_c u_c = -n_h u_h$, giving E = α'(Z) $\nu_c n_h u_h$, the ohm's law value in dense regions of a plasma [α' (Z) is Braginskii's coefficient (= 0.51 when Z = 1 , for example)].

and $E \simeq \dfrac{-1}{en_h}\ \dfrac{\partial p_h}{\partial x}$ in the corona.

In the alternate moment method[6] scheme for E we start with the time integrated momentum equations

$$j_{h,c}^{\ (m+1)*} = \frac{j_{h,c}^{\ (m)'}}{\nu_{h,c}'\Delta t} - \frac{1}{\nu_{h,c}'}\ \frac{1}{m}\left[\frac{\partial p_{h,c}^{(m)}}{\partial x} + en_{h,c}^{(m)} E^{(m+1)}\right], \qquad (8)$$

in which the (m)' currents are the old values including sources and the (m + 1)* are <u>predictions</u> for the next cycle. By setting the predicted total currents to the quasi-neutral condition $j_h(m + 1)^* + j_c(m + 1)^* = 0$, we derive

$$E^{(m+1)} = \frac{m\left(\dfrac{j_h^{(m)'}}{\nu'_h \Delta t} + \dfrac{j_c^{(m)'}}{\nu'_c \Delta t}\right) - \dfrac{1}{\nu_c}\dfrac{\partial P_c}{\partial x} - \dfrac{1}{\nu'_h}\dfrac{\partial P_h}{\partial x}}{e\left(\dfrac{n_h^{(m)}}{\nu'_h} + \dfrac{n_c^{(m)}}{\nu'_c}\right)} \qquad (9)$$

In practice, the $j_{h,c}$ obtained in any cycle are in deviance from their $j^*_{h,c}$ predictions and the $j^{(m)'}$ terms in Eq. (9) provide an additional E contribution (beyond the equlibrium value) that works to correct the deviations. The corrections can be large. To smooth their effect we have introduced an averaging procedure by which we solve Eq. (9) not for Δt but for $M\Delta t$ giving $E^{(m+M)}$. Then the E field for the next cycle is given by

$$E^{(m+1)} = (E^{(m+M)} + (M-1)E^{(m)})/M. \qquad (10)$$

The best results are obtained for M = 2. Alternatively, one could iterate the E field calculation and the particle and fluid advancement procedures each cycle until at its end the total $j^{(m+1)}$ was, indeed, zero. This would presumably be costly.

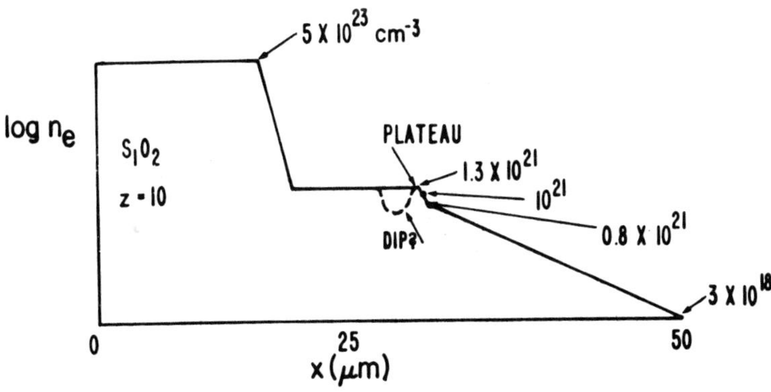

Fig. 1 Glass-foil geometry for transport simulation.

III. APPLICATIONS

The model has been used to examine electron transport in static
geometries such as that depicted in Fig. 1. Here we have a glass
foil that has expanded somewhat under laser illumination. The 1.06
μ light enters from the right. Its pondermotive force establishes
a plateau at 1.3×10^{21} cm^{-3}. Beyond the plateau the electron
density rises to the solid density value $n_t = n_h + n_c \simeq 5 \times 10^{23}$.
In all our hybrid runs the hots are emitted in a drifting Maxwellian
in a 20° half angle cone towards the laser and with $T_h \sim (I\lambda^2)^{1/3}$
$T_c^{1/3}$. As hots are created, corresponding colds are destroyed.

Figure 2 is included to demonstrate that the E field properly
establishes quasi-neutrality. Here the plateau is longer $\sim 250~\mu$,
and the high density region has been removed. The left boundary
is a mirror. Hot electrons are emitted at the site of the vertical
fiducial. In about 5 ps they exclude all the colds from the lower
density coronal region and come up to the total density. By 10 ps
they have established a nearly flat profile out to the left
boundary. In Fig. 2(b) the hot electron currents are nearly equal
and opposite to the cold currents, as required for quasi-neutrality.

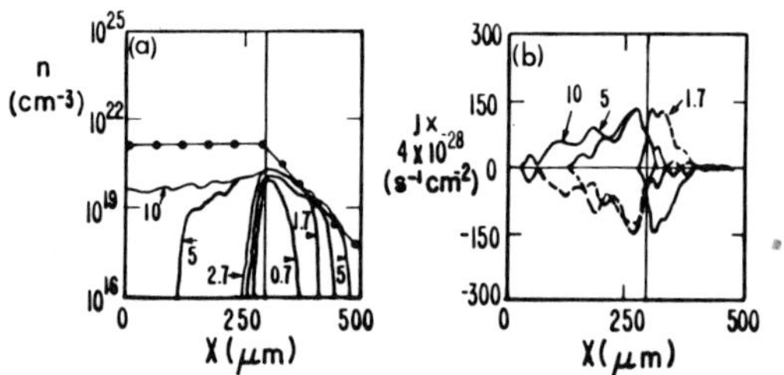

Fig. 2   (a) Evolving hot density profiles _____, background total
density Zn$_i$ _____; (b) quasi-neutral current profiles;
hot currents _ _ _ _, cold currents _____ .

The scheme has been used to investigate a number of potential
1-d inhibition mechanisms. These include: (a) return current E
field effects from classical resistivity, (b) ion-acoustic turbulent resistivity, (c) collisionless convective inhibition, and (d)
transport "choking" by troughs.

Classical Resistivity

We find that the classical resistivity inhibits the hot electrons only as long as $T_c < 10$ eV. But at plateau densities the
background goes above this temperature and "burns out" at a rate of
30 μ/ps. This is for 5×10^{15} W/cm^2 incident, 30% absorbed, and
95% of this into hot electrons with the rest into thermals--which
we shall term "standard conditions". The burn out is due princi-
pally to Joule heating by the return current. It continues at
nearly the same 30 μ/ps rate; even when the Coulomb drag and the
thermal conductivity are set to zero. So a strong thermal flux
limiter will not sustain an E field that stops the hots. Burn out
of the background is slower at higher Z and with higher density,
but, at most, we find that the hot inhibition is only equivalent
to that provided by classical electron scattering.

It is interesting that as we go to higher densities the Joule
heating is progressively less effective at burning out the resistivity. This is because the Joule heating rate goes as $\nu_c j^2/n_c =
Q_j$, which is independant of density, while it produces temperature
changes $Q_j = 3/2\, n_c \partial T_c/\partial t$. Alternatively, the drag deposition rate
goes as const. $n_c \cong Q_d$, so the drag heating survives at high densities.

It is not surprising that the classical E field fails to
stop the hots, since modern experiments indicate that they are
unhibited. A more pertinent question is, therefore, whether the
field can stop the thermals. This will be discussed below.

Ion-Acoustic Resistivity

First, we find that the calcualted ion-acoustic rate is too
low to explain the inhibition of either hots or thermals. When
the Lindman rate Eq. (4) is multiplied by 10^2, it still has no
effect. With a multiplication 3×10^3 or more the thermals become
inhibited but then so are the suprathermals. As ν_{ia} decreases the
conductivity, it also increases the electrical resistivity and the
E field, thus slowing the suprathermals, this argues against ion-
acoustic turbulence as the mechansim causing the solely observed
thermal inhibition--as does the low value for ν_{ia} found in simulation.

Convective Inhibition

Since we calculate the hydrodynamic motion of the colds, we can determine the convective effects[13] on thermal transport. In the hybrid mode of our scheme the cold temperature is governed by

$$\frac{3}{2}\frac{\partial}{\partial t}(n_c T_c) = -\frac{3}{2}\frac{\partial}{\partial x}(n_c u_c T_c) - \frac{\partial}{\partial x}(q_D + q_u) + \alpha(z)\,\nu m n_c u_c^2 + \ldots$$

(convection)   (joule heating)

$$(12a)$$

$$q_D \equiv \frac{\dfrac{-K}{\gamma(z)}\dfrac{\partial T_c}{\partial x}}{\left[1 + \left|F_D\right|/F_\ell\right]}, \qquad q_u \equiv \frac{\dfrac{3}{2}\dfrac{\beta(z)}{\hat{\gamma}(z)}n_c u_c T_c}{\left[1 + \left|F_D\right|/F_\ell\right]}, \qquad (12b)$$

in which q_D and q_u are the flux-limited heat flow from diffusion and thermelectric effects.[6] The $\gamma(z)$ and $\beta(z)$ factors are O(1) functions that can be related to Braginskii's coefficients.[9-10] Also $K \equiv 5/2\, n_c T_c/m\nu$ and $F_D \equiv \frac{K}{\gamma}\frac{\partial T_c}{\partial x}$ with $F \equiv f\, n_c T_c^{3/2}$, $f = 0.6$. The last expression gives a classical limit to the diffusion. Calcualtions have been performed with the model in its full-particle mode verifying that $f = 0.6$ is the proper choice. This will be discussed below.

Inhibition and even transport reversal can be expected when the convective and thermoelectric heat flow approach and exceed the diffusive heat flow, i.e., when

$$\left|\frac{3}{2}n_c u_c T_c + q_u\right| \gtrless \left|q_d\right|. \qquad (13)$$

By quasi-neutrality $n_c u_c \simeq -n_h u_h$, so this will occur in the presence of large suprathermal currents. The transport reversal is related to Shkarofsky's recent bi-Maxwellian results,[14] but occurs here at high relative drift speed between the components, and even in the absence of collisions. In the plateau region near the critical density in SiO_2 the mean free path for thermals heated by the standard illumination conditions is many microns, and the thermal diffusion will be, at least, classically limited. Thus, $q_D \to 0.6\, n_c v_c T_c$ where $v_c = \sqrt{T_c/m}$, and also $q_u \to 0$. Thus, Eq. (13) implies transport reversal for $\left|3/2\, n_c u_c T_c\right| > 0.6\, n_c v_c T_c$ or $\left|u_c\right| = \frac{n_h}{n_c}\left|u_h\right| > 0.4\, v_c$. Thus, very large cold drift velocities are required. These are most readily attained near critical density. Unfortunately, for inhibition, we find that these large drift velocities are associated with a high degree of Joule heating, which tends to mask the inhibition effect by heating the thermals at depths just below where the convection is acting.

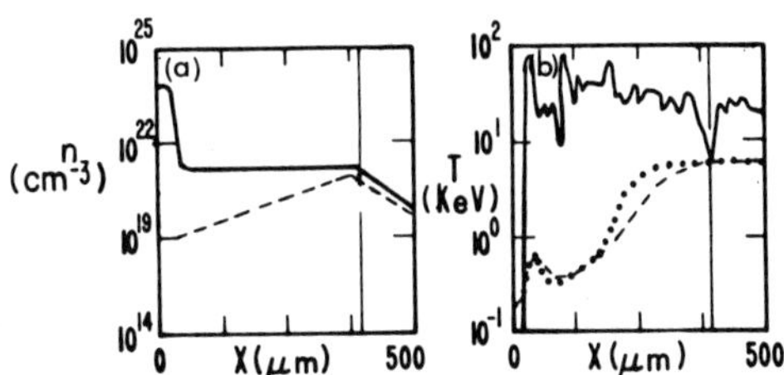

Fig. 3   (a) Density profiles: total density ____, hot electron
density __ __ __, and (b) Temperature profiles: typical
hot temperature T_h ____, T_c with convection __ __ __, T_c
without convective inhibition . . .

Figure 3 displays our calculated results for transport under the
standard illumination with and without the inclusion of the convection term in Eq. (12).   Typical of our experience from runs for
many geometries and conditions, the convective effects are weak.
Attempts to go to higher Z, so as to reduce q_D below its limit
value, have characerically resulted in less convective inhibition
due to greater Joule heating.

Density Troughs

     As an exception to these difficulties, we have found that
density troughs can intensify the convective mechanism.   Here we are
thinking of dips occuring below the critical surface to, say,
0.4 n_{crit}.   Thru such a dip the convective energy flow, $3/2\ n_c u_c T_c$,
remains unchanged, since $n_h u_h$ is unaffected by weak dips, and
quasi-neutrality prevails.   The limited diffusion $q_D\ 0.6\ n_c v_c T_c$,
however, has n_c down with the reduced $n_t = n_h + n_c = Z n_i$ in the dip
and n_c is further reduced, since n_h is fixed.   As a result, the
competition between the convective flow and q_D is more severe.
Also, the joule heating of the colds is reduced, since as n_c rises,
$v_c = v_c(T_c + u_c^2/3)$ declines.   Figure 4 shows our results under
standard illumination after 3.7 ps when the profile has a fixed
imposed dip to $n_t = 0.4\ n_{crit}$.   Clearly, the cold temperature rises
behind the dip to a value comparable to that obtained in the absence
of a dip, when the f = 0.03 limiter is imposed.

     Density dips have been predicted,[15] seen in simulation,[16] and
measured.[18]   But their duration and depth may be insufficient to
lead to the convective inhibition; nor may they exist self-consist-

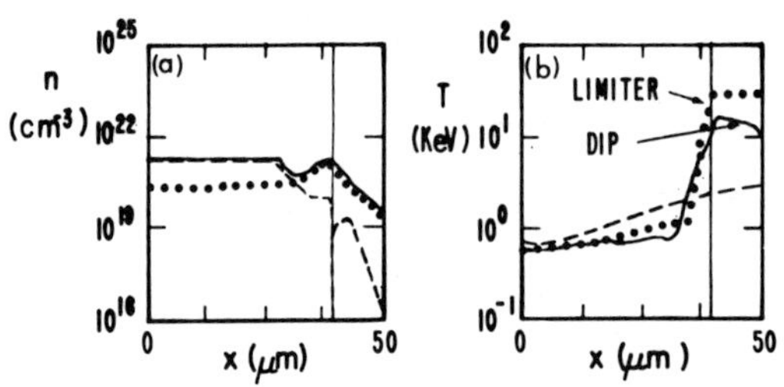

Fig. 4   (a) Density profiles including a dip: Total density Zn_i ——, n_h . . . , n_c —— —— ——.   (b) T_c profiles: No dip and classical limitations f = 0.6 —— —— ——, no dip and f = 0.3_N . . . the 0.4 n_{crit} dip and f = 0.6 ——.

ently with inhibition.   Additional work will be needed to resolve these questions.

IV.   FULL PARTICLE CALCULATIONS

     The search for a thermal inhibition mechanism has repeatedly moved us to a progressively more fundamental model. A difficulty with the hybrid model is its need for a thermal flux-limiter at low densities and high temperatures. To avoid this our scheme will do Monte Carlo transport of the thermals when run in its full-particle mode. In this mode, particle weights are decreased when new suprathermal particles are generated at critical. In other respects, the thermals particles are advanced identically as the suprathermals.

     Figure 5 gives the density and temperature profiles computed in the full-particle calculations made for the conditions leading to Fig. 4. Clearly, the two sets of results are quite similar, indirectly supporting our choice of f = 0.6 in the hybrid calculations.

     We have established that the resistive E field is too weak to significantly perturb the suprathermals, as a result of its rapid burn-out. This cannot be said of the thermals. A weak potential difference of, say, 1.5 keV would suffice to trap 500 eV thermals on the low density side of a large density interface. The Braginskii formalism is incapable of manifesting trapped electron

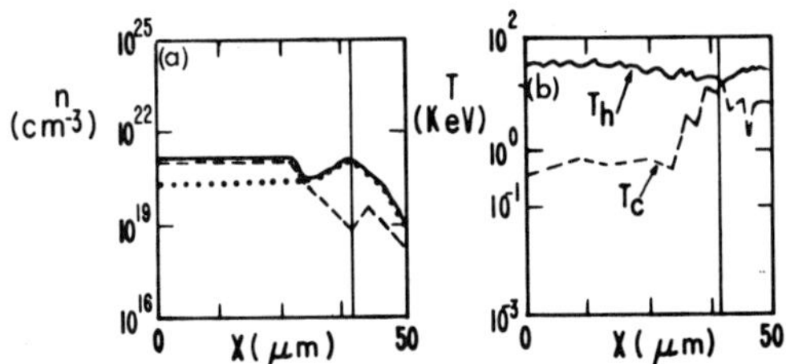

Fig. 5   (a) Density profiles computed in the full-particle mode
         for the fig.4 dip, same nomenclature, (b) the full Monte
         Carlo temperature profiles.

phenomenology, since it implicity assumes only slightly perturbed,
collision-dominated near Maxwellian distributions. Preliminary
calculations in the full-particle mode suggest that such a trapping
mechanism may lead to significant thermal inhibition.

REFERENCES

1.   J. D. Hares, J. D. Kilkenny, M. H. Key and J. G. Lunney, Phys.
     Rev. Lett. 42, 1216 (1979).
2.   W. L. Kruer, "Electron Energy Transport in Laser Produced
     Plasmas", Com. on Plasma Phys. and Contr. Fus., to be
     published.
3.   W. T. Toner, Rutherford Laboratory Annual Report March 1979,
     RL-79-036, Chap, 4.3 (unpublished).
4.   R. J. Mason, Phys. Rev. Lett. 42, 239 (1979).
5.   C. Max, private communications.
6.   R. J. Mason, "On the Monte Carlo Hybrid Modelling of Supra-
     thermal Electron Transport in Laser-produced Plasmas,"
     submitted to Phys. Fluids.
7.   R. A. Gentry, R. E. Martin and B. J. Daly, J. of Comp. phys.
     1, 87 (1966).
8.   G. Zimmerman, U. of Cal. Research Laboratory Report No. UCRL-
     74811, 1973 (unpublished).
9.   S. I. Braginskii, Rev. Plasma Phys. 1, 205 (1965).
10. C. Cranfill, private communication. Also see the appendix in
     R. S. Craxton and R. L. McCrory, "Two-dimensional Calcula-
     tions of Non-spherical Laser Fusion Implosions," submitted
     to NUCLEAR FUSION.

11. E. Lindman, J. de Physique $\underline{38}$, C6-9 (1977).
12. D. Biskamp and R. Chodura, Phys. Rev. Lett. 27, 1553 (1971),
     and D. Biskamp, R. Chodura and C. T. Dum, Phys. Rev. Lett.
     $\underline{34}$, 131 (1975).
13. R. Mason Phys. Rev. Lett. $\underline{43}$, 1795 (1979).
14. I. P. Shkarofsky, Phys. Rev. Lett. $\underline{42}$, 1342 (1979).
15. C. E. Max and C. F. Mckee, Phys. Rev. Lett. $\underline{39}$, 1336 (1977).
16. J. Virmont, R. Pellat and A. Mora, Phys. Fluids $\underline{21}$, 567 (1978),
17. R. Fedosejevs, M. D. J. Burgess, G. D. Enright and M. C.
     Richardson, submitted to Phys. Rev. Lett.

THEORETICAL ASPECTS OF PROFILE STEEPENING

IN LASER-PRODUCED PLASMAS

K.-H. Spatschek, E.W. Laedke,
M.Y. Yu, and P.K. Shukla

Fachbereich Physik, Universität Essen
D - 4300 Essen, F. R. Germany

ABSTRACT

The generation, propagation and stability of solitary waves
near the critical density are studied. First, it is shown that
due to mode conversion, solitons are created and they propagate
into the underdense region. In contrast to previous investigations,
ion inertia effects are included; the latter change the physical
behavior considerably. Secondly, for solitons of large amplitude,
electron and ion nonlinearities, space-charge fields, and relativistic mass variation can become important. Thus, the contributions of these effects are investigated. Finally, the stability
of solitons is discussed.

I. INTRODUCTION

In an inhomogeneous plasma, conversion of electromagnetic
waves into electrostatic waves enhances the field strengths of
the latter by several orders of magnitude in the vicinity of the
resonant layer, where the incident frequency matches the local
plasma frequency. The amplitude swelling is due to the reduction
of the group velocity from the velocity of light c to the electron
thermal velocity v_{te}. The enhancement factor is approximately
$(c/v_{te})^{1/2}$.

When the amplitude of the electrostatic oscillations becomes
larger, nonlinear effects can balance the dispersion of the waves
and solitary waves are formed.[1] Recently,[2,3] a theory has been
presented to explain the generation of solitons at the critical
density and to predict profile steepening by cavity formation.

However, the theory was based on the driven nonlinear Schrödinger equation[4,5] ignoring ion inertia. Since the latter is important when the solitons start moving in the presence of a density gradient, the results presented in Ref. 2 are of limited application. Thus, in the first part of this paper, we shall give an improved treatment of the problem by including ion inertia effects.

For large amplitudes, further revision of the theory is needed. Common to all observations is that the solitons are of finite amplitude (corresponding to density depressions of more than 10 percent). The interpretation of such observations thus requires more sophisticated models than those used in the small amplitude theories. The latter start from the cubic nonlinear Schrödinger equation or the Zakharov equations.[6]

In the Zakharov equations, the time-averaged wave equation for the slowly varying electric field envelope E is coupled with a <u>linear</u> ion acoustic wave equation modified by the ponderomotive force. Obviously, for finite amplitude solitons with large density variations, the plasma response is nonlinear. Therefore, the low-frequency plasma motion has been modified by including electron and ion nonlinearities.[7] Most models for subsonic finite amplitude solitons assume quasineutrality in the density response. In general, however, this assumption should be relaxed for finite amplitude solitons.[8]

Accordingly, in the second part of this paper we relax the quasi-neutrality assumption and present an improved theory of finite amplitude solitons in homogeneous plasmas.[8] The development of this theory is motivated by its many applications in laser fusion experiments where the energy density can become very large. Relativistic effects[9] shall also be included.

Finally, the stability of solitons is discussed. We present a general criterion for longitudinal stability and prove that finite amplitude solitons are stable.[10] However, as is known for the cubic nonlinear Schrödinger equation, a transverse instability occurs. The relevant k-region of the latter is calculated exactly including ion inertia effects. It is demonstrated that the cubic nonlinear Schrödinger equation overestimates the growth rate.

II. GENERATION OF SOLITONS

Consider a p-polarized electromagnetic wave obliquely incident on an inhomogeneous plasma. The component of the electric field along the density gradient can drive electrostatic density oscillations. There results a current $n\underset{\sim}{v}_0$, where $\underset{\sim}{v}_0 = e\,\underset{\sim}{E}_0 / m_e \omega_0$ is the quiver velocity of the electrons in the laser light field

$\underset{\sim}{E}_o$. This allows the electromagnetic wave to tunnel beyond the cut-off (reflection) point and arrive at the critical layer. Here, the wave frequency equals the local plasma frequency and Langmuir waves are driven resonantly. Thus, an obliquely incident electromagnetic wave creates at the critical surface enhanced density and electric field oscillations along the density gradient, having a frequency near the local plasma frequency. The power converted into the plasma oscillations is determined by the driven electric field $\underset{\sim}{E}_d = (\underset{\sim}{H} \times \hat{e}_z)$ at the resonance layer. Here, $\underset{\sim}{H}$ is the component of the laser magnetic field directed out of the plane of incidence, and $\hat{e}_z$ is the unit vector across the density gradient. In particular, we have[11,12]

$$\underset{\sim}{E}_d = \hat{e}_x \, E_o \, \phi(q) \, / \, (2\pi \, k_o \, L)^{1/2} \,, \tag{1}$$

where E_o, $k_o = 2\pi/\lambda_o$, λ_o are the free space values of the laser electric field, the wave vector, and the wavelength, respectively. The density gradient scale-length $L = n_o (dn_o / dx)^{-1}$ is assumed to be larger than the free space wavelength. The dimensionless parameter q is defined by $q = (k_o L)^{1/3} \sin\theta$, where θ is the angle of incidence. The factor $\phi(q)$ is the usual resonance function[3] and is peaked ($\phi = 1.2$) around $q \simeq 0.6$. It is of interest to note that, in the critical density region, because the electromagnetic waves are evanescent, their amplitude varies slowly on the Debye-length scale. Consequently, the laser magnetic field is approximately constant there.[13,14]

According to the linear theory,[13] the growth of the linearly transformed plasma waves can be limited by thermal convection and subsequent Landau damping after a time $t_c \simeq (L/\lambda_e)^{2/3} \, \omega_{pe}^{-1}$, where $\lambda_e = (T_e / 4\pi n_o e^2)^{1/2}$ is the electron Debye length. Near the resonance layer ($x = 0$), the amplitude of the saturated electrostatic wave is given by[2]

$$\tilde{E} = i \, \pi \, \alpha^{-2/3} \, E_d \big[G_i(0) + i \, A_i(0) \big] \leq 1.2 \, E_d \, (\omega_o L/v_{te})^{2/3} \,, \tag{2}$$

where $\alpha \equiv \lambda_e / 2L$, $v_{te} = (T_e / m_e)^{1/2}$ is the electron thermal velocity, A_i and G_i are the Airy functions. The electron plasma wave generated at the critical layer acts as a pump and couples to slow plasma motion. The dynamics of the mode-converted wave in the presence of the driving field $\tilde{\underset{\sim}{E}}_d$ is governed by[3]

$$\frac{\partial n}{\partial t} + \nabla \cdot n \underset{\sim}{v} = 0 \,, \tag{3}$$

$$\frac{\partial \underset{\sim}{v}}{\partial t} + \underset{\sim}{v} \cdot \nabla \underset{\sim}{v} = -\frac{e}{m_e} (\tilde{\underset{\sim}{E}} + \tilde{\underset{\sim}{E}}_d) - \frac{v_{te}^2}{n} \nabla n \,, \tag{4}$$

$$\nabla \cdot \tilde{\underset{\sim}{E}} = 4\pi e (n_i - n_e) \,, \tag{5}$$

where $\tilde{\underset{\sim}{E}}_d = \frac{1}{2} \underset{\sim}{E}_d \exp(-i\omega_{pe} t) + c.c.$. The other notations are standard.

The electron and ion densities vary according to

$$n_e = n_o + n^h + \delta n_e \ , \tag{6}$$

$$n_i = n_o + \delta n_i \ ,$$

whereas the velocity perturbation is

$$\underset{\sim}{v} = \underset{\sim}{v}^h + \underset{\sim}{v}_e \ , \tag{7}$$

where h denotes the fast varying quantities, and δn, $\underset{\sim}{v}_e$, $\underset{\sim}{v}_i$ are the density and velocity perturbations associated with the slow plasma motion, to be discussed later.

First, we obtain from (3) – (7) the equation for the electron plasma waves in the presence of the driving field. The result is[3]

$$\nabla \cdot \left[\frac{\partial^2 \tilde{\underset{\sim}{E}}}{\partial t^2} - 3 v_{te}^2 \nabla^2 \tilde{\underset{\sim}{E}} - \omega_o^2 \left(1 - \frac{\omega_p^2}{\omega_o^2} \right) (\tilde{\underset{\sim}{E}} + \tilde{\underset{\sim}{E}}_d) \right] = 0 \ , \tag{8}$$

where harmonic generation is excluded. Here, $\omega_p^2 = \omega_{po}^2 (1 + x/L + \delta n_e / n_o)$, $\omega_{po} = \omega_p (x = 0) \simeq \omega_o$, and a linear density profile $n = n_o(1 + x/L)$ has been assumed.

Defining the new quantity $\underset{\sim}{E} = \tilde{\underset{\sim}{E}} + \tilde{\underset{\sim}{E}}_d + c.c.$, and noting that the field is one-dimensional and electrostatic, we obtain

$$\frac{\partial^2 \underset{\sim}{E}}{\partial t^2} - 3 v_{te}^2 \nabla^2 \underset{\sim}{E} - \omega_o^2 \left(1 - \frac{\omega_p^2}{\omega_o^2} \right) \underset{\sim}{E} = \frac{1}{2} \frac{\partial^2}{\partial t^2} \left[\tilde{\underset{\sim}{E}}_d \exp(-i\omega_{po} t) + c.c. \right] . \tag{9}$$

Using a WKB ansatz, $\underset{\sim}{E} \sim \underset{\sim}{E}(x,t) \exp(-i\omega_{po} t) + c.c.$, one obtains from (9)

$$i\varepsilon \frac{\partial \underset{\sim}{E}}{\partial t} + \nabla^2 \underset{\sim}{E} - \alpha \times \underset{\sim}{E} - (n_e - 1) \underset{\sim}{E} = \underset{\sim}{E}_d \ , \tag{10}$$

where $\varepsilon = 2(m_e / 3 m_i)^{1/2}$. The slow density perturbations, enhanced by the ponderomotive force of the high-frequency waves, are described by the following equations [15]

$$\frac{\partial n_i}{\partial t} + \nabla \cdot n_i \underset{\sim}{v}_i = 0 \ , \tag{11}$$

$$\frac{\partial \underset{\sim}{v}_i}{\partial t} + \underset{\sim}{v}_i \cdot \nabla \underset{\sim}{v}_i = -\nabla \phi - \frac{T_i}{T_e} \nabla \ln n_i \ , \tag{12}$$

$$n_e \nabla |\underset{\sim}{E}|^2 = n_e \nabla \phi - \nabla n_e , \tag{13}$$

$$\frac{1}{3} \nabla^2 \phi = n_e - n_i . \tag{14}$$

Here, the following units are used: time, $\sqrt{3}/\omega_{pi}$; length, $\sqrt{3}\,\lambda_e$; potential, T_e/e; density, N_0; electric field, $(4\pi n_0 T_e)^{1/2}$; and velocity, $c_s \equiv (T_e/m_i)^{1/2}$. The ponderomotive force $\nabla |E|^2$ on the electrons is obtained by averaging the plasma motion over an electron plasma period. The ponderomotive force acting on the ions is smaller by a factor m_e/m_i and is neglected.

Equations (10) to (14) describe generation and evolution of solitons in an inhomogeneous plasma in the presence of an external driver. These equations are also appropriate for considering the stability of solitons.

III. DRIVEN SOLITONS IN INHOMOGENEOUS PLASMAS

In this section, we use the one-dimensional electrostatic approximation. Furthermore, for the initial state, or in the case of small driving fields, we assume a linearized plasma response. Then the basic equations (10) - (14) simplify to

$$i\varepsilon \frac{\partial E}{\partial t} + \frac{\partial^2 E}{\partial x^2} - (\alpha x + n)E = E_d , \tag{15}$$

$$\frac{\partial n}{\partial t} = -\frac{\partial u}{\partial x} , \tag{16}$$

$$\frac{\partial u}{\partial t} = -\frac{\partial}{\partial x}(n + E E^*) . \tag{17}$$

In previous theories,[2,3] the static approximation $n = -E E^*$ for the low-frequency response was used in (15), and it was found that a soliton would be strongly accelerated down the density gradient. Because of the neglect of ion inertia effects, that result is of limited application. We expect that due to the large ion mass (compared to that of an electron), the acceleration should be much smaller than predicted by the cubic nonlinear Schrödinger equation. In this case, the soliton stays for a longer time in the resonant region, and the energy gain through the driver might be enhanced.

Thus, the problem of profile steepening at the critical density in laser created plasmas via soliton formation has to be reconsidered.

We first formulate a very simple but powerful method which has successfully been used in similar problems.[2,16] Starting from the zeroth order constants of motion, we shall present approximate solutions of Eqs. (15) – (17).

We have the following modified conservation laws, for the plasmon number,

$$i\varepsilon \frac{\partial}{\partial t} \int E\,E^*\,dx \;=\; -E_d \int (E - E^*)dx \;, \tag{18}$$

for the total momentum,

$$\frac{\partial}{\partial t} \int \left[i\varepsilon E \frac{\partial E^*}{\partial x} + nu \right]dx \;=\; -\alpha \int E\,E^*\,dx \;, \tag{19}$$

and for the energy,

$$\int \left[\frac{\partial E}{\partial x} \frac{\partial E^*}{\partial x} + n\,E\,E^* + \frac{1}{2}\,n^2 + \frac{1}{2}\,u^2 \right]dx \tag{20}$$

$$= -\alpha \int x\,E\,E^*\,dx - E_d \int (E + E^*)\,dx + S_o \;,$$

where S_o is a constant.

In order to allow for general functional dependence in our ansatz, we also include the equations for the center of gravity of the motion,

$$i\varepsilon \frac{\partial}{\partial t} \int x\,E\,E^*\,dx \;=\; -\int \left[2E \frac{\partial E^*}{\partial x} + x\,E_d\,(E - E^*) \right]dx \;, \tag{21}$$

and for the phase ψ of E,

$$-\varepsilon \int \frac{\partial \psi}{\partial t} E\,E^*\,dx \;=\; \int \left[\frac{\partial E}{\partial x} \frac{\partial E^*}{\partial x} + (\alpha x + n)E\,E^* + \frac{1}{2}E_d\,(E + E^*) \right]dx \;. \tag{22}$$

The solutions which we shall present below will satisfy all the five moment equations.

For E, n, and u, we now make the adiabatic approximation. Adiabaticity is a good assumption as long as the time needed by sound waves to traverse the soliton is small compared with the characteristic time for changes in the system. Our results will be shown to be consistent with that assumption.

Introducing the five time-dependent parameters x_o, q, η, a and b we make the following ansatz,

$$E = \sqrt{2}(1 - \dot{x}_o^2)^{1/2}\, q\, \eta\, \text{sech}\, \eta(x - x_o)\, \exp\left[i(ax + b)\right] , \tag{23a}$$

$$n = -2\eta^2 q^2 \text{sech}^2\, \eta(x - x_o) , \tag{23b}$$

$$u = -2\eta^2 q^2 \dot{x}_o\, \text{sech}^2\, \eta(x - x_o), \tag{23c}$$

where the dot denotes derivative with respect to time. Note that this ansatz is quite general, since the velocity, amplitude, width, as well as the phase of the soliton are free.

To test our procedure and to quantify the influence of ions, we first treat the case $E_d = 0$. Inserting (23) into the moment equations (18) – (22), one gets five coupled ordinary differential equations. The latter were solved analytically. For example, the position and the amplitude of a soliton initially at rest are

$$x_o = \frac{2v^2 + 3v^4 - v^6}{C(1 - v^2)} + \kappa v^2 , \tag{24}$$

$$\eta = \eta_o / (1 - v^2) . \tag{25}$$

The relation between the velocity $v = \dot{x}_o$ and time t is

$$t = \frac{4v}{C(1 - v^2)^3} + 2\kappa v , \tag{26}$$

and the constants $C = -3\alpha/\eta_o^2$, $\kappa = -\varepsilon^2/4\alpha$, and $\eta_o = \eta(t = 0)$ have been introduced.

It is interesting to note that the amplitude-width relation ($q = 1$) as well as the functional dependence of the phase on v is not changed as compared to the cubic Schrödinger equation result.[2,3]

Fig. 1 shows the time-dependence of the velocity v. The velocity approaches the sound velocity very slowly from below. Even after several ion plasma periods, the soliton is still subsonic and ion nonlinearities can be neglected.

Fig. 2 displays the discrepancy compared to Refs. 2,3. Note that the previous investigations predict velocities of the order of the electron thermal velocity within an ion period.

The fact that v does not cross c_s (in dimensional units) has also been noted by Chukbar and Yankov[17] who used a completely different method. Therefore, we can test the present method by comparing with their results. In Fig. 3, excellent agreement is shown.

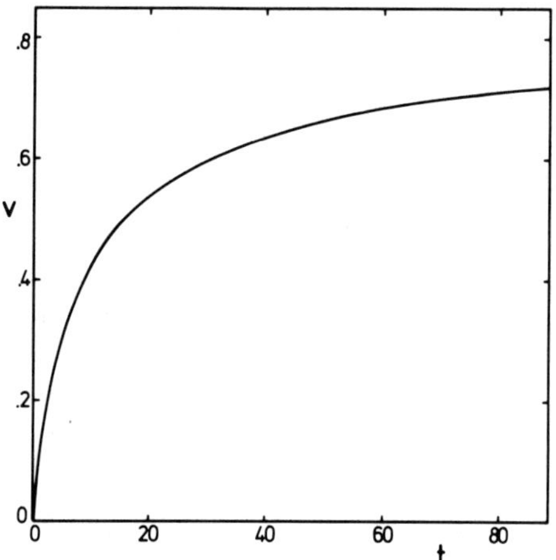

Fig. 1.   Plot of the velocity v (in units c_s) vs time t (in units $\sqrt{3}/\omega_{pi}$) of an accelerated soliton in an inhomogeneous plasma for $\alpha = 0.01$, $E_d = 0$, $\eta_o^2 = 0.1$, and $m_i / m_e = 1836$.

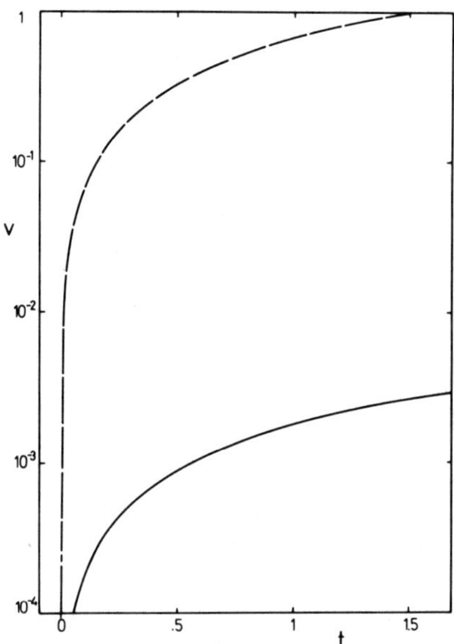

Fig. 2.   v (in units c_s) vs t (in units $\sqrt{3}/\omega_{pe}$) for the same para-- meters as in Fig. 1. The broken line is the previous result (Refs. 2,3) based on the solution of the cubic nonlinear Schrödinger equation.

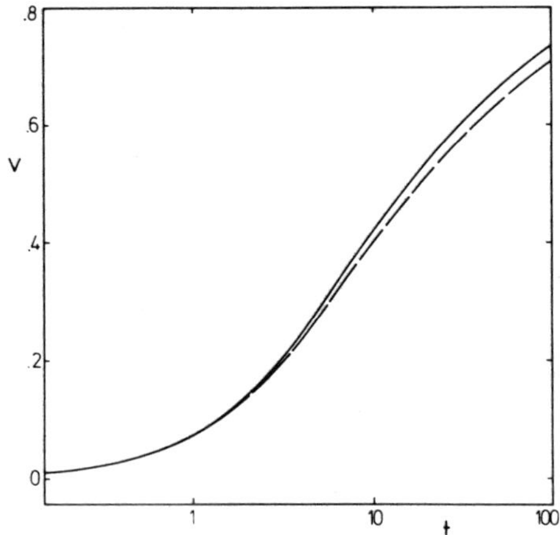

Fig. 3. v (in units c_s) vs t (in units $\sqrt{3}/\omega_{pi}$) for the same parameters as in Fig. 1. The broken line shows the result of a perturbative treatment, Ref. 17.

For applications to laser fusion devices, the computations for $E_d \neq 0$ are new and of utmost interest. We have solved the corresponding differential equations for x_o, η, q, a and b numerically. The physical picture would be as follows:[2] First, the soliton formed will have its height increasing because of the energy input due to the driver. The soliton saturates when it moves out of the resonance region due to acceleration in the density gradient. This behavior is different from the case of uniform plasma where no convective saturation can occur. However, compared to previous calculations,[2,3] ion inertia effects will drastically lower the velocity of the solitons, so that it might stay longer in the resonance region. The influence of this effect on the growth of the amplitude and decrease of the width has to be studied numerically. We investigate the time dependence of the field energy of a soliton $\left[\sim\eta(1-\dot{x}_o^2)q^2\right]$ and the width $\left[\sim\eta^{-1}\right]$. The latter gives a clear estimate of the inhomogeneity scale-length due to soliton formation at critical density.

Fig. 4 displays the dynamics of a single soliton in a highly inhomogeneous plasma. We have chosen large α to compare with previous results.[2,3] The initial value problem of the soliton formation reveals that at first the soliton energy grows linearly with time. The subsequent saturation accompanied by damped relaxation oscillations occurs when the phase shift due to nonlinearity detunes the soliton from the external driver. The growth of the soliton is limited when phase locking is broken. At this stage, the hyperbolic secant function stops the oscillations to an exponentially damped amplitude around the saturation level. In Fig. 5 we show the variation of the amplitude-width-ratio of a soliton with time.

The following conclusions can be drawn when we compare with previous results: The field energy $W = \int |E|^2 dx$ saturates at different levels and different times, depending on initial conditions. Although the time might be longer, saturation occurs much closer to the resonance point when compared with the cubic nonlinear Schrödinger result. Take for example the parameters $E_d = \sqrt{2}/10$, $\alpha = 0.2$, and $W_o = 0.05$. Then the velocity at saturation is now smaller than $\frac{1}{3} c_s$ whereas previously it was larger than v_{te}. Correspondingly, previously the soliton has travelled over several Debye lengths and thereby presumably left the resonance region before saturation could occur. In the present case, the position of the soliton was shifted less than $.1 \lambda_e$.

It is important to note that the saturation level depends on the initial values. We understand this result in the following way. The energy transfer between the external driver and the soliton is determined by a coupling coefficient which enhances the field energy density in proportion to its actual value. Then,

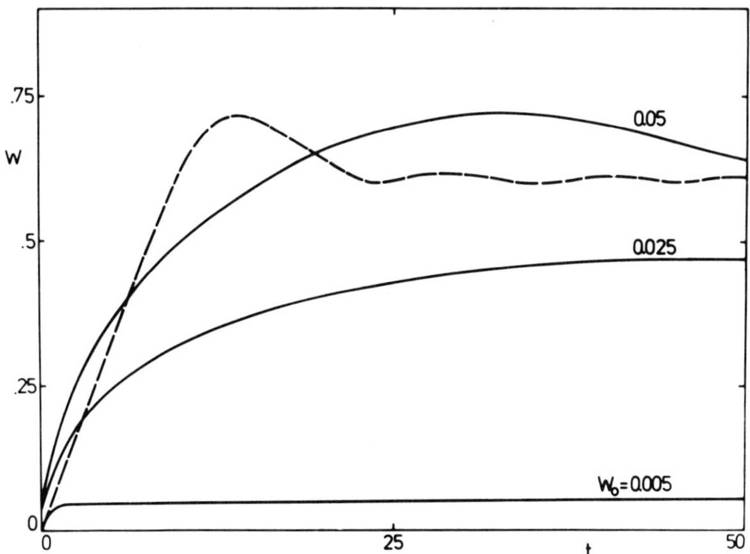

Fig. 4. Plot of the energy W (in units $96 \sqrt{3} \pi n_o T_e \lambda_e^3$) vs t (in units of ω_{pe}^{-1}) for E_d = 0.1414, α = 0.2, m_i / m_e = 1836, and various initial values W_o. The broken line is the previous result based on the solution of the cubic nonlinear Schrödinger equation.

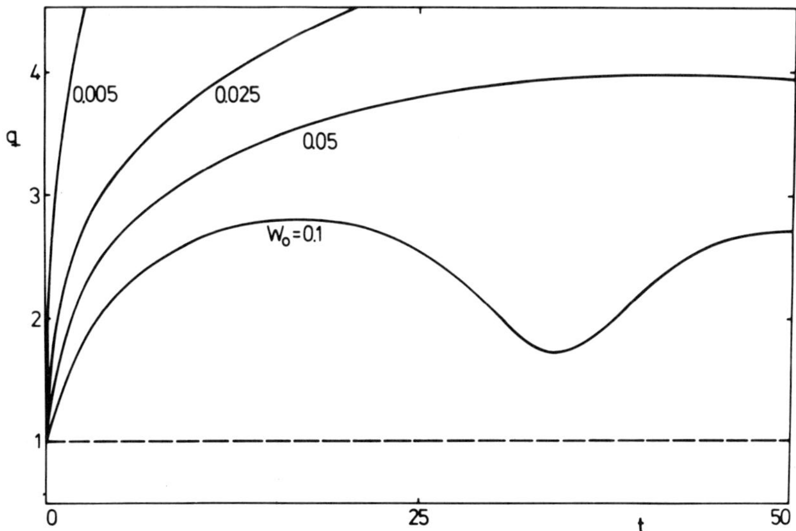

Fig. 5. Plot of the amplitude-width ratio q vs t (in units of ω_{pe}^{-1}) for the same parameters as in Fig. 4. Again, the broken line depicts the previous result, Refs. 2, 3.

however, the width of the soliton (η^{-1}) is approximately constant
in time. This new effect was detected since we have used a very
general ansatz for the soliton shape. The result is that the
driver breaks the amplitude width symmetry anticipated previously,[2,3] i.e., the width is not anymore inversely proportional to
the amplitude. On the other hand, damping sets in when $(a\,\eta^{-1}) > 1$,
and thereby the saturation level depends on the width of the soliton at $t = 0$.

IV. FINITE AMPLITUDE EFFECTS

As has been shown in the previous section, solitons can grow
to very large amplitudes. Obviously, the linear plasma response
becomes inadequate. We now study the behavior of such solitons
when they are effectively decoupled from the driver and move into
the nearly homogeneous region.

In this section, we discuss the stationary finite amplitude
solutions of the basic equations (10) – (14) for $E_d = \alpha = 0$. First,
we treat the nonrelativistic limit and point out the importance of
space charge effects which have not been included in the previous
treatment.[7]

The electron density n_e appearing in the wave equation for
electrostatic oscillations

$$i\varepsilon\,\frac{\partial E}{\partial t} + \frac{\partial^2 E}{\partial x^2} - (n_e - 1)E = 0 \;, \tag{27}$$

follows from the fluid equations for the slow motion. Under the
assumption of stationarity in a moving frame, the ion continuity
equation is

$$n_i\, u_i \;=\; -M \;, \tag{28}$$

whereas the ion and electron momentum equations are

$$\frac{1}{2}\,u_i^2 + \varphi + \frac{T_i}{T_e}\,\ln n_i \;=\; \frac{1}{2}\,M^2 \;, \tag{29}$$

$$\varphi - \ln n_e \;=\; |E|^2 \;. \tag{30}$$

We close the system with Poisson's equation

$$\frac{1}{3}\,\frac{\partial^2 \varphi}{\partial x^2} \;=\; n_e - n_i \;. \tag{31}$$

Here, $u_i = v_i - M$ is the ion velocity in the moving frame, and we
use x as the space coordinate in the moving frame.

We use the hydrodynamic approximation, anticipating weak variations in space within a Debye length. Thereby the amplitude of the solitons is also restricted since the amplitude and width are related.

Eliminating u_i from Eqs. (28) and (29), we can solve n_i in terms of φ, that is

$$n_i = \psi(\varphi) . \tag{32}$$

Inserting this and the Boltzmann distribution $n_e = \exp(\varphi - |E|^2)$ into Eq. (31), we obtain

$$\frac{1}{3}\frac{\partial^2 \varphi}{\partial x^2} = \exp(\varphi - E^2) - \psi(\varphi) , \tag{33}$$

and

$$\frac{\partial^2 E}{\partial x^2} = \eta^2 E - \left[1 - \exp(\varphi - E^2)\right]E . \tag{34}$$

We have denoted the nonlinear frequency shift by η^2 and used $E = E(x)\exp(i\eta^2 t/\varepsilon)$ for the envelope of the high frequency electrostatic field.

Equations (33) and (34) can be integrated once to yield

$$\frac{1}{2}\left(\frac{\partial E}{\partial x}\right)^2 - \frac{1}{12}\left(\frac{\partial \varphi}{\partial x}\right)^2 + \frac{1}{2}(1 - \eta^2)E^2 + \frac{1}{2}\exp(\varphi - E^2) - \frac{1}{2}\Phi(\varphi) = C . \tag{35}$$

In Eq. (35), Φ is related to ψ through $\psi = \Phi'(\varphi)$, where the prime denotes derivatives with respect to the argument. The constant C will be determined by the boundary conditions for solitary wave solutions.

We do not intend to integrate Eqs. (33) and (34) exactly although this might be possible. The reason is that within our model (as in all previous models where damping effects are neglected) the electric field energy density ($\sim 2\eta^2$) should not be too large. The larger the amplitude the narrower a soliton is, and the more effective is the interaction between the soliton and the particles.[18] We have estimated that in an argon plasma for $\eta^2 > 0.25$, a soliton interacts strongly with the background thermal distribution. Keeping this in mind, a perturbative treatment is adequate. The subsonic scaling with η^2 as a smallness parameter

$$E = \eta E_o(\eta x) + \eta^3 E_1(\eta x) + \cdots ,$$
$$\varphi = \eta^2 \varphi_o(\eta x) + \eta^4 \varphi_1(\eta x) + \cdots ,$$

is motivated by the well-known small amplitude solutions of the cubic nonlinear Schrödinger equation. Other scalings can be used in a similar way.

From Eq. (30) we find that

$$\psi'(0) \ = \ (M^2 - \alpha)^{-1} \ , \tag{36}$$

where $\alpha = T_i / T_e$. Thus, to lowest order (η^3) we obtain the following well-known solutions

$$E_o \ = \ \sqrt{2}(1 + \alpha - M^2)^{1/2} \ \text{sech} \ \eta x \ , \tag{37a}$$

$$\varphi_o \ = \ E_o^2 (\alpha - M^2) / (1 + \alpha - M^2) \ . \tag{37b}$$

To discuss the role of nonlinearities and quasi-neutrality, we now investigate the next order contributions. Note that the function $\varphi = \varphi(n_i)$ can be inverted up to arbitrary order by taking derivatives of (30). From Eq. (35), after some algebra, we get (to order η^6)

$$E_o' E_1' - E_o'' E_1 \ = \ \frac{1}{12} \varphi_o'^2 + \frac{1}{12} (1 + \alpha - 3M^2)E_o^6 / (1 + \alpha - M^2)^3 \ , \tag{38}$$

where Eqs. (33) and (34) have been used. Equation (38) is an inhomogeneous differential equation for E_1. The inhomogeneity consists of two contributions: the first term on the right hand side of Eq. (12) originates from non-neutrality while the second one corresponds to higher electron and ion nonlinearities. The electron nonlinearity is caused by the (exponential) Boltzmann distribution while the ion nonlinearities are due to convective ($M \neq 0$) and Boltzmann contributions ($\alpha \neq 0$).

By the method of variation of constant, the appropriate solution of (38) can be written in the form

$$E_1 \ = \ E_{1P} + E_{1N} \ ,$$

where E_{1P} and E_{1N} denote the contributions from Poisson's equation and (electron as well as ion) nonlinearities, respectively. After some algebra, one obtains

$$E_{1P} \ = \ -\left[2\sqrt{2}(\alpha - M^2)^2 / 3(1 + \alpha - M^2)^{1/2}\right] \text{sech} \ \eta x (1 - \text{sech}^2 \eta x), \tag{39a}$$

and

$$E_{1N} \ = \ \left[2\sqrt{2}(1 + \alpha - 3M^2)/3(1 + \alpha - M^2)^{1/2}\right] \text{sech} \ \eta x (1 - \tfrac{1}{2}\text{sech}^2 \eta x). \tag{39b}$$

To quantify the effect of space charge, we calculate the effective spatial width δ using the definition of D'Evelyn and Morales[5] (i.e., $E(\eta\delta)/\eta E_o(0) = \text{sech} \ 1$) and the ratio $R \equiv |E_{1P}|/|E_{1N}|$. The latter measures the relative importance of space charge effects, as compared with electron and ion nonlinearities. The results are given in Figs. 6 - 8.

Fig. 6 depicts a test of our perturbation analysis. Comparing with an exact numerical computation of (35), it demonstrates very high accuracy in the relevant region $\eta^2 < 0.25$.

Fig. 7 exhibits the dependence of the normalized width $\eta\delta$ on T_i / T_e for a standing soliton. We conclude that the exponential Schrödinger equation is a good model for nearly standing solitons ($M^2 \ll 1$) when $T_i \ll T_e$. On the other hand, for $T_i \simeq T_e$, as in many laboratory and astrophysical plasmas, neglect of Poisson's equation while considering finite amplitude effects is inconsistent. As is shown, space charge effects alter the amplitude-width-relation.

Fig. 8 shows the dependence of the normalized width $\eta\delta$ on Mach number for $T_i / T_e = 0$. We see that ion nonlinearities originating from the convective term are only important for $M \gtrsim 0.5$. Then, however, space charge effects can not be neglected. Especially in the region $M \simeq 0.6$, space charge effects even dominate over ion nonlinearities. This happens because at $M \approx 0.6$, the ion and electron nonlinearities tend to cancel each other. Thus, solitons in this Mach number range should be discussed by the present theory where space charge effects are included. In addition, it should be mentioned that for $M \simeq 1$, even in the small amplitude limit, nonlinearities and deviations from quasi-neutrality are of importance.[19] Therefore, we conclude that for the type of solitary waves discussed here, predictions of significant effects due to nonlinearities are only correct when space charge effects are also taken into account.

When the amplitude of the laser radiation is sufficiently large, relativistic electron mass variation introduces an additional nonlinear effect which modifies the wave propagation characteristics. In the following, we study the combined effects of the ponderomotive force and relativistic electron mass variation on the formation of solitary electromagnetic waves. Nonlinearities in the low frequency motion discussed earlier shall also be included.

The importance of the effect of relativistic mass variation on the modulational instability and solitary wave propagation can be qualitatively understood as follows. In the presence of a strong electromagnetic wave field, the electron plasma frequency $\omega_{pe} = (4\pi n e^2/m)^{1/2}$ can be modified locally in two ways. One is the usual slow density variation $n \approx 1 - \beta|\psi|^2/2(1-M^2)$ caused by the ponderomotive force, the other is the mass variation $m \approx m_o(1 + |\psi|^2/2)$ caused by the rapid oscillation of the electrons in the high frequency field. Here, $M = V/c_s$ is the Mach number of the wave packet, $\beta = c^2/v_{Te}^2$, $\psi = e \underset{\sim}{A}/m_o c^2$, n is normalized with the unperturbed density n_o, m_o is the electron rest mass, and $\underset{\sim}{A}$ is the high frequency field wave vector. The wave amplitude is assumed to be not

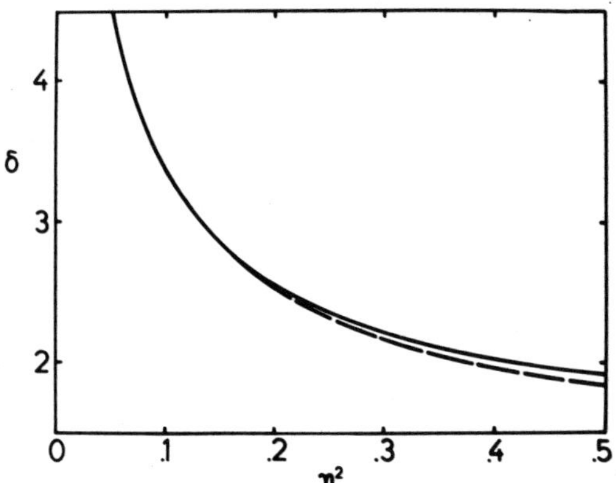

Fig. 6.   Plot of the width δ (in units $\sqrt{3}\,\lambda_e$) vs the nonlinear frequency shift η^2 for M = 0 and T_i/T_e = 0. The solid line is the exact numerical result, whereas the broken line shows the result of the perturbation analysis.

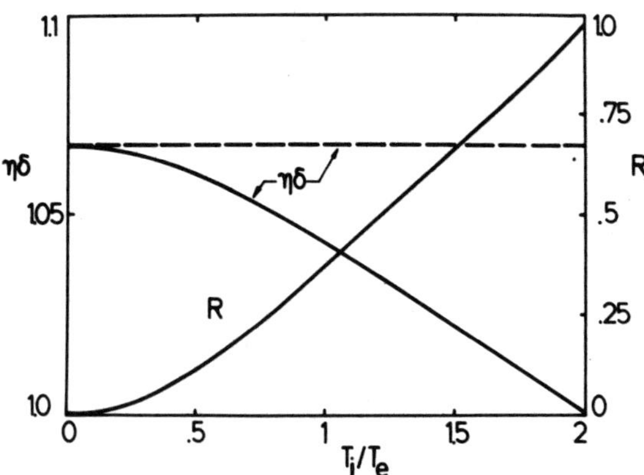

Fig. 7.   Plot of the normalized width $\eta\delta$ and the relative influence R of space charge effects vs T_i/T_e for M = 0 and η^2 = 0.1. The broken line is calculated from previous models when space charge fields are neglected.

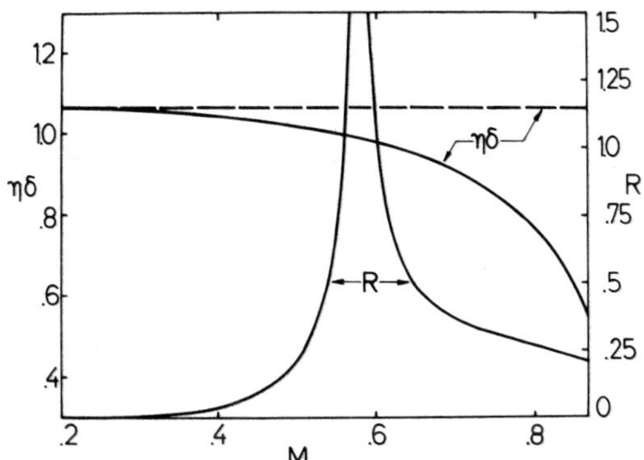

Fig. 8. $\eta\delta$ and R vs. M for $T_i/T_e = 0$ and $\eta^2 = 0.1$. The broken line depicts the normalized width resulting from the exponential Schrödinger equation.

too large. The local plasma frequency is then $\omega_{pe} = \omega_{peo}\left[1 - \beta|\psi|^2/4(1 - M^2) - |\psi|^2/4\right]$. Here, the last two terms arise from the ponderomotive force and mass variation effects, respectively. Therefore, the two effects can be of equal importance if $\beta \approx 1 - M^2$. Soliton formation is possible if the nonlinear frequency shift $\omega_{pe} - \omega_{peo}$ is negative, or $\beta|\psi|^2/(1 - M^2) + |\psi|^2 > 0$, while the growth rate of the modulational instability is proportional to the left hand side of the inequality after setting $M = 0$. Thus, the characteristic time scales associated with the two nonlinearities are of the same order if $\beta \approx 1$. Furthermore, we see that the inclusion of relativistic mass variation leads to the remarkable result that supersonic solitons can exist if $\beta < |1 - M^2|$. From the expression for the local plasma frequency, it follows that a density hump is associated with supersonic solitary waves. This happens because the nonlinear steepening effect due to mass variation dominates over the hole-digging tendency of the ponderomotive force.

To investigate the effects of relativistic mass variation in detail, we consider the propagation of a circularly polarized laser light in the form

$$\underset{\sim}{A} = A\left[\hat{x} \cos\phi + \hat{y} \sin\phi\right] , \tag{40}$$

where $\phi = \omega_o t - k_o z$. The linear dispersion relation is

$$\omega_o^2 = c^2 k_o^2 + \omega_{pe}^2 . \tag{41}$$

In Ref. 9, we have shown that the wave envelope satisfies the nonlinear Schrödinger equation

$$2i\varepsilon \frac{\partial}{\partial t}\tilde{\psi} + \beta \frac{\partial^2}{\partial z^2}\tilde{\psi} + \Delta\tilde{\psi} = \frac{n\tilde{\psi}}{(1 + |\tilde{\psi}|^2)^{1/2}} , \tag{42}$$

where $\varepsilon = (m_e/m_i)^{1/2}\omega_o/\omega_{peo}$ and $\Delta = (\omega_o^2 - c^2 k_o^2)/\omega_{pe}^2$. Here, $\tilde{\psi}$ is defined by

$$\frac{e\underset{\sim}{A}}{m_o c^2} = \underset{\sim}{\psi} = \frac{1}{2}\tilde{\psi}(t,z)(\hat{x} - i\hat{y})e^{-i\phi} , \tag{43}$$

and t is normalized by ω_{pi}, and z by λ_e.

In the moving frame $\xi = z - Mt$, we have

$$\beta \frac{\partial^2}{\partial \xi^2}\psi + \delta\psi = \frac{n\psi}{(1 + \psi^2)^{1/2}} , \tag{44}$$

where $\tilde{\psi} = \psi(\xi)\exp i\left[\theta(t) + \kappa(z)\right]$, δ is a nonlinear frequency shift, and stationarity has been assumed in the moving frame.

From the low frequency equations of motion, assuming $k_o \lambda_e \ll \beta$, we find, for localized solutions

$$(1 + \psi^2)^{1/2} = -\frac{M^2}{2\beta} \left(\frac{1}{n^2} - 1\right) - \frac{\ln n}{\beta} + 1 . \tag{45}$$

The frequency shift is given by

$$\delta = 2(1 - N + M^2 - M^2/N) / \beta \psi_m^2 . \tag{46}$$

Combining Eqs. (44) and (45), and integrating once, we obtain

$$\frac{1}{2}\left(\frac{dn}{d\xi}\right)^2 + V(n; \ \delta, \ M) = 0 , \tag{47}$$

where

$$V(n) = \frac{(n - 1 + M^2/n - M^2 + \frac{1}{2} \delta \beta \psi^2) n^6 \psi^2}{(n^2 - M^2)^2 (1 + \psi^2)} . \tag{48}$$

In Ref. 9, the potential $V(n)$ has been analyzed and solitons shown to exist. The relation between the center ψ_m of the field envelope, the center density variation N, and the Mach number M is given by the nonlinear dispersion relation

$$(1 + \psi_m^2)^{1/2} = -\frac{M^2}{2\beta} \left(\frac{1}{N^2} - 1\right) - \frac{\ln N}{\beta} + 1 . \tag{49}$$

In Fig. 9, the existence diagram of solitons is presented. The existence of supersonic solitary waves with density humps is clearly demonstrated. This study shows that relativistic effects can considerably modify the propagation characteristics of laser light. The implication of these effects on the mode conversion process is at present still not clarified, except in the small amplitude limit.

V. STABILITY ANALYSIS

We now discuss the stability properties of the generated solitons. First, to prove longitudinal stability we use the Ljapunov method. From the physical point of view, the Ljapunov theory for stability is a generalization of the physical principle that a stable state should contain minimum free energy. A few additional points are worth mentioning in this respect.

It would be unphysical to call a solitary wave solution unstable simply when the distance between the perturbed and unperturbed states grows in time. The reason is that for solitary waves, amplitude, Mach number, and nonlinear frequency shift are correlated. An amplitude perturbation will lead to a different

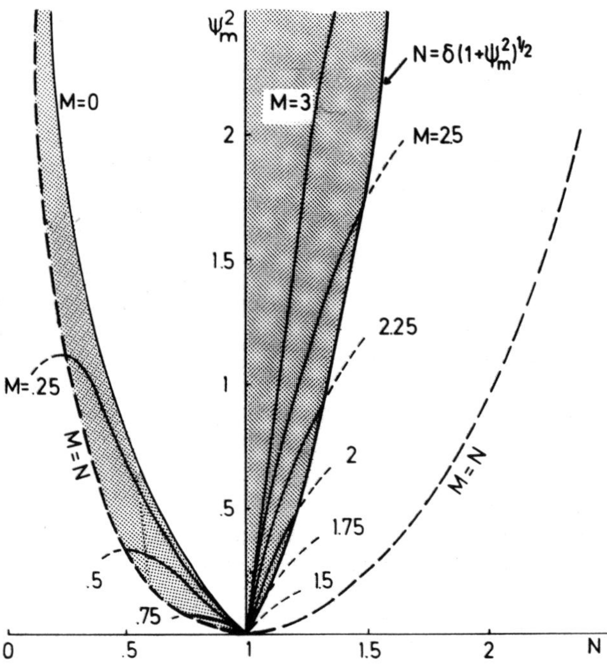

Fig. 9. Relation between the center amplitude ψ_m and the corresponding density (maximum or minimum) N for fixed β (=2) and different Mach numbers. Solitons exist only in the shaded area. The $M > 1$, $N > 1$ region exists only when relativistic effects are included.

translation velocity and phasor compared to the unperturbed soliton. To exclude effects originating from this behavior, we consider stability with respect to form. Thus, we define an invariant
set consisting of all functions which differ from the original
perturbed state by arbitrary translations, $x - x'$, and rotations
in phase, $t - t'$. The distance (metric) between a perturbed state
and the invariant set can be found by minimization with respect
to x' and t'. This defines the so called closest solitary wave
(reference state), being identical in form to the original unperturbed state. The minimization yields constraints for the stability investigation, i.e., relations between the perturbations and
the reference state.

Briefly, the stability analysis proceeds as follows. Guided
by the physical analogy with the free energy, find a functional L
with the following characteristics:

(i)   For perturbed states f from the neighborhood of the unper-
      turbed state (invariant set), upper and lower bounds for
      L(f) exist in terms of the norm;

(ii) L is not increasing, i.e., $dL/dt \leq 0$.

If such a Ljapunov functional exists, the invariant set is
stable. Physically, this conclusion is plausible: A perturbed
state, being close to the unperturbed state, will have a small
finite free energy source due to condition (i). Because of the
condition (ii), the free energy will not grow, implying (via (i)),
that the perturbed state will stay in a certain neighborhood of
the unperturbed one.

In a reversible system, a Ljapunov functional should be constructed from the constants of motion. Writing a stationary solution of the basic equations (1) - (5) in the form

$$E = G(x) \exp(i \eta^2 t/\varepsilon) \, , \tag{50}$$

we have derived the sufficient stability criterion[20]

$$\frac{\partial}{\partial \eta^2} \int dx \, G^2 > 0 \, . \tag{51}$$

We now discuss the physical implications of this criterion.
In the small amplitude limit, i.e., for the nonlinear Schrödinger
equation and the Zakharov equations, the stationary solutions

$$G = \left[2(1 - M^2)\right]^{1/2} \eta \, \mathrm{sech}\left[\eta(x - Mt)\right] \, , \tag{52}$$

obviously satisfy the stability criterion.

For finite amplitude solitons, (51) can not be used directly since the explicit analytical form of G is unknown. However, for quasi-neutral solutions G is the even and nodeless solution of the Schrödinger equation

$$-\eta^2 G + \frac{d^2}{dx^2} G + (1-n)G = 0 ,$$
(53)

and n and u follow from the algebraic relations

$$n u = -M ,$$
(54)

$$\frac{1}{2} u^2 + \ln n + G^2 = \frac{1}{2} M^2 ,$$
(55)

where u = v - M.

With the help of the last two equations the stability criterion can be rewritten as

$$\frac{\partial}{\partial \eta^2} \int \left[\frac{1}{2} M^2 (1 - n^{-2}) - \ln n \right] \left(\frac{dn}{dx} \right)^{-1} dn > 0 .$$
(56)

Integration of Eqs. (53) - (55) yields dn/dx in terms of n, N = min(n), M, and η. After integration by parts one finally obtains from (56)

$$\frac{\partial^2}{\partial (\eta^2)^2} \int_N^1 dn \left\{ -n^2 \left[1 - \eta^2 + (1 - M^2/n) \Big/ \left[M^2 (n+1) \Big/ 2n^2 \right. \right. \right.$$
$$\left. \left. \left. - (\ln n) \Big/ (n-1) \right] \right] \Big/ 3(n^2 - M^2) \right\}^{1/2} (1 - M^2/n^2) > 0 .$$
(57)

Note that for any finite amplitude soliton we can now check the stability criterion by simple integration. The explicit form of the soliton is not needed anymore. Numerical evaluation showed that all known finite amplitude solitons are longitudinally stable.

We now discuss the transverse instability of the solitary wave solution subject to perturbations governed by Eqs. (10) - (14). In general, as was the case of longitudinal stability, such an investigation should be nonlinear in the perturbations; however the nonlinear treatment confirms the linear calculation presented below.[21]

Linearizing in the form

$$\underset{\sim}{E} = (\underset{\sim}{G} + \underset{\sim}{a} + i\underset{\sim}{b}) \exp\left[i(\eta^2/\epsilon)t \right] ,$$
(58a)

$$n = n_o + n_i, \tag{58b}$$

we obtain an equation of the form[22]

$$H_+^{-1} \ddot{\underset{\sim}{\xi}} = -H_- \underset{\sim}{\xi} , \tag{59}$$

where H_+ and H_- are self-adjoint operators and H_+^{-1} is positive definite. This relation has the advantage that it is not necessary to discuss directly the higher order non-Hermitean operator $H_+ H_-$.

From (59) we obtain the maximum growth rate[10]

$$\gamma^2 = \sup_{\underset{\sim}{a},n_i} - \frac{\langle \underset{\sim}{a}|H_+ \underset{\sim}{a}\rangle + 2\langle n_i|\underset{\sim}{a}\,\underset{\sim}{G}\rangle + \frac{1}{2}\langle n_i|e^{G^2} n_i\rangle - R}{\varepsilon^2 \langle \underset{\sim}{a}|H_+^{-1} \underset{\sim}{a}\rangle + \frac{1}{2}\langle n_i|P^{-1} n_i\rangle} , \tag{60}$$

where

$$P = -\nabla \cdot \exp(-G^2)\nabla ,$$

$$R = \frac{1}{6}\langle \nabla\varphi|\nabla\varphi\rangle + \frac{1}{18}\langle \Delta\varphi|e^{G^2}\Delta\varphi\rangle \geq 0 ,$$

$$\frac{1}{3}\Delta\varphi = n_e - n_i ,$$

and n_e is implicitely given by

$$n_i = n_e - \Delta(n_e e^{G^2} + 2\underset{\sim}{a}\cdot\underset{\sim}{G}) .$$

The general expressions for H_+ and H_- were given in Ref. 10; special cases will be treated explicitly below.

We shall now discuss some interesting conclusions from (60). For simplicity, we assume quasi-neutral perturbations such that $R = 0$ and $n_i = n_e$. (When non-quasi-neutral perturbations are allowed, the actual growth rates are larger than those given below).

The result for transverse instability of the cubic nonlinear Schrödinger equation

$$i\varepsilon \frac{\partial E}{\partial t} + \nabla^2 E + |E|^2 E = 0 \tag{61}$$

as a model equation in one dimension is

$$\gamma^2 = \sup_a - \frac{\langle a|H_- a\rangle}{\varepsilon^2 \langle a|H_+^{-1} a\rangle} . \tag{62}$$

This result follows from (60) in a straightforward manner by neglecting the electron nonlinearity and the second term in the denominator of (60) (static approximation). This implies $n_i = -2aG$, and

$$H_- = H_+ - 2 G^2 , \tag{63}$$

$$H_+ = -\frac{d^2}{dx^2} + k^2 - G^2 + \eta^2 , \tag{64}$$

where G is given by

$$G(x) = \sqrt{2}\, \eta \operatorname{sech}(\eta x) . \tag{65}$$

In the small k limit, the growth rate is[23] $\gamma = 2 k\eta/\varepsilon$, whereas in the whole unstable k-region

$$3 k^2 \eta^2 (1 - k^2/k_c^2) / \varepsilon^2 \le \gamma^2 \le 4 k^2 \eta^2 (1 - k^2/k_c^2) / \varepsilon^2 \tag{66}$$

holds. The cut-off is $k_c = \sqrt{3}\,\eta$. The exact k-dependence of the growth rate is shown in Fig. 10. This result clearly shows that the order of magnitude of this growth rate (in dimensional form) is $\gamma \gtrsim \omega_{pi}$. Obviously the static ion response approximation is inapplicable. We therefore conclude that the cubic nonlinear Schrödinger equation is not a good model for considering the transverse instability of Langmuir waves.

Next, we take the dynamic ion response into account and present an analytical estimate of the growth rate for the Zakharov equations

$$i \frac{\partial E}{\partial t} - K^2 E + \frac{\partial^2}{\partial x^2} E - n E = 0 , \tag{67}$$

$$\frac{\partial^2 n}{\partial t^2} + K^2 n - \frac{\partial^2}{\partial x^2} n = \left(\frac{\partial^2}{\partial x^2} - K^2\right)|E|^2 . \tag{68}$$

Thus, in the general result (60), we formally let $\varepsilon = 1$, $\eta \equiv A$, and approximate $\exp(G^2) \simeq 1$. Note that here the time is normalized by $\sqrt{3}/\varepsilon\, \omega_{pi}$. Furthermore, $\Delta\varphi = 0$, $H_+ = -d^2/dx^2 + K^2 - G^2 + A^2$, and $G = \sqrt{2}\, A \operatorname{sech}(Ax)$. We now discuss the order of magnitude of the growth rate Γ.

First, an upper bound can be given (via a complementary variational principle)[24] in the form

$$\Gamma^2 \le \Gamma_+^2 \equiv -\frac{K^4}{2} + \left[4 K^2 Q + K^8/4\right]^{1/2} , \tag{69}$$

where $Q = <G^2|PG^2>/<G|G> = \frac{8}{15} A^4$. Note that this upper bound agrees with the growth rate given by Schmidt;[25] similarly, the expression of Valeo and Estabrook[25] is actually an upper bound. A lower bound can be easily found by inserting a specific test function. We obtain

$$\Gamma^2 \geq \Gamma_-^2 = \frac{-4\alpha - 15 K^2/8 A^4 - 20 A^2 \alpha^2/7}{15/8 A^4 K^2 + \alpha^2} \tag{70}$$

with

$$\alpha = \frac{1}{2}\left(\frac{75}{28 K^2 A^2} - \frac{15 K^2}{16 A^4}\right) - \left[\frac{1}{4}\left(\frac{75}{28 K^2 A^2} - \frac{15 K^2}{16 A^4}\right)^2 + \frac{15}{8 K^2 A^4}\right]^{1/2} .$$

For an argon plasma with $|E|^2/4\pi N_0 T_e \tilde{\simeq} 0.1$ we find $A \simeq 52$. The maximum value of Γ occurs[25] for $K = (2Q)^{1/6} \simeq 14$. At this value the upper limit is $\Gamma_+ \simeq 200$, i.e. in dimensional units $\gamma_+ \sim \frac{1}{2}\omega_{pi}$; the lower limit is $\gamma_- \sim \frac{1}{4}\omega_{pi}$.

In Fig. 11 we compare the numerical results for transverse instability for the cubic nonlinear Schrödinger equation and the Zakharov equations. Although the cut-off is the same, the growth rate is significantly reduced because of ion inertia effects.

VI. CONCLUSIONS

In this paper we have discussed the various nonlinear processes operating near the critical density in laser-produced plasmas. Ponderomotive force effects tend to dig holes (cavitons) into the density, thereby locally increasing the density gradient. Relativistic effects, however, can operate in the opposite direction. The dynamics of cavitons has been studied and it was found that ion inertia effects can be very important. Even for large amplitudes, the correct amplitude-width relations have been given. The latter is a direct measure of the effective density gradient.

Finally, for transverse perturbations, the instability region was investigated. Numerical evaluation of a very general growth rate formula allows us to obtain the wavelength for maximum growth. Because of quite short growth times, these instabilities should play a crucial role in laser-plasma interaction experiments.

The theoretical aspects discussed here do not include the very important problem of nonlinear wave particle interaction leading to particle acceleration. The latter has been recently discussed elsewhere.[26]

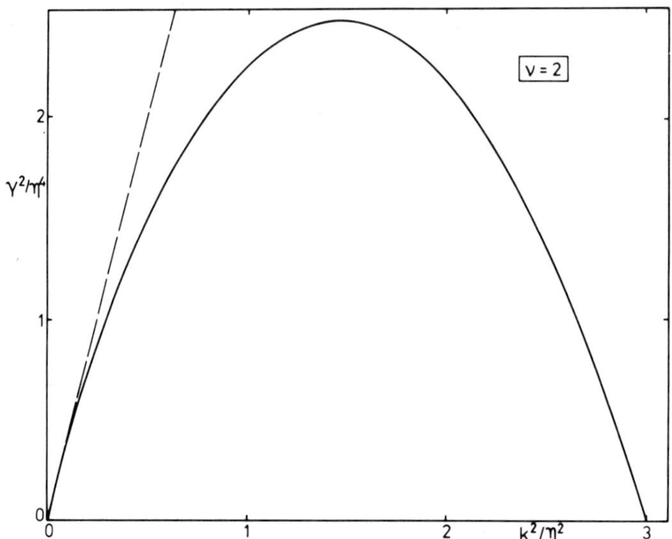

Fig. 10.   Normalized growth rate γ^2/η^4 vs k^2/η^2 for the cubic nonlinear Schrödinger equation. The broken line depicts previous results (Ref. 23) which are valid in the small k region.

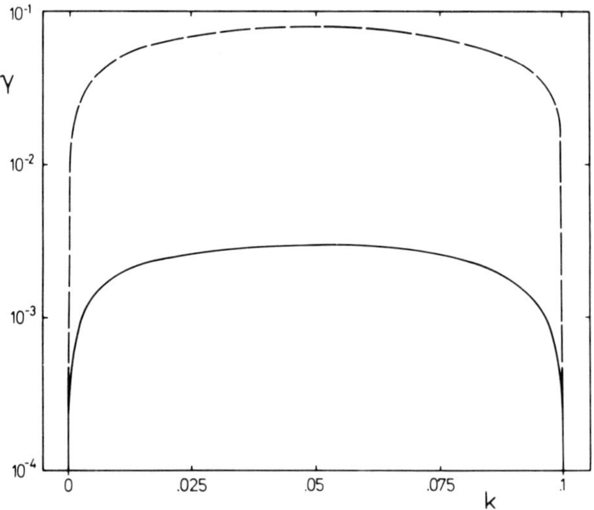

Fig. 11.   Plot of the growth rate γ (in units ω_{pe}) vs transverse wave number k (in units λ_e^{-1}) for $\eta^2 = 0.1$ and $m_i/m_e = 1836$. The solid curve is the result for the Zakharov equations whereas the broken line represents the growth rate for the cubic nonlinear Schrödinger equation.

REFERENCES

1. A. Y. Wong, _in_: "Laser Interaction and Related Plasma Pheno-
    mena", Vol. 4B, H. J. Schwarz and H. Hora, ed., Plenum
    (1977); and references therein.
2. H. H. Chen and C. S. Liu, _Phys. Rev. Lett._ 39:1147 (1977).
3. P. K. Shukla and K. H. Spatschek, _J. Plasma Phys._ 19:387
    (1978).
4. G. F. Morales and Y. C. Lee, _Phys. Rev. Lett._ 33:1016 (1974).
5. M. D'Evelyn and G. Morales, _Phys. Fluids_ 21:1977 (1978).
6. V. E. Zakharov, _Sov. Phys. JETP_ 35:908 (1972).
7. K. Lee, D. W. Forslund, J. M. Kindel, and E. L. Lindman,
    Phys. Fluids 20:51 (1977).
8. E. W. Laedke and K. H. Spatschek, _Phys. Lett. A_ (in press).
9. K. H. Spatschek, _Phys. Rev. A_ 16:2470 (1977); M. Y. Yu,
    P. K. Shukla, and K. H. Spatschek, _Phys. Rev. A_ 18:1591
    (1978).
10. E. W. Laedke and K. H. Spatschek, _Phys. Rev. Lett._ 42:1534
    (1979).
11. A. D. Piliya, _Sov. Phys. Tech. Phys._ 11:619 (1966).
12. K. G. Estabrook, E. J. Valeo, and W. Kruer, _Phys. Fluids_
    18:1151 (1975).
13. V. L. Ginzburg, "The Propagation of Electromagnetic Waves in
    Plasmas", p. 260, Pergamon (1970).
14. D. L. Kelly and A. Banos, Preprint UCLA PPG 170 (1974).
15. K. H. Spatschek, _Phys. Fluids_ 21:1032 (1978).
16. D. R. Nicholson and M. V. Goldman, _Phys. Fluids_ 19:1621
    (1976).
17. K. V. Chukbar and V. V. Yankov, _Sov. J. Plasma Phys._ 3:780
    (1977).
18. A. Y. Wong, P. Leung, and D. Eggleston, _Phys. Rev. Lett._
    39:1407 (1977).
19. K. Nishikawa, H. Hojo, K. Mima, and H. Ikezi, _Phys. Rev.
    Lett._ 33:148 (1974); V. G. Makhankov, _Phys. Lett._ 50 A:
    42 (1974).
20. E. W. Laedke and K. H. Spatschek, _Phys. Fluids_ (in press).
21. E. W. Laedke and K. H. Spatschek, _Phys. Rev. Lett._ 41:1798
    (1978).
22. E. W. Laedke and K. H. Spatschek, _J. Plasma Phys._ (in press).
23. V. E. Zakharov and A. M. Rubenchik, _Zh. Eksp. Teor. Fiz._
    65:997 (1973).
24. E. W. Laedke and K. H. Spatschek, _Phys. Lett. A_ (in press).
25. E. J. Valeo and K. G. Estabrook, _Phys. Rev. Lett._ 34:1008
    (1975); G. Schmidt, _Phys. Rev. Lett._ 34:724 (1975).
26. B. W. Boreham and H. Hora, _Phys. Rev. Lett._ 42:776 (1979).

NONLINEAR BEHAVIOR OF STIMULATED SCATTER IN LARGE UNDERDENSE
PLASMAS

W. L. Kruer and K. G. Estabrook

University of California, Lawrence Livermore Laboratory

Livermore, California 94550

ABSTRACT

Several nonlinear effects which limit Brillouin and Raman
scatter of intense light in large underdense plasmas are
examined. After briefly considering ion trapping and harmonic
generation, we focus on the self-consistent ion heating which
occurs as an integral part of the Brillouin scattering process.
In the long-term nonlinear state, the ion wave amplitude is
determined by damping on the heated ion tail which
self-consistently forms. A simple model of the scatter is
presented and compared with particle simulations. A similar model
is also applied to Raman scatter and compared with simulations.
Our calculations emphasize that modest tails on the electron
distribution function can significantly limit instabilities
involving electron plasma waves.

I. INTRODUCTION

As lasers increase in energy and the targets become larger,
longer pulse length laser light is used to irradiate them. Much

---

Work performed under the auspices of the U.S. Department of
Energy by the Lawrence Livermore Laboratory under contract number
W-7405-ENG-48.

larger underdense plasmas are then created, and the size (L) of
this plasma as seen by the light with wavelength λ_0 becomes an
important parameter. This size parameter can be crudely estimated
as

$$\frac{L}{\lambda_0} \simeq \min \frac{(v_{exp}\tau, R)}{\lambda_0} \quad . \tag{1}$$

Here v_{exp} is a typical plasma expansion speed, τ is the pulse
length of the light, and R is either the radius of the target or
the focal spot radius. Note that the characteristic size
parameter scales as τ/λ_0 and that it is sometimes set by
geometrical considerations.

The mix of coupling processes is expected to be sensitive to
the size of the underdense plasma. When $L/\lambda_0 \lesssim 10$, the
coupling is primarily determined by processes which occur at or
near the steepened, rippled critical density surface. When
$L/\lambda_0 \gtrsim 10^2$, processes such as inverse bremsstrahlung,
stimulated Brillouin and Raman scatter, and filamentation can
begin to play a sizeable role. For example, a significant
fraction of the incident light may then be scattered by the
Brillouin instability before it reaches the critical density
surface.

We here consider several nonlinear effects which limit
Brillouin and Raman back-scatter of intense light in large
underdense plasmas. After briefly considering ion trapping and
harmonic generation, we focus on the self-consistent ion heating
which occurs as an integral part of the Brillouin scattering
process. In the long-term nonlinear state, the ion wave
amplitude is determined by damping on the heated ion tail which
self-consistently forms. A simple model of the scatter is
presented and compared with particle simulations.

A similar model is also applied to Raman scatter. In this
case, we focus on the self-consistent electron heating which
occurs in this scattering process. Simulations are presented
and compared with the model. Our calculations emphasize that
modest tails on the electron distribution function can
significantly limit instabilities involving electron plasma
waves. This choking effect can be greatly enhanced by either
transport inhibition or hot electron production near the
critical density surface.

II. 1. BRILLOUIN SCATTER

Let us first consider Brillouin scatter.[1,2] This scatter
can be most simply described as the resonant decay of an
incident photon into a scattered photon plus an ion sound wave.
Hence

$$\omega_o = \omega_t + \omega_{ia} \quad , \tag{2}$$

where ω_o (ω_t) is the frequency of the incident (reflected) light wave and ω_{ia} is the frequency of the ion sound wave. As is apparent from the frequency matching conditions, this process occurs throughout the underdense plasma. In addition, since $\omega_o \gg \omega_{ia}$, nearly all of the energy of the incident light wave is transferred to the scattered wave. Hence Brillouin scatter is particularly dangerous.

It's easy to show that Brillouin scatter is potentially an important effect. Consider a light wave with electric vector E_i incident onto a uniform underdense plasma with density n_p, and let there be an ion sound wave of amplitude δn with frequency and wave number appropriate to scatter the incident wave into a backscattered light wave with electric vector Er. Using Maxwell's equations and separating out the fast time and space scale dependences, we readily obtain coupled equations for the slowly-varying amplitudes:

$$\frac{\partial E_r}{\partial x} = - \alpha \frac{\delta n}{n_p} E_i \quad , \tag{3}$$

$$\frac{\partial E_i}{\partial x} = - \alpha \frac{\delta n}{n_p} E_r \quad .$$

Here

$$\alpha = \frac{\pi}{2} \frac{1}{\lambda_o} \frac{n_p}{n_{cr}} / \left(1 - \frac{n_p}{n_{cr}}\right)^{1/2} \quad , \tag{4}$$

where λ_o is the wavelength of the incident light and n_{cr} is the critical density. These equations are readily integrated to determine the reflectivity, r. If we consider a plasma with size L and neglect $E_r(L)$ compared to $E_i(o)$, we obtain

$$r = \tanh^2 \left(\alpha \frac{\delta n}{n_p} L\right) \quad . \tag{5}$$

A simple example is very instructive. Consider an underdense plasma with $n_p = .1\, n_{cr}$, $L = 10^3 \lambda_o$, and an ion wave amplitude of only one-percent of the plasma density. Then Eq. 5 predicts that $r \simeq 80\%$. Even a small density fluctuation can lead to sizeable scatter in a large underdense plasma.

Given that Brillouin scatter can be a significant effect, it's very important to understand what nonlinear effects serve to limit the amplitude of the ion sound wave. We will first estimate the effects of ion trapping and harmonic generation and then concentrate on the important effect of self-consistent ion heating, which produces a tail which damps the ion wave. We will emphasize the nonlinear regime in which the light pressure is less than or comparable to the pressure of the underdense plasma. In the opposite regime, which obtains at very high intensity, enhanced profile steepening will clearly act to significantly reduce the scatter.[3] In order to focus on the nonlinear behavior, we will neglect the effect of gradients and simply treat the underdense plasma as an equivalent region of plasma with uniform density.

II. 2.   ION TRAPPING

Ion trapping is one effect commonly invoked to limit the ion wave amplitude. The basic idea is simple. As the ion wave amplitude increases, its potential becomes large enough to nonlinearly bring ions into resonance with the waves. Since such ions are efficiently accelerated by the wave, a strong damping results, which serves to restrict the ion wave amplitude from further increase. If the ions are cold, the trapping condition is simply $e\phi = Mv_p^2/2$, where ϕ is the potential, M the ion mass, and v_p the phase velocity of the wave. Neglecting Debye length corrections, the trapping condition corresponds to $\delta n/n_p \simeq e\phi/\Theta_e \simeq 1/2$, which is a large amplitude.

It's important to realize that even a small ion temperature significantly reduces the trapping amplitude.[4] This temperature effect is readily estimated if one assumes a so-called waterbag velocity distribution for the ions. In one-dimension, such a distribution is constant with velocity between $\pm\sqrt{3}v_i$ (v_i is the ion thermal velocity) and zero elsewhere. Since the majority of the ions in a Maxwellian distribution have velocities $\lesssim 2v_i$, the waterbag distribution gives a reasonable first approximation for the onset of strong trapping. Trapping now occurs when the fastest ion is nonlinearly brought into resonance with the wave; i.e.

$$e\phi = \frac{M}{2} (v_p - \sqrt{3}\, v_i)^2 \quad ,$$

$$\frac{\delta n}{n_p} \simeq 1/2\left(\sqrt{1 + \frac{3\Theta_i}{Z\Theta_e}} - \sqrt{\frac{3\Theta_i}{Z\Theta_e}}\right)^2 \quad , \tag{6}$$

Here Θ_i is the ion temperature, Z is the ion charge and Θ_e is the electron temperature. For $\Theta_i/Z\Theta_e = .1$, Eq.(6) predicts a fluctuation amplitude of $\delta n/n_p \simeq 0.13$. Clearly the ion temperature serves to significantly reduce the amplitude, but

note that the trapping amplitude is still of order 10%, unless
the ions are quite hot (i.e. $\Theta_i/Z\Theta_e \sim \Theta(1)$). Strong
trapping does not in general limit the fluctuation amplitude to
a small value.

II.3. HARMONIC GENERATION

Harmonic generation is another effect which acts to limit
the ion wave amplitude provided the wave number times the
electron Debye length ($k\lambda_{De}$) is small. If we neglect $k\lambda_{De}$
effects, the frequency of an ion sound wave is simply
proportional to its wavenumber. Such a wave will then steepen,
since harmonics are resonantly driven. If we consider an ion
sound wave with amplitude δn, wave number k, and frequency ω_{ik}
and compute the growth of its second harmonic by linearizing the
two-fluid equations, we obtain

$$\frac{\delta n(2k)}{n_p} = 1/2 \left[\frac{\delta n(k)}{n_p}\right]^2 \omega_{ik}t \quad . \tag{7}$$

We estimate a characteristic steepening time (t_s) by the
condition $\delta n(2k) \sim \delta n(k)$, yielding

$$\omega_{ik}t_s \sim 2/\frac{\delta n(k)}{n_p} \quad . \tag{8}$$

When $k\lambda_{De}$ corrections are included, the resonant coupling
is spoiled. In particular, we then have

$$\omega_{ik} = \frac{kV_s}{\sqrt{1 + k^2\lambda^2_{De}}} = kv_s + \Delta\omega , \tag{9}$$

where V_s is the ion sound velocity and $\Delta\omega = \omega_{ik} k^2\lambda^2_{De}/2$
for $k\lambda_{De} \ll 1$. Significant harmonic generation then at least
requires that $t_s < 1/\Delta\omega$, where $\Delta\omega$ is to be evaluated for the
second harmonic. This corresponds to the condition

$$\frac{\delta n(k)}{n_p} > 4k^2 \lambda^2_{De} \quad . \tag{10}$$

The estimate in Eq. 10 suffices to show that harmonic
generation is not very effective in the hot underdense plasmas
which are typical of recent laser-irradiated targets. For
example, let us consider the ion wave produced by Brillouin
backscatter in an underdense plasma with $n_p/n_{cr} = 1/3$ and an
electron temperature of 3 keV. Then $k^2\lambda^2_{De} \simeq .07$. Hence
significant harmonic generation would require that the ion wave
reach a very sizeable amplitude: $\delta n(k)/n_p \gtrsim 0.3$ for this
example.

II.4 ION TAIL FORMATION

Let us now take a somewhat different approach and focus on the important changes in the ion velocity distribution which are a natural consequence of the Brillouin reflection. As shown by computer simulations, the long term nonlinear state[5] is one in which the Brillouin-generated ion wave is damped by a heated ion tail, which indeed is necessary to carry off the energy deposition into the ion wave. The greater the reflectivity, the larger the ion tail, and so the larger is the ion wave damping. Hence the reflectivity is at least partially self-correcting.

A very simple model can be given to estimate the reflectivity in the nonlinear state. Since the ion tail is produced by an ion wave with phase velocity v_s, its characteristic temperature is approximately $\Theta_h = Mv_s^2$, an estimate in factor of two agreement with our simulation. The magnitude of the ion heating (the tail density, n_h) is then readily determined by balancing the energy flux deposited in the damped ion wave with the flux carried off the tail:

$$rI \frac{\omega_i k}{\omega_0} = 0.8 n_h \Theta_h \sqrt{\frac{\Theta_h}{M}} \quad , \tag{11}$$

where r is the reflectivity, I (ω_0) is the intensity (frequency) of the incident light wave, and the heat flux carried off by the ion tail has been estimated in the free-streaming limit. Having determined n_h, we can now estimate the ion wave damping as simply Landau damping on the heated ion tail; i.e. $\nu_i/\omega_i \simeq 0.8\, n_h/n_p$. One can use more sophisticated estimates, including for example resonance broadening, but this just enhances the damping.

We complete our simple model by computing the reflection from the damped ion wave. The amplitude to which the wave is driven is essentially the ponderomotive force due to the beat between the incident and reflected waves divided by the damping rate;

$$\frac{\delta n}{n_p} = \frac{Z k^2 e^2 E_i E_r}{2mM\omega_0^2 \, \nu_i \omega_i} \quad . \tag{12}$$

Here Z is the charge of the ions, ω_i (ν_i) is the ion wave frequency (energy damping rate), M(m) is the ion (electron) mass, and E_i (E_r) is the electric vector of the incident (reflected) light wave. Subsituting Eq.(12) into Eq.(3) and integrating then gives the well-known result[6] for the reflectivity r:

$$r(1-r) = B \left\{ \exp \left[Q(1-r) \right] - r \right\} \quad , \tag{13}$$

$$Q = \frac{1}{4} \frac{n_p}{n_{cr}} k_o L \left(\frac{v_{os}}{v_e} \right)^2 \Bigg/ \left[\frac{v_i}{\omega_i} \left(1 + \frac{3 \Theta_i}{Z \Theta_e} \right) \left(1 - \frac{n_p}{n_{cr}} \right) \right] \quad .$$

Here L is again the size of the plasma, n_p is the plasma density, v_{os} is the oscillation velocity of an electron in the incident light wave, v_e is the electron thermal velocity, and B is the noise level of the reflected wave at x = L normalized to the intensity of the incident light wave at x = 0.

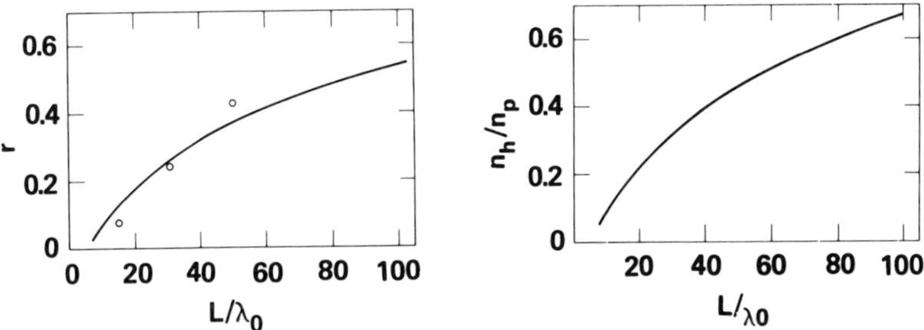

Fig. 1.   The reflection due to the Brillouin instability and the ion tail density as a function of the size of the underdense plasma.   The parameters are $(v_{os}/v_e)^2$ = 0.4, n_p/n_{cr} = 1/3, and $\Theta_i/Z \Theta_e$ = 0.2 initially. The solid lines denote the model predictions (B = 10^{-4}).   The circles denote simulation results using a hybrid code (m/M = 1/400).

The reflection is now determined as a function of plasma conditions.   As an example, we consider $(v_{os}/v_e)^2$ = 0.4 (which corresponds to $I\lambda_o^2$ = 3 x 10^{15} w-μ^2/cm^2 and Θ_e = 3 keV) and n_p/n_{cr} = 1/3 (which is a typical density, taking into account modest profile steepening near the critical density).   A noise level B = 10^{-4} and an initial ion-electron temperature ratio of 0.2 are assumed.   The model predictions for the reflectivities and tail density as a function of plasma size are shown by the solid lines in Fig. 1.   These predictions are in reasonable agreement with simulation results.   For comparison, the open circles denote the long-term Brillouin reflectivity calculated in a one-dimensional simulation code which used particle ions and fluid electrons. Note that the reflectivity is rather small ($\sim 10\%$) when $L/\lambda_0 \sim 10$, and gradually increases to a value of $\sim 50\%$ when $L/\lambda_0 \sim 100$. Note also the substantial ion tail which self-consistently forms. The mean ion temperature quickly becomes of order unity. Self-consistent ion heating is a very potent effect.

Fig. 2 compares the model predictions (solid line) with simulation results for the reflectivity as a function of plasma density, when the scale length and intensity are fixed: $L/\lambda_0 =$ 20 and $(v_{os}/v_e)^2 = 0.40$. The circles denote simulation results using a code which treats both the electrons and the ions

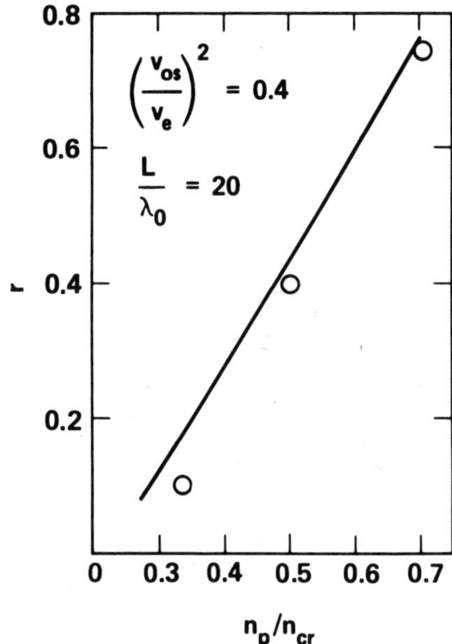

Fig. 2. The reflection due to the Brillouin instability as a function of the background plasma density. Here $(v_{os}/v_e)^2 = 0.4$, $L/\lambda_0 = 20$, and $\Theta_i/Z\Theta_e = 0.2$ initially. The solid line is the simple model prediction ($B = 10^{-4}$). The open circles denote simulation results using a particle code ($m/M = 1/100$).

as particles. For these runs, the initial ion-electron
temperature ratio was 0.2, and the ion-electron mass ratio was
100. The result at n_p/n_{cr} = 0.33 provides a point of a
comparison with the numerical results using the hybrid code
discussed in the previous figure, and the agreement is quite
reasonable. Note that the reflectivity is rather sensitive to the
density of the plasma, as expected from the ion heating model.
The fewer the ions, the larger is the heating consistent with a
given reflectivity. Hence profile steepening near the critical
density can play an important role in reducing the scatter by
limiting it to lower densities.

It's particularly instructive to also examine the reflectivity
as a function of intensity, keeping the plasma density and size
constant (n_p/n_{cr} = 1/3, L/λ_0 = 50). Figure 3 shows the

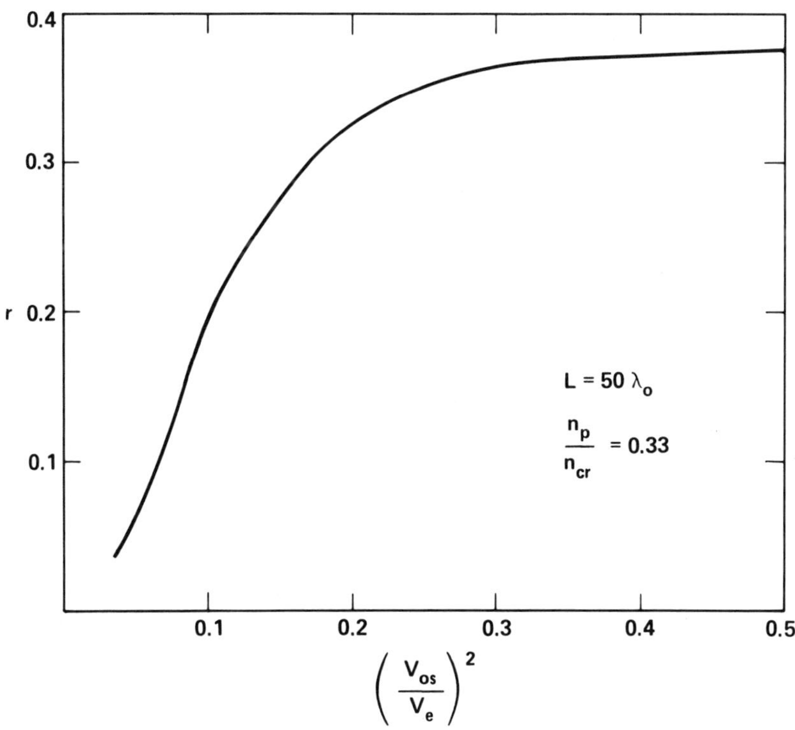

Fig. 3.   The simple model prediction for the reflection due to the
Brillouin instability as a function of $(v_{os}/v_e)^2$.
Here n_p/n_{cr} = 1/3, L/λ_0 = 50, B = 10^{-4} and
$\theta_i/Z\theta_e$ = 0.2 initially.

simple model prediction for the reflectivity as a function of $(v_{os}/v_e)^2$. Note that the reflectivity increases with $v_{os}^2 \propto I\lambda_0^2$ for low intensity but then tends to saturate as the ion tail becomes sizeable. As $I\lambda_0^2$ increases, the self-consistent ion heating increases, which acts to compensate for further increases in reflectivity.

A crude analytic estimate of the reflectivity in the saturated state is

$$r \sim 1 \, / \, \left[1 + \ell nB^{-1} \, / \, \frac{\pi}{2} \frac{L}{\lambda_0} \left(\frac{n_p}{n_{cr}} \right)^2 \right] \qquad (14)$$

Note the dependence of the reflection on the size parameter (L/λ_0) and the plasma density (n_p/n_{cr}). This result again indicates that, for a given n_p/n_{cr}, long wavelength lasers are less sensitive to Brillouin scatter at high intensity. There are simply fewer wavelengths of underdense plasma available to reflect the light.

We note that simpler ion heating models[5,7] give similar results as long as the reflectivity and ion heating are sizeable. For example, since the fraction of the ions in the heated tail is sizeable, it is convenient to approximate the ion heating as a main body heating. In this case, we simply balance the energy flux deposited into ion waves with the energy flux carried off by a heated Maxwellian distribution with density n_p. The effective temperature of this distribution is then

$$\frac{\Theta_i}{Z\Theta_e} \simeq \left[r \, \frac{n_{cr}}{n_p} \left(\frac{v_{os}}{v_e} \right)^2 \right]^{2/3} \quad .$$

Similar reflectivities are obtained. The fact that the numbers are not greatly sensitive to the details of the ion heating is another manifestation of the partially self-correcting feature of the scatter in this simple model. However, it should be noted that details of the ion heating may be significant for understanding the frequency spectrum of the reflected light. Simulations show that one develops a heated tail, and when that tail becomes heavily populated another higher energy tail is formed. The normal modes of a multi-temperature distribution differ in detail from the normal modes of a single-temperature distribution. We also note that ion tails are less sensitive to expansion cooling,[8] an effect which can compete with the ion heating in an expanding plasma.

Finally we can crudely estimate the time required for the plasma to approach this nonlinear state. The time (t_h) to produce the heated tail is

$$t_h \sim \frac{1}{2r} \frac{n_h}{n_{cr}} \frac{v_e}{v_{os}^2} \frac{L}{c_s} \quad .$$

Assuming $n_h \sim 0.2 n_p$, n_p/n_{cr} $v_e^2/v_{os}^2 = 1$, $r = 1/2$, $\Theta_e = 4$ keV and $L = 50$ λ_o gives $t_h \sim 20$ ps. Of course, in practice, it may be difficult to observe a transient state in which the ions are initially heated, since this heating can gradually accumulate as the laser pulse increases from low to high intensity. In addition, other effects such as shock heating can independently contribute to the ion heating.

II.5. THE EFFECTS OF GRADIENTS

It should be noted that gradients are not necessarily as effective in reducing Brillouin scatter in the nonlinear regime as they are in the linear regime. It's well known that gradients can greatly increase[1] the threshold intensity for the instability. The basic idea is simple. Gradients limit the region over which any given three waves can resonantly interact. Noting that the wavenumbers of the three waves (k_i, $i = 1, 2, 3,$) are now a function of position, let us define

$$K = k_1(x) - k_2(x) - k_3(x) \quad .$$

At some point $K = 0$ (i.e., the waves are resonantly coupled), but away from this point a mismatch develops which spoils the resonant coupling. Hence the three waves can resonantly interact only over some interaction region which can be estimated by

$$\int_0^{\ell int} \kappa dx \sim 1/2 \quad .$$

Taylor expanding about the point of resonance ($\kappa = \kappa(0) + \kappa^1 x$, where $\kappa(0) = 0$) then gives

$$\ell int \sim \frac{1}{\sqrt{\kappa 1}} \ll L \quad .$$

Propagation of wave energy out of this interaction region introduces an enhanced damping rate of approximately $v_{gi}/\ell int$, where v_{gi} is the group velocity of the ith wave. Hence the threshold intensity is increased.

However, well above the threshold set by gradients, the reflectivity from a given interaction region is determined nonlinearly, and one must further add up the contributions of the different interaction regions. Hence the bulk of the plasma still contributes to the net reflection. Note that nonlinearly the reflectivity is not an exponential function of L (see Fig. 1), and

hence the net contribution[9] from a number of smaller regions
need not be much less than the contribution from one larger region
(this assumes a distributed and/or broad band noise source).
Gradients will no doubt play a quantitatively significant role,
but probably not the crucial role which they often play in the
linear theory. Perhaps the most important aspect of allowing for
plasma expansion and inhomogeneity will be in properly determining
the various density regions available for scattering, since the
reflectivity depends sensitively on n_p/n_{cr}.

II.6. EXPERIMENTAL EVIDENCE: OPTIONS

Recent experiments[5, 10-18] with longer scale length plasmas
do indicate that significant Brillouin scatter is possible. In
these experiments, sizeable underdense plasmas ($L/\lambda_0 \gtrsim 30$) were
formed in various ways: by using a prepulse, a long pulse length
in the sense of τ/λ_0, or a preformed plasma. Table 1 shows a
brief summary of some of the experimental results for the back
reflection of 1.06 μ and 10.6 μ laser light. Note that a peak
reflectivity of about 50% has been observed with either
wavelength. Note also that the pulse lengths (or prepulse delays)
used in the experiments with 10.6 μ light are much longer than
those used with 1.06 μ light, reflecting the expected L/λ_0
dependence.

TABLE 1. A brief summary of some of the recent experimental
          results for the peak reflectivity (r_B) of laser
          light. For more details, see references 5, 10-18.

Institution	$\lambda_0 (\mu)$	Target	Pulse length	$I \dfrac{W\text{-}\mu^2}{cm^2}$	Peak r_B
NRL	1.06	disk	pp + 50 ps ($\sim$ 2 ns)	10^{15}-10^{16}	50% — f/2
LLL	1.06	disk, spheres	200 ps — 1 ns	3×10^{14}- 3×10^{16}	50% — f/1 15% — f/2.6
U of R	1.06	spheres	pp + 50 ps ($\sim$ 2 ns)	10^{15}-10^{16}	25% — f/2
KMS	1.06	gas jet	100 ps	10^{14}-10^{15}	40% — f/3.5
LASL	10.6	sphere	pp + 1 ns (10-35 ns)	$\sim 10^{17}$	10-50% — f/2.5
U of Alberta	10.6	gas jet (0.1 n_{cr})	35 ns	$\sim 10^{15}$	60% — f/2
UCLA	10.6	arc (0.015 n_{cr})	50 ns	$\sim 2 \times 10^{13}$	5% — f/7.5
U of Wash.	10.6	solenoid	$\sim$ 400 ns	$\sim 5 \times 10^{11}$	5% — f/6.6

Finally, if Brillouin scatter does prove to be a serious problem, there are some options to reduce the scatter. Of course, one option is to operate at lower intensity ($I\lambda_0^2$), so that the plasma is more collisional and the Brillouin instability is more weakly driven. Indeed, recent experiments[19,20] with several ns pulses of 1.06 μ light have shown a high absorption ($\sim$ 70-80%) at low intensities of $\sim 10^{13} - 10^{14}$ W/cm^2. Increasing the bandwidth and/or reducing the wavelength of the light are other effects[21] in the same direction: the plasma becomes more collisional and the instability is more weakly driven. And, of course, the complementary approach is to operate at high intensity and try to minimize the scatter by using long wavelength laser light. Many more experiments are needed to determine the intensity and wavelength window available for laser fusion applications.

III.1. RAMAN SCATTER

We now briefly consider Raman scatter. The process can be most simply characterized as the resonant decay of the incident photon into a scattered photon plus an electron plasma wave. Hence the frequency matching conditions are

$$\omega_o \rightarrow \omega_t + \omega_{pe} \quad , \tag{15}$$

where ω_{pe} is the frequency of the electron plasma wave. Since ω_{pe} is much greater than the frequency of an ion sound wave, the Raman instability is generally a much poorer scattering mechanism than is the Brillouin instability. However, it is potentially a source of high energy electrons and should not be too readily dismissed. Indeed it has been recently observed[22] at low levels in Livermore long-pulse length experiments.

III.2. ELECTRON TAIL FORMATION

Self-consistent electron heating or electron tail formation is an important mechanism for limiting the Raman scatter as shown by computer simulations. Let us then apply the same simple model used for Brillouin scatter. We envision a nonlinear state in which the electron distribution is composed of a main body plus a self-consistent tail of heated electrons[23], which in turn damps the plasma wave. The characteristic temperature Θ_h of this tail is estimated to be mv_p^2 (v_p is the phase velocity of the plasma wave), in factor of two agreement with our simulations. The tail density is determined by balancing the energy deposition into the electrons via the damped plasma wave with the heat flux carried off the electron tail. The damping of the electron wave is then estimated as the Landau damping rate on the heated tail.

The simple model is completed by computing the reflection from the damped plasma wave. The analysis is just like that for the reflection due to a damped ion wave, and results in an analogous expression for the reflectivity:

$$\Lambda(1-\Lambda) \;=\; B\left[\exp\left[x\,(1-\Lambda)\right]-\Lambda\right] \quad , \tag{16}$$

where

$$\Lambda = \frac{\omega_r}{\omega_0}\, r \quad ,$$

$$x = \frac{k_p^2 L}{8 k_r}\left(\frac{v_{os}}{c}\right)^2 \frac{\omega_{pe}}{\gamma} \quad .$$

Here ω_0 (ω_r) is the frequency of the incident (reflected) light wave, k_p (k_r) is the wave number of the plasma wave (reflected light wave), γ is the Landau damping rate, c is the velocity of light, and the other symbols have already been defined.

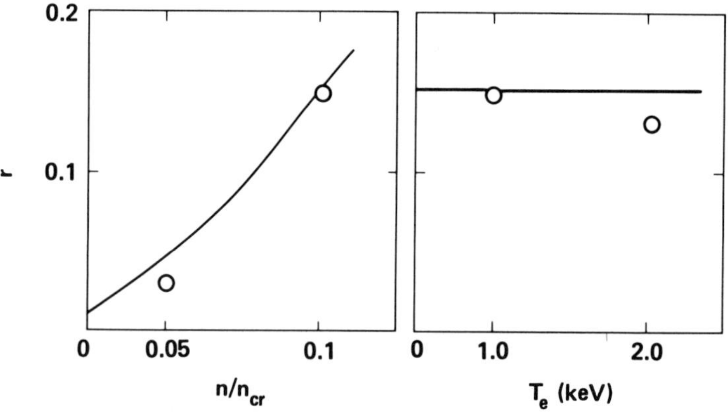

Fig. 4. The reflection due to the Raman instability as a function of the density and temperature of the background plasma. Here $I\lambda_0^2 = 2.5 \times 10^{15}$ W-μ^2/cm^2. The solid lines denote the theoretical estimates using $B = 10^{-4}$. The circles denote simulations results using a particle code (m/M = 1/100 and $Z\Theta_e/\Theta_i = 0.2$).

The simple model predictions of the Raman reflectivity are compared with sample simulation results in Fig. 4. In these simulations, a particle code was used to propagate laser light through a uniform region of low density plasma. The plasma size was $L/\lambda_0 = 127$, the incident light intensity was 2.5×10^{15}W $-\mu^2/cm^2$, and the ion-electron mass ratio was 100. Note that a Raman reflectivity of order 10% was computed, and the variation with background density and temperature were in reasonable agreement with the simple model.

The restriction of Raman scatter by electron tails has some interesting implications for experiments. This choking effect is even more important if the transport is less than the optimum value assumed in our simple model (say, due to self-generated B fields in the underdense plasma) or if critical surface heating independently provides a hot tail of heated electrons. Note also that rather large regions of plasma are needed to give a significant reflectivity, even neglecting the effect of gradients which are especially significant for Raman scatter.

III.3. A NOTE ON THE $2\omega_{pe}$ INSTABILITY

Before concluding, let us speculate that electron tail formation can also play an interesting role in the nonlinear evolution of the $2\omega_{pe}$ instability.[24] This instability corresponds to the resonant decay of an incident light wave into two electron plasma waves. It has a smaller threshold intensity than the Raman instability but only occurs over a narrow range of densities near $1/4\ n_{cr}$. There it becomes an absolute instability with a maximum growth rate of $k_0 v_{os}/4$, where k_0 is the wave number of the light wave. Both computer simulations[25] and theory[26] emphasize that ion dynamics play a crucial role in the nonlinear evolution of this instability. In the strongly-driven simulations, the unstable plasma waves beat to directly drive up ion density fluctuations, which in turn help to damp the waves, and the profile locally steepens to drive the instability towards threshold. At high intensity ($I\lambda_0^2 \sim 10^{16}$ W-μ^2/cm^2), the instability then appears to operate in bursts, which are governed by the ion dynamics.

During instability bursts, a heated electron tail is produced by the unstable plasma waves. If we consider an example from reference 25 in which $(v_{os}/c)^2 = .01$ and the background electron temperature is 10 keV, the temperature of the heated tail is about 75 keV and the average absorption is about 6%. Conservatively estimating the heat flow in a free-streaming limit, and noting that the instability is only operative about half the time then gives $n_h/n_p \gtrsim 04\%$. As a crude measure[27] of the strength of the wave particle interaction, we again take Landau damping on this tail: $\gamma_L/\omega_{pe} \sim 0.4\ n_h/n_p \simeq .016\omega_{pe}$. We note that this dissipation rate is about equal to the linear growth rate of the absolute instability, which in the fairly steep profiles is $\lesssim 1/2$ its maximum value. These simple estimates suggest that improved theories of the nonlinear evolution of the strongly-driven $2\omega_{pe}$ instability should allow for both ion dynamics and dissipative effects due to the wave-particle interaction. In addition, our estimates emphasize that rather small high energy tails ($n_h/n_{cr} \sim 1/4\ v_{os}/c$) independently produced by critical surface heating can significantly affect the $2\omega_{pe}$ decay instability.

V. SUMMARY

The size of the underdense plasma in units of the laser light wavelength is an importnat parameter which affects the mix of coupling processes. Stimulated scattering of the light in large underdense plasmas is a very real concern. We have considered several nonlinear effects which limit the Brillouin and Raman scatter. Our calculations emphasize the importance of the self-consistent changes in the velocity distribution which are produced when the electrostatic waves damp into the particles. Simple theoretical models were formulated and compared with computer simulations. The reflectivity was found to be at least partially self-correcting. As the reflectivity increases, the velocity distribution becomes more distorted which enhances the damping of the electrostatic wave and acts to reduce the reflectivity.

We emphasize that there are significant uncertainties in our attempts to quantitatively understand the level of stimulated scatter. Of course, more quantitative predictions will require a realistic treatment of the effects of gradients in the underdense plasma. The incorporation[8,28] of a model for stimulated scattering into a hydrodynamic code is being actively pursued. In general, even the zero-order plasma conditions are rather poorly known because of uncertainties in the electron transport. More experiments to diagnose the underdense plasma conditions are needed. The angular distribution[29] of the stimulated scatter also requires further investigation. Two-dimensional simulations suggest that the scatter occurs over a sizeable range of angles. Hence, the back reflection through a high f-number lens can be a gross underestimate of the level of stimulated scatter. Finally, the competition of this scatter with other processes such as filamentation is not well understood either theoretically or experimentally.

We gratefully acknowledge many discussions with B. Langdon, B. Lasinski, C. Max, and C. Randall.

REFERENCES

1. C. S. Liu, M. N. Rosenbluth, and R. B. White, Phys. Fluids 17, 1211 (1974), and many references therein.
2. D. W. Forslund, J. M. Kindel, and E. Lindman, Phys. Fluids 18, 1002, 1017 (1975), and many references therein.
3. W. L. Kruer, E. J. Valeo, and K. G. Estabrook, Phys. Rev. Letters 35, 1076 (1975).
4. J. M. Dawson, W. L. Kruer, and B. Rosen, in Dynamics of Ionized Gases, edited by M. Lighthill, I. Inai, and H. Sato (University of Tokyo Press, Tokyo, 1973), p. 47-61.
5. D. W. Phillion, W. L. Kruer, and V. C. Rupert, Phys. Rev. Letters 39, 1529 (1977); Kent Estabrook, Lawrence Livermore Laboratory UCRL-50021-76 (1977).
6. C. L. Tang, J. Appl. Physics 37, 2945 (1966).
7. J. J. Thomson and K. Mima (to be published)
8. Kent Estabrook and J. Harte, Lawrence Livermore Laboratory UCRL-82620 (1979); R. G. Evans, Rutherford Laboratory RL-79-061 (1979).
9. Private communication, C. Randall.
10. B. H. Ripin et al., Phys. Rev. Letters 39, 611 (1977).
11. M. D. Rosen et al., Phys. Fluids 22, 2020 (1979).
12. R. E. Turner and L. M. Goldman, University of Rochester preprint (1979).
13. F. J. Mayer et al., KMSF preprint U-904 (1979).
14. G. H. McCall, LASL preprint LA-UR-79-1298 (1979).
15. A. Ng et al., Phys. Rev. Letters 42, 307 (1979).
16. M. J. Herbst, C. E. Clayton, and F. F. Chen, Phys. Rev. Letters 43, 1591 (1979).
17. R. Massey, K. Berggren, and Z. Pietryzk, Phys. Rev. Letters 36, 963 (1976).
18. There are numerous other references to experiments, including L. M. Gorbunov et al., Sov. Phys. JETP Letters 27, 226 (1978); A. Maaswinkel, K. Eidmann, and R. Sigel, Phys. Rev. Letters 42, 1625 (1979); S. Nakai et al, Phys. Rev. A 17, 1133 (1978).
19. B. H. Ripin et al., Phys. Rev. Letter 43, 350 (1979).
20. K. Manes et al., Bull. Am. Phys. Soc. 24, 1011 (1979).
21. C. E. Max and K. G. Estabrook (to be published, Comments on Plasma Physics and Controlled Fusion); D. W. Forslund, Ninth Anomalous Absorption Conference, paper A-2 (1979); F. Amiranoff et al., Bull. Am. Phys. Soc. 24, 1054 (1979).
22. D. W. Phillion, Bull. Am. Phys. Soc. 24, 991 (1979).
23. See also W. M. Manheimer and H. H. Klein, Phys. Fluids 17, 1889 (1974).
24. C. S. Liu and M. N. Rosenbluth, Phys. Fluids 19, 967 (1976).
25. A. B. Langdon, B. F. Lasinski, and W. L. Kruer, Phys. Rev. Letters 43, 133 (1979).

26. H. H. Chen and C. S. Liu, Phys. Rev. Letters $\underline{39}$, 881 (1977).
27. Complementary estimates of the effective dissipation using trapping arguments give a similar expression.
28. C. J. Randall, J. J. Thomson, and K. G. Estabrook, Phys. Rev. Letters, $\underline{43}$, 924 (1979), D. G. Colombant and W. M. Manheimer, NRL preprint (1979).
29. R. Lehmberg (to be published).

BOHM-LIKE DIFFUSION IN A LASER PRODUCED PLASMA

C. Deutsch, M. M. Gombert, and H. Minoo

Physiquedes Plasmas, Université Paris XI, 91405-Orsay,
France

ABSTRACT

The Guiding Center Model is used as a starting hypothesis to
develop a three-dimensional deductive scheme for the particle transport across the Mega-Gauss magnetic fields in the underdense domain.
We make use of equilibrium properties for a two-component plasma
evaluated up to 1st order in the plasma parameter, the diffraction
corrections and the symmetry effects.   One gets $D_{\perp} \sim \alpha + (\alpha^2 + 2\beta)^{\frac{1}{2}}$
with $\alpha \sim B^{-2}$ and $\beta \sim B^{-2}$.

I.   INTRODUCTION

It is now a well-established fact [1] that in the underdense
region noncolinear density and temperature gradients can develop
easily Mega-Gauss magnetic fields according to the approximate
growth rate formula

$$\frac{\partial \vec{B}}{\partial t} \simeq - \frac{c}{en_e} \vec{\nabla} n_e \times \vec{\nabla} \left(k_B T \right) . \tag{1.1}$$

For instance for an Nd laser one gets [1] $\dot{B} \sim 7$ MG/100 psec with
$n_e \sim 10^{21}$ cm^{-3} while a CO_2 laser shows $\dot{B} \sim 17$ MG/100 psec with n_e
$\sim 10^{19}$ cm^{-3}.   The most widely discussed configurations involve the

region near the edge of a laser spot, where $\vec{\nabla}T$ points radially inward toward the axis of the laser beam, whereas $\vec{\nabla}n_e$ points axially
into the target face. The field produced is toroidal in shape, has
a scale size comparable to the spot radius, and falls to zero at the
laser beam. Similarly, discontinuities of chemical composition or

ionization state may produce very localized regions of strong $\vec{\nabla}n_e$ x
$\vec{\nabla}T$ magnetic fields. In these cases the characteristic scale sizes
of the field can be quite small, because the source is confined
nearly the width of the density jump or ionization layer. For instance, composition jumps can cause magnetic fields of several Mega-
Gauss localized in regions a few microns in diameter.

These small scale structures would be superposed on the previous toroidal field.

Other mechanisms to be considered include resonance absorption
and magnetic instability. Mega-Gauss magnetic fields possessing a
full range of scale lengths, from the scale of the laser spot down
to the submicron range, may be expected to be present in laser
plasmas. Also, many or all of these mechanisms may occur simultaneously. The total dc magnetic field can therefore be modelized
as [1]

$$\vec{B}(\vec{x}) = \vec{B}_0 + \sum_{\vec{k} \neq 0} \delta\vec{B}_k e^{i\vec{k}\cdot\vec{x}}. \qquad (1.2)$$

$\vec{B}_0$ could be taken as the large-scale toroidal field surrounding the
laser spot. $\delta\vec{B}$ then represents the combined effect of all the
other, smaller scale fields.

From the practical point of view, the most important quantity
to consider in relation with these huge B values, is the thermal
conductivity across the field lines.

We know from basic transport theory that the thermal conductivity is a many-body quantity which is far more involved than the
particle diffusion coefficient $D_\perp$ itself. Hopefully, in the present
case both quantities are nearly equivalent in the large field limit
$\Omega_i \gtrsim \omega_{pi}$. Moreover, when $\omega_{pi} > \Omega_i$, $D_\perp$ appears to be an upper limit
to the heat conductivity. So, in all cases useful information can
be derived from this more accessible quantity.

The few available pieces of numerical and experimental evidence
lend to support a heat flow intermediate between a Bohm-type $\sim B^{-1}$

and a classical one $\sim B^{-2}$. In this work, we work out a scheme introduced by Vahala [3], to treat on an equal footing the hydrodynamical
convection $\perp \vec{B}$ and the classical diffusion (mostly kinetic) along $\vec{B}$.

We wish to pay attention to the possibility of Quantum effects
such as diffraction corrections arising from the delocalization of
charges through the uncertainty principle $\left(\lambda_{ee} \geq e^2/k_B T \right)$, and also
to symmetry effects arising from the Fermi Statistics operating
within the electron component.

Although small in the corona, these Quantum effects are expected
to increase near the critical surface, and beyond it. The latter
domain could become of increased interest in relation with recent
speculations about the possibility that self-generated magnetic
fields may travel to the overdense region [4].

In contradistinction to the neoclassical approach [1] which
builds up step-by-step the transport theory, we start with a deductive point of view based upon the asymptotically exact Bohn diffusion [2,3] in the $\Omega_i \geq \omega_{pi}$ range. Furthermore, we introduce the
appropriate ingredients (finite Larmor radius, free streaming along
field lines) accounting for the specific features of the diffusion
in the Laser Fusion conditions with $\omega_{pi} > \Omega_i$.

We hope that the simple model we are proposing here may be of
interest, at least at an exploratory stage, to help in analyzing
the difficult experiments.

II. GUIDING CENTER MODEL (3D)

Starting from the Green–Kubo formalism of Linear Response
Theory, the particle diffusion coefficient reads [5]

$$D_\perp = \frac{c^2}{B^2} \int_0^\infty \langle \vec{E}_\perp(\tau) \cdot \vec{E}_\perp(0) \rangle d\tau \qquad (2.1)$$

in terms of the equilibrium canonical average of the 2-point electric field correlation. A first and nearly unavoidable approximation consists in factorizing out the average into a plasma one and
a test particle one, so that (c = speed of light)

$$D_\perp \simeq \frac{c^2}{B^2} \int_0^\infty d\tau \sum_{\vec{k}} \langle \vec{E}_{\perp,\vec{k}}(\tau) \cdot \vec{E}_{\perp,-\vec{k}}(0) \rangle \langle e^{i\vec{k}\cdot\vec{x}(\tau)} \rangle \qquad (2.2)$$

where $\vec{x}(\tau)$ denotes the test particle location. The motion transverse to the magnetic field $\vec{B}$ is that of the lines of force themselves (Guiding Center Model assumption), i.e.,

$$\vec{V} = \frac{c\vec{E}(\vec{x},t) \times \vec{B}}{B^2}. \tag{2.3}$$

A salient feature of the present derivation is that each one of the above averages can be expressed in terms of the 2-point electric correlations. So, we get

$$<\vec{E}_{\perp,\vec{k}}(\tau) \cdot \vec{E}_{\perp,-\vec{k}}(0)> = \sum_{i,j} \frac{(4\pi)^2 e_i e_j}{k^2 V^2} \frac{k_\perp^2}{k^2} <e^{-i\vec{k}\cdot\left(\vec{x}_i(\tau)-\vec{x}_j(0)\right)}> \tag{2.4}$$

and

$$<e^{i\vec{k}\cdot\vec{x}(\tau)}> \simeq e^{-\frac{c^2 k^2}{4B^2} \int_0^\tau d\tau_1 \int_0^\tau d\tau_2 <\vec{E}_\perp(\tau_1)\cdot\vec{E}_\perp(\tau_2)>}. \tag{2.5}$$

In three dimensions, the present approach amounts to modelize the particle transport across $\vec{B}$ through small increments in $\vec{r}$-space, while allowing for a parallel contribution built upon diffusion in $V_\parallel$-space, so that

$$\frac{d\vec{x}_\perp}{dt} = \frac{c\vec{E}_\perp(t) \times \vec{B}}{B^2}, \text{ and } \frac{dv_\parallel}{dt} = \frac{e}{m} E_\parallel(t). \tag{2.6}$$

Performing the particle summation in Eq. (2.4) yields

$$<\vec{E}_\perp(\tau) \cdot \vec{E}_\perp(0)> = \sum_{\vec{k},\vec{k}_\parallel \neq 0} \frac{(4\pi)^2 n_e e^2 k_\perp^2}{Vk^4} e^{-k_\parallel^2 V_i^2 \tau^2/2} H_1(k)$$

$$+ \sum_{\vec{k}_\perp} \frac{(4\pi)^2 n_e e^2}{Vk_\perp^2} H_2(k)$$

$$\cdot e^{-\frac{c^2 k_\perp^2}{2B^2} \int_0^\tau d\tau' \int_0^\tau d\tau'' <\vec{E}_\perp(\tau')\cdot\vec{E}_\perp(\tau'')>} \tag{2.7}$$

where $H_1(k)$ and $H_2(k)$ denote appropriate combinations of the electron-electron, ion-ion, and ion-electron structure factors, given as [2,6]

$$H_1(k) = e^{-k_\parallel^2 V_e^2 \tau^2/2} + e^{-k_\parallel^2 V_i^2 \tau^2/2}$$

$$+ n_e\left[-e^{-k_\parallel^2 V_e^2 \tau^2/2} S_{ee}(k) - e^{-k_\parallel^2 V_i^2 \tau^2/2} S_{ii}(k)\right.$$

$$\left. + \left(e^{-k_\parallel^2 V_e^2 \tau^2/2} + e^{-k_\parallel^2 V_i^2 \tau^2/2}\right) S_{ei}(k)\right] \tag{2.8}$$

and

$$H_2(k) = 2 + n_e\left[-S_{ee}(k) - S_{ii}(k) + 2S_{ei}(k)\right]. \tag{2.9}$$

$D_\perp$ can now be computed through the quantity [5,2,3]

$$R_\perp(\tau) = \frac{c^2}{4B^2} \int_0^\tau d\tau' \int_0^\tau d\tau'' \langle \vec{E}_\perp(\tau'' - \tau') \cdot \vec{E}_\perp(0)\rangle \tag{2.10}$$

fulfilling $\left(\varepsilon_b = (4\pi)^2 e^2 c^2 n_e / B^2 V k_D^2\right)$

$$\ddot{R}_\perp(\tau) = \varepsilon_b \sum_{k_\perp} \frac{e^{-2k_\perp^2 R_\perp(\tau)}}{2k_\perp^2} k_D^2 H_2(k)$$

$$+ \varepsilon_b \sum_{\vec{k}, \vec{k}_\parallel \neq 0} \frac{k_D^2 k_\perp^2}{2k^4} H_1(k) e^{-k_\parallel^2 V_i^2 \tau^2/2 - k_\parallel^2 D_\parallel \tau^3}. \tag{2.11}$$

In Eq. (2.11) we made use of the consistent [3,5] 3-D formulation of the Guiding Center Model through the canonical variables (x,y, z,$V_\parallel$). This way, one can include [3] free streaming of test particles along $\vec{B}$ for short times and also diffusion in $V_\parallel$-space for larger times. The time value discriminating between these two complementary behaviors is [3,7]

$$t_0 = \frac{16\sqrt{2}\pi^{9/2}}{\Lambda \ln \frac{1}{\Lambda}}, \text{ with } \Lambda = \frac{e^2}{k_B T \lambda_D} \tag{2.12}$$

obtained from an estimate of $D_\parallel = \frac{e^2}{m_i^2} \int_0^\infty d\tau \langle E_\parallel(\tau) E_\parallel(0)\rangle$.

Solving Eq. (2.11) with

$$\frac{dR_\perp(\infty)}{dt} = \frac{D_\perp}{2} \text{ and } R_\perp(0) = \frac{dR_\perp(0)}{dt} = 0 \tag{2.13}$$

one readily obtains the fundamental relations

$$\frac{D_\perp}{2} = \frac{\beta}{D_\perp} + \alpha, \quad D_\perp = \alpha + (\alpha^2 + 2\beta)^{\frac{1}{2}} \tag{2.14}$$

where

$$\beta = \frac{\varepsilon_b V^{2/3}}{2\pi} \int dk_\perp k_\perp \frac{k_D^2}{2k^4} H_2(k)_{k_\parallel=0} \tag{2.15}$$

and

$$\alpha = \frac{\varepsilon_b}{(2\pi)^3} \int_0^\infty dt \int d\vec{k} f(\vec{k},t) \tag{2.16}$$

with

$$f(\vec{k},t) = \frac{k_D^2 k_\perp^2}{2k^4} H_1(k) e^{-k_\parallel^2 V_i^2 t^2/2} .$$

Now, we make use of the generalized Debye-Hückel structure factors
[2]

$$n_e S_{ee}(k) = \frac{1}{1 + k^2 + \eta^2 k^4} + \left(a - \frac{b}{1 + k^2} + \frac{c}{(1 + k^2)^2}\right) e^{-\alpha' k^2} \tag{2.18}$$

$$n_e S_{ei}(k) = -\frac{3 + \eta^2 k^2}{1 + k^2 + \eta^2 k^2} + \frac{\eta^2}{\eta^2 k^2 + 2} + \frac{a' e^{-\alpha' k^2}}{(1 + k^2)^2} \tag{2.19}$$

and

$$n_e S_{ii}(k) = \left[1 + \frac{b}{2} \cdot \left(1 - \frac{1}{(z + 1)(1 + k^2)}\right) e^{-\alpha' k^2}\right] \frac{z^2}{1 + k^2} \tag{2.20}$$

where k is in number of k_D, with

$$\alpha' = \frac{\pi \eta^2 \ell n2}{4}, \quad a = z\rho\bar{\lambda}^3 \pi^3 (\ell n2)^{5/2}$$

$$b = \frac{2z}{z + 1} \, n_e \lambdabar^3 \pi^3 (\ell n 2)^{5/2}, \quad c = \frac{z}{(z + 1)^2} \, n_e \lambdabar^3 \pi^3 (\ell n 2)^{5/2}$$

$$a' = \frac{za}{4(z + 1)}, \quad \lambdabar = \frac{\hbar}{\sqrt{\pi m_e k_B T}}, \quad \text{and} \quad \lambdabar_{ee} = \lambdabar \sqrt{2\pi}.$$

$S_{ij}(k)$ reduces to $(k^2 + 1)^{-1}$ in the $\hbar \to 0$ limit.

The dimensionless parameters Λ, $\eta = \lambdabar_{ee}/\lambda_D$, and $4\pi\rho\lambdabar^3_{ee}$ provide, respectively, an estimate of the coupling strength, diffraction corrections, and symmetry effects.

III. RESULTS

Introducing Eqs. (2.18)–(2.20) into Eq. (2.15), one gets

$$\beta_1 = \frac{\varepsilon_b V^{2/3} k_D^2}{2\pi} \int_{2\pi\lambda_D/L}^{1} \frac{dk_\perp k_\perp}{2k^4} \, H_2(k)_{k_\parallel = 0}$$

$$= \left(\frac{ck_B T}{eB}\right)^2 \Lambda \left(\frac{\lambda_D}{L}\right) \left[\left(2 + z(\eta^2 - z)\right)\ell n\left(\frac{L}{2\pi\lambda_D}\right)^2 - \left[\frac{(1 + 6z) + 2\eta^2}{1 - \eta^2}\right]\ell n 2 \right.$$

$$+ \left[\frac{1 + 7z}{1 - \eta^2} - z\right]\eta^2 \ell n \frac{\eta^2}{\Lambda^2} + z\eta^2 \ell n 2 + z^2 + \ell n a'\left(b\left(1 - \frac{z^2}{z}\right)\right.$$

$$\left. - c - 2za'\right) - \ell n 2\left(be^{\alpha'}\left(1 - \frac{z^2}{2}\right) + c + 2za'\right)\Big] \tag{3.1}$$

if one restricts the $k_\perp$-integration to the so-called fluid limit interval $0 \le k \le k_D$. L denotes the average size of the toroidal magnetized region. Retaining the large $k_\perp$ contribution up to the Landau limit $k_B T/e^2$, provides the alternative expression

$$\beta_{\frac{1}{\Lambda}} = \left(\frac{ck_B T}{eB}\right)^2 \Lambda \left(\frac{\lambda_D}{L}\right) \left[(2 + z\eta^2 - z^2)\left(\ell n\left(\frac{L}{2\pi\lambda_D}\right)^2 + \ell n \frac{1}{\Lambda^2}\right) \right.$$

$$+ z^2 - \left[\left((1 + 6z) + z^2 \right) \ln\left(1 + \frac{1}{\Lambda^2} \right) + \eta^2 \left[z \ln 2 \right. \right.$$

$$+ \left[\frac{1 + 7z}{1 - \eta^2} - z \right] \ln \frac{\eta^2}{\Lambda^2}$$

$$\left. + 3(1 + 2z) \ln\left(1 + \frac{1}{\eta^2} \right) \right] - Cb\left(1 - \frac{z^2}{2} \right) - (c + 4za')$$

$$\left. + (C + \ln a') \left[be^{\alpha'} \left(1 - \frac{z^2}{2} \right) + c + 4za' \right] \right], \quad C = .5772. \quad (3.2)$$

These β values can be corrected for finite gyroradius effects by multiplying them with $\left(1 + \frac{\omega_{pi}}{\Omega_i} \right)^{-1}$, so that $\beta \sim \frac{1}{B^2}$ when $\Omega_i \geq \omega_{pi}$ (Bohm behavior) or $\beta \sim B^0$ for $\omega_{pi} > \Omega_i$ (plateau regime). For typical underdense conditions ($n_e \sim 10^{21}$ cm^{-3}, $k_B T \sim 500$ eV and $z = 1$) one can expect a 50% discrepancy between β_1 and $\beta \frac{1}{\Lambda}$. However, these quantities are not shifted by more than a few percent (usually 3) when $\hbar \to 0$ limit is taken out. When $L \to \infty$, it is easy to understand that the finite quantum contributions can only play a very minor role. However, even for $L/2\pi\lambda_D \sim 10$, there is a subtle compensation of diffraction and symmetry which restrains the subdominant quantum contribution to a low level.

As expected, $\hbar$-corrections are much more significant for the partly kinetic contribution α written as $\alpha \equiv \alpha_1 + \alpha_2$ with

$$\alpha_1 = \frac{\varepsilon_b V}{(2\pi)^3} k_D^2 \int_{\Lambda \ln 1/\Lambda}^{1/\Lambda} dk_\parallel \int_{2\pi\lambda_D/L}^{1/\Lambda} \frac{dk_\perp k_\perp^2}{2k^4} \int_0^\infty dt$$

$$\cdot \, e^{-k_\parallel^2 V_i^2 t^2/2} H_1(k) \qquad\qquad (3.3)$$

and

$$\alpha_2 = \frac{\varepsilon_b V}{(2\pi)^3} k_D^2 \int_{2\pi\lambda_D/L}^{\Lambda \ln 1/\Lambda} dk_\parallel \int_{2\pi\lambda_D/L}^{1/\Lambda} \frac{dk_\perp k_\perp^3}{2k^4} \int_0^\infty dt$$

$$\cdot \, e^{-k_\parallel^2 D \, t^3} H_1(k). \qquad\qquad (3.4)$$

Obviously α_1 accounts for free streaming along $\vec{B}$, while α_2 provides the adequate $V_\parallel$-diffusion term.

In contradistinction to β, the α's exhibit a very weak volume (L)-dependence, and a finite $L \to 0$ limit.

Explicitly, one gets

$$
\alpha_1 \simeq \frac{1}{16\pi^2\sqrt{2\pi}} \cdot \frac{ck_BT}{eB} \cdot \frac{\omega_{pi}}{\Omega_i}\Lambda \cdot \left[\ell n \frac{1}{\Lambda^2 \ell n \frac{1}{\Lambda}} \cdot \left\{ 1 + \frac{1}{43} \right. \right.
$$

$$
- \left[\frac{\sqrt{2}}{43} + 3z\left(\frac{\sqrt{2}}{43} + 1\right) \right] - z^2 + \eta^2 \left[z\left(1 + \frac{\sqrt{2}}{43}\right) + \frac{3}{4} - 3\left(\frac{\sqrt{2}}{43}\right. \right.
$$

$$
\left. \left. + 3z\left(\frac{\sqrt{2}}{43} + 1\right) \right] \right\} + \frac{bz^2}{4(z+1)} \cdot \left(\frac{1}{\alpha'} + 1 - C\right) - \frac{a}{43\sqrt{2}} \right] \quad (3.5)
$$

$$
+ \ell n \frac{1}{\Lambda} \cdot \left[\frac{5z^2}{4} - \eta^2\left\{\frac{1}{4} + 2z\left(1 + \frac{\sqrt{2}}{43}\right) + \frac{\sqrt{2}}{43}(1 + 3z) + 3z \right\} \right.
$$

$$
\left. + \frac{bz^2}{4(2+1)} \cdot \left(C - 1 - \frac{1}{\alpha'}\right) + \frac{a(1 - C)}{43\sqrt{2}} \right]
$$

$$
\left. + \eta^2 \ell n\eta \cdot \left[\frac{z}{4} - \frac{1}{2} + \frac{\sqrt{2}}{43} \cdot \left(\frac{z}{4} - 3z - 1 + \frac{1}{86}\right) \right] \right],
$$

and

$$
\alpha_2 \simeq \frac{1}{32\pi^2\sqrt{2\pi}} \cdot \frac{ck_BT}{eB} \cdot \frac{\omega_{pi}}{\Omega_i}\Lambda \cdot \left[2\left(\ell n \frac{1}{\Lambda^2} - \frac{1}{5}\right) + 2\ell n2 \right.
$$

$$
\left. + \left(-1 + 2\ell n\left(\Lambda \, \ell n \frac{1}{\Lambda}\right)\right) \right].
$$

$$\cdot \left(1 + \frac{1}{43}\right) + \frac{1}{2}\,\ell n\left(1 + \frac{1}{\Lambda^2}\right)\left[(1 + 6z) + 2\left(3 + \frac{4}{43}\right)\right]$$

$$+ \left[\left(\ell n(1 + \eta^2) - \ell n\left(\frac{1}{\Lambda^2} + \frac{1}{\eta^2}\right)\right)\left(-1 + \frac{1}{43}\right)\frac{\eta^2}{1 - \eta^2}\right.$$

$$+ \frac{Cbz^2}{2} + (2C + \ell n\alpha')b(1 - z^2)$$

$$\left. + \left(2a'(1 - z) - \frac{bz^2}{2(z + 1)}\right)\ell n\alpha'\right]. \tag{3.6}$$

Like $\beta\,\frac{1}{\Lambda}$, the α's exhibit characteristic kinetic terms $\sim \ell n\,\frac{1}{\Lambda}$. Numerically speaking, we need only to inject an n_e and a T_e value to implement the present computation of $D_\perp$. Such an argument may be of some interest to code makers.

REFERENCES

[1] C. E. MAX, W. M. MANHEIMER, and J. J. THOMSON – Phys. Fluids 21 (1978) 128.

[2] C. DEUTSCH – Phys. Rev. A17 (1978) 909.

[3] G. VAHALA – Phys. Fluids 16 (1973) 1876.

[4] R. PELLAT – Private Communication, November 1979.

[5] D. MONTGOMERY – Les Houches 1972 Lecture Notes, Gordon and Breach, New York (1975).

[6] C. DEUTSCH, Y. FURUTANI, and M. M. GOMBERT – Submitted to Physics Reports.

[7] Y. FURUTANI – Private Communication, July 1979.

[8] M. M. GOMBERT, C. DEUTSCH, and H. MINOO – J. Physique (Suppl.) C7 (1979) 695, and to be published.

SPECTROSCOPIC DENSITY MEASUREMENTS IN COMPRESSED PLASMAS

Hans R. Griem

Department of Physics and Astronomy
University of Maryland
College Park, MD   20742

ABSTRACT

X-ray line spectra from compressed DT plasmas seeded with,
e.g., neon or argon have been used in several laboratories to
determine densities and, to some extent, also diameters near maximum compression. Most useful are the various hydrogen and
helium-like satellites, respectively. Densities can be inferred
both from the widths of lines and from the relative intensities of
some satellites, provided the opacity is not too large. Diameters
then follow from the widths of lines having large opacities.
Similar methods are available to estimate $\rho \Delta R$ in the imploded
glass, e.g., from silicon lines. The theoretical foundations and
limits of the various methods are discussed.

INTRODUCTION

Quantitative spectroscopy of hot plasmas had served as temperature and density diagnostic over a very large plasma parameter
space, before inertial fusion research led to a substantial
increase in the plasma density achievable in the laboratory. The
major problem was therefore to extend and to apply analogous
methods in the new density regime. Such regimes had already for
some time been of theoretical interest, e.g., in radiative transfer
calculations for stellar interiors. In this case, the stellar
evolution could depend on the line contribution to the radiative
energy transfer. We now have the opportunity to confront similar
problems in the laboratory.

The present paper is mostly concerned with methods for the
determination of densities from X-ray line spectra that are

applicable up to DT densities of $\sim$10g cm^{-3}. In the next section
the theoretical foundations of these methods will be reviewed,
followed by a discussion of various applications. In the final
section an attempt is made to assess the prospects of extensions
to substantially higher densities.

THEORY

Three more-or-less related features of X-ray line spectra
depend mostly on the density of the emitting or absorbing plasma
rather than on its temperature. These features are profiles of
lines primarily broadened by electric fields produced by electrons
and ions interacting with the radiating ions, e.g., Stark
broadened lines. Second, there are relative intensities of certain lines or satellites of lines, one of which requires more than
one binary collision to be excited. Finally, there is the advance
of series limits in dense plasmas. It is closely related to line
broadening, as is the second effect in that also it is a typical
high density phenomenon involving very similar collision processes.

Stark Broadening

Because of the different time scales for electron- and ion-
produced electric fields, the following theoretical model[1] has
been rather successful in describing experimental results for
neutral atom lines and lines from singly charged ions emitted from
plasmas with temperatures in the eV range and densities up to
$\sim$10^{18} cm^{-3}. One first considers what happens to the radiator on
an intermediate time scale so that the ion-produced field acting
on the radiator is essentially static. This field splits and
shifts the energy levels according to time-independent perturbation
theory. It may also change the relative strengths of the various
radiative transitions, especially those associated with normally forbidden transitions, e.g., $(1s2s)^1S \rightarrow (1s)^2\,^1S$ in helium-like systems.

During the time of nearly constant ion-produced Stark effects,
there will in general be several electron-radiator collisions.
These cause transitions between the levels involved, thus broadening and further shifting them and also redistributing the
intensities of radiative transitions. To the extent that the
electron-radiator interaction is well approximated by the interaction between a point charge (electron) and the radiating ion
endowed only with monopole and dipole moments, also the electron
collisions are equivalent to Stark effects, albeit time-dependent.
(Note that the monopole moment is independent of the internal
state of the radiating ion, provided the electron-radiator
collisions are mostly long-range, i.e., mostly outside the excited
state Bohr orbits.) Complete Stark profiles are then computed by
averaging the electron-collision broadened quasi-statically split,
etc., profiles over the appropriate distribution of ion-produced

fields. The line shape corresponding to this model is[1]

$$L(\omega) = -\frac{1}{\pi} \text{ Re Tr} \int_0^\infty dF W(F) \{D(F)[i\omega - i\omega(F) + \phi]^{-1} \rho\} \ . \tag{1}$$

This is a generalized Lorentzian determined by the collision operator ϕ convolved with the quasi-static patterns characterized by ion-field-dependent frequencies $\omega(F)$ and field distributions $W(F)$. The trace is a sum over atomic basis states contributing to the line, whose initial populations are represented by the density matrix ρ. Finally, $D(F)$ is an operator whose matrix elements give the relative strengths of the radiative transitions. In principle, the collision term ϕ is also a function of F and, perhaps, even ρ. However, it is commonly assumed that initial populations are statistical and that ϕ is essentially field-independent. In contrast to purely Lorentzian broadening, ϕ can have nondiagonal elements and may be frequency-dependent (unified line shape). This frequency dependence is numerically important only at frequency separations larger than the electron plasma frequency.

Extensive calculations[2] for high Z ions within the confines of this model have been made for one-electron ions from C^{5+} to Si^{13+} with field strength distribution functions provided by Hooper[3] for the case that ions of the same species dominate the quasi-static broadening. (At fixed electron density, the dependence on perturbing ion charge is rather weak.[4]) The electron collision broadening was evaluated using the distorted wave method, i.e., without the usual expansions. Comparisons for Ne^{9+} with profiles calculated by Hooper and Woltz,[5] who evaluated profiles up to Ar^{17+}, suggest that for this ion deviations between calculations of the full Coulomb effects and dipole-monopole interactions are still small. Even for L_β of Si^{13+} the two calculations are in $\sim 20\%$ agreement, probably because errors from neglecting other than dipole terms are compensated by errors in cutoff parameters.[5] Similar agreement with Lee's results[6] cannot be used to argue that other than dipole interactions are not important, because he adjusted his cutoff in the electron-ion momentum transfer to get agreement with the distorted wave, full Coulomb interaction, results.[2]

Comparisons of analogous calculations with precision measurements[7] of hydrogen L_α and L_β profiles verified a shortcoming of the basic profile model that had first been suspected because of (perturbing ion-radiating atom) mass-dependent deviations near the line centers of hydrogen Balmer lines. Evidently there was considerable broadening (profile smoothing) near the line centers not allowed for in the quasi-static ion, electron-collision model. A number of procedures were therefore developed to include the effects of radiator and perturbing ion motions, of which only two appear to be

consistent with measurements of both L_α and L_β. One of these
procedures[9] replaces the quasi-static field by a model microfield
giving the correct distribution function and electric field correlation function. The other, mostly analytic, procedure[10] allows
for the rate of change of ion-produced and low-frequency collective
fields. In either case, the corrections are therefore sensitive to
correlations.

Another modification of the line broadening model has recently
been reconsidered by Lee[6] in order to allow for the pronounced
inhomogeneity of the plasma in the vicinity of a highly charged
radiating ion. In this region of space, the plasma is no longer
neutral on the average over time, and the corresponding average
interaction suggests plasma polarization shifts[1,6] and asymmetries[6]
of lines that otherwise exhibit only symmetrical linear Stark
effects. In Lee's formulation,[6] the net ionic charge also leads
to additional broadening, although it appears that the corresponding terms should largely cancel. Moreover, it is not clear how
many of the latter effects were already allowed for by the distorted wave approximation for the electron-ion scattering.[2] As to
the former effects, the problem goes beyond ordinary scattering
theory because it is concerned with electrons and perturbing ions
which are already near the radiating ion at some initial time, say,
t=0 when the ion is assumed to be in some well-defined state.

Besides the average interaction, one should also examine the
approximation concerning the distribution over initial states of
the entire radiator-plasma system. Until the corresponding
problems have been solved, any shifts and asymmetries of lines from
hydrogenic ions should not be interpreted in terms of densities in
the emitting region.

Calculations based on the Stark broadening model summarized in
Eq. (1) for helium-like ions are slightly more involved. In this
case ion-field-dependent Stark levels and the corresponding eigen-
vectors have to be calculated first by diagonalizing the Hamiltonian
for the ion in a given field, using of course empirical or
previously calculated energy levels for the zero-field matrix
elements of the radiator Hamiltonian. Such calculations were made
for many HeI lines[1] but so far only for very few ion lines, namely,
the 1^1S-3^1P, ^1D lines of Si^{12+} and Ar^{16+} by Lee[11] and Hooper and
Woltz[5] for special cases. These profiles have two maxima, like
the L_β profiles of hydrogenic ions, but the peak corresponding to
the forbidden 1^1S-3^1D transition is generally lower than that
corresponding to 1^1S-3^1P. Corrections for low frequency Stark
effects would again be most important near the minimum between
these peaks, making it shallower. Calculations for ions with more
bound electrons would be just as feasible and probably not be much
less accurate.

We finally note that resolved fine-structure of L_α lines of
hydrogenic ions leads to profiles with similar appearances than the
1^1S-3^1P, ^1D profiles of helium-like ions. This was shown
theoretically by Lee[12] for Ne^{9+} and Ar^{17+}, and should be taken
into account if electron densities are below 10^{23} cm^{-3} and 10^{25}
cm^{-3}, respectively, and central profile structures of these lines
are resolved and not obscured by opacity effects. For higher
Lyman-series lines, fine-structure is much less important.

Relative Line Intensities

For lines excited primarily by electron collisions with ground
state ions, relative intensities are mainly a function of temperature, provided radiative decay is more likely than collisional
energy transfer. For lines with relatively small radiative transition probabilities, e.g., 1^1S-2^3P intercombination lines of helium-
like ions, the latter condition is easily violated, and one then
finds a decrease of the 1^1S-2^3P to 1^1S-2^1P line ratio with
increasing density. In practice, this observation is not very
useful, both because opacity corrections must be made and because
the intercombination line may blend with lithium-like satellites
and the 1^1S-2^1S Stark component.

A similar method would involve the (integrated) intensity
ratio of L_α fine-structure components, of which the $1S_{1/2}-2P_{1/2}$
component could be more than 1/2 of the $1S_{1/2}-2P_{3/2}$ intensity.
This can occur because the $2S_{1/2}$ level may be overpopulated
relative to the other n=2 levels and because its population is
readily transferred to $2P_{1/2}$. (This possibility was evidently not
considered in Ref. 12.) Again, opacity corrections and difficulties in unambiguously separating the two components severely limit
the utility of this approach to a density measurement.

A more promising method is based on the observation of
helium[13] or lithium-like[14] satellite lines to resonance lines of
hydrogen- or helium-like ions. These satellite line intensities
are at low densities (corona model) given by[14]

$$I^z \sim \frac{A_r}{\Sigma A_a + A_r}\ (C_c N_i^{z+1} + C_e N_i^z) N_e \ , \tag{2}$$

where C_c and C_e are rate coefficients for dielectronic capture and
inner shell excitation on or in the appropriate ions with ground
state densities N_i^{z+1} and N_i^z, respectively. The first factor in
Eq. (2) is the branching ratio between the specific radiative
transition rate and the sum of all auto-ionizing and radiative
rates from the particular doubly-excited level. In this model,
dielectronic satellites have therefore the same density dependence
as the corresponding resonance line of the (z+1) times charged
ion. More importantly, dielectronic satellites for which the

capture coefficient nearly vanishes because of selection rules
are very weak in the spectrum.

At higher densities, electron-ion collisions and, at least in
principle, also ion-ion collisions will tend to transfer some
population to these inaccessible levels by changing the angular
momentum of a bound electron. For example, population will be
transferred from $2s2p\,^3P$ to $(2p)^{23}P$. This collision process is
therefore very similar to the most important contribution
(2s-2p) to the broadening of the parent line. To include these
transfer processes, coupled (steady state) rate equations must be
solved with appropriate rate coefficients. This has been done in
an approximate manner[13] for satellites to L_α of Ne^{9+} and with a
more complete theory[14] for satellites to the helium-like resonance
lines of $A\ell^{11+}$ and Si^{12+}. Usable intensity ratios were obtained
in the 10^{22} cm^{-3} – 10^{24} cm^{-3} electron density range, depending on
the ion.

It is difficult to assess the potential accuracy of this
method, because the low density values of the satellite line
intensity ratios will generally have some uncertainty, and because
the approach to the LTE limits at high densities may not be
properly modelled, not to mention the difficulties of separating
blends, etc. (line broadening).

Advance of Series Limits

Since the Stark broadening increases rapidly with increasing
principal quantum of the upper level, spectral lines of a given
series tend to merge well before the unperturbed series limit.
Inglis and Teller[15] estimated the corresponding last discrete level
by equating the mean linear Stark splitting produced by a typical
quasi-static, ion-produced field with the unperturbed separation
of adjacent hydrogenic levels of principal quantum numbers n and
n+1. This consideration leads to

$$4\,\frac{n^2}{z}\,e^2 a_o\,\bar{z}^{-1/3} N_e^{2/3} \approx \frac{z^2 e^2}{n^3 a_o} \tag{3}$$

or

$$N_e \approx \frac{z^{9/2} a_o^{-3}}{8 n^{15/2 - 1/2\bar{z}}} \approx 10^{24}\,\frac{z^{9/2}}{n^{15/2 - 1/2\bar{z}}}\,. \tag{4}$$

Here z is the nuclear charge for hydrogenic ions, or the screened
nuclear charge for helium-like ions, and $\bar{z}$ the mean charge of perturbing ions. Clearly, the numerical factor in Eq. (4) is only an
order of magnitude estimate. Uncertainties in the definition of
the last discrete level n are also a large source of errors.

It is very difficult to improve the theoretical accuracy of
this estimate. Therefore Eq. (4) is best used only for scaling
purposes. Also, it is important to remember that a reasonable
value for n can only be obtained if the last line is not too weak
or otherwise obscured for any experimental or theoretical reason.

APPLICATIONS

Almost all spectroscopic density measurements in compressed
plasmas have involved the use of Stark broadened profiles of
higher members of the resonance line series of elements added to
the DT fills, or gases replacing the DT, and from elements in the
glass microballoons. First resonance lines have generally not
been used because of their large opacities, but can after the density has been determined otherwise be used to estimate the
dimension of the emitting layer.[16] The internal consistency and
uniqueness of such ρ and ρR determinations[16,17,18] has been
questioned[11] because within given errors of theory and (time-
integrated) experiments the two spectroscopic parameters (N_e and
$N_i^Z R$) are difficult to separate. This criticism is less valid for
experiments only using relatively small admixtures of, e.g., neon
to the DT fill. A series of such experiments[19] showed that spectroscopic density measurements agreed with a careful analysis of
X-ray images within expected errors, i.e., within a factor ~ 1.5.

Work in progress in various laboratories with argon admixtures
or pure argon fills indicates that hydrogen and helium-like lines
of this element should provide density indicators at DT densities
up to ~ 10g cm^{-3}.

The other two spectroscopic methods were so far applied mainly
in Refs. 13 and 17, respectively.

OUTLOOK

To extend the range of the spectroscopic methods of density
measurement, even heavier seed elements than argon must be considered, because otherwise absorption would become prohibitive.
Clearly, this necessitates also an increase in the temperature,
conflicting with the desire for high density at given driver
energy. Be this as it may, typically the absorption coefficient
for the line radiation in the surrounding layers may decrease,
say, as z^{-6}, so that the scaling of Eq. (4) suggests even for fixed
final target radius that measurable Stark widths could still be
obtainable at $\rho \approx 10^3$g cm^{-3}, provided the fuel region is sufficiently hot for one or two electron ions to be abundant and excited.
Although the latter requirement would have to be satisfied
eventually, there may first have to be considerable efforts to
exhaust the possibilities in the argon spectrum or of various
elements in the glass, e.g., by calibrating the Inglis-Teller

relation experimentally at higher n and then using it down to n $\approx$ 4, corresponding to $N_e \approx 10^{25} cm^{-3}$. However, opacity problems and blends may well prevent the successful completion of such a scheme, whose prospects of success should therefore be investigated by a theoretical analysis and numerical modelling.

1. H. R. Griem, "Spectral Line Broadening by Plasmas," Academic Press, New York (1974).
2. H. R. Griem, M. Blaha and P. C. Kepple, Phys. Rev. A19, 2421 (1979); see also P. C. Kepple and H. R. Griem, NRL Memo. Rep. 3634 (1978).
3. R. J. Tighe and C. F. Hooper, Jr., Phys. Rev. A15, 1773 (1977).
4. J. T. O'Brien and C. F. Hooper, Jr., Phys. Rev. A5, 867 (1972).
5. C. F. Hooper, Jr. and L. A. Woltz, Univ. of Florida Tech. Rep. No. DE-ASOS-76DP40016 (1979).
6. R. W. Lee, J. Phys. B (Atom. Mol. Phys.) 12, 1129 (1979).
7. K. Grützmacher and B. Wende, Phys. Rev. A16, 243 (1977); A18, 2140 (1978).
8. W. L. Wiese, D. E. Kelleher and V. Helbig, Phys. Rev. A11, 1858 (1975).
9. J. Seidel, Z. Naturforsch. 32a, 1207 (1977).
10. H. R. Griem, Phys. Rev. A20, 606 (1979).
11. J. K. Kilkenny, R. W. Lee, M. H. Key and J. G. Lunney, Symposium on Atomic Spectroscopy, Tucson (1979).
12. R. W. Lee, Physics Letters 71A, 224 (1979).
13. J. F. Seely, Phys. Rev. Letters 42, 1606 (1979).
14. V. L. Jacobs and M. Blaha, Phys. Rev. A (in press).
15. D. R. Inglis and E. Teller, Astrophys. J. 90, 439 (1939).
16. B. Yaakobi, D. Steel, E. Thorsos, A. Hauer and B. Perry, Phys. Rev. Letters 39, 1526 (1977); B. Yaakobi et al., Phys. Rev. A19, 1247 (1979).
17. C. M. Lee and A. Hauer, Appl. Phys. Letters 33, 692 (1978).
18. M. H. Key, J. G. Lunney, J. M. Ward, R. G. Evans and P. T. Rumsby, J. Phys. B (Atom. Mol. Phys.) 12, L213 (1979).
19. K. B. Mitchell, D. B. van Hulsteyn, G. H. McCall, Ping Lee and H. R. Griem, Phys. Rev. Letters 42, 232 (1979).

STARK BROADENING OF AR^{16} LINES IN DENSE LASER PRODUCED PLASMAS

R.F. Joyce, L.A. Woltz, and C.F. Hooper, Jr.

Physics Department
University of Florida
Gainesville, Florida 32611

INTRODUCTION

Recently, the Stark broadening of X-ray lines has been used
as a diagnostic technique to determine the densities of laser-
produced plasmas. Much work has been done on hydrogenic line
shapes, but relatively little use has been made of line shapes
from heliogenic, high-z, ions. In this paper we will outline a
theory of Stark broadening for "heliogenic" lines, we distinguish
the differences between this theory and that used when analyzing
hydrogenic lines. Further, we illustrate the temperature-density
sensitivity of the first two members of the "Lyman" series of
x-ray lines of heliogenic argon.

FORMALISM

It has been conjectured[1] that for high temperature and
densities, a product of hydrogenic wavefunctions will provide a
satisfactory approximation for the heliogenic wavefunctions of
Ar XVII. However, a strict interpretation of this statement would
imply a degeneracy in the orbital quantum number, "ℓ", a degeneracy which is not justifiable. In our model, we take one electron
to be in the n=1 level and the other initially in a higher level.
The sole effect of the ground state electron (in our approximation)
will be to remove the ℓ-degeneracy. Matrix elements are calculated using hydrogenic wavefunctions for a radiator with a charge
of Z-1.

For an ion undergoing spontaneous electric dipole transitions, in the presence of a plasma, the intensity distribution of
the emitted radiation is given by the well known expression,[2]

$$I(\omega) = \int_0^\infty P(\varepsilon)J(\omega,\varepsilon)d\varepsilon \ .$$

(1)

In this equation, which uses the quasistatic approximation for the ions, $P(\varepsilon)$ is the probability distribution function for the electric microfield due to the static ions. $J(\omega,\varepsilon)$ contains the trace over the internal states of the radiating ion as well as the average over states of the perturbing electrons.

$$J(\omega,\varepsilon) = -\pi^{-1}\text{Im} \ \text{Tr}_R\{\vec{d}\cdot[\Delta\omega - e\vec{\varepsilon}\cdot\vec{R}/\hbar - H(\omega)]^{-1}\rho^{(R)}\vec{d}\}$$

(2)

$\vec{d}$ is the dipole moment operator for the radiating ion and $\Delta\omega$ is the frequency separation measured from the unperturbed transition frequency. The second term inside the square brackets contains a shift due to the ion microfield. $\vec{R}$ is the position operator for the excited electron that undergoes the radiative transition. Averaged electron broadening effects appear in $H(\omega)$.

A final step in the calculation involves the inclusion of Doppler broadening by a convolution of the Stark profile with a Doppler profile.

The calculation of $P(\varepsilon)$ has been discussed elsewhere,[3] and we will not discuss it here.

In the calculation of $J(\omega,\varepsilon)$, the interaction between the radiator and the electron perturbers is assumed to be a point-dipole interaction,

$$V_{\text{int}} = e\vec{R}\cdot\vec{\varepsilon}_e(\vec{r}).$$

(3)

$\vec{\varepsilon}_e(\vec{r})$ is the electric field at the position of the radiator due to an electron perturber located at a distance, $\vec{r}$.

In the case of the Lyman series, where there is no lower-state broadening, the relaxation theory expression for the quantity inside the square brackets of Eq. (2) becomes[4]

$$[\Delta\omega - e\vec{\varepsilon}\cdot\vec{R}/\hbar - H(\omega)]_{\mu\mu'} = \Delta\omega_{\mu\mu'} - e\vec{\varepsilon}R^z_{\mu\mu'}/\hbar - H(\omega)_{\mu\mu'}$$

(4)

$$H(\omega)_{\mu\mu'} = \hbar^{-2}\Gamma(\Delta\omega_{\mu\mu'})\sum_{\mu''}\vec{R}_{\mu\mu''}\cdot\vec{R}_{\mu''\mu'}; \ \Delta\omega_{\mu\mu'} \equiv (\omega-\omega_{\mu1})\delta_{\mu\mu'}$$

(5)

R^z is the z component of the position operator for the bound radiator electron. The sum over μ'' runs over the quantum numbers of the upper level of the desired Lyman transition. The non-degeneracy of the orbital quantum number of the upper level is evidenced in the $\Delta\omega_{\mu\mu'}$ of Eqs. (4) and (5). Other levels are displaced from the p-state level. The amount of the displacement is equal to the separation between the p-level and the other level in an unperturbed heliogenic ion. This splitting is determined externally by theory or experiment.[5,6]

All that remains is to determine $\Gamma(\Delta\omega)$, which contains the average electron-broadening effects. When the electron perturbers are treated as particles moving in the field of the charged radiator, the result for $\Gamma(\Delta\omega)$ is[4]

$$\Gamma(\Delta\omega) = -i\left(\frac{4ne^4\lambda_T^3}{3\pi^2}\right) \int_0^\infty \int_0^\infty \int_0^\infty dt\,dk_1\,dk_2 \ \exp\left[it\left(\Delta\omega + \frac{\hbar k_1^2}{2m} - \frac{\hbar k_2^2}{2m}\right)\right]$$

$$\exp\left(\frac{-\beta\hbar^2 k_1^2}{2m}\right) \ f(k_1,k_2) \tag{6}$$

where n is the electron number density and λ_T is the electron thermal wavelength; k_1 and k_2 are wave vectors of the perturbing electrons.

$$f(k_1,k_2) = k_1 k_2 (\pi/2\sqrt{3}) g_{ff}(k_1,k_2)$$

where g_{ff} is the free-free Gaunt factor. If we separate $\Gamma(\Delta\omega)$ into its real and imaginary parts, it can be shown that

$$\Gamma_{Im}(\Delta\omega>0) = -\left(\frac{2ne^4}{3}\right)\left(\frac{8\pi m}{kT}\right)^{1/2} \left(\frac{2\pi}{\sqrt{3}\theta_R}\right)$$

$$x \int_0^\infty e^{-K_1^2/\theta_R} K_1 g(K_1,X_1)\,dK_1; \quad X_1^2 = K_1^2 + |\Delta\Omega| \tag{7}$$

$$\Gamma_{Im}(\Delta\omega<0) = e^{-|\Delta\Omega|/\theta_R} \ \Gamma_{Im}(\Delta\omega>0) \tag{8}$$

where $\theta_R = 2\hbar^2 kT/me^4$, $\Delta\Omega = 2\hbar^3\Delta\omega/me^4$, and $K_1 = \hbar^2 k_1^2/me^2$.
The real part of $\Gamma(\Delta\omega)$ contributes only to the shift of the line and was neglected in this calculation.

In order to approximate the effect of electron correlations, we modify our ideal-gas result for $\Gamma_{Im}(\Delta\omega)$ such that for $\Delta\omega<\omega_p$,

$$\Gamma_{Im}(\Delta\omega<\omega_p) = \Gamma_{Im}(\omega_p) . \tag{9}$$

This procedure is suggested by Fig. 1 of Ref. 7. Changes in the line shape are relatively insensitive to variation of the cutoff frequency.

Once $\Gamma(\Delta\omega)$ has been computed, Eqs. (4) and (5) are employed in the computation of $J(\omega,\varepsilon)$ and subsequently $I(\omega)$.

Results

    Figs. (1) and (2) show α and β lines respectively for fixed
density and varying temperatures, while Figs. (3) and (4) show
the same lines for fixed temperatures and several densities.
Although the line shapes are not highly sensitive to changes in
temperature, their sensitivity to density differences shows that
line broadening may be useful as a compression diagnostic in
inertial confinement pellet implosions. Both the α and β lines
are asymmetric, as opposed to the corresponding symmetric hydrogenic lines. This asymmetry is evident in experiments.[8]

    The α line is markedly different than the hydrogenic α which
has only one peak (and perhaps two shoulders). This difference is
due to the natural splitting between the s and p states in heliogenic ions. In spite of the non-degeneracy, the states still mix
in the presence of an electric field, giving rise to the otherwise forbidden component, 2S-1S. This double peaked characteristic is also verified by experiment.[8]

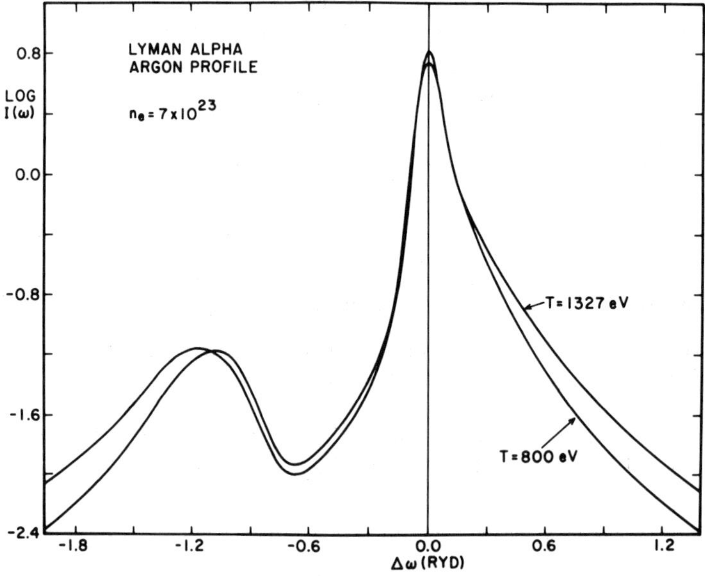

Figure 1

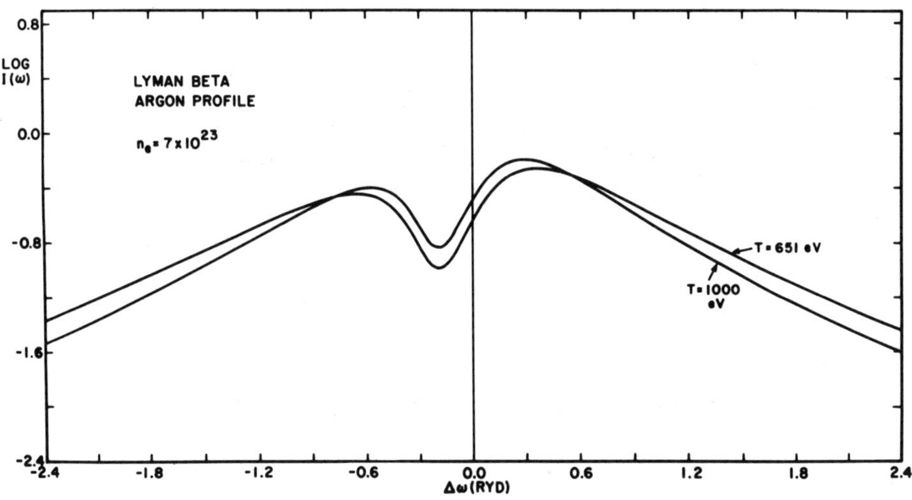

Figure 2

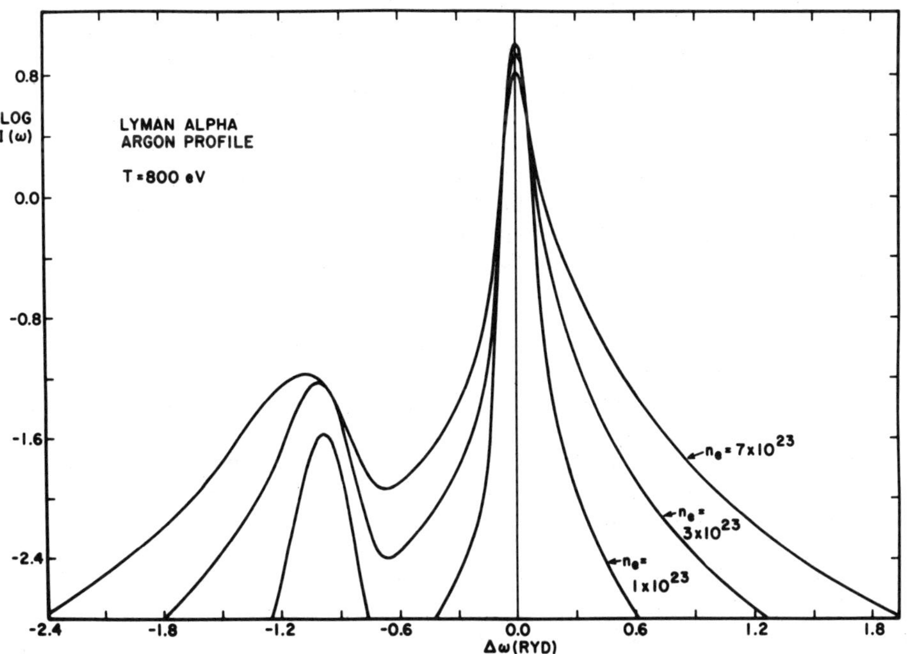

Figure 3

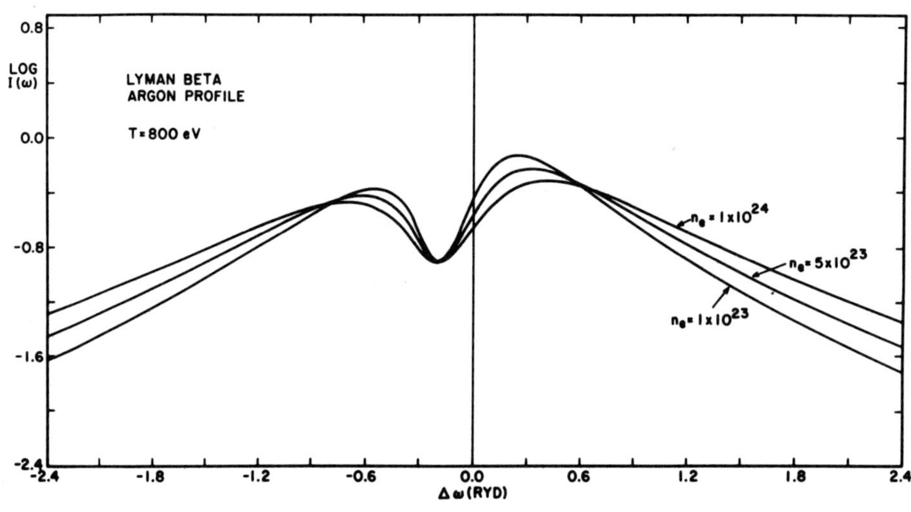

Figure 4

Summary

The ability to calculate heliogenic line shapes as well as
hydrogenic line shapes, greatly increases the possibility of using
several lines to analyze a single experimental shot. A problem
with this, however, is that the peak hydrogenic line intensity
and the corresponding peak heliogenic line intensity will, in
general, occur at different temperatures and densities that arise
during the course of the experiment. Therefore the choice of
temperature and density conditions which best fit a hydrogenic
line in a given experiment will not necessarily fit a heliogenic
line occurring in the same experiment. Time averaged profiles
may be needed to obtain consistent results.

References

1. H. R. Griem, private communication.
2. H. R. Griem, Spectral Line Broadening by Plasmas, (Academic,
       New York, 1974).
3. John T. O'Brien and C. F. Hooper, Jr., J. Quant. Spectrosc.
       Radiat. Transfer 14, 479 (1974).
4. Richard J. Tighe and C. F. Hooper, Jr., Phys. Rev. A 14, 1514
       (1976).
5. Robert E. Knight and Charles W. Scherr, Rev. Mod. Phys. 35,
       431 (1963).
6. Alan Hauer, private communication.
7. E. W. Smith, Phys. Rev. 166, 102 (1968).
8. B. Yaakobi, private communication.

AUTHOR INDEX

Numbers in paranthese following the text page numbers are reference numbers, and are included to assist in locating a reference at the end of each contribution when the author's name is not cited at the point of reference in the text.

A